Ultrafast photoinduced energy and charge transfer

Ventura, California, USA

8–10 April 2019

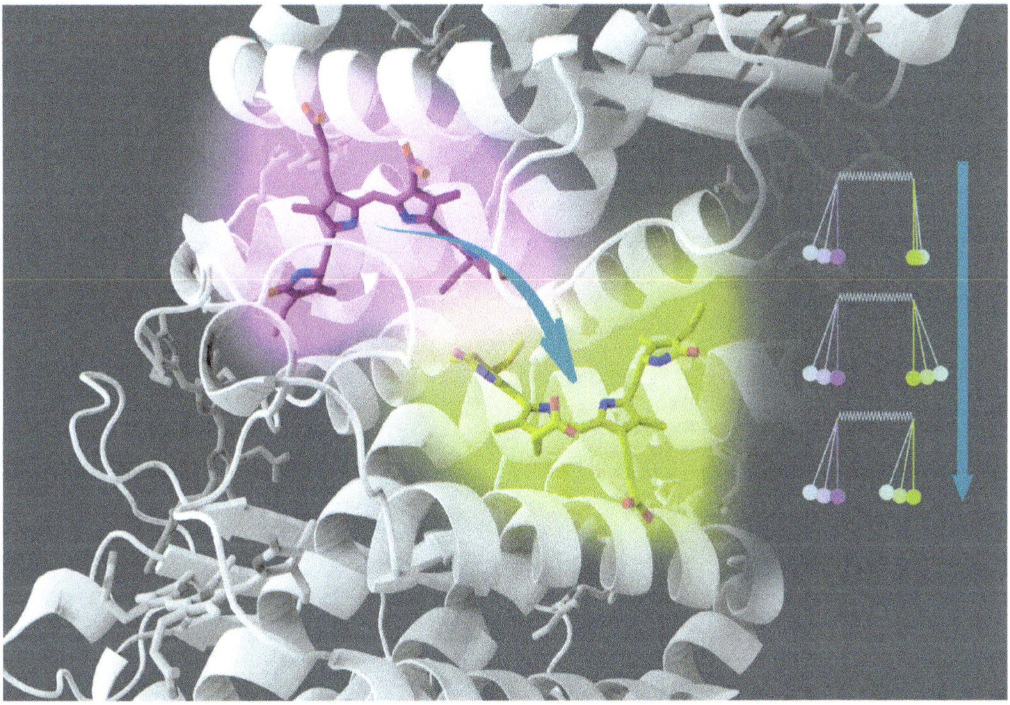

FARADAY DISCUSSIONS

Volume 216, 2019

ROYAL SOCIETY OF CHEMISTRY

The Faraday Division of the Royal Society of Chemistry, previously the Faraday Society, was founded in 1903 to promote the study of sciences lying between chemistry, physics and biology.

Faraday Discussions (Print ISSN 1359-6640, Electronic ISSN 1364-5498) is published 8 times a year by the Royal Society of Chemistry, Thomas Graham House, Science Park, Milton Road, Cambridge, UK CB4 0WF.

Volume 216 ISBN 13: 978-1-78801-673-5

2019 annual subscription price: print+electronic £1220 US $2148; electronic only £1162, US $2046. Customers in Canada will be subject to a surcharge to cover GST. Customers in the EU subscribing to the electronic version only will be charged VAT.

All orders, with cheques made payable to the Royal Society of Chemistry, should be sent to the Royal Society of Chemistry Order Department, Royal Society of Chemistry, Thomas Graham House, Science Park, Milton Road, Cambridge, CB4 0WF, UK Tel +44 (0)1223 432398; E-mail **orders@rsc.org**

If you take an institutional subscription to any Royal Society of Chemistry journal you are entitled to free, site-wide web access to that journal. You can arrange access via Internet Protocol (IP) address at **www.rsc.org/ip**

Customers should make payments by cheque in sterling payable on a UK clearing bank or in US dollars payable on a US clearing bank.

Printed in the UK

MIX
Paper from responsible sources
FSC® C013604

ROYAL SOCIETY OF CHEMISTRY | Celebrating IYPT 2019

Faraday Discussions are unique international discussion meetings that focus on rapidly developing areas of chemistry and its interfaces with other scientific disciplines.

Ultrafast photoinduced energy and charge transfer

Faraday Discussions

www.rsc.org/faraday_d

A General Discussion on Ultrafast photoinduced energy and charge transfer was held in Ventura, California, U.S.A. on the 8 th , 9 th and 10 th of April 2019.

RSC Publishing is a not-for-profit publisher and a division of the Royal Society of Chemistry. Any surplus made is used to support charitable activities aimed at advancing the chemical sciences. Full details are available from www.rsc.org

CONTENTS

ISSN 1359-6640; ISBN 978-1-78801-673-5

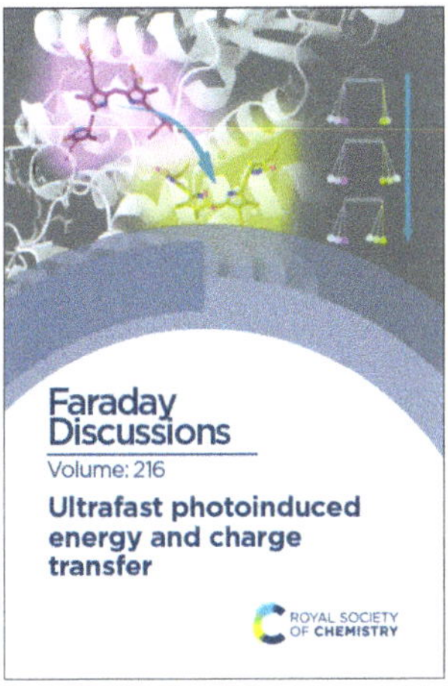

Cover
See Stefan Siwiak-Jaszek and Alexandra Olaya-Castro, Faraday Discuss., 2019, **216**, 38–56.

The image illustrates a vibronic heterodimer. As energy transfer happens from one exciton to another, some local coherent vibrations (represented as pendula) become first anti-synchronised and then in-phase synchronised.

Image reproduced by permission of Alexandra Olaya-Castro from Faraday Discuss., 2019, **216**, 38–56.

INTRODUCTORY LECTURE

PAPERS AND DISCUSSIONS

SPONSORS

Structural Dynamics

AMERICAN INSTITUTE OF PHYSICS

EGOROV SCIENTIFIC (DELMAR SYSTEMS)

KAVLI FOUNDATION

LAYERTEC GMBH

LIGHT CONVERSION USA

NATURE COMMUNICATIONS

PHASETECH SPECTROSCOPY

TELEDYNE PRINCETON INSTRUMENTS

CONCLUDING REMARKS

ADDITIONAL INFORMATION

Faraday Discussions

PAPER

Ultrafast photoinduced energy and charge transfer

Majed Chergui

Received 15th April 2019, Accepted 16th April 2019
DOI: 10.1039/c9fd00036d

After presenting the basic theoretical models of excitation energy transfer and charge transfer, I describe some of the novel experimental methods used to probe them. Finally, I discuss recent results concerning ultrafast energy and charge transfer in biological systems, in chemical systems and in photovoltaics based on sensitized transition metal oxides.

1 Introduction

Photoinduced energy and charge transfer are at the heart of how nature functions and are the basis of an impressive number of applications, both present and to come. Photosynthesis is one of the most obvious examples as it embodies a wide range of energy and charge transfer events, as shown in Fig. 1.

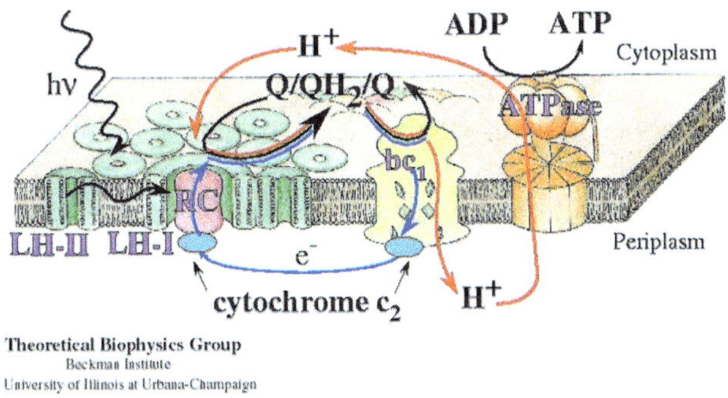

Fig. 1 The integral membrane proteins forming the photosynthetic unit and the processes triggered by light: light absorption and excitation energy transfer (black wavy arrows), electron transfer (blue arrows), proton transfer (red arrows), and ATPase (black arrows). Reproduced from ref. 1 with permission.

Ecole Polytechnique Fédérale de Lausanne, Laboratoire de Spectroscopie Ultrarapide, ISIC, Lausanne Centre for Ultrafast Science (LACUS), FSB, Station 6, CH-1015 Lausanne, Switzerland. E-mail: Majed.Chergui@epfl.ch; Fax: +41 21 6930365; Tel: +41 21 6930447 extn 0457

Light absorption excites the light harvesting complexes (LHI and LHII), which undergo excitation energy transfer (EET) to the reaction centre (RC). This then drives electron transfer (ET) between the cytochromes c_2 and the RC, and the bc1 complex. Proton transfer (PT) ensues, which triggers the ATPase. In this sequence of events, the charge (electron and proton) transfers are not directly triggered by the absorption of light. However, there are numerous examples of light-induced charge transfer processes in Nature. For example, the blue copper protein involved in electron transport in plant photosynthesis can directly undergo photoinduced charge transfer by excitation in the visible region.[2,3] Another class of system that plays a central role in converting light into the driving force for biological functions, such as ion transport and signal transduction, and is indispensable to the activity of organisms, is represented by the visual proteins.[4–8] Upon absorption of light, the embedded chromophores undergo *e.g.*, proton/electron/energy transfers in addition to isomerization, typically on extremely short timescales (100s femtoseconds). These subtle changes in the chromophore trigger a sequence of events involving higher-order structural changes in the protein exterior, leading to the final outcome of a long-lived signaling state that triggers biological functions.

Tracking the flow of energy and charge in natural systems is crucial for understanding how Nature works, but it is also the basis for an impressive variety of present and potential applications in solar energy conversion, photocatalysis, sensors and optoelectronic devices. In photovoltaics, interfacial charge transport implies a photoexcited sensitizer: a dye molecule, a semi-conducting or metallic nanoparticle, and more recently, bulk semi-conductors (such as perovskites), which transfer electrons to an electron conducting substrate material (usually a transition metal oxide, TMO), thus generating electrical current. These sensitized systems have been the focus of intense research over the past three decades.[9–11] TMOs are also important in photocatalytic applications.[12–14] Finally, charge and or energy transfers are being explored in molecular assemblies (donor–bridge–acceptor, J-/H-aggregates, *etc.*) with special emphasis on artificial photosynthesis.[15–20] These assemblies are common in biology and the most obvious example is DNA.

There are innumerable examples of ultrafast *intra*molecular charge transfer in molecules, in particular transition metal-based molecular complexes, that are the basis for several phenomena, such as electronic and spin dynamics, structural dynamics, and heat dissipation.[21–28] We will not deal with these in this article. Rather we focus on the study of *inter*molecular energy and electron transfer between a donor (D) and an acceptor (A) species that may or may not be linked by chemical bonds.

The study of these intermolecular processes has benefited from an impressive improvement in methodologies, both experimental and theoretical, over the past two decades, which I will briefly review bearing in mind that the experimental tools do not necessarily probe EET and ET in the same way, nor with the same observables. I will then present some case studies on biological systems and on sensitized TMOs.

2 Modelling excitation energy transfer

EET is the basis for fluorescence resonance energy transfer (FRET), which has become a common tool for the investigation of protein dynamics.[29,30] Förster's elegant derivation[31] begins with a Golden rule expression for the FRET transition rate. Förster theory[32,33] has been used to model EET by a quantum-mechanical coupling between the electronic transitions of a donor (D) and acceptor (A). The EET rate for a weak dipole–dipole interaction is given by:

$$k_{\mathrm{EET}} = \frac{1}{\tau_{\mathrm{EET}}} = \frac{1}{\tau_{\mathrm{D}}}\left(\frac{R_0}{r}\right)^6 = 8.79 \times 10^{23}\frac{Q_{\mathrm{D}}\kappa^2}{\tau_{\mathrm{D}}n^4r^6}J$$

where τ_{EET} is the D lifetime quenched by EET, Q_{D} is the quantum yield of the D fluorescence (in the absence of A), κ^2 is the orientation factor, τ_{D} is the pure radiative lifetime of the D fluorescence, n is the refractive index of the medium or solvent, and r is the distance between the centres of the two molecules or two dipoles (point dipole approximation) in Å. R_0 is the classical Förster distance. J is the overlap integral of the donor fluorescence and acceptor absorption. The orientation factor κ^2 is a function of R, μ_{D} and μ_{A}:

$$\kappa^2 = (\mu_{\mathrm{D}}\mu_{\mathrm{A}} - 3(\mu_{\mathrm{D}}R)(\mu_{\mathrm{A}}R))^2$$

where R is the unit vector for the D–A centre-to-centre distance, μ_{D} is the unit vector of the D dipole and μ_{A} is the unit vector of the A dipole.

Despite the success of the Förster theory for predicting EET rates and assessing conformational equilibrium configurations, in particular *via* the well-known amino-acid spectroscopic ruler tryptophan (Trp), used in FRET studies of protein dynamics, there are several situations where it fails. These have nicely been evaluated in several recent reviews.[34,35] Specifically, the theory is formulated under the assumption that the electronic coupling between D and A is very weak.[36] It also assumes that the D and A sizes are much smaller than R_{DA}, so that the point-dipole approximation can be used. Indeed, deviations from the $(1/R_{\mathrm{DA}})^6$ distance dependence of the EET rate usually occur at closer D–A separations when the point-dipole approximation fails. The latter works well for quantifying the optical spectroscopy of allowed transitions because the wavelength of light is long compared to the size of a molecule and therefore its oscillating electric field averages over details of the molecular wave function. This does not apply for close lying chromophores. Indeed, as two molecules are brought together, the Coulomb interaction of electrons and nuclei in one molecule with those in the other increases. Thus, the eigenstates of the isolated molecules are no longer eigenstates of the full Hamiltonian. This gives rise to excitonic couplings, *i.e.* interactions between excited states localized on different molecules.

Depending on the strength of the intermolecular interaction, it is still useful to think of the system as two interacting molecules rather than as a super-molecule. To do so, one expands the full Hamiltonian in the basis of the molecular states and the off-diagonal elements of that expansion are the couplings. The latter contain short- and long-range contributions. The former include exchange, overlap of donor and acceptor wave functions, and exciton transfer mediated by charge-transfer states.[37,38] Because they depend on the spatial overlap of the D and A wave functions, short-range terms decrease exponentially with distance. Beljonne *et al.*[35] and Kenny and Kassal[39] have presented the hierarchy of methods that can accurately describe EET: point-dipole approximation, extended dipole, transition charges from electrostatic potentials, transition monopole approximation,[40–42] transition density cubes,[43] and the fragment excitation difference (FED) method.[44,45]

Another deviation of the $(1/R)^6$ dependence of Förster theory is the case when electronic couplings can be mediated by molecular bridges between D to A, and when screening of the electronic couplings by the environment can also change the distance dependence.[35] Dexter EET can be viewed as double charge transfer: an electron D–A transfer and a hole A–D transfer.[46] In the case of several chromophores either with bonded or non-bonded interactions, the process is described as an exciton moving an electron–hole pair through the assembly of chromophores. This case is being investigated in a paper by Bai *et al.* (DOI: 10.1039/c9fd00007k) in this issue, which looks at interferences between different Dexter pathways.

The importance of a proper description of the couplings and rates in the strong coupling case pertains to the debate about the electronic coherence between chromophores, which is treated by Siwiak-Jaszek and Olaya-Castro, Lishchuk *et al.*, Richter and Fingerhut, and Wu *et al.* (DOI: 10.1039/c9fd00006b, DOI: 10.1039/c8fd00241j, DOI: 10.1039/c8fd00189h, and DOI: 10.1039/c8fd00190a, respectively) in this issue and to which we will come back later.

3 Modelling electron transfer

ET is described by an extraordinarily robust theory.[47] The non-adiabatic rate ($k_{\rm ET}$) for electron transfer between a donor D and an acceptor A is given by:

$$k_{\rm ET} = \frac{1}{\tau_{\rm ET}} = \frac{2\pi}{\hbar}\frac{1}{\sqrt{4\pi\lambda k_{\rm B}T}}H_{12}{}^2 \exp\left(\frac{-(\Delta G^0 + \lambda)^2}{4\lambda k_{\rm B}T}\right)$$

where $\tau_{\rm ET}$ is the lifetime of the D quenched by ET, ΔG_0 is the change in free energy of the reaction (also called the driving force), λ is the reorganization energy and H_{12} is the adiabatic electronic coupling between the initial and final electronic states. This expression applies provided nuclear motion is treated classically, a reasonable approximation in the high temperature limit, and if the electronic coupling between the ET states is weak, typically below a few tens of $\rm cm^{-1}$.[48] Goldsmith *et al.* (DOI: 10.1039/c8fd00240a) present in this issue a nice discussion of Marcus theory applied to the case of photoinduced proton-coupled-electron-transfer.

In the case of ET in biological systems, Beratan and co-workers proposed a relatively simple and successful way of estimating the electronic coupling between donor and acceptor.[49–53] This so-called pathway model assumes that ET is mediated over different pathways, by consecutive interactions between atoms connecting the donor and the acceptor. The ET events described therein assume migration of only one electron. The electronic coupling equation for ET (tunnelling through a structure-less one-dimensional barrier between localized vibronic states) in a protein is given by:

$$h_{12} = A\prod_i \varepsilon_i^{\rm C}\prod_j \varepsilon_j^{\rm H}\prod_k \varepsilon_k^{\rm S}$$

where A is the pre-factor. Segments of the pathway are characterized as covalent (C), hydrogen-bonded (H), or through-space (S), depending on whether the bonds share a common atom (C), are linked by a hydrogen bond (H), or are in van der Waals contact (S). In through-bond ET, the backscattering between bonds and all (variable) side groups are neglected. Through space decay assumes for simplicity that the noncovalent contact through which the electron tunnels is composed of only two orbitals, one on each backbone chain. Assuming two identical orbitals, the transfer matrix element connecting them is on the order of V_bS_{12}, where V_b is the orbital energy and S_{12} is the orbital overlap.[54]

Work with proteins has shown that ET rates depend exponentially on R_{DA}; a decay factor of 1.1 $Å^{-1}$ describes most systems.[55] This steep distance dependence limits single-step excited-state ET reactions to distances under 20 Å (Fig. 2), but multistep reactions can transport charges over distances of 30 Å or more.[56]

The above discussion dealt with ET in biosystems, but ET at molecule/solid or film/solid interfaces is central to the function of sensitized solar cells, involving molecular dyes,[11,57-59] metallic or semiconductor nanoparticles[60] or solid state sensitizers such as organic–inorganic perovskites.[61,62] It also plays a crucial role in charge separation in organic photovoltaics (Voss *et al.*, DOI: 10.1039/c8fd00210j).[63] ET at the molecular/solid interface is also crucial in the function of photocatalytic reactions,[12,64,65] molecular electronics,[66] photo-electrolysis,[67] and photography.[68,69] In many of these systems, the ET is extremely fast (few femtoseconds)[70-73] such that it cannot be described by a simple rate description, and quantum-chemical methods are needed, often with time-domain modelling. These methods have been presented in detail in several review papers and will not be repeated here.[74-78] Rather, we will focus on novel experimental approaches to probing such interfacial charge transfer phenomena and how they deliver further insight into the charge behaviour subsequent to the transfer.

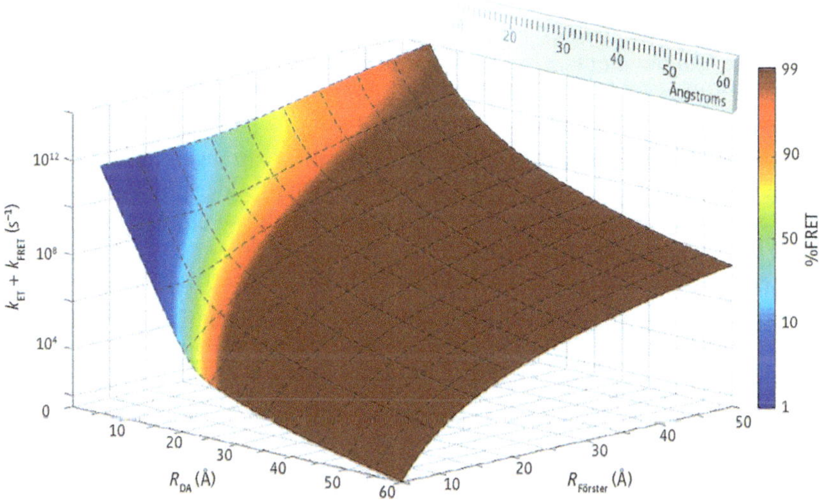

Fig. 2 Electron transfer rate (k_{ET}) and Förster resonance energy transfer rate (k_{FRET}), as a function of donor–acceptor separations (R_{DA}) and Förster radii ($R_{Förster}$). Shorter R_{DA} and $R_{Förster}$ tend to favor ET. However, beyond 20 Å FRET is expected to be dominant. Reproduced from ref. 79 with permission from the American Association for the Advancement of Science, copyright (2013).

4 Novel experimental tools

With the advent of ultrafast (fs–ps) spectroscopy at the end of the 1980s, it became possible to track in "real-time" EET and ET processes in biological[80–89] and chemical systems[90–92] and at interfaces,[11,58,60,93,94] using pump–probe techniques. Vibrational coherences generated by ultrafast EET and ET have also been reported in various systems[90,95–99] and a nice discussion of such processes and of the debates around them are presented in the paper by Wu *et al.* (DOI: 10.1039/c8fd00190a) of this issue. In general, ultrafast pump–probe spectroscopy identifies these coherences from an *a priori* knowledge of frequencies and modes, but electronic coherences that would be characterised by similar frequencies as the vibrational ones makes the two difficult to distinguish. Furthermore, interrogating EET or ET in (bio)chemical systems or at interfaces requires spectroscopic probes that can clearly distinguish the donor from the acceptor. The constant development of methods over the past two decades has offered more specific ways of tracking these phenomena. Here we mainly focus on the novel tools by briefly recalling their main features.

(A) X-ray spectroscopy

Probing on ultrafast time scales oxidation state changes that occur in the D–A pair upon ET processes can best be achieved with element-specificity using time-resolved X-ray absorption (XAS) or emission (XES) spectroscopy.[100–102] These methods are universal markers of the oxidation and spin states of the atoms. XAS probes the unoccupied density of the valence states, while XES probes the occupied density of states. Both are therefore particularly sensitive to the electron density in and occupancy of the valence orbitals, whose changes result in shifts of a specific core–shell edge absorption or emission. The absorption edge represents the ionization threshold of a given core electron. At energies higher that the edge, XAS contains modulations due to the scattering of the generated photoelectron onto the neighbour atoms, giving rise to the X-ray near-edge absorption structure (XANES) and extended X-ray absorption fine structure (EXAFS) features, which reflect the local geometry around the selected atom.[103] In a time-domain experiment, such features may change due to structural rearrangements. XES is in addition an ideal marker of the spin state of the system under investigation.[104–107] Thus, XAS and XES contain specific information about the electronic, spin and geometric structure of the elements constituting the system. In summary, because of their element-specificity and sensitivity to electron density, X-ray spectroscopies can, in principle, distinguish the donor from the acceptor in ET and EET processes.

The methodology for time resolved X-ray spectroscopy was first demonstrated in the hard X-ray range in the case of photoinduced intramolecular ET in transition metal complexes,[108–111] and it was expanded to other types of ET processes: charge-transfer-to-solvent,[112–114] interfacial ET,[115–117] and hole and electron trapping in TiO_2 (ref. 115 and 118) and ZnO.[119] Expanding the use of time-resolved XAS or XES to the soft X-rays allows the investigation of light elements, since the K-edges of elements such as C, N, O, *etc.* are found in the 300–600 eV region. This was achieved in closed cells by Huse and co-workers,[120,121] but novel liquid sample delivery schemes under vacuum,[122–124] are now opening the field.[125]

In addition to the above X-ray spectroscopies, we should also mention another core-level spectroscopy: X-ray photoelectron spectroscopy (XPS), which allows element-selective detection along with the absolute binding energies of given orbitals.[126–129] While its extension to solutions has been achieved for some years now[122,129–133] it has mostly been used for studies of ET (or electron injection) of sensitized systems, for which they are actually an ideal tool.[134-137]

(B) Deep-ultraviolet spectroscopy

Amino-acid residues in proteins and the nucleobases of DNA and RNA all have their absorption bands below 300 nm in the deep-ultraviolet (deep-UV) region. In addition, most large band-gap semi-conductors, such as TMOs or transition metal nitrides, that are of great importance in photovoltaics and photocatalysis, have their band gaps lying below 380 nm, some actually closer to 300 nm. There is therefore need for a tool that can excite and probe in this spectral region. While pumping in the deep-UV is not an issue using the third harmonic of the Ti:Sa laser, probing over large spectral windows in the deep-UV region was achieved by scanning the probe wavelength,[4,138] prior to the advent of achromatic doubling,[139] which allows the generation of a white light continuum in this spectral range.[140] This was exploited[141–143] to generate a probe continuum from 250 to 380 nm. When used with a tuneable monochromatic pump pulse over the same range, 2D deep- to near-UV TA experiments (see next section) can be performed, which enable the detection of hitherto novel ultrafast intramolecular processes in molecules,[144] and intermolecules ones in haemoproteins (see below).[145,146]

Ultrafast deep-UV TA together with UV fluorescence up-conversion spectroscopy[147-155] and deep-UV Raman spectroscopy[8,156,157] represent an ideal toolbox for the identification of ET and EET involving amino-acid residues and nucleobases of biosystems. The multidimensional extension of these spectroscopies is a powerful addition to the toolbox.

(C) Multidimensional spectroscopy

The description of the electronic couplings, the involved electronic states and the transfer rates between donor and acceptor is of the utmost importance to understand ultrafast ET and EET. In pump–probe TA spectroscopy, the pump pulse is narrow-band, while the time-delayed weaker probe pulse is a continuum. The resulting signal is the difference of the probe signal with pump and without pump. Repeating experiments with different pump wavelengths yields a 2D map (an axis for the pump energy and the other for the probe energy) corresponding to a specific time delay. This type of 2D spectroscopy has the great advantage of simplicity, but has limited temporal resolution. 2D TA spectroscopy is thus suitable when the timescales of the processes are not too short and we will present below such a case.

When couplings become stronger the spectral signatures are more complex, and in the case of biosystems congestion makes the disentangling of the various spectral signatures nearly impossible using TA spectroscopy. Only with coherent multidimensional (MD) spectroscopic techniques can one disentangle them. Indeed, MD spectroscopy allows one to excite specific states of the system and then monitor the response coming from the excited states with similar or different energies. Coherent or Fourier Transform (FT) MD spectroscopies thus provide excitation frequency, detection frequency, and time-resolved spectroscopic data. 2D FT electronic spectroscopy[158–166] has emerged as a powerful tool to disentangle highly congested spectral regions in a given system and specifically variants of the method such as 2D photon echo (2DPE) measurements, can in principle discriminate between nuclear quantum beats and electronic coherent oscillations,[161,167] which arise from very strong electronic couplings.[168,169] The underlying reason for this ability is that the data contain information about electronic coherences as well as populations.[170]

Another spectroscopy that provides detailed information about the interactions between the electronic and vibrational degrees of freedom in molecular systems is 2D electronic-vibrational (2DEV) spectroscopy, which has been witnessing increasing interest in recent years.[171–173] In 2DEV, one combines the advantages of infrared vibrational spectroscopy with visible electronic spectroscopy, which allows one to directly correlate the electronic and vibrational degrees of freedom simultaneously.

(D) Circular dichroism (CD)

CD spectroscopy measures the absorption difference between left- and right-circularly polarized light. Although, this typically represents 0.1% of the total absorption,[174,175] CD spectroscopy is a common tool in analytical biochemistry. In multichromophoric systems, the CD signal arises from coupling between the dipoles associated to each chromophore. The relative orientation and spatial arrangements of the coupled transition dipoles are encoded in the resulting CD signal, whose derivation is based on the Fermi Golden rule with electric and magnetic dipole interaction. It is proportional to the scalar product of electric and magnetic transition dipole moments.[176] Therefore, CD spectroscopy probes the same couplings as 2D FT spectroscopy, but with the advantage that the theoretical modelling is, in principle, simpler. However, in the time-domain an additional experimental difficulty arises in that, in addition to the difference between left- and right-circularly polarized light absorption, the difference between pumped and un-pumped CD spectra makes the signals extremely weak, reaching a few tenths of a milliOD at most. Nevertheless, several time-resolved CD set-ups have been developed, but very few in the deep-UV region where the important chromophores of biology absorb.[177–179] My group has recently implemented such an experiment as an extension of our 2D deep-UV set-up.[142,143] The CD experiment employs a photoelastic modulator to achieve shot-to-shot polarization switching of a 20 kHz pulse train of broadband fs deep- to near-UV (250–370 nm) pulses. The sequence of alternating left- and right-circularly polarized probe pulses is used in a pump–probe scheme with shot-to-shot dispersive detection, allowing the acquisition of broad-band CD spectra of ground- and excited-state species. A sensitivity of $<2 \times 10^{-5}$ OD ($\approx$0.7 mdeg) is achieved.[180] This set-up allows detecting chirality changes of molecular complexes and assemblies of great relevance to biology, such as DNA.

5 Scientific questions

In the following section, I will dwell on some of the scientific subjects relevant to this Faraday Discussion. This section is by no means exhaustive and it does not reflect the wealth of novel insights that are presented at the meeting. I refer the reader to the respective articles in this volume and their appended discussions.

(A) Electron *versus* excitation energy transfer in proteins

The amino-acid residue tryptophan (Trp) is commonly used as the "molecular ruler"[181] in studies of protein dynamics because it is known to undergo dipole–dipole EET to nearby acceptor chromophores *via* the FRET process. The strong distance dependence of the FRET rate allows one to extract, from the Trp decay times, information about the protein structure and its changes. Because of its low oxidation potential, Trp is also involved in multistep ET processes in proteins and DNA.[56,84,182–188] Excited Trp fluorescence quenching has also been attributed to ET to Cu atoms in azurin.[189]

Trp fluorescence decay in myoglobins (Mbs) has repeatedly been interpreted as being due to EET to the haem.[154,190,191] We recently used 2D deep-UV and visible TA, and ultrafast fluorescence up-conversion, to investigate the decay of Trp fluorescence in Mbs,[145,146] and in ferric and ferrous cytochrome c (Cyt c).[192,193]

Mb has two tryptophan residues: Trp7 and Trp14, which are located at distances of 21.2 Å and 15.2 Å (centre to centre), respectively, from the haem group (Fig. 3). Their fluorescence decay times are, in all forms of Mb (ferric or ferrous, and ligated or deoxy), typically 20–30 ps for Trp14 and 110–140 ps for Trp7, compared to the ~2–3 ns lifetime of Trp in water.

Our 2D-UV TA studies of ferric Mbs (MbCN and metMb) revealed that Trp14 undergoes both EET and ET to the haem,[145] with comparable yields. How coupled EET and ET are in this case is unclear (the issue of coupling is discussed in this issue by M. Richter and B. P. Fingerhut *et al.* DOI: 10.1039/c8fd00189h). On the other hand, the more distant Trp7 undergoes exclusively EET to the haem. It was also found that the transferred electron ends up on the iron atom, yielding a ferrous porphyrin. Extending the study to the unligated ferrous deoxyMb form, it was found that Trp14-to-haem ET yields a low-valence porphyrin anion

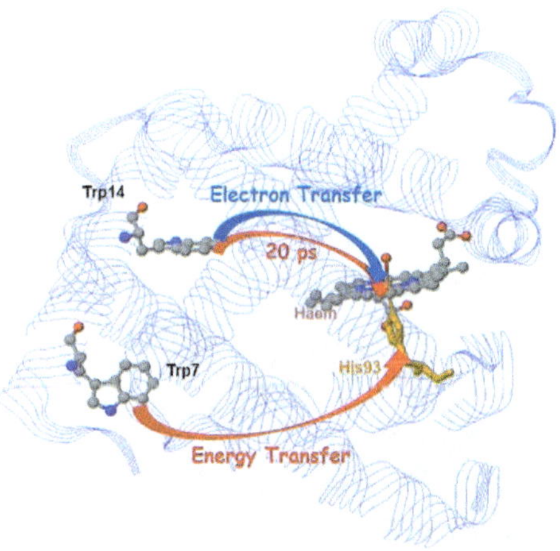

Fig. 3 The main result of our 2D deep-UV and visible TA studies of myoglobins. While Trp7 fully decays *via* FRET to the haem (red arrow), Trp14 undergoes at comparable yields both electron transfer (blue arrow) and energy transfer.[145,146]

radical.[146] In order to investigate these processes in ferrous ligated Mbs (MbNO, MbCO), we used IR TA because the dissociation–recombination kinetics of the diatomic ligand, in particular in the case of NO (~10–20 ps and ~150–200 ps), occurs on similar time scales to the decay of the two Trps. The IR probe data yields several new insights into the outcome of the ET, which were not detected in the visible probe experiments.[145] In particular, in MbCN and MbCO the transferred electron density is distributed between the iron and the porphyrin, *i.e.* it does not fully reduce the Fe atom in the ferric MbCN.[194]

We proposed that the ET proceeds from Trp14 *via* the leucine 69 (Leu69) and valine 68 (Val68) amino acid residues to the haem.[146] This hypothesis was recently supported by theoretical modelling[195] based on DFT calculations and on the Beratan model.[49,53]

As already mentioned, ET rates depend exponentially on R_{DA} (Fig. 2). The above results obtained on Trp7 and Trp14 decay in Mbs fully fall within these trends. The more efficient ET, overwhelming the EET for smaller distances would be expected in the case with ferric and ferrous Cytochrome c (Cyt c), since their single Trp59 residue lies ~9 Å from the haem and its centre-of-mass is located almost in the plane of the porphyrin, while its indole plane forms an angle of ~70–80° with the latter (Fig. 4).

We performed ultrafast UV fluorescence up-conversion experiments[193] and TA spectroscopy on both systems,[192] and found that the Trp fluorescence decays on extremely short time scales: 350 fs for ferrous Cyt c, and 770 fs for the ferric form. These values are orders of magnitude shorter than those in Mbs, implying much more efficient ET and/or EET. However, our TA studies[192] did not point to an ET, in particular for ferric Cyt c, where the reduction of the haem would have yielded the ferrous form, as was observed in the case of Trp-to-haem ET in ferric Mbs.[145] The very short Trp decay times hint to a relatively strong interaction with the haem. In the case of bacteriorhodopsin, excitation of retinal yielded a strong response of the nearby Trp residues in the form of a Stark shift.[4,5,196] These results had been modelled in terms of an excitonic coupling of retinal with the two nearest lying Trps. We believe that for Cyt c, a similar excitonic coupling occurs at the excitation energy of Trp, due to a resonance with the N bands of the haem. The differences between ferric and ferrous Cyt c may be attributed to the different

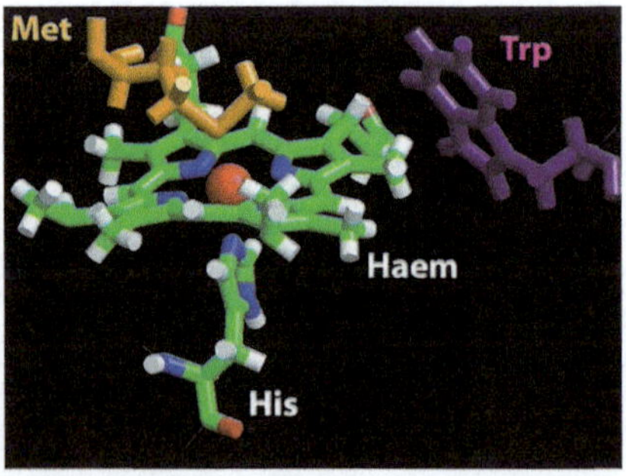

Fig. 4 Scheme of the haem pocket in cytochrome c, with the methionine (Met) and histidine amino-acid residues bound to the Fe atom (red) and the tryptophan near the haem.

relative orientations and distances between the Trp and the haem groups in the two redox states.

The strong coupling case where excitonic pairs are formed as in the above examples, or excitons in larger assemblies of chromophores, *e.g.* the photosynthetic system (Lishchuk *et al.*, DOI: 10.1039/c8fd00241j; Richter and Fingerhut, DOI: 10.1039/c8fd00189h; Hsieh *et al.*, DOI: 10.1039/c8fd00205c), organic materials (Mroczek *et al.*, DOI: 10.1039/c8fd00182k), or molecular crystals (Wang *et al.*, DOI: 10.1039/c8fd00157j), is related to the hotly debated issue of electronic *vs.* vibrational coherences in the photosynthesis, nicely presented by Wu *et al.* (DOI: 10.1039/c8fd00190a) in this issue. As far as Trps are concerned, an interesting case of strong coupling and its consequences was recently simulated and discussed by Kurian and co-workers in the case of tubulin.[197,198] Microtubules are biological protein polymers with critical and diverse functions. They share some similarities with photosynthetic antenna complexes, particularly in the ordered arrangement of photoactive molecules with large transition dipole moments, which applies to Trps. While the biological role of photoexcitation of microtubules is still an open question, using positions (in their native configuration), dipole orientations from realistic models (Fig. 5) and a Hamiltonian describing the light–matter interactions, Kurian and co-workers showed that the Trps exhibit a superradiant lowest excitonic state, which represents an excitation fully extended on the chromophore lattice. Furthermore, such a superradiant state

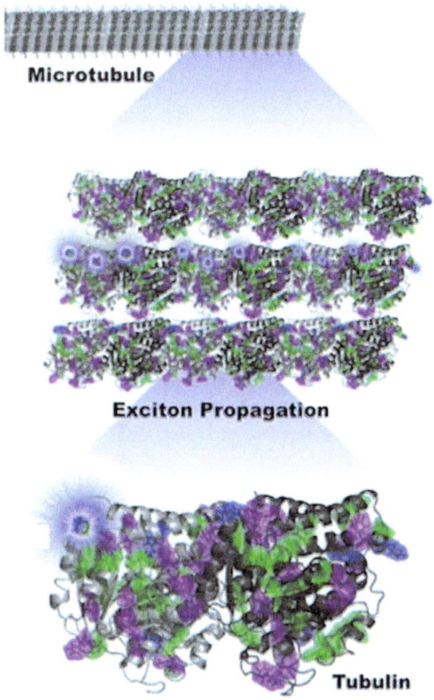

Fig. 5 In microtubules, tubulins form regular networks. Each tubulin contains 8 tryptophans and 34 tyrosines. Upon absorption of deep-UV light, energy transfer occurs along the neuronal microtubules and the exciton propagation extends on the order of dendritic length scales and beyond, according to ref. 197. Reproduced from ref. 197 with permission from Elsevier, copyright 2017.

emerges as a result of super-transfer coupling between the lowest exciton states of the microtubule. These cooperative effects (superradiance and supertransfer) may induce ultra-efficient photoexcitation absorption and could enhance excitonic energy transfer in microtubules over long distances under physiological conditions.

Another system in which excitons are known to occur but whose coherence length, *i.e.* the number of monomer units it involves, is still a matter of debate is DNA (in single- or double-stranded form).[199,20] However, in contrast to microtubules, DNA is very flexible and this property is critical for its function as a carrier of genetic information, for binding by transcription factors (TF) and for its organization into chromatin. Therefore, probing DNA conformational dynamics will allow understanding to how DNA allostery, modulated by DNA base modifications, affects TF binding. They also directly impact on the coherence length of the excitons and are the origin of the uncertainty on its value. In return, the dynamics of the excitons can report on the conformational evolution of DNA strands (single or double). However, the excited state dynamics of DNA still raise fundamental questions, which are hotly debated. What is the nature of the absorbing state in a single strand? The examination of the absorption spectrum of the monomer and 20-mer (Fig. 6) suggests formation of Frenkel excitons in the 20-mer, which was supported by quantum chemical calculations.[201] However, Nogueira *et al.*[202] recently argued that the differences between the features of the 20-mer spectrum are due to long-range perturbations of the monomer spectrum.

Another important question that needs to be addressed is; does the initially excited state remain unchanged or does it evolve into other species, *i.e.* if monomers are excited, do they evolve into excitons delocalized on 2 bases or

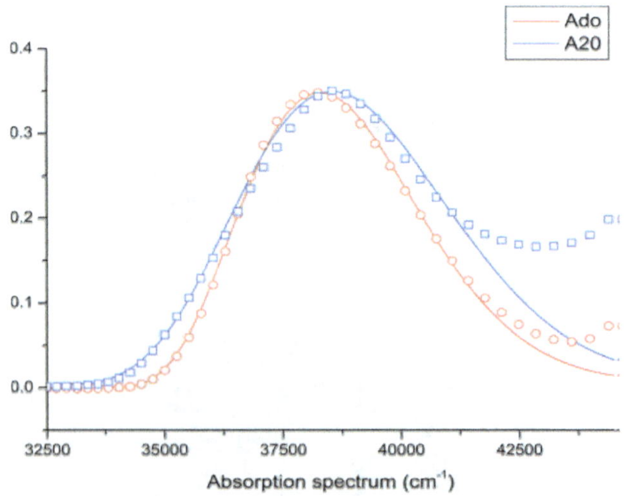

Fig. 6 Experimental (symbols) absorption spectra in aqueous solution of single base adenosine (Ado) and of 20-mer adenine strands in aqueous solution. The 20-mer spectrum shows a lower energy red wing and a blue shifted maximum. The calculated spectra (solid lines) were derived from quantum chemical calculations described in ref. 201. Reproduced from ref. 201 with permission from the American Chemical Society, copyright 2007.

more? Finally, assuming we have excitons, how do they trap to form what Kohler and co-workers[199] define as excimers and exciplexes?

Despite a rich literature using a variety of experimental tools (single wavelength deep-UV[199,200] and broadband visible[203,204] TA, fluorescence up-conversion,[147,150,151,205,206] 2D spectroscopy[207–209] and liquid phase ultrafast photoelectron spectroscopy[210,211]), many of these questions are still open. The studies by Fiebig and co-workers using a UV visible (320 500 nm) continuum probe on monomeric and stacked DNA bases of variable number (from dimer to 18-mer) already delivered some rich insight. From the evolution of the signal at 430 nm as a function of the number of bases (Fig. 7), they concluded that the delocalization length of the exciton is between 3 and 4. My group has recently carried out broadband deep-UV (240–360 nm) probe studies upon 266 nm of the adenosine monomer, dimer and 20-mer and 2D deep-UV spectroscopy, which will soon be published. The interesting feature of these studies is that we can also detect the ground state bleach signal of the species, allowing a more complete visualization of the evolution of the photoexcited species.

The advent of coherent 2D UV spectroscopies[140,208,212–214] should pave the way to finer studies of the dynamics of DNA and, more generally, biological systems.[209,215–218]

(B) Chemical systems

Chemical reactions in solution often imply an exchange of charges, in particular an ET. This not only concerns the two reactants, which may or may not be linked by a bridge, but also the solvation shell around them is not just a spectator but reacts to the charge exchange, which creates a new field of forces, by rearranging itself in such a way as to minimize the free energy. If in addition the reactant is photoexcited, then intramolecular relaxation processes may occur therein, which

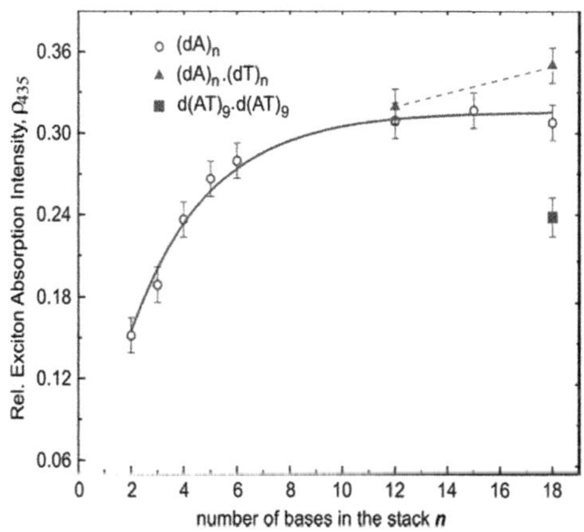

Fig. 7 Ratio of the spectral intensity (open circles) between the transient absorption of different strains of adenosine, associated to the exciton (at 435 nm) and to the monomer (at 330 nm), 3 ps after excitation, as a function of the stack length n, according to ref. 204. The fit of the curve shows a 1/e delocalization length of the exciton of $d \approx 3-4$ bases. The values for the A:T duplexes are connected with a dotted line. Reproduced from ref. 204 with permission from the National Academy of Science, copyright 2007.

can take place on timescales comparable to those of ET and solvation dynamics, making these difficult to disentangle. This is why unambiguous markers of the oxidation state, the solvation dynamics and the intramolecular relaxation are needed. The latter two phenomena have already been dealt with rather well by ultrafast 2D spectroscopies that can disentangle the homogeneous and inhomogeneous contributions to a spectral feature and their evolution with time. The detection of an oxidised or reduced species is more difficult due to the fact that their spectral features are different to those of the neutral form, and conclusions about charge transfer processes rely on an *a priori* knowledge of the system, as was done in ref. 145.

As already mentioned, XAS and XES are universal markers of the oxidation state.[101] Of course, this is the case when the whole charge density is localised on a given element. This is often the case with metal complexes and several examples exist of ps XAS detection of intramolecular electron transfer in transition metal complexes.[108–110] The extension of such experiments to ET in molecular dyads has been achieved by Canton *et al.*[219] who combined XES and X-ray scattering (XRS) with femtosecond resolution and characterized the non-equilibrated electron transfer (ET) dynamics in a bimetallic complex consisting of a light harvesting, ruthenium (Ru)-based complex linked to an optically dark cobalt (Co) complex that acts as an electron sink. The bridge between them mediates the ultrafast ET. This system is characteristic of the wide class of synthetic and natural photocatalysts, for which studies by various methods have barely started: a combination of a UV and IR pump aimed at vibrational control of ET and an IR probe,[17,220] 2D electronic spectroscopy,[221] and 2DEV.[171,173] Canton and co-workers could thus detect the dynamics of electron departure from the donor, the transit time *via* the bridge, the arrival to the acceptor, and, finally, the formation time of the high spin state in the latter (Fig. 8). This work showed the great advantages of X-ray probes in studies of ultrafast photoinduced ET: (i) the unambiguous identification of the

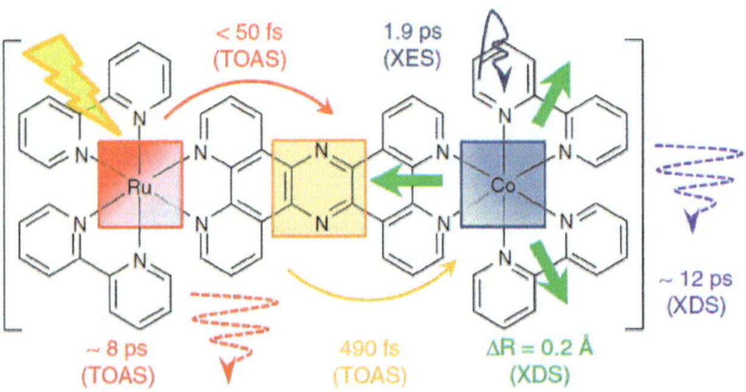

Fig. 8 Schematic representation of the results obtained by Canton *et al.*[219] on the electron transfer in the bimetallic RuCo complex $[(bpy)_2Ru^{II}(tpphz)^1Co^{III}(bpy)_2]^{5+}$ (with bpy = bipyridine, tpphz = tetrapyridophenazine) using transient optical absorption spectroscopy (TOAS), fs X-ray emission spectroscopy (XES) and X-ray scattering (XDS). Excitation of the Ru complex transfers an electron to the Co-complex *via* the bridge (red and yellow arrows). The reduction and high spin state of the Co-complex is detected by XES, while its structural rearrangement and solvent relaxation is detected by XDS. Reproduced from ref. 219 with permission.

donor and acceptor; (ii) the detection of the optically silent acceptor; (iii) the identification of its high spin electronic state.

Other examples of intermolecular electron transfer processes detected by time-resolved X-ray spectroscopy are those involving charge-transfer-to-solvent states, for which a typical example is aqueous iodide. These states are metastable states of the solute–solvent system and for that reason they exist only in polar solvents.[222] Our ps and fs X-ray studies of photoexcited aqueous iodide showed the appearance of neutral iodine, but most interestingly revealed the change of the solvation shell structure from hydrophilic to hydrophobic after electron abstraction.[112,113,223] This process requires 2–3 ps to be completed. Further studies[224] using fluorescence up-conversion allowed us to detect the departure of the electron from the CTTS state, since it fluoresces as long as the excited state wave function still has an overlap with the ground state one. These studies showed that the decay of the CTTS states (between 50 and 450 fs) is highly inhomogeneous due to the infinite number of initial configurations of iodide in the labile solvation shell. Another system exhibiting CTTS states, aqueous $[Fe(CN)_6]^{2+}$, which we investigated using time-resolved X-ray spectroscopy, did not yield any ultrafast fluorescence hinting to a possible intramolecular channel for the electron departure.[225]

Finally, I would like to dwell on our recent development of ultrafast broadband (260–360 nm) deep-UV circular dichroism (CD) spectroscopy, which although mainly implemented to tackle biological systems, *e.g.* DNA, we first demonstrated on a simple model system, ruthenium-trisbipyridine ($[Ru(bpy)_3]^{2+}$) in water undergoing a metal-to-ligand-charge transfer (MLCT) upon photoexcitation. This molecule is chiral in the ground state and the dipoles of the 3 bpy ligands are coupled and have their ligand-centred π–π^* bands centred at 292 nm. MLCT excitation transfers an electron to one of the bpy ligands, breaking the chiral character of the molecule. Our experiments exhibited a very clear response of the two enantiomers,[180] paving the way to the study of biological systems. As a start, we investigated thiolated dipeptides as the model units of peptide chains. The coupling of the two thioamide units could not be resolved by achiral 2D-UV spectroscopy, however it gives rise to a pronounced bisignate CD spectrum. The time-resolved CD experiments reveal the weakening of this coupling upon 266 nm excitation, which releases conformational constraints.[226] These results show that direct local probing of fast backbone conformational change *via* CD is possible. Furthermore, site-selective thio-substitution promises a more detailed probing of conformational dynamics in peptides and proteins.

(C) Photovoltaics and photocatalysis

Transition metal complexes are highly studied candidates for photocatalysis and photovolatics in so-called dye sensitized solar cells (DSSCs).[227,228] In DSSCs, the most popular dyes are ruthenium-polypyridine complexes, adsorbed onto a transition metal oxide (TMO) substrate (TiO_2 or ZnO). The ruthenium complexes serve as visible light harvesters and upon excitation of their singlet metal-to-ligand-charge-transfer (^{1}MLCT) state, they undergo ultrafast electronic relaxation and intersystem crossing to the ^{3}MLCT state,[229,230] and simultaneously to these, electron injection into the conduction band (CB) of the TMO occurs. This process has been probed using ultrafast spectroscopy in the visible to the THz range.[72,94,231–236]

However, these probes are only sensitive to the free carriers (*i.e.* the Drude electrons in the CB of the material), cannot detect trapped charges (important for catalysis) and are not material specific (the response of Drude electrons is not specific to the material). In the past decade my group has developed several tools to interrogate the fate of charge carriers (both electrons and holes) in TMOs, in particular upon charge injection.[73,115,118,119,237,238]

We implemented a picosecond optical pump/XAS probe in the case of ruthenium dye sensitized anatase TiO_2 nanoparticles (NPs) in solution, probing both photooxidation of the dye and the reduction of Ti atoms at defects due to electrons injected into the substrate. We also compared these results to the case of direct band gap excitation of the bare NPs. These were chosen to reproduce the average grain size of mesoporous TiO_2 films used in DSSCs. Previously, Katz *et al.*[116,239] had carried out an Fe K-edge absorption study of iron oxide nanoparticles, upon electron injection from adsorbed organic dyes. By comparing their transients with simulated ones, they concluded that the reduced metal sites formed small polarons on a 100 ps time scale. Our ps XAS studies on bare NPs showed that upon deep-UV band gap excitation, the electron gets trapped at defects characterised by pentacoordination of the Ti atoms reducing them from 4+ to 3+.[115] Further, fs XAS studies showed that the electron is trapped in ~100 fs, but the structural relaxation of the oxygen cage around it requires a longer time of about 330 fs.[118,240] This ultrafast electron trapping suggests that the electron barely travels in the lattice and is trapped at or near the unit cell where it was created. Given that in anatase TiO_2 NPs the shell region is defect-rich, we conclude that most of the trapped electrons are found therein. However, it is obvious that not all electrons are trapped. When conducting the experiment on dye-sensitized NPs, the XAS spectral signatures were somewhat different to the bare NP case, which we attributed to the electron being trapped in the outer

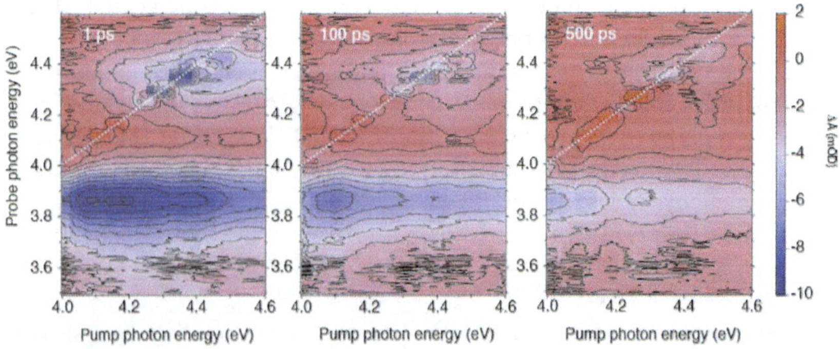

Fig. 9 2D deep-UV TA maps (pump and probe photon energy) of a colloidal solution of anatase TiO_2 NPs at room temperature. The spectral response is displayed at three different time delays between pump and probe: 1 ps, 100 ps, and 500 ps. The time resolution is 150 fs. The signal is dominated by a bleach over the entire probe region, with two bands appearing at 3.85 eV and 4.4 eV, which are not distinguished in the steady-state absorption spectrum. The signal is due to a combination of phase space filling or bleach, which attenuates the absorption, and long-range Coulomb screening, which broadens it. Reproduced from ref. 244 with permission from the American Chemical Society, copyright 2018.

surface region of the NPs. We also carried out similar studies on the amorphous[115] and rutile[237] polymorphs of TiO_2, yielding the same trends as in anatase.

ZnO is another very promising material for photovoltaics and photocatalysis, as well as many other applications. In this material, the Zn atoms have a d^{10} electronic structure, therefore they cannot trap electrons as in TiO_2, yet when we carried out ps XAS on ZnO NPs, we found a very pronounced response at the Zn K-edge. At the same time, the ps X-ray emission spectra did not hint (as expected) to a change of electronic structure of the Zn atoms. In fact, their strong K-edge response is due to the fact that the material contains singly (positively) charged oxygen vacancies. Upon photoexcitation, free holes are created, which migrate and get trapped at these vacancies to turn them into doubly charged ones. This process induces a dramatic displacement of the four surrounding Zn atoms,[241] which explains why the Zn K-edge is so strongly affected. Thus, one charge can affect several atoms simultaneously. The additional interesting aspect of this work is that it allows us to detect for the first time selectively the hole in the VB of a TMO. Indeed, the VB is dominated by the oxygen orbitals in all TMOs, and detecting them using XAS or XES implies working in the soft X-ray range at the O K-edge. This requires a solid sample in vacuum, while here we detect them in solution.

As already mentioned, the high contrast achieved by X-ray spectroscopy upon the oxidation state change of a given element implies that for the TMOs only localised charges can unambiguously be detected using X-ray spectroscopy, but not the free ones. A method is needed that can detect the free charges, while being element-specific. 2D deep-UV TA spectroscopy turned out to be an extremely valuable tool for the characterisation of charge carrier dynamics in TMOs. We recently carried out a study of anatase TiO_2 NPs, and revealed hitherto unknown features in the TA, as can be seen in Fig. 9.[242] These features are missing in the steady state absorption spectrum but show up here because of the suppression of

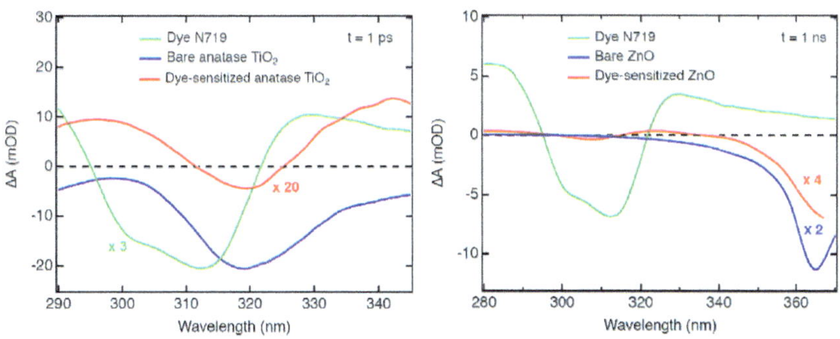

Fig. 10 Spectral responses of the ruthenium dye sensitizer in solution upon 550 nm excitation (green), the bare NPs of the substrate material (blue) upon above band-gap excitation in the deep-UV region and of the sensitized material upon 550 nm excitation. In the case of the bare TiO_2 NPs, the signal is identical to that in Fig. 9. For ZnO NPs, the signal is entirely due to a bleach due to the filling of the conduction band. In the sensitized NPs, the same signatures appear as in the bare case, except for an offset in the case of TiO_2, due to the contribution of Drude electrons, and a shift in the case ZnO, which will be discussed elsewhere. Reproduced from ref. 238 with permission from the American Chemical Society, copyright 2017.

scattering in the difference spectra (pumped minus unpumped). Interestingly, the same features are present in the transient reflectivity spectra of single crystals of anatase TiO_2, pointing to a bulk like origin. Further analysis using a wide range of experimental (spectroscopic ellipsometry and angle-resolved photoemission) and theory established that the first feature at 3.85 eV is due to a strongly bound exciton (binding energy $\sim$ 150 meV) intermediate between a Frenkel and a Wannier exciton. However, most remarkable is that this exciton is confined to a 2D plane of the 3D lattice. Furthermore, we found that it is robust against defects, temperature and strain.[243] This finding prompted us to propose the use of excitonic transitions as a substrate-specific probe of electron injection from a sensitizer. Indeed, injecting electrons into the CB should affect the excitonic transitions by reducing the transition probability (phase-space filling) or broadening it by Coulomb screening.

Fig. 10 shows the results obtained for a ruthenium-dye adsorbed on anatase TiO_2 NPs and ZnO NPs. In both cases, the response at the energy of the first exciton is very clear and its kinetic behaviour reproduces what was obtained from the detection of Drude electrons using other methods in other spectral ranges (THz, IR and Visible).[72,245] However the deep-UV detection teaches us something new about what the effect of the Drude electrons is in the CB of the material. Indeed, in TiO_2 the excitonic response is mainly due to Coulomb screening, while in ZnO it is due to phase space filling. These responses have implications on the functioning of the material as a photovoltaic device. This work also suggests an avenue for performing 2D visible/deep-UV spectroscopy of charge injection, as it has been suggested that the injection efficiency is excitation energy-dependent.[246]

6 Outlook

In this article, I have attempted to review a number of recent results we and others have obtained using novel ultrafast spectroscopic tools (X-ray spectroscopy, 2D deep-UV, *etc.*), casting them in the context of this Faraday Discussion. What, in my humble opinion, emerges are a number of points regarding both methodologies and concepts. These are far from being exhaustive. As far as methodologies are concerned, there is no doubt that the emergence of 2D spectroscopies in their various forms, but more specifically electronic ones, represent a crucial turning point in the study of EET and ET. Extending them to shorter wavelengths is a promising avenue that will yield new insights, especially when they will be implemented as core-level spectroscopies.[218,247,248] A complement to these is offered by ultrafast circular dichroism spectroscopy, which in the optical domain has the advantage of a relatively simple theoretical modelling. CD spectroscopy offers in addition access to structural information, while working in the optical domain. For biosystems in physiological solutions, this is unique as the only alternative is X-ray solution scattering, which requires heavy simulations for the interpretation of results.

In terms of concepts, there has been no doubt about vibrational coherence in biosystems, and in the case of electronic coherence, these systems exhibit quantum mechanical properties such as delocalization and super-radiance. Excitons are key objects in the capture of light, its transport in neutral form and its conversion to electrical charges. What we are now seeing is that in addition, excitons can serve as probes of conformational dynamics, *e.g.* in DNA.

Interfacial charge transfer is central to electrochemistry, photocatalysis and to photovoltaics based on sensitized systems and on organic materials. The way it is governed by the nature of the donor and the acceptor is still not rationalised and much needs to be done yet in this respect. Finally, and somewhat related, proton coupled electron transfer, which is essential in enzymatic activities and in artificial synthesis to allow the system to recover, is an area where intense work is needed if any of the proposed schemes for solar energy conversion can "see the light".

Conflicts of interest

There no conflicts of interest to declare.

Acknowledgements

I am grateful to all my co-workers for the studies reported in this paper. Several of these projects have been supported by the Swiss NSF *via* the NCCR MUST. The results on transition metal oxides were supported by the ERC Advanced grant DYNAMOX.

References

1 http://www.ks.uiuc.edu/Research/quantum_biology/LHmembrane_stamped.gif.
2 M. A. Webb, C. M. Kwong and G. R. Loppnow, *J. Phys. Chem. B*, 1997, **101**, 5062–5069.
3 E. Fraga and G. Loppnow, *J. Phys. Chem. B*, 1998, **102**, 7659–7665.
4 S. Schenkl, F. van Mourik, G. van der Zwan, S. Haacke and M. Chergui, *Science*, 2005, **309**, 917–920.
5 J. Leonard, E. Portuondo-Campa, A. Cannizzo, F. van Mourik, G. van der Zwan, J. Tittor, S. Haacke and M. Chergui, *Proc. Natl. Acad. Sci. U. S. A.*, 2009, **106**, 7718–7723.
6 Y. Sudo, M. Mizuno, Z. Wei, S. Takeuchi, T. Tahara and Y. Mizutani, *J. Phys. Chem. B*, 2014, **118**, 1510–1518.
7 S. Tahara, S. Takeuchi, R. Abe-Yoshizumi, K. Inoue, H. Ohtani, H. Kandori and T. Tahara, *J. Phys. Chem. Lett.*, 2015, **6**, 4481–4486.
8 H. Kuramochi, S. Takeuchi, K. Yonezawa, H. Kamikubo, M. Kataoka and T. Tahara, *Nat. Chem.*, 2017, **9**, 660.
9 A. Hagfeldt, G. Boschloo, L. C. Sun, L. Kloo and H. Pettersson, *Chem. Rev.*, 2010, **110**, 6595–6663.
10 D. Gust, T. A. Moore and A. L. Moore, *Acc. Chem. Res.*, 2009, **42**, 1890–1898.
11 M. K. Nazeeruddin, E. Baranoff and M. Gratzel, *Sol. Energy*, 2011, **85**, 1172–1178.
12 A. J. Esswein and D. G. Nocera, *Chem. Rev.*, 2007, **107**, 4022–4047.
13 A. Fujishima, X. T. Zhang and D. A. Tryk, *Surf. Sci. Rep.*, 2008, **63**, 515–582.
14 K. Nakata and A. Fujishima, *J. Photochem. Photobiol., C*, 2012, **13**, 169–189.
15 L. De Cola and P. Belser, *Coord. Chem. Rev.*, 1998, **177**, 301–346.
16 K. Ohta, M. Yang and G. R. Fleming, *J. Chem. Phys.*, 2001, **115**, 7609–7621.

17 P. A. Scattergood, M. Delor, I. V. Sazanovich, M. Towrie and J. A. Weinstein, *Faraday Discuss.*, 2015, **185**, 69–86.
18 S. Verma, A. Ghosh, A. Das and H. N. Ghosh, *J. Phys. Chem. B*, 2010, **114**, 8327–8334.
19 A. Schubert, V. Settels, W. Liu, F. Würthner, C. Meier, R. F. Fink, S. Schindlbeck, S. Lochbrunner, B. Engels and V. Engel, *J. Phys. Chem. Lett.*, 2013, **4**, 792–796.
20 P. Parkinson, C. Muller, N. Stingelin, M. B. Johnston and L. M. Herz, *J. Phys. Chem. Lett.*, 2010, **1**, 2788–2792.
21 A. Vlček Jr, J. Víchová and F. Hartl, *Coord. Chem. Rev.*, 1994, **132**, 167–174.
22 D. J. Stufkens, M. P. Aarnts, J. Nijhoff, B. D. Rossenaar and A. Vlcek, *Coord. Chem. Rev.*, 1998, **171**, 93–105.
23 A. Vlcek, *Coord. Chem. Rev.*, 2000, **200**, 933–977.
24 J. K. McCusker, *Acc. Chem. Res.*, 2003, **36**, 876–887.
25 E. A. Juban, A. L. Smeigh, J. E. Monat and J. K. McCusker, *Coord. Chem. Rev.*, 2006, **250**, 1783–1791.
26 A. Cannizzo, C. J. Milne, C. Consani, W. Gawelda, C. Bressler, F. van Mourik and M. Chergui, *Coord. Chem. Rev.*, 2010, **254**, 2677–2686.
27 M. Chergui, *Dalton Trans.*, 2012, **41**, 13022–13029.
28 M. Chergui, *Acc. Chem. Res.*, 2015, **48**, 801–808.
29 K. Truong and M. Ikura, *Curr. Opin. Struct. Biol.*, 2001, **11**, 573–578.
30 B. Schuler, A. Soranno, H. Hofmann and D. Nettels, *Annu. Rev. Biophys.*, 2016, **45**, 207–231.
31 T. Förster, *Ann. Phys.*, 1948, **437**, 55–75.
32 G. D. Scholes, *Annu. Rev. Phys. Chem.*, 2003, **54**, 57–87.
33 B. Masters, *Eur. Phys. J. H*, 2014, **39**, 87–139.
34 A. Muñoz-Losa, C. Curutchet, B. P. Krueger, L. R. Hartsell and B. Mennucci, *Biophys. J.*, 2009, **96**, 4779–4788.
35 D. Beljonne, C. Curutchet, G. D. Scholes and R. J. Silbey, *J. Phys. Chem. B*, 2009, **113**, 6583–6599.
36 A. Agabekyan and A. Melikyan, *Opt. Spektrosk.*, 1972, **32**, 153–157.
37 R. D. Harcourt, K. P. Ghiggino, G. D. Scholes and S. Speiser, *J. Chem. Phys.*, 1996, **105**, 1897–1901.
38 K. J. Fujimoto, *J. Chem. Phys.*, 2012, **137**, 034101.
39 E. P. Kenny and I. Kassal, *J. Phys. Chem. B*, 2015, **120**, 25–32.
40 C. Weiss, *J. Mol. Spectrosc.*, 1972, **44**, 37–80.
41 J. C. Chang, *J. Chem. Phys.*, 1977, **67**, 3901–3909.
42 K. Sauer, R. J. Cogdell, S. M. Prince, A. Freer, N. W. Isaacs and H. Scheer, *Photochem. Photobiol.*, 1996, **64**, 564–576.
43 B. P. Krueger, G. D. Scholes and G. R. Fleming, *J. Phys. Chem. B*, 1998, **102**, 5378–5386.
44 C.-P. Hsu, Z.-Q. You and H.-C. Chen, *J. Phys. Chem. C*, 2008, **112**, 1204–1212.
45 C.-P. Hsu, *Acc. Chem. Res.*, 2009, **42**, 509–518.
46 S. S. Skourtis, C. Liu, P. Antoniou, A. M. Virshup and D. N. Beratan, *Proc. Natl. Acad. Sci. U. S. A.*, 2016, **113**, 8115–8120.
47 R. A. Marcus and N. Sutin, *Biochim. Biophys. Acta, Rev. Bioenerg.*, 1985, **811**, 265–322.
48 A. de La Lande, N. Gillet, S. Chen and D. R. Salahub, *Arch. Biochem. Biophys.*, 2015, **582**, 28–41.

49 D. Beratan, J. Betts and J. Onuchic, *Science*, 1991, **252**, 1285–1288.

50 J. N. Onuchic and D. N. Beratan, *J. Chem. Phys.*, 1990, **92**, 722–733.

51 J. N. Onuchic, D. N. Beratan, J. R. Winkler and H. B. Gray, *Annu. Rev. Biophys. Biomol. Struct.*, 1992, **21**, 349–377.

52 J. R. Winkler, H. B. Gray, T. R. Prytkova, I. V. Kurnikov and D. N. Beratan, *Bioelectronics: From Theory to Applications*, 2005, pp. 15–33.

53 I. A. Balabin, X. Hu and D. N. Beratan, *J. Comput. Chem.*, 2012, **33**, 906–910.

54 D. N. Beratan, J. N. Onuchic and J. Hopfield, *J. Chem. Phys.*, 1987, **86**, 4488–4498.

55 H. B. Gray and J. R. Winkler, *Biochim. Biophys. Acta, Bioenerg.*, 2010, **1797**, 1563–1572.

56 J. J. Warren, M. E. Ener, A. Vlček Jr, J. R. Winkler and H. B. Gray, *Coord. Chem. Rev.*, 2012, **256**, 2478–2487.

57 A. Hagfeldt and M. Gratzel, *Acc. Chem. Res.*, 2000, **33**, 269–277.

58 M. K. Nazeeruddin, S. M. Zakeeruddin, J. J. Lagref, P. Liska, P. Comte, C. Barolo, G. Viscardi, K. Schenk and M. Graetzel, *Coord. Chem. Rev.*, 2004, **248**, 1317–1328.

59 C. S. Ponseca, P. Chábera, J. Uhlig, P. Persson and V. Sundström, *Chem. Rev.*, 2017, **117**, 10940–11024.

60 Y. Yang and T. Q. Lian, *Coord. Chem. Rev.*, 2014, **263**, 229–238.

61 M. A. Green, A. Ho-Baillie and H. J. Snaith, *Nat. Photonics*, 2014, **8**, 506.

62 N.-G. Park, *Mater. Today*, 2015, **18**, 65–72.

63 J. Teuscher, J. C. Brauer, A. Stepanov, A. Solano, A. Boziki, M. Chergui, J. P. Wolf, U. Rothlisberger, N. Banerji and J. E. Moser, *Struct. Dyn.*, 2017, **4**, 061503.

64 A. L. Linsebigler, G. Q. Lu and J. T. Yates, *Chem. Rev.*, 1995, **95**, 735–758.

65 K. J. Young, L. A. Martini, R. L. Milot, R. C. Snoeberger, V. S. Batista, C. A. Schmuttenmaer, R. H. Crabtree and G. W. Brudvig, *Coord. Chem. Rev.*, 2012, **256**, 2503–2520.

66 C. Lindstrom and X.-Y. Zhu, *Chem. Rev.*, 2006, **106**, 4281–4300.

67 I. R. Gould, J. R. Lenhard, A. A. Muenter, S. A. Godleski and S. Farid, *J. Am. Chem. Soc.*, 2000, **122**, 11934–11943.

68 D. Liu, G. L. Hug and P. V. Kamat, *J. Phys. Chem.*, 1995, **99**, 16768–16775.

69 N. S. Lewis, *J. Electroanal. Chem.*, 2001, **508**, 1–10.

70 J. Schnadt, P. A. Bruhwiler, L. Patthey, J. N. O'Shea, S. Sodergren, M. Odelius, R. Ahuja, O. Karis, M. Bassler, P. Persson, H. Siegbahn, S. Lunell and N. Martensson, *Nature*, 2002, **418**, 620–623.

71 R. Huber, J.-E. Moser, M. Grätzel and J. Wachtveitl, *J. Phys. Chem. B*, 2002, **106**, 6494–6499.

72 G. Benko, J. Kallioinen, J. E. I. Korppi-Tommola, A. P. Yartsev and V. Sundstrom, *J. Am. Chem. Soc.*, 2002, **124**, 489–493.

73 O. Bram, A. Cannizzo and M. Chergui, *Phys. Chem. Chem. Phys.*, 2012, **14**, 7934–7937.

74 W. R. Duncan and O. V. Prezhdo, *Annu. Rev. Phys. Chem.*, 2007, **58**, 143–184.

75 J. L. Bredas, J. E. Norton, J. Cornil and V. Coropceanu, *Acc. Chem. Res.*, 2009, **42**, 1691–1699.

76 F. De Angelis, S. Fantacci, E. Mosconi, M. K. Nazeeruddin and M. Gratzel, *J. Phys. Chem. C*, 2011, **115**, 8825–8831.

77 C. F. A. Negre, R. L. Milot, L. A. Martini, W. D. Ding, R. H. Crabtree, C. A. Schmuttenmaer and V. S. Batista, *J. Phys. Chem. C*, 2013, **117**, 24462–24470.
78 J. Li, H. Li, P. Winget and J.-L. Brédas, *J. Phys. Chem. C*, 2015, **119**, 18843–18858.
79 J. R. Winkler, *Science*, 2013, **339**, 1530–1531.
80 J. W. Petrich, J. W. Longworth and G. R. Fleming, *Biochemistry*, 1987, **26**, 2711–2722.
81 J. W. Petrich, C. Poyart and J. Martin, *Biochemistry*, 1988, **27**, 4049–4060.
82 J. E. Hansen, S. J. Rosenthal and G. R. Fleming, *J. Phys. Chem.*, 1992, **96**, 3034–3040.
83 J. L. Martin and M. H. Vos, *Annu. Rev. Biophys. Biomol. Struct.*, 1992, **21**, 199–222.
84 C. Aubert, M. H. Vos, P. Mathis, A. P. Eker and K. Brettel, *Nature*, 2000, **405**, 586.
85 M. Byrdin, V. Sartor, A. P. M. Eker, M. H. Vos, C. Aubert, K. Brettel and P. Mathis, *Biochim. Biophys. Acta, Bioenerg.*, 2004, **1655**, 64–70.
86 T. Polivka and V. Sundstrom, *Chem. Rev.*, 2004, **104**, 2021–2071.
87 V. Sundstrom, *Annu. Rev. Phys. Chem.*, 2008, **59**, 53–77.
88 Y.-C. Cheng and G. R. Fleming, *Annu. Rev. Phys. Chem.*, 2009, **60**, 241–262.
89 O. Farver and I. Pecht, *Coord. Chem. Rev.*, 2011, **255**, 757–773.
90 K. Wynne, G. D. Reid and R. M. Hochstrasser, *J. Chem. Phys.*, 1996, **105**, 2287–2297.
91 N. Mataga and H. Miyasaka, Electron Transfer and Exciplex Chemistry, In *Advances in Chemical Physics*, ed. I. Prigogine and S. A. Rice, 1999, DOI: 10.1002/9780470141663.ch8.
92 A. H. Zewail, *Angew. Chem., Int. Ed.*, 2000, **39**, 2587–2631.
93 N. A. Anderson and T. Lian, *Coord. Chem. Rev.*, 2004, **248**, 1231–1246.
94 R. Katoh, A. Furube, A. V. Barzykin, H. Arakawa and M. Tachiya, *Coord. Chem. Rev.*, 2004, **248**, 1195–1213.
95 M. H. Vos, M. R. Jones, C. N. Hunter, J. Breton, J.-C. Lambry and J.-L. Martin, *Biochemistry*, 1994, **33**, 6750–6757.
96 C. Zimmermann, F. Willig, S. Ramakrishna, B. Burfeindt, B. Pettinger, R. Eichberger and W. Storck, *J. Phys. Chem. B*, 2001, **105**, 9245–9253.
97 Y. Song, S. N. Clafton, R. D. Pensack, T. W. Kee and G. D. Scholes, *Nat. Commun.*, 2014, **5**, 4933.
98 S. M. Falke, C. A. Rozzi, D. Brida, M. Maiuri, M. Amato, E. Sommer, A. De Sio, A. Rubio, G. Cerullo and E. Molinari, *Science*, 2014, **344**, 1001–1005.
99 R. Huber, L. Dworak, J. E. Moser, M. Grätzel and J. Wachtveitl, *J. Phys. Chem. C*, 2016, **120**, 8534–8539.
100 C. J. Milne, T. J. Penfold and M. Chergui, *Coord. Chem. Rev.*, 2014, **277**, 44–68.
101 M. Chergui, *Struct. Dyn.*, 2016, **3**, 031001.
102 R. Abela, P. Beaud, J. A. van Bokhoven, M. Chergui, T. Feurer, J. Haase, G. Ingold, S. L. Johnson, G. Knopp, H. Lemke, C. J. Milne, B. Pedrini, P. Radi, G. Schertler, J. Standfuss, U. Staub and L. Patthey, *Struct. Dyn.*, 2017, **4**, 061602.
103 M. Chergui, *Acta Crystallogr., Sect. A: Found. Crystallogr.*, 2010, **66**, 229–239.
104 U. Bergmann and P. Glatzel, *Photosynth. Res.*, 2009, **102**, 255–266.

105 W. K. Zhang, R. Alonso-Mori, U. Bergmann, C. Bressler, M. Chollet, A. Galler, W. Gawelda, R. G. Hadt, R. W. Hartsock, T. Kroll, K. S. Kjaer, K. Kubicek, H. T. Lemke, H. Y. W. Liang, D. A. Meyer, M. M. Nielsen, C. Purser, J. S. Robinson, E. I. Solomon, Z. Sun, D. Sokaras, T. B. van Driel, G. Vanko, T. C. Weng, D. L. Zhu and K. J. Gaffney, *Nature*, 2014, **509**, 345–348.
106 G. Vanko, A. Bordage, P. Glatzel, E. Gallo, M. Rovezzi, W. Gawelda, A. Galler, C. Bressler, G. Doumy, A. M. March, E. P. Kanter, L. Young, S. H. Southworth, S. E. Canton, J. Uhlig, G. Smolentsev, V. Sundstrom, K. Haldrup, T. B. van Driel, M. M. Nielsen, K. S. Kjaer and H. T. Lemke, *J. Electron Spectrosc. Relat. Phenom.*, 2013, **188**, 166–171.
107 G. Vanko, T. Neisius, G. Molnar, F. Renz, S. Karpati, A. Shukla and F. M. F. de Groot, *J. Phys. Chem. B*, 2006, **110**, 11647–11653.
108 M. Saes, C. Bressler, R. Abela, D. Grolimund, S. L. Johnson, P. A. Heimann and M. Chergui, *Phys. Rev. Lett.*, 2003, **90**, 047403.
109 W. Gawelda, M. Johnson, F. M. F. de Groot, R. Abela, C. Bressler and M. Chergui, *J. Am. Chem. Soc.*, 2006, **128**, 5001–5009.
110 A. El Nahhas, R. M. van der Veen, T. J. Penfold, V. T. Pham, F. A. Lima, R. Abela, A. M. Blanco-Rodriguez, S. Zalis, A. Vlcek, I. Tavernelli, U. Rothlisberger, C. J. Milne and M. Chergui, *J. Phys. Chem. A*, 2013, **117**, 361–369.
111 T. J. Penfold, S. Karlsson, G. Capano, F. A. Lima, J. Rittmann, M. Reinhard, M. H. Rittmann-Frank, O. Braem, E. Baranoff, R. Abela, I. Tavernelli, U. Rothlisberger, C. J. Milne and M. Chergui, *J. Phys. Chem. A*, 2013, **117**, 4591–4601.
112 V. T. Pham, W. Gawelda, Y. Zaushytsin, M. Kaiser, D. Grolimund, S. Johnson, R. Abela, C. Bressler and M. Chergui, *J. Am. Chem. Soc.*, 2007, **129**, 1530–1531.
113 V. T. Pham, T. J. Penfold, R. M. van der Veen, F. Lima, A. El Nahhas, S. L. Johnson, P. Beaud, R. Abela, C. Bressler, I. Tavernelli, C. J. Milne and M. Chergui, *J. Am. Chem. Soc.*, 2011, **133**, 12740–12748.
114 C. G. Elles, I. A. Shkrob, R. A. Crowell, D. A. Arms and E. C. Landahl, *J. Chem. Phys.*, 2008, **128**, 061102.
115 M. H. Rittmann-Frank, C. J. Milne, J. Rittmann, M. Reinhard, T. J. Penfold and M. Chergui, *Angew. Chem., Int. Ed.*, 2014, **53**, 5858–5862.
116 J. E. Katz, B. Gilbert, X. Y. Zhang, K. Attenkofer, R. W. Falcone and G. A. Waychunas, *J. Phys. Chem. Lett.*, 2010, **1**, 1372–1376.
117 J. Huang, O. Buyukcakir, M. W. Mara, A. Coskun, N. M. Dimitrijevic, G. Barin, O. Kokhan, A. B. Stickrath, R. Ruppert, D. M. Tiede, J. F. Stoddart, J. P. Sauvage and L. X. Chen, *Angew. Chem., Int. Ed.*, 2012, **51**, 12711–12715.
118 F. G. Santomauro, A. Lubcke, J. Rittmann, E. Baldini, A. Ferrer, M. Silatani, P. Zimmermann, S. Grubel, J. A. Johnson, S. O. Mariager, P. Beaud, D. Grolimund, C. Borca, G. Ingold, S. L. Johnson and M. Chergui, *Sci. Rep.*, 2015, **5**, 14834.
119 T. J. Penfold, J. Szlachetko, F. G. Santomauro, A. Britz, W. Gawelda, G. Doumy, A. M. March, S. H. Southworth, J. Rittmann, R. Abela, M. Chergui and C. J. Milne, *Nat. Commun.*, 2018, **9**, 478.
120 N. Huse, H. Cho, K. Hong, L. Jamula, F. M. F. de Groot, T. K. Kim, J. K. McCusker and R. W. Schoenlein, *J. Phys. Chem. Lett.*, 2011, **2**, 880–884.
121 K. Hong, H. Cho, R. W. Schoenlein, T. K. Kim and N. Huse, *Acc. Chem. Res.*, 2015, **48**, 2957–2966.

122 B. Winter and M. Faubel, *Chem. Rev.*, 2006, **106**, 1176–1211.
123 M. Ekimova, W. Quevedo, M. Faubel, P. Wernet and E. T. J. Nibbering, *Struct. Dyn.*, 2015, **2**, 054301.
124 J. D. Koralek, J. B. Kim, P. Brůža, C. B. Curry, Z. Chen, H. A. Bechtel, A. A. Cordones, P. Sperling, S. Toleikis and J. F. Kern, *Nat. Commun.*, 2018, **9**, 1353.
125 C. Kleine, M. Ekimova, G. Goldsztejn, S. Raabe, C. Strüber, J. Ludwig, S. Yarlagadda, S. Eisebitt, M. J. Vrakking and T. Elsaesser, *J. Phys. Chem. Lett.*, 2018, **10**, 52–58.
126 U. Gelius and K. Siegbahn, *Faraday Discuss.*, 1972, **54**, 257–268.
127 H. O. G. Siegbahn, *Prog. Colloid Polym. Sci.*, 1992, **88**, 146–153.
128 K. Siegbahn, *Philos. Trans. R. Soc., A*, 1970, **268**, 33–39.
129 M. Faubel, K. R. Siefermann, Y. Liu and B. Abel, *Acc. Chem. Res.*, 2012, **45**, 120–130.
130 J. Ojeda, C. A. Arrell, J. Grilj, F. Frassetto, L. Mewes, H. Zhang, F. van Mourik, L. Poletto and M. Chergui, *Struct. Dyn.*, 2016, **3**, 023602.
131 J. Ojeda, C. A. Arrell, L. Longetti, M. Chergui and J. Helbing, *Phys. Chem. Chem. Phys.*, 2017, **19**, 17052–17062.
132 L. Gallmann, I. Jordan, H. J. Worner, L. Castiglioni, M. Hengsberger, J. Osterwalder, C. A. Arrell, M. Chergui, E. Liberatore, U. Rothlisberger and U. Keller, *Struct. Dyn.*, 2017, **4**, 061502.
133 J. Hummert, G. Reitsma, N. Mayer, E. Ikonnikov, M. Eckstein and O. Kornilov, *J. Phys. Chem. Lett.*, 2018, **9**, 6649–6655.
134 K. R. Siefermann, C. D. Pemmaraju, S. Neppl, A. Shavorskiy, A. A. Cordones, J. Vura-Weis, D. S. Slaughter, F. P. Sturm, F. Weise, H. Bluhm, M. L. Strader, H. Cho, M.-F. Lin, C. Bacellar, C. Khurmi, J. Guo, G. Coslovich, J. S. Robinson, R. A. Kaindl, R. W. Schoenlein, A. Belkacem, D. M. Neumark, S. R. Leone, D. Nordlund, H. Ogasawara, O. Krupin, J. J. Turner, W. F. Schlotter, M. R. Holmes, M. Messerschmidt, M. P. Minitti, S. Gul, J. Z. Zhang, N. Huse, D. Prendergast and O. Gessner, *J. Phys. Chem. Lett.*, 2014, **5**, 2753–2759.
135 O. Gessner and M. Gühr, *Acc. Chem. Res.*, 2016, **49**, 138–145.
136 T. Arion, S. Neppl, F. Roth, A. Shavorskiy, H. Bluhm, Z. Hussain, O. Gessner and W. Eberhardt, *Appl. Phys. Lett.*, 2015, **106**, 121602.
137 S. Neppl, J. Mahl, A. S. Tremsin, B. Rude, R. Qiao, W. Yang, J. Guo and O. Gessner, *Faraday Discuss.*, 2016, **194**, 659–682.
138 A. S. Mereshchenko, P. K. Olshin, O. S. Myasnikova, M. S. Panov, V. A. Kochemirovsky, M. Y. Skripkin, P. N. Moroz, M. Zamkov and A. N. Tarnovsky, *J. Phys. Chem. A*, 2016, **120**, 1833–1844.
139 P. Baum, S. Lochbrunner and E. Riedle, *Opt. Lett.*, 2004, **29**, 1686–1688.
140 A. Cannizzo, *Phys. Chem. Chem. Phys.*, 2012, **14**, 11205–11223.
141 C. Consani, M. Premont-Schwarz, A. ElNahhas, C. Bressler, F. van Mourik, A. Cannizzo and M. Chergui, *Angew. Chem., Int. Ed.*, 2009, **48**, 7184–7187.
142 G. Aubock, C. Consani, R. Monni, A. Cannizzo, F. van Mourik and M. Chergui, *Rev. Sci. Instrum.*, 2012, **83**, 093105.
143 G. Aubock, C. Consani, F. van Mourik and M. Chergui, *Opt. Lett.*, 2012, **37**, 2337–2339.
144 M. Reinhard, G. Aubock, N. A. Besley, I. P. Clark, G. M. Greetham, M. W. D. Hanson-Heine, R. Horvath, T. S. Murphy, T. J. Penfold,

M. Towrie, M. W. George and M. Chergui, *J. Am. Chem. Soc.*, 2017, **139**, 7335–7347.

145 C. Consani, G. Aubock, F. van Mourik and M. Chergui, *Science*, 2013, **339**, 1586–1589.

146 R. Monni, A. Al Haddad, F. van Mourik, G. Auböck and M. Chergui, *Proc. Natl. Acad. Sci. U. S. A.*, 2015, **112**, 5602–5606.

147 D. Onidas, D. Markovitsi, S. Marguet, A. Sharonov and T. Gustavsson, *J. Phys. Chem. B*, 2002, **106**, 11367–11374.

148 T. Gustavsson, A. Sharonov, D. Onidas and D. Markovitsi, *Chem. Phys. Lett.*, 2002, **356**, 49–54.

149 T. Gustavsson, N. Sarkar, E. Lazzarotto, D. Markovitsi and R. Improta, *Chem. Phys. Lett.*, 2006, **429**, 551–557.

150 T. Gustavsson, A. Banyasz, R. Improta and D. Markovitsi, 2011.

151 W.-M. Kwok, C. Ma and D. L. Phillips, *J. Am. Chem. Soc.*, 2006, **128**, 11894–11905.

152 W. M. Kwok, C. Ma, M. W. George, D. C. Grills, P. Matousek, A. W. Parker, D. Phillips, W. T. Toner and M. Towrie, *Phys. Chem. Chem. Phys.*, 2003, **5**, 1043–1050.

153 W. Qiu, T. Li, L. Zhang, Y. Yang, Y.-T. Kao, L. Wang and D. Zhong, *Chem. Phys.*, 2008, **350**, 154–164.

154 J. A. Stevens, J. J. Link, Y. T. Kao, C. Zang, L. J. Wang and D. P. Zhong, *J. Phys. Chem. B*, 2010, **114**, 1498–1505.

155 A. Cannizzo, O. Bram, G. Zgrablic, A. Tortschanoff, A. A. Oskouei, F. van Mourik and M. Chergui, *Opt. Lett.*, 2007, **32**, 3555–3557.

156 G. Balakrishnan, C. L. Weeks, M. Ibrahim, A. V. Soldatova and T. G. Spiro, *Curr. Opin. Struct. Biol.*, 2008, **18**, 623–629.

157 H. Kuramochi, T. Fujisawa, S. Takeuchi and T. Tahara, *Chem. Phys. Lett.*, 2017, **683**, 543–546.

158 P. Hamm, M. Lim, W. F. DeGrado and R. M. Hochstrasser, *J. Chem. Phys.*, 2000, **112**, 1907–1916.

159 D. M. Jonas, *Annu. Rev. Phys. Chem.*, 2003, **54**, 425–463.

160 T. Brixner, I. V. Stiopkin and G. R. Fleming, *Opt. Lett.*, 2004, **29**, 884–886.

161 T. Brixner, J. Stenger, H. M. Vaswani, M. Cho, R. E. Blankenship and G. R. Fleming, *Nature*, 2005, **434**, 625–628.

162 E. L. Read, G. S. Engel, T. R. Calhoun, T. Mancal, T. K. Ahn, R. E. Blankenship and G. R. Fleming, *Proc. Natl. Acad. Sci. U. S. A.*, 2007, **104**, 14203–14208.

163 P. Hamm, J. Helbing and J. Bredenbeck, *Annu. Rev. Phys. Chem.*, 2008, **59**, 291–317.

164 P. Hamm and M. T. Zanni, *Concepts and methods of 2d infrared spectroscopy*, Cambridge University Press, Cambridge, New York, 2011.

165 A. Chenu and G. D. Scholes, *Annu. Rev. Phys. Chem.*, 2015, **66**, 69–96.

166 K. El Hage, S. Brickel, S. Hermelin, G. Gaulier, C. Schmidt, L. Bonacina, S. C. van Keulen, S. Bhattacharyya, M. Chergui, P. Hamm, U. Rothlisberger, J. P. Wolf and M. Meuwly, *Struct. Dyn.*, 2018, **5**, 061507.

167 D. Zigmantas, E. L. Read, T. Mancal, T. Brixner, A. T. Gardiner, R. J. Cogdell and G. R. Fleming, *Proc. Natl. Acad. Sci. U. S. A.*, 2006, **103**, 12672–12677.

168 A. V. Pisliakov, T. Mancal and G. R. Fleming, *J. Chem. Phys.*, 2006, **124**, 234505.

169 G. S. Engel, T. R. Calhoun, E. L. Read, T. K. Ahn, T. Mancal, Y. C. Cheng, R. E. Blankenship and G. R. Fleming, *Nature*, 2007, **446**, 782–786.
170 V. Butkus, A. Gelzinis, R. Augulis, A. Gall, C. Buchel, B. Robert, D. Zigmantas, L. Valkunas and D. Abramavicius, *J. Chem. Phys.*, 2015, **142**, 212414.
171 T. L. Courtney, Z. W. Fox, K. M. Slenkamp and M. Khalil, *J. Chem. Phys.*, 2015, **143**, 154201.
172 H. Dong, N. H. Lewis, T. A. Oliver and G. R. Fleming, *J. Chem. Phys.*, 2015, **142**, 174201.
173 T. L. Courtney, Z. W. Fox, L. Estergreen and M. Khalil, *J. Phys. Chem. Lett.*, 2015, **6**, 1286–1292.
174 N. Berova, K. j. Nakanishi and R. Woody, *Circular dichroism: principles and applications*, Wiley-VCH, New York, 2nd edn, 2000.
175 N. Berova, L. Di Bari and G. Pescitelli, *Chem. Soc. Rev.*, 2007, **36**, 914–931.
176 B. Nordén, A. Rodger and T. Dafforn, *Linear dichroism and circular dichroism*, The Royal Society of Chemistry, 2010.
177 T. Dartigalongue and F. Hache, *Chem. Phys. Lett.*, 2005, **415**, 313–316.
178 C. Niezborala and F. Hache, *J. Phys. Chem. A*, 2007, **111**, 7732–7735.
179 T. Dartigalongue, C. Niezborala and F. Hache, *Phys. Chem. Chem. Phys.*, 2007, **9**, 1611–1615.
180 M. Oppermann, B. Bauer, T. Rossi, F. Zinna, J. Helbing, J. Lacour and M. Chergui, *Optica*, 2019, **6**, 56–60.
181 J. R. Lakowicz, *Principles of fluorescence spectroscopy*, Springer, New York, 3rd edn, 2006.
182 S. V. Jovanovic, A. Harriman and M. G. Simic, *J. Phys. Chem.*, 1986, **90**, 1935–1939.
183 C. Shih, A. K. Museth, M. Abrahamsson, A. M. Blanco-Rodriguez, A. J. Di Bilio, J. Sudhamsu, B. R. Crane, K. L. Ronayne, M. Towrie and A. Vlček, *Science*, 2008, **320**, 1760–1762.
184 H. B. Gray and J. R. Winkler, *Proc. Natl. Acad. Sci. U. S. A.*, 2005, **102**, 3534–3539.
185 M. Cordes and B. Giese, *Chem. Soc. Rev.*, 2009, **38**, 892–901.
186 K. Takematsu, P. Pospíšil, M. Pižl, M. Towrie, J. Heyda, S. Záliš, J. T. Kaiser, J. R. Winkler, H. B. Gray and A. Vlček, *J. Phys. Chem. B*, 2019, **123**, 1578–1591.
187 K. Takematsu, H. R. Williamson, P. Nikolovski, J. T. Kaiser, Y. Sheng, P. Pospíšil, M. Towrie, J. Heyda, D. Hollas, S. Záliš, H. B. Gray, A. Vlček and J. R. Winkler, *ACS Cent. Sci.*, 2019, **5**, 192–200.
188 B. R. Crane, A. J. Di Bilio, J. R. Winkler and H. B. Gray, *J. Am. Chem. Soc.*, 2001, **123**, 11623–11631.
189 J. W. Petrich, J. W. Longworth and G. R. Fleming, *Biochemistry*, 1987, **26**, 2711–2722.
190 E. R. Henry and R. M. Hochstrasser, *Biophys. J.*, 1986, **49**, A60.
191 S. M. Janes, G. Holtom, P. Ascenzi, M. Brunori and R. M. Hochstrasser, *Biophys. J.*, 1987, **51**, 653–660.
192 C. Consani, O. Bram, F. van Mourik, A. Cannizzo and M. Chergui, *Chem. Phys.*, 2012, **396**, 108–115.
193 O. Bram, C. Consani, A. Cannizzo and M. Chergui, *J. Phys. Chem. B*, 2011, **115**, 13723–13730.
194 R. Monni, R. Horwarth, X.-Z. Sun, D. Tiemessen, Y. Zhao, M. Towrie, M. George and M. Chergui, *J. Am. Chem. Soc.*, 2019, submitted.

195 C. J. Suess, J. D. Hirst and N. A. Besley, *J. Comput. Chem.*, 2017, **38**, 1495–1502.
196 J. Briand, J. Léonard and S. Haacke, *J. Opt*, 2010, **12**, 084004.
197 P. Kurian, T. Obisesan and T. J. Craddock, *J. Photochem. Photobiol., B*, 2017, **175**, 109–124.
198 G. L. Celardo, M. Angeli, P. Kurian and T. Craddock, *New J. Phys.*, 2018, **21**, 023005.
199 C. T. Middleton, K. de La Harpe, C. Su, Y. K. Law, C. E. Crespo-Hernandez and B. Kohler, *Annu. Rev. Phys. Chem.*, 2009, **60**, 217–239.
200 B. Kohler, in *Ultrafast Dynamics at the Nanoscale*, Pan Stanford, 2017, pp. 21–82.
201 L. Hu, Y. Zhao, F. Wang, G. Chen, C. Ma, W.-M. Kwok and D. L. Phillips, *J. Phys. Chem. B*, 2007, **111**, 11812–11816.
202 J. J. Nogueira, F. Plasser and L. González, *Chem. Sci.*, 2017, **8**, 5682–5691.
203 A. Trifonov, M. Raytchev, I. Buchvarov, M. Rist, J. Barbaric, H. A. Wagenknecht and T. Fiebig, *J. Phys. Chem. B*, 2005, **109**, 19490–19495.
204 I. Buchvarov, Q. Wang, M. Raytchev, A. Trifonov and T. Fiebig, *Proc. Natl. Acad. Sci. U. S. A.*, 2007, **104**, 4794–4797.
205 D. Onidas, T. Gustavsson, E. Lazzarotto and D. Markovitsi, *J. Phys. Chem. B*, 2007, **111**, 9644–9650.
206 W.-M. Kwok, C. Ma and D. L. Phillips, *J. Phys. Chem. B*, 2009, **113**, 11527–11534.
207 A. Nenov, J. Segarra-Martí, A. Giussani, I. Conti, I. Rivalta, E. Dumont, V. K. Jaiswal, S. F. Altavilla, S. Mukamel and M. Garavelli, *Faraday Discuss.*, 2015, **177**, 345–362.
208 V. I. Prokhorenko, A. Picchiotti, M. Pola, A. G. Dijkstra and R. D. Miller, *J. Phys. Chem. Lett.*, 2016, **7**, 4445–4450.
209 J. Segarra-Martí, V. K. Jaiswal, A. J. Pepino, A. Giussani, A. Nenov, S. Mukamel, M. Garavelli and I. Rivalta, *Faraday Discuss.*, 2018, **207**, 233–250.
210 F. Buchner, B. Heggen, H.-H. Ritze, W. Thiel and A. Lübcke, *Phys. Chem. Chem. Phys.*, 2015, **17**, 31978–31987.
211 F. Buchner, H.-H. Ritze, J. Lahl and A. Lübcke, *Phys. Chem. Chem. Phys.*, 2013, **15**, 11402–11408.
212 B. A. West and A. M. Moran, *J. Phys. Chem. Lett.*, 2012, **3**, 2575–2581.
213 Q. Li, A. Giussani, J. Segarra-Martí, A. Nenov, I. Rivalta, A. A. Voityuk, S. Mukamel, D. Roca-Sanjuán, M. Garavelli and L. Blancafort, *Chem.–Eur. J.*, 2016, **22**, 7497–7507.
214 M. Chergui, *J. Chem. Phys.*, 2019, **150**, 070901.
215 J. Jiang and S. Mukamel, *Angew. Chem., Int. Ed.*, 2010, **49**, 9666–9669.
216 D. Abramavicius, J. Jiang, B. M. Bulheller, J. D. Hirst and S. Mukamel, *J. Am. Chem. Soc.*, 2010, **132**, 7769–7775.
217 J. Jiang and S. Mukamel, *Phys. Chem. Chem. Phys.*, 2011, **13**, 2394–2400.
218 W. Hua, S. Oesterling, J. D. Biggs, Y. Zhang, H. Ando, R. de Vivie-Riedle, B. P. Fingerhut and S. Mukamel, *Struct. Dyn.*, 2016, **3**, 023601.
219 S. E. Canton, K. S. Kjaer, G. Vanko, T. B. van Driel, S. I. Adachi, A. Bordage, C. Bressler, P. Chabera, M. Christensen, A. O. Dohn, A. Galler, W. Gawelda, D. Gosztola, K. Haldrup, T. Harlang, Y. Z. Liu, K. B. Moller, Z. Nemeth, S. Nozawa, M. Papai, T. Sato, T. Sato, K. Suarez-Alcantara, T. Togashi, K. Tono, J. Uhlig, D. A. Vithanage, K. Warnmark, M. Yabashi, J. X. Zhang, V. Sundstrom and M. M. Nielsen, *Nat. Commun.*, 2015, **6**, 6359.

220 M. Delor, T. Keane, P. A. Scattergood, I. V. Sazanovich, G. M. Greetham, M. Towrie, A. J. Meijer and J. A. Weinstein, *Nat. Chem.*, 2015, **7**, 689.
221 O. Bixner, V. Lukeš, T. Mančal, J. Hauer, F. Milota, M. Fischer, I. Pugliesi, M. Bradler, W. Schmid and E. Riedle, *J. Chem. Phys.*, 2012, **136**, 204503.
222 X. Y. Chen and S. E. Bradforth, *Annu. Rev. Phys. Chem.*, 2008, **59**, 203–231.
223 V. T. Pham, I. Tavernelli, C. J. Milne, R. M. van der Veen, P. D'Angelo, C. Bressler and M. Chergui, *Chem. Phys.*, 2010, **371**, 24–29.
224 F. Messina, O. Bram, A. Cannizzo and M. Chergui, *Nat. Commun.*, 2013, **4**, 2119.
225 M. Reinhard, T. J. Penfold, F. A. Lima, J. Rittmann, M. H. Rittmann-Frank, R. Abela, I. Tavernelli, U. Rothlisberger, C. J. Milne and M. Chergui, *Struct. Dyn.*, 2014, **1**, 024901.
226 M. Oppermann, J. Spekowius, B. Bauer, R. Pfister, M. Chergui and J. Helbing, *J. Phys. Chem. Lett.*, 2019, **10**, 2700–2705.
227 M. Gratzel, *Nature*, 2001, **414**, 338–344.
228 M. Gratzel, *Acc. Chem. Res.*, 2009, **42**, 1788–1798.
229 A. Cannizzo, F. van Mourik, W. Gawelda, G. Zgrablic, C. Bressler and M. Chergui, *Angew. Chem., Int. Ed.*, 2006, **45**, 3174–3176.
230 O. Bräm, F. Messina, A. El-Zohry, A. Cannizzo and M. Chergui, *Chem. Phys.*, 2012, **393**, 51–57.
231 J. Kallioinen, V. Lehtovuori, P. Myllyperkio and J. Korppi-Tommola, *Chem. Phys. Lett.*, 2001, **340**, 217–221.
232 R. Huber, J. E. Moser, M. Gratzel and J. Wachtveitl, *J. Phys. Chem. B*, 2002, **106**, 6494–6499.
233 J. Kallioinen, G. Benko, P. Myllyperkio, L. Khriachtchev, B. Skarman, R. Wallenberg, M. Tuomikoski, J. Korppi-Tommola, V. Sundstrom and A. P. Yartsev, *J. Phys. Chem. B*, 2004, **108**, 6365–6373.
234 J. Kallioinen, G. Benko, V. Sundstrom, J. E. I. Korppi-Tommola and A. P. Yartsev, *J. Phys. Chem. B*, 2002, **106**, 4396–4404.
235 G. Benko, P. Myllyperkio, J. Pan, A. P. Yartsev and V. Sundstrom, *J. Am. Chem. Soc.*, 2003, **125**, 1118–1119.
236 P. Myllyperkio, G. Benko, J. Korppi-Tommola, A. P. Yartsev and V. Sundstrom, *Phys. Chem. Chem. Phys.*, 2008, **10**, 996–1002.
237 J. Budarz, F. G. Santomauro, M. H. Rittmann-Frank, C. J. Milne, T. Huthwelker, D. Grolimund, J. Rittmann, D. Kinschel, T. Rossi and M. Chergui, *Chimia*, 2017, **71**, 768–772.
238 E. Baldini, T. Palmieri, T. Rossi, M. Oppermann, E. Pomarico, G. Auböck and M. Chergui, *J. Am. Chem. Soc.*, 2017, **139**, 11584–11589.
239 J. E. Katz, X. Y. Zhang, K. Attenkofer, K. W. Chapman, C. Frandsen, P. Zarzycki, K. M. Rosso, R. W. Falcone, G. A. Waychunas and B. Gilbert, *Science*, 2012, **337**, 1200–1203.
240 Y. Obara, H. Ito, T. Ito, N. Kurahashi, S. Thürmer, H. Tanaka, T. Katayama, T. Togashi, S. Owada, Y.-i. Yamamoto, S. Karashima, J. Nishitani, M. Yabashi, T. Suzuki and K. Misawa, *Struct. Dyn.*, 2017, **4**, 044033.
241 A. Janotti and C. G. Van de Walle, *Phys. Rev. B: Condens. Matter Mater. Phys.*, 2007, **76**, 165202.
242 E. Baldini, L. Chiodo, A. Dominguez, M. Palummo, S. Moser, M. Yazdi-Rizi, G. Aubock, B. P. P. Mallett, H. Berger, A. Magrez, C. Bernhard, M. Grioni, A. Rubio and M. Chergui, *Nat. Commun.*, 2017, **8**, 13.

243 E. Baldini, T. Palmieri, A. Dominguez, P. Ruello, A. Rubio and M. Chergui, *Nano Lett.*, 2018, **18**, 5007–5014.

244 E. Baldini, T. Palmieri, E. Pomarico, G. Aubock and M. Chergui, *ACS Photonics*, 2018, **5**, 1241–1249.

245 H. Nemec, J. Rochford, O. Taratula, E. Galoppini, P. Kuzel, T. Polivka, A. Yartsev and V. Sundstrom, *Phys. Rev. Lett.*, 2010, **104**, 197401.

246 J. E. Moser, M. Wolf, F. Lenzmann and M. Gratzel, *Z. Phys. Chem.*, 1999, **212**, 85–92.

247 S. Mukamel, D. Healion, Y. Zhang and J. D. Biggs, *Annu. Rev. Phys. Chem.*, 2013, **64**, 101–127.

248 D. Healion, Y. Zhang, J. D. Biggs, W. Hua and S. Mukamel, *Struct. Dyn.*, 2014, **1**, 014101.

Faraday Discussions

PAPER

Transient synchronisation and quantum coherence in a bio-inspired vibronic dimer

Stefan Siwiak-Jaszek* and Alexandra Olaya-Castro *

Received 13th January 2019, Accepted 17th January 2019
DOI: 10.1039/c9fd00006b

Synchronisation is a collective phenomenon widely investigated in classical oscillators and, more recently, in quantum systems. However it remains unclear what features distinguish synchronous behaviour in these two scenarios. Recent works have shown that investigating synchronisation dynamics in open quantum systems can give insight into this issue. Here we study transient synchronisation in a bio-inspired vibronic dimer, where electronic excitation dynamics is mediated by coherent interactions with intramolecular vibrational modes. We show that the synchronisation dynamics of local mode displacements exhibit a rich behaviour which arises directly from the distinct time-evolutions of different vibronic quantum coherences. Furthermore, our study shows that coherent energy transport in this bio-inspired system is concomitant with the emergence of positive synchronisation between mode displacements. Our work provides further understanding of the relations between quantum coherence and synchronisation in open quantum systems and suggests an interesting role for coherence in biomolecules, that of promoting synchronisation of vibrational motions driven out of thermal equilibrium.

1 Introduction

Synchronisation can be broadly defined as the adjustment of rhythms of oscillating objects due to their weak interaction.[1] This description corresponds to numerous processes throughout the natural world that occur on a wide range of length and time scales. This is especially true in living systems, where synchronisation is common and is often closely related to biological function.[2,3] On the metre scale, the synchronous flashing of male fireflies aids each individual in reproduction[2] and on the micrometre scale, cardiac cells synchronise contractions with their neighbours.[3] The question of whether this phenomenon persists on even smaller length scales such as that of individual biomolecules (nanometre) and time scales such as the relaxation times of intramolecular motions (picoseconds) has not been investigated. On these length and time scales quantum phenomena cannot be neglected and synchronisation, if it occurs, could exhibit features that have no equivalence in the classical regime.[4–9]

Department of Physics and Astronomy, University College London, London WC1E 6BT, UK. E-mail: Stefan.siwiak-jaszek.11@ucl.ac.uk; a.olaya@ucl.ac.uk

An interesting biophysical scenario in which one can explore synchronisation in the quantum regime, and at the same time investigate its possible relations to coherence and biological function, is during electronic excitation transport in photosynthetic complexes,[10,11] where coherent dynamics has been observed lasting a few hundred of femtoseconds.[12–17] While there remains some controversy on such observations,[18] the leading hypothesis on the mechanism underlying long-lived coherent dynamics in these biophysical systems is the quantum mechanical exchange of energy between excitonic and vibrational degrees of freedom.[16,17,19–26] Hence there is a widespread interest in understanding the intertwined dynamics of electronic and vibrational motions during energy transfer. In this context, it is timely and interesting to investigate whether molecular vibrations may be synchronised during ultra-fast energy transfer processes in photosynthetic complexes and if such synchronisation happens, what its relations to coherence and function could be.

The form of synchronisation that may emerge in the ultra-fast energy transfer scenario should be fully described as transient, emphasising that synchronous behaviour occurs before relaxation to the ground state, and spontaneous, meaning that synchronisation happens solely due to the interactions within the quantum system considered and not due to the influence of an external fixed-frequency driving force.[27] This differs from other efforts to define synchronisation in self-sustained quantum oscillators.[4,8]

Transient spontaneous quantum synchronisation – referred from hereon as synchronisation – has been recently investigated in a range of open quantum systems.[27–31] In the simplest case of coupled two-level systems (TLSs), it has been shown that synchronisation cannot occur in the presence of dephasing channels alone and dissipation appears to be essential.[28] In quantum harmonic oscillator (QHO) networks, adjusting the coupling between oscillators can create asymptotically synchronised states that avoid dissipation altogether;[31] and in hybrid TLS-QHO systems the form of coupling of QHOs to TLSs can induce and control synchronisation between QHOs with a well-defined phase difference.[30] From these works it can be concluded that the specific forms of the decoherence channels and of the interactions among quantum subsystems play a determinant role in reaching synchronisation. However exactly the interplay between coherent dynamics and decoherence enables synchronisation in a particular time scale, or how coherences may relate to synchronisation, is not fully understood. Here, we show that the investigation of synchronisation in a bio-inspired vibronic dimer provides interesting insights on these open questions.

Specifically, we investigate the synchronisation of molecular motions during electronic energy transport in a prototype photosynthetic vibronic dimer where two local excited electronic states interact with each other and with local quasi-coherent harmonic intramolecular motions. The harmonic motions are subject to local dissipation into thermal baths while the electronic subsystem undergoes pure dephasing. We provide an intuitive explanation of the mechanism behind synchronisation by analysing the eigenstate coherences that influence the oscillatory patterns of the local displacements being synchronised. Our results indicate that interferences between vibronic coherences can result in negative and positive synchronisation. The periods of negative and positive synchronisation reflect a rich and distinctive interplay between coherence and decoherence mechanisms at different time-scales. Furthermore, we show that a faster onset of synchronisation is correlated with a larger degree of coherent excitation transport. From a biological standpoint, our study suggests local-mode synchronisation may be present during energy transfer in some photosynthetic pigment–protein complexes at physiological temperatures.

2 Modelling synchronisation in exciton-vibration dimers

In the following sections we describe the model and methods used. In Section 2.1 we introduce the Hamiltonian for the exciton-vibration dimer model that can represent dimers in a variety of photosynthetic pigment–protein complexes. In Section 2.2 we describe the Markovian master equation that phenomenologically describes the open quantum system dynamics of the exciton-vibration dimer and briefly discuss our numerical methods used to solve it. In Section 2.3 we describe the measure used to quantify synchronisation and discuss its limitations.

2.1 The exciton-vibration dimer

Photosynthetic light-harvesting proteins exhibit complex excitation transfer (ET) dynamics due to an overlap of energy scales in electronic and vibrational degrees of freedom and the timescales of their associated coherent and incoherent processes.[10] Experiments and theory suggest that these complexes are capable of sustaining quantum coherence at room temperature that last a few hundred of femtoseconds, thereby suggesting a functional role for coherent dynamics.[10] Here we consider a prototype light-harvesting unit, which we called an exciton-vibration dimer model (Fig. 1), formed by a pair of chromophores whose local electronic excitations interact with quasi-coherent vibrational modes.[19,23]

Let the chromophores have single excited states $|e_{i=1,2}\rangle$ of energy $e_{i=1,2}$ interacting *via* dipole–dipole coupling of strength V and each locally coupled to a quantised intramolecular vibration of energy $\omega_{i=1,2}$ with strength $g_{i=1,2}$. The system has a total Hamiltonian of the form: $H = H_{\mathrm{el}} + H_{\mathrm{vib}} + H_{\mathrm{el-vib}}$. The electronic Hamiltonian reads

$$H_{\mathrm{el}} = e_1|e_1\rangle\langle e_1| + e_2|e_2\rangle\langle e_2| + V(|e_2\rangle\langle e_1| + |e_1\rangle\langle e_2|) \tag{1}$$

and its eigenstates are delocalised electronic states known as excitons $|E_{d=1,2}\rangle$ with energies:

$$\begin{aligned} E_1 &= \frac{1}{2}\left(e_1 + e_2 - \sqrt{\Delta e^2 + 4V^2}\right) \\ E_2 &= \frac{1}{2}\left(e_1 + e_2 + \sqrt{\Delta e^2 + 4V^2}\right), \end{aligned} \tag{2}$$

where $\Delta e = e_2 - e_1$. The unitary rotation $U(\theta)$ which diagonalises H_{el} to create an excitonic Hamiltonian $H_{\mathrm{exc}} = U(\theta)\ H_{\mathrm{el}}\ U^{\dagger}\ (\theta) = E_1|E_1\rangle\langle E_1| + E_2|E_2\ \rangle\langle E_2|$ has the form:[32]

$$U = \begin{pmatrix} \cos\theta & \sin\theta \\ -\sin\theta & \cos\theta \end{pmatrix} \tag{3}$$

where $\theta = \frac{1}{2}\arctan(2V/|\Delta e|)$ is referred to as the mixing angle and is an effective measure of delocalisation of the electronic sub-system or the exciton size.

The bare vibrational Hamiltonian reads:

$$H_{\mathrm{vib}} = \omega_1 b_1^\dagger b_1 + \omega_2 b_2^\dagger b_2: \tag{4}$$

where $b_{i=1,2}^\dagger(b_{i=1,2})$ are the creation (annihilation) operators for the modes. The eigenstates of H_{vib} are Fock states given by the tensor product of eigenstates of mode 1 and mode 2: $|n_1\rangle \otimes |n_2\rangle$.

The interaction Hamiltonian reads $H_{\mathrm{el-vib}} = g_1|e_1\rangle\langle e_1|(b_1 + b_1^\dagger) + g_2|e_2\rangle\langle e_2|(b_2 + b_2^\dagger)$ where we have assumed a linear coupling between mode position and electronic states.[33,34] We define the electronic operators in the exciton basis as $\Theta_i = U(\theta)\ |e_i\rangle\langle e_i|\ U^\dagger\ (\theta)$ and insert into $H_{\mathrm{el-vib}}$ to give:

$$H_{\mathrm{exc-vib}} = g_1\Theta_1(b_1 + b_1^\dagger) + g_2\Theta_2(b_2 + b_2^\dagger) \tag{5}$$

The final exciton-vibration Hamiltonian is then

$$H = E_1|E_1\rangle\langle E_1| + E_2|E_2\rangle\langle E_2| + \omega_1 b_1^\dagger b_1 + \omega_2 b_2^\dagger b_2 + g_1\Theta_1 X_1 + g_2\Theta_2 X_2: \tag{6}$$

where we have introduced the position operator for each mode $X_{i=1,2} = b_i + b_i^\dagger$. In the joint basis of exciton and local vibrational states, the eigenstates of H can be written as:

$$\begin{aligned}|\psi_\mathrm{j}\rangle &= \sum_{\mathrm{d}=1,2}\alpha_\mathrm{d}|\mathrm{E_d}\rangle \otimes \sum_{\mathrm{n_1}=1}^{\mathrm{M}}\beta_{\mathrm{n_1}}|\mathrm{n_1}\rangle \otimes \sum_{\mathrm{n_1}=1}^{\mathrm{M}}\gamma_{\mathrm{n_2}}|\mathrm{n_2}\rangle \\ &= \sum_{d,n_1,n_2} c(d,n_1,n_2)|E_d,n_1,n_2\rangle.\end{aligned} \tag{7}$$

To obtain convergent dynamics we account for a maximum occupation $M = 8$ in each mode. Eigenstates $|\psi_j\rangle$ are ordered in ascending energy.

In this paper we investigate synchronisation of local vibration displacements assuming equal frequencies $\omega_1 = \omega_2 = \omega$ and identical coupling strengths $g_1 = g_2 = g$. We also consider the regime of weak electronic coupling where $\Delta E \approx \omega > g > V$, which is characteristic of chromophore pairs present in a variety of light-harvesting proteins.[13,19,20,24,35–38] In this regime excitons are not fully delocalised and excitonic energies are close to the local energies. The resultant quasi-localised nature of eigenstates gives validity to our analysis of synchronisation of local mode displacements.

2.2 Open quantum system model

For simplicity, we assume Markovian relaxation processes described in the Lindblad form:

$$\dot{\rho}(t) = -i[H,\rho(t)] + D_{\mathrm{deph}}[\rho(t)] + D_{\mathrm{th}}[\rho(t)], \tag{8}$$

where $\rho(t)$ is the exciton-vibration density matrix and the Lindblad-form superoperators $D_\mathrm{v}[\rho] = \Gamma_\mathrm{v}\left(O_\mathrm{v}\rho O_\mathrm{v}^\dagger - \frac{1}{2}\rho O_\mathrm{v}^\dagger O_\mathrm{v} - \frac{1}{2}O_\mathrm{v}^\dagger O_\mathrm{v}\rho\right)$ for the system transition operator O_V at rate Γ_V. We assume local pure dephasing processes on the

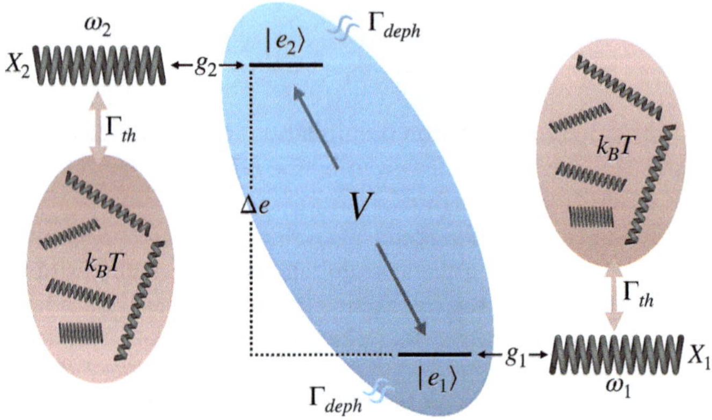

Fig. 1 Schematic diagram of the exciton-vibration dimer. Two chromophores (subscripts 1 and 2) with single excited states $|e_j\rangle$ interact *via* dipole–dipole coupling of strength V. Each electronic state is coupled linearly with strength g_i to a harmonic mode of energy ω_i. Electronic sub-system (blue oval) experiences local pure dephasing at rate Γ_{deph}, each mode dissipates into separate thermal baths (red ovals) of temperature k_BT at rate Γ_{th}.

electronic sub-system[39,40] with transition operators $|e_1\rangle\langle e_1|$ and $|e_2\rangle\langle e_2|$ at equal rates of $\Gamma_{\text{deph}} = [0.1 \text{ ps}]^{-1}$ such that exciton coherence decays in approximately 0.5 ps as inferred from experimental evidence of algal photosynthetic protein PC645.[20] Each mode is assumed to undergo relaxation[40] due to thermal reservoirs at temperature 298 K (207.1 cm^{-1}), with associated transition operators b_1 and b_2 at rate $\Gamma_{\text{th}}(1 + B)$ and jump operators $b_1^\dagger$ and $b_2^\dagger$ at rate $\Gamma_{\text{th}}B$. Here $B = \left(e^{\frac{\omega}{k_BT}} - 1\right)^{-1}$ is the mean number of quanta in a thermally occupied mode of frequency ω and $\Gamma_{\text{th}} = [1 \text{ ps}]^{-1}$ is the mode thermal dissipation rate. Table 1 summarises the parameters considered.

To proceed numerically we linearise the master equation:

$$|\dot{\rho}(t)\rangle\rangle = \mathscr{L}\,|\rho(t)\rangle, \tag{9}$$

where $\mathscr{L}$ is the Liouvillian superoperator and $|\rho(t)\rangle\rangle$ are flattened density matrices. We solve this ordinary differential equation in Python 3 with packages NumPy and SciPy and author generated scripts.

Since we are interested in the process of synchronisation during energy transfer, we fix the initial electronic state to be the higher energy excitonic state $|E_2\rangle$ and assume that both intramolecular modes are initially in thermal equilibrium with their respective baths. This results in the initial state:

Table 1 Parameters used for numerical calculations representing the central dimer in the cryptophyte antennae PE545 (PEB$_{50/61}$).[19,38] Units in spectroscopic wavenumbers cm^{-1} except final two columns

Δe	V	ω	g	k_BT	Γ_{th}	Γ_{deph}
1042	92	1111	267.1	207.1	$[1 \text{ ps}]^{-1}$	$[0.1 \text{ ps}]^{-1}$

$$\rho(0) = |E_2\rangle\langle E_2| \otimes \rho_1^{\mathrm{th}} \otimes \rho_2^{\mathrm{th}}: \tag{10}$$

where $\rho_i^{\mathrm{th}} = \sum_{n_i} P_{n_i}|n_i\rangle\langle n_i|$ and $P_{n_i} = \left(1 - e^{\frac{-\omega}{k_B T}}\right) e^{\frac{-n_i\omega}{k_B T}}$. In analogy of classical studies of synchronisation, here we focus on the expected values of the position operators X_i.

2.3 Measuring synchronisation

To quantify synchronisation between signals we adopt a commonly used measure based on the Pearson correlation factor.[41] Defined generally for any two time dependent functions $f_1(t)$ and $f_2(t)$:

$$C_{f_1,f_2}(t|\Delta t) = \frac{\int_t^{t+\Delta t} \delta f_1 \delta f_2 \mathrm{d}t}{\left(\int_t^{t+\Delta t} \delta {f_1}^2 \mathrm{d}t \int_t^{t+\Delta t} \delta {f_2}^2 \mathrm{d}t\right)^{1/2}} \tag{11}$$

where $\delta f = f - \bar{f}, \bar{f} = \frac{1}{\Delta t}\int_t^{t+\Delta t} f(t')\mathrm{d}t'$ is a time average and Δt is the averaging window. This function returns a continuous value in the range of -1 to 1 corresponding to a phase shift between f_1 and f_2 of π to 0. We choose this function for its time dependent quantitative measure of synchronisation and for its wide use in the quantum synchronisation literature.[28,29,31,42] An analysis of different synchronisation measures[41] states that the Pearson correlation factor returns a value of 1 for positive synchronisation (in-phase), -1 for negative synchronisation (π out of phase) and 0 for asynchrony. Here however we note that, by an appropriate choice of Δt to be as close as possible to time period of the dominant frequency in $f_{1,2}$, the synchronisation measure in fact indicates the phase difference between two oscillating signals. Therefore any constant value of $C_{f_1,f_2}(t)$ corresponds to a constant phase between the signals *e.g.* $C_{f_1,f_2}(t) = 0$ corresponds to a phase difference of $\pi/2$ at time t. A characterisation of this relationship is presented in Fig. 2 where the value of the synchronisation function is plotted as a function of constant phase difference, ϕ, between two perfect sinusoids. Based on this, we define our condition for synchronisation as a constant value in $C_{f_1,f_2}(t)$ over time.

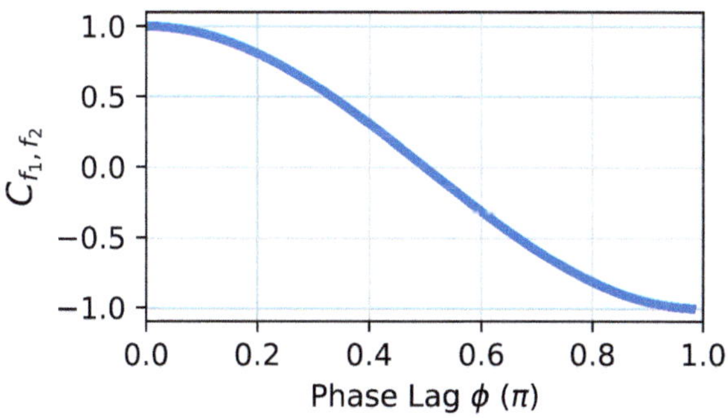

Fig. 2 Value of synchronisation measure C_{f_1,f_2} for two identical sinusoids as a function of their phase lag ϕ. $f_1 = \sin(at)$, $f_2 = \sin(at + \phi)$, and $\Delta t = 1/a$ yield $C_{f_1,f_2} = \cos(\phi)$.

3 Results

In the following sections we present and explain the main findings of this paper. In Section 3.1 we introduce the exciton-vibration coherences description of synchronisation by considering the purely coherent dynamics of the dimer. In Section 3.2 we show that when dissipative processes are included, spontaneous synchronisation of the displacements of intramolecular modes in exciton-vibration dimers emerges and it is accompanied by a negatively synchronised transient. We illustrate how this synchronisation can be understood as the dominance of a specific exciton-vibration coherence over a set of competing coherences that contribute to oscillatory dynamics in the displacement of the modes. Finally, we provide a qualitative explanation for the dominance of one coherence over others. In Section 3.3 we demonstrate a correlation between coherent energy transfer and the time taken to reach positive synchronisation.

3.1 Closed system dynamics and the conditions for synchronisation

In order to understand the dynamical emergence of synchronisation between the positions of intramolecular modes we begin by exploring the time dependence of the expected values of the position operators, $\langle X_i(t)\rangle$.

In the basis of system eigenstates, the density matrix of the exciton-vibration systems reads $\rho(t) = \sum_{j,k}\rho_{jk}(t)|\psi_j\rangle\langle\psi_k|$ where $|\psi_j\rangle$ are the exciton-vibration eigenstates and $\rho_{jk}(t) = \langle\psi_j|\rho(t)|\psi_k\rangle$. The expected value $\langle X_i(t)\rangle$ becomes

$$\begin{aligned}\langle X_i(t)\rangle &= Tr\{X_i\rho(t)\}\\ &= \sum_l\langle\psi_l|\left(X_i\sum_{j,k}\rho_{jk}(t)|\psi_j\rangle\langle\psi_k|\right)|\psi_1\rangle\\ &= \sum_{j,k}\rho_{jk}(t)\langle\psi_k|X_i|\psi_j\rangle\\ &= \sum_{j,k}\rho_{jk}(t)X_{i,kj}\end{aligned} \tag{12}$$

where $X_{i,kj} = \langle\psi_k|X_i|\psi_j\rangle$. Eqn (12) indicates that the time-evolution of local displacements is given by the matrix elements $\rho_{jk}(t)$, yet the ability to exhibit synchronised behaviour depends critically on the form of $X_{i,kj}$ as it is the only source of difference between $\langle X_1(t)\rangle$ and $\langle X_2(t)\rangle$.

To explore the consequences of eqn (12) in more detail we analyse whether synchronisation can occur in the closed quantum system whose dynamics is solely given by the system Hamiltonian H. Let us denote $\rho^H(t)$ the density matrix of the close system, which becomes:

$$\rho^H(t) = \sum_{j,k}\rho_{jk}(0)e^{i\Omega_{kj}t}|\psi_j\rangle\langle\psi_k| \tag{13}$$

Here $\rho_{jk}(0) = \langle\psi_j|\rho(0)|\psi_k\rangle$ are the populations ($j = k$) and coherences ($j \neq k$) of the initial state while $\Omega_{kj} = \epsilon_k - \epsilon_j$ where ϵ_j are the eigenenergies of H. One can therefore see that in close quantum dynamics only coherences will contribute oscillatory components, with specific frequencies Ω_{kj}, to the dynamics of $\langle X_i(t)\rangle$.

To exhibit synchronisation we require oscillations in $\langle X_1(t)\rangle$ to align in frequency and phase with $\langle X_2(t)\rangle$ for an extended period of time. Using eqn (12) and (13) we require the equality:

$$\sum_{j,k}\rho_{jk}(0)e^{i\Omega_{jk}t}X_{1,kj} = \sum_{j,k}\rho_{jk}(0)e^{i\Omega_{jk}t}X_{2,kj} \tag{14}$$

for the oscillating components $j \neq k$. Eqn (14) is trivially true if the initial state is an eigenstate of H. However this cannot be considered synchronised as there will not be time evolution at all. Hence, eqn (14) can only be satisfied (and be non-zero) if the initial state $\rho(0)$ is such that $\rho_{jk}(0)$ is zero when $X_{1,kj} \neq X_{2,kj}$. The only $\rho(0)$ that can fulfil these criteria contains either a single (or specific combination of) eigenstate coherence(s) $|\psi_j\rangle\langle\psi_k|$.

For the system under consideration, we identify the seven largest position matrix elements $X_{i,kj}$ contributing to the position dynamics and present their associated exciton-vibration coherences $|\psi_j\rangle\langle\psi_k|$ and frequencies, Ω_{kj}, in Table 2. Examining the values of $X_{i,kj}$ in Table 2 shows that, in the parameter regime considered, *i.e.* $\omega_1 = \omega_2$ (see also Section 4 for more details), the position matrix elements fall into two distinct groups: those for which $X_{1,kj} = X_{2,kj}$ and those for which $X_{1,kj} = -X_{2,kj}$. For the latter set, the amplitude scaling of -1 results in a phase factor of π between $\langle X_1(t)\rangle$ and $\langle X_2(t)\rangle$. For example a $\rho(0)$ consisting of $|\psi_1\rangle\langle\psi_5|$ and $|\psi_3\rangle\langle\psi_7|$ and no other coherences result in oscillations of frequency 1111.0 cm^{-1} with no phase separation between $\langle X_1(t)\rangle$ and $\langle X_2(t)\rangle$. Similarly, a combination of only $|\psi_1\rangle\langle\psi_4|$ and $|\psi_3\rangle\langle\psi_8|$ would result in oscillations between 1102.6 cm^{-1} and 1109.2 cm^{-1} except with a constant phase separation of π. Both of these scenarios can be identified as synchronised as there is a constant phase difference over time: in the former displacements are positively synchronised and in the latter they are negatively synchronised. A combination of the two groups of coherences however would create interferences yielding a cyclic phase change between $\langle X_i(t)\rangle$.

Whilst it is possible to find an initial state in which the dynamics of $\langle X_i(t)\rangle$ evolve in a synchronised way, this cannot be classed as spontaneous synchronisation as it has not emerged from an initially non-synchronised state. Furthermore, in the closed system evolution, the ratio of these coherences is determined only by the initial state $\rho(0)$ thereby fixing the frequency composition and indicating there is no emergence of synchronisation. A change in the ratio of coherences can only be induced by dissipation. Therefore we conclude that

Table 2 Seven largest amplitude exciton-vibration coherences. Top row is the associated oscillation frequency from eqn (12). The remaining rows are the matrix elements corresponding to coherence $|\psi_j\rangle\langle\psi_k|$ of different operators which represent: the coupling to position of mode 1, coupling to position of mode 2, coupling to inter-exciton coherence, and coupling to ground state of both modes

	$\lvert\psi_0\rangle\langle\psi_2\rvert$	$\lvert\psi_0\rangle\lvert\psi_3\rangle$	$\lvert\psi_1\rangle\langle\psi_4\rvert$	$\lvert\psi_1\rangle\langle\psi_5\rvert$	$\lvert\psi_3\rangle\langle\psi_7\rvert$	$\lvert\psi_3\rangle\langle\psi_8\rvert$	$\lvert\psi_1\rangle\langle\psi_3\rvert$
Ω_{kj} (cm^{-1})	1111.0	1125.0	1102.6	1111.0	1111.0	1119.2	81.0
$\langle\psi_k\lvert X_1\rvert\psi_j\rangle$	0.707	−0.637	0.767	0.707	0.707	−0.935	−0.174
$\langle\psi_k\lvert X_2\rvert\psi_j\rangle$	0.707	0.637	−0.767	0.707	0.707	0.935	0.174
$\langle\psi_k\lvert\sigma_x\rvert\psi_j\rangle$	0.000	0.385	0.340	0.000	0.000	0.384	0.196
$\langle\psi_k\lvert(\lvert 0\rangle\langle 0\rvert_1 \otimes \lvert 0\rangle\langle 0\rvert_2)\rvert\psi_j\rangle$	0.161	−0.144	−0.131	0.133	0.032	0.026	−0.351

synchronisation cannot emerge in the closed quantum system. This analysis concurs with previous studies of coupled TLSs which show that synchronisation cannot occur in the presence of dephasing alone and some energy loss is required.[28] In the presence of dissipation we would expect one (or a set of) coherence(s) to emerge with a significantly larger amplitude than the others allowing it to dominate the dynamics of $\langle X_i(t)\rangle$ and produce a constant phase difference in $\langle X_i(t)\rangle$.

Fig. 3a displays the numerical results of evolution of $\langle X_i(t)\rangle$ and the correlation measure $C_{\langle X_1\rangle\langle X_2\rangle}(t)$ in the closed system with initial state $\rho(0)$ (eqn (10)). As expected, we see a large range of frequency oscillations in $\langle X_i(t)\rangle$. The value of $C_{\langle X_1\rangle\langle X_2\rangle}(t)$ changes in a cyclic pattern, indicating a continuous change in phase between the oscillations and a clear difference in frequency compositions. The frequency components in each $\langle X_i(t)\rangle$ can be resolved by taking the real part of the Fourier Transform (FT) which we present in Fig. 3b. We note here how the frequencies present are exactly those in Table 2 and that the frequencies that correspond to negative synchronisation can be clearly seen as those which have opposite sign in the FT.

Fig. 3c displays the short-time dynamics of the real parts of the five dominant exciton-vibration coherences, weighted by their associated position matrix

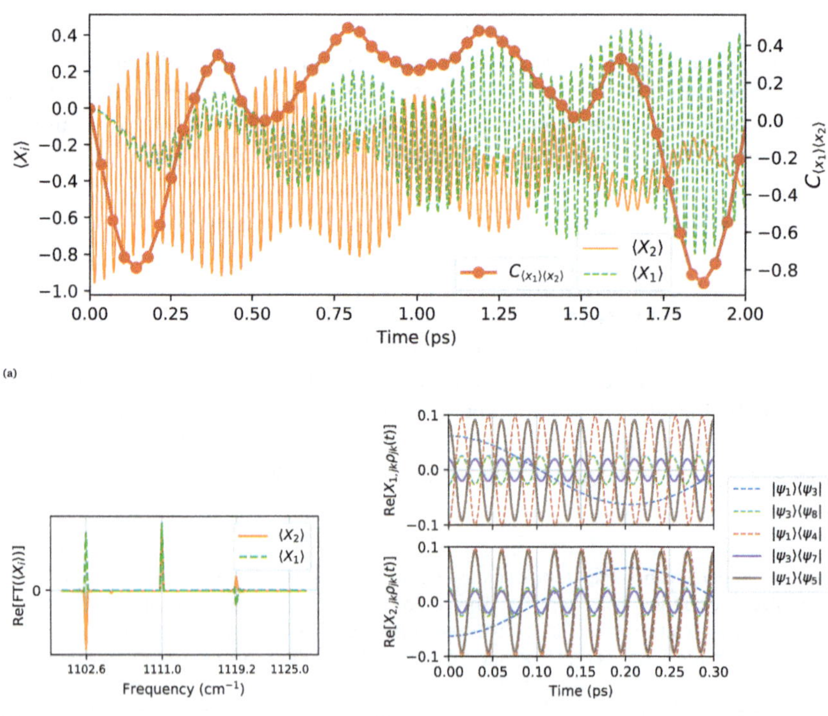

Fig. 3 Coherent evolution of excited vibronic dimer. Initial state is eqn (10) and relevant parameters are listed in Table 1 (closed system evolution so no dephasing or dissipation). (a) Evolution of expectation value of intramolecular mode displacements $\langle X_i(t)\rangle$ and their synchronisation $C_{\langle X_1\rangle,\langle X_2\rangle}(t)$. (b) Real part of Fourier transform of full time $\langle X_i(t)\rangle$. (c) Evolution of the real part of exciton-vibration coherences scaled by their coupling to $\langle X_1(t)\rangle$ [TOP] and $\langle X_2(t)\rangle$ [BOTTOM].

elements, dominating the evolution of $\langle X_i(t)\rangle$ (see eqn (12)). Interference between these coherences generate the overall $\langle X_i(t)\rangle$ signals. Coherences $|\psi_1\rangle\langle\psi_5|$ and $|\psi_3\rangle\langle\psi_7|$ (bold lines) have identical frequencies and remain in phase throughout, whereas coherences $|\psi_1\rangle\langle\psi_4|$ and $|\psi_3\rangle\langle\psi_7|$ (dotted lines) begin to accumulate a phase difference due to their differing frequencies. The phase shift over time manifests as an oscillation in $\langle X_i(t)\rangle$ at a frequency equal to the differences between the pairs of Ω_{kj} involved. These are 8 cm^{-1} (time period of 4.2 ps) and 17 cm^{-1} (time period of 1.9 ps) which explains the approximate 2 ps periodicity seen in Fig. 3a. The oscillation of period 0.4 ps is due to interference with low frequency coherence $|\psi_1\rangle\langle\psi_3|$ (dotted line slowly changing). It is clear that no single coherence dominates the dynamics and that synchronisation does not emerge from the chosen initial state, in the closed system evolution.

3.2 Open system dynamics and the emergence of synchronisation

To understand how synchronisation can emerge we must understand the form of $\rho_{kj}(t)$ in the open quantum system. Solving eqn (8) for an element j,k results in a set of coupled differential equations:

$$\dot{\rho}_{jk}(t) = i\Omega_{jk}\rho_{jk}(t) + \sum_{\alpha,\alpha'} R_{j,k,\alpha,\alpha'}\rho_{\alpha,\alpha'}(t): \tag{15}$$

where $R_{j,k,\alpha,\alpha'}$ is commonly known as the Redfield tensor which captures the dependence of each $\rho_{j,k}$ on all other matrix elements as induced by the local dissipators. In this case we cannot say that each coherence $|\psi_j\rangle\langle\psi_k|(t)$ consists of only one single oscillation frequency Ω_{jk}. However our numerical results show that the oscillatory dynamics of each of the seven coherences in Table 2 are dominated by the coherent component and that the latter terms in eqn (15) contribute mainly decaying dynamics. It is this decay that allows a change in the ratio of coherences and the potential for synchronisation to emerge.

Fig. 4 reports the synchronisation of $\langle X_1(t)\rangle$ and $\langle X_2(t)\rangle$ in the first two picoseconds of evolution. Inspection of the fast oscillations in the displacements reveal an almost π phase difference between $\langle X_i(t)\rangle$ at 0.15 ps and exactly in phase oscillations after 1 ps. This observation is captured quantitatively with $C_{\langle X_1\rangle,\langle X_2\rangle}(t)$

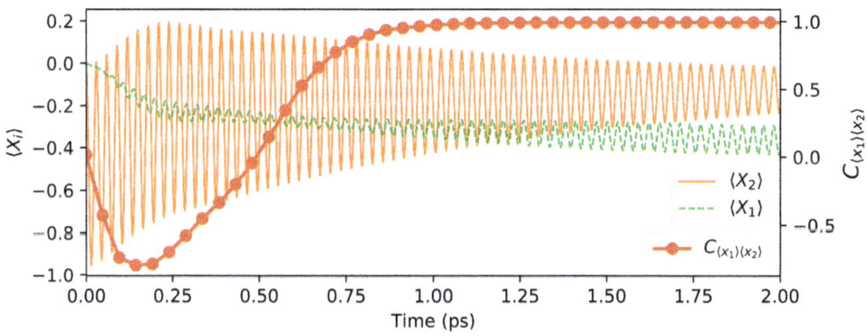

Fig. 4 Evolution of the expectation value of intra-molecular mode displacements $\langle X_i(t)\rangle$ and their synchronisation $C_{\langle X_1\rangle,\langle X_2\rangle}(t)$ in the open system evolution of an excited vibronic dimer as described in Section 2. Initial state is eqn (10) and relevant parameters are listed in Table 1.

dipping to −0.75 for negative synchronisation at 0.15 ps and then up to 1 for positive synchronisation at 1 ps. This figure shows that synchronisation indeed occurs in the displacements of intramolecular modes of exciton-vibration dimers during the energy transfer process. To understand the underlying mechanism we perform an analysis similar to the previous section.

Fig. 5 displays the FT of $\langle X_i(t)\rangle$ at two time instances at 0.15 ps and 1.50 ps. Synchronisation can be seen again in this figure as the change in frequency distribution between the two time points. At 0.15 ps the FT resembles that of the coherent case in Fig. 3b which indicates at this time coherent dynamics are dominating. The presence of negatively synchronised frequency 1102.6 cm^{-1} at a magnitude comparable to positively synchronised frequency 1111.0 cm^{-1} results in an interference and change in phase. This is reflected in $C_{\langle X_1\rangle,\langle X_2\rangle}(t)$ in Fig. 4 at early times 0–1 ps where the measure is continuously changing. At 1.5 ps the dominant frequency in both $\langle X_1(t)\rangle$ and $\langle X_2(t)\rangle$ become 1111.0 cm^{-1} and the synchronisation measure becomes 1.

This FT picture does not allow us to understand why the frequency composition of $\langle X_i(t)\rangle$ changes over time. To do so we must consider the evolution of the underlying coherence dynamics which we present in Fig. 6. In Fig. 6a we present the first 0.30 ps of evolution where the two signals follow the same phase curve towards being negatively synchronised. As expected, in this short-time scale, the dominant frequency in the oscillation of each coherence is Ω_{jk} as can be evidenced by the similarity of Fig. 6a to that of the purely coherent evolution. In the longer-time scale shown in Fig. 6b, this picture changes and the two signals appear much more similar. A change in the amplitude of the coherences has occurred and the dominant contribution is now the three positively synchronised coherences $|\psi_0\rangle\langle\psi_2|$, $|\psi_1\rangle\langle\psi_5|$ and $|\psi_3\rangle\langle\psi_7|$ (solid lines). These coherences constructively interfere to dominate the oscillations seen in $\langle X_i(t)\rangle$ and therefore a value of 1 in $C_{\langle X_1\rangle,\langle X_2\rangle}(t)$. We conclude that the dynamic change in the ratio of coherences over time determines the emergence of synchronisation.

In previous studies[28] the mechanism for synchronisation has been related to a difference in the decay rates of eigenmodes of the Liouvillian superoperator, $\mathscr{L}$.

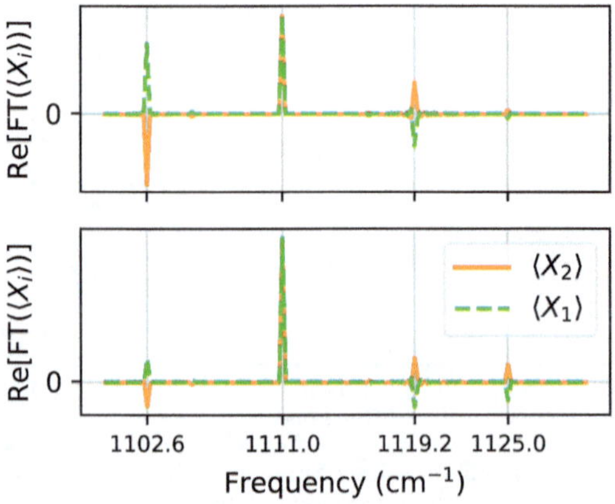

Fig. 5 Real part of Fourier transform of $\langle X_i\rangle$ at 0.15 ps [TOP] and 1.50 ps [BOTTOM] during open system evolution.

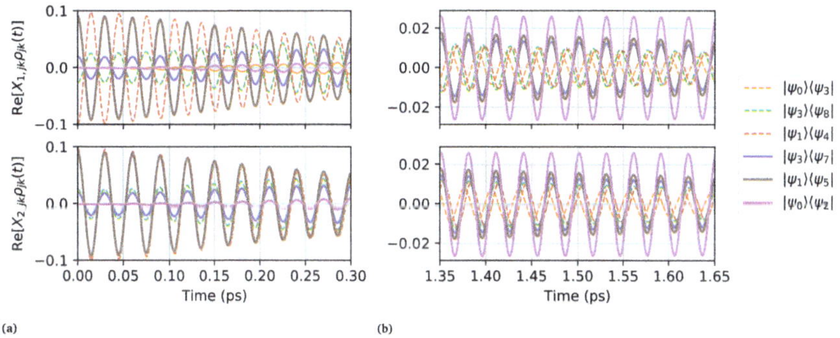

Fig. 6 Evolution of the real part of exciton-vibration coherences in the open system scaled by their coupling to $\langle X_1(t)\rangle$ [TOP] and $\langle X_2(t)\rangle$ [BOTTOM] at times (a) 0–0.30 ps and (b) 1.35–1.65 ps.

The reasoning is that synchronisation occurs when one normal mode of the system, that is equally coupled to the operators of interest, significantly outlives the other normal modes such that the operators are synchronised until steady state is reached. Recently this explanation has been consolidated analytically with an exact treatment of a single dissipating qubit coupled to a probe qubit.[5] In the original case these normal modes are found by diagonalising $\mathscr{L}$ and finding the conjugate pair of eigenvectors that have eigenvalues with real parts closest to zero and which couple significantly to the desired operator. Applying this process here results in an eigenmode of the Liouvillian that consists almost entirely of exciton-vibration coherence $|\psi_0\rangle\langle\psi_2|$. This analysis corroborates with our results which we present by plotting the absolute value of each coherence in Fig. 7. We find coherence $|\psi_0\rangle\langle\psi_2|$ is indeed longest lived. The Liouvillian eigenmode analysis provides a straight forward prediction on the emergence of synchronisation at

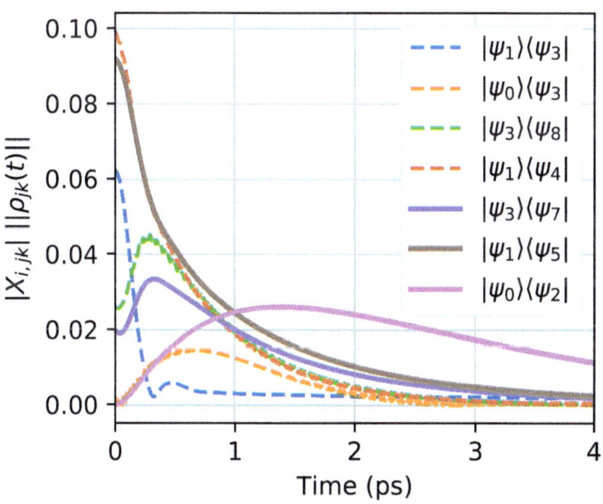

Fig. 7 Complex magnitude of exciton-vibration coherences scaled by the absolute value of their corresponding position matrix element in the open quantum system evolution.

long times, but it does not facilitate an understanding of the early transient synchronisation dynamics. Neither does it provide understanding of why certain eigenmodes survive longer than others. The tracking of coherences we present in this paper give us insight into these early transients. In addition it allows us to give a qualitative explanation of why certain coherences survive longer in the presence of dissipation and dephasing.

We expect pure dephasing to result in exponential decays in excitonic coherences and thermal dissipation to cause exponential decays in the population of modes.[40] In the final two rows of Table 2 we consider the matrix elements of an operator relating to excitonic coherences only *i.e.* $\sigma_x = |E_1\rangle\langle E_2| + |E_2\rangle\langle E_1|$, and the ground-state vibration projector operator $|0_1\rangle\langle 0_1| \otimes |0_2\rangle\langle 0_2|$, for the seven largest coherences contributing to the dynamics of $\langle X_i(t)\rangle$. Exciton-vibration coherences that result in a large matrix element for σ_x will be more affected by the fast electronic dephasing. Similarly exciton-vibration coherences that result in the largest values for the projector of the ground vibrational states would last the longest due to thermal dissipation operating in a longer time scale and preferentially populating such ground states. Out of the set of coherences considered in Table 2, the one that has the lowest exciton coherence component, and the largest ground-state for the vibrational component is $|\psi_0\rangle\langle\psi_2|$, which we find indeed to be the longest lived. This explanation can be consolidated with a numerical test. By reversing the dephasing and thermal dissipation rates (results not shown), we would expect the eigenstate coherence containing a larger excitonic coherence component would survive longest. Indeed we find that in this case the coherence $|\psi_0\rangle\langle\psi_1|$, which has a much larger excitonic coherence element $\langle\psi_0|\sigma_x|\psi_1\rangle = -0.8572$, would be the longest lived coherence and would lead to long lived negative synchronisation.

In summary, we have shown how the interplay between quantum exciton-vibration dynamics and the different noise sources considered lead to a rich synchronisation dynamics of local modes mapping in a direct way the evolution of quantum coherences in the system.

3.3 The role of coherent energy transfer in synchronisation of intramolecular modes

Thus far we have described the coherent and dissipative contributions to the dynamics of exciton-vibration coherences and therefore to the dynamics of synchronisation. Presently we turn our attention to the electronic energy transfer dynamics in the light-harvesting unit considered, its coherent character and their relation to synchronisation of local displacements.

One of the most important features of the prototype system considered is that efficient ET is aided by a resonance in energy between the exciton energy splitting and an energy quanta of the local intramolecular vibrations.[19,23,26] This resonance results in a range of vibronic eigenstates that are close in energy, yet in the picture of the exciton-vibration states $|E_d, n_1, n_2\rangle$ have different weights. When coherently evolving from the initial state considered – with the excitonic system being initially in the highest-energy exciton state – joined vibrational transfer pathways are open and population is coherently transferred to exciton-vibration states involving the lowest lying exciton.

We can intuitively understand how ET is essential for synchronisation of local displacements in the prototype dimers studied by recalling both the quasi-localised nature of the excitons for the parameter regime considered and the local nature of the electronic–vibration interactions. Trivially, ET between excitons must occur for the intramolecular modes to become effectively coupled, exchange energy and synchronise. However the precise relations between the degree of coherent electronic ET and synchronisation is less obvious. In our synchronisation analysis so far we observe that the ET period (0–0.5 ps) is concomitant with the negatively synchronised transient of Fig. 4. It appears that during this energy transfer period the displacements of the modes tend towards a negatively synchronised oscillation. This suggests a signature of coherent ET could be found in a measure of synchronisation of local mode displacements, which we discuss further in Section 4.

To investigate the relationship between coherent ET and synchronisation quantitatively, we compare scenarios in which the resonance condition between exciton energy splitting and mode energy quanta is kept fixed but the degree of delocalisation of excitons is increased such that coherent ET transfer is enhanced. We also analyse the case in which the frequency of the mode is detuned from the exciton energy splitting to illustrate the fundamental role of such energy matching condition both for ET and for synchronisation.

The energy difference between exciton splitting and vibrational energies, *i.e.* $\Delta = \Delta E - \omega$; the coupling strength between local vibrational and electronic degrees of freedom g; and the exciton size or delocalisation θ which depends on the ratio $\eta = 2V/|\Delta e|$ all influence the coherent character of ET. An approximate indicator of the degree of coherent ET is derived[43] from the transition probability between the two exciton-vibration states dominating electronic ET in our prototype dimer. The indicator is an estimate of the maximum amplitude A for the population oscillations and is given by

$$A = \frac{1}{1 + \left(\frac{\Delta}{2g\sin(2\theta)}\right)^2}. \tag{16}$$

The parameter regime used throughout this paper (see Section 2) corresponds to $A = 0.76$ and its associated exciton population and synchronisation dynamics are presented in Fig. 8c. For comparison, we consider the case in which the vibrational modes are detuned from the exciton energy splitting and have frequencies equal to 1500 cm^{-1} (a mode at this frequency is also present in PE545 (ref. 36)) such that $A = 0.04$. Hence, ET is overall decreased and slowed down (in comparison to our reference case) as illustrated in Fig. 8e. We also consider the regime in which coherent transport is increased by increasing the exciton delocalisation, *i.e.* $\frac{2V}{\Delta E} = 0.5$, but keeping fixed the energy matching conditions. In this case we have $A = 0.95$ *i.e.* enhanced coherent energy transport as shown in Fig. 8a. Alongside the different exciton population dynamics shown in Fig. 8a, c and e, we immediately notice a correlation with synchronisation times, with positive synchronisation being reached in around 0.5 ps in the case of the largest degree of coherent ET (Fig. 8a), around 1 ps in the reference case (Fig. 8c) and not stabilising at all in the detuned regime (Fig. 8e). This correlation is also seen in

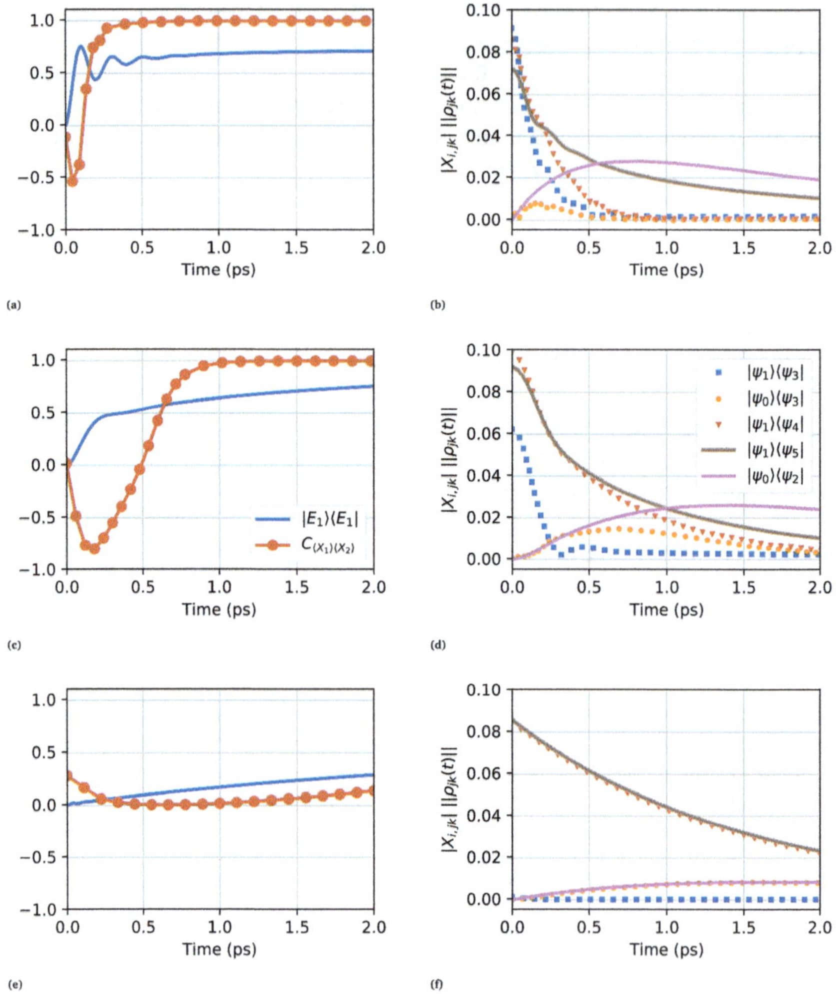

Fig. 8 Comparison of population transfer and synchronisation of $\langle X_i(t)\rangle$ for three different transport regimes with their corresponding exciton-vibration coherence magnitudes. (c) and (d) Central PEB dimer in PE545, parameters in Table 1 corresponding to $A = 0.76$; (a) and (b) modified parameters $2V/\Delta E = 0.5$, $A = 0.95$; (e) and (f) modified parameters $\omega = 1500^{-1}$, $A = 0.04$.

Fig. 8b, d and f where we plot the evolution of the relevant exciton-vibration coherences for the synchronisation of displacements (two have been removed for clarity). In the regime of the largest coherent ET as seen in Fig. 8b, negatively synchronised coherences (plotted with shape markers) decay faster than in the regime of Fig. 8d. This results in the positively synchronised coherences (solid lines) dominating from an earlier time and hence the earlier emergence of positive synchronisation. In the case of reduced ET shown in Fig. 8f we observe that positively and negatively synchronised coherences have almost identical amplitude throughout the evolution and therefore neither can dominate. Small differences are amplified and interferences prevent synchronisation in the time scale considered. Positive synchronisation only emerges at around 10 ps (not shown) when positive synchronised coherences finally outlive the negative synchronised ones however by this time the oscillation amplitudes have nearly decayed to zero.

Furthermore we identify that the eigenstate coherence $|\psi_1\rangle\langle\psi_3|$ is an additional indicator of the magnitude of coherent ET. In the three regimes considered its amplitude appears correlated with the amplitude of coherent ET. This can be understood by considering the composition of the states involved and their energy splitting. The eigenstates $|\psi_1\rangle$ and $|\psi_3\rangle$ have an energy splitting of only 81 cm^{-1} $\ll \omega$ which indicates resonance energy transfer between these states. In the basis of $|E_d,\ n_1,\ n_2\rangle$, these eigenstates have the following approximate compositions:

$$\begin{aligned}|\psi_1\rangle &\approx 0.3|E_1\rangle(|01\rangle - |10\rangle) + 0.9|E_200\rangle - 0.2|E_201\rangle, |\psi_3\rangle \\ &\approx 0.6|E_1\rangle(|01\rangle - |10\rangle) - 0.4|E_200\rangle + 0.1|E_201\rangle.\end{aligned} \quad (17)$$

We can see that they have different amplitudes on each exciton state. This points towards this coherence as being part of the mechanism for ET.

Our analyses point towards a clear positive correlation between coherent transport and the time it takes for synchronisation to emerge. This is of particular interest as it suggests that a level of control of synchronisation can be achieved by adjusting coherent exciton-vibration interactions without changing the dissipation or dephasing rates. The relationship we observe can be understood as follows: parameter changes that increase ET, effectively increase the coupling between the subsystems, allowing a faster exchange of energy between modes. This results in the electronic subsystem "overshooting" its equilibrium position and ET oscillations occurring. This accelerates the damping of exciton coherence and in turn the decays of negatively synchronised exciton-vibration coherences, thus allowing the system to reach a state dominated by positive synchronised coherences in a shorter time scale.

4 Discussions and conclusion

Before concluding we discuss some of the implications and limitations of our results.

The results of Sections 3.2 and 3.3 indicate negative synchronisation is concomitant with coherent energy transport. Moreover, our analysis of the exciton-vibration coherences that are negatively synchronised showed that they all have a large component of excitonic coherence. This raises the question: can we conclude that the negative synchronisation of local displacements is a signature of the survival of excitonic coherence? A dip towards negative synchronisation persists in all different initial preparations of exciton state (except for an initial state equal to the steady state), and the dip becomes less negative as coherent exciton population transport was inhibited. Additionally, as reported in Section 3.2, if dephasing rates are much slower than thermal relaxation *i.e.* excitonic coherence is longer lived, we find that the length of this negative synchronisation period extends to the steady state. Altogether this suggests that, for the system considered, a negatively synchronised transient in the positions of intramolecular modes can indeed be a signature of excitonic coherence and coherent energy transfer.

The parameter regime studied throughout this paper is for the special case of $\omega_1 = \omega_2$. This restriction allowed us to focus on the complex relationship between synchronisation, coherence and dissipation. However, it also raises some interesting points. Firstly, a classical view of two coupled oscillators with $\omega_1 = \omega_2$ might lead one to expect synchronisation to trivially always occur. We have shown that in the quantum setting, due to the exciton-vibration nature of the complex, the frequencies at which the local displacements may oscillate are not equal and, in fact, change over time. This effect cannot be thought of classically. On the other hand, the relationship between ET and synchronisation can be thought of as analogous to how increasing the coupling strength between two classical oscillators allows them to synchronise faster. Our intramolecular modes are coupled to local electronic states that exchange energy through electronic coupling therefore increasing electronic coupling, and the associated increase in ET, leads to a stronger effective coupling between the modes.

Secondly, as mentioned in Section 3.2, the exciton-vibration coherences we study have equal or opposite values in $X_{i,kj}$. We can explain this result by examining our Hamiltonian for differences between X_1 and X_2. Whilst $\omega_1 = \omega_2$, swapping X_i would have no effect on H_{vib}. A difference only enters in $H_{\text{exc-vib}}$ where swapping would result in a -1 scaling of the excitonic coherences *i.e.* $\langle E_2|\Theta_1|E_1\rangle = -\langle E_2|\Theta_2|E_1\rangle$ hence we only see this positive and negative synchronisation. If $\omega_1 \neq \omega_2$, H_{vib} is no longer symmetric upon exchange of X_i. We expect this detuning would result in a wide range of $X_{i,kj}$ and therefore contributions of different phases to the dynamics.

To conclude, we have predicted the transient spontaneous synchronisation of the position of intramolecular modes on neighbouring molecules in a bio-inspired vibronic dimer. Synchronisation has not been investigated in a hybrid quantum system where excitonic and vibrational coherence overlap in such a way. We have presented an understanding of the mechanism for synchronisation as the survival of specific exciton-vibration coherences, and detailed how coherences are selected for by dissipation. This analysis may provide a perspective from which we can understand synchronisation in other dissipating hybrid quantum systems such as larger multi-chromophore systems. We showed that both dissipative and coherent dynamics play an important role in the formation of synchronisation in these systems. Coherent ET is correlated with the time scale in which synchronisation is achieved. Dissipation is required for the decay of exciton-vibration coherences to allow one to dominate and can control the form of synchronisation that occurs. The parameter regime that most closely resembles a real photosynthetic dimer appears to be one where there is a balance between coherent and dissipative dynamics in which synchronisation can emerge before the steady state is reached. Our work highlights a possible role for exciton-vibrational coherence in biomolecular complexes, that of supporting synchronisation of vibrational motions that are driven out of thermal equilibrium.

5 Conflicts of interest

There are no conflicts to declare.

References

1 A. Pikovsky, M. Rosenblum and J. J. Kurths, *Synchronisation. A Universal Concept in Nonlinear Sciences*, Cambridge University Press, 2001, p. 411.
2 J. Buck and E. Buck, *Sci. Am.*, 1976, **234**, 74–85.
3 I. Nitsan, S. Drori, Y. E. Lewis, S. Cohen and S. Tzlil, *Nat. Phys.*, 2016, **12**, 472.
4 N. Lörch, S. E. Nigg, A. Nunnenkamp, R. P. Tiwari and C. Bruder, *Phys. Rev. Lett.*, 2017, **118**, 243602.
5 G. L. Giorgi, F. Galve and R. Zambrini, *Phys. Rev. A*, 2016, **94**, 052121.
6 H. Qiu, R. Zambrini, A. Polls, J. Martorell and B. Juliá-Díaz, *Phys. Rev. A*, 2015, **92**, 043619.
7 M. R. Hush, W. Li, S. Genway, I. Lesanovsky and A. D. Armour, *Phys. Rev. A*, 2015, **91**, 061401.
8 S. Walter, A. Nunnenkamp and C. Bruder, *Phys. Rev. Lett.*, 2014, **112**, 094102.
9 T. E. Lee and H. R. Sadeghpour, *Phys. Rev. Lett.*, 2013, **111**, 234101.
10 F. Fassioli, R. Dinshaw, P. C. Arpin and G. D. Scholes, *J. R. Soc., Interface*, 2014, **11**, 20130901.
11 G. D. Scholes, G. R. Fleming, L. X. Chen, A. Aspuru-Guzik, A. Buchleitner, D. F. Coker, G. S. Engel, R. van Grondelle, A. Ishizaki, D. M. Jonas, J. S. Lundeen, J. K. McCusker, S. Mukamel, J. P. Ogilvie, A. Olaya-Castro, M. A. Ratner, F. C. Spano, K. B. Whaley and X. Zhu, *Nature*, 2017, **543**, 647.
12 G. S. Engel, T. R. Calhoun, E. L. Read, T.-K. Ahn, T. Mančal, Y.-C. Cheng, R. E. Blankenship and G. R. Fleming, *Nature*, 2007, **446**, 782.
13 E. Collini, C. Y. Wong, K. E. Wilk, P. M. G. Curmi, P. Brumer and G. D. Scholes, *Nature*, 2010, **463**, 644–647.
14 E. Harel and G. S. Engel, *Proc. Natl. Acad. Sci. U. S. A.*, 2011, **109**, 706–711.
15 G. Panitchayangkoon, D. V. Voronine, D. Abramavicius, J. R. Caram, N. H. C. Lewis, S. Mukamel and G. S. Engel, *Proc. Natl. Acad. Sci. U. S. A.*, 2011, **108**, 20908–20912.
16 E. Romero, R. Augulis, V. I. Novoderezhkin, M. Ferretti, J. Thieme, D. Zigmantas and R. van Grondelle, *Nat. Phys.*, 2014, **10**, 676–682.
17 F. D. Fuller, J. Pan, A. Gelzinis, V. Butkus, S. S. Senlik, D. E. Wilcox, C. F. Yocum, L. Valkunas, D. Abramavicius and J. P. Ogilvie, *Nature Chemistry*, 2014, **6**, 706–711.
18 H.-G. Duan, V. I. Prokhorenko, R. J. Cogdell, K. Ashraf, A. L. Stevens, M. Thorwart and R. J. D. Miller, *Proc. Natl. Acad. Sci.*, 2017, **114**, 8493–8498.
19 A. Kolli, E. J. O'Reilly, G. D. Scholes and A. Olaya-Castro, *J. Chem. Phys.*, 2012, **137**, 174109.
20 G. H. Richards, K. E. Wilk, P. M. G. Curmi, H. M. Quiney and J. A. Davis, *J. Phys. Chem. Lett.*, 2012, **3**, 272–277.
21 A. W. Chin, J. Prior, R. Rosenbach, F. Caycedo-Soler, S. F. Huelga and M. B. Plenio, *Nat. Phys.*, 2013, **9**, 113.
22 S. Huelga and M. Plenio, *Contemp. Phys.*, 2013, **54**, 181–207.
23 E. J. O'Reilly and A. Olaya-Castro, *Nature Communications*, 2014, **5**, 3012.

24 F. Novelli, A. Nazir, G. H. Richards, A. Roozbeh, K. E. Wilk, P. M. G. Curmi and J. A. Davis, *J. Phys. Chem. Lett.*, 2015, **6**, 4573–4580.
25 H. Chen, X. Wang, A.-P. Fang and H.-R. Li, *Chin. Phys. B*, 2016, **25**, 098201.
26 J. C. Dean, T. Mirkovic, Z. S. Toa, D. G. Oblinsky and G. D. Scholes, *Chemistry*, 2016, **1**, 858–872.
27 G. L. Giorgi, F. Galve, G. Manzano, P. Colet and R. Zambrini, *Phys. Rev. A*, 2012, **85**, 052101.
28 G. L. Giorgi, F. Plastina, G. Francica and R. Zambrini, *Phys. Rev. A*, 2013, **88**, 042115.
29 C. Benedetti, F. Galve, A. Mandarino, M. G. A. Paris and R. Zambrini, *Phys. Rev. A*, 2016, **94**, 052118.
30 B. Militello, H. Nakazato and A. Napoli, *Phys. Rev. A*, 2017, **96**, 023862.
31 G. Manzano, F. Galve, G. L. Giorgi, E. Hernández-García and R. Zambrini, *Sci. Rep.*, 2013, **3**, 1439.
32 V. May and O. Kühn, *Charge and Energy Transfer Dynamics in Molecular Systems*, Wiley-VCH Verlag GmbH & Co. KGaA, Weinheim, Germany, 2011, pp. 9–66.
33 T. Renger, *Phys. Rev. Lett.*, 2004, **93**, 188101.
34 G. C. Schatz and M. A. Ratner, *Quantum Mechanics in Chemistry*, Dover Publications, 2002, ch. 10, pp. 231–261.
35 J. M. Womick and A. M. Moran, *J. Phys. Chem. B*, 2011, **115**, 1347–1356.
36 L. Viani, M. Corbella, C. Curutchet, E. J. O'Reilly, A. Olaya-Castro and B. Mennucci, *Phys. Chem. Chem. Phys.*, 2014, **16**, 16302–16311.
37 A. B. Doust, C. N. J. Marai, S. J. Harrop, K. E. Wilk, P. M. G. Curmi and G. D. Scholes, *J. Mol. Biol.*, 2004, **344**, 135–153.
38 V. I. Novoderezhkin, A. B. Doust, C. Curutchet, G. D. Scholes and R. Van Grondelle, *Biophys. J.*, 2010, **99**, 344–352.
39 H. Haken and G. Strobl, *Zeitschrift für Physik*, 1973, **262**, 135–148.
40 H.-P. Breuer and F. Petruccione, *The Theory of Open Quantum Systems*, Oxford University Press, 2002, p. 636.
41 F. Galve, G. Luca Giorgi and R. Zambrini, *Lectures on General Quantum Correlations and their Applications*, Springer, Cham, 2017, pp. 393–420.
42 B. Bellomo, G. L. Giorgi, G. M. Palma and R. Zambrini, *Phys. Rev. A*, 2017, **95**, 043807.
43 E. J. O'Reilly, PhD thesis, UCL (University College London), 2014.

Faraday Discussions

PAPER

Turning the challenge of quantum biology on its head: biological control of quantum optical systems

Anna Lishchuk,[a] Cvetelin Vasilev,[b] Matthew P. Johnson,[b] C. Neil Hunter,[b] Päivi Törmä[c] and Graham J. Leggett*[a]

Received 16th December 2018, Accepted 24th January 2019
DOI: 10.1039/c8fd00241j

When light-harvesting complex II (LHCII), isolated from spinach, is adsorbed onto arrays of gold nanostructures formed by interferometric lithography, a pronounced splitting of the plasmon band is observed that is attributable to strong coupling of the localised surface plasmon resonance to excitons in the pigment–protein complex. The system is modelled as coupled harmonic oscillators, yielding an exciton energy of 2.24 ± 0.02 eV. Analysis of dispersion curves yields a Rabi energy of 0.25 eV. Extinction spectra of the strongly coupled system yield a resonance at 1.43 eV that varies as a function of the density of nanostructures in the array. The enhanced intensity of this feature is attributed to strong plasmon–exciton coupling. Comparison of data for a large number of light-harvesting complexes indicates that by control of the protein structure and/or pigment compliment it is possible to manipulate the strength of plasmon–exciton coupling. In strongly coupled systems, ultra-fast exchange of energy occurs between pigment molecules: coherent coupling between non-local excitons can be manipulated *via* selection of the protein structure enabling the observation of transitions that are not seen in the weak coupling regime. Synthetic biology thus provides a means to control quantum-optical interactions in the strong coupling regime.

Introduction

Organic semiconductors, produced from earth-abundant elements *via* low-energy pathways, are attractive for the sustainable production of devices and materials for many new and emerging technologies, including consumer electronics, solar energy capture, quantum computing, quantum communications and photo-catalysis. However, molecular systems are intrinsically disordered; dephasing rates are thus high,[1,2] and excitons are transported *via* incoherent hopping

[a]*Department of Chemistry, University of Sheffield, Brook Hill, Sheffield S3 7HF, UK. E-mail: Graham.Leggett@sheffield.ac.uk*

[b]*Department of Molecular Biology and Biotechnology, University of Sheffield, Western Bank, Sheffield S10 2TN, UK*

[c]*Department of Applied Physics, Aalto University, School of Science, P.O. Box 15100, 00076 Aalto, Finland*

processes and have small diffusion lengths, typically ~10 nm but rising to ~50 nm in exceptional cases.[2] Design rules for the efficient transport of excitons across long distances are lacking, placing significant constraints on device architecture and impeding the development of these technologies.[2]

The importance of coherence as a design concept for molecular photonic materials has thus been recognised.[3–5] Theory[6–9] and measurements by ultra-fast spectroscopy[10–14] have suggested that quantum coherent excitations of multiple pigment molecules in photosynthetic light-harvesting complexes (LHCs) facilitate efficient energy transfer, leading to the delocalisation of excitation around individual complexes and also to more efficient transfer of energy between complexes *via* Förster resonance energy transfer (FRET).[15] There has thus been a resurgence of interest in quantum biology, and it has been suggested that photosynthetic mechanisms provide a model for the design of molecular photonic structures to achieve efficient transport of excitons.[5,16]

However, the idea remains controversial.[17,18] For example, Miller and coworkers recently argued that decoherence rates in light harvesting proteins are so fast under physiological conditions that electronic coherence could not contribute to photosynthesis.[19] Moreover, while it is thought that intra-membrane FRET transfers are efficient and enable excitonic transport across distances of at least several 10s of nm, a direct experimental determination of the exciton diffusion length is still lacking.

Here we describe a different approach based on the strong coupling of plasmon modes to excitons in light harvesting proteins. In contrast to the electronic coherence posited to occur in LHCs, strong plasmon–exciton coupling does not rely upon achieving a superposition of excitonic wavefunctions; instead, pigment molecules exchange energy coherently *via* a confined electromagnetic mode.

Plasmons are collective oscillations of surface electrons.[20,21] Their characteristic frequencies are typically in the visible region of the electromagnetic spectrum, and resonant coupling of a plasmon to incident electromagnetic radiation yields a surface plasmon polariton. When the plasmon is formed at the surface of a nanostructure the polariton mode is confined – it is a localised surface plasmon resonance (LSPR). In strong plasmon–exciton coupling,[22–27] light and matter states exchange energy faster than their respective decay channels, giving rise to new quasiparticles called *plasmon–exciton polaritons* ("plexcitons") in which the electronic states of the plasmon and exciton are mixed to form hybrid light-matter states.[27] The coupling is a collective phenomenon: the plasmon is hybridised to an array of emitters.[28] The properties of the plexcitons are thus determined by the arrangement of the emitters as well as the properties of the plasmon and exciton. An important consequence of this is that all of the emitters coupled to a particular nanostructure – which may be 200–300 nm in size – are coherent.[27]

Recently we reported the first example of strong coupling of a plasmon mode to a biological molecule.[28] Gold nanostructure arrays were found to exhibit surface plasmon resonances that split after attachment of light harvesting complexes 1 and 2 (LH1 and LH2) from purple bacteria. The splitting was attributed to strong coupling between the localized surface plasmon resonances and excitons in the light-harvesting complexes, and the coupling was modelled as coupled harmonic oscillators.

Here we show that plasmon modes are strongly coupled to excitons in plant light-harvesting complexes. Furthermore we show that there is enhanced evanescent coupling between nanostructures in arrays of gold nanostructures as a result of this strong coupling. Data from a wide range of natural and synthetic light-harvesting complexes demonstrate that the energies of plexcitonic states can be controlled *via* selection of the protein structure and the spatial arrangement of emitters, and that coherent ultra-fast exchange of energy between non-local pigments occurs *via* the plasmon mode.

Experimental

Materials and chemicals

Microscope coverslip slides (22 mm × 50 mm, no. 1.5 thickness) were obtained from Menzel-Gläser, Germany. Gold wire (99.997% trace metals basis) and chromium chips (99.5% trace metals basis) used for the thermal evaporation were purchased from Sigma-Aldrich. The 30% hydrogen peroxide solution and 95% concentrated sulfuric acid used for preparation of the piranha solution were supplied by VWR Chemicals, UK. For preparation of the gold etchant solution, 32% ammonia solution, HPLC purity ethanol, and cysteamine, obtained from Sigma-Aldrich, were used. 1-Octadecanethiol (98%) and HEPES were also obtained from Sigma-Aldrich. Chemicals used for the gold nanostructures functionalization, *i.e.* 11-amino-1-undecanethiol hydrochloride, glutaraldehyde (25%), *Nα,Nα*-bis(carboxymethyl)-L-lysine trifluoroacetate salt (AB-NTA), and nickel sulfate were supplied by Sigma-Aldrich. All chemicals were used as received.

Fabrication of gold nanostructures

All glassware, *i.e.* the microscope coverslip slides and vials used, were cleaned initially by submersion in piranha solution, a mixture of hydrogen peroxide and concentrated sulfuric acid in the ratio 3 : 7, for 40–60 min, until the solution has stopped bubbling and cooled down to room temperature. The glassware was rinsed thoroughly with deionized water and sonicated for 10 min before being placed in the oven (*ca.* 90 °C) to dry.

Gold substrates were prepared by evaporating a 3–5 nm thick chromium film followed by a 20–22 nm (unless otherwise is stated) thick gold layer. Chromium and gold were both deposited by thermal evaporation using an Edwards Auto 306 bell jar vacuum coating system under pressure of 8×10^{-7} mbar. Evaporation rates of 0.1 nm s^{-1} for Cr and 0.1–0.2 nm s^{-1} for Au were used. It should be noted that the above-stated thickness values were taken from the evaporator QCM thickness monitor. They may differ (by up to 8%) from the actual thickness values, which were determined later on by spectroscopic ellipsometry.

Chromium/gold coated glass slides were immersed in a 1 mM solution of 1-octadecanethiol (ODT) in ethanol for at least 24 h to form closely packed self-assembled monolayers (SAMs). SAMs of ODT on gold were photopatterned by interferometric lithography (IL) using a Lloyd's mirror two-beam interferometer in conjunction with the frequency-doubled argon ion laser emitting at 244 nm (Innova FreD 300C, Coherent, UK). The angle between the mirror and the sample in the interferometer was 30 ± 2.5°. Samples were patterned using IL with a dose of 34 J cm^{-2}. Subsequently, samples were rotated by different angles on the sample stage and exposed again, to a dose of 20 J cm^{-2}.[29]

Photopatterned ODT monolayers on gold were etched by immersion in 2 mM cysteamine with an added 8% v/v of ammonia in HPLC ethanol. After etching, the samples were then rinsed with ethanol, dried under a steam of nitrogen and annealed in a chamber furnace (Carbolite, UK) at 500–550 °C for 60–90 min. The heating rate was *ca.* 7 °C min^{-1} and the annealed samples were left to cool in air to room temperature. Highly crystalline structures and strong plasmon bands were observed in extinction spectra after annealing.

Samples were cleaned for re-use by immersion in piranha solution (which was allowed to cool down to room temperature) for 5–7 min, washed thoroughly with deionized water and blown dry with nitrogen.

Surface functionalization and protein adsorption

Arrays of gold nanostructures were functionalized with 11-amino-1-undecanethiol (AUT) by immersion in a 2 mM solution of the adsorbate in ethanol for 18 h, washed with ethanol and dried with nitrogen. The samples were then immersed in LHCII in 20 mM HEPES buffer, 0.03% BDDM, pH 7.5. The surfaces were immersed in the protein/buffer solution overnight in a humid chamber in a fridge. Finally, the samples were then gently washed with HEPES buffer and deionized water and dried under a steam of nitrogen.

Characterization

Morphology of the as-fabricated and annealed gold nanostructures was determined with atomic force microscopy (AFM). AFM images were acquired in air using a Nanoscope Multimode 8 atomic force microscope (Bruker, Germany) operated in a tapping mode. Tapping mode probes used were OTESPA-R3 model (Bruker), with a resonance frequency of *ca.* 300 kHz and a nominal tip radius of 7 nm. Image analysis was performed with the Bruker NanoScope Analysis (v.1.5) software.

UV-visible absorption spectra at normal incidence were recorded in air using a Cary50 spectrophotometer (Agilent Technologies, USA). The wavelength scan range was 350–850 nm (unless otherwise stated). The samples were placed in a special holder enabling absorption measurements of the same spot on the sample during all experimental stages.

Results and discussion

Extinction spectra

LHCII is the most abundant antenna protein in the photosynthetic apparatus of higher plants, and its primary function is to funnel energy into the photosystem II reaction centre. It is a trimeric protein,[30] containing 7–8 chlorophyll (Chl) *a*, 5–6 Chl *b* and 3–4 carotenoids (Crt),[31] the latter being a mixture of lutein, neoxanthin and violaxanthin. Fig. 1 shows the extinction spectrum for LHCII isolated from spinach and dissolved in buffer (green trace). The Chl *a* and Chl *b* Qy transitions are observed at 1.83 and 1.90 eV, respectively. The Chl *b* Soret band is observed at 2.84 eV, and a broad feature peaking at 2.63 eV results from overlapping bands due to the Chl Soret transitions and the $S_0 \rightarrow S_2$ transitions in the Crt.

Macroscopically extended ($\sim$1 cm^2) arrays of gold nanostructures were fabricated using interferometric lithography in a double-exposure process, as

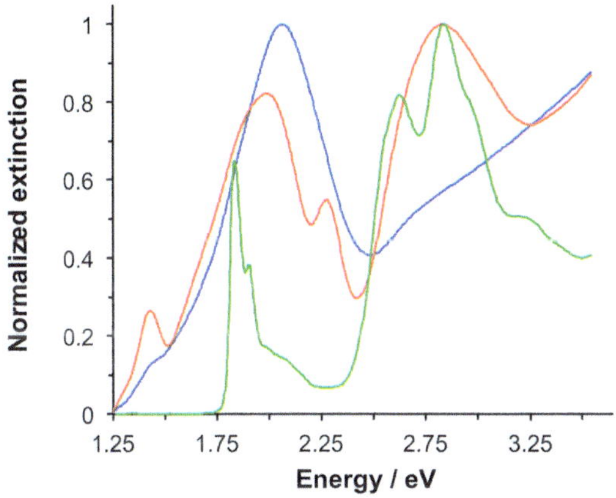

Fig. 1 Normalized extinction spectra of LHCII in buffer solution (green), clean gold nanostructures (blue) and gold nanostructures after adsorption of LHCII (red).

described previously.[32] An advantage of fabricating nanostructures over large areas is that spectroscopic measurements can be made with a simple bench-top spectrophotometer with illumination at low light intensities. The blue trace in Fig. 1 is the extinction spectrum of an array of clean gold nanostructures. The nanostructures were approximately disc-shaped with a height of 60 ± 10 nm and diameter 140.5 ± 17.5 nm at a pitch of 296 ± 16 nm. A strong feature is observed at 2.06 eV corresponding to the localised surface plasmon resonance (LSPR) of the gold nanostructures. The LSPR energy, E_{LSPR}, can be controlled by varying the angle 2θ between the sample and mirror in the spectrophotometer, the angle of rotation between exposures and the etch conditions.

Fig. 1 also shows an extinction spectrum of the same array after adsorption of a monolayer of light-harvesting complex II (LHCII) (red trace). It can be seen that after adsorption of LHCII onto the gold nanostructures there is a dramatic change in the spectrum. The plasmon mode is split to yield a broad feature at 1.98 eV and a narrow feature at 2.27 eV. This type of splitting is characteristic of a type of asymmetric Fano resonance in which a broad mode (the plasmon mode) is coupled to a narrow one (the exciton).

Modelling

Gallinet and Martin provided a first-principles analysis of strong plasmon–exciton coupling.[33] They demonstrated that in the case of a broad resonance coupled to a narrow one, the coupling may be modelled as coupled harmonic oscillators. In our previous work we described the application of such a model to the strong coupling of LSPRs to excitons in bacterial light-harvesting complexes. The same approach was used here to model the extinction spectra obtained after adsorption of LHCII onto arrays of metal nanoparticles. Fig. 2 shows the region of the extinction spectrum from 1.7 to 2.35 eV, containing the plasmon band (red symbols) together with a spectrum fitted using our coupled oscillator model. It can be seen that the fit is very good. The model yields an exciton energy of 2.22 ± 0.01 eV and a coupling strength of 0.27 ± 0.015 eV. These data are consistent with strong plasmon–exciton coupling.

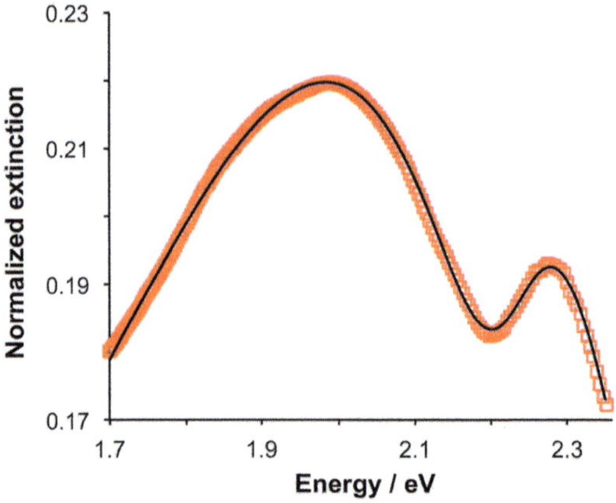

Fig. 2 Extinction spectrum showing the plasmon band at higher resolution after adsorption of LHCII onto gold nanostructure arrays (red symbols) and a fitted spectrum obtained by modelling the system as coupled harmonic oscillators (black line).

To test this hypothesis further, measurements were made for a series of arrays of nanostructures with different LSPR energies after adsorption of LHCII. The spectra were modelled and the energies of the upper and lower polariton branches of the coupled system were determined. The data are shown in Fig. 3. The data were fitted to yield dispersion curves using the relationship:[34]

$$E_{\text{plexiton}}^{\text{UB,LB}}(\hbar\omega_{\text{LSPR}}) = \frac{\hbar\omega_{\text{LSPR}} + \hbar\omega_{\text{mol}}}{2} \pm \frac{1}{2}\sqrt{(\hbar\Omega_{\text{R}})^2 + (\hbar\omega_{\text{LSPR}} - \hbar\omega_{\text{mol}})^2} \quad (1)$$

where $\hbar\omega_{\text{LSPR}}$ and $\hbar\omega_{\text{mol}}$ are the energies of the uncoupled LSPR and exciton, and $\hbar\Omega_{\text{R}}$ is the Rabi splitting, the separation between the upper (UB) and lower (LB) polariton branches at resonance ($\omega_{\text{LSPR}} = \omega_{\text{mol}}$). The Rabi splitting (the coupling energy) is determined as the difference between the energies of the upper and lower polariton branches at resonance, when $\hbar\omega_{\text{LSPR}} = \hbar\omega_{\text{mol}}$. It was not possible

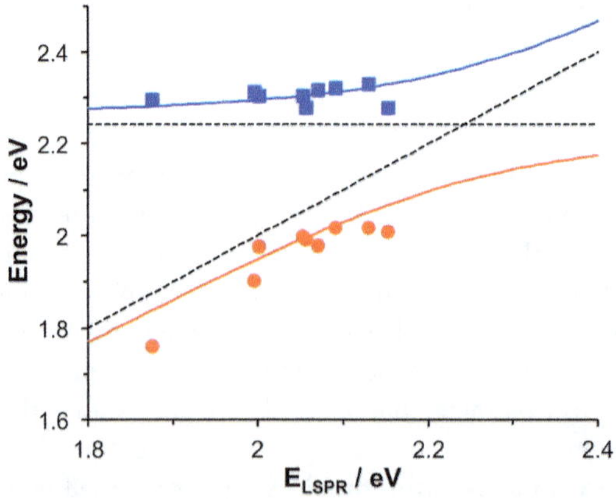

Fig. 3 Dispersion curves for the plexcitonic states determined from experimental data (circles and squares) together with curves fitted using eqn (1). The dotted lines represent the energies of the uncoupled exciton and LSPR states.

to fabricate arrays of nanostructures with LSPR energies greater than E_{mol} (2.24 eV), thus the dispersion curves were fitted using only data for which $E_{LSPR} < E_{mol}$. However, it is still possible to estimate the Rabi energy as 0.25 eV. Using this value we can test whether the system has entered the strong coupling regime. There are a number of criteria for this. One widely used measure is $\hbar\Omega_R \geq \sqrt{\gamma_{mol}\gamma_{LSPR}}$,[27] where γ_{LSPR} and γ_{mol} are the linewidths of the uncoupled LSPR and exciton states. In the present case, γ_{LSPR} is ~ 0.6 eV and γ_{mol} is ~ 0.1 eV, hence the Rabi splitting should be greater than 0.24 eV, a condition that is satisfied here.

Modelling of the spectra yields the exciton energy E_{mol}. Fig. 4 shows the variation in E_{mol} with the plasmon energy. As expected, the value of E_{mol} remains invariant within experimental error at a mean value of 2.24 eV. However, this value does not match the energy of any of the main transitions in the LHCII pigment molecules. In our previous work on bacterial photosynthetic proteins, the calculated value of E_{mol} was found to be equal to that of the Crt $S_0 \rightarrow S_2$ transition for Crt-containing LHCs, and that of the bacteriochlorophyll (BChl) Qx transition for a Crt-free mutant of LH1. In the present study the value of E_{mol} lies close to the energy of the Crt $S_0 \rightarrow S_2$ transitions between 2.5 and 2.6 eV, but is smaller in magnitude. We hypothesise that this reflects the fact that the plasmon mode couples strongly to an ensemble of emitters.

The coupling energy depends on the square root of the density of excitons, but is proportional to the transition dipole moment:[27]

$$E_C = \sqrt{\frac{\mu^2 E_{mol}{}^2}{\varepsilon_0 \varepsilon_b E_{LSPR}} \frac{N}{V_{LSPR}}} \tag{2}$$

In LH2, there are 3 BChl for each Crt and the Crt $S_0 \rightarrow S_2$ transition dipole moment is ~10 × that of the BChl Qx transition dipole moment. Given that the coupling energy is $E_C \propto \sqrt{N/V}$ but that $E_C \propto \boldsymbol{\mu}$, it seems reasonable that plasmon–exciton coupling for LH2 is dominated by coupling to the Crt $S_0 \rightarrow S_2$ transition. However, in LHCII there are ~4 Chl for each Crt and, moreover, the transition dipole moment for the $S_0 \rightarrow S_2$ transition in lutein is ~3× the magnitude of that

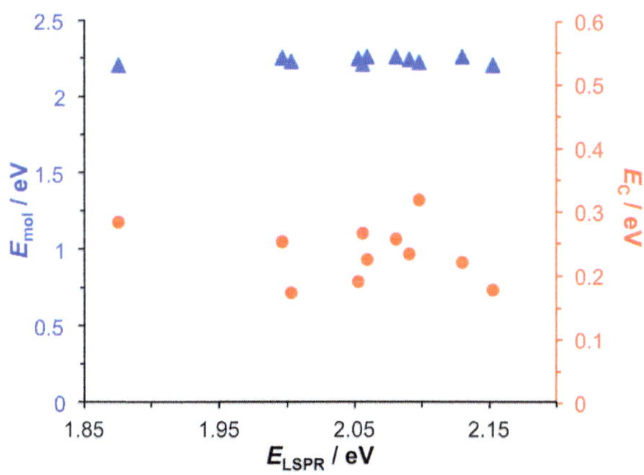

Fig. 4 Variation in the exciton energy (triangles) and scaled coupling energy (circles) as a function of the LSPR energy for a monolayer of LHCII attached to gold nanostructures.

for the Qy transition dipole moments of the Chl.[35] Thus the situation is more complicated and one might not expect the coupling to simply be dominated by coupling to the Crt $S_0 \rightarrow S_2$ transition. Our current model is unable to separate the couplings to different excitons, but instead models the ensemble behaviour. In the future it is expected that more sophisticated models may be capable of analysing the separate contributions to the plasmon–exciton coupling.

In the coupled harmonic oscillator model, the coupling constant g has the dimensions of frequency squared. When scaled to be expressed in units of energy, the coupling constant is G and the coupling energy (equal to the splitting between the normal modes) is $E_C = G/E_{LSPR}$, where E_{LSPR} is the energy of the LSPR. The variation in E_C with E_{LSPR} is shown in Fig. 4 (red circles). It can be seen that as the LSPR energy is decreased below 2.24 eV, the coupling gradually increases to reach values close to 0.3 eV. This behaviour is consistent with that predicted using eqn (2).

Plasmonic coupling in arrays

After adsorption of LHCII onto gold nanostructure arrays, features are observed at 2.83 eV and 1.43 eV in addition to the features at 1.98 eV and 2.27 eV. The feature at 2.83 eV is of unknown origin. Although the energy of this feature is close to that of the Soret transition in the protein, a similar feature was observed at the same energy in extinction spectra of bacterial LHCs and synthetic light-harvesting maquette proteins, which have Soret transitions at significantly different energies. Thus, because it appears to be independent of the exciton energies in the LHCs, it seems unlikely that this feature is due to strong plasmon–exciton coupling. Its origin is currently unclear.

The feature at 1.43 eV is smaller, although distinct. A very small shoulder is just visible at this energy in the spectrum of the clean gold arrays, suggesting that this feature results from a process that is present in the clean gold nanostructure arrays but which is enhanced by strong plasmon–exciton coupling. Coupling between nanostructures is known to yield resonances at low energies. To examine whether the feature at 1.43 eV was associated with coupling between nanostructures, its area was measured and plotted as a function of the nanostructure density (Fig. 5). It was found that the intensity of the feature at 1.43 eV increased with increasing array density but that the relationship was non-linear, increasing slowly at high densities ($8–18 \times 10^{12}$ m^{-2}) but changing more rapidly at lower densities. As the density of nanostructures increases, the amount of analyte increases, and it is expected that extinction will increase, but in a linear fashion. Moreover, the feature at 1.43 eV increases relative to the intensity of the plasmon mode as a function of array density.

The non-linearity in Fig. 5 suggests that the feature at 1.43 eV is associated with distance-dependent coupling between nanostructures. Determination of the mechanism of this coupling will be a subject for further research.

Discussion

These data may now be combined with results from previous studies of strong coupling of plasmon modes to bacterial light-harvesting complexes[28] and

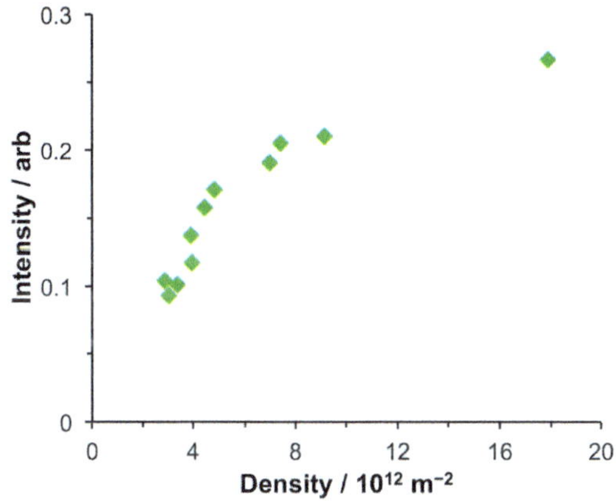

Fig. 5 Variation in the intensity of the feature at 1.43 eV in the extinction spectra of nanostructure arrays coupled to LHCII as a function of the density of nanostructures in the array.

synthetic maquette proteins[36] to begin to assemble a detailed picture of the way that protein structure may be used to control strong plasmon–exciton coupling.

The pigment complement of light harvesting complexes determines the coupling strength

In strong plasmon–exciton coupling, the energies of the resulting plexcitonic states are determined by the energy of the LSPR (which is controllable, *via* modification of the lithographic process),[32] the energy and transition dipole moment of the exciton, and the organisation of the excitons. Because of their exquisitely controlled architectures, light-harvesting complexes provide powerful model systems within which to explore strong plasmon–exciton coupling.

Fig. 6 combines data from all three studies completed to date. It shows the mean coupling energy E_C determined from fitting the extinction spectra for

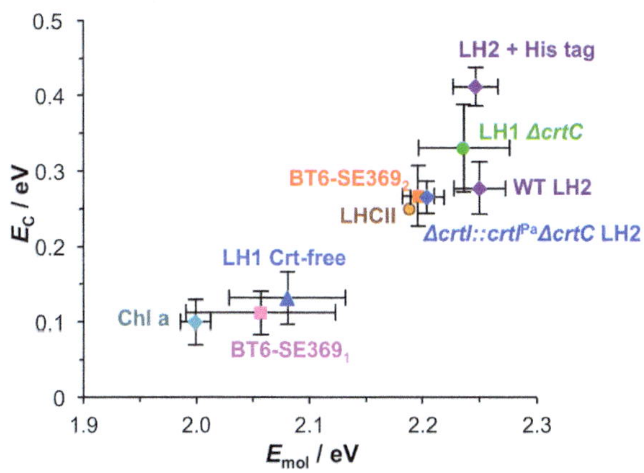

Fig. 6 Mean coupling energy E_C as a function of the exciton energy E_{mol} for a variety of light-harvesting complexes and for self-assembled monolayers derivatised by attachment of chlorophyll *a*. For LHCII, the error bars are similar in size to the symbol used.

a number of strongly coupled systems as a function of the exciton energy E_{mol}. The data are clustered into two broad groups. Three points lie just outside the threshold for the strong coupling regime: a monolayer derivatised with Chl *a*, BT6 maquettes containing a single binding site for a chlorin (BT6-SE3691), and "blue" LH1, a carotenoid-free mutant of LH1 from *R. sphaeroides* all yield coupling energies that are significant but not sufficiently large to be said to have entered the strong coupling regime. In all cases the pigment is a chlorin (or bacteriochlorin) with a comparatively small transition dipole moment, and in all cases the exciton density is low ($\sim 10^{17}$ m^{-2}).

A second group of points exhibits larger coupling energies. For each of these proteins, the coupling energy is large enough to yield strong splitting of the plasmon mode. This group includes wild-type (WT) LH2, in which the Crt is spheroidenone, and the *ΔcrtI::crtI^{Pa}ΔcrtC* LH2 mutant, which has the same structure as WT-LH2 save that spheroidenone has been replaced by lycopene. In the case of these proteins, the Crt transition dipole moment is aligned perpendicular to the surface of the gold nanostructure to which they are attached, meaning that it lies in the direction of the electric field associated with the LSPR. The Qx transition dipole moment lies in this direction, but its transition dipole moment is much smaller and the extinction spectra appear to be dominated by strong coupling of plasmon modes to the Crt $S_0 \rightarrow S_2$ transition, yielding clear differences in the splitting for different mutants (Fig. 7). The Qy transition dipole moment lies orthogonal to the LSPR field direction and it does not couple to the plasmon mode. In blue LH2 it is the Qx transition dipole moment that couples to the LSPR.

The coupling energy is still larger for the *ΔcrtC* mutant of LH1, in which the Crt is neurosporene which has a larger transition energy than those of spheroidenone and lycopene.[28] The Crt in LHCII have energies closer to that of the $S_0 \rightarrow S_2$ transition in neurosporene than the other mutants of LH2, so that the coupling energy measured here for LHCII appears slightly small. This is probably a consequence of the different structure of the protein: LH2 has a pronounced circular symmetry and the transition dipole moments of the Crt will consequently lie close to the direction of the LSPR field, but the structure of LHCII is more complex and the transition dipole moments of neither the Crt nor the Chl lie in a single direction. This will inevitably reduce the coupling strength.

His-tagged LH2 yields the largest coupling energy measured in our studies of light harvesting systems, at 0.41 eV, significantly larger than the value obtained for WT LH2. This is attributed to the strong binding of His-tagged proteins to nanostructures that are functionalised with nitrilotriacetic acid (NTA), which is expected to lead to a higher surface coverage of immobilised LH2 and thus a higher density of excitons within the plasmon mode volume.

Synthetic biology enables the control of ultra-fast non-local exchange of energy in strongly coupled systems

In our previous studies we demonstrated that for bacterial light harvesting complexes and synthetic maquette proteins the coupling strength varied with the square root of the density of proteins at the gold surface. These observations reflect the fact that in strong plasmon–exciton coupling, an LSPR couples to an ensemble of emitters. Although we have successfully used a simple classical

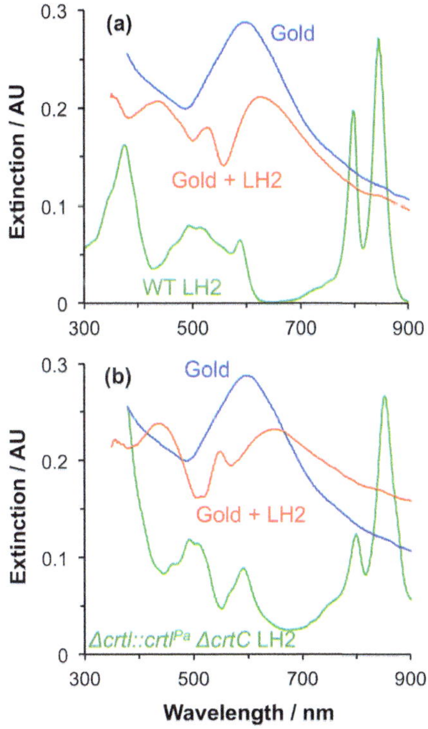

Fig. 7 Extinction spectra for arrays of gold nanostructures before (blue) and after (red) attachment of (a) WT LH2 and (b) the *Δcrtl::crtl*^{Pa}*ΔcrtC* mutant of LH2 from *R. sphaeroides*. Absorption spectra of the proteins in solution are shown in green.

model to analyse our strongly coupled systems, this dependence of the coupling strength on the organisation of the excitons within the plasmon mode volume is a reflection of the quantum optical character of strong plasmon–exciton coupling. The strongly coupled system effectively consists of macroscopically extended states in which the LSPR is coupled coherently to all excitons within the mode volume of each nanostructure. This allows ultra-fast exchange of energy between non-local excitons.

Evidence for this comes from the unexpectedly large coupling energy of the synthetic light-harvesting maquette protein BT6-SE369$_2$ (0.27 eV) when attached to gold nanostructures (Fig. 6). This protein contains two binding sites for synthetic SE369 chlorins. The neighbouring data points in Fig. 6 are all for proteins that contain Crt, which have larger transition dipole moments. However, BT6-SE369$_2$ contains no carotenoids. Moreover, its two chlorin binding sites are separated by >2 nm; at such separations dipole coupling is weak. Consistent with this, the absorption spectrum of BT6-SE369$_2$ when collected in solution is indistinguishable from that of the protein BT6-SE369$_1$ that contains only a single chlorin binding site (Fig. 8a and b, green traces). However, when attached to gold nanostructures, BT6-SE369$_1$ yields a coupling energy of only 0.11 eV, less than half that of BT6-SE369$_2$ and similar in magnitude to the coupling energies calculated for the *ΔcrtC* mutant of LH1 and Chl *a* functionalised surfaces.

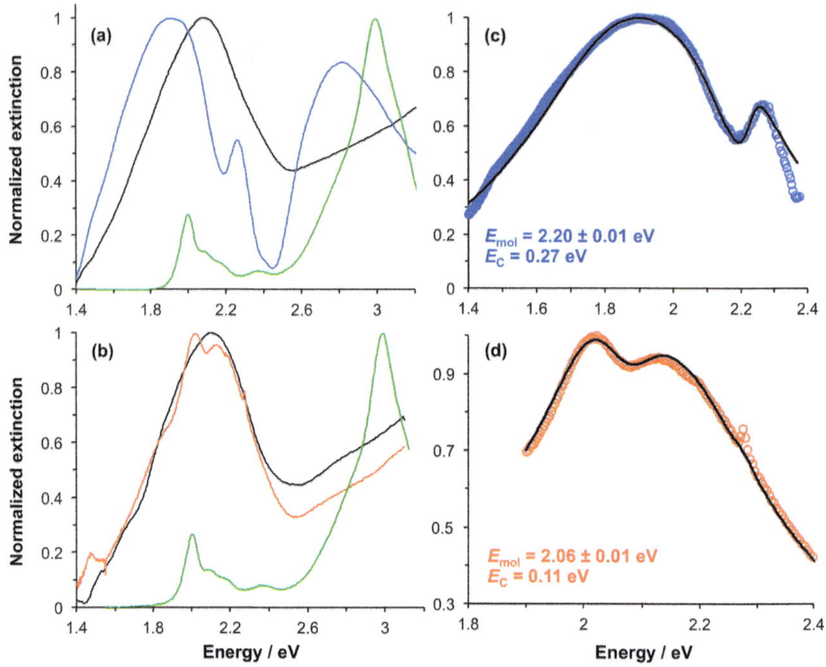

Fig. 8 (a) Extinction spectra for the clean gold nanostructure array (black), BT6-SE369$_2$ in buffer (green) and nanostructures derivatised by attachment of BT6-SE369$_2$ (blue). (b) Extinction spectra for clean gold nanostructure array (black), BT6-SE369$_1$ in buffer (green) and nanostructures derivatised by attachment of BT6-SE369$_1$ (red). (c) Experimental data (blue) and fitted spectrum (black) for gold nanostructures derivatised by attachment of BT6-SE369$_2$. (d) Experimental data (red) and fitted spectrum (black) for gold nanostructures derivatised by attachment of BT6-SE369$_1$.

Modelling of the extinction spectra yields an exciton energy for BT6-SE369$_1$ of 2.06 $\pm$ 0.07 eV, close to that of the Qy transition in the protein. For BT6-SE369$_2$, however, an exciton energy of 2.20 $\pm$ 0.01 eV is obtained, intermediate between the energies of the Qx and Qy transitions of the chlorin. A transition with this energy is not observed in the absorption spectrum of either protein.

We hypothesise that these surprising observations are explained by strong coupling of the LSPR to a dimer state not observed under weak coupling. Transition dipole moments in aggregates of pigment molecules may couple to form J-dimers (or aggregates) leading to a red shift in the exciton energy,[37] or H-dimers (or aggregates) leading to a blue shift[38] (as shown in Fig. 9). However, as noted above, the chlorins in BT6-SE369$_2$ are too far apart for effective dipole coupling. We hypothesise that in the strongly coupled system, the dimer state results from ultra-fast exchange of energy between non-local pigment molecules *via* the plasmon mode. The observation of this dipole coupling is thus a consequence of the coherence that is intrinsic to strongly coupled systems.

The blue shift in the exciton energy that results from strong coupling of the plasmon mode to excitons in BT6-SE369$_2$ suggests the formation of H-dimers. Such coupling of the chlorins would also yield an increase in the transition dipole moment, which, combined with the larger exciton energy, would account

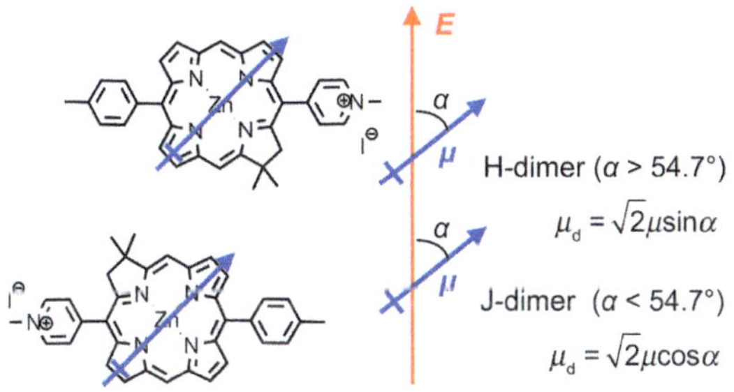

Fig. 9 Possible alignment of pairs of SE369 chlorins in maquettes. The blue arrow represents the Qy transition dipole moment, and the red arrow the direction of the field associated with the surface plasmon mode.

for the much larger coupling energy determined for the two-chlorin maquette. However, the use of an H-dimer model here may represent an over-simplification of what may be a more complex coupling mechanism: given that in strong plasmon–exciton coupling the plasmon mode couples to an array of emitters, it is indeed plausible that the observed couplings involve chlorins in different proteins. These data, combined with the close similarity of the absorption spectra acquired for the two proteins under weak coupling conditions, provide evidence that by changing the structure of the maquette it was possible to manipulate the coherent non-local transfer of energy in the strongly coupled system.

Conclusions

Plasmon modes are strongly coupled to excitons in light-harvesting complexes from plants and bacteria. The strong coupling regime is reached when the plasmon mode and the excitons in pigment molecules exchange energy faster than their respective decay channels. By manipulating the structures of these pigment–protein complexes, the organisation of excitons within the plasmon mode volume can be controlled precisely, enabling the properties of the coupled states to be manipulated through the ultra-fast exchange of energy *via* the plasmon mode. Non-local coupling between excitons can be manipulated *via* the design of synthetic proteins and through the control of protein organisation at the surfaces of plasmonic nanostructures. Strong plasmon–exciton coupling achieves coherent transport of energy across distances of at least 100s of nm, but this can be extended by manipulation of plasmonic coupling mechanisms in extended arrays of plasmonic nanostructures.

Conflicts of interest

There are no conflicts to declare.

Acknowledgements

The authors are grateful to the EPSRC (Programme Grant EP/I012060/1 and equipment grant EP/M028437/1) for Financial Support. C. V., M. P. J. and C. N. H. were supported by research grant BB/M000265/1 from the Biotechnology and

Biological Sciences Research Council (UK) and Advanced Award 338895 from the European Research Council.

Notes and references

1 C. J. Bardeen, *Annu. Rev. Phys. Chem.*, 2014, **65**, 127–148.
2 G. J. Hedley, A. Ruseckas and I. D. W. Samuel, *Chem. Rev.*, 2017, **117**, 796–837.
3 A. Chenu and G. D. Scholes, *Annu. Rev. Phys. Chem.*, 2015, **66**, 69–96.
4 J.-L. Bredas, E. H. Sargent and G. D. Scholes, *Nat. Mater.*, 2017, **16**, 35–44.
5 G. D. Scholes, G. R. Fleming, L. X. Chen, A. Aspuru-Guzik, A. Buchleitner, D. F. Coker, G. S. Engel, R. van Grondelle, A. Ishizaki, D. M. Jonas, J. S. Lundeen, J. K. McCusker, S. Mukamel, J. P. Ogilvie, A. Olaya-Castro, M. A. Ratner, F. C. Spano, K. B. Whaley and X. Zhu, *Nature*, 2017, **543**, 647.
6 M. G. Cory, M. C. Zerner, X. Hu and K. Schulten, *J. Phys. Chem. B*, 1998, **102**, 7640–7650.
7 A. Damjanović, T. Ritz and K. Schulten, *Phys. Rev. E: Stat. Phys., Plasmas, Fluids, Relat. Interdiscip. Top.*, 1999, **59**, 3293–3311.
8 M. L. Cartron, J. D. Olsen, M. Sener, P. J. Jackson, A. A. Brindley, P. Qian, M. J. Dickman, G. J. Leggett, K. Schulten and C. Neil Hunter, *Biochim. Biophys. Acta, Bioenerg.*, 2014, **1837**, 1769–1780.
9 M. Sener, J. Strumpfer, A. Singharoy, C. N. Hunter and K. Schulten, *eLife*, 2016, **5**, e09541.
10 H. Lee, Y.-C. Cheng and G. R. Fleming, *Science*, 2007, **316**, 1462–1465.
11 G. S. Engel, T. R. Calhoun, E. L. Read, T.-K. Ahn, T. Mancal, Y.-C. Cheng, R. E. Blankenship and G. R. Fleming, *Nature*, 2007, **446**, 782–786.
12 E. Collini, C. Y. Wong, K. E. Wilk, P. M. G. Curmi, P. Brumer and G. D. Scholes, *Nature*, 2010, **463**, 644–647.
13 P. D. Dahlberg, G. J. Norris, C. Wang, S. Viswanathan, V. P. Singh and G. S. Engel, *J. Chem. Phys.*, 2015, **143**, 101101.
14 P. D. Dahlberg, P.-C. Ting, S. C. Massey, M. A. Allodi, E. C. Martin, C. N. Hunter and G. S. Engel, *Nat. Commun.*, 2017, **8**, 988.
15 J. Strümpfer, M. Şener and K. Schulten, *J. Phys. Chem. Lett.*, 2012, **3**, 536–542.
16 F. Fassioli, R. Dinshaw, P. C. Arpin and G. D. Scholes, *J. R. Soc., Interface*, 2014, **11**, 20130901.
17 A. Halpin, P. J. M. Johnson, R. Tempelaar, R. S. Murphy, J. Knoester, T. L. C. Jansen and R. J. D. Miller, *Nat. Chem.*, 2014, **6**, 196–201.
18 D. M. Wilkins and N. S. Dattani, *J. Chem. Theory Comput.*, 2015, **11**, 3411–3419.
19 H.-G. Duan, V. I. Prokhorenko, R. J. Cogdell, K. Ashraf, A. L. Stevens, M. Thorwart and R. J. D. Miller, *Proc. Natl. Acad. Sci. U. S. A.*, 2017, **114**, 8493–8498.
20 T. W. Ebbesen, H. J. Lezec, H. F. Ghaemi, T. Thio and P. A. Wolff, *Nature*, 1998, **391**, 667–669.
21 W. L. Barnes, A. Dereux and T. W. Ebbesen, *Nature*, 2003, **424**, 824–830.
22 J. Dintinger, S. Klein, F. Bustos, W. L. Barnes and T. W. Ebbesen, *Phys. Rev. B: Condens. Matter Mater. Phys.*, 2005, **71**, 035424.
23 Y. Sugawara, T. A. Kelf, J. J. Baumberg, M. E. Abdelsalam and P. N. Bartlett, *Phys. Rev. Lett.*, 2006, **97**, 266808.
24 S. Aberra Guebrou, C. Symonds, E. Homeyer, J. C. Plenet, Y. N. Gartstein, V. M. Agranovich and J. Bellessa, *Phys. Rev. Lett.*, 2012, **108**, 066401.

25 L. Shi, T. K. Hakala, H. T. Rekola, J. P. Martikainen, R. J. Moerland and P. Törmä, *Phys. Rev. Lett.*, 2014, **112**, 153002.
26 A. I. Väkeväinen, R. J. Moerland, H. T. Rekola, A. P. Eskelinen, J. P. Martikainen, D. H. Kim and P. Törmä, *Nano Lett.*, 2014, **14**, 1721–1727.
27 P. Törmä and W. L. Barnes, *Rep. Prog. Phys.*, 2015, **78**, 013901.
28 A. Tsargorodska, M. L. Cartron, C. Vasilev, G. Kodali, O. A. Mass, J. J. Baumberg, P. L. Dutton, C. N. Hunter, P. Törmä and G. J. Leggett, *Nano Lett.*, 2016, **16**, 6850–6856.
29 S. R. J. Brueck, *Proc. IEEE*, 2005, **93**, 1704–1721.
30 X. Wei, X. Su, P. Cao, X. Liu, W. Chang, M. Li, X. Zhang and Z. Liu, *Nature*, 2016, **534**, 69–74.
31 C. C. Gradinaru, I. H. M. van Stokkum, A. A. Pascal, R. van Grondelle and H. van Amerongen, *J. Phys. Chem. B*, 2000, **104**, 9330–9342.
32 A. Tsargorodska, O. El Zubir, B. Darroch, M. I. L. Cartron, T. Basova, C. N. Hunter, A. V. Nabok and G. J. Leggett, *ACS Nano*, 2014, **8**, 7858–7869.
33 B. Gallinet and O. J. F. Martin, *Phys. Rev. B: Condens. Matter Mater. Phys.*, 2011, **83**, 235427.
34 A. E. Schlather, N. Large, A. S. Urban, P. Nordlander and N. J. Halas, *Nano Lett.*, 2013, **13**, 3281–3286.
35 P. López-Tarifa, N. Liguori, N. van den Heuvel, R. Croce and L. Visscher, *Phys. Chem. Chem. Phys.*, 2017, **19**, 18311–18320.
36 A. Lishchuk, G. Kodali, J. A. Mancini, M. Broadbent, B. Darroch, O. A. Mass, A. Nabok, P. L. Dutton, C. N. Hunter, P. Törmä and G. J. Leggett, *Nanoscale*, 2018, **10**, 13064–13073.
37 S. Shoji, T. Ogawa, T. Hashishin, S. Ogasawara, H. Watanabe, H. Usami and H. Tamiaki, Nanotubes of Biomimetic Supramolecules Constructed by Synthetic Metal Chlorophyll Derivatives, *Nano Lett.*, 2016, **16**, 3650–3654.
38 B. Z. Packard, D. T. Toptygin, A. Komoriya and L. Brand, *Proc. Natl. Acad. Sci. U. S. A.*, 1996, **93**, 11640–11645.

Faraday Discussions

PAPER

Coupled excitation energy and charge transfer dynamics in reaction centre inspired model systems

Martin Richter† and Benjamin P. Fingerhut *

Received 19th November 2018, Accepted 14th January 2019

DOI: 10.1039/c8fd00189h

Functional operation conditions of reaction centre core complexes require the tight coupling of exciton states to concomitant charge separation. Rigorous theoretical treatment of such integrated excitation energy transfer (EET) and charge transfer (CT) dynamics is particularly challenging due to (i) appreciable system sizes, (ii) inter-site and system-bath couplings of similar magnitude that render the Born–Markov approximation invalid, (iii) substantial reorganization energies of CT states, and (iv) the presence of complex structured spectral densities due to vibrational modes of the surroundings. We present numerical simulations on bacterial reaction centre (bRC) inspired model systems that utilize the recently developed MACGIC-iQUAPI method [Richter *et al.*, *J. Chem. Phys.*, 2017, **146**, 214101]. The simulations demonstrate that the method provides a rigorous framework for the investigation of such integrated EET–CT dynamics. First, the applicability of the MACGIC-iQUAPI method is explored for a transition from monotonically decaying to oscillatory system-bath influence coefficients, a behavior inherently imposed by structured bath spectral densities. Tightly coupled EET and CT dynamics is further addressed for an excitonic subsystem that resembles strong coupling of special pair states and serves as donor towards a generic bridge–acceptor system. By solving the dissipative quantum dynamics of such bRC inspired model systems, the quenching of excitonic coherence on the hundreds of femtoseconds timescale is explored *via* a variation of the bridge state energetics, resembling a continuous transition from sequential to superexchange mediated CT regimes. Further, the simulations explore the influence of resonant vibrational modes on the quenching of excitonic coherence *via* CT. The results reveal a moderate influence of vibrational mode on charge separation dynamics in regimes of biologically relevant EET and CT dynamics.

Max-Born-Institut für Nichtlineare Optik und Kurzzeitspektroskopie, 12489 Berlin, Germany. E-mail: fingerhut@mbi-berlin.de

† Present address: Institut für Physikalische Chemie, Friedrich-Schiller-Universität Jena, 07743 Jena, Germany.

1 Introduction

Reaction centre core complexes can be considered as highly efficient nano-machines where excitonic excitation energy, collected by the light-harvesting unit of a photosynthetic apparatus, is trapped and converted *via* electron transfer (ET) reactions into a charge separated state of polaronic character. This process proceeds with astonishing, near unit quantum efficiency on the hundreds of femtosecond to few picosecond timescale *via* a coupling of exciton states to concomitant multi-step charge transfer.[1–10] In bacterial reaction centres (bRCs) two adjacent bacteriochlorophyll (B) molecules form a so-called special pair that gives rise to a strongly coupled pair of exciton states (P^+/P^-, Fig. 1(a)). The lower energy exciton state P^- acts as primary electron donor for charge separation, which at room temperature proceeds unidirectionally and sequentially *via* a primary charge separated state P^+ $B_A^-H_A$ and secondary separated state P^+ $B_AH_A^-$ of the active A-branch.[5,11] The energetics of charge separated states[12] is subtly balanced, thus preventing thermal repopulation of the initially excited state while at the same time assuring high quantum efficiency and avoiding futile charge recombination over molecular distances.[8,13,14]

Due to congestion of spectrally overlapping chlorin cofactors an observation of real time exciton- and charge transfer dynamics is challenging. bRCs offer the advantage of reduced congestion and better spectral resolution with particular prospects for multi-dimensional electronic spectroscopy (2DES) to resolve coupled EET and CT dynamics *via* a spreading of information content into the 2D plane.[11,15–18] While the kinetics of CT in bRCs has been resolved and at room temperature is well represented by a two-step sequential CT cascade,[6,8,11] the functional impact of (electronic/vibrational) coherences on CT dynamics remains debated.[16,19–22] The frequencies of identified vibrational modes match frequency differences between electronic excitons and CT states. Accordingly, electronic-vibrational resonances were predicted to represent an important design principle for the high quantum efficiency in primary charge separation. Nevertheless, the increased fluctuations of CT states reflected in the larger reorganization energies and their coupling to exciton states is hardly explored. Recent investigations employing second order perturbation theory reported robustness and insensitivity of charge separation with respect to vibrational modes in the parameter regime of the photosystem II reaction centre.[23] Further, the importance of coherent vibronic transport as a design principle was recently questioned.[24]

Rigorous theoretical treatment of tightly integrated EET and CT dynamics is particularly challenging due to appreciable system sizes and the fact that inter-site and system-bath couplings of photosynthetic proteins are found to be of similar magnitudes, which renders theoretical treatments that rely on the Born–Markov approximation invalid. Such finite environment relaxation times are imposed by the presence of a sluggish protein environment. Vibrational modes further induce resonances in the spectral density function. The arising complex environments impose system-bath correlation times that can become comparable to the ultrafast, sub-ps timescale of EET and CT dynamics. Accordingly, charge separation can potentially arise from a non-equilibrium state. Excitonic quantum coherence has further been shown to be decisively affected by the presence of

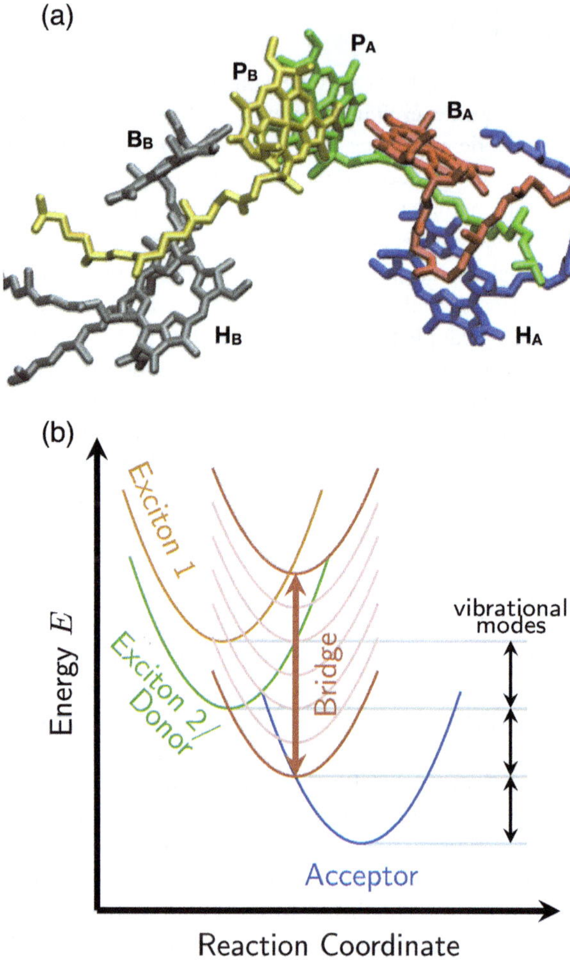

Fig. 1 Schematic representation of the bacterial reaction centre (bRC) inspired model system. (a) Structural arrangement of bacteriochlorophylls (B) and bacteriopheophytines (H) in the bacterial reaction centre *B. viridis*. Redox cofactors of the active A-branch and considered in the bRC-inspired model are highlighted with colors, cofactors of the inactive B-branch are indicated in grey (pdb: 1PRC[27]). (b) two exciton states (orange and green) are coupled to a bridge state (red) and an acceptor CT state (blue). A tuning of the bridge state energy allows a switch from sequential to superexchange regimes of CT.

vibrational modes.[25] Additionally, for CT states the coupling to the environment along the reaction coordinate is increased, imposing the necessity for a large number of auxiliary density operators and challenges for convergence in numerically exact hierarchical equation of motion (HEOM) simulations.[26] Together, such scenarios potentially give rise to complex, nonexponential dynamics.

Theoretical investigations of integrated EET–CT dynamics are of particular relevance beyond the investigated bRC model. The fundamental understanding of artificial photochemical devices subject to aforementioned scenarios is important for, for example, hot exciton dissociation in organic photovoltaic cells[28] and ultrafast ET in oligothiophene–fullerene heterojunctions.[29] Exciton coupled CT thus constitutes the basic mechanism for the utilization of solar energy. While isolated CT and EET dynamics have been extensively studied, the mechanisms leading to the irreversibility of tightly coupled EET-CT processes have only recently been addressed.[30]

Here, we present numerical simulations on bRC inspired model systems that realize tightly coupled EET and CT dynamics. The remainder of the paper is outlined as follows: in Section 2, the employed methodology is presented where the path integral based iQUAPI method (Section 2.1), as well as the recently developed MACGIC-iQUAPI approach (Section 2.2) are summarized. Section 2.3 introduces the spectral densities employed in this work. Results are presented in Section 3 where we successively construct the bRC inspired model: first, we establish the numerical accuracy of the MACGIC-iQUAPI method for the treatment of dissipative quantum dynamics subject to an environment that is composed of a structured spectral density (Section 3.1); further, we investigate the non-equilibrium dynamics of model systems of increasing complexity. Section 3.2 describes coupled EET and CT dynamics in a bRC inspired three-state model, and Section 3.3 presents the dynamics of a pair of strongly coupled exciton states that act as donor towards a bridge–acceptor system. Upon varying the bridge state energy, a continuous interpolation between regimes of sequential and superexchange CT dynamics is realized. The effects of vibrational modes on the dynamics are explored.

2 Methodology

We partition the entire system, composed of many degrees of freedom, into a low-dimensional quantum system of interest H_s and the environment acting as the heatbath H_b.[31] Both are coupled *via* the system-bath interaction H_{sb}

$$H = H_s + H_b + H_{sb}. \tag{1}$$

Assuming a harmonic bath, composed of oscillators with masses m, frequencies ω, positions x and conjugate momenta p, and a bilinear system-bath coupling c_j to the collective system coordinate s of the quantum system, the Hamiltonian of the Caldeira–Legett model takes the form[31,32]

$$H = H_s + \sum_j \left[\frac{p_j^2}{2m_j} + \frac{1}{2} m_j \omega_j^2 \left(x_j - \frac{c_j s}{m_j \omega_j^2} \right)^2 \right]. \tag{2}$$

The system Hamiltonian

$$\begin{aligned} H_s = & \sum_{n=1}^{N^{\mathrm{EET}}} \varepsilon_n |e_n\rangle\langle e_n| + \sum_{n=1}^{N^{\mathrm{EET}-1}} \left(V_{n,n+1} |e_n\rangle\langle e_{n+1}| + \mathrm{h.c.} \right) \\ & + \sum_{m=N^{\mathrm{EET}}+1}^{M} \varepsilon_m |c_m\rangle\langle c_m| + \sum_{m=N^{\mathrm{EET}}+1}^{M-1} \left(t_{m,m+1} |c_m\rangle\langle c_{m+1}| + \mathrm{h.c.} \right) \\ & + \left(t_{e,c} |e_{N^{\mathrm{EET}}}\rangle\langle c_{N^{\mathrm{EET}}+1}| + \mathrm{h.c.} \right) \end{aligned} \tag{3}$$

describes an excitonic system composed of N^{EET} states $|e_n\rangle$ with energy ε_n and excitonic coupling $V_{n,m}$ (first and second term) and N^{CT} CT states $|c_m\rangle$ with energy

ε_m and electronic coupling $t_{m,m+1}$ due to wavefunction overlap (third and fourth term) with $M = N^{\text{EET}} + N^{\text{CT}}$ being the total number of states. The EET–CT coupling element $t_{e,c}$ mediates the transfer from EET to CT subsystem.

All relevant information about the environment is contained in the spectral density function (*cf.* Section 2.3)

$$J(\omega) = \frac{\pi}{2}\sum_j \frac{c_j{}^2}{m_j\omega_j}\delta(\omega - \omega_j). \tag{4}$$

Time-dependent state populations and coherences are evaluated *via* the reduced density matrix $\tilde{\rho}(t)$ that is obtained upon taking the trace over the bath degrees of freedom

$$\tilde{\rho}(t) = \text{Tr}_\text{b}[\text{e}^{-iHt/\hbar}\rho(0)\text{e}^{iHt/\hbar}]. \tag{5}$$

2.1 Iterative quasi-adiabatic propagator path integral (iQUAPI)

The time evolution of the reduced density matrix $\tilde{\rho}(t)$ (eqn (5)) can be expressed in discretized path integral representation by employing a factorized initial density matrix of system and environment and introducing discrete value representation (DVR) of system coordinates s[33]

$$\begin{aligned}\left\langle u_{k_N^+}\middle|\tilde{\rho}(t)\middle|u_{k_N^-}\right\rangle = &\sum_{k_0^+}^{M}\sum_{k_1^+}^{M}\cdots\sum_{k_{N-1}^+}^{M}\sum_{k_0^-}^{M}\sum_{k_1^-}^{M}\cdots\sum_{k_{N-1}^-}^{M} \\ &\times S\left(s_{k_0^\pm},\ldots,s_{k_N^\pm};t\right)I\left(s_{k_0^\pm},\ldots,s_{k_N^\pm};t\right),\end{aligned} \tag{6}$$

with $[s_{k_j^+},\ j = 0,\ldots,N]$ and $[s_{k_j^-},\ j = 0,\ldots,N]$ denoting discretized path segments of forward and backward paths, respectively, at time $j\Delta t$ with N being the number of time steps with step size $\Delta t = t/N$. The contribution of each path arises from a time-local short time propagator S

$$\begin{aligned}S\left(s_{k_0^\pm},\ldots,s_{k_N^\pm};t\right) = &\left\langle u_{k_0^+}\middle|\tilde{\rho}(0)\middle|u_{k_0^-}\right\rangle \\ &\times\prod_{j=1}^{N} S_{k_j}\left(s_{k_j^\pm},s_{k_{j-1}^\pm}\right)\end{aligned} \tag{7}$$

with

$$S_{k_j}\left(s_{k_j^\pm},s_{k_{j-1}^\pm}\right) = \left\langle u_{k_j^+}\middle|\text{e}^{-iH_\text{s}\Delta t/\hbar}\middle|u_{k_{j-1}^+}\right\rangle \times \left\langle u_{k_{j-1}^-}\middle|\text{e}^{iH_\text{s}\Delta t/\hbar}\middle|u_{k_j^-}\right\rangle \tag{8}$$

and the time-nonlocal Feynman–Vernon influence functional

$$\begin{aligned}I\left(s_{k_0^\pm},\ldots,s_{k_N^\pm};t\right) &= \prod_{j=0}^{N}\prod_{j'=0}^{j} I_{k_jk_{j'}}\left(s_{k_j^\pm},s_{k_{j'}^\pm}\right) \\ &= \exp\left\{-\frac{1}{\hbar}\sum_\alpha\sum_{j=0}^{N}\sum_{j'=0}^{j}\left(s_{k_j^+,\alpha} - s_{k_j^-,\alpha}\right)\times\left(\eta_{jj'}^{(\alpha)}s_{k_{j'}^+,\alpha} - \eta_{jj'}^{(\alpha)*}s_{k_{j'}^-,\alpha}\right)\right\}.\end{aligned} \tag{9}$$

Here $\eta_{jj'}$ denotes influence functional coefficients with timelag $\Delta t(j - j')$ that represent the discretized analog of the bath response function.[33] Eqn (9) accounts for multiple independent baths by extending lattice coordinates s and influence functional coefficients η into multiple dimensions α (ref. 34) (explicit expressions for $\eta_{jj'}$ are given in the appendix (eqn (21)–(26))). Taking into account the finite memory time $\tau_M = \Delta k_{max}\Delta t$ of condensed phase systems allows us to construct the propagator functional iteratively (iterative quasi-adiabatic propagator path integral – iQUAPI).[35–37]

The iQUAPI propagator (eqn (7)–(9)) can be re-factorized[38] into a history term

$$\mathcal{H}(\mathcal{P}_i) = \left\langle u_{k_0^+} \middle| \tilde{\rho}(0) \middle| u_{k_0^-} \right\rangle \prod_{j=1}^{N-1} S_{k_j}\left(s_{k_j^\pm}, s_{k_{j-1}^\pm}\right) \times \prod_{j=0}^{N-1} \prod_{j'=0}^{j} I_{k_j k_{j'}}\left(s_{k_j^\pm}, s_{k_{j'}^\pm}\right) \tag{10}$$

that accounts for interactions between all $N-1$ past time points and a propagator term

$$\mathcal{P}\left(\mathcal{P}_i, s_{k_N^\pm}\right) = S_{k_N}\left(s_{k_N^\pm}, s_{k_{N-1}^\pm}\right) \prod_{j'=0}^{N} I_{k_N k_{j'}}\left(s_{k_N^\pm}, s_{k_{j'}^\pm}\right) \tag{11}$$

that accounts for interactions of the current and previous time steps. The re-factorized propagator for L paths p

$$\left\langle u_{k_N^+} \middle| \tilde{\rho}(t) \middle| u_{k_N^-} \right\rangle = \sum_{i=1}^{L} \mathcal{H}(\mathcal{P}_i)\mathcal{P}(\mathcal{P}_i, k_N^\pm) \tag{12}$$

allows for on-the-fly path selection by taking into account the finite memory time $\tau_M = \Delta k_{max}\Delta t$ of condensed phase systems. Information on all segments of Feynman paths (of fixed length τ_M) is used for propagation, employing the history and propagator terms. In on-the-fly path selection[38] Feynman paths that coincide in every path segment are merged. The maximum number of conceivable paths is defined by the path length ($M^{2\Delta k_{max}}$, with Δk_{max} being the number of time steps contained in memory time τ_M).

Iterative propagation of system dynamics can be further improved by filtering contributing paths according to their quantum amplitude/weight (*i.e.* absolute value of the summand of eqn (6)) using a filter threshold θ.[38–40] The on-the-fly filtered propagator functional (OFPF) algorithm preserves important properties, *i.e.*, convergence to numerical exact results can be rigorously obtained for $\tau_M \to \infty$, $\Delta t \to 0$ and $\theta \to 0$.

2.2 Mask assisted coarse graining of influence coefficients (MACGIC)

Recently, we introduced an algorithmic extension of the OFPF method that employs mask assisted coarse graining of influence coefficients (MACGIC) for a reduction of considered paths during propagation.[41] Within the OFPF algorithm, information about all segments of the Feynman path of size Δk_{max} is contained. In the MACGIC-iQUAPI algorithm path segments are classified into two distinct categories *via* an optimized mask function (*cf.* Fig. 2). Path segments selected by the mask define category 1 while the intermediate path segments of category 2 are discarded (*cf.* Fig. 2, blue and orange path segments, respectively).

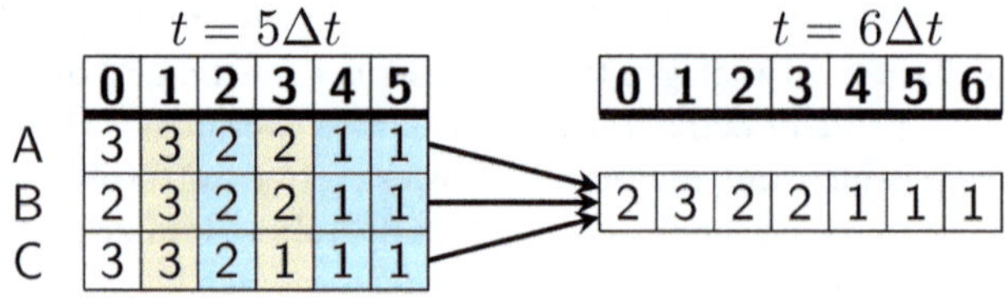

Fig. 2 Merging of indistinguishable Feynman paths in the MACGIC scheme: three exemplary forward paths A, B, and C are shown with number of states $M = 3$ and memory time steps $\Delta k_{\max} = 5$ as denoted by colored path segments. Blue segments are the coarse grained path segments selected by the mask function and employed in the merging procedure (category 1); orange path segments are discarded during path merging (category 2). Although Feynman paths A and B are distinguishable from path C at time $t = 3\Delta t$, all three coincide for the coarse grained path segments defined by the mask function. The merging into a single effective Feynman path that is employed for propagation reduces computational effort.

Feynman paths that coincide in all path segments of category 1 are merged, just as in OFPF simulations. Propagation of the propagator tensor is performed with the reduced number of Feynman paths that contain the dynamic information of both categories along the entire path. Using a sparse mask of k_{eff} path segments (category 1), the number of conceivable paths is reduced to $M^{2k_{\text{eff}}}$, compared to $M^{2\Delta k_{\max}}$ in the OFPF algorithm. We could demonstrate, that careful optimization of the coarse grained mask allows for a lowering of computational effort of up to several orders of magnitude while preserving the accuracy of the simulations. The MACGIC-iQUAPI method was tested on a set of benchmark systems covering isolated CT or EET dynamics and spectral densities of Drude–Lorentz and ohmic type.

2.3 Spectral density $J(\omega)$

Spectral densities of ohmic and Drude–Lorentz type are both characterized by a monotonic decay of the absolute value of influence functional coefficients $|\eta_{jj'}|$ with increasing time lag/memory time. In contrast, structured spectral densities containing one (or more) peaks, result in a complex decaying behaviour of $|\eta_{jj'}|$. A straightforward way to include such behaviour in simulations is the use of i shifted Drude–Lorentz peaks

$$J(\omega)^{\text{DL}} = \sum_i \lambda_i \gamma_i \omega \left(\frac{1}{{\gamma_i}^2 + (\omega + \omega_{0,i})^2} + \frac{1}{{\gamma_i}^2 + (\omega - \omega_{0,i})^2} \right). \tag{13}$$

Here, λ is the reorganization energy, γ determines the peak width and therefore directly affects the memory time τ_{M} and ω_0 defines the peak position. Fig. 3(a) illustrates the effect of changing ω_0 from 0 to 400 cm^{-1} for a single resonance peak (eqn (13), $i = 1$).

For typical realizations of photosynthetic proteins, the spectral density of chromophores within the environment is composed of a broad background and discrete peaks that arise from molecular vibrations (*cf.* Fig. 3(d)). We constructed a prototypical spectral density as the sum of vibrations modeled by Drude–Lorentz peaks and a log-normal background

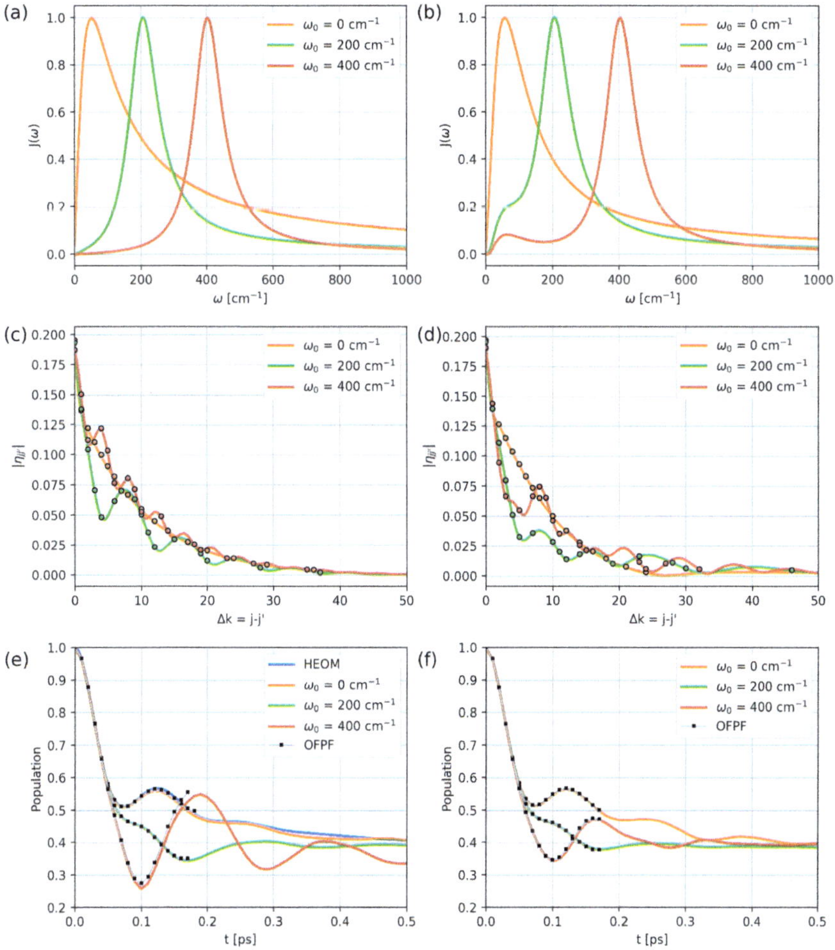

Fig. 3 Excitation energy transfer (EET) in the presence of vibrational structure: normalized spectral densities of (a) Drude–Lorentz type $J(\omega)^{\text{DL}}$ (eqn (13)) and (b) as the sum of a Drude–Lorentz peak (eqn (13)) and a log-normal background (eqn (14)) for varying resonance frequency ω_0. (c and d) Effect of spectral density on Feynman–Vernon influence functional coefficients $\eta_{jj'}$ for the spectral densities shown in (a) and (b), respectively. (e and f) Non-equilibrium population dynamics of the initially populated state in an asymmetric two-level system employing the different spectral densities of (a) and (b) as indicated. For $\omega_0 = 0\ \text{cm}^{-1}$ benchmarking to numerical exact HEOM results confirms the accuracy of MACGIC-iQUAPI simulations. Black symbols: OFPF short time propagations serve as further benchmark of convergence of the MACGIC-iQUAPI method for $\omega_0 > 0\ \text{cm}^{-1}$. Converged MACGIC-iQUAPI propagations employed a propagation tensor of size $L \approx 176 \times 10^6$ where L denotes the number of Feynman paths (eqn (12)), simulation parameters are given in Table 1.

$$J(\omega)^{\text{BG}} = \frac{\pi S \omega}{\sqrt{2\pi}\sigma} \mathrm{e}^{-[\ln(\omega/\omega_c)]^2} \Big/ 2\sigma^2, \tag{14}$$

which shows the functional form of experimentally determined spectral densities of bacteriochlorophylls in pigment–protein complexes.[42] Here ω_c denotes the cutoff frequency, σ is the standard deviation and S is the dimensionless Huang–

Table 1 Parameter settings of MACGIC-iQUAPI simulations

M	τ_M (fs)	Δt (fs)	k_{max}	k_{eff}	θ	L_{MACGIC}	Figure	ω_0 (cm^{-1})
2	480	10	48	16	5×10^{-9}	1.76×10^8	3e	0
2	480	10	48	16	5×10^{-9}	3.74×10^7	3e	200
2	480	10	48	16	5×10^{-9}	8.70×10^7	3e	400
2	480	10	48	48	1×10^{-10}	$>2.5 \times 10^9$	3e	0
2	480	10	48	48	1×10^{-10}	$>2.5 \times 10^9$	3e	200
2	480	10	48	48	1×10^{-10}	$>2.5 \times 10^9$	3e	400
2	480	10	48	16	1×10^{-9}	5.58×10^8	3f	0
2	480	10	48	16	1×10^{-9}	4.69×10^8	3f	200
2	480	10	48	16	1×10^{-9}	5.48×10^7	3f	400
2	480	10	48	48	1×10^{-9}	$>5.67 \times 10^8$	3f	0
2	480	10	48	48	1×10^{-9}	$>5.67 \times 10^8$	3f	200
2	480	10	48	48	1×10^{-9}	$>5.67 \times 10^8$	3f	400
3	480	1.76	272	10	1×10^{-9}	5.59×10^6	4b	0
3	480	1.76	272	10	1×10^{-9}	6.58×10^6	4b	500
4	455	1.09	416	10	1×10^{-9}	2.41×10^6	5a	0
4	455	1.09	416	10	1×10^{-9}	6.50×10^6	5a	400
4	455	1.09	416	10	1×10^{-9}	3.92×10^6	5a	0[a]
4	455	1.09	416	10	1×10^{-9}	6.50×10^6	5a	400[a]
4	455	1.09	416	12	1×10^{-8}	2.70×10^7	5a	0[b]
4	455	1.09	416	12	1×10^{-8}	2.71×10^7	5a	0[c]
4	455	1.09	416	12	1×10^{-8}	2.80×10^7	5a	0[d]

[a] $t_{3,4} = 31.3$ cm^{-1}. [b] $t_{3,4} = 51.3$ cm^{-1}. [c] $t_{3,4} = 71.3$ cm^{-1}. [d] $t_{3,4} = 91.3$ cm^{-1}.

Rhys factor $S = \lambda/(\omega_c \exp(0.5\sigma^2))$ with coupling strength λ. Compared to a spectral density of Drude–Lorentz form, the log-normal background has non-algebraic superohmic behavior at low frequencies.[43]

3 Results and discussion

In the following, numerical results of integrated EET–CT dynamics employing the MACGIC-iQUAPI method are presented. We begin by establishing the accuracy of the method for structured spectral densities in an asymmetric two-level system in the regime of EET (Section 3.1). Further, integrated EET and CT dynamics are investigated for bRC inspired model systems of increasing complexity (Sections 3.2 and 3.3), where the influence of structured spectral densities on the CT dynamics is investigated.

3.1 Excitation energy transfer in the presence of vibrational structure

In order to numerically demonstrate applicability of the MACGIC-iQUAPI method for structured spectral densities we start by discussing the dynamics in a prototypical, asymmetric two-level system[44] where each level is coupled to its own independent bath ($T = 300$ K). The system Hamiltonian (in cm^{-1}) is given by

$$H_S^{EET} = \sum_{n=1}^{N^{EET}=2} \varepsilon_n |e_n\rangle\langle e_n| + \sum_{n=1}^{1} (V_{n,n+1}|e_n\rangle\langle e_{n+1}| + \text{h.c.}) = \begin{pmatrix} 100. & 100. \\ 100. & 0. \end{pmatrix}. \tag{15}$$

We start by considering a single peak Drude–Lorentz spectral density (eqn (13), $i = 1$, Fig. 3(a)) with $\gamma = 53.08\ \text{cm}^{-1}$ ($\gamma = 10\ \text{ps}^{-1}$) that imposes substantial non-Markovian behavior in the dynamics. The resulting memory time $\tau_M = 480$ fs is equal for different realizations of $J(\omega)$ with varying ω_0 and necessitates for $\Delta k_{\max} = 48$ to account for the long-time decay of system bath correlations ($\Delta t = 10$ fs, Fig. 3(c)). For increasing $\omega_0 > 0$ the decay of influence functional coefficients $\eta_{jj'}$ is qualitatively altered from a monotonic decay to oscillatory behaviour.

Fig. 3(e) presents dynamics obtained with the MACGIC-iQUAPI method.[41] Simulations are compared to non-perturbative, numerical exact HEOM method of Tanimura and Kubo[45,46] that is employed for reference simulations with spectral densities of Drude–Lorentz form ($\omega_0 = 0\ \text{cm}^{-1}$, yellow and blue line, respectively). HEOM simulations have been performed with the PHI program[47] and converged with a high hierarchy level $L = 9$ and $K = 10$ Matsubara terms ($\approx 5.9 \times 10^6$ auxiliary density matrices). OFPF simulations[38] were performed for short time propagations, *i.e.*, the occurrence of the first oscillatory feature, that provide a further benchmark for convergence of the MACGIC-iQUAPI method. Due to the exponentially increasing number of paths within $\tau_M = \Delta k_{\max} \times \Delta t = 480$ fs short time propagations are limited to $\approx$160–170 fs on state-of-the-art compute nodes.

The MACGIC-iQUAPI method allows us to quantitatively reproduce benchmark HEOM and OFPF results, in particular the first coherent feature around 100–150 fs. For $\omega_0 = 200\ \text{cm}^{-1}$ and $\omega_0 = 400\ \text{cm}^{-1}$ (green and red lines, respectively) such coherent dynamics appears more pronounced and longer lived compared to the $\omega_0 = 0\ \text{cm}^{-1}$ case. The increased amplitude of coherences matches nearly quantitatively the OFPF results. The temporal shift of minima and maxima positions of the coherent feature points to a renormalizations of the eigenbasis of the system Hamiltonian in the presence of the bath, beyond perturbative treatments in the weak coupling limit. The presence of a structured spectral density function has been demonstrated to enhance the non-equilibrium quantum dynamics and extend electronic coherence.[25] The longer lived coherence arises from the interactions of excitons and the nuclear vibrations of the environment thus sustaining coherent oscillatory EET.

We further consider spectral densities composed of a broad log-normal background and one distinct Drude–Lorentz peak ($J(\omega) = J(\omega)^{\text{BG}} + J(\omega)^{\text{DL}}$, eqn (13) and (14), $i = 1$, Fig. 3(b)) that is particularly relevant for the modeling of spectral densities of bacteriochlorophylls in pigment–protein complexes. Numerical values of parameters are $S = 0.3$, $\sigma = 0.7$, $\omega_c = 38\ \text{cm}^{-1}$ (*cf.* ref. 42). Both contributions to $J(\omega)$ are weighted to give a total coupling strength of exciton states to the environment $\lambda = \lambda_{\text{BG}} + \lambda_{\text{DL}} = 25\ \text{cm}^{-1} + 75\ \text{cm}^{-1}$. During memory time $\tau_M = 480$ fs ($\Delta k_{\max} = 48$) system-bath correlations decay non-monotonously already for the $\omega_0 = 0\ \text{cm}^{-1}$ case (orange line Fig. 3(d)) where the weak oscillatory behaviour is imposed by the log-normal background. Such nontrivial decay of $|\eta_{jj'}|$ is further enhanced for $\omega_0 > 0\ \text{cm}^{-1}$.

Non-equilibrium dynamics simulations employing MACGIC-iQUAPI and OFPF methods show excellent agreement for the coherent oscillatory feature around 100–150 fs (*cf.* solid lines and symbols in Fig. 3(f)). Already for $\omega_0 = 0\ \text{cm}^{-1}$ due to the more complex temporal behaviour of $|\eta_{jj'}|$ (orange lines in Fig. 3(e) and (f)). For $\omega_0 = 200\ \text{cm}^{-1}$ and $\omega_0 = 400\ \text{cm}^{-1}$ (green and red line, respectively) the admixture of multiple nuclear frequency components decreases the amplitude of the coherent oscillation compared to the dynamics observed with a single peak

Drude–Lorentz spectral density. We note that dynamics simulations for spectral densities with a log-normal background and an additional Drude–Lorentz peak pose challenges to the mask optimization algorithm developed in ref. 41. For reliable masks and accurate dynamics we performed manual comparison to dynamics simulated with the OFPF method, showing possibilities for future algorithmic improvements of the mask optimization procedure in the MACGIC algorithm.

3.2 Excitation energy and charge transfer dynamics in a bRC-inspired three-state model

A bRC inspired model was constructed by coupling a dimeric exciton system that resembles special pair states P^+ and P^- to a single charge transfer state. Hamiltonian parameters reflect bRC *B. viridis* and were derived from ref. 15 where an isoenergetic, symmetric two-level system was considered in a particle hole basis with particularly strong excitonic coupling of the special pair

$$\begin{aligned} H_s = & \sum_{n=1}^{N^{\mathrm{EET}}=2} \varepsilon_n |e_n\rangle\langle e_n| + \sum_{n=1}^{1} (V_{n,n+1}|e_n\rangle\langle e_{n+1}| + \mathrm{h.c.}) \\ & + \sum_{m=N^{\mathrm{EET}}+1}^{M=3} \varepsilon_m |c_m\rangle\langle c_m| \\ & + (t_{e,c}|e_{N^{\mathrm{EET}}}\rangle\langle c_{N^{\mathrm{EET}}+1}| + \mathrm{h.c.}) \\ = & \begin{pmatrix} 11000 + \dfrac{\Delta\varepsilon}{2} & V & 0 \\ V & 11000 - \dfrac{\Delta\varepsilon}{2} & 32.7 \\ 0 & 32.7 & 9634.8 \end{pmatrix}. \end{aligned} \tag{16}$$

For $\Delta\varepsilon = 0$ and $V = 852\ \mathrm{cm}^{-1}$ (ref. 15) states 1 and 2 give rise to exciton states of the special pair (P^+/P^-) subject to maximal excitonic mixing of states, whereas state 3 resembles the primary charge separated state of the A-branch $P^+\ B_A{}^-H_A$. *Via* unitary transformation

$$U = \begin{pmatrix} \cos\phi & -\sin\phi \\ \sin\phi & \cos\phi \end{pmatrix}, \tag{17}$$

the two-level excitonic sub-system can be rotated into an asymmetric realization ($\Delta\varepsilon \neq 0$) without affecting P^+ and P^- eigenstates. Assuming $\phi = 30°$, we obtain $\Delta\varepsilon = 852\ \mathrm{cm}^{-1}$ and $V = 738\ \mathrm{cm}^{-1}$.

The total 12 state Hamiltonian of ref. 15 has been shown to reproduce the temperature dependent linear absorption spectrum of bRC *B. viridis*[48] with high accuracy. Modeled special pair exciton states are located at $P^+ = 11\,852\ \mathrm{cm}^{-1}$ and $P^- = 10\,148\ \mathrm{cm}^{-1}$ (corresponding to the weak and stronger Q_Y absorption at ≈850 and 980 nm, respectively). For $\phi = 30°$ (eqn (17)) the reduced excitonic coupling resembles *ab initio* findings on excitonic coupling and further accounts for asymmetry of the local electrostatic environment imposed by the protein.[49–52] Note that the energy of the P^+ state closely resembles recent reports of the state in

Rb. capsulatus[11] where reduced energetic splitting ($\approx$650 cm^{-1}) of P$^-$ and P$^+$ states is observed.

The simulations on the bRC-inspired three-state model (eqn (16)) assume that all states of the Hamiltonian are diagonally coupled to their own independent heatbath (T = 300 K), and fluctuations are characterized by the spectral densities described by eqn (13). We employ two Drude–Lorentz resonances, the first resembling a broad background ($\omega_0 = 0$, $\lambda_{BG} = \lambda_1 = 25$ cm^{-1}, $\gamma_{BG} = \gamma_1 = 53.08$ cm^{-1}), and the second peak accounting for a distinct mode with variable ω_0 ($\lambda_2 = 75$ cm^{-1}, $\gamma_2 = 53.08$ cm^{-1}; Fig. 4(a)). Accordingly, aggregated coupling strength of exciton states to the environment is $\lambda = \lambda_{BG} + \lambda_2 = 100$ cm^{-1}, whereas the coupling of the CT state with its environment is assumed to be stronger due to its polaronic character ($\lambda = 500$ cm^{-1}).[12,53] Note that the finite dephasing timescale γ substantially affects excitonic dynamics during memory time τ_M. The appreciable system-bath coupling strength together with the finite timescale of environment relaxation ($\tau_M = \Delta k_{max} \times \Delta t = 480$ fs) can not be accounted for in perturbative approaches, like Redfield theory. The numerical exact HEOM method formally provides access but the large coupling strength requires to increase the number of auxiliary density operators and a high hierarchy level for convergence.[26]

Fig. 4(b) presents the non-equilibrium dynamics for two different realizations of the spectral density (solid: $\omega_0 = 0$ cm^{-1}; dashed: $\omega_0 = 500$). In both cases coherent oscillations of exciton state population can be identified that are damped on a timescale of a few hundred femtoseconds. The population is then transferred to the low energy CT state on the $\approx$2 ps timescale. Coherent oscillations appear slightly more pronounced for the spectral density containing vibrational structure ($\omega_0 = 500$, dashed lines), resembling the findings of Section 3.1. Noticeably, the resonance frequency in the spectral density mostly affects equilibration among exciton states 1 and 2 while the depopulation of lower energy exciton state 2 (P$^-$) is unaffected. Such effects of vibrational structure on exciton equilibration dynamics could be even enhanced for models of bRC *Rb. sphaeroides*[54] and *Rb. capsulatus*[11] where the reduced energetic splitting of P$^-$ and P$^+$ states ($\approx$780–650 cm^{-1}) closely matches the investigated vibrational resonance frequency ω_0 in the spectral density (*cf.* Section 3.1). We further observe slightly accelerated population of the CT state (*cf.* dashed and solid red lines in Fig. 4(b)) and the details of spectral density can moderately affect population transfer from the excitonic subsystem to the CT trap state *via* the resonance frequency ω_0.

Quantitative insight into the dynamics for varying ω_0 was obtained by fitting the decay of the initially populated exciton state 1 (P$^+$) to a function of the form:

$$f_{\rm Ex1}(t) = A_{\rm EET}\cdot\cos(\omega_{\rm EET}\cdot t)\cdot e^{-t/\tau_{EET}} + (1 - A_{\rm EET} - A_{\rm Ex1})\cdot e^{-t/\tau_{\rm Ex1}} + A_{Ex1}. \quad (18)$$

Here, damping of coherence, represented by a cosine function with frequency $\omega_{\rm EET}$ and amplitude $A_{\rm EET}$, is reflected in the lifetime $\tau_{\rm EET}$. The exponential lifetime $\tau_{\rm Ex1}$ represents population decay of the exciton state Ex1. The CT state dynamics were characterized by fitting the rise with the exponential function

$$f_{\rm CT}(t) = N\cdot\left(1 - e^{-t/\tau_{\rm CT}}\right), \quad (19)$$

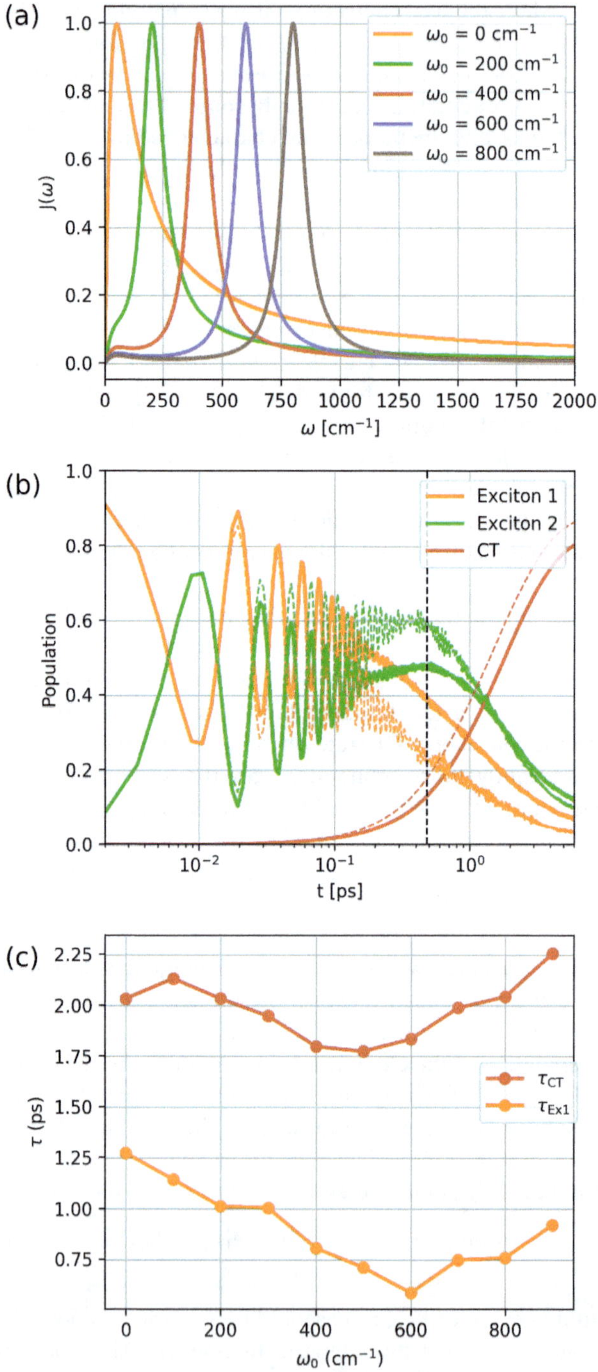

Fig. 4 Excitation energy and charge transfer dynamics in a bRC-inspired three-state model: (a) spectral densities composed of a background peak and a shifted mode located at ω_0 (eqn (13)) employed for simulations of bRC model systems; (b) non-equilibrium dynamics of a bRC inspired three-state model (solid lines – $\omega_0 = 0$ cm^{-1}, dashed lines – $\omega_0 = 500$ cm^{-1}); (c) lifetimes of exciton state Ex1 τ_{Ex1} and charge transfer τ_{CT} (eqn (18) and (19)) for varying resonance frequency ω_0. The minimum in τ_{CT} (ω_0 = 400–600 cm^{-1})

with τ_{CT} being the CT rise time. Eqn (18) and (19) assume exponential kinetics of CT reactions. We find that after the decay of initially coherent population oscillations the exponential fits provide a reasonable description of population dynamics, especially for >1 ps, while for short times minor deviations due to non-exponential dynamics are observed (with typical $\approx$10% and 5% deviation for τ_{Ex1} and τ_{CT}, respectively).

Fig. 4(c) shows τ_{CT} for varying values of the spectral density resonance frequency ω_0. We find that the decay time constants vary substantially (almost by a factor of 2) in the range 0.59 to 1.28 ps. The fastest depopulation of Ex1 is observed for $\omega_0 \approx 600$ cm^{-1}. In contrast, damping of P^+/P^- coherence occurs largely independent of ω_0 with a uniform time constant $\tau_{EET} \approx$ 50–70 fs, in agreement with experimental reports of excitonic equilibration in bRC.[11,50] The CT timescale τ_{CT} (eqn (19)) shows a minimal lifetime for $\omega_0 \approx 500$ cm^{-1}, slightly shifted to lower resonance frequencies and varies by about 15% ($\tau = 2.03$–1.78 ps). Here, the minimum in τ_{CT} qualitatively coincides with the energy gap of Ex1 state and the CT state that is 516 cm^{-1} upon diagonalization of eqn (16). As the CT process proceeds optimally, *i.e.*, non-activated ($-\Delta E \approx \lambda$), only weak dependences on ω_0 is expected.[3] Such findings of insensitivity of CT dynamics are in line with recent investigations[23] of the promoting role of vibrational modes on charge separation in the photosystem II RC that report a moderate influence for realistic reorganization energies >300 cm^{-1}. The second-order perturbative treatment employed in ref. 23 assumes instantaneous medium relaxation and charge separation from an equilibrated donor state. Here we numerically demonstrate that robustness of charge separation with respect to intramolecular vibrations is preserved in bRC if finite timescales of medium relaxation and charge separation from a coherent non-equilibrium state is taken into account.

In summary, we find that vibrations of the environment, here considered *via* a resonance frequency ω_0 in the spectral density function $J(\omega)$, can assist the charge separation process and lead to a moderate acceleration of population of the CT state even in the native, non-activated parameter regime of bRC. In principle such acceleration of dynamics can arise from two origins: (1) *via* a vibrational resonance that bridges the energy gap between the excitonic subsystem and the accepting manifold of CT states and (2) in an indirect way *via* affecting the coherence properties of the excitonic subsystem (*cf.* Section 3.1). In the current model we find that for a resonance frequency that approximately bridges the gap between excitonic subsystem and accepting CT states ($\omega_0 = 400$–600 cm^{-1}) most efficient transfer to CT states occurs and the moderate acceleration in τ_{CT} can be directly understood from energy differences of the model Hamiltonian (eqn (16)). The more pronounced dependence of τ_{Ex1} on ω_0 is less obvious due to the larger energy splitting of exciton states ($\Delta E_{12} = 1704$ cm^{-1} upon diagonalization of eqn (16)). Multi-quantal transitions in resonance frequency ω_0 provide an explanation that agrees with the minimum position of τ_{CT} in the range 500–600 cm^{-1} but other origins due to subtle effects on coherent exciton dynamics *via* ω_0 can not be ruled out. As the depopulation of exciton state

coincides with the energy gap between the lower exciton and charge separated state in the adiabatic reference Hamiltonian (eqn (16)). Simulation parameters are summarized in Table 1.

2 (P^-) is largely unaffected (*cf.* Fig. 4), a primary coherent origin of the accelerated dynamics within excitonic subsystem is suggested.

3.3 Coupled excitation energy and charge transfer dynamics in an excitonic donor–bridge–acceptor model

Further, an excitonic donor–bridge–acceptor tetrad was considered where the excitonic dimer P^+/P^- is coupled to charge transfer states $P^+\ B_A^-H_A$ and $P^+\ B_AH_A^-$ (*cf.* Fig. 1). Such a realization of downstream CT reactions is an eminent requirement for efficient primary charge separation within a photochemical unit in order to avoid thermal repopulation of initially excited state together with futile charge recombination.[8,13] The system Hamiltonian (in cm^{-1}) is given by

$$
\begin{aligned}
H_s = & \sum_{n=1}^{N^{EET}=2} \varepsilon_n |e_n\rangle\langle e_n| + \sum_{n=1}^{1} (V_{n,n+1}|e_n\rangle\langle e_{n+1}| + \text{h.c.}) \\
& + \sum_{m=N^{EET}+1}^{M=4} \varepsilon_m |c_m\rangle\langle c_m| + \sum_{m=N^{EET}+1}^{3} (t_{m,m+1}|c_m\rangle\langle c_{m+1}| + \text{h.c.}) \\
& + (t_{e,c}|e_{N^{EET}}\rangle\langle c_{N^{EET}+1}| + \text{h.c.}) \\
= & \begin{pmatrix} 11\,000 + \frac{\Delta\varepsilon}{2} & V & 0 & 0 \\ V & 11\,000 - \frac{\Delta\varepsilon}{2} & 32.7 & 0 \\ 0 & 32.7 & E_{\text{Bridge}} & 31.3 \\ 0 & 0 & 31.3 & 8478 \end{pmatrix}.
\end{aligned} \tag{20}
$$

As before asymmetric configuration of special pair excitons was assumed ($\phi = 30°$, eqn (17), $\Delta\varepsilon = 852\ cm^{-1}$ and $V = 738\ cm^{-1}$, *cf.* Section 3.2). In eqn (20) states 1 and 2 give rise to exciton states P^+/P^- of the special pair, state 3 corresponds to the primary charge separated bridge state $P^+\ B_A^-H_A$ and state 4 resembles the secondary charge separated state of the A-branch ($P^+\ B_AH_A^-$). The energetics of bRC *B. viridis* is approximately reproduced for $E_{\text{Bridge}} = 9634.8\ cm^{-1}$ where sequential CT is expected to be dominant at room temperature.[9,12]

For the system-bath interaction we assume that both exciton states are coupled to their own independent heatbath ($T = 300$ K), characterized by a dual peak spectral density (eqn (13)), *i.e.*, a Lorentzian background ($\lambda_{BG} = \lambda_1 = 25\ cm^{-1}$, $\gamma_{BG} = \gamma_1 = 53.08\ cm^{-1}$) together with vibrational structure around ω_0 ($\lambda_2 = 75\ cm^{-1}$, $\gamma_2 = 53.08\ cm^{-1}$). Aggregated coupling strength of exciton states to the environment is $\lambda = \lambda_{BG} + \lambda_2 = 100\ cm^{-1}$. Due to polaron character of CT states, substantial reorganization energies λ arise for ET reactions. We consider CT along a linear chain[53,55] where the primary CT reaction $P^*\ B_AH_A \rightarrow P^+\ B_A^-H_A$ is characterized by a reorganization energy $\lambda = 500\ cm^{-1}$ (DVR coordinate $d = 2.24$) and the secondary ET step $P^+\ B_A^-H_A \rightarrow P^+\ B_AH_A^-$ is characterized by a reorganization energy $\lambda = 1157\ cm^{-1}$ (DVR coordinate $d = 5.64$).[12,56] Reorganization energies approximately match the excitonic donor–bridge energy gap, as well as the bridge–acceptor energy gap, respectively, upon diagonalization of the Hamiltonian (eqn (20), non-activated CT reactions with $-\Delta E \approx \lambda$).[3,6,9,12]

Fig. 5(a) presents the non-equilibrium dynamics of the excitonic donor–bridge–acceptor model in the sequential regime of CT reactions ($E_{\text{Bridge}} = 9634.8\ \text{cm}^{-1}$). The dynamics appears initially coherent among exciton states 1 and 2 where oscillations are damped on a timescale of several hundred femtoseconds, followed by sequential transfer *via* the bridge–acceptor system on the few picosecond timescale. The simulated sequential CT dynamics compares reasonably to reported population dynamics of bRC. The maximum intermediate population of the bridge state $P^+ B_A^- H_A$ is $\approx$30% and slightly higher than respective values derived from experiment.[6,8] Such differences can be remedied by investigating the coupling dependence of the secondary CT reaction ($P^+ B_A^- H_A \rightarrow P^+ B_A H_A^-$) on the dynamics (dashed dotted lines in Fig. 5(a)). We find that for an increased coupling $t_{3,4} = 51.3–71.3\ \text{cm}^{-1}$ (*cf.* eqn (20)) the maximum intermediate population of the bridge state closely matches the experimentally derived population dynamics ($\approx$18% at 1.5 ps) reflecting a two-step sequential process with faster, sub-ps secondary ET reaction.[6,8,11,16]

In Fig. 5(b) the numerical efficiency of the MACGIC-QUAPI algorithm is analyzed. The number of employed Feynman paths selected by on-the-fly path filtering in MACGIC-QUAPI initially increases to a maximum value of $\approx$3.9 million. The maximum number of paths is selected after $\approx$550 fs where weak residual excitonic coherence prevails and, in parallel, population starts to

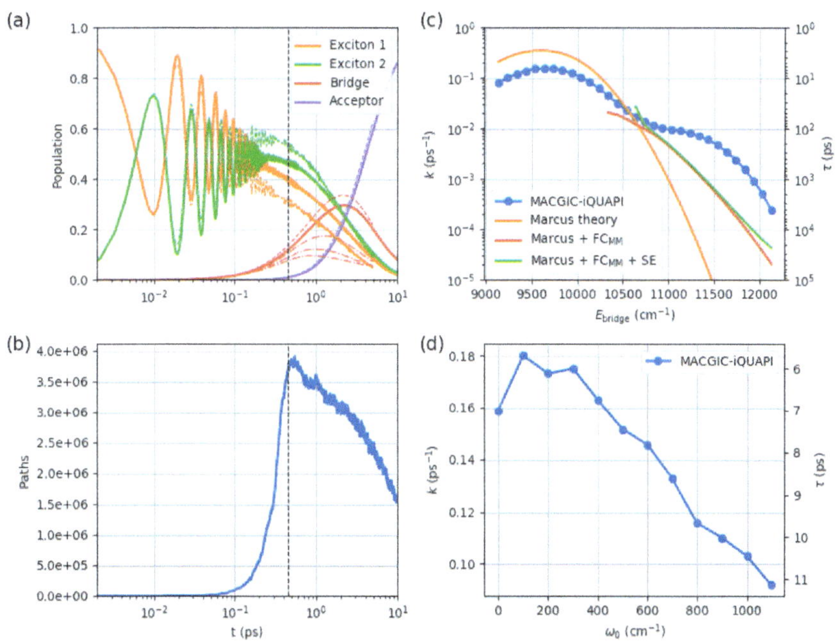

Fig. 5 Excitation energy and charge transfer dynamics in a bRC-inspired excitonic donor–bridge–acceptor model (*cf.* Fig. 1): (a) non-equilibrium population dynamics (solid lines – $\omega_0 = 0\ \text{cm}^{-1}$, dashed lines – $\omega_0 = 400\ \text{cm}^{-1}$), dashed-dotted lines denote the bridge state population for $t_{3,4} = [51.3, 71.3, 91.3]\ \text{cm}^{-1}$ (*cf.* eqn (20)); (b) number of Feynman paths used for propagation of system dynamics; (c) dependence of the bridge-to-acceptor transfer rate on the bridge state energy; (d) dependence of the bridge-to-acceptor transfer rate on spectral density resonance frequency ω_0. Simulation parameters are summarized in Table 1.

distribute among CT states. Later on, the number paths decreases upon the trapping of population in the acceptor state demonstrating the on-the-fly capabilities of the algorithm.

Fig. 5(c) demonstrates the dependence of the bridge-to-acceptor transfer rate on the bridge state energy. Transfer rates and characteristic lifetimes are evaluated *via* fitting the rise of acceptor state population to eqn (19). The MACGIC-QUAPI results are compared to transfer rates employing Marcus theory in an incoherent (three-state) donor–bridge–acceptor model that inherently assumes an equilibrated donor state (see Appendix for details). For both methods, maximal transfer rate is observed for $-\Delta E \approx \lambda$ ($E_{\text{Bridge}} \approx 9600\ \text{cm}^{-1}$) but respective rates differ by approximately a factor of 2 due to finite timescale of the bath reorganization in the numerically more elaborate path integral treatment. For $-\Delta E > \lambda$ parabolic dependence of the rate on driving force is observed. For $E_{\text{Bridge}} > 10\,300\ \text{cm}^{-1}$ ($-\Delta E + \lambda > 700\ \text{cm}^{-1}$) Marcus theory loses validity due to (i) the absence of vibrational excitations of intramolecular modes and (ii) the absence of superexchange contributions to the transfer rate.[3] Such deficiencies can to some extend qualitatively be recovered by considering a multi-mode picture in Franck–Condon factors (*cf.* appendix, eqn (31)) and first-order superexchange correction to the transfer rate. We note that such perturbative treatment is only valid for $E_{\text{Bridge}} > 11\,300\ \text{cm}^{-1}$ ($-\Delta E + \lambda > 1700\ \text{cm}^{-1}$) *i.e.* in the absence of donor and bridge state resonances. The non-perturbative MACGIC-iQUAPI method describes population transfer to the acceptor for all bridge energies without the emergence of divergences and can further account for coherent excitation of donor states, *i.e.*, charge separation from non-equilibrium states. The plateau-like behaviour of the transfer rates for $10\,500\ \text{cm}^{-1} < E_{\text{Bridge}} < 11\,500\ \text{cm}^{-1}$ resembles the findings of HEOM simulations[57] and represents the transition to a parameter regime where an appreciable contribution to the total transfer rate *via* vibrational excitations of intramolecular modes and a superexchange mechanism arise. In agreement with ref. 58, we find that superexchange contributions to the transfer rate become relevant for $E_{\text{Bridge}} > 11\,700\ \text{cm}^{-1}$ ($-\Delta E + \lambda > 2100\ \text{cm}^{-1}$, *cf.* the red and green lines in Fig. 5 (cs)).

Fig. 5(d) analyzes the effect of spectral density resonance frequency ω_0 (*cf.* Fig. 3(a)) on the bridge-to-acceptor transfer rate in the sequential CT regime ($E_{\text{Bridge}} = 9634.8\ \text{cm}^{-1}$); transfer rates have been evaluated by fitting eqn (19) to the rise of acceptor state population evaluated on MACGIC-iQUAPI level of theory. We find that the bridge-to-acceptor transfer rate is only weakly affected by ω_0, in particular in the range $\omega_0 = [0\text{–}400]\ \text{cm}^{-1}$. For $\omega_0 > 400\ \text{cm}^{-1}$ the transfer rate moderately decreases, corresponding to a range of lifetime $\tau_{\text{CT}} = 7\text{–}10$ ps. Such behaviour arises from two effects that together lead to an insensitivity of the bridge-to-acceptor transfer rate on details of the spectral density. First, the bridge-to-acceptor CT reaction proceeds optimally, *i.e.*, non-activated ($-\Delta E \approx \lambda$) and only weak dependence on ω_0 is expected (*cf.* Section 3.2). Second, the variation of ω_0 moderately increases the population of the CT bridge state ($\omega_0 = 400\ \text{cm}^{-1}$, *cf.* dashed red line in Fig. 5). For $\omega_0 > 500\ \text{cm}^{-1}$ the intermediate population of the CT bridge state is reduced again due to detuning from the exciton P^--to-bridge energy gap. Such reduced intermediate CT bridge state decelerates the transfer rate to the acceptor state.

The presented model simulations of coupled EET–CT population dynamics in the excitonic donor–bridge–acceptor setup is characterized by substantial driving force and system-bath coupling as the reaction coordinate progresses from the EET to CT subsystem. The finite timescale of environment relaxation ($\tau_M = \Delta k_{max} \times \Delta t = 480$ fs, *cf.* dashed line in Fig. 5) is in agreement with reported energy gap fluctuation correlation functions of sequential CT processes that suggest strong electrostatic coupling of the protein matrix to the electron transfer.[56,59,60] The bRC inspired model thus covers extended, complex dynamics ranging from initially coherent oscillatory dynamics to the incoherent trapping of population within low energy charge separated states where medium reorganization occurs on comparable timescales to system dynamics. Note that due to the appreciable driving force of sequential CT steps the thermal repopulation of exciton donor states is largely suppressed and equilibrated charge separation proceeds highly efficient. In particular, equilibrated long-time population of the acceptor state 4 is >99%. In contrast, the acceptor state in the three-state model (Section 3.2) is populated with 92% while 8% are distributed over P^+/P^- states of the special pair.

4 Summary and conclusions

We have presented the step-by-step construction of an excitonic donor–bridge–acceptor model that is inspired by the available parameters and dynamics of bRC. The effects of structure in the spectral density function on population dynamics and persistence of coherent dynamics has been benchmarked and evaluated with the MACGIC-iQUAPI method. The results suggest a moderate impact of vibrational modes on transfer dynamics and a decoupling of timescales and of coherent EET and CT dynamics. Accordingly, the classification of CT reactions as non-adiabatic[48] provides a reasonable description of the charge separation dynamics in bRC at room temperature where equilibration of excitonic coherence precedes CT. Our findings are in agreement with conclusions of ref. 23 and suggest an emerging picture where intramolecular vibrations assure the robustness of optimal CT reactions. Vibrational modes of the spectral density function mediate resonance conditions for optimal, *i.e.*, non-activated CT reactions within photosynthetic reaction centre complexes and allow for insensitivity to various realizations of static disorder. Under such optimum conditions the effect of vibrational modes on transfer dynamics due to a mixing of electronic and vibrational states is minor.

5 Appendix

5.1 Explicit expressions of influence functional coefficients $\eta_{jj'}$ (ref. 36):

If $0 < j' < j < N$:

$$\eta_{jj'} = \frac{2}{\pi}\int_{-\infty}^{\infty} \mathrm{d}\omega \frac{J(\omega)\mathrm{e}^{\beta\hbar\omega/2}}{\omega^2 \sinh(\beta\hbar\omega/2)} \sin^2(\omega\Delta t/2)\mathrm{e}^{-i\omega\Delta t(j-j')} \tag{21}$$

If $0 < j' = j < N$:

$$\eta_{jj} = \frac{1}{2\pi}\int_{-\infty}^{\infty} \mathrm{d}\omega \frac{J(\omega)\mathrm{e}^{\beta\hbar\omega/2}}{\omega^2 \sinh(\beta\hbar\omega/2)}\left(1 - \mathrm{e}^{-i\omega\Delta t}\right) \tag{22}$$

If $j = j' = 0$ or $j = j' = N$:

$$\eta_{00} = \eta_{NN} = \frac{1}{2\pi}\int_{-\infty}^{\infty} \mathrm{d}\omega \frac{J(\omega)\mathrm{e}^{\beta\hbar\omega/2}}{\omega^2 \sinh(\beta\hbar\omega/2)}\left(1 - \mathrm{e}^{-i\omega\Delta t/2}\right) \tag{23}$$

If $0 = j' < j < N$:

$$\eta_{j0} = \frac{2}{\pi}\int_{-\infty}^{\infty} \mathrm{d}\omega \frac{J(\omega)\mathrm{e}^{\beta\hbar\omega/2}}{\omega^2 \sinh(\beta\hbar\omega/2)} \sin(\omega\Delta t/4) \times \sin(\omega\Delta t/2)\mathrm{e}^{-i\omega(j\Delta t - \Delta t/4)} \tag{24}$$

If $0 < j' < j = N$:

$$\eta_{Nj'} = \frac{2}{\pi}\int_{-\infty}^{\infty} \mathrm{d}\omega \frac{J(\omega)\mathrm{e}^{\beta\hbar\omega/2}}{\omega^2 \sinh(\beta\hbar\omega/2)} \sin(\omega\Delta t/4) \times \sin(\omega\Delta t/2)\mathrm{e}^{-i\omega(t - j'\Delta t - \Delta t/4)} \tag{25}$$

If $j' = 0$ and $j = N$:

$$\eta_{N0} = \frac{2}{\pi}\int_{-\infty}^{\infty} \mathrm{d}\omega \frac{J(\omega)\mathrm{e}^{\beta\hbar\omega/2}}{\omega^2 \sinh(\beta\hbar\omega/2)} \sin^2(\omega\Delta t/4)\mathrm{e}^{-i\omega(t - \Delta t/2)} \tag{26}$$

5.2 Marcus theory

Second order perturbation theory provides an incoherent description of CT dynamics in a donor–bridge–acceptor model and lacks non-equilibrium effects, and quantum phenomena such as coherent exciton dynamics or superexchange. CT resembling the dynamics of bRC was further simulated in a 3 state donor–bridge–acceptor model that omits details of exciton states; the reduced Hamiltonian is

$$H_{\text{Marcus}} = \begin{pmatrix} E_\text{D} & V_\text{DB} & V_\text{SE} \\ V_\text{DB} & E_\text{B} & V_\text{BA} \\ V_\text{SE} & V_\text{BA} & E_\text{A} \end{pmatrix} \tag{27}$$

with donor state energy $E_\text{D} = 10\,148\ \text{cm}^{-1}$ corresponding to the lower exciton state P^-, acceptor state energy $E_\text{A} = 8478\ \text{cm}^{-1}$ and the bridge energy was tuned from 9100 to 12 100 cm^{-1}. As in eqn (20) donor–bridge coupling is $V_\text{DB} = 32.7\ \text{cm}^{-1}$, and bridge–acceptor coupling is $V_\text{BA} = 31.3\ \text{cm}^{-1}$. Considering the first-order superexchange correction, V_SE is defined as

$$V_\text{SE} = \frac{V_\text{DB} V_\text{BA}}{E_\text{B} - E_\text{D}}. \tag{28}$$

Marcus theory rate constants were calculated for reorganization energies ($\lambda_\text{DB} = 500\ \text{cm}^{-1}$, $\lambda_\text{BA} = 1157\ \text{cm}^{-1}$, $\lambda_\text{DA} = 3181\ \text{cm}^{-1}$), free energy ($\Delta G_\text{DB} = E_\text{B} - E_\text{D}$, $\Delta G_\text{BA} = E_\text{A} - E_\text{B}$, $\Delta G_\text{DA} = E_\text{A} - E_\text{D}$) and interstate couplings (V_DB, V_BA, V_SE) according to:

$$k_{ij} = \sqrt{\frac{4\pi^3}{h^2 \lambda_{ij} k_\text{B} T}} V_{ij}{}^2 \mathrm{e}^{-\frac{\left(\Delta G_{ij} + \lambda_{ij}\right)^2}{4\lambda_{ij} k_\text{B} T}}, \tag{29}$$

with temperature $T = 300$ K and Boltzmann constant k_B. Rate constants for the back reaction were derived from detailed balance condition

$$k_{ji} = k_{ij} e^{-\frac{E_2 - E_1}{k_B T}}. \tag{30}$$

Fig. 5(c) additionally presents rate constants accounting for a multi-mode picture in Franck–Condon factors (high-temperature limit).[3,8] Here, an effective resonant mode with frequency ω_H affects transfer dynamics and the rate constant is given by:

$$k_{ij} = \frac{2\pi}{\hbar} V_{ij}{}^2 \frac{1}{\sqrt{4\pi\lambda_{ij} k_B T}} e^{-S_H} \sum_{n_i=0}^{\infty} \frac{S_H^{n_i}}{n_i!} e^{-\frac{\left(\Delta G_{ij} + \lambda_{ij} + n_i \hbar\omega_H\right)^2}{4\lambda_{ij} k_B T}}, \tag{31}$$

with the dimensionless coupling strength $S_H = 2$, resembling polar environments, and $\omega_H = 500\ \text{cm}^{-1}$.

Conflicts of interest

There are no conflicts to declare.

Acknowledgements

The authors would like to thank Wolfgang Zinth for stimulating and fruitful discussions. B. P. F. gratefully acknowledges support through the DFG within the Emmy Noether Programme (Grant No. FI 2034/1-1).

References

1 C. C. Moser, J. M. Keske, K. Warncke, R. S. Farid and P. L. Dutton, *Nature*, 1992, **355**, 796–802.
2 C. K. Chan, T. J. DiMagno, L. X. Chen, J. R. Norris and G. R. Fleming, *Proc. Natl. Acad. Sci. U. S. A.*, 1991, **88**, 11202–11206.
3 M. Bixon and J. Jortner, *Electron Transfer—from Isolated Molecules to Biomolecules*, Wiley-Blackwell, 2007, pp. 35–202.
4 W. Holzapfel, U. Finkele, W. Kaiser, D. Oesterhelt, H. Scheer, H. U. Stilz and W. Zinth, *Proc. Natl. Acad. Sci. U. S. A.*, 1990, **87**, 5168.
5 T. Arlt, S. Schmidt, W. Kaiser, C. Lauterwasser, M. Meyer, H. Scheer and W. Zinth, *Proc. Natl. Acad. Sci. U. S. A.*, 1993, **90**, 11757.
6 W. Zinth and J. Wachtveitl, *ChemPhysChem*, 2005, **6**, 871–880.
7 A. R. Holzwarth, M. G. Müller, M. Reus, M. Nowaczyk, J. Sander and M. Rögner, *Proc. Natl. Acad. Sci. U. S. A.*, 2006, **103**, 6895.
8 B. P. Fingerhut, W. Zinth and R. de Vivie-Riedle, *Chem. Phys. Lett.*, 2008, **466**, 209–213.
9 W. W. Parson and A. Warshel, in *Mechanism of Charge Separation in Purple Bacterial Reaction Centers*, ed. C. N. Hunter, F. Daldal, M. C. Thurnauer and J. T. Beatty, Springer Netherlands, Dordrecht, 2009, pp. 355–377.
10 M. G. Müller, C. Slavov, R. Luthra, K. E. Redding and A. R. Holzwarth, *Proc. Natl. Acad. Sci. U. S. A.*, 2010, **107**, 4123.

11 A. Niedringhaus, V. R. Policht, R. Sechrist, A. Konar, P. D. Laible, D. F. Bocian, D. Holten, C. Kirmaier and J. P. Ogilvie, *Proc. Natl. Acad. Sci. U. S. A.*, 2018, **115**, 3563–3568.
12 M. Bixon, J. Jortner and M. E. Michel-Beyerle, *Chem. Phys.*, 1995, **197**, 389–404.
13 A. Warshel and D. W. Schlosser, *Proc. Natl. Acad. Sci. U. S. A.*, 1981, **78**, 5564.
14 B. P. Fingerhut, W. Zinth and R. de Vivie-Riedle, *Phys. Chem. Chem. Phys.*, 2010, **12**, 422–432.
15 B. P. Fingerhut and S. Mukamel, *J. Phys. Chem. Lett.*, 2012, **3**, 1798–1805.
16 F. Ma, E. Romero, M. R. Jones, V. I. Novoderezhkin and R. van Grondelle, *J. Phys. Chem. Lett.*, 2018, 1827–1832.
17 H.-G. Duan, V. I. Prokhorenko, E. Wientjes, R. Croce, M. Thorwart and R. J. D. Miller, *Sci. Rep.*, 2017, **7**, 12347.
18 E. Romero, J. Prior, A. W. Chin, S. E. Morgan, V. I. Novoderezhkin, M. B. Plenio and R. van Grondelle, *Sci. Rep.*, 2017, **7**, 2890.
19 M. H. Vos, F. Rappaport, J.-C. Lambry, J. Breton and J.-L. Martin, *Nature*, 1993, **363**, 320–325.
20 S. Spörlein, W. Zinth and J. Wachtveitl, *J. Phys. Chem. B*, 1998, **102**, 7492–7496.
21 V. I. Novoderezhkin, A. G. Yakovlev, R. van Grondelle and V. A. Shuvalov, *J. Phys. Chem. B*, 2004, **108**, 7445–7457.
22 D. Paleček, P. Edlund, S. Westenhoff and D. Zigmantas, *Sci. Adv.*, 2017, **3**, e1603141.
23 Y. Fujihashi, M. Higashi and A. Ishizaki, *J. Phys. Chem. Lett.*, 2018, **9**, 4921–4929.
24 D. I. G. Bennett, P. Malý, C. Kreisbeck, R. van Grondelle and A. Aspuru-Guzik, *J. Phys. Chem. Lett.*, 2018, **9**, 2665–2670.
25 A. W. Chin, J. Prior, R. Rosenbach, F. Caycedo-Soler, S. F. Huelga and M. B. Plenio, *Nat. Phys.*, 2013, **9**, 113–118.
26 T. Kramer, M. Rodríguez and Y. Zelinskyy, *J. Phys. Chem. B*, 2017, **121**, 463–470.
27 J. Deisenhofer, O. Epp, I. Sinning and H. Michel, *J. Mol. Biol.*, 1995, **246**, 429–457.
28 Y. Yan, L. Song and Q. Shi, *J. Chem. Phys.*, 2018, **148**, 084109.
29 E. Mangaud, C. Meier and M. Desouter-Lecomte, *Chem. Phys.*, 2017, **494**, 90–102.
30 S. Sakamoto and Y. Tanimura, *J. Phys. Chem. Lett.*, 2017, **8**, 5390–5394.
31 U. Weiss, *Quantum Dissipative Systems*, World Scientific, 3rd edn, 2008.
32 I. de Vega and D. Alonso, *Rev. Mod. Phys.*, 2017, **89**, 015001.
33 N. Makri, *J. Math. Phys.*, 1995, **36**, 2430–2457.
34 P. Nalbach, J. Eckel and M. Thorwart, *New J. Phys.*, 2010, **12**, 065043.
35 D. E. Makarov and N. Makri, *Chem. Phys. Lett.*, 1994, **221**, 482–491.
36 N. Makri and D. E. Makarov, *J. Chem. Phys.*, 1995, **102**, 4600–4610.
37 N. Makri and D. E. Makarov, *J. Chem. Phys.*, 1995, **102**, 4611–4618.
38 E. Sim, *J. Chem. Phys.*, 2001, **115**, 4450–4456.
39 E. Sim and N. Makri, *Chem. Phys. Lett.*, 1996, **249**, 224–230.
40 E. Sim and N. Makri, *Comput. Phys. Commun.*, 1997, **99**, 335–354.
41 M. Richter and B. P. Fingerhut, *J. Chem. Phys.*, 2017, **146**, 214101.
42 A. Kell, X. Feng, M. Reppert and R. Jankowiak, *J. Phys. Chem. B*, 2013, **117**, 7317–7323.
43 G. Ritschel and A. Eisfeld, *J. Chem. Phys.*, 2014, **141**, 094101.
44 A. Ishizaki and G. R. Fleming, *J. Chem. Phys.*, 2009, **130**, 234111.

45 Y. Tanimura and R. Kubo, *J. Phys. Soc. Jpn.*, 1989, **58**, 101–114.
46 Y. Tanimura, *Phys. Rev. A: At., Mol., Opt. Phys.*, 1990, **41**, 6676–6687.
47 J. Strümpfer and K. Schulten, *J. Chem. Theory Comput.*, 2012, **8**, 2808–2816.
48 P. Huppmann, S. Spörlein, M. Bibikova, D. Oesterhelt, J. Wachtveitl and W. Zinth, *J. Phys. Chem. A*, 2003, **107**, 8302–8309.
49 A. Warshel, *Proc. Natl. Acad. Sci. U. S. A.*, 1980, **77**, 3105.
50 D. C. Arnett, C. C. Moser, P. L. Dutton and N. F. Scherer, *J. Phys. Chem. B*, 1999, **103**, 2014–2032.
51 M. E.-A. Madjet, F. Müh and T. Renger, *J. Phys. Chem. B*, 2009, **113**, 12603–12614.
52 F. Zheng, M. Jin, T. Mančal and Y. Zhao, *J. Phys. Chem. B*, 2016, **120**, 10046–10058.
53 E. Sim and N. Makri, *J. Phys. Chem. B*, 1997, **101**, 5446–5458.
54 X. J. Jordanides, G. D. Scholes and G. R. Fleming, *J. Phys. Chem. B*, 2001, **105**, 1652–1669.
55 N. Makri, E. Sim, D. E. Makarov and M. Topaler, *Proc. Natl. Acad. Sci. U. S. A.*, 1996, **93**, 3926.
56 H. Treutlein, K. Schulten, A. T. Brünger, M. Karplus, J. Deisenhofer and H. Michel, *Proc. Natl. Acad. Sci. U. S. A.*, 1992, **89**, 75.
57 M. Tanaka and Y. Tanimura, *J. Chem. Phys.*, 2010, **132**, 214502.
58 E. Sim, *J. Phys. Chem. B*, 2004, **108**, 19093–19095.
59 M. Marchi, J. N. Gehlen, D. Chandler and M. Newton, *J. Am. Chem. Soc.*, 1993, **115**, 4178–4190.
60 W. W. Parson and A. Warshel, in *Theoretical Analyses of Electron-Transfer Reactions*, ed. R. E. Blankenship, M. T. Madigan and C. E. Bauer, Springer Netherlands, Dordrecht, 1995, pp. 559–575.

Faraday Discussions

PAPER

A theoretical study on the dynamics of light harvesting in the dimeric photosystem II core complex: regulation and robustness of energy transfer pathways†

Shou-Ting Hsieh, [a] Lu Zhang,[b] De-Wei Ye,[a] Xuhui Huang *[c] and Yuan-Chung Cheng *[a]

Received 25th November 2018, Accepted 22nd January 2019

DOI: 10.1039/c8fd00205c

Here we present our theoretical investigations into the light reaction in the dimeric photosystem II (PSII) core complex. An effective model for excitation energy transfer (EET) and primary charge separation (CS) in the PSII core complex was developed, with model parameters constructed based on molecular dynamics (MD) simulation data. Compared to experimental results, we demonstrated that this model faithfully reproduces the absorption spectra of the RC and core light-harvesting complexes (CP43 and CP47) as well as the full EET dynamics among the chromophores in the PSII core complex. We then applied master equation simulations and network analysis to investigate detailed EET plus CS dynamics in the system, allowing us to identify key EET pathways and produce a coarse-grained cluster model for the light reaction in the dimeric PSII core complex. We show that non-equilibrium energy transfer channels play important roles in the efficient light harvesting process and that multiple EET pathways exist between subunits of PSII to ensure the robustness of light harvesting in the system. Furthermore, we revealed that inter-monomer energy transfer dominated by the coupling between the two CLA625 molecules enables efficient energy exchange between two CP47s in the dimeric PSII core complex, which leads to significant energy pooling in the CP47 domain during the light reaction. Our study provides a blueprint for the design of light harvesting in the PSII core and show that a structure-based approach using molecular dynamics simulations and quantum chemistry calculations can be effectively utilized to elucidate the dynamics of light harvesting in complex photosynthetic systems.

[a]*Department of Chemistry, National Taiwan University, Taipei City, Taiwan. E-mail: yuanchung@ntu.edu.tw*

[b]*State Key Laboratory of Structural Chemistry, Fujian Institute of Research on the Structure of Matter, Chinese Academy of Sciences Fuzhou, Fujian, CN 350002, China*

[c]*Department of Chemistry, Institute for Advance Study and School of Science, Hong Kong University of Science and Technology, Kowloon, Hong Kong. E-mail: xuhuihuang@ust.hk*

† Electronic supplementary information (ESI) available. See DOI: 10.1039/c8fd00205c

1 Introduction

Photosynthetic systems utilize complex structures composed of pigment–protein complexes (PPCs) to harvest sunlight and convert the energy into chemical potentials that can be used to drive subsequent chemical reactions.[1–4] In particular, oxygenic photosynthetic organisms employ the photosystem II (PSII) supercomplex in their photosynthetic machinery.[5–9] The PSII supercomplex utilizes several hundreds of chlorophylls to perform light harvesting and charge separation at almost unity quantum efficiency. This remarkable process, called the light reaction, is at the core of photosynthesis, yet many aspects of its fundamental molecular mechanisms remain elusive due to the vast complexity of photosynthetic systems.[7,8]

Previous studies indicated that the PSII supercomplex exhibits a variety of different forms in natural photosynthesis[10–15] that nevertheless share the same basic functional unit, the PSII core complex (PSII-cc). The atomistic structure of the PSII-cc is available in high resolution.[6,9,15,16] The PSII core complex consists of several pigment–protein complexes assembled in a dimeric structure with near C2 symmetry. In each monomer, it consists of two antenna complexes (CP43 and CP47) and one reaction center (RC). The location of the antenna complexes and reaction centers are shown in Fig. 1. In the past few decades, the excitation energy transfer (EET) and charge separation in PSII-cc have been investigated extensively using various spectroscopic techniques.[10,17–25] In addition, with the availability of the high-resolution structures, theoretical models with detailed descriptions of the dynamics of light harvesting in the PSII-cc have also been constructed.[10,23,24,26–29]

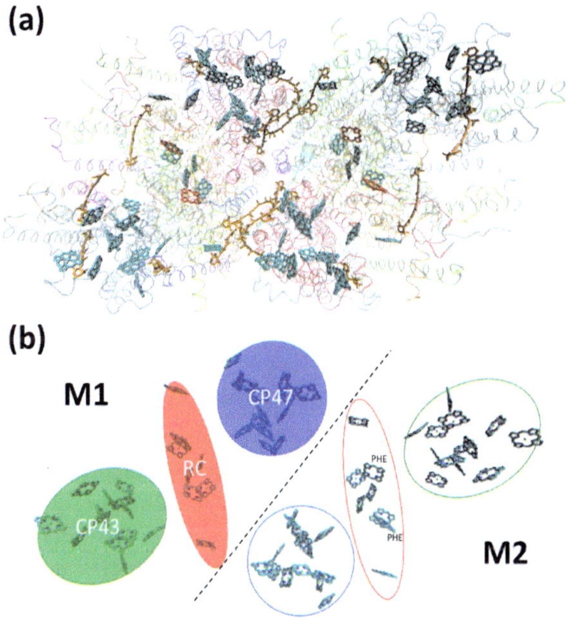

Fig. 1 Pigment arrangement in the dimeric PSII-cc. (a) The top view of the PSII core complex. The illustration is based on the crystal structure from *T. vulcanus* at a resolution of 1.9 Å (PDB: 3ARC). (b) The chlorophyll arrangement in the PSII-cc. The chlorin rings of chlorophyll a and pheophytin molecules are shown. Furthermore, we depict the two monomers (M1 and M2) as well as the respective subunits (CP43, RC, and CP47).

The accumulated data have allowed the elucidation of many mechanistic aspects of the light reaction in PSII, yet some key questions regarding the energy landscapes, the relative time scales of EET and charge separation processes, and energy regulation in PSII remain to be answered.[7,8,30,31] In particular, previous theoretical studies have largely overlooked the dimeric nature of the PSII-cc and almost exclusively focus on dynamics within a single monomeric unit. It remains unclear, however, whether the dimeric form serves only a structural role to stabilize the assembly on the thylakoid membrane, or it also provides a functional role to offer more advantages in light harvesting. To this end, an accurate molecular-level understanding of the dynamics of light harvesting in the dimeric PSII-cc is crucial for revealing the design principles underlying the efficient light reaction in the system.

In recent years, molecular dynamics (MD) simulations have been utilized as powerful tools to elucidate detailed pigment–protein interactions crucial in the mechanistic of light harvesting in PPCs.[29,32–35] Here, we follow the MD simulation approach published previously[36] to obtain key parameters that govern EET dynamics in the dimeric PSII-cc system. Our goal is to construct an accurate effective model for light harvesting in the dimeric PSII-cc system. This model would allow us to focus on the role of the inter-monomer EET pathways and to elucidate whether the dimeric structure of PSII-cc contributes to the function of the system.

2 Theoretical methods

2.1 Effective Frenkel exciton model

For a pigment–protein complex, it is conventional to adopt the Frenkel exciton model to describe the photo-excitations in the system.[1,2,37] The Frenkel exciton Hamiltonian for a system with N pigments (sites) is given by

$$H_s = \sum_{n=1}^{N} E_n|n\rangle\langle n| + \sum_{n\neq m} J_{nm}|n\rangle\langle m|, \tag{1}$$

where $|n\rangle$ represents a local excitation on the nth pigment, E_n is the transition energy (site energy) of $|n\rangle$, and J_{nm} is the excitonic coupling between $|n\rangle$ and $|m\rangle$. The effective Hamiltonian describes electronic excitations in a pigment–protein complex, and in this work we determine the site energies and excitonic couplings based on structures obtained in MD simulations.

2.2 Molecular dynamics simulation and excitonic parameters

Our effective excitonic model for EET in the PSII-cc is constructed based on MD simulations of the PSII-cc embedded in a lipid-bilayer membrane. Details about the structure model and MD simulations were published previously,[36] and here we only present essential information in order to make this work self-contained. The structure model is based on the crystal structure of PSII-cc from *T. vulcanus*.[16] To capture the dynamical effects, 30 snapshots sampled with equal time intervals (1 ns) from the last 5 ns of five independent 20 ns equilibrium MD runs at 300 K were used as initial conditions to carry out equilibrium MD simulations, resulting in an ensemble of 30 50 ps MD trajectories for the PSII-cc system. Instantaneous structures in these MD trajectories were saved every 4 fs to obtain structure dynamics of the PSII-cc in high temporal resolution.

To estimate site energies based on the MD trajectories, the charge density coupled (CDC) method[27,38,39] is used. In this approach, the time-dependent site energy shift of a pigment in a PPC is calculated by considering Coulomb interactions between the pigment and the background atomic partial charges of the protein, water, and all other molecules. Thus, for the nth site

$$\begin{aligned}\Delta E_n(t) &= \frac{1}{\varepsilon_{\text{eff}}}\sum_i\sum_j\left(\frac{q_{\text{e},i}^n q_j^{\text{bg}}}{\left|r_i^n(t)-r_j^{\text{bg}}(t)\right|}-\frac{q_{\text{g},i}^n q_j^{\text{bg}}}{\left|r_i^n(t)-r_j^{\text{bg}}(t)\right|}\right)\\ &= \frac{1}{\varepsilon_{\text{eff}}}\sum_i\sum_j\frac{\Delta q_i^n q_j^{\text{bg}}}{\left|r_i^n(t)-r_j^{\text{bg}}(t)\right|},\end{aligned} \tag{2}$$

where i runs over all heavy atoms on the nth site, and j includes all the charged atoms in the background. ε_{eff} is an effective dielectric constant, $q_{\text{e},i}^n$ and $q_{\text{g},i}^n$ are the atomic partial charges on atom i for the excited state and ground state, respectively, of the nth pigment. $r_i^n(t)$ and $r_j^{\text{bg}}(t)$ are the instantaneous positions of atom i and atom j, respectively. Δq_i^n is the charge difference due to electronic excitation, $\Delta q_i^n = q_{\text{e},i}^n - q_{\text{g},i}^n$. Following the treatment detailed in ref. 36, the atomic charges for the excited state and ground state of the pigment are obtained by fitting to TDDFT electrostatic potential distributions, and the background partial charges are obtained from the AMBER03 force field.[40,41] We apply eqn (2) to calculate site energy shifts for all the structural snapshots sampled by the MD simulations, resulting in a total of 30 50 ps trajectories of site energy shifts for each site. We then calculate the site energy of each site by averaging over all the MD snapshots:

$$E_n = E_0 + f\langle\overline{\Delta E_n(t)}\rangle, \tag{3}$$

where E_0 is an empirical total energy shift that depends on the type of pigments, $\overline{A(t)}$ denotes the time average of the observable $A(t)$ over a single trajectory, and $\langle\cdot\rangle$ denotes the average over all the trajectories. f is an empirical scaling factor used to compensate for the errors in the CDC scheme, and it will be determined by comparing to experimental data.

We applied the transition charge from electrostatic potential (TrEsp)[42] method as described in ref. 36 to calculate excitonic couplings between excitations on each pair of pigments. Our results indicate that the structure fluctuations at 300 K do not significantly affect the excitonic couplings. In general, the couplings do not fluctuate more than 5% from the values calculated using the crystal structure. Thus, in order to simplify our model, we choose to neglect the time-dependence of the excitonic couplings by using the excitonic couplings calculated from the crystal structure of PSII-cc. In addition, since most exciton states in PSII-cc are localized at pigments no greater than 30 Å apart, we consider the excitonic coupling only if the center-to-center distance between the two pigments is less than 30 Å in order to avoid unphysical delocalization among pigments with a large spatial separation due to accidental energy degeneracy.

2.3 System-bath interactions from MD data

The description of quantum dynamics for excitons in PPCs is well established.[4,27,37,43–45] The standard Hamiltonian used to study energy transfer in a PPC contains three parts:

$$H = H_s + H_b + H_{sb}, \tag{4}$$

where H_s is the system Hamiltonian (eqn (1)), H_b is the bath Hamiltonian, and H_{sb} describes the system–bath interactions. The bath Hamiltonian is represented by a set of harmonic oscillators:

$$H_b = \sum_\alpha \hbar\omega_\alpha \left(b_\alpha^\dagger b_\alpha + \frac{1}{2} \right), \tag{5}$$

where $\hbar\omega_\alpha$ is the energy of the vibrational quanta in the αth bath mode. The system-bath Hamiltonian is described by

$$H_{sb} = \sum_\alpha \sum_n \hbar\omega_\alpha g_{\alpha,n} \left(b_\alpha^\dagger + b_\alpha \right) |n\rangle\langle n|, \tag{6}$$

where $g_{\alpha,n}$ is the system–bath coupling constant between $|n\rangle$ and the αth bath mode. We also assume that the fluctuations of site energies are linearly coupled to the displacement of nuclei from their equilibrium positions. Therefore, the system–bath couplings lead to fluctuations in the site energies of the pigments, which can be calculated from the MD data.

To evaluate the system–bath couplings from the MD data we calculate the time-correlation function (TCF) of the energy gap fluctuations for each pigment in the PSII-cc system using the time-dependent site energy shifts (eqn (2)) obtained from the MD trajectories. For each trajectory, we average the site energy shift within a 10 ps window to yield the time-averaged shift $\overline{\Delta E_n(t)}$, and then evaluate the site-energy fluctuations around this average:

$$\delta E_n(t) = \Delta E_n(t) - \overline{\Delta E_n(t)}. \tag{7}$$

Then, we can calculate the TCF for the site-energy fluctuations on the nth site as:

$$\begin{aligned} C_n(\tau) &= \langle \overline{\delta E_n(\tau)\delta E_n(0)} \rangle \\ &= \frac{1}{T}\int_0^T \langle \delta E_n(\tau + t)\delta E_n(t) \rangle \mathrm{d}t, \end{aligned} \tag{8}$$

where the window of time average is $T = 10$ ps. We emphasize that the division into 10 ps windows allows us to separate the fast dynamical disorder from the slow static disorder, therefore $C_n(\tau)$ calculated here would not be significantly affected by inadequate sampling of slow protein dynamics in the MD simulation.

In principle, one can calculate the TCF of site energy fluctuations for each site and model the full PSII-cc system using all of the MD details of dynamical fluctuations. Nevertheless, we further assume that the system–bath couplings of different sites are independent and identical. As a result, we use the same averaged quantum TCF for all of the sites to describe the system–bath couplings.

$$C_{MD}(\tau) = \frac{1}{M}\sum_n C_n(\tau), \tag{9}$$

where M is the total number of sites in the system.

The classical TCF calculated based on MD data does not satisfy the detailed balance condition. Therefore, we adopt a harmonic quantum correction to calculate the quantum time-correlation function for the site-energy fluctuations.[46,47] The quantum correction is performed in the frequency domain. Given a classical TCF, $C_{\mathrm{MD}}(\tau)$, we can calculate its Fourier transform:

$$\tilde{C}_{\mathrm{MD}}(\omega) = \int_{-\infty}^{\infty} \mathrm{e}^{i\omega\tau} C_{\mathrm{MD}}(\tau)\mathrm{d}\tau, \tag{10}$$

where ω is the angular frequency. Then, the quantum TCF in the frequency domain can be obtained by multiplying the classical TCF with an harmonic quantum correction factor:

$$\tilde{C}_{\mathrm{qc}}(\omega) = \frac{\beta\hbar\omega}{1 - \mathrm{e}^{\beta\hbar\omega}} \tilde{C}_{\mathrm{MD}}(\omega), \tag{11}$$

where $\beta = \dfrac{1}{k_{\mathrm{b}}T}$ is the inverse temperature. Finally, the inverse Fourier transform is applied to $\tilde{C}_{\mathrm{qc}}(\omega)$ to obtain the quantum TCF in the time domain, $C_{\mathrm{qc}}(\tau)$.

2.4 Simulation of the absorption spectrum and EET dynamics

The above-mentioned system-base Hamiltonian for a pigment–protein complex can be used to calculate EET dynamics and spectra. To this end, we diagonalize H_{s} in order to obtain the exciton states. The exciton state $|\alpha\rangle$ satisfies $H_{\mathrm{s}}\,|\alpha\rangle = E_\alpha\,|\alpha|$, where E_α is the exciton energy of $|\alpha\rangle$, and can be represented as a linear combination of localized excitations, $|\alpha\rangle = \sum_n c_n^\alpha |n\rangle$, where c_n^α is the coefficient for the excitation on site n.

To simulate the linear absorption spectrum and the EET dynamics, we adopt the modified-Redfield approach.[43,48,49] We define the TCF of energy fluctuations in the exciton basis:

$$C_{\alpha\beta\gamma\delta}(\tau) = \sum_n c_n^\alpha c_n^\beta c_n^\gamma c_n^\delta C_{\mathrm{qc}}(\tau), \tag{12}$$

where $C_{\mathrm{qc}}\,(\tau)$ is the quantum TCF for site-energy fluctuations calculated from the MD data. Furthermore, we define the line shape function in the exciton basis,

$$g_{\alpha\beta\gamma\delta}(t) = \int_0^t \mathrm{d}\tau C_{\alpha\beta\gamma\delta}(\tau)(t-\tau), \tag{13}$$

and the reorganization energy function,

$$\lambda_{\alpha\beta\gamma\delta} = -\int_0^\infty \mathrm{Im}(C_{\alpha\beta\gamma\delta}(\tau))\mathrm{d}\tau. \tag{14}$$

Using the exciton Hamiltonian and the quantum TCF, the optical density (OD) of linear absorption can be calculated:

$$\mathrm{OD}(\omega) = \omega\sum_\alpha {\mu_\alpha}^2 \mathrm{Re}\int_0^\infty \mathrm{d}t\mathrm{e}^{i(\omega-\omega_\alpha)t - g_{\alpha\alpha\alpha\alpha}(t) - R_{\alpha\alpha\alpha\alpha}t}, \tag{15}$$

$$R_{\alpha\alpha\alpha\alpha} = -\sum_{\beta\neq\alpha} R_{\beta\beta\alpha\alpha}, \tag{16}$$

where $\mu_\alpha = \sum_n c_n^\alpha \mu_n$ and $\omega_\alpha = E_\alpha/\hbar$ denote the transition dipole moment and frequency of the αth exciton state, respectively. $R_{\beta\beta\alpha\alpha}$ is the transition rate constant from the αth exciton state to the βth exciton state based on the modified Redfield theory:

$$R_{\beta\beta\alpha\alpha} = -2\text{Re}\int_0^\infty \text{d}t\hat{W}(\omega_{\beta\alpha},t)\left\{\begin{array}{l}\ddot{g}_{\beta\alpha\alpha\beta}(t)-\{\dot{g}_{\alpha\beta\alpha\alpha}(t)-\dot{g}_{\alpha\beta\beta\beta}(t)+2i\lambda_{\alpha\beta\alpha\alpha}\}\\ \times\{\dot{g}_{\alpha\alpha\beta\alpha}(t)-\dot{g}_{\beta\beta\beta\alpha}(t)+2i\lambda_{\alpha\alpha\beta\alpha}\}\end{array}\right\}, \quad (17)$$

where

$$\hat{W}(\omega_{\beta\alpha},t) = \exp\{-i\omega_{\beta-\alpha}t - g_{\alpha\alpha\alpha\alpha}(t) - g_{\beta\beta\beta\beta}(t) + 2g_{\alpha\alpha\beta\beta} + 2i(\lambda_{\alpha\alpha\beta\beta} - \lambda_{\alpha\alpha\alpha\alpha})t\}$$

and

$$\omega_{\beta\alpha} = \omega_\beta - \omega_\alpha. \quad (18)$$

The rate constants given in eqn (16) and (17) can also be used to construct a master equation describing the full EET dynamics in the exciton states of the PSII-cc. In addition to energy transfer, we phenomenologically include fluorescence decay with $k_{FL} = 1\ \text{ns}^{-1}$ for all exciton states, and the two charge separation pathways for primary charge separation in each RC are also included in our rate matrix. The two pathways are the P_{D1} pathway and the Chl_{D1} pathway, with rate constants of 0.33 ps^{-1} and 1.43 ps^{-1}, respectively. We can then propagate the master equation with given initial populations on the exciton states to obtain the full EET dynamics.

3 Results and discussions

3.1 Quantum time correlation function

We have performed a MD simulation of the PSII-cc system and calculated the TCF for site energy fluctuations using the MD data. To justify our choice of using a single TCF to describe system–bath couplings of all of the chlorophyll a molecules (CLAs) and pheophytins (PHEs), we have separately calculated TCFs for CLAs in different subunits of the PSII-cc (Fig. S1†), and the results confirmed that different CLAs do not show significantly different site energy fluctuations. Fig. 2 shows the averaged quantum TCF of site energy fluctuations ($C_{qc}(t)$) in PSII-cc. The TCF decays rapidly, showing a correlation time of ~50 fs, which is in agreement with the time scales of bath relaxation in empirical TCFs normally used for photosynthetic PPCs. However, the variance of fluctuations ($C_{qc}(0)$) is only ~830 cm^{-2}, which is more than an order of magnitude smaller than the normal empirical TCFs used for PPCs at 300 K. We thus conjecture that while the classical force fields properly describe the molecular dynamics and time scales of site energy fluctuations, the magnitude of the fluctuations were described poorly by the atomic partial charges, which is not surprising since these partial charges were not fitted to describe couplings to the excited state of the pigments. Therefore, we propose to use a scaled quantum TCF in order to correct for the deficiency in the partial charge model:

$$C_{qc}^{s}(\tau) = f^2 C_{qc}(\tau), \quad (19)$$

where f is a scaling factor. This approach is also consistent with the scaling factor proposed for site energy shifts (eqn (3)). The scaled quantum TCF will be used to

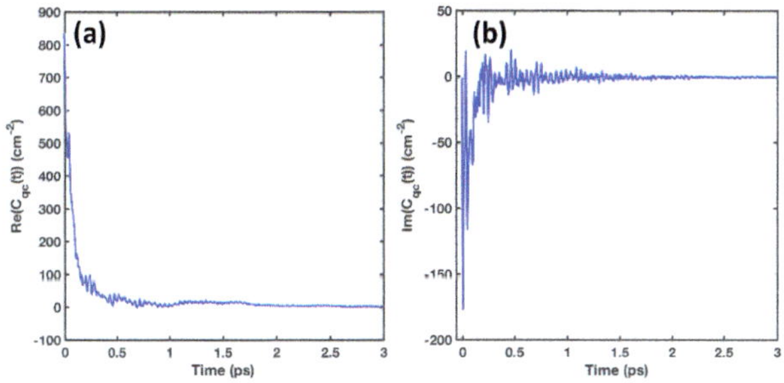

Fig. 2 The averaged quantum time-correlation function of site energy fluctuations calculated based on the PSII-cc MD trajectory. (a) The real part of the quantum time-correlation function. (b) The imaginary part of the quantum time-correlation function.

calculate spectra and EET dynamics presented in this work, and the scaling factor will be determined by comparing to experimental spectral data.

3.2 Linear absorption spectra

We first focus on the linear absorption spectrum of the RC of the M1 monomer. In our spectral simulation study, we found that the scaling factor f, the transition energy in vacuum E_0, the two pheophytin site energies, the site energies of the special pair CLAs and the excitonic coupling between the special pair CLAs have to be adjusted in order to achieve a reasonable fit to the experimental spectrum. These fitting parameters are well justified because they are precisely where the CDC scheme for site energy shifts and the TrESP approach for excitonic couplings are expected to perform poorly. Moreover, since the simulation time of each MD trajectory is only 50 ps, the inhomogeneous broadening is not captured by the MD trajectories. Therefore, we apply an additional diagonal Gaussian disorder with $\sigma_{\text{static}} = 70\ \text{cm}^{-1}$ to capture the static disorder effects.[23,26] Fig. 3 shows the comparison between the optimized theoretical absorption spectrum and the experimental spectrum for the RC in 300 K. There, we found that a scaling factor of $f = 5$ together with the MD structure-based data yield an excellent fit to the absorption spectrum. The resulting exciton Hamiltonian is shown in the RC block in Table S1.† Fig. 4 shows the labeling of the pigments used in this work.

We also simulated the absorption spectra for the CP43 and CP47 subunits of the PSII-cc (Fig. 5). In these two light-harvesting complexes, we found that if we use the site energies obtained from the MD structures and a scaling factor of $f = 5$ for the quantum TCF, the experimental spectra can be nicely reproduced.

Overall, we have demonstrated that the MD structures together with the CDC scheme for site energy shifts and the TrESP approach for excitonic couplings can be used to obtain effective models of PSII-cc that reasonably reproduce the linear spectra of each subunit in the PSII-cc. The universal scaling factor $f = 5$ for the TCF of site energy fluctuations confirm our conjecture that the MD trajectory captures the essential dynamics of site energy fluctuations. Note that the comparison between the experimental spectrum and the theoretical spectrum is utilized as a means to correct for the effective electronic screening and local field effects in the scaling factor f, and we do not consider the agreements unequivocally prove that the MD structure and the CDC scheme yields correct site energy

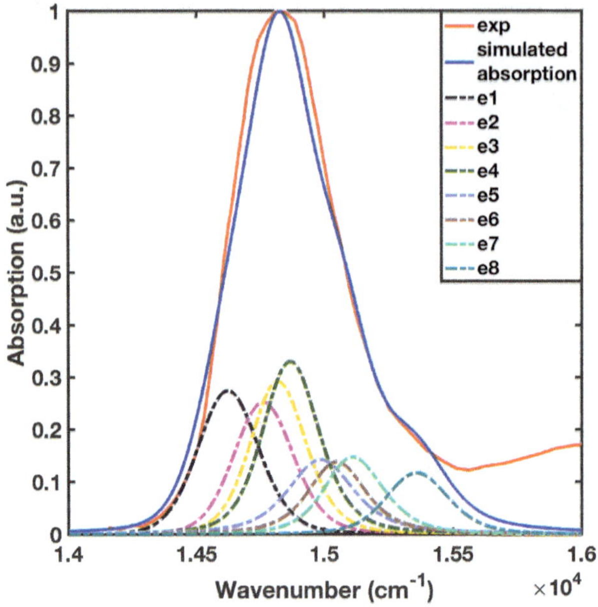

Fig. 3 The comparison of the simulated and experimental absorption spectra of the PSII RC. The red curve depicts the experimental result at 297 K.[50] The blue peak is the simulated spectrum based on our PSII-cc exciton model. Curves depicted with dashed lines are contributions from each individual exciton states.

values. Instead, our goal is mainly to show that the MD structure-based models can be made consistent with available experimental data with a minimal number of adjustable parameters. To further and unbiasedly validate the model, it is necessary to compare it to additional spectral data and available time-resolved spectroscopic measurements. For PSII-cc, that would present a formidable challenge since clean experimental data that could be directly compared to simulations are still not available. As a result, we choose to not carry out a detailed comparative study. Instead, we would go on to carry out simulations of EET dynamics, and see how far the MD data could take us. Intriguingly, we found that

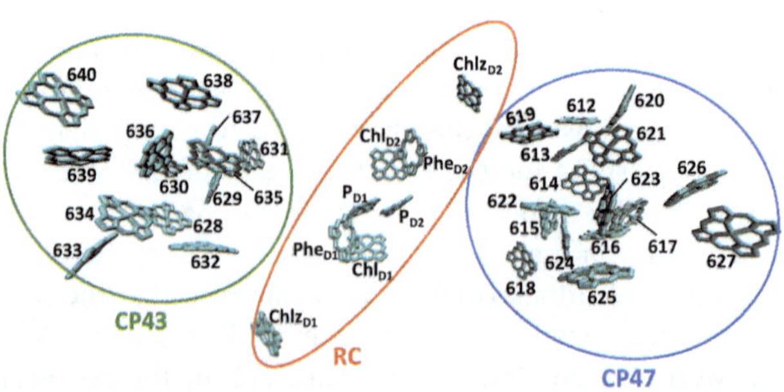

Fig. 4 Labeling of pigments in a monomer of the PSII-cc. The blue ellipse represents the location of the CP47 antenna complex, and the red ellipse represents the location of the RC. The green ellipse represents the location of the CP43 antenna complex.

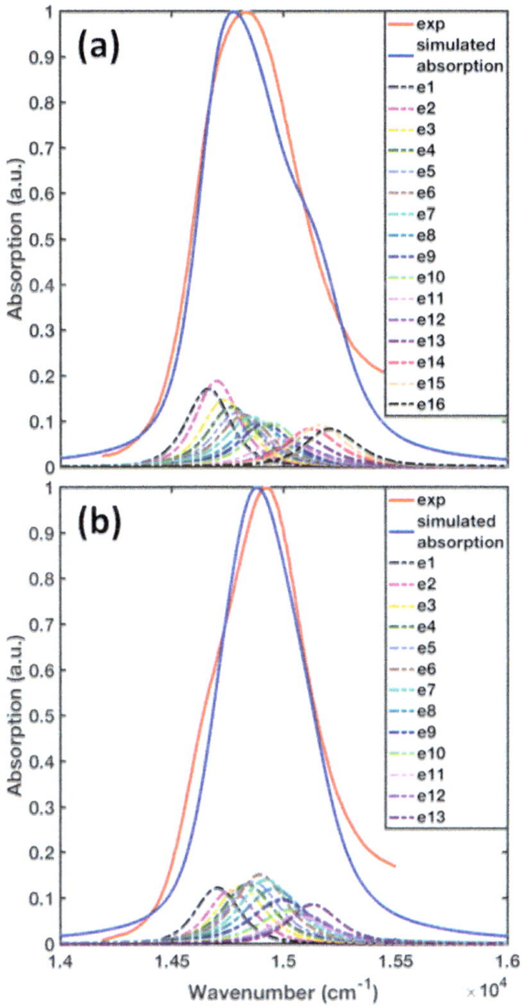

Fig. 5 Simulated absorption spectra of CP47 and CP43. (a) The simulated absorption spectrum of CP47 compared to the experimental spectrum. (b) The simulated absorption spectrum of CP43 compared to the experimental spectrum. The experimental spectra at 297 K was taken from ref. 51. Contributions from each individual exciton states are depicted in colored dashed lines.

the MD based model could semi-quantitatively reproduce most of the results presented in previous studies, and furthermore allowed us to explore new aspects of the dynamics of light harvesting in the PSII-cc.

3.3 Exciton states in the PSII-cc

Before we present simulations of EET dynamics in the PSII-cc, we examine the model Hamiltonian. The full excitonic Hamiltonian for the PSII-cc system is presented in Tables S2 and S3.† We see that the Hamiltonian exhibits a blocked form, where significant couplings only exist between CLAs in the same subunit. Moreover, this model Hamiltonian shows that the lowest energy site is Chl_{D1} in the RC, which is in agreement with previous work.[36] Aside from Chl_{D1}, several CLAs in the CP47 subunit represent a group of low energy sites in the PSII-cc, causing the averaged site energy of the CP47 to be significantly lower than that of CP43 and the RC. Again, this is in agreement with previous experimental observations and theoretical calculations.[23,26,30,31] We note that the assignments of low-energy sites in CP47 are still controversial,[23,30,52] yet the overall key features in our model Hamiltonian are consistent with previously published effective models for the PSII-cc.

The exciton states that diagonalize our model Hamiltonian are illustrated in Fig. S2.† To quantify the degree of delocalization, we calculated the delocalization length (Fig. S3†) as well as the spatial delocalization distance (Fig. S4†) for all the exciton states. We found that most exciton states exhibit a delocalization length between 1–3 across a distance shorter than 15 Å. The average delocalization length for the 74 exciton states is 2.3, meaning that most of the exciton states are delocalized among 2 CLAs. The average spatial delocalization distance for the 74 exciton states is 9.5 Å, whereas the average distance between nearest-neighbor CLAs is 10.6 Å. Therefore, the spatial delocalization shows that in general excitons are delocalized among nearest-neighbor CLAs, which confirms the average value of the delocalization length.

Our data also revealed that exciton states e12 and e13 have especially large values of spatial delocalization. These states are contributed by $Chlz_{D2}$ and CLA627 in the CP47 of the M2. Interestingly, these two CLAs do not strongly couple to each other. The delocalization is due to excitonic couplings between these pigments and CLA621 and CLA619. In this case, EET from $Chlz_{D2}$ to CP47 is mediated by a cluster of CLAs.

3.4 EET dynamics in the PSII-cc

Our MD-structure based exciton model and scaled quantum TCF allows us to simulate full EET dynamics in the PSII-cc. Previous studies have established that excitation energy from the peripheral light-harvesting complexes could enter the PSII-cc though two pathways in the PSII supercomplex:[13–15] one into CP43 and the

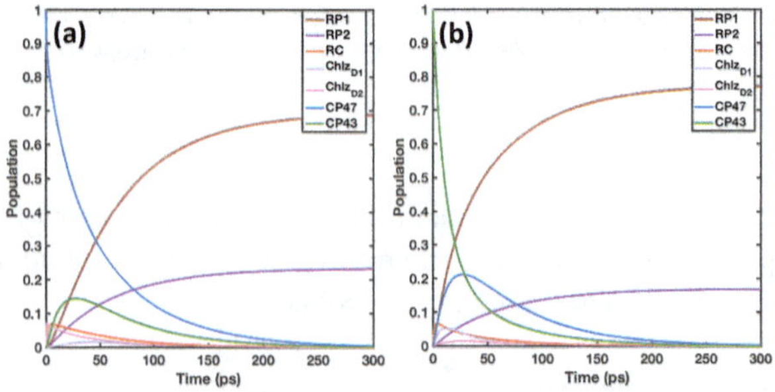

Fig. 6 EET dynamics in the M1 monomer of the PSII-cc. (a) EET population dynamics when CLAs in the CP47 subunit are uniformly excited initially. We depict aggregated population for each subunit as a function of time. In addition, we show the populations of the excited $Chlz_{D1}$ and $Chlz_{D2}$, respectively. The populations that reach the two radial-pair states (RP1 and RP2) are shown as well. (b) EET population dynamics when CLAs in the CP43 subunit are uniformly excited initially.

other into CP47. Therefore, we will focus on EET dynamics with either CP43 or CP47 excited initially. Furthermore, to investigate the efficiency of light harvesting, fluorescence decay of CLAs and the charge separation pathways leading to the two radical-pair states, RP1 and RP2, respectively, are also included in all the subsequent dynamical simulations (Tables S3 and S4†).

We first study the EET dynamics in a single monomer (M1) of the PSII-cc by inspecting the aggregated excitation population in each subunit of the M1 monomer as a function of time. Fig. 6a shows the EET dynamics when the exciton states in CP47 are initially excited. In this case, the aggregated population of CP47 decays in a 50 ps time scale, and at the time scale the two radical pair states are populated, showing rapid EET from CP47 to the RC and the subsequent charge separation dynamics. The quantum efficiency of charge separation in this case is about 0.93, which is also in agreement with previous studies. Noticeably, transient population of CP43 is observed, and it reaches a maximum of about 0.15 at ~30 ps. This EET from CP47 to CP43 occurs presumably through the RC because CP47 and CP43 are not adjacent in space and all excitonic couplings between CP47 and CP43 pigments are effectively zero. This indicates that regardless of the much faster (~1 ps) charge separation dynamics in the RC, EET between the two antenna complexes could still occur rapidly. Thus, EET dynamics in the PSII-cc shows significant non-equilibrium effects and a multiple-pathway nature, which can not be described by a hopping rate from the antenna to the RC.

In Fig. 6b, we show EET dynamics in the M1 monomer when the exciton states in CP43 are initially excited. Similar to the case when CP47 is initially excited, rapid charge separation and EET to CP47 are both observed. The time scale of population decay in CP43 is ~20 ps, slightly faster than the population decay in CP47, which could be a result of CP47 containing the low-energy CLAs. The quantum efficiency of charge separation in this case is about 0.94.

To investigate the role the peripheral chlorophylls ($Chlz_{D1}$ and $Chlz_{D2}$) play in EET to the core of RC, we also plot the population of excitation on $Chlz_{D1}$ and $Chlz_{D2}$, respectively, in Fig. 6. We observe that the population decay on either $Chlz_{D1}$ or $Chlz_{D2}$ does not cause a corresponding population increase on the RC. Therefore, excitation energy transfer from CP43 and CP47 to RC can be achieved without passing though either Chlz sites. In addition, the EET dynamics shown in Fig. 6 indicate that $Chlz_{D2}$ is more closely connected to CP47, whereas $Chlz_{D1}$ is more closely connected to CP43. This is also in agreement with previous theoretical studies.[23,26]

It is interesting to compare our simulation results with experiments. Generally speaking, our model predicts 30–50 ps EET from CP47 and CP43 to the RC and 15 ps EET from RC to the antenna complexes (Tables S3 and S4†). Note that EET in the PSII-cc does not proceed through a single energy funnel, therefore these timescales should be considered as coarse-grained averaged values over a range of multiple pathways. Nevertheless, these numbers are in reasonable agreement with EET parameters obtained in previous combined theoretical–experimental studies,[23,26] despite the differences in the site energies and couplings between our models. Moreover, in the recent 2D electronic spectroscopic study by Pan *et al.* at 77 K,[53] multiple time scales of energy relaxation dynamics were resolved by applying global data analysis on 2D spectra. Although the ultrafast sub-ps energy transfer dynamics reported in this 2D study are intentionally ignored in our study because of the setup of the initial conditions, the 7 ps and 36 ps relaxation

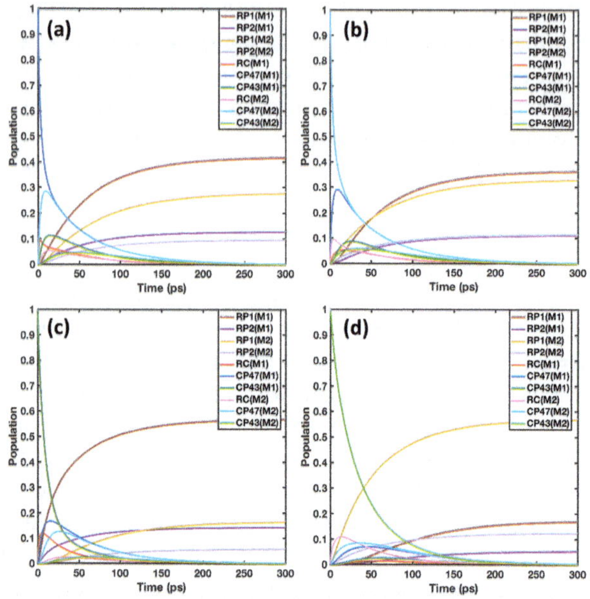

Fig. 7 EET dynamics in the dimeric PSII-cc. Here we show EET dynamics in a full dimeric PSII-cc with different initial excitation conditions. (a) EET dynamics with initial excitation at CP47 of M1. The quantum efficiency of charge separation: RP1(M1) is 0.42; RP2(M1) is 0.13; RP1(M2) is 0.28; RP2(M2) is 0.10. Total QE = 0.93. (b) EET dynamics with initial excitation at CP47 of M2. The quantum efficiency of charge separation: RP1(M1) is 0.37; RP2(M1) is 0.11; RP1(M2) is 0.34; RP2(M2) is 0.12. Total QE = 0.94. (c) EET dynamics with initial excitation at CP43 of M1. The quantum efficiency of charge separation: RP1(M1) is 0.57; RP2(M1) is 0.15; RP1(M2) is 0.17; RP2(M2) is 0.06. Total QE = 0.95. (d) EET dynamics with initial excitation at CP43 of the M2. The quantum efficiency of charge separation: RP1(M1) is 0.17; RP2(M1) is 0.05; RP1(M2) is 0.58; RP2(M2) is 0.13. Total QE = 0.93.

kinetics are in good agreement with the <10 ps intra-subunit relaxation and 30–50 ps CP43 to RC EET, respectively. Therefore, our simulation results show a good level of consistency with previous experimental data. We should emphasize that direct comparison between theoretical model and experimental data is highly non-trivial in this case because of the disordered nature of the PSII-cc system and the sensitivity of the EET dynamics to the experimental conditions (sample preparation, temperature, form of the aggregations… *etc.*) More detailed comparisons of the model parameters and EET kinetics between our model with previous theoretical and experimental studies is a work in progress and will be published later.

The EET dynamics in the dimeric PSII-cc system is investigated by performing simulations with four different initial conditions (Fig. 7). In Fig. 7a, we show the EET dynamics when initially the CP47 of the M1 monomer is excited. Following photo-excitation, the excited population in the CP47 of the M1 decreases rapidly, and at the same time the population of excitation in the CP47 of the M2 increases. Excitation energy in the two CP47s reach an equal value at ~15 ps, which indicates that excitation energy can equilibrate rapidly among the two CP47 subunits. In addition, the energy also rapidly flows to the RC of the M1 as well as the CP43 of the M1 within a 15 ps time scale. Remarkably, although the CP47 of the M1 is initially excited, a significant portion of the excitation energy ends up in the radical-pair states in the M2. This indicates that inter-monomer EET enables excitation energy to be utilized in both RCs. Moreover, Fig. 7b shows the EET dynamics when initially the CP47 of the M2 monomer is excited. The dynamics are qualitatively the same as those shown in Fig. 7a (when the CP47 of the M1 is initially excited), confirming that rapid EET between the two CP47s allows a quasi-equilibrium of excitations among the two CP47s to be established before EET to the RCs. It is interesting to note that rapid energy pooling to the CP47 subunits can also be observed in previous studies of EET dynamics in the PSII super-complex,[24,28] however the significance of this phenomenon was not explicitly investigated before.

In contrast, when initially the CP43 in one of the monomers is excited, most of the excitation energy goes to the nearest RC and the population dynamics on the two monomers show a significant difference (Fig. 7c and d). Therefore it is confirmed that the inter-monomer EET in the dimeric PSII-cc must occur through the two CP47 subunits.

Croce and co-workers investigated PSII samples with different peripheral antenna sizes.[54] The results show that the quantum efficiency of charge separation in the PSII often exceeds 0.84, whereas in the PSII with smaller antenna size, the quantum efficiency can be as high as 0.91. Because excitation energies in PSII samples with larger antenna would need more time to reach a RC, it is reasonable that the quantum efficiency increases as the size of antenna decreases. Our theoretical simulations predict that the total quantum efficiency of charge separation in the PSII-cc is about 0.93, in good agreement to the experimental data.

3.5 Inter-monomer EET between CP47s

The rapid inter-monomer EET between the two CP47s revealed in our full EET dynamics of the dimeric PSII-cc system deserves a more detailed investigation. A recent study clearly showed the existence of inter monomer EET in dimeric PSII core,[55] yet the detailed molecular mechanism and time scales of such process is still unclear. Published structure-based models for EET in the PSII-cc mostly overlooked the inter-monomer EET because the two CP47s are well separated spatially, with the

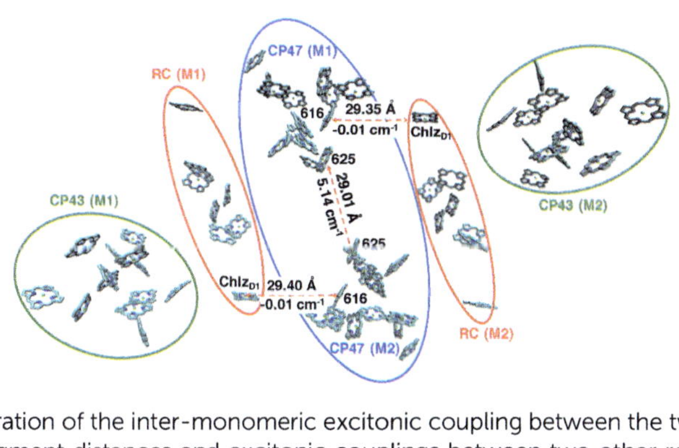

Fig. 8 Illustration of the inter-monomeric excitonic coupling between the two CLA625s. The inter-pigment distances and excitonic couplings between two other representative pairs of CLAs are also shown.

closest inter-CP47 CLAs located at a distance of 29.0 Å apart from each other. So how could excitation energy flow so rapidly between the two CP47s? In Fig. 8, we show the inter-CLA distances as well as the excitonic couplings calculated in this work for three representative CLA pairs that have a CLA–CLA separation of about 29.0 Å. It is clear that the coupling between the two CLA625 sites, which is located in the interface of the two CP47s, is anomalously large. The value of $J_{\mathrm{CLA625(M1)–CLA625(M2)}}$ is two orders of magnitude greater than those of the other two pairs. Clearly, the large $J_{\mathrm{CLA625(M1)–CLA625(M2)}}$ plays an important role in the efficient inter-monomer EET between CP47s. A close inspection indicates that the large coupling is a consequence of the near parallel transition dipoles of the two CLA625 molecules. This favorable molecular orientation is due to the C2 symmetry of the PSII-cc system. As a result, the two CLA625s serve as an efficient energy bridge that conducts excitation energy rapidly between the CP47s. Furthermore, we found that CLA625 is also located closely to several CLAs within the same CP47 subunit (Fig. 4). In particular, CLA624 and CLA623 are strongly coupled to CLA625. $J_{\mathrm{CLA625–CLA624}}$ is about $-59\ \mathrm{cm}^{-1}$, and $J_{\mathrm{CLA625–CLA623}}$ is about $119\ \mathrm{cm}^{-1}$. These values indicate that CLA625 is well connected to CLAs within the same CP47, which also facilitates the efficient EET between the two CP47s.

To demonstrate the importance of CLA625s in the inter-monomer EET of the PSII-cc, we investigated the dynamics of EET in a mutant PSII-cc, in which the two CLA625 sites are deleted from the model Hamiltonian. The EET dynamics for this mutant PSII-cc are presented in Fig. S5.† Clearly, when the two CLA625s are deleted, the inter-monomer EET can no longer be observed, and the EET dynamics resemble those observed in the monomeric PSII-cc (Fig. 6). We conclude that the EET pathway between the two CLA625 sites plays a crucial role in the inter-monomer EET in the PSII-cc. Nevertheless, the mutant PSII-cc still exhibits a high quantum efficiency for light harvesting since the two monomers could function independently.

3.6 EET in the PSII-cc with a closed RC

The significance of inter-monomer EET in the light harvesting of the PSII-cc is not obvious in normal circumstances. We conjecture that the inter-monomer EET could play a significant role in maintaining the efficiency of light harvesting when

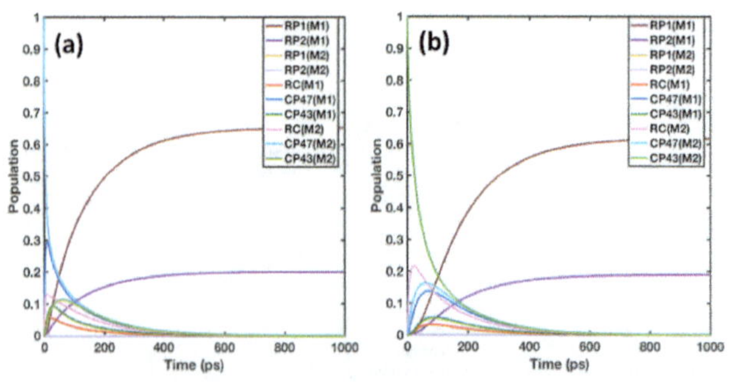

Fig. 9 EET dynamics in PSII-cc when the RC of the M2 is closed. (a) EET dynamics with initial excitation at CP47 of M2. The quantum efficiency of charge separation: RP1(M1) is 0.65; RP2(M1) is 0.20. Total QE = 0.85. (b) EET dynamics with initial excitation at CP43 of M2. The quantum efficiency of charge separation: RP1(M1) is 0.62; RP2(M1) is 0.20. Total QE = 0.82.

one of the RCs is closed. Under ambient conditions for photosynthesis, it is highly likely that one of the RCs in a dimeric PSII-cc is oxidized and only one RC remains active for charge separation, due to the vast mismatch of the time scales between light harvesting (<1 ns) and the recovery of the neutral ground state of the special pair (>1 μs).[1] To mimic the dynamics when one of the RCs is closed, we simulated EET dynamics of the PSII-cc system while removing the two radical-pair generation pathways in the RC of the M2 monomer. In this case, excitation energy can flow into and out of the RC of M2, but charge separation can not take place in the closed RC of M2.

The EET dynamics in the PSII-cc with the RC of M2 closed are shown in Fig. 9. There, we examine the worst-case scenarios when the core light-harvesting complex belonging to the M2 monomer is initially excited. Because the RC of M2 is closed, we expect that the quantum efficiency of charge separation would be very low if the EET is limited to a single monomer. Nevertheless, EET dynamics shown in Fig. 9 exhibit rapid EET to the CP47 of M2, followed by inter-monomer EET to the CP47 of the M1, and then finally reach the RC of M1, leading to successful charge separation in M1. The quantum efficiency is still greater than 0.8. Our simulations indicate that even with one of the RCs closed, the dimeric PSII-cc still maintains a high quantum efficiency. Thus the inter-monomer EET enhances the robustness of the PSII-cc.

To elucidate the importance of the two CLA625s in the robustness of the light harvesting in the PSII system, we simulated EET dynamics in the mutant PSII-cc (without the CLA625s) with the RC of M2 closed. Fig. 10 shows the simulated EET dynamics when both CP47s are initially populated. The EET dynamics within the M1 monomer (Fig. 10a) are not very different from the corresponding case in a normal monomeric PSII-cc (Fig. 6a), because the RC in M1 is still open. In contrast, the EET dynamics within the M2 monomer are drastically different due to the closed RC. The fluorescence decay dominates the longtime dynamics in M2. As a result, the mutant PSII-cc with the RC of M2 closed shows a significantly reduced total quantum efficiency of ∼0.47. Therefore, we have clearly demonstrated that the inter-monomeric EET in the PSII-cc mediated by the two CLA625s play an important role in the robustness of the light reaction in the PSII system.

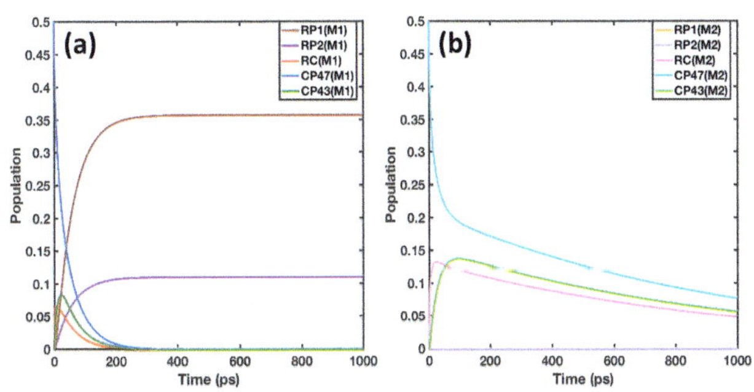

Fig. 10 EET dynamics in mutant PSII-cc with the RC of the M2 closed. (a) EET dynamics in the M1 monomer. (b) EET dynamics in the M2 monomer. The initial condition is that both CP47s are equally populated. The quantum efficiency of charge separation: RP1(M1) is 0.36; RP2(M1) is 0.11.

3.7 Minimum-cut network analysis

To further elucidate the EET dynamics in the PSII-cc system we carried out an analysis of the EET network in the PSII-cc based on a minimum cut approach for network clustering.[56–58] A minimum cut in a flow network is the division that separates the network into two clusters of nodes, a source cluster and a target cluster, and minimizes the total flow from the source to the target.[56] The concept is in line with the idea of finding bottlenecks in EET networks and we have successfully adapted the minimum cut method and applied it to study energy transfer in several photosynthetic pigment–protein complexes.[59]

To apply the minimum cut approach to the PSII-cc system, we consider EET network in the PSII-cc. In this network, nodes comprised of exciton states and edges weighted by the calculated rates are used as a weighted network representation for the EET dynamics. We then apply the minimum cut algorithm to locate the bottleneck in the EET network and divide the network into two clusters. When this procedure is applied recursively, a binary minimum-cut tree is obtained.

Fig. S6† shows the minimum-cut tree of the PSII-cc EET network. The minimum-cut tree depicts the source cluster and the target cluster for each cut as well as the corresponding maximal flow. These cuts represent bottlenecks of EET dynamics within the PSII-cc system, and the maximum flows are the sum of the rate constants from the source cluster to the target cluster. The maximum-cut tree also enables us to develop a efficient kinetic clustering method that can be used to generate effective coarse-grained model for EET dynamics. We can start from the bottom of the tree and regroup clusters when maximum flows between them is greater than a selected cut-off value to yield a clustering model for the network.[57,58] Furthermore, we can control the number of clusters in the reduced model by varying the cut-off value. Once the clustering is determined, we can calculate a coarse-grained (CG) effective energy transfer rate between each two clusters:

$$R_{TS} = \sum_{\beta \in T} \sum_{\alpha \in S} k_{\beta\alpha} \frac{\mathrm{e}^{-E_\alpha/k_\mathrm{B}T}}{Z_S}, \tag{20}$$

where T and S denote the target cluster and the source cluster, respectively. R_{TS} is the effective EET rate from S to T, E_α is the energy of the exciton state $|\alpha\rangle$, $k_{\beta\alpha}$ is the rate constant of EET from $|\alpha\rangle$ to $|\beta\rangle$, and $Z_S = \sum_{\alpha \in S} \mathrm{e}^{-E_\alpha/k_\mathrm{B}T}$ is a partition function. Eqn (20) assumes rapid thermal equilibrium of excitation energy among the exciton states in the source cluster, therefore the effective rate is a sum with Boltzmann weights. To validate the effectiveness of the CG scheme, we can simulate EET dynamics using the CG model and compare the results to those calculated from the full dynamics. A quantitative comparison can be achieved using the integrated absolute population difference between the CG dynamics and the exact dynamics as an error function for a given CG model.

In Fig. S7a,† we plot the integrated absolute population difference for CG models constructed with different numbers of clusters. Clearly, the 7-cluster model yields an excellent description of the EET dynamics, while still keeping the number of clusters in the model minimal. Fig. S7b† shows the excellent agreement between the CG dynamics calculated using the 7-cluster model and the dynamics from the full simulation. Fig. 11 depicts the 7-cluster CG model as well

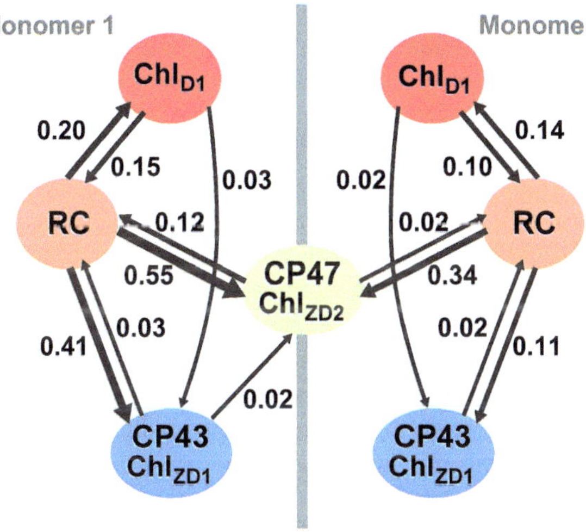

Fig. 11 The 7-cluster kinetic CG model for the PSII-cc obtained from a minimum-cut tree based cluster scheme. The effective inter-cluster rates are also labeled in the network model (unit: ps^{-1}).

as the effective inter-cluster EET rates. It is interesting to note that this CG model basically follows the subunit divisions in the PSII-cc system. However, the minimum-cut scheme places Chlz$_{D1}$ in the CP43 cluster and Chlz$_{D1}$ in the CP47 cluster, respectively, which is in agreement with our dynamical simulations (Fig. 6) and previous structure-based model for EET in the PSII.[23,26] The minimum-cut clustering scheme also successfully identified the lowest energy Chl$_{D1}$ sites and separates them from the RC. Moreover, the two CP47s are considered as a large cluster in the CG model, reflecting the rapid inter-monomer EET through CLA625s discussed above. The CG model also indicates that the two PSII core monomers are very similar but not exactly the same, which is a result of the sampling of heterogeneity by the MD simulation.

The CG model for EET in the PSII-cc presented in Fig. 11 suggests that the dimeric PSII-cc is specifically designed to perform light harvesting with high efficiency and robustness. The two CP47s, locating in the center of the PSII-cc structure, could serve as an effective energy pool that collects energy efficiently from the antenna. In this regard, the fact that the CP47 complex contains several low energy sites could be a design feature to facilitate directional and rapid energy transfer by providing an energy gradient. In addition, since CP47 is lower in energy compared to the RC, the energetics also favor energy pooling in CP47, which could be beneficial when the RC is closed. The dimeric form of the PSII-cc also provides the extra redundancy such that the excitation energy could be utilized even when one of the RCs is closed. It seems that nature adopted this design of pooling excitation energy into the CP47s of PSII-cc to enable efficient light harvesting as well as the mitigation of the mismatch in the time scales of light harvesting and the recovery of a closed RC.

Very recently, the entire pigment network and punitive energy transfer pathways in the PSII supercomplex have been elucidated based on high-resolution structural models.[9,13,15] On the basis of these structural data, it was proposed that energy transfer from the major light-harvesting complexes to CP29 and then to CP47 is likely the dominant pathway for conducting energy from the outer antenna to the PSII RC. It is interesting to note that our CG model indicates that the PSII-cc with CP47 pooling the excitation energy is well positioned to take advantage of this EET pathway.

4 Conclusions

In this work, we present a structure-based effective model for the dimeric PSII core complex. The model was constructed based on MD simulations of the PSII core complex in combination with the charge density coupled method for site energies and the transition charge from electrostatic potential approach for excitonic couplings. We demonstrated that with proper separation of dynamical and static disorders and rescaling of the site energy fluctuations, system–bath couplings evaluated by our method describes the absorption spectrum of CP47, CP43 and the RC at 297 K as well as the full excitation energy transfer dynamics among the 74 pigments in the PSII core complex.

Simulation of EET dynamics in the monomer of PSII core complex shows that our model describes highly efficient light harvesting in the system. Sites important for EET from CP43 and CP47 to the RC can be identified, and we found that the Chlz sites do not mediate EET into the RC. In addition, we observe that EET between two antenna complexes takes place through the RC, despite the rapid charge separation dynamics in the RC. Thus, EET dynamics in the PSII core complex exhibit strong dynamical heterogeneity and is conducted by multiple pathways. We conclude that the EET dynamics can not be describe by a single hopping rate and a reasonable description of light harvesting in the complex must take into account the detailed non-equilibrium dynamics. In this regard, the long-lasting debate regarding the validity of the exciton radical pair equilibrium model and the transfer-to-the-trap limited model in the PSII[7,8,10,26] is partially a result of the overly-simplified nature of gross kinetic models. When the detailed geometry and disordered energy landscape of the PSII-cc are considered, the dynamics become intrinsically heterogeneous and can not be described by a single hopping time.

In the EET dynamics of dimeric PSII core complex, we reveal that the two CP47s can mediate rapid EET between the two monomers, and the two CLA625 sites in the CP47s play crucial role in the inter-monomer EET. The relatively strong excitonic coupling between the two CLA625s is achieved by the parallel orientation of the two molecules due to the C2 symmetry of the core complex. We showed that after removing the two CLA625s in the system Hamiltonian, the inter-monomer EET is totally suppressed.

We further investigate EET dynamics in the PSII core complex when one of the RCs is closed. In normal PSII core complex, the closed RC does not significantly affect the overall quantum efficiency of the system, since rapid inter-monomer EET can transfer the excitation energy to be utilized by the RC that is still open. However, the quantum efficiency is drastically reduced when the inter-monomer EET is suppressed. We conclude that the dimeric form of the PSII core complex could significantly enhance the robustness of the system, which suggests that the evolution did not make the dimeric PSII core complex just for a structural reason.

Finally, we applied a minimum-cut clustering method to analyze the complex EET network of the 74 exciton states in the PSII core complex. The network analysis approach produces a 7-cluster coarse-grained model that captures the key EET pathways in the PSII core complex. Furthermore, the coarse-grained model provides a clear blueprint for the design of the PSII core complex. The CP47 subunits, with their favorable energetics and location, serve as a energy pool to collect and distribute the excitation energy to both RCs within the dimer. We suggest that this design allows the high efficiency and robustness of light harvesting in the PSII system, especially when the large time-scale mismatch between the dynamics of light harvesting and the recovery of the radical-pair state back to the neutral state is considered. Furthermore, we suspect that the energy pooling in the CP47s could also contribute to the regulation of excess energy *via* non-photochemical quenching.[60]

Our model shows that the method based on MD simulations and quantum chemistry calculations can be effectively utilized to elucidate the dynamics of light harvesting in complex photosynthetic complexes. Furthermore, the minimum cut based clustering method provides a powerful approach to generate accurate CG models for EET dynamics in large EET networks. This framework should be applicable to general organic molecular aggregates and useful for the design of efficient light-harvesting materials. We believe this approach is likely to bring significant breakthrough towards the understanding of geometrical factors and functional design principles of photosynthetic light reactions.

Conflicts of interest

There are no conflicts to declare.

Acknowledgements

YCC thanks the Ministry of Science and Technology, Taiwan (Grant No. NSC 105-2113-M-002-012), and Center for Quantum Science and Engineering (Subproject: 103R891401) for financial support. We are grateful to the Computer and Information Networking Center, National Taiwan University for the support of high-performance computing facilities. XH thanks the support of the Hong Kong Research Grant Council (HKUST C6009-15G, 16305817, 16302214, 16304215, 16318816, AoE/P-705/16, M-HKUST601/13, and T13-607/12R), King Abdullah University of Science and Technology (KAUST) Office of Sponsored Research (OSR) (OSR-2016-CRG5-3007), Shenzhen Science and Technology Innovation Committee (JCYJ20170413173837121), Innovation and Technology Commission (ITCPD/17-9 and ITC-CNERC14SC01), and National Natural Science Foundation of China (11275022, 11635002). LZ thanks the support of the National Natural Science Foundation of China (21733007 and 21803071), and the National 1000 Youth Talents Program of China.

Notes and references

1 R. E. Blankenship, *Molecular mechanisms of photosynthesis*, Blackwell Science Ltd, Oxford, UK, 2008, pp. 1–321.
2 Y.-C. Cheng and G. R. Fleming, *Annu. Rev. Phys. Chem.*, 2009, **60**, 241–262.
3 N. Nelson and W. Junge, *Annu. Rev. Biochem.*, 2015, **84**, 659–683.

4 T. Mirkovic, E. E. Ostroumov, J. M. Anna, R. Van Grondelle, Govindjee and G. D. Scholes, *Chem. Rev.*, 2017, **117**, 249–293.

5 S. Vasil'ev, P. Orth, A. Zouni, T. G. Owens and D. Bruce, *Proc. Natl. Acad. Sci. U. S. A.*, 2001, **98**, 8602–8607.

6 J. Nield and J. Barber, *Biochim. Biophys. Acta, Bioenerg.*, 2006, **1757**, 353–361.

7 R. Croce and H. van Amerongen, *J. Photochem. Photobiol., B*, 2011, **104**, 142–153.

8 H. van Amerongen and R. Croce, *Photosynth. Res.*, 2013, **116**, 251–263.

9 X. Wei, X. Su, P. Cao, X. Liu, W. Chang, M. Li, X. Zhang and Z. Liu, *Nature*, 2016, **534**, 69–74.

10 K. Broess, G. Trinkunas, C. D. Van Der Weij-De Wit, J. P. Dekker, A. Van Hoek and H. Van Amerongen, *Biophys. J.*, 2006, **91**, 3776–3786.

11 R. Danielsson, M. Suorsa, V. Paakkarinen, P.-A. Albertsson, S. Styring, E.-M. Aro and F. Mamedov, *J. Biol. Chem.*, 2006, **281**, 14241–14249.

12 S. Caffarri, R. Kouril, S. Kereiche, E. J. Boekema and R. Croce, *EMBO J.*, 2009, **28**, 3052–3063.

13 X. Su, J. Ma, X. Wei, P. Cao, D. Zhu, W. Chang, Z. Liu, X. Zhang and M. Li, *Science*, 2017, **357**, 815–820.

14 L. S. van Bezouwen, S. Caffarri, R. S. Kale, R. Kouril, A.-M. W. H. Thunnissen, G. T. Oostergetel and E. J. Boekema, *Nat. Plants*, 2017, **3**, 17080.

15 P. Cao, X. Su, X. Pan, Z. Liu, W. Chang and M. Li, *Biochim. Biophys. Acta, Bioenerg.*, 2018, **1859**, 633–644.

16 Y. Umena, K. Kawakami, J. R. Shen and N. Kamiya, *Nature*, 2011, **473**, 55–60.

17 J. P. Dekker and R. Van Grondelle, *Photosynth. Res.*, 2000, **63**, 195–208.

18 L. Konermann and A. R. Holzwarth, *Biochemistry*, 1996, **35**, 829–842.

19 M. Groot, J. Dekker, R. van Grondelle, F. D. Hartog and S. Völker, *J. Phys. Chem.*, 1996, **3654**, 11488–11495.

20 V. I. Novoderezhkin, E. G. Andrizhiyevskaya, J. P. Dekker and R. Van Grondelle, *Biophys. J.*, 2005, **89**, 1464–1481.

21 V. I. Novoderezhkin, E. Romero, J. P. Dekker and R. van Grondelle, *ChemPhysChem*, 2011, **12**, 681–688.

22 E. Romero, I. H. M. Van Stokkum, V. I. Novoderezhkin, J. P. Dekker and R. Van Grondelle, *Biochemistry*, 2010, **49**, 4300–4307.

23 Y. Shibata, S. Nishi, K. Kawakami, J. R. Shen and T. Renger, *J. Am. Chem. Soc.*, 2013, **135**, 6903–6914.

24 D. I. Bennett, K. Amarnath and G. R. Fleming, *J. Am. Chem. Soc.*, 2013, **135**, 9164–9173.

25 R. N. Frese, M. Germano, F. L. de Weerd, I. H. M. van Stokkum, A. Y. Shkuropatov, V. a. Shuvalov, H. J. van Gorkom, R. van Grondelle and J. P. Dekker, *Biochemistry*, 2003, **42**, 9205–9213.

26 G. Raszewski and T. Renger, *J. Am. Chem. Soc.*, 2008, **130**, 4431–4446.

27 T. Renger, M. E.-A. Madjet, M. S. Am Busch, J. Adolphs and F. Müh, *Photosynth. Res.*, 2013, **116**, 367–388.

28 C. Kreisbeck and A. Aspuru-Guzik, *Chem. Sci.*, 2016, **7**, 4174–4183.

29 C. Curutchet and B. Mennucci, *Chem. Rev.*, 2017, **117**, 294–343.

30 J. Hall, T. Renger, F. Müh, R. Picorel and E. Krausz, *Biochim. Biophys. Acta*, 2016, **1857**, 1580–1593.

31 F. Müh, M. Ploeckinger and T. Renger, *J. Phys. Chem. Lett.*, 2017, **8**, 850–858.

32 C. Olbrich, T. L. C. Jansen, J. Liebers, M. Aghtar, J. Struempfer, K. Schulten, J. Knoester and U. Kleinekathoefer, *J. Phys. Chem. B*, 2011, **115**, 8609–8621.
33 Y. Jing, R. Zheng, H.-X. Li and Q. Shi, *J. Phys. Chem. B*, 2012, **116**, 1164–1171.
34 K. Ogata, T. Yuki, M. Hatakeyama, W. Uchida and S. Nakamura, *J. Am. Chem. Soc.*, 2013, **135**, 15670–15673.
35 J. Huh, S. K. Saikin, J. C. Brookes, S. Valleau, T. Fujita and A. Aspuru-Guzik, *J. Am. Chem. Soc.*, 2014, **136**, 2048–2057.
36 L. Zhang, D. A. Silva, H. Zhang, A. Yue, Y. Yan and X. Huang, *Nat. Commun.*, 2014, **5**, 1–9.
37 T. Renger, V. May and O. Ku, *Phys. Rep.*, 2001, **343**, 137–254.
38 F. Müh, M. E.-A. Madjet, J. Adolphs, A. Abdurahman, B. Rabenstein, H. Ishikita, E.-W. Knapp and T. Renger, *Proc. Natl. Acad. Sci. U. S. A.*, 2007, **104**, 16862–16867.
39 J. Adolphs, F. Muh, M.-a. Madjet, M. Schmidt am Busch and T. Renger, *J. Am. Chem. Soc.*, 2010, **132**, 3331–3343.
40 J. W. Ponder and D. A. Case, *Adv. Protein Chem.*, 2003, **66**, 27–85.
41 L. Zhang, D.-A. Silva, Y. Yan and X. Huang, *J. Comput. Chem.*, 2012, **33**, 1969–1980.
42 J. Adolphs, F. Müh, M. E. A. Madjet and T. Renger, *Photosynth. Res.*, 2008, **95**, 197–209.
43 M. Yang and G. R. Fleming, *Chem. Phys.*, 2002, **275**, 355–372.
44 V. I. Novoderezhkin and R. van Grondelle, *Phys. Chem. Chem. Phys.*, 2010, **12**, 7352–7365.
45 T. Renger and F. Müh, *Phys. Chem. Chem. Phys.*, 2013, **15**, 3348–3371.
46 S. Egorov and J. L. Skinner, *Chem. Phys. Lett.*, 1998, **293**, 469–476.
47 S. Valleau, A. Eisfeld and A. Aspuru-Guzik, *J. Chem. Phys.*, 2012, **137**, 224103.
48 V. I. Novoderezhkin, D. Rutkauskas and R. van Grondelle, *Biophys. J.*, 2006, **90**, 2890–2902.
49 Y.-H. Hwang-Fu, W. Chen and Y.-C. Cheng, *Chem. Phys.*, 2015, **447**, 46–53.
50 G. Gatzen, M. G. Müller, K. Griebenow and A. R. Holzwarth, *J. Phys. Chem.*, 1996, **100**, 7269–7278.
51 M. Boehm, E. Romero, V. Reisinger, J. Yu, J. Komenda and L. A. Eichacker, *J. Biol. Chem.*, 2011, **286**, 14812–14819.
52 M. Jassas, T. Reinot, A. Kell and R. Jankowiak, *J. Phys. Chem. B*, 2017, **121**, 4364–4378.
53 J. Pan, A. Gelzinis, V. ChoroĄ̊ajev, M. Vengris, S. S. Senlik, J.-R. Shen, L. Valkunas, D. Abramavicius and J. P. Ogilvie, *Phys. Chem. Chem. Phys.*, 2017, **19**, 15356–15367.
54 E. Wientjes, H. van Amerongen and R. Croce, *J. Phys. Chem. B*, 2013, **117**, 11200–11208.
55 Y. Yoneda, T. Katayama, Y. Nagasawa, H. Miyasaka and Y. Umena, *J. Am. Chem. Soc.*, 2016, **138**, 11599–11605.
56 L. R. Ford and D. R. Fulkerson, *Can. J. Math.*, 1956, **8**, 399–404.
57 G. W. Flake, R. E. Tarjan and K. Tsioutsiouliklis, *Internet Math*, 2011, **1**, 385–408.
58 R. Görke, T. Hartmann and D. Wagner, *J. Graph. Algorithms Appl.*, 2012, **16**, 411–446.
59 D.-W. Ye, W.-H. Tseng and Y.-C. Cheng, in preparation.
60 M. Yokono, R. Nagao, T. Tomo and S. Akimoto, *Biochim. Biophys. Acta, Bioenerg.*, 2015, **1847**, 1274–1282.

Faraday Discussions

PAPER

Two-dimensional electronic vibrational spectroscopy and ultrafast excitonic and vibronic photosynthetic energy transfer

Eric C. Wu,[ab] Eric A. Arsenault,[a] Pallavi Bhattacharyya,[abd] Nicholas H. C. Lewis[c] and Graham R. Fleming*[abd]

Received 19th November 2018, Accepted 22nd January 2019
DOI: 10.1039/c8fd00190a

Two-dimensional electronic-vibrational (2DEV) spectroscopy is a new coherent spectroscopic technique, which shows considerable promise for unravelling complex molecular dynamics. In this Discussion we describe an application to the energy transfer pathway in the major light harvesting protein, LHCII, providing new data on the center line slopes (CLS) of the spectral peaks. The CLS provides information that appears unique to the 2DEV method. We then outline a general approach to calculating 2DEV spectra which is valid for strongly and weakly coupled molecular systems. We conclude with some prospects for the future development of 2DEV spectroscopy and its theoretical analysis.

Introduction

The timescale of individual energy transfer steps in the light harvesting antennae of plants, algae, and photosynthetic bacteria has been the subject of speculation and a spur to experimentalists and theorists alike for over 70 years. Once it was realized that antenna chlorophylls (Chls) outnumbered the primary charge separation centers of Photosystem II by about 300 : 1, and that the quantum efficiency of transfer from the antenna at low light levels could approach 95%, it was apparent that individual energy transfer steps must be very fast given the known fluorescence lifetime of Chl. In 1964 Duysens suggested that individual steps might take 100 fs [1] while Robinson talked of exciton "spreading" as rapidly as 50 fs.[2] Such timescales were far beyond the experimental methods available in the 1960s but once mode locked lasers with few ps duration pulses became available, attention rapidly turned to photosynthetic energy transport. Early

[a]Department of Chemistry, University of California, Berkeley 94720, USA. E-mail: grfleming@lbl.gov
[b]Molecular Biophysics and Integrated Bioimaging Division, Lawrence Berkeley National Lab, Berkeley, California 94720, USA
[c]James Franck Institute, University of Chicago, Illinois 60637, USA
[d]Kavli Energy Nanosciences Institute at Berkeley, Berkeley, California 94720, USA

measurements with low repetition rate (10^{-2} Hz!) lasers were severely compromised by the very high intensities used. With the advent of high repetition rate dye and Ti:Sapphire lasers, precise measurements began to be possible, notably done by Gillbro, Sundström, and van Grondelle,[3] by Holzwarth and coworkers,[4] and by Fleming and coworkers.[5] By 1989 Eads *et al.* were able to roughly measure the Chl *b* to Chl *a* transfer timescale in the plant LHCII complex using fluorescence upconversion.[6] Although there were earlier hints that Förster theory based on molecule to molecule hopping was not adequate, it was the measurements of B800 to B850 transfer in the bacterial light harvesting complex 2 (LH2) protein that provided the stimulus to develop new models. It proved impossible to provide B800 to B850 transfer by simple Forster theory and simultaneously and independently, Sumi[7] and Scholes and Fleming[8] developed Generalized Förster Theory that takes account of delocalized multi-chromophore states.

Photon echo spectroscopies began to be used in the 1990s to study photosynthetic energy transfer, and theoretical methods based on a modified form of Redfield theory were developed.[9–11] In 2005 Brixner *et al.* showed two-dimensional electronic spectra (2DES) of the bacterial FMO light harvesting complex.[12] There were hints of oscillatory behavior in the initial data but it was the 2007 paper by Engel *et al.* that clearly showed long-lived oscillations or beats.[13] This was followed by observations of long-lived oscillations in an alga by Scholes and coworkers[14] and temperature dependent 2DES studies of FMO by Engel and coworkers.[15] Engel *et al.* initially interpreted the oscillations as arising from electronic coherences between excitonic states.[13] This led to a huge surge in theoretical interest and continuing discussion of the oscillatory features as being electronic, vibronic, or vibrational in origin. Two important contributions were the development of the Hierarchy Equations of Motion approach for formally exact calculation of energy transfer,[16] and the insightful contribution of Jonas and coworkers who showed that the longest lived oscillations were very likely from vibrational wavepackets on the ground electronic state.[17] Debate continues on the relative importance of vibronic contributions to the overall rate of transfer, compared to that from purely electronic interactions. A salutary point was made by Fujihashi *et al.* who showed that while oscillatory signals may be prominent in 2DES, they may have little impact on the rates of population transfer.[18]

Considering which experimental approach might be most suited to unravelling the role of vibrations in ultrafast energy transfer led us to develop a new form of 2D spectroscopy – two-dimensional electronic-vibrational (2DEV) spectroscopy.[19] In 2DEV spectroscopy two UV-vis pulses interact with the system of interest, followed by an IR pulse and finally the signal IR pulse is emitted by the system. Fig. 1 shows a schematic of the relationship of 2D infrared, 2D electronic and 2DEV spectroscopies.

2DEV spectroscopy correlates the evolution of the electronic and nuclear degrees of freedom, and thus holds promise to provide insight into the complex question of the dynamics of mixed electronic–vibrational states in natural photosynthetic light harvesting. For systems with small to moderate electronic coupling 2DEV spectroscopy also provides a window into the spatial location of excitation, since Chl vibration frequencies are sensitive to binding site, but very little dependent on whether the Chl is electronically excited or not.[20] In addition to the peak frequencies and amplitudes, the evolution of the peak shape in a 2DEV spectrum provides a novel picture of the dynamics.[21] We characterize the

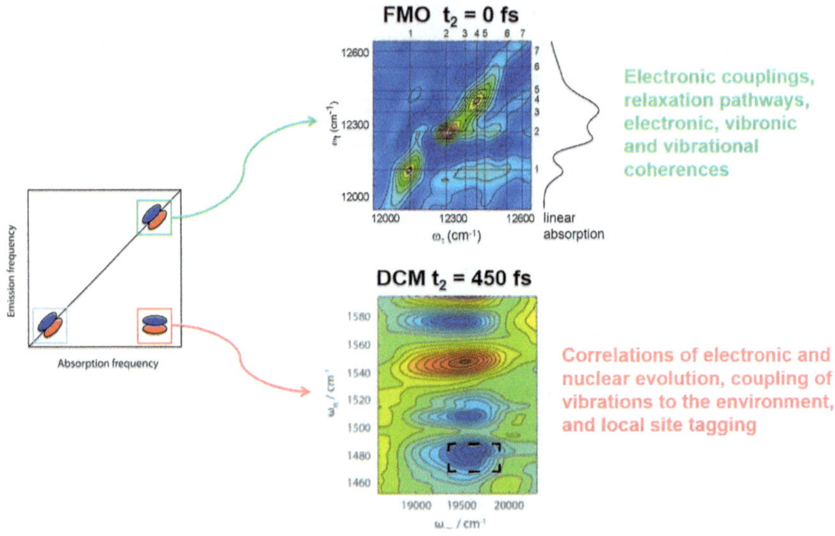

Fig. 1 Two-dimensional coherent spectroscopies. Left: the positions of 2D infrared spectroscopy (lower left), 2D electronic spectroscopy (upper right), and 2D electronic-vibration (2DEV) spectroscopy (lower right) in a two-dimensional frequency plot. Right upper: the 2D electronic spectrum of the FMO light harvesting complex at waiting time (t_2) of zero.[12] The diagonal peaks show the exciton absorption bands, the off-diagonal (cross) peaks show coupling and, at later waiting times, relaxation pathways. Right lower: 2DEV spectrum of DCM dye at $t_2 = 450$ fs.[21] There are no diagonal peaks. Excited state (absorptive) peaks are colored blue and ground state bleaches red. The box around the lower excited state peak is centered on the position of that peak at $t_2 = 0$ fs, and shows shifts in both spectral dimensions by $t_2 = 450$ fs.

peak shape by a single variable, the center line slope (CLS). In this paper we describe 2DEV spectroscopic results for the major light harvesting protein of green plants, LHCII. We then describe a new theoretical description of 2DEV spectra applicable to both weakly and strongly coupled vibronic systems. When the electronic and vibrational states become mixed, creating vibronic energy levels, the number of pathways contributing to the 2DEV signal increases substantially. By separating the response into decoherence and population relaxation contributions we show how to select the significant pathways contributing to the signal and simplify the analysis of the 2DEV spectrum.

LHCII 2DEV spectra and SVD analysis

The major pigment-protein complex utilized by plants and green algae to collect sunlight is called LHCII. About half the chlorophyll (Chl) on earth is found in this protein and the antenna systems of plants consist largely of arrays of LHCII surrounding Photosystem II supercomplexes where water splitting occurs. LHCII is generally found in the trimeric form with each monomer containing 8 Chl *a* and 6 Chl *b*. Fig. 2 (top panels) shows the 2DEV spectrum of LHCII at 77 K as a function of waiting time, t_2.

The features in the 2DEV spectra of isolated Chl *a* and Chl *b*[22] and that of LHCII[23] have been assigned and discussed in previous works, but for clarity we

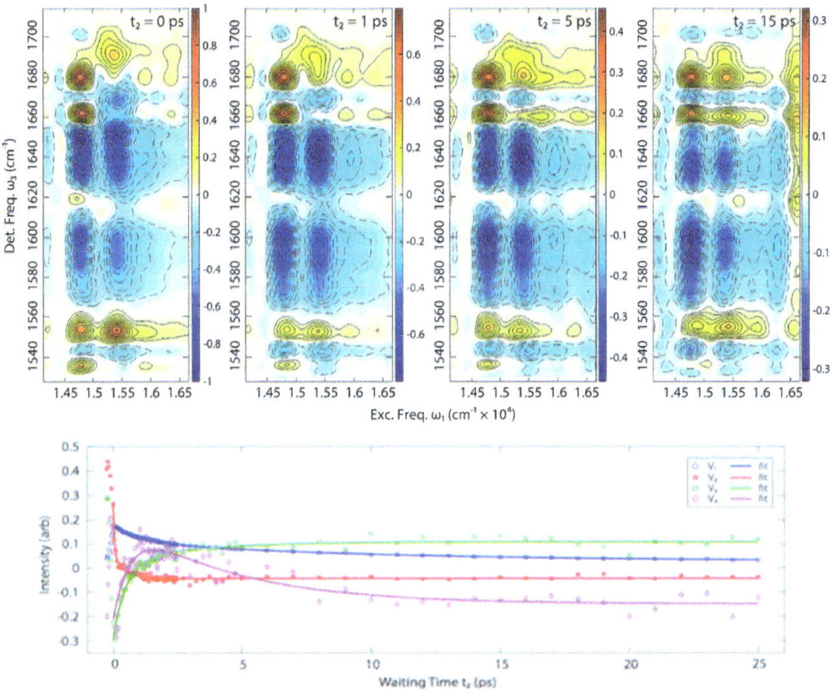

Fig. 2 2DEV spectra of the LHCII light harvesting complex at four waiting times (upper panels)[23] and singular value decomposition (SVD) into four components (lower panel). The results of fits to the four singular vectors are given in Table 1. Adapted with permission from *J. Phys. Chem. Lett.*, 7, 4197–4206. Copyright 2016 American Chemical Society.

will briefly summarize the main conclusions here. These assignments are based on the assumption that the vibrations remain local to either monomers or strongly coupled dimers or trimers of Chl. First, we will focus on the excitation bands, which are straightforward to assign since the transitions also appear in the linear electronic absorption spectrum of LHCII. The 14 800 cm^{-1} absorption band has been assigned to electronic states associated with the Q_y transition of Chl *a*.[24,25] However, as will be discussed below, the long-lived shoulder at 1690 cm^{-1} suggests that some long-lived population on Chl *b* contributes to this band as well. The excitation band at 15 300 cm^{-1} has been assigned to a combination of Chl *b* transitions with additional contributions from higher energy Chl *a* states, such as the Q_x transition.[24,25] The shoulder above 16 000 cm^{-1} is likely due to the Chl *b* Q_x transition and possibly higher energy vibronic excitations. To assign peaks along the detection axis, assignments from previous work on Chl *a* and Chl *b* were used as references.[22] The bleach at 1690 cm^{-1} in the 15 400 cm^{-1} band was attributed to the five-coordinated, rather than six-coordinated, Chl *b* species as seen in the spectrum of isolated Chl *b* in solution. The 1690 cm^{-1} bleach is also present in the 14 800 cm^{-1} band, but it appears slightly obscured as a shoulder of the 1680 cm^{-1} peak. The 1660 cm^{-1} and 1680 cm^{-1} bleaches present in the 14 800 cm^{-1} band are likely from the same transition, possibly the GSB of C=O stretch of Chl *a*, because their intensities have the same dynamics. However, the appearance of two bleaches, rather than one, was attributed to overlap with an absorption at 1670 cm^{-1}. A peak at 1670 cm^{-1} is also present in the 15 400 cm^{-1} band. The origin of the 1670 cm^{-1} transition remains unknown, as it does not appear in the spectra of isolated Chl *a* or Chl *b*. The bleaches and absorptions at 1555 cm^{-1} and 1545 cm^{-1}, respectively, were assigned to C=C stretches of the Chlorin rings. It should be noted that these peaks appear blue-shifted relative to the C=C stretches observed in the spectra of isolated Chl *a* and Chl *b* due to pigment-protein interactions. The most pronounced features along the detection axis are the two broad absorptions centered at 1640 cm^{-1} and 1590 cm^{-1}. These broad absorptions were assigned to excited state absorptions (ESAs) of C=O stretches, where overlap with another feature, a ground state bleach in this case, results in the observation of two absorption features. The broadness of the C=O ESA is due to the different protein environments for the individual sites, therefore, the different dynamics between the two bands can provide information on population transfer between these sites in LHCII.

With spectral assignments as a starting point, we can now focus on the peak dynamics. The spectrum at a waiting time, t_2, of zero represents the distribution of population initially created by the laser excitation, *i.e.* the probe spectrum at a specific excitation frequency is given by the vibrational spectrum of the sites that are initially populated by the visible excitation at that excitation frequency. Therefore, the differences along the detection axis between the bands at 15 300 cm^{-1} and 14 800 cm^{-1} represent the differences between the sites that were excited. As waiting time increases, the two excitation bands become more similar due to the relaxation of the excited state population. A comparison of the spectral evolution along the detection axis of the two excitation bands then reflects both the mutual and distinct population relaxation pathways of the initial population created by the visible pump pulse. At the longest waiting time, $t_2 = 15$ ps, the probe spectrum along each excitation band should then be similar to the vibrational spectrum of the lowest energy state. At 77 K, this means that the equilibrated population resides in the trimer states composed of a610, a611, and a612 (the Chl numbering system is taken from Liu *et al.*[26]). Upon comparison of the two excitation bands, in combination with the previously discussed spectral assignments, some indication of population and energy transfer between the sites in LHCII can be observed. For example, the 15 400 cm^{-1} band unambiguously shows the transfer of population from Chl *b* to Chl *a*. This is because the intensity of the 1690 cm^{-1} bleach, which as a reminder, has been assigned to Chl *b*, decreases while the intensities at the 1660 cm^{-1} and 1680 cm^{-1} bleaches, both assigned to Chl *a*, increase. Inspection of the two broad C=O ESAs also reveals population transfer, in this case from the 1640 cm^{-1} peak, assigned Chl *b*, to the 1590 cm^{-1} peak of Chl *a*.

To further study the dynamics, singular value decomposition (SVD) was performed on the spectrum. SVD breaks the spectrum into different spectral surfaces and dynamical contributions weighted by the singular values. SVD analysis gives the corresponding four spectral components that are not dominated by noise, shown in Fig. 3. The first four singular vectors are shown in the lower panel of Fig. 2. Table 1 collects the results of biexponential fits to these curves.

The first component, the spectral surface and corresponding dynamical component with the largest singular value, is what best describes the 2DEV spectrum. The second component is then the surface and dynamics that best describes the 2DEV spectrum once the first component is removed. The first

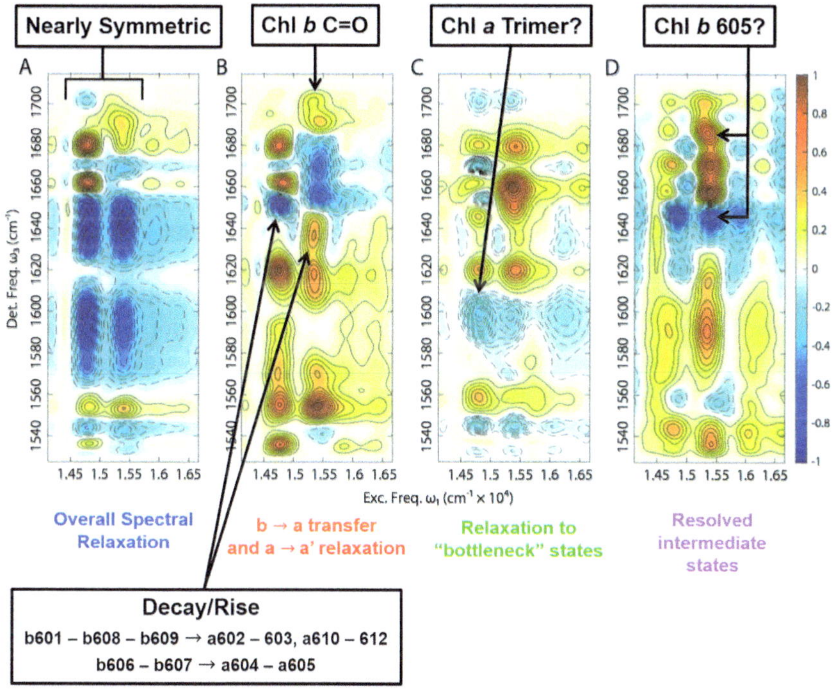

Fig. 3 SVD spectral components corresponding to the four singular vectors (A–D) in Fig. 2.[23] Assignments of a number of the spectral features are shown on the figure while overall relaxation pathways are indicated below. Note the nearly symmetric appearance of the first panel, indicating overall relaxation.[23] Adapted with permission from *J. Phys. Chem. Lett.*, 7, 4197–4206. Copyright 2016 American Chemical Society.

spectral surface is very similar to the 2DEV spectrum at waiting time of 1 ps. This indicates that the changes in the spectrum as a function of waiting time are minor and the dynamics of the first component reflect the overall decay of the spectrum as the population returns from the excited state to ground state. The second spectral component, which is notably the only component beside the first to have a peak at 1690 cm^{-1}, shows the relaxation both from Chl *b* to Chl *a* and within Chl *a*. In the 15 400 cm^{-1} band, the second spectral component has a positive peak at 1690 cm^{-1} and two negative peaks at 1680 and 1660 cm^{-1}. This solidifies the earlier claim of the observation of population transfer from Chl *b* to Chl *a*.

Table 1 Biexpontial fits to the four singular vectors in Fig. 2

Fit to $V_i(t) = A_1e^{-t/\tau_1} + A_2e^{-t/\tau_2} + A_3$

	A_1	τ_1 (ps)	A_2	τ_2 (ps)	
V_1	0.07	1.13	0.08	9.97	Overall relaxation of spectrum
V_2	0.22	0.05	0.09	0.50	b → a transfer a → a′ relaxation
V_3	−0.30	0.46	−0.10	2.90	Equilibration to "bottleneck" states
V_4	0.45	0.67	0.39	4.21	Resolved intermediates along the relaxation pathway

Moreover, in the second spectral component, the differences in the C=O ESA region, 1570 cm^{-1} to 1650 cm^{-1}, between the two excitation bands reveal the different relaxation pathways for these bands. Starting with the 14 800 cm^{-1} band, the negative peak at 1650 cm^{-1} and the positive peak at 1620 cm^{-1} indicate that the population relaxes directly from 1650 cm^{-1} to 1620 cm^{-1}. On the other hand, for the 15 300 cm^{-1} band, the additional positive peak at 1640 cm^{-1} indicates a completely different or an additional relaxation pathway. The dynamics belonging to the second spectral surface, which again has been attributed to Chl *b* to Chl *a* relaxation, was best fit with a biexponential function with time constants of 50 fs and 0.5 ps. The dynamics of the third component, which rises then plateaus, suggest that the third spectral surface represents the lowest energy state, possibly the Chl *a* trimer. Finally, the fourth component, exhibiting a rise followed by a decay, suggest that the fourth spectral surface may be attributed to intermediate species. While it is clear that 2DEV spectroscopy can provide remarkable insight into the dynamics of complex systems it is also clear that further theory and modeling is needed to fully assign the abundance of observed transitions and to unveil the wealth of information contained within these experiments.

LHCII center line slope data and analysis

Information that may be unique to 2DEV spectra can be obtained *via* analysis of the center line slope (CLS) dynamics of the spectral features. The CLS describes the correlation between initial electronic coherence created by the first visible pulse and the final vibrational coherence created by the IR probe pulse. The CLS can be positive, negative, or zero. A non-zero center line slope indicates that there is a correlation between the initial electronic and final vibrational transitions. For LHCII, prior to calculating the CLSs, we did not expect to see any correlations, especially long-lived correlations, due to the rapid energy and population transfer in LHCII. Instead, we expected that population and energy transfer would scramble and ultimately destroy any correlation between electronic and vibrational coherences. However, contrary to intuition, CLSs with a rich variety of dynamics were observed, as can been seen in Fig. 4.

The CLSs for some peaks were not shown for one of two reasons: either the CLS dynamics were similar to those of other peaks for which the CLS dynamics were already shown or the CLSs were zero throughout the entire waiting time. Some CLSs decay monotonically and rapidly, such as the 1670 cm^{-1} ESA, and some decay monotonically and slowly, such as the 1590 cm^{-1} ESA and the 1680 cm^{-1} GSB. The CLS of the 1670 cm^{-1} ESA decays with a time constant of 0.9 ps, therefore the decay of the CLS is likely due to the population of Chl *b* to Chl *a* destroying the correlation. The CLS of the 1590 cm^{-1} ESA of the 14 800 cm^{-1} band decays very slowly and appears to plateau. Since the spectrum is measured at 77 K, the long-lived correlation may be due to a static inhomogeneous contribution, due to the frozen protein environment. On the other hand, the CLS of the 1590 cm^{-1} ESA of the 15 300 cm^{-1} band is zero for all waiting times. This could indicate that the static contribution to the CLS was lost due to population transfer. In addition to the CLSs that decay monotonically, some CLSs exhibit sign changes from negative to positive, such as the 1660 cm^{-1} GSB and the 1635 cm^{-1} ESA, and some decrease from zero to negative values, then decay back to zero,

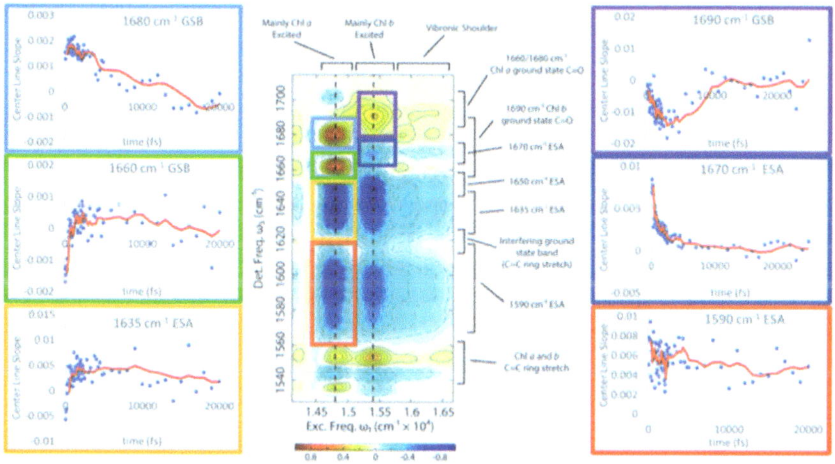

Fig. 4 Center line slopes (CLS) for the peaks indicated in the central panel by colored boxes. Note the wide variation in sign and time dependence of the CLS for different spectral features.

such as the 1690 cm^{-1} GSB. While the underlying causes of the observed CLS dynamics remain to be detailed, a previous 2DEV study on triphenylmethane dyes[27] argued that a change in the sign of CLS could indicate a change in the overall molecular dipole moment due to changes in the electronic structure. The fast initial rise in the CLSs of the 1660 cm^{-1} GSB and the 1635 cm^{-1} ESA can be fit with time constants of 0.4 ps and 0.2 ps, suggesting that the initial rise could be due to large changes in the overall dipole direction due to population relaxation from Chl *b* to Chl *a*. Moreover, it is interesting to note that even though the 1680 cm^{-1} GSB and the 1660 cm^{-1} GSB have identical dynamics in terms of peak intensities, these features exhibit quite different CLS dynamics. The distinction between the peak intensity and the CLS dynamics could simply be due to overlap with the 1670 cm^{-1} ESA, or more intriguingly, it could indicate that these two peaks are from the same exciton, but different sites. For example, the 1680 cm^{-1} and the 1660 cm^{-1} GSBs come from C=O stretches on different Chls in different environments, which result in the CLSs having different signs. Finally, some CLSs appear to have oscillations during the first ps, which could be caused by vibronic coherences, which is an area of particular interest in the development of theoretical models. However, to confirm such a claim, more waiting times must be sampled. Ultimately, the full interpretation of CLS dynamics is contingent on further theoretical development.

Theoretical modeling of 2DEV spectra

To theoretically evaluate nonlinear spectra, we can proceed in two different ways: (a) calculate the net nonlinear polarization: nonperturbative approaches[28–30] are generally used to theoretically evaluate polarization, however, this approach offers limited insights and requires extensive post-processing[31] to extract the spectroscopic signal and, (b) evaluate the Nonlinear Response Functions (NRFs),[32] relevant to the pulse sequence: NRFs are mathematically expressed in terms of the transition dipole moments describing the field-matter interactions at the different times of the spectroscopy and offer more readily apparent physical insight, than comes from calculating the polarization. Numerical methods, for instance, the Liouville Hierarchical Equation of Motion method,[33] are employed to evaluate the NRFs. However, such numerical methods are computationally expensive and scale poorly with the system dimensionality.

To circumvent this problem, we propose a near-analytical theoretical approach to evaluate NRFs for 2DEV spectroscopy.[46] The principal approximation we employ to achieve analyticity is that non-adiabatic transitions among the system states could be treated perturbatively, which implies that the system resides close to its adiabatic limit. The principal goal is not to obtain exact results, for which we need extensive numerical simulations, but to obtain an understanding of how 2DEV spectroscopy can help to unravel the mechanism of excitation energy transfer in light-harvesting complexes, both spatially and temporally. Analyticity ensures more expedient calculations and provides clear insights. The key feature of the near-analytical approach we introduce for 2DEV spectroscopy is a unitary transformation, analogous to the one used for the non-adiabaticity theory in chemical dynamics.[34]

For our initial calculations, we consider a model dimer system, with an IR mode, of angular frequency ω ($\omega = \omega_A = \omega_B$, $\hbar = 1$), local to each of the monomers (monomers labeled as A and B, respectively). We consider the IR modes on A and B to be uncoupled. The electronic coupling is given as $\langle A|H|B\rangle = \langle B|H|A\rangle = J$. The site basis, given as $|X\ \nu_\mu^A\nu_\nu^B\rangle$, therefore, should reflect information about both the electronic and IR modes. Here, $X \in \{g, A, B\}$, where g labels the electronic state where both A and B are in their respective ground states, A labels the state where only A is electronically excited and B labels the state where only B is electronically excited. We do not consider the doubly-excited electronic state, which is not important in 2DEV spectroscopy. The superscript on ν in $|X\ \nu_\mu^A\nu_\nu^B\rangle$ denotes that the IR mode is local to A or B, and $\{\mu,\nu\} \in \{g0, g1, e0', e1'\}$. Here, g and e denote the ground and first excited electronic states, respectively, for the given monomer. 0 and 0′ are the ground vibrational states on the ground and first excited electronic states, respectively. Similarly, 1 and 1′ are the first excited vibrational states on the ground and first excited electronic states, respectively. For simplicity, in the site basis, the electronic wavefunctions are considered to be independent of the nuclear wavefunctions, *i.e.*, $|X\ \nu_\mu^A\nu_\nu^B\rangle \approx |X\rangle \otimes |\nu_\mu^A\rangle \otimes |\nu_\nu^B\rangle$. The vibrational overlap between the nuclear wavefunctions on the ground and excited electronic states, for a given monomer, are expressed in terms of the Huang Rhys factors σ:

$$\left\langle \nu_{g0}^A \middle| \nu_{e0'}^A \right\rangle = e^{-\frac{\sigma}{2}}; -\left\langle \nu_{e0'}^A \middle| \nu_{g1}^A \right\rangle \approx \left\langle \nu_{g0}^A \middle| \nu_{e1'}^A \right\rangle = \sqrt{\sigma} e^{-\frac{\sigma}{2}}; \left\langle \nu_{g1}^A \middle| \nu_{e1'}^A \right\rangle = (1-\sigma) e^{-\frac{\sigma}{2}}. \quad (1)$$

Again, the ground and excited electronic states are coupled only radiatively. Consequently, for the dynamics during the Ground State Bleaching (GSB) pathway, we work with the system Hamiltonian in the ground electronic state manifold, $H_{sys,g}$, formed by the basis states, $|g\ \nu_{g0}^A\nu_{g0}^B\rangle$, $|g\ \nu_{g1}^A\nu_{g0}^B\rangle$ and $|g\ \nu_{g0}^A\nu_{g1}^B\rangle$, respectively. The ground electronic state is considered to be the zero of energy.

$$H_{sys,g} = \begin{pmatrix} 0 & 0 & 0 \\ 0 & \omega & 0 \\ 0 & 0 & \omega \end{pmatrix} \quad (2)$$

For the dynamics during the Excited State Absorption (ESA) pathway, we work with the system Hamiltonian in the excited electronic state manifold, $H_{\text{sys,ex}}$, formed by the basis states $|\mathrm{A}\nu_{\mathrm{e0'}}^{\mathrm{A}}\nu_{\mathrm{g0}}^{\mathrm{B}}\rangle$, $|\mathrm{A}\nu_{\mathrm{e1'}}^{\mathrm{A}}\nu_{\mathrm{g0}}^{\mathrm{B}}\rangle$, $|\mathrm{A}\ \nu_{\mathrm{e0'}}^{\mathrm{A}}\nu_{\mathrm{g1}}^{\mathrm{B}}\rangle$, $|\mathrm{B}\ \nu_{\mathrm{g0}}^{\mathrm{A}}\nu_{\mathrm{e0'}}^{\mathrm{B}}\rangle$, $|\mathrm{B}\ \nu_{\mathrm{g0}}^{\mathrm{A}}\nu_{\mathrm{e1'}}^{\mathrm{B}}\rangle$, and $|\mathrm{B}\ \nu_{\mathrm{g1}}^{\mathrm{A}}\nu_{\mathrm{e0'}}^{\mathrm{B}}\rangle$, respectively.

$$H_{\text{sys,ex}} = \begin{pmatrix} E_{\mathrm{A}} & 0 & 0 & J\mathrm{e}^{-\sigma} & J\sqrt{\sigma}\mathrm{e}^{-\sigma} & -J\sqrt{\sigma}\mathrm{e}^{-\sigma} \\ 0 & E_{\mathrm{A}}+\omega' & 0 & J\sqrt{\sigma}\mathrm{e}^{-\sigma} & J\sigma\mathrm{e}^{-\sigma} & J(1-\sigma)\mathrm{e}^{-\sigma} \\ 0 & 0 & E_{\mathrm{A}}+\omega & -J\sqrt{\sigma}\mathrm{e}^{-\sigma} & J(1-\sigma)\mathrm{e}^{-\sigma} & J\sigma\mathrm{e}^{-\sigma} \\ J\mathrm{e}^{-\sigma} & J\sqrt{\sigma}\mathrm{e}^{-\sigma} & -J\sqrt{\sigma}\mathrm{e}^{-\sigma} & E_{\mathrm{B}} & 0 & 0 \\ J\sqrt{\sigma}\mathrm{e}^{-\sigma} & J\sigma\mathrm{e}^{-\sigma} & J(1-\sigma)\mathrm{e}^{-\sigma} & 0 & E_{\mathrm{B}}+\omega' & 0 \\ -J\sqrt{\sigma}\mathrm{e}^{-\sigma} & J(1-\sigma)\mathrm{e}^{-\sigma} & J\sigma\mathrm{e}^{-\sigma} & 0 & 0 & E_{\mathrm{B}}+\omega \end{pmatrix} \quad (3)$$

Here, the displaced IR mode in the excited electronic state has a shifted frequency ω'. The first excited electronic states of A and B have energies, E_{A} and E_{B}, respectively. The system interacts with the environment/bath/phonons, considered to be a collection of harmonic oscillators. Therefore, the total Hamiltonian including the system, bath and the system–bath coupling, approximated to be linear in the position coordinates of the phonon harmonic oscillators, is given as:

$$H_{\mathrm{g/ex}} = \sum_i (E_i + Q_i)|i\rangle\langle i| + \sum_{i,j,i\neq j} V_{ij}|i\rangle\langle j| + H_{\mathrm{ph}} \quad (4)$$

where

$$H_{\mathrm{ph}} = \sum_{i,k}\left(\frac{{p_{ik}}^2}{2m_{ik}} + \frac{1}{2}m_{ik}{\omega_{ik}}^2{q_{ik}}^2\right), \quad (5)$$

where i runs over 3 basis states, as discussed above, for H_{g} and 6 basis states for H_{ex}. V_{ij} describes the off-diagonal coupling between the system states, k labels the harmonic oscillator. The system–bath coupling is defined as $Q_i = \sum_k m_{ik}\nu_{ik}q_{ik}$. q_{ik}, m_{ik}, ν_{ik}, ω_{ik} and p_{ik} are the position, mass, coupling strength, angular frequency and momentum, respectively, of the kth oscillator pertaining to the ith system state. H_{g} is diagonal with respect to its basis and therefore, easy to work with. H_{ex}, on the other hand, when diagonalized, is given as:

$$H_{\mathrm{ex}} = \sum_m \varepsilon_m(\boldsymbol{Q})|m(\boldsymbol{Q})\rangle\langle m(\boldsymbol{Q})| + H_{\mathrm{ph}}, \quad (6)$$

where $|m(\boldsymbol{Q})\rangle$ denotes the adiabatic basis. The adiabatic basis is tedious due to its dependence on $\boldsymbol{Q}$, the system-bath coupling vector. Also, it does not commute with the momenta in H_{ph}. We, therefore, introduce a unitary transformation, $U(\boldsymbol{Q})$, which transforms from the adiabatic basis $|m(\boldsymbol{Q})\rangle$ to the stationary adiabatic basis $|m(\boldsymbol{Q}=0)\rangle$, to be denoted as $|m\rangle$.[35–37]

The 2DEV spectra can be theoretically evaluated from the relevant response functions, using Feynman diagrams. As an example, we will discuss in detail the calculation for the ESA rephasing pathway shown in Fig. 5; the other pathways are evaluated in an analogous fashion. To ease the notation, g_0 will now denote the

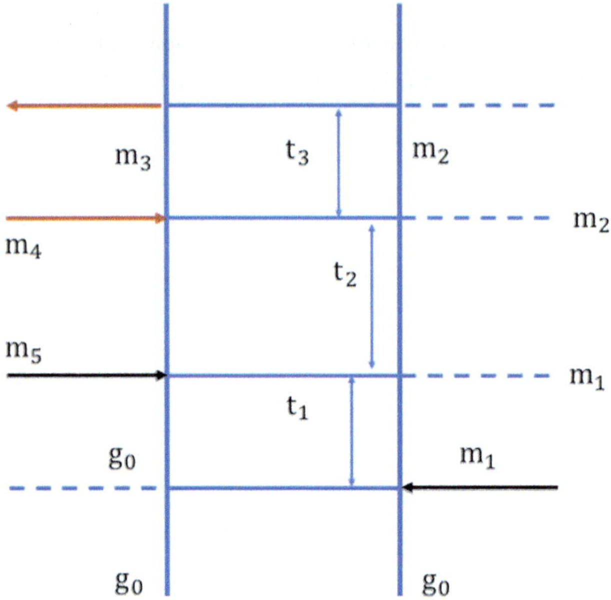

Fig. 5 Feynman diagram for an excited state rephasing pathway. Electronic interactions with the light are shown as black arrows, infrared interactions as red arrows. m_i denotes a vibronic level.

state $\left|g\ v_{g0}^{A} v_{g0}^{B}\right\rangle$. The m_i states, where i is a positive integer, are used to denote the stationary adiabatic basis states. Despite the different definition, the stationary adiabatic basis states have same energies as the eigenstates obtained by diagonalizing the system Hamiltonian in the electronic excited state manifold, $H_{sys,ex}$. The ESA rephasing response function can be written as:

$$R_{ESA}(t_3,t_2,t_1) = -\langle \mu_{el}(0)\mu_{vib}(t_1+t_2+t_3)\mu_{vib}(t_1+t_2)\mu_{el}(t_1)\rho(0)\rangle, \tag{7}$$

where $\rho(0) = |g_0\rangle\langle g_0|\rho_{ph}(0)$. $\mu_{el/vib}(t)$ is the transition dipole operator in the Heisenberg picture at time t, defined as $\mu_{el/vib}(t) = e^{iHt}\mu_{el/vib}e^{-iHt}$, where $H = H_g + H_{ex}$. Therefore,

$$R_{ESA}(t_3,t_2,t_1) = -\mathrm{Tr}_{ph}\{\langle g_0|\mu_{el}e^{iH(t_1+t_2+t_3)}\mu_{vib}e^{-iHt_3}\mu_{vib}e^{-iHt_2}\mu_{el}e^{-iHt_1}|g_0\rangle\rho_{ph}(0)\}. \tag{8}$$

Introducing the unitary transformation and the stationary adiabatic basis,[35–37] we have

$$R_{ESA}(t_3,\ t_2,t_1) = -\sum_{\{m\}}\mu_{el,g_0m_1}\mu_{vib,m_2m_3}\mu_{vib,m_3m_4}\mu_{el,m_5g_0}e^{-iH_{ph}t_1}F_{m_1m_2m_3m_4m_5}(t_3,\ t_2,t_1), \tag{9}$$

where

$$F_{m_1m_2m_3m_4m_5}(t_3,t_2,t_1) \approx \mathrm{Tr}_{ph}\{\langle m_1|e^{i\tilde{H}_{ex}t_1}|m_1\rangle\langle m_1|e^{i\tilde{H}_{ex}t_2}|m_2\rangle \langle m_2|e^{i\tilde{H}_{ex}t_3}|m_2\rangle \times \langle m_3|e^{-i\tilde{H}_{ex}t_3}|m_3\rangle\langle m_4|e^{-i\tilde{H}_{ex}t_2}|m_5\rangle\rho_{ph}(0)\}. \tag{10}$$

Here, $\bar{H}_{\text{ex}} = \bar{H}_0 + \bar{H}_{\text{na}}$, where $\overline{H}_0 = \sum_m \varepsilon_m(\boldsymbol{Q})|m\rangle\langle m| + H_{\text{ph}}$ and $\bar{H}_{\text{na}}$ is the perturbative term.[35,36] We have approximated above that $\langle g_0|\mu|m_i(\boldsymbol{Q})\rangle \approx \langle g_0|\mu|m_i\rangle$. In the 2DEV experiments, t_1 and t_3 are small, so we can neglect population relaxation during these times. The dynamics time t_2 could be small or large — if it is large compared to phonon relaxation time, population relaxation needs to be accounted for. If t_2 is small, we can neglect population relaxation, therefore, $m_1 = m_2$ and $m_4 = m_5$. We now make the approximation that, since decoherence and incoherent population relaxation are largely independent processes, we can decouple them at the lowest order. Introducing $\bar{H}_0$[35,36] and working in the interaction picture, we obtain:

$$
\begin{aligned}
&F_{m_1m_2m_3m_4m_5}(t_3,\ t_2, t_1)\\
&\approx \mathrm{Tr}_{\text{ph}}\left\{\begin{array}{l}\left(\hat{T}^{\dagger}\mathrm{e}^{i\int_0^{t_1}\mathrm{d}t'\varepsilon_{m_1}(\boldsymbol{Q}(t'))}\right)\left(\hat{T}^{\dagger}\mathrm{e}^{i\int_0^{t_2}\mathrm{d}t'\varepsilon_{m_2}(\boldsymbol{Q}(t'))}\right)\left(\hat{T}^{\dagger}\mathrm{e}^{i\int_0^{t_3}\mathrm{d}t'\varepsilon_{m_2}(\boldsymbol{Q}(t'))}\right)\\ \times\left(\hat{T}\mathrm{e}^{-i\int_0^{t_3}\mathrm{d}t'\varepsilon_{m_3}(\boldsymbol{Q}(t'))}\right)\left(\hat{T}\mathrm{e}^{-i\int_0^{t_2}\mathrm{d}t'\varepsilon_{m_4}(\boldsymbol{Q}(t'))}\right)\rho_{\text{ph}}(0)\end{array}\right\}\mathrm{e}^{iH_{\text{ph}}t_1}\\
&\times \mathrm{Tr}_{\text{ph}}\left\{\langle m_1|U_{\text{I}}^{\dagger}(t_2)|m_2\rangle\rho_{\text{ph}}(0)\langle m_4|U_{\text{I}}(t_2)|m_5\rangle\right\},
\end{aligned}
\tag{11}
$$

where $\hat{T}^{\dagger}\mathrm{e}^{i\int_0^{t}\mathrm{d}t'\varepsilon_{m_i}(\boldsymbol{Q}(t'))} = \mathrm{e}^{i\varepsilon_{m_i}t}\left(\hat{T}^{\dagger}\mathrm{e}^{i\int_0^{t}\mathrm{d}t'\nabla_{\boldsymbol{Q}}\varepsilon_{m_i}(\boldsymbol{Q}(t'))\cdot\boldsymbol{Q}(t')}\right)$, and the phonon coupling to the m_i state in $\boldsymbol{Q}$ causes decoherence, evaluated almost exactly for an ohmic or Debye spectral density for the environment, while $U_{\text{I}}(t) = \mathrm{e}^{i\bar{H}_0 t}\mathrm{e}^{-i\bar{H}_{\text{ex}}t}$, and $\mathrm{Tr}_{\text{ph}}\{\langle m_1|U_{\text{I}}^{\dagger}(t_2)|m_2\rangle\rho_{\text{ph}}(0)\langle m_4|U_{\text{I}}(t_2)|m_5\rangle\}$ results in population relaxation and is treated perturbatively by a Markovian master equation.[35,36] This is justified since the system is assumed to stay close to its adiabatic limit and therefore, non-adiabatic transitions leading to population relaxation among the stationary adiabatic states are treated perturbatively.

Results and discussion

For the ESA rephasing pathway, as seen in Fig. 5, we will have 6^5 possible pathways, since m_i, where $i \in \{1,2,3,4,5\}$, could be any of the 6 states obtained by diagonalizing $H_{\text{sys.ex}}$. Since we are seeking mechanistic insights into the dynamics, it would be valuable to look at each of these individual pathways and the respective contributions. However, performing calculations for 6^5 pathways is tedious and despite the near-analyticity that the method promises, time-consuming. We exploit the separability of the transformed Hamiltonian into decoherence and population relaxation. The population relaxation rates,[35,36] which are given as,

$$
\Gamma_{\text{mn}} = 2\pi\frac{1}{\mathrm{e}^{\beta\omega_{\text{mn}}}-1}\sum_j J_j(\omega_{\text{mn}})(\langle \text{n}|j\rangle\langle j|\text{m}\rangle)^2, \tag{12}
$$

and $\Gamma_{\text{nm}} = \mathrm{e}^{\beta\omega_{\text{mn}}}\Gamma_{\text{mn}}$, where m, n label the stationary adiabatic states, j labels the site state, $J_j(\omega)$ is the spectral density for the jth site and $\omega_{\text{mn}} = \varepsilon_{\text{m}}(\mathbf{0}) - \varepsilon_{\text{n}}(\mathbf{0})$, $\omega_{\text{mn}} > 0$, allow us to select the significant pathways; we discard those that have very small rates. Also, we can neglect pathways that have small values for the transition dipole moments. Using the two tools, we are able to significantly reduce the number of pathways we need to compute in order to obtain the spectroscopic signal.

In our calculations, we choose $\mu_{el} = 1$ for both the monomers A and B, in the site basis. Also, the values for the vibrational transition dipole moments μ_{vib}, in the site basis are considered to be unity. $E_A = 12\,000\ cm^{-1}$, $E_B = 12\,450\ cm^{-1}$, $J = 250\ cm^{-1}$, $\omega = 1030\ cm^{-1}$, $\omega' = 950\ cm^{-1}$, $\sigma = 0.0025$ and $T = 77$ K. The ratio of the displacements for the IR mode in the ground and excited vibrational states in the electronic excited state with respect to the ground electronic state are given as $\alpha_0 = 0.6$ and $\alpha_1 = 1.6$, respectively.[21] The environment is modeled by the Debye spectral density, with $\lambda_{el} = 35\ cm^{-1}$, $\lambda_{vib} = 5\ cm^{-1}$ and $\omega_c = 53.3\ cm^{-1}$, for both electronic and vibrational degrees of freedom. The eigenstates formed by diagonalizing $H_{sys.ex}$ are shown in Fig. 6. Table 2 shows the vibrational transition dipole moments (TDM) for the eigenstates, we enumerate only the ones that have sizable values.

The total ESA rephasing signals at $t_2 = 0$ fs and $t_2 = 625$ fs are shown in Fig. 7(a) and (b), respectively. Since each pathway can be computed separately, we are able to see that the signal arises mostly from the populations at excitons 5(12 560 cm^{-1}) and 6(11 890 cm^{-1}), the "electronic-only" eigenstates of $H_{sys.ex}$. Coherences do not contribute to the signal for these parameters.

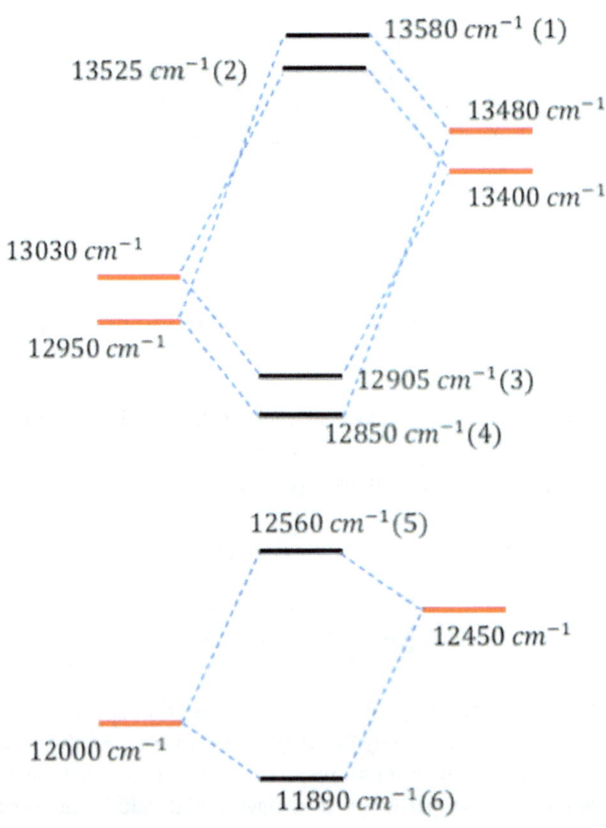

Fig. 6 Site states (red) and vibronic level (black) along with their energies for the dimer model described in the text. The numbers in brackets indicate the vibronic excitons from highest to lowest energy.

Table 2 Vibrational transition moments for the excited states of the dimer model

	Vibrational TDM					
	1	2	3	4	5	6
1					−1.0045	
2					−0.9903	
3						−0.9809
4						−1.0166
5	−1.0045	−0.9903				
6			−0.9809	−1.0166		

At long times, population relaxation becomes important. Therefore, at $t_2 = 625$ fs, the signal consists of populations that (i) have started and stayed at the same exciton, Fig. 7(c) shows these contributions from excitons 5 and 6, and (ii) have relaxed from one exciton to another, downhill relaxation being more predominant. Fig. 7(d) shows the relaxing population from 5 to 6. These contributions add up to give the total signal at $t_2 = 625$ fs in Fig. 7(b).

Also, unlike 2DES spectroscopy, the surviving (diagonal peaks in 2DES) and relaxing (off-diagonal peaks in 2DES) populations show up at roughly the same peak positions in 2DEV spectra. For instance, for exciton 5, the signal appears at $\{\varepsilon_5,(\varepsilon_1 - \varepsilon_5)\}$ and $\{\varepsilon_5,(\varepsilon_2 - \varepsilon_5)\}$ for surviving population and at $\{\varepsilon_5,(\varepsilon_4 - \varepsilon_6)\}$ and $\{\varepsilon_5,(\varepsilon_3 - \varepsilon_6)\}$ for relaxing population from 5 to 6. Here, $(\varepsilon_2 - \varepsilon_5)$ in the surviving pathway and $(\varepsilon_4 - \varepsilon_6)$ in the relaxing pathway have similar values. Also, $(\varepsilon_1 - \varepsilon_5)$ and $(\varepsilon_3 - \varepsilon_6)$ have similar values. For exciton 6, the signal appears at $\{\varepsilon_6,(\varepsilon_3 - \varepsilon_6)\}$ and $\{\varepsilon_6,(\varepsilon_4 - \varepsilon_6)\}$ for the surviving population.

Fig. 8 shows the total rephasing signal, obtained by summing the contributions from the Ground State Bleaching (GSB) and Excited State Absorption (ESA)

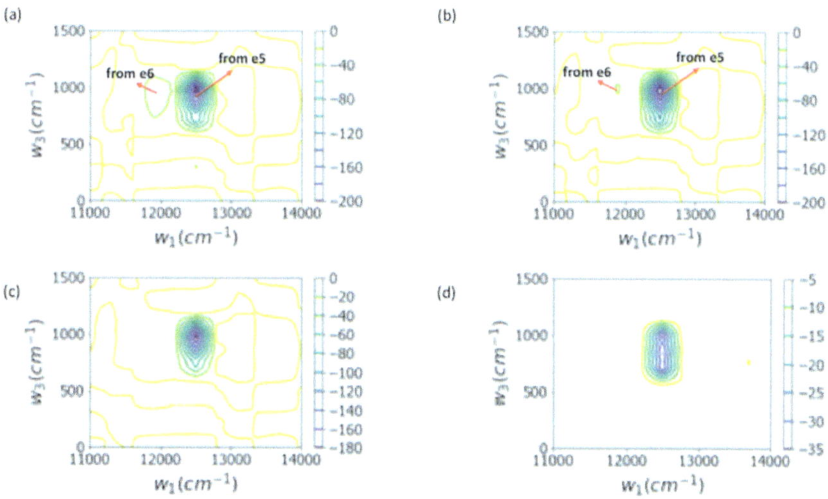

Fig. 7 2DEV spectra for ESA pathway for the model dimer system. (a) and (b) Total rephasing contribution at waiting times of 0 fs and 625 fs, respectively. (c) Population that has stayed at the initial excitons after 625 fs, and (d) relaxing population from exciton 5 to exciton 6 at 625 fs.

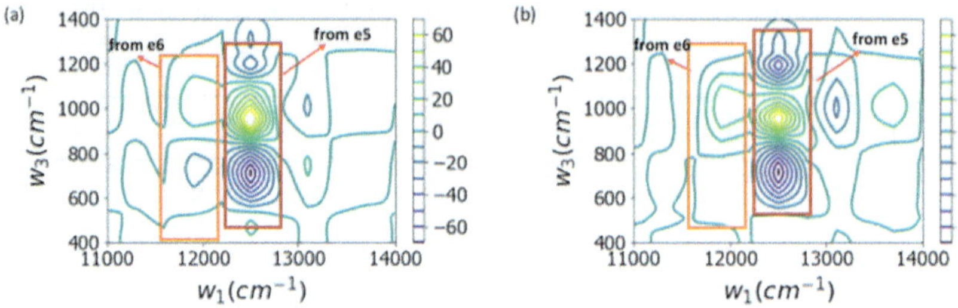

Fig. 8 Total rephasing signal (GSB + ESA pathways) at (a) $t_2 = 0$ fs, (b) $t_2 = 625$ fs.

pathways, at (a) $t_2 = 0$ fs, and (b) $t_2 = 625$ fs. The positive signal stems from GSB pathway and the negative signal from the ESA pathway. Also, the signal is strongest at exciton 5, with faint contours at exciton 6.

It is important to note that for the model dimer system with each monomer possessing a single vibration in both ground and excited electronic state, the six-exciton picture is accurate for all coupling strengths. However, for weakly coupled systems, specifically in the Förster limit, a number of vibrational transitions in the excited state become rigorously forbidden as they involve "cross molecule" transitions, for example, $|Y\ \nu_{g0}^{X}\nu_{e0'}^{Y}\rangle \rightarrow |Y\ \nu_{g1}^{X}\nu_{e0'}^{Y}\rangle$, where $\{X,Y\} \in \{A,B\}$. In this case the vibrational transitions observed in 2DEV spectra revert to the more intuitive flags indicating the location of electronic excitation.

Summary and conclusions

2DEV spectroscopy, though still in its development stage, clearly demonstrates considerable promise for the study of complex molecular dynamics such as energy, electron, coupled proton-electron transfer, as well as molecular dynamics involving large structural changes.[27] The enhanced spectral resolution through the vibrational axis provides insights into electronic structural evolution, the spatial location of excitation in energy transfer complexes, and, through the information contained in the center line slope, provides a unique window into the correlation of electronic and vibrational evolution.[21] Further development of theory and modeling of the CLS dynamics is clearly needed to expose the full potential of this new quantity, but it has already proved very useful in describing conical intersection dynamics of triphenyl methane dyes.[27]

A key motivation for developing the technique was its expected sensitivity to vibronic coherences, which have recently been suggested to play a key role in primary photosynthetic charge separation in Photosystem II,[38–41] and in organic photovoltaic systems.[42–45] This required development of theoretical techniques to calculate 2DEV spectra for both strongly and weakly coupled systems. The application of our approach to weakly coupled (localized vibrations) is described above. Application to strongly coupled systems are described in a recent publication.[46]

Conflicts of interest

There are no conflicts to declare.

Acknowledgements

This work was supported by the U.S. Department of Energy, Office of Science, Basic Energy Sciences, Chemical Sciences, Geosciences, and Biosciences Division under Field Work Proposal 449A. G. R. F. thanks Magdalen College, Oxford for a visiting Fellowship, and the Department of Physical and Theoretical Chemistry, Oxford for their hospitality during the writing of this paper. E. A. A. acknowledges the support of the Berkeley Fellowship.

References

1 L. N. M. Duysens, *Prog. Biophys. Mol. Biol.*, 1964, **14**, 1–104.
2 G. W. Robinson, *Brookhaven Symposium in Biology*, 1966, **19**, 16–48.
3 H. Bergström, V. Sundström, R. van Grondelle, E. Åkesson and T. Gillbro, *Biochim. Biophys. Acta Bioenerg.*, 1986, **852**, 279–287.
4 M. G. Müller, K. Griebenow and A. R. Holzwarth, *Biochim. Biophys. Acta Bioenerg.*, 1991, **1098**, 1–12.
5 T. Joo, Y. Jia, J.-Y. Yu, D. M. Jonas and G. R. Fleming, *J. Phys. Chem.*, 1996, **100**, 2399–2409.
6 D. D. Eads, E. W. Castner, R. S. Alberte, L. Mets and G. R. Fleming, *J. Phys. Chem.*, 1989, **93**, 8271–8275.
7 H. Sumi, *J. Phys. Chem. B*, 1999, **103**, 252–260.
8 G. D. Scholes and G. R. Fleming, *J. Phys. Chem. B*, 2000, **104**, 1854–1868.
9 R. Agarwal, B. P. Krueger, G. D. Scholes, M. Yang, J. Yom, L. Mets and G. R. Fleming, *J. Phys. Chem. B*, 2000, **104**, 2908–2918.
10 R. Agarwal, M. Yang, Q.-H. Xu and G. R. Fleming, *J. Phys. Chem. B*, 2001, **105**, 1887–1894.
11 M. Yang and G. R. Fleming, *Chem. Phys.*, 2002, **275**, 355–372.
12 T. Brixner, J. Stenger, H. M. Vaswani, M. Cho, R. E. Blankenship and G. R. Fleming, *Nature*, 2005, **434**, 625–628.
13 G. S. Engel, T. R. Calhoun, E. L. Read, T.-K. Ahn, T. Mančal, Y.-C. Cheng, R. E. Blankenship and G. R. Fleming, *Nature*, 2007, **446**, 782–786.
14 J. C. Dean, T. Mirkovic, Z. S. D. Toa, D. G. Oblinsky and G. D. Scholes, *Chem*, 2016, **1**, 858–872.
15 G. Panitchayangkoon, D. Hayes, K. A. Fransted, J. R. Caram, E. Harel, J. Wen, R. E. Blankenship and G. S. Engel, *Proc. Natl. Acad. Sci. U. S. A.*, 2010, **107**, 12766–12770.
16 A. Ishizaki and G. R. Fleming, *J. Chem. Phys.*, 2009, **130**, 234111.
17 V. Tiwari, W. K. Peters and D. M. Jonas, *Proc. Natl. Acad. Sci. U. S. A.*, 2013, **110**, 1203–1208.
18 Y. Fujihashi, G. R. Fleming and A. Ishizaki, *J. Chem. Phys.*, 2015, **142**, 212403.
19 T. A. A. Oliver, N. H. C. Lewis and G. R. Fleming, *Proc. Natl. Acad. Sci. U. S. A.*, 2014, **111**, 10061–10066.
20 J. K. Gillie, G. J. Small and J. H. Golbeck, *J. Phys. Chem.*, 1989, **93**, 1620–1627.
21 N. H. C. Lewis, H. Dong, T. A. A. Oliver and G. R. Fleming, *J. Chem. Phys.*, 2015, **142**, 174202.
22 N. H. C. Lewis and G. R. Fleming, *J. Phys. Chem. Lett.*, 2016, **7**, 831–837.
23 N. H. C. Lewis, N. L. Gruenke, T. A. A. Oliver, M. Ballottari, R. Bassi and G. R. Fleming, *J. Phys. Chem. Lett.*, 2016, **7**, 4197–4206.

24 V. Novoderezhkin, A. Marin and R. van Grondelle, *Phys. Chem. Chem. Phys.*, 2011, **13**, 17093.

25 G. S. Schlau-Cohen, T. R. Calhoun, N. S. Ginsberg, E. L. Read, M. Ballottari, R. Bassi, R. van Grondelle and G. R. Fleming, *J. Phys. Chem. B*, 2009, **113**, 15352–15363.

26 Z. Liu, H. Yan, K. Wang, T. Kuang, J. Zhang, L. Gui, X. An and W. Chang, *Nature*, 2004, **428**, 287–292.

27 E. C.-K. Wu, Q. Ge, E. A. Arsenault, N. H. C. Lewis, N. L. Gruenke, M. J. Head-Gordon and G. R. Fleming, *Phys. Chem. Chem. Phys.*, DOI: 10.1039/C8CP05264F.

28 T. Mančal, A. V. Pisliakov and G. R. Fleming, *J. Chem. Phys.*, 2006, **124**, 234504.

29 A. V. Pisliakov, T. Mančal and G. R. Fleming, *J. Chem. Phys.*, 2006, **124**, 234505.

30 B. J. Ka and E. Geva, *J. Chem. Phys.*, 2006, **125**, 214501.

31 L. Seidner, G. Stock and W. Domcke, *J. Chem. Phys.*, 1995, **103**, 3998–4011.

32 M. Cho, *J. Chem. Phys.*, 2001, **115**, 4424–4437.

33 L. Chen, R. Zheng, Y. Jing and Q. Shi, *J. Chem. Phys.*, 2011, **134**, 194508.

34 M. Baer, *Beyond Born-Oppenheimer*, John Wiley & Sons, Inc., Hoboken, NJ, USA, 2006.

35 P. Bhattacharyya and K. L. Sebastian, *Phys. Rev. E*, 2013, **87**, 062712.

36 P. Bhattacharyya and K. L. Sebastian, *J. Phys. Chem. A*, 2013, **117**, 8806–8813.

37 P. Bhattacharyya and N. Ananth, arXiv:1705.00738.

38 E. Romero, R. Augulis, V. I. Novoderezhkin, M. Ferretti, J. Thieme, D. Zigmantas and R. van Grondelle, *Nat. Phys.*, 2014, **10**, 676–682.

39 V. I. Novoderezhkin, E. Romero, J. Prior and R. van Grondelle, *Phys. Chem. Chem. Phys.*, 2017, **19**, 5195–5208.

40 F. D. Fuller, J. Pan, A. Gelzinis, V. Butkus, S. S. Senlik, D. E. Wilcox, C. F. Yocum, L. Valkunas, D. Abramavicius and J. P. Ogilvie, *Nat. Chem.*, 2014, **6**, 706–711.

41 G. D. Scholes, G. R. Fleming, L. X. Chen, A. Aspuru-Guzik, A. Buchleitner, D. F. Coker, G. S. Engel, R. van Grondelle, A. Ishizaki, D. M. Jonas, J. S. Lundeen, J. K. McCusker, S. Mukamel, J. P. Ogilvie, A. Olaya-Castro, M. A. Ratner, F. C. Spano, K. B. Whaley and X. Zhu, *Nature*, 2017, **543**, 647–656.

42 A. De Sio and C. Lienau, *Phys. Chem. Chem. Phys.*, 2017, **19**, 18813–18830.

43 S. M. Falke, C. A. Rozzi, D. Brida, M. Maiuri, M. Amato, E. Sommer, A. De Sio, A. Rubio, G. Cerullo, E. Molinari and C. Lienau, *Science.*, 2014, **344**, 1001–1005.

44 Y. Song, S. N. Clafton, R. D. Pensack, T. W. Kee and G. D. Scholes, *Nat. Commun.*, 2014, **5**, 4933.

45 Y. Song, C. Hellmann, N. Stingelin and G. D. Scholes, *J. Chem. Phys.*, 2015, **142**, 212410.

46 P. Bhattacharyya and G. R. Fleming, *J. Phys. Chem. Lett.*, 2019, **10**, 2081–2089.

Faraday Discussions

DISCUSSIONS

Energy and charge-transfer in natural photosynthesis: general discussion

Shuming Bai, Vytautas Balevicius, Jr., Eric Bittner, Yuan-Chung Cheng, Majed Chergui, Jeffrey Cina, Natércia das Neves Rodrigues, Animesh Datta, Jahan Dawlaty, Amro Dodin, Benjamin Fingerhut, Graham Fleming, Naomi Ginsberg, Sharon Hammes Schiffer, Vanessa Huxter, Bern Kohler, Yumin Lee, Graham Leggett, Andrew Marcus, Karen Morenz, Jennifer Ogilvie, Alexandra Olaya-Castro, Thomas A. A. Oliver, Minjung Son, Yin Song and Vasilios Stavros

DOI: 10.1039/C9FD90027F

Andrew Marcus opened a discussion of the introductory lecture by Majed Chergui: My question pertains to the theoretical interpretation of the interesting time-resolved circular dichroism (CD) experiments on DNA that you presented. The determination of nucleic acid base conformation from CD data is a fairly complex and challenging problem, which has been studied for many years. In an oligonucleotide with many bases, the CD spectrum depends on the numerous non-degenerate couplings between the many transition dipole moments of the bases, which in turn depends on their relative distances and orientations (*i.e.* base conformation). In general, there is no unique mapping between a given CD spectrum and a specific oligonucleotide conformation. How might this issue be addressed for the structural interpretation of time-resolved CD experiments on oligonucleotides?

Majed Chergui responded: This is a very good question and I agree with you that the modelling of CD spectra is far from trivial for systems like nucleic acid bases or even peptides. What I showed in my talk aimed at starting to resolve this problem, at least in the case of peptide chains, using thioamide substitution. We investigated a dipeptide whose two carbonyl oxygen atoms were replaced by sulfur, so that the strong lowest-lying pi–pi* transitions are shifted to the red. This not only brings them in the range of our time-resolved CD set-up but more importantly, it offers a site specific labelling of the peptidic chains.

Similar substitutions are possible with DNA strands and they are already used in either fluorescence or CD studies, *e.g.* 2-aminopurine dinucleotides, which we are planning to use.

From a more general perspective, there are still huge challenges to be overcome in order to make time-resolved CD a reliable and easy to implement method:

extending the spectral range, developing labels that can be incorporated into chains and, very importantly, developing the theory. However, I see this as a very promising avenue because there are not many experimental tools that allow you to look into the conformational changes of systems such as peptides or DNA/RNA (that consist of light atoms) in their physiological media and with the adequate temporal resolution. The only other method I can think of is time-resolved X-ray solution scattering, which I believe is extremely challenging with such systems and no less prone to heavy modeling. Furthermore, one needs to carry out the X-ray experiments at large scale installations (synchrotrons and X-ray free electron lasers), whereas time-resolved deep-UV CD can be performed in the lab.

Bern Kohler commented: You drew a contrast between excitons in the dinucleoside monophosphate $(dA)_2$ and in the 20-mer $(dA)_{20}$. Because base stacking is only weakly stabilized at room temperature, domains of well-stacked bases are interrupted by one or more unstacked bases with the result that the average number of bases in a stacked domain can be much less than the total number of residues in a strand.[1,2] For this reason, it would be very interesting to study strands of intermediate length as the maximum extent of excitonic delocalization could be reached for an oligonucleotide of length much less than 20.

1 C. Su, C. T. Middleton and B. Kohler, *J. Phys. Chem. B*, 2012, **116**, 10266–10274.
2 B. Kohler, *Excited States of Single-Stranded DNA Revealed by Femtosecond Transient Absorption Spectroscopy. In Ultrafast Dynamics at the Nanoscale: Biomolecules and Supramolecular Assemblies*, ed. I. Burghardt and S. Haacke, Pan Stanford Publishing Pvt Ltd, Singapore, 2017, pp. 3–64.

Eric Bittner opened the discussion of the paper by Alexandra Olaya-Castro: In your paper, you present a fully quantum model of a system driven by a fully quantum ensemble of oscillators. However, it would seem that a classical bath model would also produce synchronization simply by symmetry. At what point does it matter if the bath is quantum or classical and does this really matter for realistic biological systems?

Alexandra Olaya-Castro replied: The model assumes a quantum description for the excited states of each chromophore and for the local vibrational motions interacting with each chromophore. The influence of the remaining the thermal environment is assumed to be Markovian. As we discussed during the presentation, synchronicity of the local vibrational motions is preceded by a transient where motions are anti-synchronised. This corresponds to the dynamics as a collective relative motion that assists the energy transfer. We have previously shown that such a collective mode exhibits non-classical thermal fluctuations, that is fluctuations that are narrower than Poissonian and can only be described *via* a quasiprobability distribution that exhibit negativities, thereby eliminating any classical description.[1] We have also shown that a classical description of such high energy vibrations will not appropriately describe the directionality of the energy transfer.

1 E. O'Reilly and A. Olaya-Castro, *Nat. Commun.*, 2014, **5**, 3012.

Eric Bittner asked: The exciton/phonon model you present is much like that used to study molecular dimers in J- and H-aggregate systems. As such is it not anticipated that the synchronization between oscillators is simply reflecting that coupling? In essence, the steady-states of the system must always be in symmetric or anti-symmetric irreducible representations. Furthermore, is it possible to construct a phase diagram in terms of the coupling parameters, exciton gap, and phonon lineshape?

Alexandra Olaya-Castro replied: Our model differs from H- and J- aggregates as in this case the on-site energies are not identical. Most importantly, the synchronisation dynamics that we describe here are not between the electronic states of the heterodimer we consider but between vibrational displacements of local quasi-coherent modes coupled to each local excited state. The vibrational modes do not interact directly, yet their synchronisation dynamics are mediated by the interaction between electronic states. The emergence of anti-synchronisation and synchronisation of such displacements does reflect the formation of collective vibrational motions during energy transfer. The vibrational motions are assumed to be of equal frequency and locally coupled to each electronic state, with the same strength. Hence, in the short time regime, a collective anti-correlated relative motion is formed and drives energy transfer dynamics.[1,2] Given that vibrational coherence lasts longer, in the long time regime, the centre-of-mass correlated collective motion survives and leads to synchronised motions.

1 E. J. O'Reilly and A. Olaya-Castro, *Nat. Commun.*, 2014, **5**, 3012.
2 V. Tiwari, *et al.*, *Proc. Natl. Acad. Sci. U. S. A.*, 2013, **110(4)**, 1203–1208.

Jeffrey Cina asked: In some famous measurements of polarized fluorescence up-conversion from LH-1 in the middle nineties[1] vibrational quantum beats in the perpendicular channel were shown to appear in-phase with those in the parallel channel. These results suggested that coherent intra-chromophore vibrations in the energy-transfer acceptors were somehow synchronized with those in the donor chromophore. Graham Fleming and I later found an explanation in a surface-crossing picture of electronic excitation transfer.[2] In this description, it is seen that in a homodimer the crossing seam between donor-excited and acceptor-excited electronic states passes through the Franck–Condon points of both surfaces. Consequently, when a nuclear wave packet is launched by short-pulse excitation to the donor potential, energy-transfer is most efficient each time it returns to this point, and the Franck–Condon excited nuclear wave packet in the acceptor is therefore synchronized with that in the donor. Does this sound like a phenomenon that is related to the synchronization revealed in your paper?

1 S. E. Bradforth, R. Jimenez, F. van Mourik, R. van Grondelle and G. R. Fleming, *J. Phys. Chem.* 1995, **99**, 16179.
2 J. A. Cina and G. R. Fleming, *J. Phys. Chem. A*, 2004, **108**, 11196–11208.

Alexandra Olaya-Castro responded: This is a very interesting observation. Thank you for pointing out these works and it is fantastic to see that there is a direct experimental approach that will allow us to confirm our predictions of negative and positive synchronisation of Franck–Condon active vibrations during energy transfer. We have shown that as excitonic states are more delocalised, the period of negative synchronisation is shorter and vibrational displacements reached positive synchronisation faster. We therefore expect that in the case of a homodimer, with fully delocalised excitonic states, positive synchronisation will dominate all the way through.

Animesh Datta queried: Can we prove that the synchronisation effect studied here is uniquely quantum? Coupled harmonic oscillators, which the vibrations are, such as pendulums can be synchronised classically as well. Here the mediator is a two-level system. Is that what makes this quantum?

Alexandra Olaya-Castro replied: The mediators here are two collective electronic states (not just a two level system). As we have discussed in ref. 1, it is the exchange of exactly one quantum of energy between these collective electronic states and the emergent collective vibrational states that leads to truly quantum behaviour with a non-classical analogue for the vibrational motions in this system. In other words, it is the formation of collective vibronic coherences and their decoherence mechanisms that leads to the synchronisation dynamics observed. Different from the classical case, the oscillatory pattern of the observables being synchronised emerges due to quantum interference among different coherences, and not by simply natural oscillating frequencies of the oscillators involved. These are all features indicating that the spontaneous synchronisation dynamics described in our manuscript is uniquely quantum.

1 E. J. O'Reilly and A. Olaya-Castro, *Nat. Commun.*, 2014, 5, 3012.

Vanessa Huxter asked: Can you comment on the physical interpretation of negative or positive synchronization?

Alexandra Olaya-Castro answered: Thanks for this question. Negative synchronisation indicates that as energy transfer happens the oscillating displacement of one site is anti-correlated with the ones of these other sites. The message this gives is that a collective relative displacement mode has been activated during exciton energy transfer. This relative mode promotes both quantum superposition of the two excitons and coherent exciton population transfer in the short time-regime. In our case, electronic dephasing is faster than vibrational dephasing, then, as time evolves, the phases of these anti-correlated motions get destroyed due to electronic dephasing and the complementary collective mode (centre of mass motion) is activated *via* exciton population transfer, which carries information on the surviving local vibrational phase thereby promoting synchronisation.

Jennifer Ogilvie commented: In your paper you find that negative synchronization occurs simultaneously with coherent energy transfer and that there is a positive correlation between coherent transport and the time it takes for positive synchronized coherence to emerge. Your vibronic dimer model is related to the work of Tiwari and Jonas that also considers a dimer model with an intramolecular vibration common to both monomers and resonant in frequency to an electronic energy gap in the Fenna-Matthews Olson complex.[1] Are the positive and negatively synchronized coherences in your paper related to Tiwari and Jonas's correlated and anti-correlated vibrations, of which they propose that the latter can influence energy transfer?

1 V. Tiwari, W. K. Peters and D. M. Jones, *Proc. Natl. Acad. Sci. U. S. A.*, 2013, **110(4)**, 1203–1208.

Alexandra Olaya-Castro responded: Thanks for this great question. The answer is yes. Our negative and positive synchronised motions are consistent with the anti-correlated and correlated vibrational motions discussed by Tiwari and Jonas[1] and with the non-classical behaviour of such collective relative motions that we discuss in ref. 2. Our main contribution in this case is that we quantify and follow the dynamics of such synchronisation while establishing a direct relation to the coherence dynamics.

1 V. Tiwari, W. K. Peters and D. M. Jones, *Proc. Natl. Acad. Sci. U. S. A.*, 2013, **110(4)**, 1203–1208.
2 E. J. O'Reilly and A. Olaya-Castro, *Nat. Commun.*, 2014, **5**, 3012.

Amro Dodin said: The energy eigenstate analysis in this paper provides a great deal of insight into how environment-induced relaxation drives synchronization. Can this mechanism be complementarily thought of using excited state forces? In this picture, when the exciton is in an in-phase superposition between the two sites (Fig. 1(A)) the excitonic coupling would produce an attractive force acting on the two oscillators and an out-of phase superposition (Fig. 1(B)) would produce a repulsive force. Would it be correct, in this case, to think of the exciton dynamics providing an oscillatory driving force on the two oscillators as the mechanism for synchronization? In that case, does negative synchronization cause enhanced electron transfer rates by bringing the two sites into closer proximity, increasing their coupling?

Alexandra Olaya-Castro answered: Vibrational motions induce fluctuations of the on-site energies. Hence, if the on-site energies are very different (as it is in our case), negative synchronised displacements will bring electronic levels into resonance and therefore promote energy transfer from one site to another. So your picture of bringing sites on close proximity is correct as long as we understand is proximity in energy and not in space. It is correct to think of the electronic interactions as the driving force of the vibrational negative and positive synchronisation. And it is important to bear in mind that synchronisation is a collective phenomenon. The electronic interactions mediate the collective response of the oscillators. Negative synchronisation is associated with the

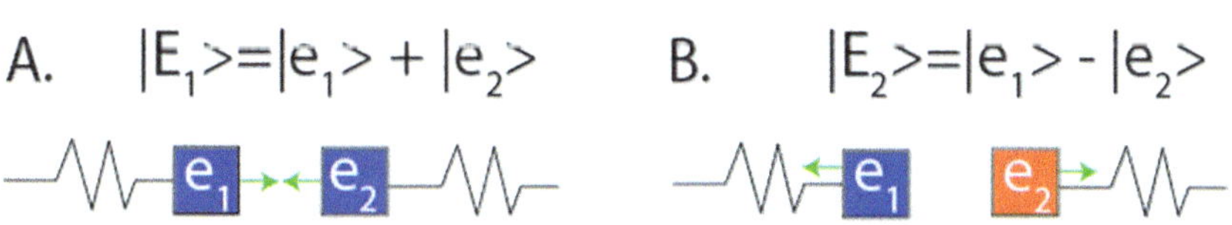

Fig. 1 Potential mechanism for synchronisation using excited state forces. (A) shows an attractive force acting on two oscillators due to an in-phase superposition and (B) shows a repulsive force acting on the two oscillators due to an out-of-phase superposition.

emergence of collective relative displacement motions while positive synchronisation will be associated to the emergence of a collective centre of mass motion, just as in the case of two coupled oscillators. So instead of the idea of a repulsive and attractive force, the picture of relative and centre of mass motion is more appropriate.

Yin Song asked: How does the preparation of the initial state affect the synchronization and energy transfer rate? Does this study also suggest that there is an approach to regulate energy transfer using selective laser pulse excitation?

Alexandra Olaya-Castro replied: The preparation of the initial state affects both the synchronisation dynamics of local displacements and the electronic energy transfer, as discussed in Sections 3.1 of our paper (DOI: 10.1039/c9fd00006b) and therefore, yes, this suggests selective laser pulse excitations can be used to regulate energy transfer. If the initial state is such that energy will be transferred from one quasi-localised exciton to another one *via* the mode, then short time negative synchronised motions will precede positive synchronisation of local vibrational displacements, as we have explained. If the initial state is such that now one has a population distribution between the excitons (not single excited states) and the vibrational motions start out in thermal equilibrium, this will naturally result in slow population dynamics as the joint exciton-vibration system is closer to its thermal equilibrium distribution. In this case one will still expect some transient dynamics for the displacements with a very fast emergence of positive synchronisation. In a separate work developed by one of my master's students, we have shown that if the initial states of the local vibrations are coherent states (instead of thermal states) energy transfer from one exciton to the other is enhanced (in comparison to the thermal vibrations). In this case we expect rich synchronisation dynamics as well. In short, selective laser pulse excitation can create different initial vibrational states which, in combination with an approach that selectively excites the highest exciton state, offers a way to regulate energy transfer.

Karen Morenz questioned: I understand that with increased coupling you get increased coherent energy transfer and so you reach synchrony sooner, but I'm wondering, with the way that you've defined synchrony as a constant phase difference between the two states *i.e.* a constant Pearson correlation value, what is the mathematical definition of perfect asynchrony? Or is there a way to quantify the extent of synchrony or asynchrony?

Alexandra Olaya-Castro responded: This is an important question as it touches on the conceptual differences between defining perfect synchronous or asynchronous states in the quantum regime. We have only investigated synchronisation of one observable. But, does one need to measure every observable and therefore the entire density matrix representing the quantum state to assert perfect synchronisation?

But in the spirit of your question (having in mind the classical picture synchronisation), we have considered using the rate of change of the Pearson correlation to quantify the 'degree of synchronicity'. This would have a nice zero value for any constant phase difference. However, I don't imagine there would be a value for a maximum asynchrony as two signals could be made to be arbitrarily different.

Animesh Datta addressed Alexandra Olaya-Castro: Related to my previous question, synchronicity can be attained between classical coupled pendulums. So can we make, as the paper says, the 'negative synchronicity' something stronger than a 'signature of excitonic coherence and coherent energy transfer'?

In other words is negative synchronicity necessary and sufficient for excitonic quantum coherence? A separate question: This work (such as Fig. 8 in 10.1039/c9fd00006b) shows correlations between population transfer and synchronicity. Is this statistically significant? This can be shown by calculating the *p*-value. We did a similar study for a related problem in ref. 1.

1 G. C. Knee, P. Rowe, L. D. Smith, A. Troisi and A. Datta, *J. Phys. Chem. Lett.*, 2017, **8**(10), 2328–2333.

Alexandra Olaya-Castro responded: I will address the question:

"Is negative synchronicity necessary and sufficient for excitonic quantum coherence?" For the initial state considered, *i.e.* the system is in a separable state with the electronic system in the highest exciton state and the quasi-coherent vibrations in thermal states, negative synchronisation of local displacements emerges due to the formation of a collective vibronic coherence that involves the two excitonic states (quasi-localised in different chromophores) and different local vibrational states. Such vibronic coherence will lead necessarily to a non-zero excitonic coherence. Notice that all of this is satisfied when the energy quanta of the identical vibrations commensurate with the inter-exciton energy gap.

Jahan Dawlaty commented: This question is not on our paper. However, I find the topic very interesting and have a quick comment to add. Two classical oscillators can and do couple with a classical coupling term. Franck–Condon active modes that are excited as wave-packets behave as classical oscillators, at least for a short length of time. The coupling term between two chromophores is dipole–dipole interaction (which is classical in nature). Therefore, it would be interesting to isolate the classical and quantum manifestations of the reported synchronicity.

Tom Oliver asked: Does synchronisation between the two chromophores only occur for the case that the thermal bath fluctuations surrounding each molecule are correlated?

Alexandra Olaya-Castro said: I am not sure what you mean. The thermal baths and the vibrations in question are assumed local and, to begin with, uncorrelated. What we show is that, due to the electronic interaction between local excited states, these specific, local vibrational motions can become anti-synchronised or synchronised depending on the local decoherence and dissipation mechanisms taking place at different time scales.

Jeffrey Cina remarked: If I understand correctly, your calculations proceed by transferring to the excited electronic state the quiescent nuclear wave function from the electronic ground state. Have you considered turning things around by using external sources to drive coherent nuclear motion in the electronic ground state prior to short-pulse electronic excitation? The thought would be that doing so might allow the inter-mode synchronization you investigate to serve as a means of influencing the time-course of electronic energy-transfer dynamics, or excited-state dynamics more generally. Jason Biggs and I recently explored some simple consequences of vibrational pre-excitation for electronic excitation transfer.[1]

1 J. D. Biggs and J. A. Cina, *J. Phys. Chem. A*, 2012, **116**, 1683–1693.

Alexandra Olaya-Castro responded: We have not yet considered this scenario but it certainly sounds like an interesting avenue to approach. Thanks for the suggestion.

Graham Fleming commented: In light of your demonstration of vibrational synchronization in a vibronic dimer the 2DVE (2D vibrational-electronic) spectroscopy developed by Munira Khalil might be very interesting.[1] Here the vibrational degree of freedom is excited first, followed by the electronic excitation.

1 T. L. Courtney, Z. W. Fox, K. M. Slenkamp and M. Khalil, *J. Chem. Phys.*, 2015, **143**, 153201.

Alexandra Olaya-Castro responded: One of our motivations to investigate observables of the vibrations, came precisely from knowing there was an experimental technique that can probe concomitantly the interplay between electronic and vibrational motions such as the 2D vibrational-electronic spectroscopy. We expect that the non-thermal equilibrium states that are formed *via* excitation of the vibrational degree of freedom will affect the electronic dynamics for as long as the vibrations affected have energies comparable to the excitonic gap. I definitely think it will be extraordinary to use 2DEV to demonstrate without ambiguity the emergence of negative and positive vibrational synchronisation during exited state dynamics.

Natercia das Neves Rodrigues communicated: Your studies describe synchronisation between different systems, *i.e.* a "through-space" coherence between the vibrational states of different molecular units—is this correct? However, the vibrational states of each one of these individual molecular units may themselves couple to each other, so that there is intramolecular coherent energy dissipation. I was wondering, then, how this potential intramolecular coherence may affect the intermolecular case (can energy get trapped in molecular unit 1 if there is coherence, unable to then transfer to molecular unit 2, for example?) and, if so, how do you account for these effects in your calculations?

Alexandra Olaya-Castro communicated in reply: Yes, our work describes the transient synchronisation dynamics of some specific vibrations located at different chromophores, so it is correct to think of these as correlations between the oscillatory patterns of displacements of different molecular units. Notice that a key fact in our systems is that the energy of the vibrations involved commensurate with the exciton energy splitting. In other words, the mechanism of synchronisation involves direct energy exchange of energy between excitons and the vibrational states. This is important, as it means that other vibrations which do not satisfy those energy conditions will not be involved in such energy exchange. These vibrations can however affect both the energy transfer dynamics and will effectively broaden the mode in question. The way we account for these in our calculations is that we consider those as part of the thermal bath inducing incoherent transitions in the system. If the coupling of single excitation states to those local vibrations is very different and very strong, the emergence of positive synchronisation will be delayed, if it happens at all.

Jahan Dawlaty opened discussion of the paper by Graham Leggett: Adding a layer of dielectric to the metal can shift the plamson frequency. Is that accounted for in the assigning the splitting?

Graham Leggett replied: Adding a dielectric layer to the surface of a plasmonic nanoparticle changes the local refractive index, and causes a small shift in the position of the plasmon band in the extinction spectrum. However, that is a very different phenomenon from the one described in our paper; we do not see a small shift, but rather a dramatic splitting of the plasmon mode. The observation of splitting demands a different explanation than simply the modification of the local refractive index. The type of splitting observed here is a type of asymmetric Fano resonance, and it is explained by a hybridisation of the confined optical mode (the localised surface plasmon resonance) with an ensemble of excitons (the chlorins and carotenoids in the light-harvesting complexes).

Jahan Dawlaty asked: The coupling between plasmons and the molecular excitons on the overlayer is likely not describable by dipole–dipole coupling. However, an electrostatic approach may be used to estimate that coupling from simpler arguments. Can one justify the 0.25 eV electronic coupling in this way?

Graham Leggett responded: Strong plasmon–exciton coupling is a quantum optical phenomenon in which a localised electromagnetic mode is coupled to an ensemble of quantum emitters. In our case, the confined light mode is a localised surface plasmon resonance, and the emitters are pigment molecules in light-harvesting proteins. There are a number of approaches to the modelling of such interactions. In a particularly important paper, Gallinet and Martin performed a first-principles analysis of strong coupling, but then demonstrated that in the special case where a broad mode (the LSPR in our system) is coupled to a narrow mode (the transition in the pigment), the predicted spectrum is also obtained by modelling the system as coupled harmonic oscillators.[1] The fact that our spectra are effectively modelled using this approach is evidence for strong coupling.

However, we can apply a number of other tests for whether we are really in the strong coupling regime. The most basic is that the coupling energy is greater than or equal to the square root of the products of the linewidths of the coupled modes. In Fig. 3 of 10.1039/c8fd00241j the data do seem to suggest an avoided crossing. Unfortunately it was not possible to fabricate arrays with LSPR energies greater than the exciton energy, making the figure look a little lop-sided. However, the energy difference between the upper and lower polariton branches for the plasmonic array with the LSPR of highest energy was approximately 0.25 eV, suggesting that there was indeed an avoided crossing with a Rabi energy large enough for us to be in the strong coupling regime.

1 B. Gallinet and O. J. F. Martin, *Phys. Rev. B: Condens. Matter Mater. Phys.*, 2011, **83**, 235427.

Tom Oliver remarked: I presume that wild-type LH2 and LH1 proteins preferentially orient onto the gold nanostructures with their rings aligned parallel to the gold surface, however, is any alignment inferred for wild-type LHCII? Would the lack of LHCII protein orientation with respect to the electric field associated with the localised surface plasmon resonance explain (at least in part) the smaller bulk coupling energy for LHCII compared to LH2 and LH1?

Graham Leggett replied: The gold nanoparticles in our arrays are functionalised with a self-assembled monolayer of alkylthiolates. In particular, the monolayer is derivatised with nitrilotriacetic acid (NTA) terminal groups, which after complexation with Ni^{2+}, enable us to bind His-tagged proteins site-specifically. For proteins that cannot be His-tagged (*e.g.* LH1) there is still expected to be some control of orientation, because the NTA functionalised surfaces tend to be negatively charged, with the consequence that binding to the cationic face of the protein predominates. The fact that naturally occurring light-harvesting complexes are membrane proteins is helpful in this respect, because one face contains the C-termini, and the other the N-termini, meaning that they posses an intrinsic anisotropy of interaction with the surface.

Andrew Marcus questioned: Based on the various light harvesting complexes that you have examined, can you say what the effects of structural inhomogeneity are, if any, on the plasmon–exciton coupling strength? Is the inhomogeneity due mostly to disorder of the attachment sites, or rather internal to the protein scaffolds?

Graham Leggett answered: An important type of inhomogeneity is found in the orientations of the transition dipole moments. Of course in LHCII, the pigment complement is diverse, and the transition dipole moments have a variety of orientations. However, in LH2 from bacterial antenna complexes, there is important aniosotropy. In particular, the transition dipole moments for the bacteriochlorophyll Qx transitions and the carotenoid S0 to S2 transitions are aligned perpendicular to the membrane, or the gold surface in our nanostructures, *i.e.* they lie in the electric field direction. In contrast, the Qy transitions are orthogonal to the Qx transitions. It is notable that we see strong coupling of the plasmon mode to the Qx transitions but not to the Qy transitions. Thus the introduction of the plasmon field as a reference direction leads to a sensitivity to orientation in the excitons.

There is likely to be some disorder in the protein attachment sites at the surface, but it is hard to gauge the significance of this. For His-tagged proteins measurements of the amount of adsorbate using spectroscopic ellispsometry are consistent with the formation of close-packed monolayers, and in such systems the disorder in the attachment sites relative to the plasmon field direction is not likely to be significant. However, it is hard to be categorical because these are difficult systems to characterise at the level of spatial resolution really required to answer the question unequivocally.

Jennifer Ogilvie said: It is very interesting in LHCII that you extract an exciton energy of 2.24 eV which does not match any of the main exciton transitions in the LHCII pigment molecules. In 2DIR studies of vibrational polaritons there have been reports of the observation of "dark states" of molecules that can be accessed when the vibrational states of the molecule are strongly coupled to a cavity.[1] Could this be a similar effect and if so do you have any ideas of what the dark states could be?

1 B. Xiang *et al.*, *Proc. Natl. Acad. Sci. U. S. A.*, 2018, **115**(19), 4845–4850.

Graham Leggett replied: This is a very interesting question, and one that we did not consider. If there was coupling to "dark" states then it might be very interesting. It is physically feasible that this could be the case, but, bearing in mind the comparative simplicity of our current model, I would be reluctant to want to conclude that nevertheless. Our current model does not allow us to include multiple different excitons in the coupled system, which we know is the case for LHCII because of its diverse complement of caroteinoids. In LH2, where there is a single carotenoid, and a large difference between the transition dipole moments of the bacteriochlorophyll Qx and carotenoid S2 transitions, it is reasonable to assume that coupling to the S2 transition may dominate the strongly coupled system. However, for LHCII the situation is much less clear because of its diverse pigment complement. It would be useful to improve the model so that we could address such systems in a more meaningful way. Then we might be able to give a better answer to the question!

Vasilios Stavros asked: Have you considered examining the influence of plasmonic resonances from other metal surfaces on the substrate?

Graham Leggett responded: This is an interesting question. To date we have predominantly worked with arrays of gold nanoparticles. However, we have recently done some work with arrays of aluminium nanostructures fabricated using interferometric lithography. Al is an attractive material because it is cheap and abundant, but the fabrication presents greater challenges. In preliminary experiments we appear to have demonstrated strong coupling of Al plasmon modes to light harvesting complexes, but this work is currently in its early stages. Certainly there are other materials that are plasmonically active and some of these may also be suitable for use in studies of strong coupling of plasmon modes to excitons in light harvesting systems.

Naomi Ginsberg asked: What is the impact of the relative size of the gold particles and the light harvesting complexes on top of them?

Graham Leggett replied: The morphology of the gold nanostructures is important, because it determines the plasmon energy. The size is important – the larger the structures the smaller the energy of the plasmon mode – but their geometries are also important (needle-like nanostructures have longitudinal and transverse modes). In addition, the packing of nanostructures can be important because of the possibility for coupling between nanostructures at short range (see Fig. 5 of DOI: 10.1039/c8fd00241j). The interferometric lithography technique (from ref. 32 of DOI: 10.1039/c8fd00241j) that we use to fabricate our arrays (each of which covers nearly 1 cm^2) allows us to vary these parameters systematically, and we have built up a library of hundreds of arrays with different characteristics.

Minjung Son queried: In the paper (DOI: 10.1039/c8fd00241j), you state that the coupling strength was dependent on the density of the bacterial light-harvesting and synthetic maquette proteins. Have you tried the same experiment for the case of LHCII? Furthermore, LHCII proteins are known to exhibit strong protein–protein interactions by self-aggregation when multiple LHCIIs are closely packed together, which likely happens at a high protein concentration (*i.e.*, density). Have you considered the possibility of these protein–protein interactions might be interfering with your signal of interest, *i.e.* the LHCII-gold nanostructure interaction? Do you think you would be able to tell that by looking at the changes in the UV/Vis spectra?

Graham Leggett responded: This is a very interesting question. In strong plasmon–exciton coupling the plasmon mode couples to an ensemble of emitters, and the coupling energy is thus proportional to the square root of the density of excitons (eqn 2 in our paper (DOI: 10.1039/c8fd00241j)).

We did not model the coupling as a function of fractional coverage for LHCII, unfortunately. It would certainly have been good to do so. However, these measurements are very hard to do, because to be sure that spectral changes are due solely to the change in coverage, it is best to re-use the same plasmonic array. We have done this successfully for LH2 and maquettes by cleaning the array between each adsorption using piranha solution (a mixture of concentrated sulfuric acid and hydrogen peroxide) but it takes its toll on the arrays and we lost a number along the way! Having been through this process for other proteins, we are now quite confident of our model, and so would now feel that a good fit of the experimental spectrum to the calculated one is sufficient evidence that strong coupling is observed. But in an ideal world we would certainly have done a study of coverage-dependence.

We did not consider LHCII aggregation at the time we carried out the work presented. I do not have a detailed understanding of the aggregation phenomena described, but note from a quick survey of the literature that these are predominantly phenomena that have been studied in bulk phases (in solution) and that aggregation is thought to be strongly influenced by the nature and concentration of lipid in solution. It is important to note that the conditions at the surface are probably very different. There is a substantial literature on protein adsorption and based on our knowledge of this phenomenon, and our ellipsometric data, we believe that LHCII forms close-packed monolayers at the gold surface. One might argue that our samples consist of macroscopic aggregates, therefore! This may in fact be rather important. My understanding is that aggregation is thought to alter the rates of energy transfer processes in the protein; I do not know whether this would have an effect on plasmon–exciton coupling. There are some suggestions in the literature that new transitions may be observed in aggregates; it might be that this would have an effect, if these transitions coupled to the plasmon mode.

Our measurements do not currently have any time resolution, so we cannot measure affects related to changes in energy transfer rates at present. However, changes in the energies of transitions are potentially detectable. A distinguishing feature of our measurements is that the directionality of the plasmon field (perpendicular to the gold surface) means that couplings in that direction have particular importance. Thus aggregation effects would become evident if they led to changes of transition energies or transition dipole moments able to couple to the plasmon field. Perhaps with a more sophisticated model than our current two-oscillator model we might be able to explore that better. It would certainly be interesting to know whether there was any evidence for coverage-dependent changes in the transition energies. However, at the moment, I do not think I can give a more detailed answer to the question, interesting though it is.

Bern Kohler remarked: You write that the exciton density is low (10^{17} m^{-2}), yet this value appears to be roughly equal to the surface density of carotenoids in your sample. How can every carotenoid molecule be excited by the low light intensities used in your experiments?

Graham Leggett replied: A very important characteristic of strong plasmon–exciton coupling is that the confined optical mode (the plasmon) couples collectively to an ensemble of excitons to form a type of quasiparticle called a plasmon–exciton polariton. Theoreticians talk about the excited state being delocalised across the nanoparticle/exciton ensemble.

The consequence of all of this is that the coupling depends on the organisation of excitons within the plasmon mode volume; for example the coupling energy depends on the square root of the density of excitons (eqn 2 in the paper (DOI: 10.1039/c8fd00241j)). As shown in our earlier Nano Letters paper,[1] a reduction in the fractional coverage of proteins at the surface from 1.0 to 0.6 leads to a visible reduction in the splitting of the plasmon mode in the extinction spectrum.

As you say, the light intensities in our experiment are very low compared to those used in the other types of experiment described at this meeting. However, I'd argue that this is not a problem because of the physics of strong plasmon–exciton coupling. If the coupling was not a quantum optical phenomenon, then we might need a photon for every carotenoid. But in our plasmon–exciton polaritons, we do not need a photon for every carotenoid because the plexcitons are delocalised modes.

The delocalised nature of the excitation in these plasmon–exciton polaritons is what gives rise to their coherence, and makes them such attractive prospects for the design of new kinds of photonic materials and devices.

1 A. Tsargorodska, M. L. Cartron, C. Vasilev, G. Kodali, O. A Mass, J. J Baumberg, P L. Dutton, C N. Hunter, P. Törmä and G. J. Leggett, *Nano Lett.*, 2016, **16**(11), 6850–68456.

Jeffrey Cina opened a discussion of the paper by Benjamin Fingerhut: Is your one-dimensional plot of exciton-state energies and bridge and electron-acceptor states meant to be merely schematic? I would have imagined that the reaction coordinates for the excitonic states and those serving as bridge and acceptor for electron transfer would be distinct.

Benjamin Fingerhut replied: The one-dimensional plot of exciton-state energies and bridge/electron-acceptor states (Figure 2 in the talk) is schematic in order to highlight the tight integration of exciton-states and charge transfer states in bRC. In the performed simulations a multidimensional system is employed. In particular, we consider a 3-dimensional system, where the excitonic states are coupled to an independent heat bath each along a single dimension. The lower exciton state P− serves as donor state towards charge transfer states (bridge: $P^+ B_A^-$, acceptor: $P^+ H_A^-$) that are displaced along the collective coordinate of electron transfer reactions, which is treated as the third dimension. The position along this collective coordinate affects the reorganization energy of respective the charge transfer reaction (see Section 3.2 of the main text (DOI: 10.1039/c8fd00189h) for DVR coordinate values).

Alexandra Olaya-Castro commented: During your presentation you argue the idea that for photosynthetic reaction centres, including vibrational motions *via* narrow features in spectral density did not necessarily result in an improved performance in terms of energy transfer efficiency. Could you elaborate on the physical underlying mechanisms supporting your conclusions? Your conclusions are consistent with the analysis we have done for one of the energy transfer pathways in PSIIRC where we demonstrate that coupling to such local vibrational motions does not necessarily results in an improve performance in terms of increased rate of population transfer.[1] We argue the main reason is the asymmetry of the system-bath coupling for excitation states and for charge transfer states. The much stronger coupling between charge transfer states and such local vibrations guarantees directional and step-wise charge transfer process absolutely necessary for reactions centres, but this doesn't mean it accelerates the process. This is different from what is expected for light-harvesting complexes with no charge transfer states.

1 R. Stones *et al.*, *Chem. Sci.*, 2017, **8**, 6871–6880.

Benjamin Fingerhut replied: I would like to clarify that our simulations show that the performance of energy transfer efficiency within the electronic coherence time (memory time) can indeed be affected and manipulated by the details of the spectral density. This is demonstrated in Fig. 4b and 5a of the paper (*cf.* dashed and solid green and orange lines) where the resonance frequency ω_0 shows profound impact on excitonic dynamics. In contrast, we argue that upon strong coupling of states to the environment (*e.g.* due to the polaronic character of charge transfer states) the impact on directional and irreversible trapping dynamics appears to be moderate in the investigated parameter regime of bRC. The physical mechanism for such different behavior is rooted in the separation of timescale of excitonic and charge transfer dynamics. The latter proceeds largely outside the memory time and consequently phase relations between pathways are evaded.

Compared to the findings of Stones *et al.*,[1] we show that the coupled excitation energy-charge transfer dynamics is particularly stable with respect to details of the spectral density even upon coupling to non-equilibrium exciton states, in particular with respect to the low (∼3) ps timescale of charge transfer between the special pair and accessory bacteriochlorophyll A. Such findings are consistent with a picture where the strong coupling between charge transfer states and local vibrations guarantees directional and step-wise charge transfer with moderate accelerating impact on the dynamics.

1 R. Stones *et al.*, *Chem. Sci.*, 2017, **8**, 6871–6880.

Shuming Bai commented: The parameters of orange, red and green lines in Fig. 5(c) of the paper (DOI: 10.1039/c8fd00189h) are not clear. In the Marcus theory method, is it true that only V_SE from superexchange through bridge is considered, while $V = 738\ cm^{-1}$ in eqn (20) is not included? Since the reason of the invalidity of Marcus theory here is clarified, is it possible to modify the original Marcus theory for Donor–Bridge–Acceptor systems, or it is beyond the limits?

Benjamin Fingerhut responded: In descriptions of population dynamics with Marcus theory, charge transfer from an equilibrated donor state is inherently assumed, which precludes the coupling to the coherently evolving special pair exciton states. We thus employ a simplified three-state Donor–Bridge–Acceptor model system, as described in the Appendix (Sec. 5.2 Marcus theory) with the Hamiltonian given in eqn 27 (see DOI: 10.1039/c8fd00189h). As correctly noted, this model does not include the strongly coupled excitonic states ($V = 738\ cm^{-1}$) but a single donor state that is coupled to the bridge *via* the direct coupling element V_{DB} and to the acceptor *via* the superexchange coupling element V_{SE}. In Fig. 5(c) different 'flavors' of Marcus theory are compared to results obtained with the MACGIC-iQUAPI method.[1] Employing the celebrated Marcus rate formula (eqn 29, $V_{SE} = 0$), quadratic dependence of transfer rate with driving force is obtained (orange line in Fig 5b), which deviates qualitatively from the MACGIC-iQUAPI results for bridge energies $E_{Bridge} > 10700\ cm^{-1}$. Accounting for a multi-mode picture in Franck–Condon factors (eqn 31, $V_{SE} = 0$) improves on the qualitative agreement with the MACGIC-iQUAPI results for bridge energies $11700\ cm^{-1} > E_{Bridge} > 10700\ cm^{-1}$ (red line in Fig 5b). The incoherent description of CT dynamics employing different 'flavors' of Marcus theory thus allows us to demonstrate the relevant contributions to the transfer rate as demonstrated for the first-order super-exchange correction to the transfer rate (green line in Fig 5b) that only becomes relevant for $E_{Bridge} > 11700\ cm^{-1}$ (eqn 31, $V_{SE} > 0$).

A modification of Marcus theory to rigorously account for a coupling of populations and coherences in non-equilibrium population dynamics has to the best of our knowledge not yet been realized, apart from elaborate numerical methods like MACGIC-iQUAPI or HEOM. An intermediate and numerically more efficient level of theory is given by the generalized Redfield theory[2] that reduces to the second order perturbation theory of Marcus-type treatments in the respective limits, but the time evolution of populations and coherences evolves decoupled, precluding the description of, *e.g.*, coherence to population transfer that is rigorously accounted for by the MACGIC-iQUAPI method.

1 M. Richter *et al.*, *J. Chem. Phys.*, 2017, **146**, 214101.
2 W. Zhang *et al.*, *J. Chem. Phys.*, 1998, **108**, 7763–7774.

Jennifer Ogilvie commented: You cite the recent work of Ishizaki and coworkers wherein they explore the effect of electronic-vibrational resonance on the charge separation in a model dimer inspired by the Photosystem II reaction center.[1] Can you comment on how your study compares to this one? How does it differ in its approach, the approximations that are made and the parameters that are chosen?

1 Y. Fujihashi, M. Higashi and A. Ishizaki, *J. Phys. Chem. Lett.*, 2018, **9**, 4921–4929.

Benjamin Fingerhut replied: Charge separation in the model dimer inspired by the PSII RC as investigated by Ishizaki and coworkers (ref. 23 in DOI: 10.1039/c8fd00189h) is described with fixed couplings on the order of 70 cm^{-1} and reorganization energies in a range 100–500 cm^{-1}, parameters derived from TDDFT calculations. The resulting population transfer dynamics thus appears as incoherent, justifying a treatment with second order perturbation theory. Herein instantaneous medium relaxation and charge separation from an equilibrated donor state is inherently assumed. In the current work, we numerically investigate the impact of excitonic coherence and intramolecular vibrations on charge separation dynamics. The explored parameter regime (reorganization energies > 500 cm^{-1}, couplings > 30–700 cm^{-1}) goes beyond the validity of perturbation theory and covers a regime of coherent oscillatory population dynamics within exciton states. The investigated bRC inspired model thus covers extended dynamics that connects the initially coherent dynamics to an incoherent trapping of population within low energy charge separated states. Moreover, medium reorganization is considered to occur on comparable timescales to system dynamics. We find that distinct vibrational modes can substantially affect the population dynamics among strongly coupled exciton states but a moderate effect on the charge separation process is found. In summary, our results suggest a moderate impact of vibrational modes on transfer dynamics and a decoupling of timescales and of coherent EET and CT dynamics even if a finite timescale of medium relaxation is taken into account and charge separation from a coherent non-equilibrium state is considered. In this view, molecular vibrations primarily assure the robustness of optimal CT reactions in bRC, in agreement with conclusions from Ishizaki and coworkers for charge separation in PSII.

Yuan-Chung Cheng commented: I think this work using a reliable and accurate method to look into the coupled excitation energy transfer-charge transfer dynamics is very enlightening and important for clarifying several important issues in the literature. I am a little bit surprised by the result that the high frequency vibrational mode only has a minor effect on the charge transfer rates. So, I am wondering if the energy resonance effect has been fully explored. What are the respective energy gaps between the donor and acceptor states of the two charge transfer processes? Were the dynamics affected by the vibrational mode as the mode frequency matches the energy gap? I also notice that the highest frequency examined in Fig. 5(d) of DOI: 10.1039/c8fd00189h is around 1100 cm^{-1}, why not try higher frequency modes? In pi-conjugated systems the often strongly coupled carbon–carbon stretching mode is around 1400–1600 cm^{-1}. It will be nice if some comments on the resonance effect can be provided.

Benjamin Fingerhut replied: Frankly, we were similarly surprised by the finding that resonant vibrational modes (represented *via* structured spectral density function centered around ω_0) show only moderate impact on the charge transfer dynamics. The adiabatic energy gap of the low energy exciton state P– and primary CT state is 517 cm^{-1} and thus well within the investigated range of vibrational mode resonance frequencies ω_0. The resulting impact on charge transfer dynamics is found to be moderate (*cf.* solid and dashed red lines in Fig.

4b of the paper). For the secondary ET step the adiabatic energy gap is ~1100 cm^{-1} and thus also comparable to the investigated resonance frequency ω_0. Here, similar conclusions apply for the impact on dynamics apply (*cf.* solid and dashed violet lines in Fig. 5a of the paper). In contrast, for the excitonic system (P− and P+ special pair states, $\Delta E = 1704$ cm^{-1} upon diagonalization of eqn (16)) we find pronounced effects of the resonance frequency ω_0 on the dynamics already in the investigated 0–1100 cm^{-1} range, properly described with the nonperturbative MACGIC-iQUAPI method.

Sharon Hammes-Schiffer asked: Does this work provide any insight into the functional asymmetry that leads to unidirectional electron transfer in photosynthetic reaction centers?

Benjamin Fingerhut responded: The presented model of exciton coupled charge transfer accounts for electron transfer along the active A-branch of bRC in order to demonstrate the general applicability of the MACGIC-iQUAPI algorithm to both regimes of excitation energy transfer (intermediate system-bath coupling strength) and charge transfer dynamics (strong system-bath coupling strength) subject to structured spectral density functions. Furthermore, model parameters are reliably available only for the A-branch with substantial uncertainty for the B-branch.

The favorable numerical scaling of the MAGCIC-QUAPI algorithm in principle allows for an extension of the model to system sizes that account for the inactive B-branch. The current work thus establishes a benchmark for further investigations. The established methodology should allow to establish boundary conditions for unidirectional charge separation originating from the functional and structural asymmetry of bRC.

Jennifer Ogilvie said: Can you comment on how sensitive your conclusions are to the parameters of your model? For example, you employ a Huang Rhys factor that is considerably larger than many other treatments such as the recent work by Ishizaki.[1]In addition, the separation you use between upper and lower exciton states of the special pair is considerably larger than we recently found from our two dimensional electronic spectroscopy experiments on Rhodobacter *Capsulatus.*[2] Your spectral densities are also quite simple, containing a single vibrational mode that is relatively broad. How might these various parameter choices affect the generality of your conclusions?

1 Y. Fujihashi, M. Higashi and A. Ishizaki, *J. Phys. Chem. Lett.*, 2018, **9**, 4921–4929.
2 A. Niedringhaus *et al.*, *Proc. Natl. Acad. Sci. U. S. A.*, 2018, **115**(14), 3563–3568.

Benjamin Fingerhut responded: The employed model parameters intend to reflect the bRC B. *viridis*. The current system Hamiltonian is derived from ref. 1 that was shown to reproduce the temperature dependent linear absorption spectrum of bRC B. *viridis* with high accuracy. Modeled special pair exciton states are found at P+ = 11 852 cm^{-1} and P− = 10 148 cm^{-1}, corresponding to the weak and strong Q_Y absorption observed at 850 and 980 nm, respectively. Other bRC, like Rb. *capsulatus*[2] and Rb. *Sphaeroides* are characterized by smaller P+/P− energy gaps. We note that the energy of the P+ state closely resembles recent reports of the P+ state in Rb. *Capsulatus*[2] in the direct vicinity of the BA absorption band. The parameters of the current bRC B. *viridis* model can be adopted to other bRC, in particular the reduced excitonic splitting can be addressed efficiently using the MACGIC-iQUAPI method. For such reduced excitonic splitting, vibrations could have enhanced impact on excitonic equilibration dynamics as the reduced energetic splitting of P– and P+ states ($\approx$ 780–650 cm^{-1}) closely matches the vibrational resonance frequency ω_0 in the spectral density. However, as charge separation proceeds *via* couplings on the order of 30–60 cm^{-1} we do not expect that the charge separation dynamics is significantly altered by EET dynamics.

Concerning the coupling strength to the environment (as reflected in the magnitude of Huang Rhys factors) our bRC model employs established parameters from experimental mutation studies with $-\Delta E \approx \lambda = 500$ cm^{-1} for the initial charge transfer reaction P B $\rightarrow$ P^+ B_A^- in bRC.[3] The substantial driving force of the sequential CT steps allows to realize efficient (> 90%) charge separation, which is a prerequisite to suppressed thermal repopulation of exciton donor states (see also ref. 4–6).

Given that sequential charge transfer at room temperature proceeds with highly conserved dynamics in the various bRC, we do not expect the generality of our conclusions to be affected when changing the bRC B. *viridis* Hamiltonian parameters to *e.g.* Rb. *capsulatus* or Rb. *Sphaeroides* (see also the answer to Jennifer Ogilvie's previous question).

1 B. Fingerhut *et al.*, *J. Phys. Chem. Lett.*, 2012, **3**, 1798–1805.
2 A. Niedringhaus *et al.*, *Proc. Natl. Acad. Sci. U. S. A.*, 2018, **115**(14), 3563–3568.
3 M. Bixon *et al.*, *Chem. Phys.*, 1995, **197**, 389–404.
4 A. Warshel and D. W Schlosser, *Proc. Natl. Acad. Sci. U. S. A.*, 1981, **78**, 5564, 1981.
5 B. Fingerhut *et al.*, *Chem. Phys. Lett.*, 2008, **466**, 209–213.
6 B. Fingerhut *et al.*, *Phys. Chem. Chem. Phys.*, 2010, **12**, 422–432.

Vasilios Stavros addressed Benjamin Fingerhut and Alexandra Olaya-Castro: Your calculations assume (as far as I can tell) that the excitation is instantaneous. Can you comment on the importance of the pulse profile on the dynamics and how easy (or difficult) it would be to include the pulse profile in your calculations?

Alexandra Olaya-Castro answered: We are interested in investigating spontaneous and transient synchronisation of vibrational displacements while excitation energy transfer takes place. Hence, the experimental scenario we consider is indeed that in which a particular initial state is created and allowed to evolve over time. In this case we assume the initial state is formed instantaneously and is such that only the highest exciton state is populated. Introducing pulsed excitation will not be difficult for the simplified model we have considered and it will effectively be the same as considering a different initial state for the system of interest. We expect the specific synchronisation dynamics (*i.e.* their evolution in time) will naturally be different, yet the mechanisms underlying the emergence of negative or positive synchronisation will still be as we have described in our manuscript: the formation and decay of specific collective vibronic coherences.

Benjamin Fingerhut responded: The excitation is indeed assumed to be instantaneous occurring *via* a delta excitation pulse. Including realistic (Gaussian) pulse shapes in our simulations could occur *via* two complementary approaches: (i) *via* a perturbative approach in a response function formalism and (ii) by directly including the matter–electric field interaction (within dipole approximation) into the system Hamiltonian. Within approach (i) convolution of dynamics with realistic pulse profiles is accounted for, see ref. 1 for a recent overview. Approach (ii) further accounts for phase sensitivity of the excitation process and opens the prospects of coherent control *via* affecting off-diagonal coherences and strong field effects. This approach requires the implementation of the QUAPI algorithm with a time-dependent Hamiltonian, to be diagonalized under the action of the electric field in every time step. While such developments are in principle feasible, important properties of the QUAPI algorithm are lost (*e.g.* norm conservation that provides an important indicator of convergence) and a substantially higher numerical effort is expected due to the required reduction in time step size. We would like to note that the simulated, oscillatory dynamics of exciton states is readily observed with current laser technology as demonstrated in pump probe studies of exciton coherences with high temporal resolution[2] and more recently *via* oscillation in 2D signals.[3,4]

1 M. Kowalewski *et al.*, *Chem. Rev.* 2017, **117**, 12165–12226.
2 D. C. Arnett *et al.*, *J. Phys. Chem. B*, 1999, **103**, 2014–2032.
3 A. Niedringhaus *et al.*, *Proc. Natl. Acad. Sci. U. S. A.*, 2018, **115**(14), 3563–3568.
4 F. Ma *et al.*, *J. Phys. Chem. Lett.*, 2018, **9**, 1827–1832.

Graham Fleming commented: The question of whether quantum coherence is important in systems driven incoherently by natural sunlight is very challenging. Analysis of photon statistics is an attractive method of investigating this question as demonstrated by Prof. Olaya-Castro's paper (DOI: 10.1039/c9fd00006b) on g2(t, ω1,ω2) in a vibronic system.

Benjamin Fingerhut responded: I would like to add that spectroscopy with entangled biphoton states[1] increases the information content and the prospects of control over observables, paving a route towards addressing questions about the importance of quantum coherence. For example, the unique temporal and spectral features of entangled photons may be employed for nonlinear spectroscopy in order to reveal multiexciton properties in the bacterial reaction centre. The absorption of photon pairs with short entanglement time allows suppression of population transport in the intermediate single-exciton manifold, thus allowing the manipulation of the distribution of two-exciton states. Here, the quantum nature of light is essential for achieving such a degree of control. While classical light is fundamentally limited by the frequency-time uncertainty, entangled photons provide independent temporal and spectral characteristics not subjected to this uncertainty.

1 F. Schlawin *et al. Nat. Commun.*, 2013, **4**, 1782.

Jennifer Ogilvie said: I think theory has a critical role to play in guiding the experiments that can test the functional relevance of electronic-vibrational resonance to photosynthetic energy transfer and charge separation. Do you have ideas for any 'smoking gun' experiments, or further simulations for that matter, that could definitively prove or disprove the functional relevance?

Benjamin Fingerhut answered: In order to demonstrate functional relevance of electronic-vibrational resonances spectroscopic measurements in my opinion should aim at successively expanding on the information content of the observed signals (see also ref. 1 for an overview). Near future challenges are the improvement in pulse bandwidth to cover the energy scale of special pair exciton states and/or two-color experiments with improved (broadband) probing towards the near-IR region. Such experiments address the direct probing of $P^+ B_A^-$ and $P^+ H_A^-$ charge separated states with possibly imprinted coherence effects in the transient signals.

The use of polarized pulse sequences further provides a route to increase contrast (for prospects of ambitious experiments relying on spectroscopy with entangled biphoton states see answer to the previous question). The combination of parallel and orthogonal polarized pulse sequences (as suggested in ref. 2) provides increased selectivity towards the orientation of dipole moments in RC. The signal contribution of parallel-aligned exciton transitions can be largely suppressed, with the prospect of highlighting weaker signal features and allowing us to dissect the kinetic components of charge separation from EET and possible (coherent) coupling of both.

1 M. Kowalewski *et al. Chem. Rev.*, 2017, **117**, 12165–12226.
2 B. Fingerhut *et al. J. Phys. Chem. Lett.*, 2012, **3**, 1798–1805.

Alexandra Olaya-Castro asked: Have you carried out your analysis for the different transfer pathways that are argued to exist in PSIIRC?

Benjamin Fingerhut replied: We have not yet performed simulations for the different transfer pathways proposed in PSIIRC but plan to do so in the near future. The current study on bRC model Hamiltonians serves to establish the methodological groundwork for such investigations, *i.e.*, the treatment of structured spectral densities and non-equilibrium exciton coupled charge separation with the MACGIC-iQUAPI method.

Alexandra Olaya-Castro commented: One recurrent subject of discussion over the years has been whether the conclusions we are drawing from experiments done with a laser applied to the situation in which these biomolecules are illuminated with incoherent sunlight as happens in physiological conditions. It is about time for the community to achieve conceptual and experimental clarity on the issue so that we can move on! It seems to me that the whole discussion in the field comes mostly from our inability, and sometimes unwillingness, to listen to each other's scientific arguments carefully.

My take on the issue is the following. It is clear that coherent superpositions are formed when illuminating with lasers and, therefore, experiments are probing how those superpositions evolve in time. What we are after is a clear understanding of the mechanisms that support such coherent dynamics because such coherent mechanisms will definitely play a role under incoherent illumination. Under incoherent input and output of excitation one expects non-thermal equilibrium steady states to be formed, and how far these states are from thermal equilibrium will depend fully on the coherent mechanism that may be present.

Jennifer Ogilvie said: I would like to comment that I think experiments on model systems such as dimers, in which electronic energy gaps and vibrational modes can be independently varied and more easily modeled than photosynthetic complexes, will be important to ultimately answer the question of the relevance of electronic-vibrational resonance to photosynthetic function.

Graham Fleming opened the discussion of the paper by Yuan-Chung Cheng: How does the cutting method you describe relate to the systematic (and unique) reduction of the rate matrix into kinetic domains described by Mino Yang?[1]

1 M. Yang *et al.*, *J. Chem. Phys*, 2003, **119**, 5614.

Yuan-Chung Cheng replied: The kinetic domain approach proposed by Mino Yang and yourself focus on the trapping time of each site and iteratively constructs a kinetic domain by identifying the sites that are closest to the trap site(s). Therefore, the method is based on finding the most rapid pathways towards the traps. On the other hand, the minimum-cut approach used in our work is based on a philosophy of finding the bottleneck and applying time-scale separation to construct coarse-grained cluster models. In our experience, this approach produced better coarse-grained models compared to methods that focus on rapid pathways. In my opinion, these various coarse-graining methods might contain complementary information and it would be interesting to carry out a comparative study for complex photosynthetic energy transfer networks.

Benjamin Fingerhut said: An empirical scaling factor is applied to the time correlation function in order to derive a scaled time correlation function (eqn 19 in DOI: 10.1039/c8fd00205c). Could you elaborate on the procedure used to obtain the scaling factor *f*?

Yuan-Chung Cheng responded: A scaling factor of 5 was applied to the site-energy fluctuations calculated from the charge-density-coupling method and MD structures in order to reproduce the absorption spectra of the PSII-cc subunits. Basically we tested different values of the scaling factor, and found that scaling the energy shift by a factor of 5 and the energy-gap fluctuation time-correlation function by a factor of 5^2 best produces the spectra shown in Fig. 3 and 5 of the paper (DOI: 10.1039/c9fd00006b). The rationale of applying the scaling factor is that the site-energy shifts are calculated from partial charges of the protein residues and waters defined in the AMBER forcefield, however the classical forcefield was never meant to be used to simulate spectroscopic data. Nevertheless, the protein dynamics should be captured reasonably well by the AMBER forcefield. Therefore, we choose to neglect the magnitude of the charge-density-coupling interactions by introducing the scaling factor, whereas the information on the protein dynamics is kept in the time-dependent profile of the time-correlation function. Indeed, our results yields site-energy fluctuations with $\sim$30 cm^{-1} standard deviation at the room temperature, which is too small. After scaling, the standard deviation for energy fluctuation is $\sim$150 cm^{-1}, which is in the expected scale of the thermal energy and better reproduces the spectra as well as the dynamics.

Benjamin Fingerhut queried: The empirical scaling factor *f* of the time correlation function accounts for deficiencies of both, the employed partial charge model and the simulation method of the optical density (relying on modified Redfield theory, eqn 15–18 in DOI: 10.1039/c9fd00006b) in an average fashion. Could you comment on the magnitude of both sources of error and discuss a strategy to disentangle the two?

Yuan-Chung Cheng answered: I think this is a good point. In our study, we believe the modified Redfield method yields accurate dynamics, so the larger error comes from the spectral density. In fact, I do not think it is possible to fully disentangle the uncertainties in the spectral density (Hamiltonian) and the errors in the methods used to carry out the dynamical simulations (the modified-Redfield approach in this study). In particular, whenever a fitting to empirical data is used to determine the parameters in the model, the resulting Hamiltonian and the dynamical method should be bound together, *i.e.* the parameters best describe the system when the same dynamical method is used. There, using the same parameters with a higher-order or even numerically exact dynamical method does not mean the results would be a better representation of the real system. At least it is more likely that the results will deviate more from the experiments. Hence, fitting to determine empirical parameters in a consistent level of theory is important. Having said that, we still hope to reduce the ambiguities as much as possible. That is why we perform this study to utilize MD simulations in our parametrization of the Hamiltonian. Our results also serve the purpose of verifying the approximations in conventional models used to describe energy transport in photosynthesis, and so far we conclude that the typical assumptions such as identical independent harmonic baths and diagonal system-bath couplings do work reasonably well.

Vytautas Balevicius Jr. asked: In the paper you mention that the time-correlation function (TCF) of the energy gap fluctuations is calculated within a 10 ps window of the MD trajectory, which allows the TCF to capture the fast/dynamical aspects of fluctuation as opposed to the static disorder. However, the motivation for the specific number (10 ps) is presented in a somewhat vague manner. The question is then: is there a specific measure to help extract the duration value of 10 ps, or is this number purely empirical and based on attempts to obtain TCFs within different windows?

Yuan-Chung Cheng replied: The 10 ps window was chosen based on the relevance of the timescale for population transfer in photosynthetic energy conduction, *i.e.* 95% quantum yield means a total mean trapping time of about 50 ps. We did experiments with different numbers ranging from 5 ps to 100 ps with our MD data, and we found that the resulting TCFs were only slightly different and the 10 ps window consistently yields results that have the static part of the fluctuations eliminated.

Jennifer Ogilvie said: Renger and coworkers propose that the energy funnel in the PSII CC is not a simple one in which the lowest energy states are located in the reaction center.[1] In particular they propose a specific low energy site in CP47 that could be important for photoprotection. Can you comment on how your work differs in this regard and what the origin of the differences might be?

1 Y. Shibata *et al.*, *J. Am. Chem. Soc.*, 2013, **135**(18), 6903–6914.

Yuan-Chung Cheng responded: Our model is in general in agreement with Renger's model, although the low-energy site in CP47 in our model is different from that in Renger's model. We also emphasize the complex nature of the energy funnel in the PSII-cc and hence the function of the combined CP47 subunit for energy pooling and to enhance the robustness of the dimeric PSII-cc. Actually, our MD data also show significant energy fluctuations in long times that scramble the order of energies of the sites across the PSII-cc, and as a result, the different low-energy sites proposed by Renger and coworkers can be explained as different realizations of the system. We are currently investigating the details of the fluctuations captured by the MD simulations and their implications on the energy transfer dynamics. Regarding the implications for NPQ, there are many speculations on the molecular mechanisms and explicit energy quenching sites, but we do not think our data so far would allow us to comfortably comment on this. We will work towards building a full PSII supercomplex microscopic model in order to test various NPQ hypothesis.

Vanessa Huxter questioned: How does the system change when both reaction centers are closed?

Yuan-Chung Cheng answered: I do not know about any structure changes accompanying the closing down of both reaction centers. In terms of the change in energy transfer dynamics, the dimeric PSII core complex is only a small portion of the larger photosynthetic apparatus on the thylakoid membrane. It is known that excitation energy can visit more than one reaction center *via* long-range transport across PSII supercomplexes. So, I suspect when both reaction centers are closed, the excitation energy could reach an open reaction center in a different supercomplex, although in a much longer timescale and lower quantum efficiency.

Graham Fleming said: To study the regulation of light harvesting *via* non-photochemical quenching requires a much larger region of the thylakoid membrane than a single super complex to be modeled. We made a model of a 200 $\times$ 200 nm patch of the membrane and find that the quantity that is being controlled is the diffusion length.[1] In fact any combination of the density of quenchers and intrinsic quenching rate that produces the same diffusion length gives the same fluorescence lifetime.

The classic experiments of Joliot and Joliot[2] show the high degree of connectivity in the membrane and our model reproduces the hyperbolic shape of their plot of photochemical yield *vs.* fraction of open reaction centers extremely well.

1 G. Fleming *et al.*, *Proc. Natl. Acad. Sci. U. S. A.*, 2019, **116**, 3385–3390.
2 A. Joliot and P. Joliot, *Comptes Redues*, 1964, **258**, 4622–4625.

Yuan-Chung Cheng answered: Thank you very much for the informative comment. I also agree that our current model is too small and a much larger area of the photosynthetic membrane must be studied in order to realistically describe the regulation of light harvesting *via* non-photochemical quenching. Your theoretical model (and the works by you and Doran Bennett[1–3]) is still the state-of-the-art and the only one that provides a full perspective on the energy regulation problem. We are working on constructing models for the full PSII supercomplex and beyond, and hopefully I will be able to report our results and comparisons to your model in the near future.

1 D. I. G. Bennett, K. Amarnath and G. R. Fleming, *J. Am. Chem. Soc.*, 2013, **135**, 9164.
2 K. Amarnath, D. I. G. Bennett, A. R. Schneider and G. R. Fleming, *Proc. Natl. Acad. Sci. U. S. A.*, 2016, **113**, 1156.
3 D. I. G. Bennett, G. R. Fleming and K. Amarnath, *Proc. Natl. Acad. Sci. U. S. A.*, 2018, **115**, E9523.

Jahan Dawlaty asked: Given that circular dichroism experiments are likely to be more information rich compared to linear absorption, is it possible to simulate those experiments?

Yuan-Chung Cheng responded: Yes, we plan to simulate circular dichroism and other spectroscopic signals (fluorescence, 2D, *etc.*) in future work in order to better verify the Hamiltonian obtained from the MD trajectories. Note that we did compare our Hamiltonian with that calculated by Renger and coworkers.[1] For the most part, our site energies and couplings are consistent with each other. Renger and coworkers have calculated the CD spectrum and shown that their theoretical results are in good agreement with experiments. So we believe our model should also capture the CD spectrum of the PSII-cc system.

1 Y. Shibata *et al.*, *J. Am. Chem. Soc.*, 2013, **135**(18), 6903–6914.

Tom Oliver commented: How would the molecular re-arrangements associated with various forms of non-photochemical quenching, for example the xanthophyll cycle,[1] re-organisation of pigment protein complexes[2] or state transitions,[3] affect the excitonic energy landscape and thus the electronic energy transfer rates and pooling onto CP47?

1 B. Demmig-Adams and W.W. Adams, *Trends Plant Sci.,* 1996, **1**, 21.
2 A.V. Ruban *et al.*, *Nature*, 2007, **450**, 575.
3 J. Minagawa, *Biochim. Biophys. Acta Bioenerg.,* 2011, **1807**, 897.

Yuan-Chung Cheng replied: I think this is a fair and important question, but at this point, I do not think we have an answer. The phenomena related to NPQ occur in a bigger length scale and timescale, and we will need to, at the minimum, build a model of the PSII supercomplex in order to have any data meaningful for NPQ. We are currently working on that.

Amro Dodin asked: Has this study looked at differences in energy transfer pathways for different realizations of the system? In particular, are these dominant pathways (*e.g.* in Fig 11 of DOI: 10.1039/c9fd00006b) conserved for different realizations of static disorder?

Yuan-Chung Cheng answered: This is an excellent question. In the present work, we have focused on the time-averaged Hamiltonian that should give a better overall picture of the energy transfer pathways. Nevertheless, the MD simulation results do provide us the opportunity to investigate energy transfer pathways for different realizations of the PSII-cc system. This further analysis is currently in progress and will be published in a separate paper. Generally speaking, the dominant pathways presented in the 7-cluster model are very conservative (robust) against static disorders, but the magnitude of the rates do vary significantly across different realizations. I want to point out that the data presented in Fig. 11 in the current work (DOI: 10.1039/c9fd00006b) are already indicative of this. The two monomeric subunits (M1 and M2) are symmetric, but the rates belonging to the same pathways on different subunits shown in Fig. 11 of the paper can be significantly different, meaning the static disorders do affect the magnitudes of the rates.

Yumin Lee said: Thank you for the great presentation and work. I just wonder why does this work exclude the ultrafast relaxation component with a sub-ps ultrafast timescale which was previously observed in 2D-ES measurement. Do you think this component is not related to this work?

Yuan-Chung Cheng responded: I think they are related, but at this point, we intend to first focus on the general behavior of the system by studying the time-averaged Hamiltonian. Note that we do explicitly include the spectral density representing the averaged dynamical fluctuations obtained from the MD data in our dynamics and spectral calculations. There a bath correlation of $\sim$100 fs in the dynamical fluctuations is observed. So I do believe some of the relaxation dynamics observed in 2DES is described in our simulations. Our MD data also have the influence of structural relaxations and high-frequency vibrational modes, which we will include in our dynamical simulations in future work. We also plan to simulate 2DES spectra and have a more solid comparison with the experimental observations.

Naomi Ginsberg asked: Although I appreciate that it is generally expected based on bulk measurements that an exciton may make multiple 'trips' to a reaction centre prior to quenching, could you explain the logic behind the statement that the rate of quenching and concentration of quenchers (reaction centres) can simply trade off because they are multiplied together and the exciton diffusion length is large?

Graham Fleming responded: First, the Photosystem II antenna is highly connected as shown by the classic experiments of Joliot and Joliot[1] in which the photochemical yield was measured as a function of the fraction of open reaction centers. The dependence is hyperbolic - in other words the yield is greater than 50% of the value when all RCs are open when 50% of the RCs are closed. This can only arise from a high degree of connectivity. Our model[2] reproduces this curve quantitatively, suggesting we are capturing the connectivity adequately. Our model is sufficiently detailed to enable us to see what is the longest time and excitation spends on a specific chlorophyll on LHCII. The median dwell time on Chl 610-612 is 2.8 ps. As long as the intrinsic quenching time is longer than 10 ps a 2D Diffusion picture holds and for quenching times between 10 ps and 100 ps we find that a wide range of quenching times and quencher densities fit the experimental fluorescence decay curves. We further find that any combination of these two quantities that fit a given decay curve at a specific degree of quenching produce the same diffusion length for the excitation. In other words the single quantity that is controlled is the diffusion length. This arises because the quenching is weak (multiple visits before quenching) combined with the very large diffusion length in the thylakoid membrane (50 nm for dark adapted thylakoids). If the quenching rate were much faster than the inverse of 10 ps then this picture would not hold as well.

1 A. Juliot and P. Juliot, *Comptes Rendus*, 1964, **258**, 4622–4625.
2 D. I. G. Bennett *et al. Proc. Natl. Acad. Sci. U. S. A.*, 2018 **115**, E9523–E9531.

Alexandra Olaya-Castro commented: The complexity of simulating with atomistic level the full photosynthetic membrane is too great and probably unnecessary. Can we identify a minimal functional unit that will suffice to understand the energy transfer and conversion in photosynthetic membrane? For instance, in ref. 1 we investigate the performance of a minimal membrane subarea containing the fewest possible LH1s, such that both the LH2:LH1 ratio and the typical local core environment observed in the larger membrane are maintained. Remarkably, the energy transfer performance (quantum yield, excitation life-time, fluorescence yield) of this minimal subunit compare very well with the energy transfer performance in the full membrane. Will this be a relevant unit to investigate at the atomic level? What insight are we expecting to gain form such coarse-grained description?

1 F. Fassioli *et al.*, *Biophys Journal*, 2009, **97**, P2464–P2473.

Andrew Marcus opened the discussion of the paper by Graham Fleming: My question pertains to the central line slope analysis of your 2D electronic vibrational spectra of LHCII. You mentioned that the long-lived correlation that you observe between an initially excited electronic coherence and a subsequently excited vibrational coherence defies expectations due to the rapid energy and population transfer in LHCII. Does this long-lived correlation imply that there is a quantum coherent pathway that connects the initial electronic coherence to the final vibrational coherence, or instead could this correlation be due to a pathway of classical incoherent processes that are, nevertheless, statistically meaningful?

Graham Fleming answered: I believe that the reciprocal nature of the Center Line Slopes (CLS) in donor acceptor systems arises in part from the mixing of the levels, but as you suggest there may also be a contribution coming from a bias in energy transfer rates within the distributions of the two states. It is true, in our model calculations, that the magnitude of the CLS is strongly influenced by the magnitude of the reorganization energy.

Tom Oliver said: In your calculated two-dimensional electronic-vibrational spectra, the centre line slope appears to oscillate as a function of the waiting time- what does this physically correspond to?

Graham Fleming replied: The oscillation frequency is at the energy gap of the heterodimer v=0 levels.

Tom Oliver said: In the 2DEV experiments do you see any evidence for anharmonic coupling between the low-frequency wavepackets initiated with your short visible pump pulses and the high frequency vibrational modes probed in the mid-infrared region?

Graham Fleming answered: We have not yet modeled spectra with this type of coupling in detail but preliminary calculations in Nick Lewis's thesis suggest such coupling distorts the line shape of the higher frequency modes in the 2DEV spectrum

Jeffrey Cina queried: In her poster, my student Alexis Kiessling demonstrates that at least in the case of two-dimensional electronic spectroscopy, it can be helpful to plot the data as time-*versus*-time, rather than frequency-*versus*-frequency. This approach helps to expose the underlying nuclear wave-packet dynamics. Has your group considered using time-*versus*-time plots for your two-dimensional electronic-vibrational interferograms?

Graham Fleming responded: This is a good idea. Thanks for the suggestion.

Jennifer Ogilvie said: You mentioned in your paper that partial motivation for developing the 2DEV approach was to gain insight into the role of mixed electronic-vibrational states in energy transfer and charge separation. Many of the vibrational modes that have been suggested to have possible functional importance are in the THz regime. Do you envision extending the method to employ a THz probe? In addition, what specific signatures in 2DEV spectra do you think will be particularly illuminating for addressing the possible functional role of mixed electronic-vibrational states in photosynthesis?

Graham Fleming answered: Because the phase differences are only imposed on the visible pulses, provided appropriate sources and detectors are available, the 2DEV method could be used in the THz region. Having an effective way to distinguish vibronic coherences from ultrafast population transfer peaks will likely be important in unraveling the complex question of vibronic *vs.* electronic *vs.* vibration coherences.

Jahan Dawlaty asked: Can you comment on the similarities and differences between the 2D-EV and femtosecond stimulated Raman scattering (FSRS)? What type of information can be obtained from 2D-EV that can not be retrieved from FSRS?

Graham Fleming responded: This is an interesting question but not one I have considered in any detail. I suspect some of the time resolution issues in FSRS are less of a problem in 2DEV. We are currently working on a detailed comparison of 2DES and 2DEV.

Alexandra Olaya-Castro remarked: Two-dimensional (2D) electronic as well as 2D electronic-vibrational spectroscopy have allowed us to gain a wealth of information on the ultrafast excitation dynamics in light-harvesting systems but since the signals are congested their interpretation is not entirely clear. Which other techniques do you think will help us gain complementary insight into the quantum dynamics of complex molecular systems?

Graham Fleming replied: Experimental and complementary theoretical studies of the photon statistics from light harvesting complexes have the potential to open new windows into the dynamics. It may be possible to determine if an incoherently excited system evolves coherently, whether a complex system can generate its own coherent dynamics, and what the difference (if any) is between microscopic and macroscopic quantum yields. It has also been suggested by Mukamel and coworkers[1] that combined high time and spectral resolution may be achieved in congested spectra with rapid relaxation rates through the use of correlations between entangled photon pairs, perhaps used in conjunction with "classical" ultrashort light pulses.

And a second area, which may also involve photon statistics, is the use of micro cavities to study polariton spectra and dynamics.

1 K. E. Dorfman, F. Schlawin and S. Mukamel, *Rev. Mod. Phys.* 2016, **88**, 045008.

Animesh Datta commented: It seems that temporal correlations in the emitted light in the form of g^2 measurements may be the way to resolving quantumness in EET. Do we have any estimate of how much statistics this would take, given that we are seeking two-photon coincidences? And how does that compare to nonlinear spectroscopy statistics?

Graham Fleming responded: This is an important experimental question to which we do not yet have a complete answer.

Yin Song asked: In 2DES, rephasing and non-rephasing spectra are often used to separate signals from different pathways, *e.g.* differentiating electronic and vibrational coherences in complex systems. Is there a similar protocol that can be applied to 2DEV to shed light on the interplay of electronic and vibrational states and corresponding dynamics?

Graham Fleming responded: Because of the frequency difference between the infrared and visible pulses the rephasing effect in 2DEV spectra is not large, see ref. 1.

1 H. Dong *et al.*, *J. Chem. Phys.*, 2015, **142**, 17429.

Karen Morenz remarked: You seemed to be concerned that when doing a second order correlation experiment with a pulsed laser, you would be limited in time resolution by the detector deadtime. The way that people usually get around this is by using a beam splitter and two detectors; this wastes 50% of the data when both coincident photons go to the same detector within the detector deadtime, but the other half of the time they go to two different detectors, and so you can still build up statistics over time, and you are not limited by the detector deadtime. Perhaps I didn't quite grasp the concern with time resolution.

Graham Fleming answered: There is an issue here to do with whether one is thinking of single molecule or ensemble measurements. If the latter I suspect that the g(2) needs to be measured one photon at a time (with an entangled pair source) so that the sample only ever has one photon in it at once. This then means one would like to run the experiment at the maximum possible repetition rate. This was the point I was making.

Jennifer Ogilvie asked: Can you comment on the potential of your approach in other frequency regimes? For example, 2D UV excitation, combined with vibrational probing could provide an exciting new probe of protein and nucleic acid structure and dynamics.

Graham Fleming replied: Yes there is great potential for the method in protein and nucleic acid studies.

Naomi Ginsberg addressed Graham Fleming and Alexandra Olaya-Castro :How does one go about 'scaling up' from a model dimer system in an experiment that employs pair of correlated photons as a probe?

Alexandra Olaya-Castro replied: I find the question unclear but I can make the following comment. The possibility of probing ultrafast dynamics with quantum light (entangled photons) was mentioned. Given that entangled photons offer different frequency and time resolutions, this can provide an advantage for isolating excited state features in two-dimensional spectroscopy, for instance. Systematic studies showing the specific advantaged offered are needed.

Graham Fleming responded: Multiple systems have be discussed in the context of cavities.[1,2]

1 J. Yuen-Zhou and V. M. Menon, *Proc. Natl. Acad. Sci. U. S. A.*, 2019, **116**, 5214.
2 Z. Zhang *et al. J. Chem. Phys.*, 2018, **148**, 074302.

Conflicts of interest

There are no conflicts to declare.

Faraday Discussions

PAPER

Systematic control of the rate of singlet fission within 6,13-diphenylpentacene aggregates with PbS quantum dot templates†

Chen Wang, [a] Mohamad S. Kodaimati,[b] Shichen Lian[b] and Emily A. Weiss *[b]

Received 24th October 2018, Accepted 20th December 2018
DOI: 10.1039/c8fd00157j

Lead chalcogenide quantum dots (QDs) are promising acceptors for photovoltaic devices that harness the singlet fission (SF) mechanism. The rate of singlet fission of polyacenes in the presence of QDs is a critical parameter in determining the performance of such devices. The present study demonstrates that the rates of SF in a pentacene derivative, 6,13-diphenylanthracene (DPP), are modulated by forming coaggregates with PbS QDs in aqueous dispersions. PbS QDs generally accelerate SF within DPP aggregates, and the extent of acceleration depends on the size of the QD. The average rate of SF increases from 0.074 ps^{-1} for DPP-only aggregates to 0.37 ps^{-1} within DPP-D co-aggregates for QDs with radius 2.2 nm, whereas co-aggregation with the smallest (r = 1.6 nm) and largest (r = 2.7 nm) QDs we tried only slightly change the SF rate. The rate variation is associated with (i) the density of surface ligands, which is influenced by the faceting of the PbS surface, and (ii) the local dielectric constant for DPP. To accelerate SF, the ligands should be dense enough to provide sufficient affinity for DPP aggregates and effectively perturb the perpendicular alignment of DPP monomers within aggregates to increase the intermolecular coupling that promotes SF, but should not be too dense so as to form a low dielectric environment that disfavors SF. The study suggests that it is critical to consider the influence of the microenvironment of the QD surface on photophysical processes when fabricating QD/organic hybrid devices.

Introduction

The paper describes the control of the rate of singlet fission (SF) within aggregates of 5,13-diphenylpentacene (DPP) adsorbed to the surfaces of colloidal PbS

[a]*Department of Chemistry and Biochemistry, City University of New York, Queens College, 65-30 Kissena Blvd., Flushing, NY, 11367, USA*

[b]*Department of Chemistry, Northwestern University, 2145 Sheridan Rd., Evanston, IL 60208-3113, USA. E-mail: e-weiss@northwestern.edu*

† Electronic supplementary information (ESI) available. See DOI: 10.1039/c8fd00157j

quantum dots (QDs) that are also capped with poly(ethylene glycol)thiolate (PEGT), in water. Singlet fission is a photophysical process that produces two triplet excited states from one singlet excited state. The SF mechanism is a possible route to increasing the efficiency of organic photovoltaic devices beyond the Shockley–Queisser limit.[1–4] Polyacene materials are promising candidates for devices that employ the singlet fission mechanism,[5,6] and the dynamics of singlet fission of polyacenes and their derivatives have been studied in detail, both in isolated aggregates[7–19] and aggregates adsorbed to gold nanoparticles.[20] The rate of the SF process is a crucial parameter for designing photovoltaic devices utilizing this mechanism; this rate should be maximized, not only to ensure the efficient generation of triplet excitons, but also to outcompete direct charge separation from the singlet excited state.[21,22] It is known that the alignment of SF chromophores within the SF material determines their intermolecular coupling, which plays a crucial role in determining the rate of formation of the SF precursor state, a correlated triplet pair $^1(T_1T_1)$, and in determining the rate of dissociation of the two triplet excitons.[5,21,23–26] The dielectric environment of the organic aggregate can also affect the SF rate,[27] because SF is mediated by a charge transfer intermediate.[28,29] Systematic control of the SF rate on the surface of a particle like a colloidal QD, where the properties of the QD scaffold are controllable through both size and surface chemistry, provides insight into the structural parameters that influence SF, and the efficiency of subsequent transfer of triplet excitons[30,31] or photoelectrons[32,33] to a QD, which could serve as a complementary active material in a solar cell.

This work demonstrates that the surface of a QD and it solubilizing ligands serve as a template for the assembly of DPP, as it has for other organic dye molecules.[34–38] Here, we show that this scaffolding effect influences the dynamics of singlet fission within adsorbed DPP aggregates, to a degree that depends on the size of the QD. We prepared PbS QD/DPP co-aggregates, with QDs of six different radii, in aqueous solution, and characterized the sizes and shapes of the co-aggregates with dynamic light scattering (DLS), and transmission electron microscopy (TEM). We evaluated the rate of the SF process of DPP in these co-aggregates using transient absorption (TA) spectroscopy, and found that, for most sizes of QDs, co-aggregation of DPP with QDs accelerates singlet fission. In particular, when DPP is co-aggregated with QDs with a radius of 2.2 nm, the singlet fission rate is 5-fold faster than that in the pure DPP aggregate, but co-aggregation with the very large (r = 2.7 nm) and very small (r = 1.6 nm) QDs does not substantially affect the SF rate.

Experimental

Preparation of poly(ethylene glycol)thiolate-(PEGT) capped PbS quantum dots

We synthesized PbS QDs capped with oleate using a protocol adapted from Hines *et al.*[39] (see the ESI†). To prepare water-soluble PbS QDs, we exchanged the oleate ligands for PEGT (MW = 1000, Sigma-Aldrich). We dissolved oleate-capped QDs in 0.3 ml $CHCl_3$ to form a 50 μM solution. PEGT ligands were dissolved in methanol in a concentration of 100 mg ml^{-1}. Different amounts of PEGT were added to the QD dispersion for different radii of QDs to ensure minimum but sufficient amounts of PEGT in the ligand exchange process; specifically: for r = 1.6 nm, 250 eq.; r = 1.8 nm, 300 eq.; r = 2.0 nm, 400 eq.; r = 2.2 nm, 500 eq.; r =

2.4 nm, 600 eq.; $r = 2.7$ nm, 800 eq. per QD. We allowed the QD/PEGT mixtures to react for at least 10 h, and precipitated the QDs by adding 6 ml of a mixture of $CHCl_3$/ethanol/hexanes (1 : 1 : 10, v/v/v) and centrifuging at 8000 rpm for 5 min. We washed the pellet with the same solvent mixture and recollected the QDs by centrifuging to remove extra PEGT ligands. The wash/centrifugation cycle was repeated twice. The final pellet was dried under N_2 flow, and redispersed in 3 ml of water. The sample solution was filtered using syringe filters with a pore size of ≤0.22 μm, diluted to 1 μM, and purged with $Ar_{(g)}$ for 10 min before being moved into an N_2-filled box.

Preparation of QD–DPP assemblies

We prepared the QD–DPP assemblies in an N_2-filled box. We dissolved DPP in acetonitrile to form a nearly saturated solution (~0.5 mM), and added 0.2 ml of DPP solution to 1.8 ml of 1 μM QDs $_{(aq)}$ while stirring vigorously. The samples were sonicated for 1 hour before use.

Transient absorption spectroscopy

Details of the TA setup can be found elsewhere.[40] Briefly, fs-TA experiments were conducted using a commercial system (Ultrafast Systems LLC, Helios) powered by a Ti:sapphire amplifier (Spectra-Physics, Solstice). The 800 nm, 2.5 mJ, 100 fs pulse generated by the amplifier was split into two to generate pump and probe beams. The pump beam was directed through an optical parametric amplifier (Light Conversion, TOPAS-C) with sum-frequency generation to produce pump pulses at wavelengths of 600 and 970 nm. The probe beam containing a small portion of the amplifier output was directed through a 3 mm thick sapphire window to generate a white light continuum probe in the spectral region. The pump and the probe were focused and overlapped at the sample, which was contained in a 2 mm quartz cuvette. During data collection, all samples were stirred continuously. The instrument response function (IRF) of the TA setup was determined by measuring the full-width-at-half-maximum of the optical Kerr effect response of the acetonitrile/water (10%, v/v) solution in the same quartz cuvette. The IRF was 200 fs for the TA experiment.

Results and discussion

Fig. 1A shows that mixing DPP with PbS QDs does not affect the electronic transitions of the QDs. The first excitonic band of the QDs, at 1220 nm, does not shift or broaden after mixing with DPP. The enhancement of the absorption at wavelengths below 500 nm is primarily due to the absorption of DPP but also has contribution from scattering of the aggregates discussed below.

Fig. 1B shows the spectra of disaggregated DPP ("DPP Mon"), aggregated DPP with no QDs ("DPP Agg"), and DPP in mixtures with QDs ($r = 2.2$ nm) ("QD + DPP"), after the spectra of the QDs have been subtracted (see the ESI†). The formation of aggregates of DPP results in a 530 cm^{-1} bathochromic shift of its absorption spectrum. The shift is primarily due to dispersive interactions of the excitons in DPP aggregates, as reported for other molecular aggregate systems.[41] The shift within DPP-only aggregates is not however attributable to J-aggregation because: (1) there is no significant change in the vibronic progression upon the

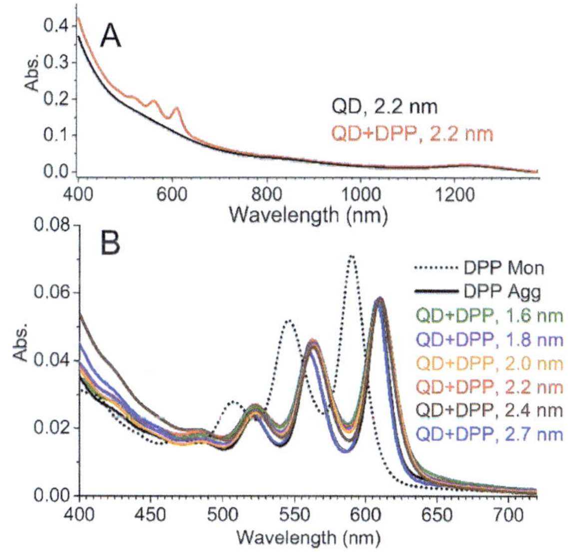

Fig. 1 (A) Ground state absorption spectra of PbS QDs with r = 2.2 nm and their assemblies with DPP in 10% (v/v) acetonitrile/water. (B) Absorption spectra of the DPP monomer in acetonitrile, DPP aggregates in water, and DPP in the presence of PbS QDs of six different radii. The spectra of DPP within the "QD–DPP" samples were obtained by subtracting the absorption spectra of the PbS QDs from the co-aggregate spectra, see the ESI.†

formation of the aggregate – that is, in the aggregate spectrum, the relative intensities of the (0–0) (~610 nm) and (0–1) (~560 nm) transitions are very similar to those of the ~590 and ~550 nm peaks in the monomer spectrum[42] – and (2) crystallography data of DPP samples with similar spectral features show that adjacent DPP molecules are aligned perpendicularly, which prevents exciton coupling between their transition dipoles.[21]

There are subtle differences among the absorption spectra of the QD–DPP samples depending on the radius of the QDs. The spectrum for the sample with the largest QDs ($r = 2.7$ nm) overlays with that of the DPP aggregate without QDs. For QDs with radii of 1.6 nm, 1.8 nm, 2.0 nm, and 2.2 nm, in the QD–DPP co-aggregate spectra there is a 120-to-160 cm^{-1} broadening of the 0–0 vibronic band, and a ~100 cm^{-1} bathochromic shift of the 0–1 vibronic band, compared to that of the pure DPP aggregate. The spectrum of the mixture of DPP with r = 2.4 nm QDs appears to be intermediate between those of the undisturbed DPP aggregate and the QD–DPP co-aggregate.

These perturbations of the DPP aggregate spectra by the QDs is most likely due to interaction between DPP and the PEGT ligands of the QDs, which can disturb the perpendicular packing of DPP molecules, and thereby induce weak exciton coupling within the DPP aggregates. Such coupling would cause a bathochromic shift and heterogeneous broadening of the absorption peaks. The $r = 2.7$ nm QDs do not appear to cause such a perturbation.

We supported the aforementioned hypothesis with transmission electron micrographs (TEM) of the QDs and the QD–DPP mixtures (Fig. 2) with four sizes of

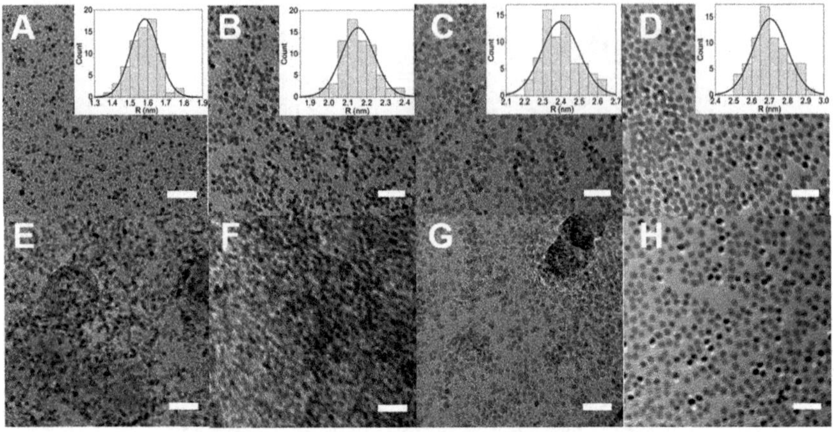

Fig. 2 TEM images of QDs of four different radii (A–D) and their assemblies with DPP (E–H). The radii of PbS QDs, determined by fitting the sizing histograms for 70 particles with Gaussian functions (insets), are 1.6 ± 0.2 nm (A and E), 2.2 ± 0.2 nm (B and F), 2.4 ± 0.2 nm (C and G) and 2.7 ± 0.2 nm (D and H). Scale bars in all images indicate 20 nm.

QDs. When there is no DPP present, the QDs are well-dispersed, or only form small clusters, on the TEM grids (Fig. 2A–D). When the 1.6 nm, 2.2 nm, and 2.4 nm QDs are mixed with DPP in solution and deposited on the TEM grid, however, we observe large QD agglomerates with blurry profiles for individual nanoparticles (Fig. 2E–G). Although we cannot capture images of the DPP assemblies because of the low contrast of the organic material in comparison with the QDs, these images indicate the formation of co-aggregated DPP and QDs, which decreases the contrast and focus of the image. The sizes of the QD–DPP aggregates are irregular, varying from 30 to 80 nm in diameter, which agrees with the results of DLS measurements (see the ESI†). The only exception is the largest QDs ($r = 2.7$ nm), which do not form co-aggregates with DPP (Fig. 2H). Given that the absorption spectrum of the QD (2.7 nm)-DPP sample is the only one that overlays with that of the DPP (aggregate)-only sample, while the absorption spectra of the QD (1.6 nm, 2.2 nm, and 2.4 nm)-DPP assemblies are shifted and broadened, we can confirm that these spectral changes are due to co-aggregation of the QDs and DPP.

The transient absorption spectra of DPP-only have three main features (Fig. 3A and B): (i) a relatively broad absorption signal at 460 nm corresponding to the S_1 state formed directly after photoexcitation at 610 nm, which has a lifetime of 13.4 ps; (ii) two narrow absorption bands peaked at 490 and 523 nm that grow in as the S_1 signal decays (with the same rate) corresponding to the T_1 state; and (iii) weak ground state bleach (GSB) signals at 565 and 610 nm, which appear within the IRF of the laser, and undergo a second growth with a time constant of 14 ps. The coordination of the decay of the S_1 (13–14 ps)$^{-1}$ signal with the generation of the T_1 signal and the second growth of the GSB signals unambiguously suggest a singlet fission mechanism, which agrees with previous reports on thin films of DPP.[21] Additionally, the 523 nm peak for T_1 has an additional growth process, with a time constant of 150 ps, after the S_1 signal has decayed completely (the 490 nm T_1 signal overlaps strongly with the S_1 signal such that it is difficult to

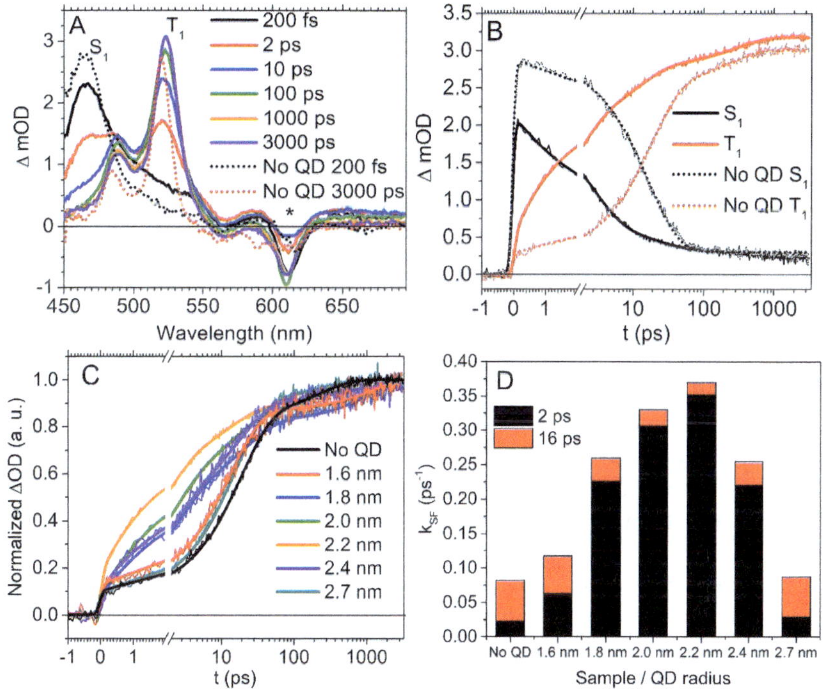

Fig. 3 (A) Selected fs-TA spectra of the QD–DPP ($r = 2.2$ nm) assembly at a series of time delays after excitation at 610 nm showing the decay of S_1 of DPP and the rise of T_1 of DPP. The asterisk indicates the spectral region affected by pump scatter. The dotted lines are the representative spectra of the S_1 and T_1 states of DPP-agg without QDs. (B) The dynamics of SF of DPP with and without QDs. Solid lines: kinetic traces extracted from the TA spectra in (A) at 467 nm (the S_1 absorption of DPP) and 523 nm (the T_1 absorption of DPP), after subtracting the photoinduced absorption (PIA) background from QDs, see the ESI.† Dotted lines: kinetics of DPP aggregates without QDs. The kinetic traces are fit with multi-exponential functions convolved with a Gaussian IRF. (C) Normalized kinetic traces extracted from the TA spectrum of DPP or the QD–DPP assemblies at 523 nm (the T_1 absorption of DPP). The legend indicates the radii of the QDs in the mixtures. The kinetic traces are fit with the convolution of multiple-exponential functions and a Gaussian IRF; two time constants (2 ps and 16 ps) that correspond to the SF process are shared over fits to all kinetic traces and their relative amplitudes are varied. (D) Amplitude-averaged SF rate constants of DPP only and QD–DPP assemblies with various QD radii. The colored segments indicate the relative contributions of 2 ps (black) and 16 ps (red) components of the SF process, obtained from the fits in (C).

track dynamics at 490 nm). Based on previous reports, we assign this rate constant to the dissociation of the triplet pair state, $^1(T_1T_1)$, after the initial singlet fission.[17,26,43–45] The complete SF mechanism can be written as eqn (1),

$$S_1 + S_0 \xleftrightarrow{\tau_1=2.7-13.4\ \text{ps}} {}^1(T_1T_1) \xleftrightarrow{\tau_2=150-310\ \text{ps}} T_1 + T_1 \tag{1}$$

where the rate constants shown are for the range of samples we examined with and without QDs (*vide infra*).

When we pump the QD (2.2 nm)/DPP sample at 610 nm within the TA experiment, both DPP and the PbS QDs are photoexcited. TA spectra at a series of delay times after photoexcitation are shown in Fig. 3A. Within the spectral region between 480 and 700 nm, PbS QDs have a broad, featureless TA background underneath the DPP signals. A bleach signal caused by the depletion of a higher exciton state appears between 470 and 450 nm, and overlaps with the S_1 signal of DPP (see the ESI†). By comparing the TA spectra of the mixtures with the TA spectra of DPP-only, we determined that, at wavelengths >640 nm, the spectrum only has contributions from the photo-induced absorption (PIA) of PbS QDs. Furthermore, when we excite the mixture at 970 nm, where only PbS QDs absorb, the dynamics of the PIA are the same as if we excited the sample at 610 nm. We can therefore remove the QD PIA background from the raw spectra of QD/DPP by subtracting off from these spectra the spectra collected on the same sample with 970 nm excitation, after appropriately scaling the PIA signals of QDs. A pair of isolated kinetic traces of the S_1 and T_1 states of DPP with QDs ($r = 2.2$ nm) obtained with this procedure are plotted in Fig. 3B (see the ESI† for details). We note that, with DPP present, the PIA signal of PbS QDs ($r = 2.2$ nm) decays with time-constants of 11 and 120 ps, primarily due to hole transfer from the valence band of PbS to DPP. Hole transfer from PbS QDs to pentacene derivatives has been reported previously.[32,46]

We did not observe either triplet-to-triplet energy transfer from QDs to DPP, or the formation of the triplet exciton of DPP following the hole transfer, when PbS QDs were excited within the co-aggregates. These processes are either too slow and therefore outside the temporal window of our TA experiment,[46] or require specific surface sites associated with ligand binding to occur, as suggested by Roberts and coauthors.[47] We also did not observe electron transfer from either the S_1 or T_1 states of DPP to the QDs. It is likely that the fast SF process out-competes electron transfer from S_1, as previously observed for gold nanoparticles,[20] and that the T_1 state of DPP does not have enough energy to donate an electron to the conduction band of the QDs.

Fig. 3A shows the TA spectrum of the QD–DPP ($r = 2.2$ nm) co-aggregate at a series of delay times after pumping at 610 nm. Fig. 3B shows kinetic traces extracted from this spectrum at the peak of the S_1 and T_1 features of DPP. The spectra and kinetics for the QD–DPP samples with QD of all other radii are in the ESI.† The S_1 TA feature of DPP is broader and undergoes a significantly faster decay when 2.2 nm QDs are present than in DPP-only samples (Fig. 3A and B). The growth of the T_1 signals accelerates commensurately. We globally fit the two kinetic traces in Fig. 3B by sharing two time constants between the two fits, allowing the amplitudes to float. We obtain $\tau_1 = 1.8$ ps, and $\tau_2 = 14$ ps. 57% of the amplitude of the triplet signal grows with the time constant τ_1, which corresponds to accelerated SF due to the formation of DPP/QD assemblies. Based on comparison with the DPP-only dynamics, τ_2 corresponds to the SF process in undisturbed DPP-agg (where the molecules are still perpendicularly aligned). The presence of two different SF rates within the same sample of DPP-aggregates implies at least two types of SF sites or local environments, which has been reported in rubrene assemblies.[15,19] A third kinetic component with $\tau_3 = 310$ ps is needed to fit the slow growth of the T_1 signal, and corresponds to the dissociation of T_1 excitons from the triplet pair state, $^1(T_1T_1)$.

Fig. 3C shows that the radius of the QDs influences the degree to which SF is accelerated within the DPP aggregates. We determine the relationship between the size of the QD and the rate of SF by fixing the two fastest time constants for growth of the T_1 signal to 2 ps and 16 ps (allowing their relative amplitudes to float), and using those two time constants to globally fit the T_1 kinetic traces for all samples. Fig. 3D shows the amplitude-averaged SF rate constants (also listed in Table 1) for samples made with each of six QD sizes, and the breakdown (with respect to amplitude) of the average rate constants into the two fixed components. The most prominent acceleration of SF occurs in the co-aggregates formed with $r = 2.2$ nm QDs, for which the overall averaged SF rate is 5-fold faster than that in DPP-only aggregates. The smallest (r = 1.6 nm) and largest QDs (2.7 nm) only marginally increase the rate of SF.

The comparison of ground state absorption, TEM and TA data indicates that the acceleration of SF in DPP is related to the formation of DPP/QD co-aggregates and subsequent disturbance of the perpendicular packing of DPP molecules. The undisturbed DPP crystal structure enables minimal intermolecular coupling between the SF precursor and the S_1 excited state.[21] Any displacement of DPP molecules from this perpendicular alignment will yield a more favorable overlap between HOMO and LUMO among different DPP molecules, and thereby accelerate SF.[2,25,48]

We believe the SF behavior of DPP is not strongly affected by the largest QDs because, for large PbS QDs, the dominant facet of the nanocrystal is (100), which has a low density of surface ligands and therefore a low affinity for DPP.[49] We used quantitative NMR to confirm that, indeed, the 2.7 nm QDs have the smallest average coverage of PEGT (see Table 1 and the ESI†). However, affinity for DPP cannot be the only factor in the size-dependence of this effect because (i) the smallest QDs ($r = 1.6$ nm) with the highest surface ligand density (3.6 nm^{-2}) also form co-aggregates with DPP with minimal effect on SF; and (ii) according to the ground state absorption spectra, the strength of the intermolecular coupling within the co-aggregates of DPP and QDs with $r \leq 2.2$ nm is nearly constant, but the SF rates vary by a factor of three for this set of samples.

We therefore propose that the QDs' ability to modulate the rate of SF also depends on the dielectric environment at the surface of the QD. It is known that the rate of SF, which can be regarded as a double electron transfer process, is accelerated in a more polar environment where charge-separated intermediates are stabilized.[28,29] For the smallest QDs, the predominant (111) facets densely covered with surface ligands (Table 1) favor the formation of co-aggregates, but also surround the DPP with PEG, which has a low dielectric constant ($\varepsilon = \sim 10$) compared to water ($\varepsilon = 80.4$). The effect on SF from the enhancement of intermolecular coupling is therefore counteracted by the decrease in local dielectric constant. Additional evidence that as the QDs get smaller, the DPP is increasingly separated from the QD by PEG is in the comparison of hole transfer rates from the QD to DPP. The rate of this charge transfer process strongly depends on the distance between the DPP and the surface of the QDs. The smallest ($r = 1.6$ nm)

Table 1 NMR-measured surface density of PEGT on PbS QDs of different sizes, and corresponding SF rates within QD–DPP co-aggregates

QD radius (nm)	2.7	2.4	2.2	2.0	1.8	1.6
Density (PEG per nm^2)	1.9	2.2	2.5	2.9	3.3	3.6
SF rates (ps^{-1})	0.088	0.25	0.37	0.33	0.26	0.12

QDs have a slower rate of hole transfer than the $r = 2.2$ nm QDs, despite a greater driving force for the process (see the ESI†). This result implies that DPP is closer and more electronically coupled to the 2.2 nm QD than to the 1.6 nm QD. PbS has a significantly higher dielectric constant ($\varepsilon = 169$) than water and can better stabilize the charge transfer-like SF precursor. We therefore believe that the 2.2 nm QDs have the optimal surface density of PEG to have sufficient affinity for DPP to disturb its aggregate structure, and a high enough local dielectric constant to stabilize the SF precursor. These two factors contribute to a factor-of-five increase in the rate of SF.

The local electric field of the photoinduced charge separated state, $QD^{\bullet-}|DPP^{+}$, formed by hole transfer, could, in principle, affect the SF rate by stabilizing the SF intermediate. We do not believe this effect is significant because: (i) this type of influence would only occur when a $QD^{\bullet-}|DPP^{+}$ state appears close to the SF site, which is very low probability considering our low photon flux (corresponding to <15% of PbS QDs photoexcited per pulse); (ii) the hole transfer process observed in QD/DPP co-aggregates is on a similar or slower time scale (11 and 120 ps) than SF; and (3) charge-separated states would also accelerate the decay of proximate triplet excitons in DPP (a known problem for photovoltaic devices trying to harness SF). We do not however observe faster decay of the T_1 state of DPP within DPP/QD co-aggregates than with DPP only.

Conclusions

We demonstrate systematic control of the SF rate within DPP aggregates by adsorbing those aggregates on PbS QDs of different sizes. The rate of SF can be accelerated by a factor of five using QDs with the optimal size, $r = 2.2$ nm. The acceleration of SF in DPP-agg can be attributed to two factors: (i) the disturbance of the perpendicular molecular packing of DPP, which enhances the intermolecular coupling; and (ii) the high local dielectric environment provided by PbS, which stabilizes the SF precursor. Our work suggests that careful and systematic manipulation of the QD surface environment can affect the excitonic behaviour of the aggregates of adsorbed organic molecules. This work informs strategies for designing SF-promoted photovoltaic and optoelectronic devices based on PbS (or other types of) QDs and polyacene derivatives.

Conflicts of interest

There are no conflicts to declare.

Acknowledgements

This material is primarily based upon work supported by the Air Force Office of Scientific Research, under AFOSR award No. FA-9550-17-1-0271 (synthesis of QD–DPP complexes, stead-state and time-resolved optical measurements), as part of the Center for Bio-Inspired Energy Science, an Energy Frontier Research Center funded by the U.S. Department of Energy (DOE), Office of Science, Basic Energy Sciences (BES), under Award # DE-SC0000989 (TEM measurements), and by the National Science Foundation under Award # CHE-1664184 (NMR measurements).

References

1 M. C. Hanna and A. J. Nozik, *J. Appl. Phys.*, 2006, **100**(7), 0745101–0745108.

2 M. B. Smith and J. Michl, *Chem. Rev.*, 2010, **110**(11), 6891–6936.

3 M. J. Y. Tayebjee, A. A. Gray-Weale and T. W. Schmidt, *J. Phys. Chem. Lett.*, 2012, **3**(19), 2749–2754.

4 C. A. Nelson, N. R. Monahan and X. Y. Zhu, *Energy Environ. Sci.*, 2013, **6**(12), 3508–3519.

5 E. C. Greyson, B. R. Stepp, X. Chen, A. F. Schwerin, I. Paci, M. B. Smith, A. Akdag, J. C. Johnson, A. J. Nozik, J. Michl and M. A. Ratner, *J. Phys. Chem. B*, 2009, **114**(45), 14223–14232.

6 I. Paci, J. C. Johnson, X. D. Chen, G. Rana, D. Popovic, D. E. David, A. J. Nozik, M. A. Ratner and J. Michl, *J. Am. Chem. Soc.*, 2006, **128**(51), 16546–16553.

7 P. M. Zimmerman, Z. Zhiyong and C. B. Musgrave, *Nat. Chem.*, 2010, **2**(8), 24–28.

8 W. L. Chan, M. Ligges, A. Jailaubekov, L. Kaake, L. Miaja-Avila and X. Y. Zhu, *Science*, 2011, **334**(6062), 1541–1545.

9 M. W. B. Wilson, A. Rao, J. Clark, R. S. S. Kumar, D. Brida, G. Cerullo and R. H. Friend, *J. Am. Chem. Soc.*, 2011, **133**(31), 11830–11833.

10 P. M. Zimmerman, F. Bell, D. Casanova and M. Head-Gordon, *J. Am. Chem. Soc.*, 2011, **133**(49), 19944–19952.

11 J. J. Burdett and C. J. Bardeen, *J. Am. Chem. Soc.*, 2012, **134**(20), 8597–8607.

12 N. R. Monahan, D. Sun, H. Tamura, K. W. Williams, B. Xu, Y. Zhong, B. Kumar, C. Nuckolls, A. R. Harutyunyan, G. Chen, H.-L. Dai, D. Beljonne, Y. Rao and X. Y. Zhu, *Nat. Chem.*, 2016, **9**, 341.

13 L. Ma, K. K. Zhang, C. Kloc, H. D. Sun, M. E. Michel-Beyerle and G. G. Gurzadyan, *Phys. Chem. Chem. Phys.*, 2013, **14**(23), 8307–8312.

14 J. J. Burdett and C. J. Bardeen, *Acc. Chem. Res.*, 2013, **46**(6), 1312–1320.

15 S. T. Roberts, R. E. McAnally, J. N. Mastron, D. H. Webber, M. T. Whited, R. L. Brutchey, M. E. Thompson and S. E. Bradforth, *J. Am. Chem. Soc.*, 2012, **134**(14), 6388–6400.

16 A. A. Bakulin, S. E. Morgan, T. B. Kehoe, M. W. B. Wilson, A. W. Chin, D. Zigmantas, D. Egorova and A. Rao, *Nat. Chem.*, 2016, **8**(1), 16–23.

17 B. S. Basel, J. Zirzlmeier, C. Hetzer, B. T. Phelan, M. D. Krzyaniak, S. R. Reddy, P. B. Coto, N. E. Horwitz, R. M. Young, F. J. White, F. Hampel, T. Clark, M. Thoss, R. R. Tykwinski, M. R. Wasielewski and D. M. Guldi, *Nat. Commun.*, 2017, **8**, 15171.

18 A. J. Musser, M. Liebel, C. Schnedermann, T. Wende, T. B. Kehoe, A. Rao and P. Kukura, *Nat. Phys.*, 2015, **11**(4), 352–357.

19 J. N. Mastron, S. T. Roberts, R. E. McAnally, M. E. Thompson and S. E. Bradforth, *J. Phys. Chem. B*, 2013, **117**(49), 15519–15526.

20 D. Kato, H. Sakai, N. V. Tkachenko and T. Hasobe, *Angew. Chem.*, 2016, **128**(17), 5316–5320.

21 S. R. Yost, J. Lee, M. W. B. Wilson, T. Wu, D. P. McMahon, R. R. Parkhurst, N. J. Thompson, D. N. Congreve, A. Rao, K. Johnson, M. Y. Sfeir, M. G. Bawendi, T. M. Swager, R. H. Friend, M. A. Baldo and T. Van Voorhis, *Nat. Chem.*, 2014, **6**(6), 492–497.

22 C. Ramanan, A. L. Smeigh, J. E. Anthony, T. J. Marks and M. R. Wasielewski, *J. Am. Chem. Soc.*, 2012, **134**(1), 386–397.

23 E. C. Greyson, J. Vura-Weis, J. Michl and M. A. Ratner, *J. Phys. Chem. B*, 2010, **114**(45), 14168–14177.

24 C. Wang, M. Angelella, C. H. Kuo and M. J. Tauber, *Proc. SPIE: Phys. Chem. Interf. Nanomater.*, vol. XI, 2012, pp. 8459051–8459113.

25 R. D. Pensack, A. Tilley, C. Grieco, G. Purdum, E. Ostroumov, D. B. Granger, D. Oblinksy, J. Dean, G. Doucette, J. Asbury, Y.-L. Loo, D. Seferos, J. E. Anthony and G. Scholes, *Chem. Sci.*, 2018, **9**(29), 6240–6259.

26 R. D. Pensack, C. Grieco, G. E. Purdum, S. M. Mazza, A. J. Tilley, E. E. Ostroumov, D. S. Seferos, Y.-L. Loo, J. B. Asbury, J. E. Anthony and G. D. Scholes, *Mater. Horiz.*, 2017, **4**(5), 915–923.

27 C. M. Mauck, P. E. Hartnett, Y. L. Wu, C. E. Miller, T. J. Marks and M. R. Wasielewski, *Chem. Mater.*, 2017, **29**(16), 6810–6817.

28 S. Hart, W. R. Silva and R. R. Frontiera, *Chem. Sci.*, 2018, **9**(5), 1242–1250.

29 B. S. Basel, J. Zirzlmeier, C. Hetzer, S. R. Reddy, B. T. Phelan, M. D. Krzyaniak, M. K. Volland, P. B. Coto, R. M. Young, T. Clark, M. Thoss, R. R. Tykwinski, M. R. Wasielewski and D. M. Guldi, *Chem*, 2018, **4**(5), 1092–1111.

30 N. J. Thompson, M. W. B. Wilson, D. N. Congreve, P. R. Brown, J. M. Scherer, T. S. Bischof, M. Wu, N. Geva, M. Welborn, T. V. Voorhis, V. Bulović, M. G. Bawendi and M. A. Baldo, *Nat. Mater.*, 2014, **13**(11), 1039–1043.

31 M. Tabachnyk, B. Ehrler, S. Gélinas, M. L. Böhm, B. J. Walker, K. P. Musselman, N. C. Greenham, R. H. Friend and A. Rao, *Nat. Mater.*, 2014, **13**(11), 1033–1038.

32 B. Ehrler, M. W. B. Wilson, A. Rao, R. H. Friend and N. C. Greenham, *Nano Lett.*, 2012, **12**(2), 1053–1057.

33 P. J. Jadhav, P. R. Brown, N. Thompson, B. Wunsch, A. Mohanty, S. R. Yost, E. Hontz, T. Van Voorhis, M. G. Bawendi, V. Bulovic and M. A. Baldo, *Adv. Mater.*, 2012, **24**(46), 6169–6174.

34 B. J. Walker, G. P. Nair, L. F. Marshall, V. Bulović and M. G. Bawendi, *J. Am. Chem. Soc.*, 2009, **131**(28), 9624–9625.

35 J. E. Halpert, J. R. Tischler, G. Nair, B. J. Walker, W. Liu, V. Bulovic and M. G. Bawendi, *J. Phys. Chem. A*, 2009, **113**(23), 9986–9992.

36 E. A. McArthur, J. M. Godbe, D. B. Tice and E. A. Weiss, *J. Phys. Chem. A*, 2012, **116**(10), 6136–6142.

37 C. Wang and E. A. Weiss, *J. Am. Chem. Soc.*, 2016, **138**(30), 9557–9564.

38 D. Savateeva, D. Melnikau, V. Lesnyak, N. Gaponik and Y. P. Rakovich, *J. Mater. Chem.*, 2012, **22**(21), 10816–10820.

39 M. A. Hines and G. D. Scholes, *Adv. Mater.*, 2003, **15**(21), 1844–1849.

40 E. A. McArthur, A. J. Morris-Cohen, K. E. Knowles and E. A. Weiss, *J. Phys. Chem. B*, 2010, **114**(45), 14514–14520.

41 F. C. Spano, *Acc. Chem. Res.*, 2010, **43**(3), 429–439.

42 J. Knoester and F. C. Spano, *Theory of pump-probe spectroscopy of molecular J-aggregates*, in Host Publication, World Scientific Publishing, 1996, pp. 111–160.

43 R. D. Pensack, E. E. Ostroumov, A. J. Tilley, S. Mazza, C. Grieco, K. J. Thorley, J. B. Asbury, D. S. Seferos, J. E. Anthony and G. D. Scholes, *J. Phys. Chem. Lett.*, 2016, **7**(13), 2370–2375.

44 R. D. Pensack, A. J. Tilley, S. R. Parkin, T. S. Lee, M. M. Payne, D. Gao, A. A. Jahnke, D. G. Oblinsky, P.-F. Li, J. E. Anthony, D. S. Seferos and G. D. Scholes, *J. Am. Chem. Soc.*, 2015, **137**(21), 6790–6803.
45 B. D. Folie, J. B. Haber, S. Refaely-Abramson, J. B. Neaton and N. S. Ginsberg, *J. Am. Chem. Soc.*, 2018, **140**(6), 2326–2335.
46 S. Garakyaraghi, C. Mongin, D. B. Granger, J. E. Anthony and F. N. Castellano, *J. Phys. Chem. Lett.*, 2017, 1458–1463.
47 J. A. Bender, E. K. Raulerson, X. Li, T. Goldzak, P. Xia, T. Van Voorhis, M. L. Tang and S. T. Roberts, *J. Am. Chem. Soc.*, 2018, **140**(24), 7543–7553.
48 M. J. Y. Tayebjee, K. N. Schwarz, R. W. MacQueen, M. Dvořák, A. W. C. Lam, K. P. Ghiggino, D. R. McCamey, T. W. Schmidt and G. J. Conibeer, *J. Phys. Chem. A*, 2016, **120**(1), 157–165.
49 H. Choi, J.-H. Ko, Y.-H. Kim and S. Jeong, *J. Am. Chem. Soc.*, 2013, **135**(14), 5278–5281.

Faraday Discussions

PAPER

Light harvesting and energy transfer in a porphyrin-based metal organic framework†

Shaunak M. Shaikh,[a] Arnab Chakraborty,[a] James Alatis,[a] Meng Cai,[a] Evgeny Danilov[b] and Amanda J. Morris*[a]

Received 20th November 2018, Accepted 18th December 2018
DOI: 10.1039/c8fd00194d

We present the synthesis and photophysical characterization of a water stable PCN-223(freebase) metal organic framework (MOF) constructed from *meso*-tetrakis(4-carboxyphenyl)porphyrin (TCPP). The photophysical properties of the synthesized crystalline material were studied using a wide range of steady-state and time-resolved spectroscopic techniques. Quenching experiments performed on TCPP and PCN-223 demonstrated that the extent and the rate of quenching in the MOF is significantly higher than the monomeric ligand. Based on these results, we propose that upon photo-excitation, the singlet excitation energy migrates across neutral TCPP linkers until it is quenched by a N-protonated TCPP linker. The N-protonated linkers act as trap states that deactivate the excited state to the ground state. Variable temperature measurements aided in understanding the mechanism of singlet–singlet energy transfer in the PCN-223 MOF. The rate of energy transfer and the total exciton hopping distance in PCN-223 were calculated in order to quantify the energy transfer characteristics of PCN-223. Nanosecond transient absorption spectroscopy was used to study the triplet excited state photophysics in both the free ligand and PCN-223 MOF. Furthermore, femtosecond transient absorption spectroscopy was employed to get a better understanding of the photophysical processes taking place in the ligand and MOF on ultrafast timescales. Efficient energy transfer (Förster radius = 54.5 Å) accompanied with long distance exciton hopping (173 Å) was obtained for the PCN-223 MOF.

Introduction

One of the most important components of artificial photosynthesis assemblies are the antennae, which collect solar energy and direct it towards the reaction centers. A multi-chromophoric array with energy cascade can direct sequential

[a]*Department of Chemistry, Virginia Tech, Blacksburg, Virginia, 24060, USA. E-mail: ajmorris@vt.edu*
[b]*Department of Chemistry, North Carolina State University, Raleigh, North Carolina, 27695-8204, USA*

† Electronic supplementary information (ESI) available: Thermogravimetric analysis, nitrogen gas adsorption, spectral overlap calculation, steady state and nanosecond TA spectrum. See DOI: 10.1039/c8fd00194d

photoexcited energy flow and perform the function of light harvesting antenna assemblies.[1] Multiporphyrin arrays have been studied extensively with the aim of constructing antenna assemblies that mimic natural photosynthetic systems in terms of the efficiency of excitation energy transfer (EET). Subtle changes in the structural parameters, such as connectivity, distance, and orientation between porphyrin units in the array, can have strong implications on the nature of interchromophoric interactions, and the rates and efficiencies of EET.[2] Highlighted by highly-ordered crystal structures and synthetic tunability *via* crystal engineering, metal organic frameworks (MOFs) allow for precise control of distances and angles between chromophores and their alignment by judicious choice of ligands and metal nodes.[3–6] Porphyrin based MOFs are therefore ideal candidates to study EET as a function of structural parameters. Such studies will aid in the design of porphyrin-based architectures that are conducive for energy transfer. Before examining the role of MOF structure in EET, it is necessary to first understand the mechanistic aspects of EET and the factors that determine the efficiency of EET in porphyrin-based MOFs. To address these issues, we probed the energy transfer characteristics of PCN-223(fb), a Zr-MOF based on *meso*-tetrakis(4-carboxyphenyl)porphyrin (TCPP).

Zr-TCPP MOFs have been studied extensively due to their exceptional chemical stability under harsh experimental conditions and their ability to exhibit a variety of functionalities, like catalysis, light harvesting, gas-storage and sensing.[4,7–12] PCN-223, in particular, has a very unique structure (Fig. 1).[10] It consists of unprecedented D_{6h} symmetric $[Zr_6O_4(OH)_4]^{12+}$ nodes connected to 12 TCPP linkers, representing the first (4,12)-connected MOF with the "shp" topology. The closely-packed structure of PCN-223 supports a high density of chromophores that can simultaneously absorb light and participate in the energy transfer process. The closest distance between two porphyrin struts is 10.71 Å (see ESI, Fig. S7†). PCN-223 also manifests a small porphyrin–porphyrin torsional angle (~55°) that facilitates interchromophoric electronic coupling between TCPP units.[13] Based on these merits, PCN-223 qualifies to serve as a model system to explore the EET mechanism in porphyrin-based MOFs.

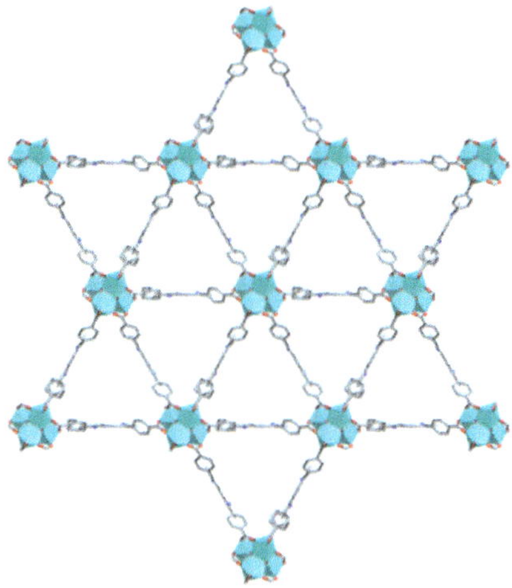

Fig. 1 View of PCN-223(fb) along the *c* axis with uniform triangular 1D channels.

Herein, we present synthesis, structural characterization and photophysical characterization of a PCN-223 MOF constructed from the free base TCPP ligand. The effects of pH and temperature on the excited state properties of PCN-223(fb) were investigated and compared with those of the ligand. The variable temperature measurements revealed the incoherent nature of excitation energy transfer (EET) in PCN-223. The excitation energy transfer was investigated by studying the pH dependence of photoluminescence (PL) quenching. The study revealed the presence of static quenching both in the ligand and in the MOF. Additionally, the study revealed that the PL quenching was greatly enhanced in the MOF as compared to the free ligand, and the efficiency of quenching increased from 51% in the ligand to 93% in the MOF. A Förster energy transfer model was used to estimate the rate of exciton hopping (k_{hop}) and the exciton hopping time in PCN-223. Nanosecond transient absorption spectroscopy was used to characterize the nature of the non-emissive triplet state. The study revealed the presence of a long-lived triplet state extending beyond 200 μs and having the characteristic signature of a TCPP based triplet state. Femtosecond transient absorption spectroscopy was employed to characterize the ultrafast processes taking place in TCPP and PCN-223. Kinetic analysis of the ultrafast data of TCPP and PCN-223 revealed the presence of three distinct time components that correspond to: (a) solvent-induced vibrational reorganization of excitation energy, (b) vibrational cooling, and (c) fluorescence. Our study revealed the presence of efficient excitation energy transfer and long-distance exciton hopping (100 Å (1D), 141 Å (2D), and 173 Å (3D)) in the PCN-223 MOF.

Experimental section

Materials

Meso-tetracarboxyphenylporphyrin (>97%) was purchased from Frontier Scientific and was used without further purification. Zirconium chloride (anhydrous, ≥99.5%) was purchased from Sigma-Aldrich. Dimethylformamide (DMF, spectrophotometric grade, ≥99.9%) was purchased from Fisher chemical. Propionic acid (PA, 99%) was purchased from Alfa Aesar.

Synthesis of PCN-223(fb)

10 mg of H_2TCPP (1.3×10^{-5} moles) and 7 mg of $ZrCl_4$ (3×10^{-5} moles) were added to 10 ml DMF and ultrasonically dissolved in a 6-dram vial. 2 ml propionic acid was added to the vial and the vial was sonicated for 15 minutes to get a homogeneous reaction mixture. The vial was placed in an oven set at 120 °C for 16 hours. After allowing them to cool down to room temperature, the resultant MOF powder was collected using centrifugation. It was washed 3 times with DMF and then soaked in ethanol for 3 days with fresh ethanol replacement every day. The MOFs were dried at room temperature and then activated by heating at 100 °C under vacuum.

Powder X-ray diffraction and scanning electron microscopy (PXRD)

A 600 W Rigaku MiniFlex powder diffractometer with a CuKα (0.15418 nm) radiation source was used, with a sweeping range of 2–25° in continuous scanning mode. PXRD traces were collected in 0.05° increments at a scanning rate of 0.2° min^{-1}.

Scanning electron microscopy (SEM)

SEM samples were prepared by suspending MOF powders in ethanol with sonication. The resulting suspensions were drop-cast on precut glass slides. After drying, the glass slides were mounted on SEM sample pegs with the help of double-sided copper tape. The sides of the glass slides and the platform of sample peg were coated with conductive carbon paint purchased from Electron Microscopy Sciences. A LEO (Zeiss) 1550 field-emission scanning electron microscope, equipped with an in-lens detector, operating at 5.0 kV, was used to obtain high-resolution images of the MOF particles.

Thermogravimetric analysis (TGA)

A Q-series thermogravimetric analyzer from TA Instruments was used to assess the thermal stability of the MOFs. Samples weighing ~3–5 mg were placed on a platinum pan and heated under air at a rate of 5 °C min^{-1} over the temperature range of 25–800 °C.

Gas sorption isotherms

The N_2 adsorption measurements were conducted using a Micromeritics 3Flex instrument. A 6 mm large bulb sample cell was used to hold the samples and was degassed under vacuum at a temperature of 100 °C for 24 h. The surface area of the MOFs was determined from the N_2 adsorption isotherms at 77 K by fitting the adsorption data within the 0.05–0.3 P/P_0 pressure range to the BET equation.

Diffuse absorption spectroscopy

The diffuse absorption spectra of TCPP and PCN-223(fb) were obtained using an Agilent Technologies 8453 UV-vis diode array spectrophotometer (1 nm resolution) where the sample compartment was replaced with an integration sphere. The powder samples were diluted by mixing with $BaSO_4$.

Steady-state emission spectroscopy and time-resolved emission lifetimes

The steady-state emission spectra were obtained using a QuantaMaster Model QM-200-4E emission spectrophotometer from Photon Technology, Inc. (PTI). The excitation light source was a 75 W Xe arc lamp (Newport). The detector was a thermoelectrically cooled Hamamatsu 1527 photomultiplier tube (PMT). Emission traces were analyzed using Origin 9.0. Time-resolved fluorescence lifetimes were obtained *via* the time-correlated single photon counting technique (TCSPC) with the same QuantaMaster Model QM-200-4E emission spectrophotometer from Photon Technology, Inc. (PTI) equipped with a 415 nm LED and a Becker & Hickl GmbH PMH-100 PMT detector with time resolution of <220 ps FWHM. Florescence lifetime decays were analyzed with the help of Origin 9.0.

Nanosecond transient absorption spectroscopy

Transient absorption difference spectra and kinetic traces were collected with a LP 980 laser flash photolysis system (Edinburgh Instruments) equipped with a PMT detector (R928, Hamamatsu). The excitation source was a 532 nm Nd:YAG laser (Spectra-Physics-Quanta-Ray Lab) operating at 1 Hz. The laser system was also equipped with an image intensified CCD (ICCD) camera detector. Triplet lifetime decays were analyzed with the help of Origin 9.0.

Femtosecond transient absorption spectroscopy

Time-resolved transient absorption measurements were performed at the Imaging and Kinetic Spectroscopy (IMAKS) Laboratory, Department of Chemistry, NCSU (North Carolina State University). A mode-locked Ti:sapphire laser (Coherent Libra, 800 nm, 1 kHz repetition rate, 100 fs, 4 mJ per pulse) was used as the main light source. The output from the laser was split into the pump beam and probe beam. The pump beam was directed into the parametric amplifier (Coherent OPerA Solo) to generate the 400 nm excitation. The probe beam was delayed in a 6 ns optical delay stage and then focused into a CaF_2 crystal for white light continuum generation between 340 nm and 750 nm. The pump beam was then focused on the sample (into an 800 μm spot) and overlapped with the probe beam (~200 μm). The relative polarizations of the pump beam and the probe beam were set at the magic angle of 54.7°. The ground state absorption spectra of the samples were measured before and after the measurements to ensure that there is no degradation. Transient kinetics were analyzed using the fitting routines available in Origin 9.0.

Results and discussion

Synthesis of PCN-223(fb) was achieved by following a procedure that gives highly crystalline, phase-pure powder. Briefly, 3×10^{-5} moles of $ZrCl_4$ and 1.3×10^{-5} moles of TCPP were dissolved in 10 ml DMF along with 2.5×10^{-2} moles of propionic acid as the modulator. The mixture was sonicated for 15 minutes then the vial was placed in an oven and heated at 120 °C for 16 h. PCN-223(fb) MOF powders were characterized using PXRD (Fig. 2a) and SEM (Fig. 2b). Comparison of the PXRD pattern obtained from synthesis to the simulated pattern from single crystal XRD data indicated high phase purity. The SEM image shows small bean-shaped particles that are characteristic of the PCN-223 MOF. All MOF particles are morphologically identical, which further verifies the phase purity. In addition, thermal stability and surface area of as-prepared PCN-223(fb) were studied with the help of thermogravimetric analysis and N_2 adsorption isotherms. The BET

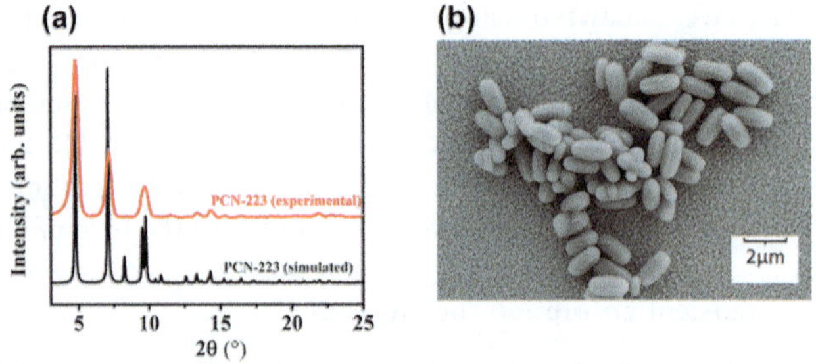

Fig. 2 (a) PXRD characterization of PCN-223(fb) (b) SEM image of PCN-223(fb).

surface area of the MOF obtained from the gas adsorption data agrees well with the literature (see ESI, Fig. S1 and S2†).[10]

To examine how the ground and excited-state properties of TCPP are affected upon coordination into the MOF structure, the absorption spectra of TCPP and PCN-223(fb) were compared (Fig. 3). The electronic absorption spectrum of TCPP consists of two distinct regions. The first involves the transition from the ground state to the second excited state ($S_0 \rightarrow S_2$) and the corresponding band is called the Soret or B band. The second region consists of a weak transition to the first excited state ($S_0 \rightarrow S_1$) in the range of 500–750 nm (the Q bands). PCN-223(fb) displays an absorption spectrum similar to that of TCPP with a sharp Soret band and four Q bands. The Soret band of PCN-223 is blue shifted by 13 nm relative to the ligand, which is attributed to structural changes that TCPP undergoes as it is incorporated in the MOF. Twisting of phenyl rings and changes in the macrocyclic ring planarity of TCPP may be responsible for increasing the electronic transition energy-gap (S_0–S_2), which causes the blue shift. The peak positions and the peak intensities of Q bands of PCN-223 match those of TCPP (see ESI, Table 1†), suggesting that the energy gap between ground state and first excited state of TCPP is relatively unaffected upon incorporation in the MOF.

The acid-base properties of porphyrins in aqueous solutions provide important information about their reactivity, aromaticity, tautomerization mechanisms and stereochemistry.[14–16] The central macrocycle of porphyrins exhibits an amphoteric behavior and can exist in either neutral (free-base), N-protonated, or deprotonated form (Fig. 4). Mono-protonation of the free-base form induces nonplanar distortions in the porphyrin structure, which makes the second protonation more favorable.[17,18] The first protonation step is immediately followed by the second step, producing the porphyrin dication, while the monoprotonated species is present in very small amounts at any given time. To determine the pK_a values corresponding to the successive protonation steps, TCPP solution (10^{-6} M) and MOF suspension in water were titrated against 0.01 M NaOH solution (see ESI, Fig. S3 and S4†). The pK_a values are reported in Fig. 4.

Varying solution pH can shift the protonation–deprotonation equilibrium of the macrocycle in favour of a particular form, which has been shown to greatly influence the photophysics of the porphyrin molecule.[18] Non-planarity induced in the porphyrin structure due to protonation of the nitrogen atoms breaks up their π-electron conjugated double-bond system. Loss of conjugation promotes non-

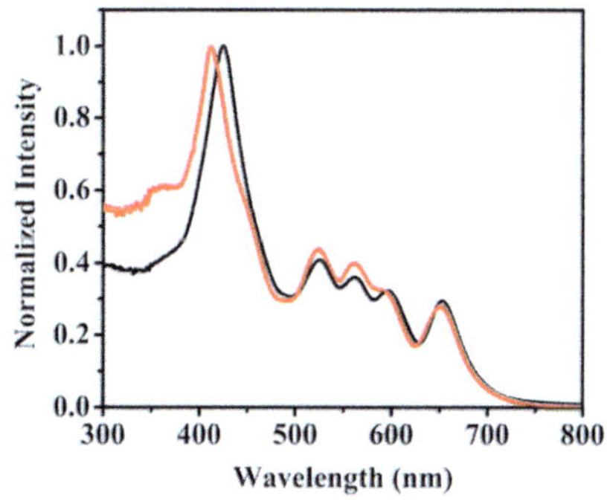

Fig. 3 Diffuse absorption spectra of TCPP (black) and PCN-223(fb) (red).

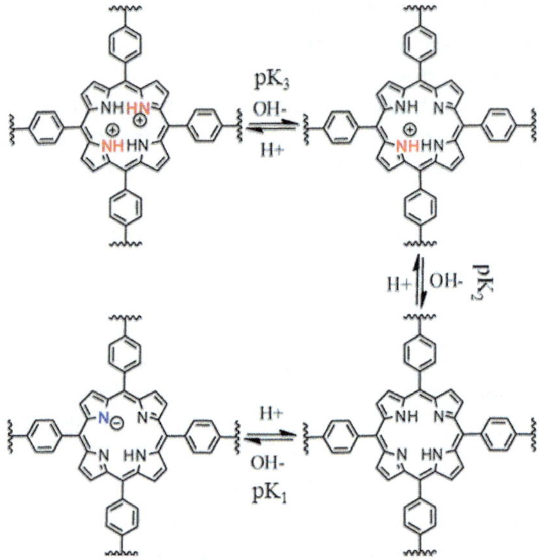

Fig. 4 Protonation and deprotonation processes of TCPP in acidic and basic media.[8] pK_1, pK_2 and pK_3 represent the pK_a values associated with these processes. The first and second protonation processes are almost indistinguishable, such that $pK_2 \approx pK_3$. For TCPP, $pK_2 \approx pK_3 = 3.2$, and for PCN-223 $pK_2 \approx pK_3 = 3.75$ (see ESI, Fig. S3 and S4†).

radiative relaxation of the excited state, resulting in significant fluorescence quenching and short fluorescence lifetimes.[19] To investigate the effects of pH variation on the fluorescence properties of TCPP and PCN-223(fb), their steady-state emission spectra were measured in an experimental pH range from 3.5 to 8.5 ($\lambda_{\text{excitation}} = 415$ nm). Given that PCN-223(fb) is stable in aqueous environments with pH values ranging from 0 to 10,[10] the experimental pH range is suitable for investigating the photophysics of the MOF without loss in crystallinity. The peak positions and intensities of fluorescence spectra of TCPP and PCN-223(fb) were found to be strongly correlated with the pH of the solution/suspension. The fluorescence spectra of the neutral form of TCPP and PCN-223(fb) ($5.5 < \text{pH} < 8.5$) displayed a sharp band at 645 nm ($Q(0,0)$) and a relatively weaker band at 720 nm ($Q(0,1)$) (Fig. 5a and b). Structural changes induced in the porphyrin macrocycle due to protonation of central nitrogen atoms cause the band at 645 nm to broaden and red shift by ~35 nm. As a result, the fluorescence spectrum appears to be a single broad band with a maximum at 680 nm (Fig. 5c).

Temperature dependence of fluorescence lifetimes

The temperature dependence of the fluorescence decay rates (k_{obs}) of an emissive MOF can give insight into the mechanism of energy transfer.[20] At room temperature, there are various vibrational degrees of freedom that allow the excited state to fully relax between energy transfer events. However, at low temperatures (~77 K) some of the vibrational degrees of freedom are frozen out. If the electronic interactions between the linkers in a MOF are very strong (strong coupling regime), then at low temperatures the rate of energy transfer can exceed that of

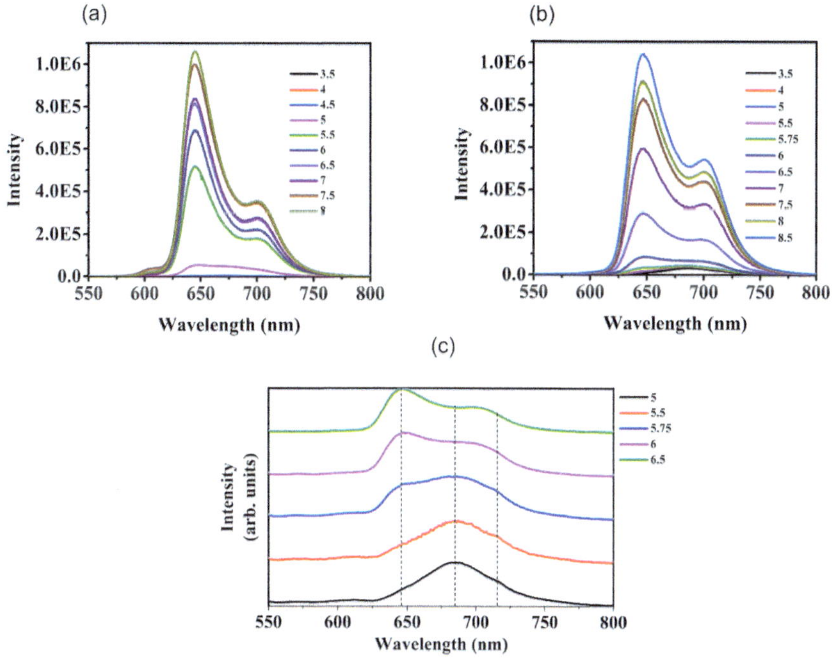

Fig. 5 (a) Fluorescence spectra of TCPP in water (10^{-6} M) as a function of pH. (b) Fluorescence spectra of PCN-223(fb) in water as a function of pH. (c) Spectral evolution of PCN-223(fb) in a pH range from 5 to 6.5.

vibrational relaxation. In such a scenario, the excitation energy can move as an exciton that is delocalized over the whole system.[1,21] The MOF should behave as an "aggregate" and the energy transfer process is termed as "coherent". In contrast, if the electronic interactions between the linkers in a MOF are weak (weak coupling regime), then vibrational relaxation dominates at all temperatures. At a given time "t", the excitation energy remains localized on a linker. The MOF behaves as a "monomer" and the excitation energy transfer is termed as "incoherent".

To determine whether PCN-223(fb) belongs to the strong coupling regime or the weak coupling regime, the temperature-dependent fluorescence decay kinetics of TCPP and PCN-223(fb) were compared.[22] The fluorescence decay rates of TCPP and PCN-223(fb) were obtained at temperatures ranging from 77 K to 333 K (Fig. 6a and b). To plot the temperature dependence curves, the experimental decay rates were fit to the following equation:

$$k_{\text{obs}} = k_0 + k_1 \mathrm{e}^{-\Delta E/k_{\text{B}}T}$$

where k_0 is the temperature independent term and the Arrhenius term describes the temperature dependence of k_{obs}.[23,24] The parameters extracted from the fits are reported in Table 1. The pre-exponential factor or the frequency factor (k_1) and the activation energy for transitioning from the ground state to the excited state (ΔE) of TCPP and PCN-223(fb) are very similar, suggesting that the temperature-dependent fluorescence behavior of the MOF is comparable to that of

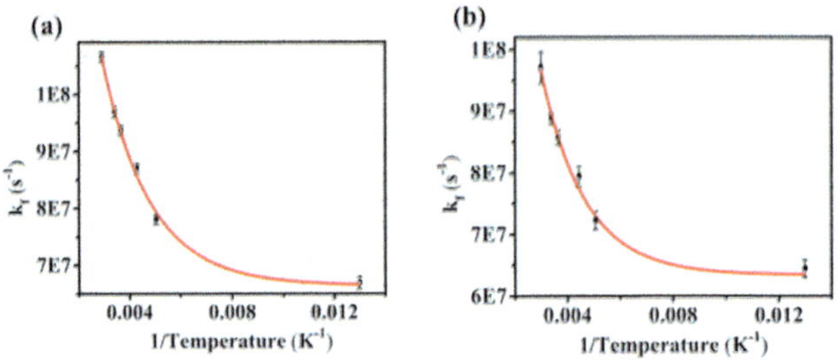

Fig. 6 (a) Temperature dependence of the observed emission decay rates of TCPP in a MeOH–EtOH (3 : 1 v/v) mixture (pH ≈ 7). (b) Temperature dependence of the observed emission decay rates of a PCN-223(fb) suspension (3 : 1 v/v) mixture (pH ≈ 7).

monomeric TCPP units. This result implies that PCN-223(fb) belongs to the weak coupling regime. Therefore, we propose an incoherent mechanism for EET in the PCN-223(fb) MOF.

Photoluminescence quenching

The energy transfer efficiency of porphyrin-based MOFs can be evaluated by relating the concentration of quencher to the extent of photoluminescence quenching.[25,26] Typically, quenching experiments on MOFs are carried out in the presence of an external quencher on its surface. The pH-dependence of the fluorescence intensity in the PCN-223 MOF shows that N-protonated porphyrins can potentially act as an internal quencher. Due to the presence of this internal quenching, no additional quencher was used in this study. The relationship between the quencher concentration and extent of quenching is provided by the Stern–Volmer equation,

$$I_0/I = 1 + K_{SV}[Q]$$

where I_0 is the fluorescence intensity in the absence of quencher, I is the fluorescence intensity at a particular concentration of quencher, K_{SV} is the Stern–Volmer quenching constant, and [Q] is the quencher concentration. Fig. 7 shows the modified Stern–Volmer plot for TCPP and PCN-223, where I_0/I is plotted against that ratio of N-protonated and neutral ligand concentration (see ESI, Section 5†).

The quenching rate k_q can be calculated by dividing K_{SV} by the fluorescence lifetime of donor in the absence of quencher (τ_0). The Stern–Volmer rate constants and quenching rate constants of TCPP and PCN-223(fb) are provided in Table 2. The quenching rate in PCN-223 is relatively higher than in TCPP (by an order of 10), which is attributed to energy migration from neutral TCPP units to N-

Table 1 Experimental parameters associated with TCPP and PCN-223(fb)

	k_0 (s^{-1}) (×10^7)	k_1 (s^{-1}) (×10^7)	ΔE (J per molecule) (×10^{-21})
TCPP	6.7 ± 0.1	4 ± 0.1	7.3 ± 0.1
PCN-223	6.4 ± 0.2	3.3 ± 0.6	8.3 ± 0.5

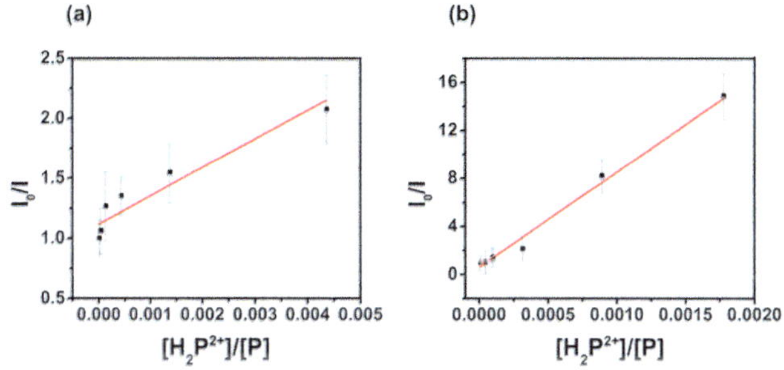

Fig. 7 (a) Modified Stern–Volmer plot of TCPP in the pH range from 5.5 to 8.5 (b) Modified Stern–Volmer plot of PCN-223(fb) in the pH range from 5.5 to 8.5.

protonated TCPP units. Although a dramatic decrease in photoluminescence intensity of MOFs was observed on lowering the pH, time resolved photoluminescence decay measurements revealed that the lifetime of the MOF was not quenched.

Energy transfer efficiency in PCN-223(fb)

Excitation energy transfer primarily occurs through two coupling mechanisms: the Dexter exchange mechanism and the Förster dipole–dipole mechanism.[4,26–30] The Dexter mechanism requires the presence of electronic communication between the donor and acceptor *via* orbital overlap.[31,32] Since the orbital overlap between adjacent porphyrin struts in Zr-based MOFs with porphyrin linkers is poor, the Dexter mechanism may not be suitable to describe energy transfer in these MOFs.[20] The Förster mechanism (FRET), on the other hand, adequately describes the "through space" energy migration in these MOFs.

The rate and efficiency of FRET is dependent on three physical parameters. These parameters are (1) the degree of overlap between donor emission and acceptor absorption (spectral overlap, J), (2) the relative geometric orientation of donor and acceptor within the framework, and (3) the distance between the donor and acceptor in the framework.[33] The FRET efficiency of porphyrin-based MOFs has been previously quantified by performing photoluminescence quenching experiments.[20,28,34] A qualitative analysis of FRET efficiency was performed by relating the extent of quenching (Φ_q) to the concentration of quencher (N-protonated porphyrin). Φ_q is defined as,

$$\Phi_q = 1 - I_q/I_0$$

where I_q and I_0 are the PL intensities of the donor in the presence and absence of the quencher.[35] The concentration of N-protonated TCPP in solution and in the

Table 2 Stern–Volmer rate constant and quenching rate constants of TCPP and PCN-223(fb)

	K_{SV} (M^{-1})	k_q (M^{-1} s^{-1})
TCPP	237.68	2.71×10^{10}
PCN-223(fb)	7298.96	8.06×10^{11}

MOF at a given pH can be found using the Henderson–Hasselbalch equation. The concentration of N-protonated TCPP in solution and in the MOF is ~1% at pH 5.5 and 6 respectively. The emission intensity of monomeric TCPP solution and in the MOF at pH 5.5 and 6 was set as I_q. The emission intensity at pH 8 was set as I_0. The Φ_q for TCPP and PCN-223 was estimated to be 93% and 51% respectively. The significantly higher magnitude of Φ_q in the MOF complements the results of PL quenching in the previous section. The quenching observed in the MOF can be considered as a combination of "self-quenching" and quenching *via* energy transfer from neutral porphyrins to N-protonated porphyrins.

The rate of energy transfer defined by Förster model is given by,

$$k_{\mathrm{ET}} = \frac{1}{\tau_0}\left(\frac{R_0}{r}\right)^6$$

where τ_0 is the lifetime of donor in the absence of acceptor, r is the donor–acceptor distance and R_0 is the Förster radius.[21] R_0 is defined as the distance at which the energy transfer efficiency is 50%, and is dependent on the fluorescence quantum yield of donor in the absence of acceptor (Φ_D) and the overlap integral (J).

$$R_0 = \left(\frac{9000(\ln 10)K^2\Phi_D J}{128\pi^5 N_A \eta^4}\right)^{1/6}$$

where K^2 is a geometric parameter that describes the relative orientation between the transition dipole moments of donor and acceptor, N_A is Avogadro's number, and n is the refractive index of the surrounding media. Based on the above equation, the Förster radius for PCN-223 was found to be 54.5 Å ($J = 5.344 \times 10^{-14}$ M^{-1} cm^3, see ESI, Fig. S6†). Quantum yield was obtained using relative method (0.27) (see ESI, Section 8†). The quantum yield of TCPP in methanol was obtained from the literature.[36] The spectral overlap integral of TCPP previously reported in literature is 2.49×10^{-15} M^{-1} cm^3.[37] The rate of exciton hopping (k_{hop}) and exciton hopping time were estimated to be 1.9×10^{12} s^{-1} and 0.53 ps respectively. These values are broadly consistent with those reported by Son *et al.* for other porphyrin-based MOFs.[25] The total exciton hopping distance was obtained from,

$$R_{\mathrm{hop}} = \sqrt{m\frac{D_m}{R_0}}$$

where D_m is the exciton diffusion coefficient.[36–40] Long distance exciton hopping corresponding to 100 Å for one-dimensional energy transfer, 141 Å for two-dimensional energy transfer, and 173 Å for three-dimensional energy transfer was estimated. Since there is limited information about the singlet exciton diffusion coefficient of porphyrin-based MOFs, the exciton diffusion coefficient of TPPS nanotubes (95×10^{-6} m^2 s^{-1}) was used for the calculation of exciton hopping distance.[36]

Nanosecond transient absorption spectroscopy

Thus far, we have investigated the singlet excited states of TCPP and PCN-223(fb). Nanosecond transient absorption (nsTA) spectroscopy was used as an additional tool to get further insight into the excited state photophysics of TCPP and PCN-223(fb). With the help of this technique, the triplet excited state of TCPP and

PCN-223(fb) were studied. Nanosecond transient absorption difference spectra were acquired for a TCPP solution and PCN-223(fb) suspension in water at pH 8 (in the nanosecond to microsecond time domain, Fig. 8a and b). The resulting difference spectra for both TCPP and PCN-223(fb) were almost identical, having an intense ground state bleach at ~420 nm followed by an excited state absorption band centred near 470 nm. The study also revealed that the nature of excited state did not change between nanosecond and microsecond time domains (see ESI, Fig. S8†). The transient absorption decay of TCPP at 470 nm exhibited single-exponential kinetics with a lifetime of 260 ± 7 µs (Fig. 8c). The triplet lifetime of TCPP was in good agreement with the previously reported studies.[41] PCN-223(fb) also exhibited a mono-exponential decay with a lifetime of 201 ± 27 µs (Fig. 8d). Due to the rigidity of the MOF structure, the rate of non-radiative deactivation pathways is expected to decrease, which in turn should lead to an increase in the lifetime of TCPP incorporated in PCN-223(fb). Contrastingly, the lifetime of PCN-223(fb) was found to be relatively shorter as compared to the ligand. The shorter triplet lifetime of PCN-223(fb) may be attributed to energy transfer pathways within the donor–acceptor framework. One of the possible pathways could potentially be energy transfer to the trap states such as N-protonated porphyrins and structural defects in the MOF. In order to further investigate the role of

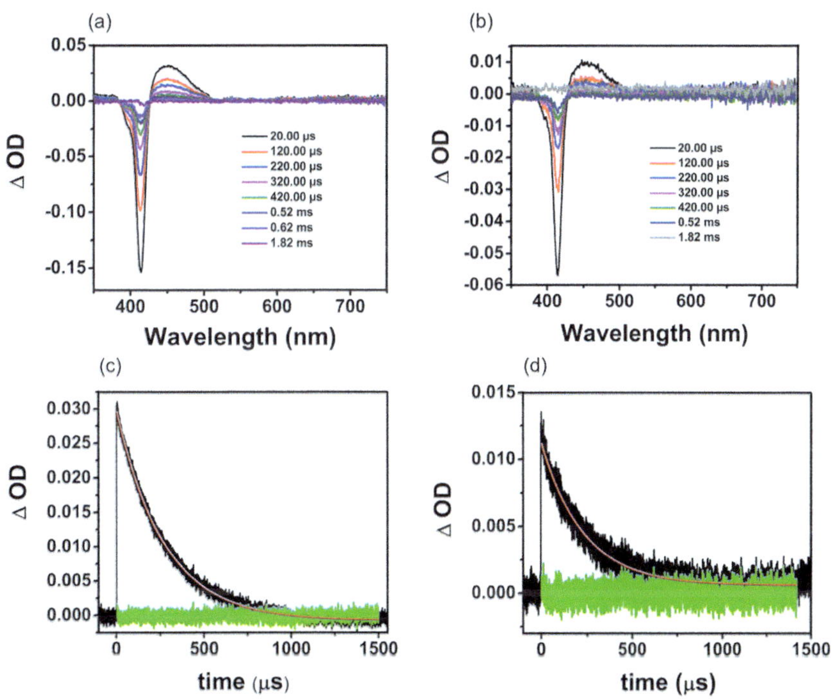

Fig. 8 Transient absorption difference spectra (mapping) of (a) TCPP and (b) PCN-223(fb) measured in degassed water at room temperature following 532 nm pulsed laser excitation (4–5 mJ per pulse, 5–7 ns FWHM). Both difference spectra represent an average of 30 transients. Fitted transient absorption decay of (c) TCPP and (d) PCN-223(fb) along with the residuals (green coloured). Both the transient absorption kinetic measurements were probed at 470 nm.

N-protonated porphyrins as quenchers, nanosecond transient absorption measurements were attempted at different pH values. However, the solubility of TCPP in water dramatically decreases on lowering the pH, resulting in very weak transient signal. Consequently, nsTA measurements were not successful at low pH.

Femtosecond transient absorption spectroscopy

To further map the excited state trajectory across all available timescales, ultrafast transient absorption (fsTA) studies on TCPP and PCN-223 were performed with 400 nm excitation and probe wavelengths that ranged across the UV-visible region. The fsTA difference spectra of both TCPP and PCN-223(fb) consists of a ground state bleach at 415 nm followed by an excited state absorption band starting from 430 nm. Composite spectral features that can be attributed to ground state bleach (from the Q band absorption) and excited state absorption can be observed in the 500–650 nm range. The evolution of their transients is presented in Fig. 9a and b. Over the course of the first 500 fs, the ground state bleach at 415 nm and the excited state absorbance centred near 470 nm appear. As time progresses, no change is observed in the positions of the bands. The features in the fsTA difference spectrum of TCPP and PCN-223 qualitatively match those of the nsTA difference spectrum (see ESI, Fig. S9a and b†), signifying that the nature of excited state in both the ligand and the MOF did not change as we moved from the femtosecond to microsecond time domain.

The transient kinetics of TCPP were probed at 418 nm and 506 nm, and analysed using single and multi-exponential models (see ESI, Fig. S10a†). The kinetic trace at 418 nm was best fit to a triexponential model. The lifetimes determined from the fit are provided in Table 3. Following previous assignments provided in the literature,[42–47] the shortest component (τ_1) is assigned to a solvent-induced process that causes vibrational reorganization of excitation energy in the S_1 state of TCPP (solvent reorganization). Elastic collisions between the ligand

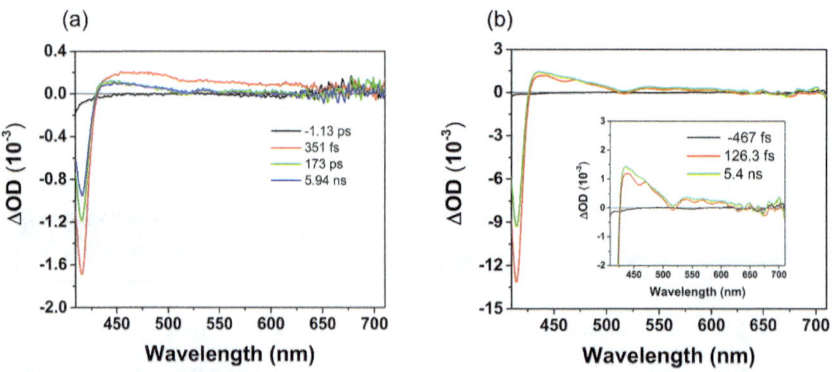

Fig. 9 Transient absorption difference spectra of (a)TCPP (in a 1 : 1 (v/v) water–ethanol mixture (~10^{-6} M)) at pH-8 following 400 nm pulsed laser excitation (140 fs FWHM), and (b) PCN-223 in water at pH-8 following 400 nm pulsed laser excitation (140 fs FWHM). The inset shows an expanded view of the excited state absorption band. Experimental delay times are indicated in the legend.

Table 3 Time constants obtained for TCPP in a 1 : 1 water–ethanol mixture at selected probe wavelengths (pump wavelength = 400 nm)

Wavelength (nm)	τ_1	τ_2	τ_3
418 nm	2 ± 1 ps	57 ± 22 ps	>1 ns
506 nm	51 ± 26 ps	—	—

Table 4 Time constants obtained for PCN-223 in water at selected probe wavelengths (pump wavelength = 400 nm)

Wavelength (nm)	τ_1	τ_2	τ_3
348 nm	5 ± 2 ps	115 ± 22 ps	>1 ns
510 nm	123 ± 29 ps	—	—

and solvent molecules are believed to be responsible for this process. The second component (τ_2) is attributed to vibrational relaxation of excited state S_1 due to interaction with solvent molecules (vibrational cooling). The third component is too long to be measured accurately with our apparatus and is attributed to fluorescence in TCPP.[42] The trace at 506 nm exhibited single exponential decay kinetics with a lifetime consistent with the vibrational cooling component of excited state relaxation.

In the case of the PCN-223 MOF, the kinetics were probed at 348 nm and 510 nm (see ESI, Fig. S10b†). Kinetic analysis of the ultrafast data revealed the presence of three distinct time constants (Table 4). Based on a comparison between the time constants of TCPP and PCN-223, the first component (τ_1) is assigned to solvent reorganization. The second component (τ_2) is assigned to vibrational cooling in the MOF. The third component ($\tau_3 > 1$ ns) is assigned to fluorescence in the MOF.

It is worth noting that the lifetime corresponding to solvent reorganization and vibrational cooling in the MOF is higher than the free ligand. Such an observation can be explained by three potential hypotheses. The first is that immobilization of TCPP within the rigid structure of MOF decreases the number of vibrational modes that can couple to this cooling process. It has been previously reported that immobilization of chromophores in MOFs decreases the rate of non-radiative excited state decay, leading to longer lifetimes.[48,49] The second is that the solvent molecules coupled to the vibrational cooling process have less accessibility to the porphyrin units in the MOF. The third hypothesis attributes the observed differences in the lifetimes between the TCPP and MOF samples to the different solvent systems used to measure each. Ultrafast TA measurements on TCPP samples were conducted in a 1 : 1 (v/v) mixture of water and ethanol. Ethanol was added to improve the solubility of TCPP in water, which in turn improves its absorbance. On the other hand, fsTA measurements on MOF samples were conducted in water (without ethanol). The difference in solvent polarities may be responsible for the higher lifetimes in the MOF.

Conclusions

In summary, the PCN-223 MOF was synthesized from free base TCPP and was thoroughly characterized. The photophysical properties of the synthesized MOF and the free ligand were explored extensively using various steady state and time-resolved spectroscopic techniques. The pH dependence of the fluorescence intensity of TCPP and PCN-223(fb) was investigated. Temperature dependence of the observed fluorescence decay rates of PCN-223(fb) was comparable to that of monomeric TCPP units, suggesting that the interchromophoric interactions between TCPP linkers in the MOF belong to the weak coupling regime. Therefore, an incoherent, hopping type mechanism was proposed for EET in PCN-223(fb). Internal fluorescence quenching experiments demonstrate that the singlet-state of PCN-223(fb) (in suspension) is quenched to a greater extent than the ligand (in solution). PCN-223(fb) demonstrated a significantly higher extent of quenching ($\Phi_q = 93\%$) as compared to monomeric TCPP solution ($\Phi_q = 51\%$), at similar concentrations of quencher. The enhanced quenching in the MOF is attributed to energy transfer from the freebase porphyrin linkers to N-protonated porphyrin linkers. The rate of exciton hopping and exciton hopping time in PCN-223 were estimated using the Förster energy transfer model. These values were found to be consistent with other porphyrin-based MOFs. Femtosecond and nanosecond transient absorption studies were conducted on free ligand and MOF to characterize their excited state properties. Nanosecond transient absorption decays showed that the triplet lifetime of the MOF is relatively shorter than that of the ligand, which may be due to triplet–triplet energy transfer to N-protonated porphyrins. Structural defects may also act as trap states and quench the triplet excited state of the MOF. The enhanced exciton migration distance (173 Å in 3D) in PCN-223(fb) MOF suggests that this type of chromophoric system could potentially have useful applications as light-harvesters in various solar energy conversion devices.

Conflicts of interest

The authors declare no conflicts of interest in preparation of this article.

Acknowledgements

This material is based upon work supported by the Department of Energy under Grant DE-SC0012445.

Notes and references

1 T. Renger, *Photosynth. Res.*, 2009, **102**(2), 471–485.
2 J. Yang, M.-C. Yoon, H. Yoo, P. Kim and D. Kim, *Chem. Soc. Rev.*, 2012, **41**(14), 4808–4826.
3 Y. Li, H. Xu, S. Ouyang and J. Ye, *Phys. Chem. Chem. Phys.*, 2016, **18**(11), 7563–7572.
4 M. C. So, G. P. Wiederrecht, J. E. Mondloch, J. T. Hupp and O. K. Farha, *Chem. Commun.*, 2015, **51**(17), 3501–3510.

5 H. Furukawa, K. E. Cordova, M. O'Keeffe and O. M. Yaghi, *Science*, 2013, **341**, 974.
6 A. Kuc, A. Enyashin and G. Seifert, *J. Phys. Chem. B*, 2007, **111**(28), 8179–8186.
7 D. Feng, Z.-Y. Gu, J.-R. Li, H.-L. Jiang, Z. Wei and H.-C. Zhou, *Angew. Chem., Int. Ed.*, 2012, **51**(41), 10307–10310.
8 H.-L. Jiang, D. Feng, K. Wang, Z.-Y. Gu, Z. Wei, Y.-P. Chen and H.-C. Zhou, *J. Am. Chem. Soc.*, 2013, **135**(37), 13934–13938.
9 W. Morris, B. Volosskiy, S. Demir, F. Gándara, P. L. McGrier, H. Furukawa, D. Cascio, J. F. Stoddart and O. M. Yaghi, *Inorg. Chem.*, 2012, **51**(12), 6443–6445.
10 D. Feng, Z.-Y. Gu, Y.-P. Chen, J. Park, Z. Wei, Y. Sun, M. Bosch, S. Yuan and H.-C. Zhou, *J. Am. Chem. Soc.*, 2014, **136**(51), 17714–17717.
11 D. Feng, H.-L. Jiang, Y.-P. Chen, Z.-Y. Gu, Z. Wei and H.-C. Zhou, *Inorg. Chem.*, 2013, **52**(21), 12661–12667.
12 P. Deria, D. A. Gómez-Gualdrón, I. Hod, R. Q. Snurr, J. T. Hupp and O. K. Farha, *J. Am. Chem. Soc.*, 2016, **138**(43), 14449–14457.
13 P. Deria, J. Yu, R. P. Balaraman, J. Mashni and S. N. White, *Chem. Commun.*, 2016, **52**(88), 13031–13034.
14 P. G. Seybold and M. Gouterman, *J. Mol. Spectrosc.*, 1969, **31**(1–13), 1–13.
15 M. Biesaga, K. Pyrzyńska and M. Trojanowicz, *Talanta*, 2000, **51**(2), 209–224.
16 P.-C. Lo, X. Leng and D. K. P. Ng, *Coord. Chem. Rev.*, 2007, **251**(17), 2334–2353.
17 A. B. Rudine, B. D. DelFatti and C. C. Wamser, *J. Org. Chem.*, 2013, **78**(12), 6040–6049.
18 P. H. Hynninen, *J. Chem. Soc., Perkin Trans. 2*, 1991, (5), 669–678.
19 M. M. Kruk, A. S. Starukhin and W. Maes, *Macroheterocycles*, 2011, **4**(2), 69–79.
20 J. Zhu, S. Shaikh, N. J. Mayhall and A. J. Morris, in *Elaboration and Applications of Metal–Organic Frameworks; Series on Chemistry, Energy and the Environment*, WORLD SCIENTIFIC, 2017, vol. 2, pp. 581–654.
21 B. Valeur and M. N. Berberan-Santos, *Excitation Energy Transfer*, 2012.
22 U. De Rossi, S. Dähne, U. Gomez and H. Port, *Chem. Phys. Lett.*, 1998, **287**(3), 395–402.
23 M. Abrahamsson, H.-C. Becker, L. Hammarström, C. Bonnefous, C. Chamchoumis and R. P. Thummel, *Inorg. Chem.*, 2007, **46**(24), 10354–10364.
24 M. Abrahamsson, H.-C. Becker and L. Hammarström, *Dalt. Trans.*, 2017, **46**(39), 13314–13321.
25 H. J. Son, S. Jin, S. Patwardhan, S. J. Wezenberg, N. C. Jeong, M. So, C. E. Wilmer, A. A. Sarjeant, G. C. Schatz, R. Q. Snurr, O. K. Farha, G. P. Wiederrecht and J. T. Hupp, *J. Am. Chem. Soc.*, 2013, **135**(2), 862–869.
26 S. Patwardhan, S. Jin, H.-J. Son and G. C. Schatz, *Mater. Res. Soc. Symp. Proc.*, 2013, **1539**, 22–27.
27 H.-J. Son, S. Jin, S. Patwardhan, S. J. Wezenberg, N. C. Jeong, M. So, C. E. Wilmer, A. A. Sarjeant, G. C. Schatz, R. Q. Snurr, O. K. Farha, G. P. Wiederrecht and J. T. Hupp, *J. Am. Chem. Soc.*, 2013, **135**(2), 862–869.
28 C. Y. Lee, O. K. Farha, B. J. Hong, A. A. Sarjeant, S. T. Nguyen and J. T. Hupp, *J. Am. Chem. Soc.*, 2011, **133**(40), 15858–15861.
29 C. A. Kent, B. P. Mehl, L. Ma, J. M. Papanikolas, T. J. Meyer and W. Lin, *J. Am. Chem. Soc.*, 2010, **132**(37), 12767–12769.
30 Q. Zhang, C. Zhang, L. Cao, Z. Wang, B. An, Z. Lin, R. Huang, Z. Zhang, C. Wang and W. Lin, *J. Am. Chem. Soc.*, 2016, **138**(16), 5308–5315.

31 D. L. Dexter, *J. Chem. Phys.*, 1953, **21**(5), 836.

32 J. Lin, X. Hu, P. Zhang, A. Van Rynbach, D. N. Beratan, C. A. Kent, B. P. Mehl, J. M. Papanikolas, T. J. Meyer, W. Lin, S. S. Skourtis and M. Constantinou, *J. Phys. Chem. C*, 2013, **117**(43), 22250–22259.

33 D. L. Andrews, *Chem. Phys.*, 1989, **135**(2), 195–201.

34 E. Maligaspe, T. Kumpulainen, H. Lemmetyinen, N. V. Tkachenko, N. K. Subbaiyan, M. E. Zandler and F. D'Souza, *J. Phys. Chem. A*, 2010, **114**(1), 268–277.

35 I. Majoul, Y. Jia and R. Duden, *Practical Fluorescence Resonance Energy Transfer or Molecular Nanobioscopy of Living Cells*, Springer, Boston, MA, 2006.

36 T. Kim, S. Ham, S. H. Lee, Y. Hong and D. Kim, *Nanoscale*, 2018, **10**(35), 16438–16446.

37 L. D. A. Siebbeles, A. Huijser and T. J. Savenije, *J. Mater. Chem.*, 2009, **19**(34), 6067–6072.

38 W. A. Maza, A. J. Haring, S. R. Ahrenholtz, C. C. Epley, S. Y. Lin and A. Morris, *J. Chem. Sci.*, 2016, **7**(1), 719–727.

39 M. Kaushal, A. L. Ortiz, J. A. Kassel, N. Hall, T. D. Lee, G. Singh and M. G. Walter, *J. Mater. Chem. C*, 2016, **4**(24), 5602–5609.

40 O. V. Mikhnenko, P. W. M. Blom and T.-Q. Nguyen, *Energy Environ. Sci.*, 2015, **8**(7), 1867–1888.

41 A. Kathiravan, R. Renganathan and S. Anandan, *J. Colloid Interface Sci.*, 2010, **348**(2), 642–648.

42 J. S. Baskin, H.-Z. Yu and A. H. Zewail, *J. Phys. Chem. A*, 2002, **106**(42), 9837–9844.

43 H.-Z. Yu, J. S. Baskin and A. H. Zewail, *J. Phys. Chem. A*, 2002, **106**(42), 9845–9854.

44 R. Kumble, S. Palese, V. S.-Y. Lin, M. J. Therien and R. M. Hochstrasser, *J. Am. Chem. Soc.*, 1998, **120**(44), 11489–11498.

45 X. Li, C. Gong, G. G. Gurzadyan, M. F. Gelin, J. Liu and L. Sun, *J. Phys. Chem. C*, 2018, **122**(1), 50–61.

46 S. Sorgues, L. Poisson, K. Raffael, L. Krim, B. Soep and N. Shafizadeh, *J. Chem. Phys.*, 2006, **124**(11), 114302.

47 J. Rodriguez, C. Kirmaier and D. Holten, *J. Chem. Phys.*, 1991, **94**(9), 6020–6029.

48 W. A. Maza, R. Padilla and A. J. Morris, *J. Am. Chem. Soc.*, 2015, **137**(25), 8161–8168.

49 W. A. Maza and A. J. Morris, *J. Phys. Chem. C*, 2014, **118**(17), 8803–8817.

Faraday Discussions

PAPER

Band-selective dynamics in charge-transfer excited iron carbene complexes

Pavel Chábera,[a] Lisa A. Fredin,[b] Kasper S. Kjær,[a] Nils W. Rosemann,[a] Linnea Lindh,[a] Om Prakash,[c] Yizhu Liu,[c] Kenneth Wärnmark,[c] Jens Uhlig,[a] Villy Sundström,[a] Arkady Yartsev[a] and Petter Persson *[b]

Received 7th December 2018, Accepted 18th February 2019

DOI: 10.1039/c8fd00232k

Ultrafast dynamics of photoinduced charge transfer processes in light-harvesting systems based on Earth-abundant transition metal complexes are of current interest for the development of molecular devices for solar energy conversion applications. A combination of ultrafast spectroscopy and first principles quantum chemical calculations of a recently synthesized iron carbene complex is used to elucidate the ultrafast excited state evolution processes in these systems with particular emphasis on investigating the underlying reasons why these complexes show promise in terms of significantly extended lifetimes of charge transfer excited states. Together, our results challenge the traditional excited state landscape for iron-based light harvesting transition metal complexes through radically different ground and excited state properties in alternative oxidation states. This includes intriguing indications of rich band-selective excited state dynamics on ultrafast timescales that are interpreted in terms of excitation energy dependence for excitations into a manifold of charge-transfer states. Some implications of the observed excited state properties and photoinduced dynamics for the utilization of iron carbene complexes for solar energy conversion applications are finally discussed.

Introduction

Characterization of ultrafast excited-state dynamics is crucial for understanding and rationally improving the performance of a wide range of emerging solar energy conversion technologies, including molecular systems for both photovoltaic and solar fuels applications.[1] Many molecular-based approaches utilize the favourable photoelectrochemical properties of transition metal complexes

[a]Division of Chemical Physics, Department of Chemistry, Lund University, Box 124, SE-22100 Lund, Sweden

[b]Division of Theoretical Chemistry, Department of Chemistry, Lund University, Box 124, SE-22100 Lund, Sweden. E-mail: Petter.Persson@teokem.lu.se

[c]Center for Analysis and Synthesis (CAS), Department of Chemistry, Lund University, Box 124, SE-22100 Lund, Sweden

(TMCs) that have been developed both for light-harvesting and photocatalytic functions.[2] TMCs can, in particular, utilize charge-transfer (CT) excited states as a first step towards generating high-energy CT states. The intrinsic capabilities of many transition metal complexes to sustain a range of oxidation states also make them suitable as central components for electrochemical and photo-electrochemical oxidation and reduction reactions as a key part of many photo-catalytic reaction schemes.[3]

Octahedral second and third row transition metal complexes sporting $n\mathrm{d}^6$ (with $n = 4$ or 5) electronic configurations on the metal have maintained a central role for many light-harvesting and molecular photochemistry applications for several decades, owing in large part to the large ligand field splitting (LFS) in which the low-spin $n\mathrm{d}^6$ electronic ground state configuration is stabilized. This has yielded many complexes with long-lived high-energy metal-to-ligand charge transfer (MLCT) excited states that have been widely used for photophysical and photochemical applications.[2] This particularly includes the famous family of Ru^{II} polypyridyl complexes such as $Ru(bpy)_3$ derivatives used extensively as photo-sensitizers, *e.g.* in molecular photovoltaics.[4,5] Other prominent examples of photoactive d^6 complexes are based on the third-row species Re^{I}, Os^{II}, and Pt^{III} which are also interesting for light-emitting applications.[6]

It has, however, increasingly been recognized that it is desirable for environ-mentally friendly and sustainable large-scale solar energy applications to replace rare and expensive second and third row transition metal complexes with more Earth-abundant alternatives.[7–11] This has prompted significant interest in devel-oping first row transition metal complexes with good photophysical and photo-chemical properties to ideally rival and replace those from the more well-established second and third row transition metals. This has led to the emer-gence in recent years of several interesting photoactive first row transition metal complexes of, for example, copper, chromium, and iron.[12–16] As the lighter and abundant congener of ruthenium and osmium, iron looks *prima facie* to be an obvious candidate to develop photofunctional first row transition metal complexes. Progress in this direction has, however, until very recently been largely hampered by poor excited state properties, including a preponderance for substantial and ultrafast excited state energy losses,[17,18] with high-energy metal-to-ligand charge-transfer excited states in iron polypyridyl complexes typically decaying on a few picoseconds or faster timescale.[19,20] This has remained rather discouraging compared to the abundance of ruthenium polypyridyl complexes maintaining high excited state energies for hundreds of nanoseconds or longer.[2]

The photophysics and photochemistry of many first-row transition metal complexes, including traditional octahedral Fe^{II} polypyridyl complexes, suffer severely from an intrinsically weaker ligand field splitting for 3d complexes compared to their 4d and 5d counter-parts.[21] Even the favourable $3d^6$ configura-tion in octahedral Fe^{II} complexes is typically not sufficient to avoid serious problems with ultrafast deactivation of high energy excited singlet and triplet $^{1/3}$MLCT states to low-energy high-spin (quintet) metal centred (^{5}MC) scavenger states.[22] Complexes with low-energy high-spin MC states are of some considerable interest in themselves as spin cross-over (SCO) complexes, and they are investi-gated in the particular context of photoinduced formation as light-induced excited state spin transition (LIESST) systems.[23] This has prompted several investigations to elucidate the ultrafast dynamics and excited state cascade of photoexcited iron complexes.[17] Recently, this has also been complemented by ultrafast X-ray spectroscopic investigations capable of probing local structure and spin dynamics involving the iron centre.[24–26] Experimental progress together with increasingly advanced quantum chemical calculations and dynamics simulations have yielded a wealth of valuable insight into the ultrafast early deactivation dynamics in such complexes.[27–31]

In the last five years, a new ligand motif featuring strongly sigma-donating *N*-heterocyclic carbenes (NHCs) has been realized synthetically and demonstrated through a combination of time-resolved optical characterization and calculations to achieve dramatically improved excited state properties of iron complexes.[32,33] This has provided a promising path for development also of iron-based photosensitizers for a variety of photoinduced electron transfer applications. Initial synthesis and characterization of octahedral Fe^{II} complexes with up to four carbene ligands yielded excited charge-transfer state lifetimes of up to a few tens of picoseconds;[34–36] significantly longer than the typical ultrafast decay of Fe-polypyridyl LIESST complexes on a few ps or faster timescale, though still quite short compared to *e.g.* many champion Ru^{II} photosensitizers. This excited state lifetime improvement to a few tens of ps excited state lifetimes was sufficient for a recent proof-of-principle demonstration of efficient photoinduced interfacial electron injection from an anchored iron carbene photosensitizer to a semiconducting TiO_2.[37]

Another step towards more broadly useful iron-based photosensitizers was recently achieved through a synthetic design that involved full octahedral coordination using a *hexa*-carbene motif in a $Fe(btz)_3$ complex with the chemical structure shown in Fig. 1. Interestingly, this complex was found to be sufficiently stable in both oxidation states II and III to allow both ground state absorption (Fig. 1b) as well as excited state dynamics characterization. Fe^{II} and Fe^{III} correspond to $3d^6$ and $3d^5$ electronic configurations or the iron centre, respectively.[38,39] The $Fe^{III}(btz)_3$ complex was first reported in 2017 as a quite rare case of a $3d^5$ complex exhibiting interesting photophysics, with a record ~100 ps charge-transfer excited state lifetime involving a ^{2}LMCT state as illustrated in Fig. 2.[38] Spin-allowed decay to the doublet ground state (^{2}GS) also showed the first persistent room temperature photoluminescence (PL) in the visible part of the

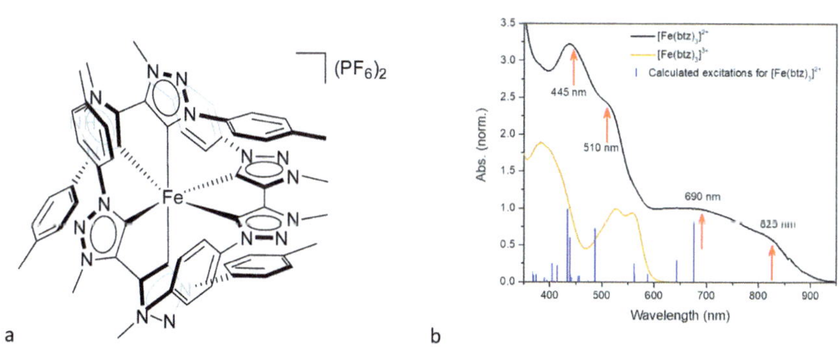

Fig. 1 (a) Chemical structure of the $Fe^{II}(btz)_3$ complex, and (b) steady state spectra of $Fe^{II}(btz)_3$ and $Fe^{III}(btz)_3$ in acetonitrile. The red arrows indicate the excitation wavelengths for transient absorption experiments.

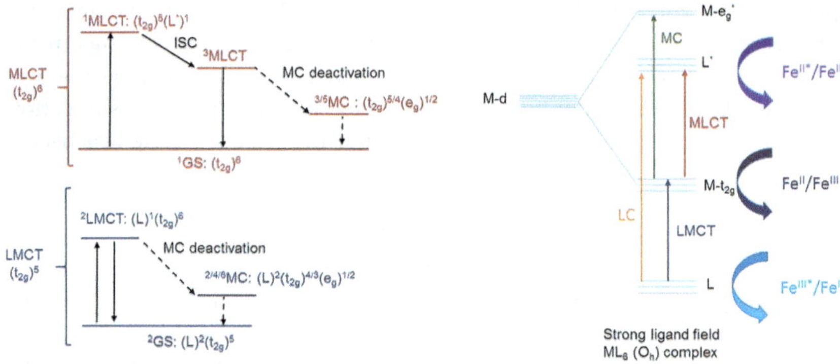

Fig. 2 Simplified Jablonski-type diagram for low-spin Fe^{II} $3d^5$ and Fe^{III} $3d^6$ complexes (left), together with a schematic illustration of key electronic levels and photoexcitation processes involved in the photophysics (right). Curved arrows indicate opportunities to drive subsequent bimolecular or interfacial electron transfer processes (far right).

electromagnetic spectrum from such an iron complex, albeit with a rather low PL quantum yield ($\Phi_{PL} \sim 10^{-4}$).[38] The corresponding Fe^{II} species $Fe^{II}(btz)_3$ was subsequently also shown to have a long room-temperature excited state lifetime of several hundred picoseconds (τ ~500 ps), following a more conventional MLCT excitation scheme for a $3d^6$ complex (Fig. 2).[39] A new hexa-carbene Fe^{III} complex was furthermore recently shown to feature enhanced photoluminescence from an excited CT state with a nanosecond lifetime, as well as being capable of reductive and oxidative bimolecular electron transfer reactions from the photoexcited ^{2}LMCT state.[40] The extension of the excited state lifetimes of the MLCT and LMCT charge transfer states for iron complexes into the hundreds of picosecond range and longer thus open up a range of photophysical and photochemical applications for iron carbene complexes.

It is, in this context, quite remarkable that the same strategy of increasing the σ-donation of the ligand surrounding the iron centre was effective for both Fe^{II} and Fe^{III} as shown in the case of $Fe(btz)_3$. It is therefore interesting to develop a more comprehensive comparison of the excited state dynamics of these two types of excited state cascades. In order to achieve this, we here report new results, in particular pertaining to dynamics of $Fe^{II}(btz)_3$ following excitations at different excitation energies, together with some computational results detailing electronic structure properties in these photosensitizers. Together with recent time-resolved and computational results that we have reported for these complexes, this allows us to present a more comprehensive discussion focused on the ultrafast excited state dynamics of this novel type of promising photosensitizer.

Methods

Experimental

The experimental setup for recording broadband transient absorption spectra was based on a Ti:Sapphire amplified laser system (Spitfire XP Pro, Spectra Physics) operating at a 1 kHz repetition rate, producing ~80 fs pulses centered at 796 nm. The pump beam was tuned by an optical parametric amplifier (TOPAS C,

Light Conversion) to excite the sample at various wavelengths (445 nm, 510 nm, 690 nm, and 825 nm) with typical fluency ~3×10^{15} photons per pulse per cm^2 and maximum fluency not exceeding 1×10^{16} photons per pulse per cm^2 which is well below absorption saturation density ~5×10^{17} photons per cm^2. A power dependency check was also performed by varying the excitation intensity by a factor of 25 (100 to 2500 μW) with no deviation of the observed ΔA signal from linear behaviour. A super-continuum white-light was used as probe, generated by focusing the NIR signal from a TOPAS C into a 5 mm sapphire plate. The desired timing between excitation and probe pulses was achieved using a computer-controlled delay line (Aerotech, 10 ns). The pump and the probe beams were overlapped on the sample with their relative polarization set to the magic-angle (54.7°) by a Berek polarization compensator placed in the pump beam path. The sample was placed in a quartz 1 mm path length cuvette with an automated sample mover to avoid sample photodamage, which was also checked by measuring the absorption spectra of the sample before and after each experiment. The probe and reference beams were collimated on the entrance aperture of a prism-based, double-beam spectrograph, and detected by a double diode-array detection system (Pascher Instruments). Correction of the data as well as fitting of the transient absorption datasets was carried out using both an in-house written Python routine and the analysis software DAFit (Pascher Instruments) and the individual kinetics were fitted using the Origin software.

Computational

First principles quantum chemical calculations have been performed at the Density Functional Theory (DFT) and time-dependent DFT (TD-DFT) levels of theory for ground- and excited-state properties, respectively. The presented computational results were obtained using the B3LYP* modification[41] of the standard B3LYP hybrid density functional[42] and employing standard 6-311G basis sets with polarization functions,[43] as well as a polarizable continuum solvation model for acetonitrile.[44] All calculations were performed using the Gaussian 09 program package.[45]

Results

In this results section, we first consider the linear absorption spectra of both $Fe^{II}(btz)_3$ and $Fe^{III}(btz)_3$ as a basis for the subsequent discussions. Next, new time-resolved optical spectroscopy results for the excited state dynamics of $Fe(btz)_3$ are presented. In particular, we present new transient absorption (TA) results for the $Fe^{II}(btz)_3$ complex in acetonitrile following excitations at different energies. Finally, absorption spectra are complemented by a presentation of selected results from quantum chemical calculations of both Fe^{II} and Fe^{II} oxidation states of this complex. Together, this forms a basis for a comparative analysis of the excited state dynamics of this complex in the complementary Fe^{II} (3d^6) and Fe^{III} (3d^5) oxidation states presented in the subsequent discussion below.

Linear absorption

The linear absorption spectra of $Fe^{II}(btz)_3$ and $Fe^{III}(btz)_3$, shown in Fig. 1b, clearly demonstrate that, in contrast to many other TMCs, this new class of Fe-carbene

TMCs exhibit two distinct CT absorption bands on the red side of the strong ligand-based absorption below ~350 nm. There are, furthermore, indications of several sub-band features of vibrational or electronic nature that appear as shoulders or double peaks in the CT regions of the absorption spectra. Additionally, the blue CT band in $Fe^{II}(btz)_3$ has at least twice as strong maximum absorption strength compared to the red band for both oxidation states. The measured CT absorption bands agree qualitatively with the calculated singlet–singlet TD-DFT spectrum for $Fe^{II}(btz)_3$ presented as sticks for comparison.

Transient absorption

Transient absorption data has been recorded for the $Fe^{II}(btz)_3$ complex as shown in Fig. 3. In particular, this includes transient data following excitations at 445 nm, 510 nm, 690 nm, and 825 nm, complementing the published data for excitation at 800 nm at key points in the Fe(II) absorption spectrum (Fig. 1).[39] The excitations into different parts of the visible absorption spectrum probe the

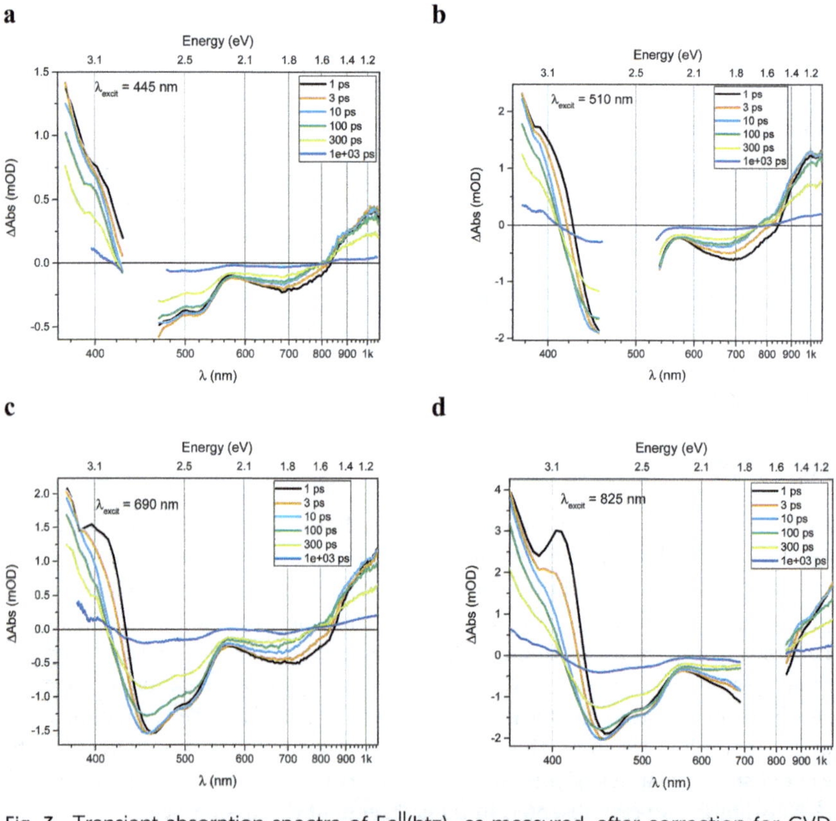

Fig. 3 Transient absorption spectra of $Fe^{II}(btz)_3$ as measured, after correction for GVD (chirp) and background subtraction. Spectra were recorded following excitations at (a) 445 nm, (b) 510 nm, (c) 690 nm and (d) 825 nm, with excitation densities of 2.4×10^{15}, 3.7×10^{15}, 3.3×10^{15} and 1.1×10^{16} photons per pulse per cm^2, respectively. Excitation scatter region is omitted for clarity. For all panels the x-axis (wavelength) is in reciprocal values (linear steps in energy).

dynamics following excitations into CT states of different electronic nature and with different initial excess energies compared to the absorption threshold.

Several main features are recognizable throughout as relatively persistent features that survive the full dynamics, also for the different excitation energies. TA dynamics are brought out further in the analysis shown in Fig. 4, which compares the spectral shape of the TA at 1 and 20 ps for all excitation energies after accounting for the excitation densities.

The main spectral features for all the four excitation energies are similar to what was recently outlined for 800 nm excitation in our paper introducing this complex.[39] Briefly, this covers broad ground state bleach (GSB) features similar to an inverted absorption spectrum in the *ca* 450–800 nm region of the spectrum, together with strong excited state absorption (ESA) in the blue part of the spectrum giving a positive signal above *ca* 425 nm as well as some ESA in the red part of the spectrum that turns positive beyond the GSB region for wavelengths longer than *ca* 800 nm. The blue ESA band is apparently relatively strong, as the positive TA signal is observed in the spectral range where one would expect a pronounced bleach signal based on the linear absorption spectrum. Clearly, the overlapping ESA band must have a sharp rise towards the blue to overshoot the expected bleach shoulder, and instead give rise to a strong net positive TA signal.

A closer look at the spectral features in Fig. 3 and 4 reveals some noticeable similarities and differences in terms of the detailed shape and evolution of these TA spectra. The 510 nm and 690 nm mid-CT band spectra are, in particular, very similar throughout. The TA results for all the excitations furthermore show a qualitatively similar GSB to the TA for the recently reported energy excitation at 800 nm. There is a noticeable reduction in the intensity of the ESA feature at ~408 nm at early timescales for the higher energy excitations compared to the corresponding features for the lowest excitation energy at 825 nm. There is, however, little further spectral evolution after the first 20 ps, as evident for excitations at all four excitation energies. Thus, this behaviour is very similar for the higher energy excitations compared to what has already been established for excitation at 800 nm.[39] It is interesting to note that although similar, the main

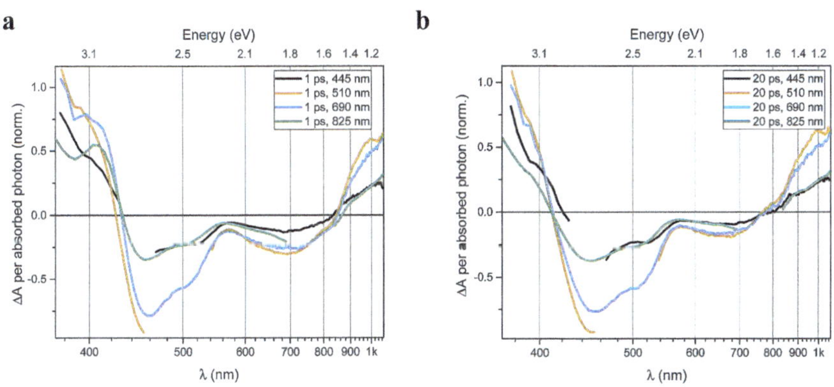

Fig. 4 Transient absorption spectra of $Fe^{II}(btz)_3$ at (a) 1 ps, and (b) 20 ps normalized by the number of absorbed photons per cm^2 at each respective excitation wavelength. The *x*-axis (wavelength) is in reciprocal values (linear steps in energy).

features of the differential spectra appear to have excitation energy dependent amplitudes.

As a final step in the analysis of the excitation energy dependent dynamics, we have investigated different aspects of the multi-exponential kinetics for different combinations of pump and probe wavelengths in further detail as shown in Fig. 5, together with an associated summary of selected key kinetics components listed in Table 1. The single kinetic analysis is focused on identifying specific decay component in the TA data. As some of the decay components only contribute significantly at some probe wavelengths it is not always possible to fit all the time regions in one kinetic with current S/N ratio in order to receive satisfactory fits. Thus only selected unambiguous components are listed in Table 1. It can be generally noted that the dynamics in several places show signs of an ultrafast component on a sub-ps timescale that is potentially interesting in terms of ISC *etc*, but in all transient spectra and kinetics discussed in this study, we do not discuss further the very early dynamics that occur during the first few 100s of fs due to rather strong contribution of the coherent and solvent related experimental artefacts. For excitation energies into both main charge-transfer bands, the major part of the decay originates from a long-term component with $\tau \sim 540$ ps, as *e.g.* clearly evident for the 1000 nm kinetics for all excitation energies, and overall in good agreement with what we recently reported for excitation at 800 nm. There is also clear evidence for a ~ 3 ps component at selected spectral regions in the more comprehensive TA obtained here, as seen *e.g.* for the 420 nm probes with different excitation energies presented in panels Fig. 5c and d; again in broad agreement with the results in our initial study based on 800 nm excitation. We also identify an additional new kinetic component with $\tau \sim 50$ ps as part of the early dynamics. This component is most clearly pronounced in the displayed TA data for the 408 nm kinetics following 825 nm excitation while appearing gradually less pronounced for higher energy excitations. Overall, the kinetics measured at different excitations thus look qualitatively similar, but the contribution of the different decay components is quantitatively distinct. For example, all kinetics look very similar as probed at 1000 nm, while there are clear differences in the kinetics probed at 408 nm for the different excitation energies. Some implications of the rich dynamics observed here is further considered in the discussion below.

Electronic structure

Quantum chemical calculations performed at the DFT and TD-DFT level of theory provide information about the electronic states involved in the excited state cascade. Some key results were presented for both $Fe^{III}(btz)_3$ and $Fe^{II}(btz)_3$ already as part of the initial characterization of the respective species and their excited state properties.[38,39] Selected computational results, including some new information, are presented here, to put results from both species on common ground to facilitate a common discussion about the similarities and differences in fundamental electronic structure properties pertinent for understanding the excited state dynamics.

First, it can be noted that the change in oxidation state from Fe^{II} to Fe^{III} has a profound influence both on the ground state electronic structure and the excited state manifold. The Fe^{II} carbene complexes are characterized as a regular

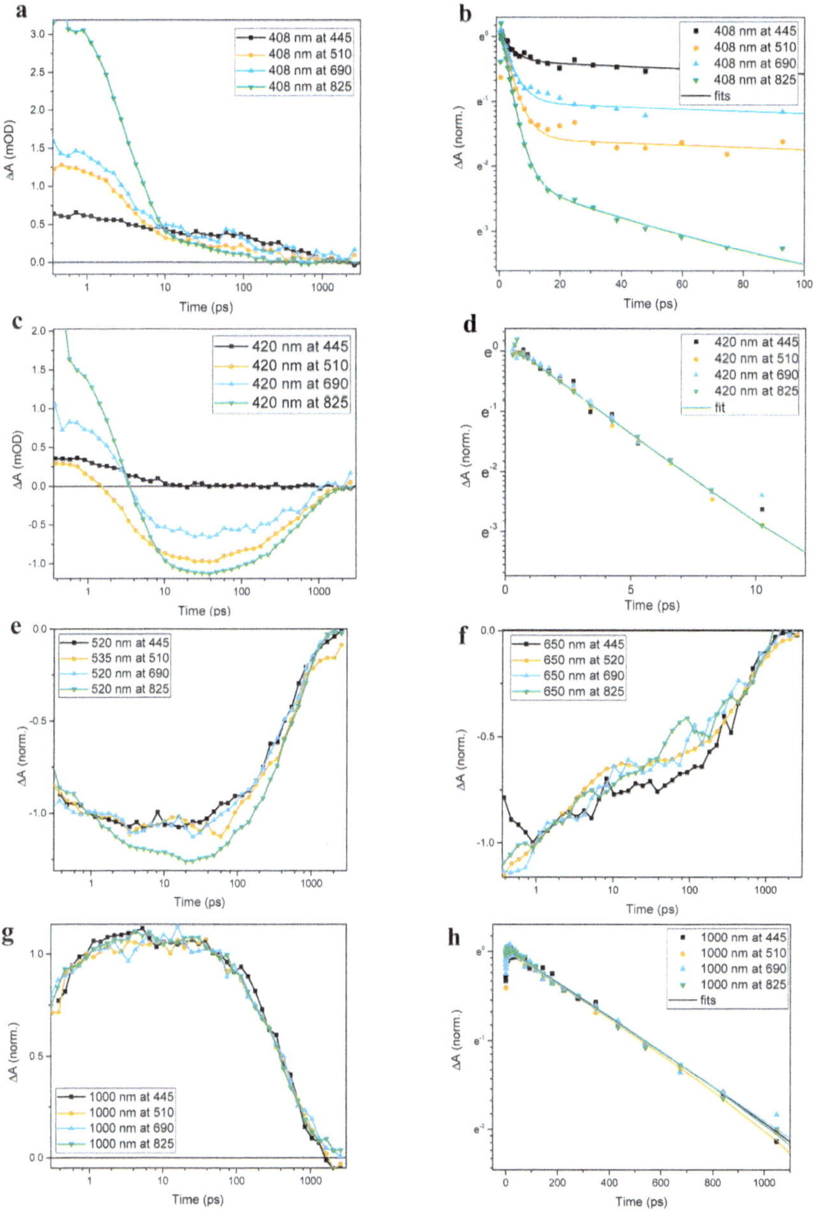

Fig. 5 Comparison of $Fe^{II}(btz)_3$ kinetics following excitations at different wavelengths. Results are shown for kinetics recorded at (a and b) 408 nm, (c and d) 420 nm, (e) 520 nm, (f) 650 nm, and (g and h) 1000 nm following excitations at 445 nm, 510 nm, 690 nm, and 825 nm. Solid lines represent multi-exponential fits. Note that graphs in panels (a), (c), (e), (f) and (g) are plotted in ΔA(lin)/time (log) scale while graphs in panels (b), (d), and (h) are shown in ΔA(ln)/time (lin) scale. All kinetics except for panels (a) and (c) have been normalised for visual clarity.

low-spin $3d^6$ species, *i.e.* with an electronic configuration of $(t_{2g})^6$ that is similar *e.g.* to many well-known Ru^{II} polypyridyl complexes.[2] This features a singlet ground state and $^{1/3}$MLCT excited states. For traditional Fe^{II} complexes with

Table 1 Key kinetic components from multiexponential fits of TA data for different excitation and probe wavelengths

$\lambda_{excitation}$ (nm)	λ_{probe} (nm)	τ_1 (ps)	τ_2 (ps)	τ_3 (ps)
445, 510, 690, 825	1000			540 ± 35
445, 510, 690, 825	420	3.2 ± 0.10		
445, 510, 690	408	3.8 ± 0.81		540 ± 35
825	408	3.5 ± 0.36	54.0 ± 5.67	540 ± 35

weaker LFS, initial excitation into MLCT excited states have been extensively discussed to decay to low-energy metal centred triplet and quintet states (^{3}MC and ^{5}MC) on ultrafast timescales.[17,21,22,27,46] This is typically followed by slower ground state recovery from meta-stable MC states on the timescale of hundreds of picosecond or slower.[47] Dramatically different excited state dynamics has instead been observed for Fe^{II} carbene complexes, including assignment of extended excited state lifetimes of ^{3}MLCT states of ~10 ps or longer, and followed by fast ground state recovery on similar timescales. The changes in excited state dynamics are largely attributed to a strong destabilization of the metal centred states in the iron carbene complexes due to the strong sigma-donating nature of the carbenes.[14] This altered photophysics also allowed a recent demonstration of efficient electron injection into a TiO_2 substrate.[37] The electronic structure for $Fe^{II}(btz)_3$ features a characteristic ^{3}MLCT-like spin density for the photofunctional charge-transfer state (Fig. 6a).

The electronic structure properties of $Fe^{III}(btz)_3$ can be understood in simple terms as originating from a one-electron oxidation of the $Fe^{II}(btz)_3$ complex in which one electron is removed from the iron center so that it takes on what could recently been characterized as a low-spin $3d^5$ electronic configuration, *i.e.* with a $(t_{2g})^5$ population on iron in quasi-octahedral symmetry notation (Fig. 2). From this follows that the ground state is a doublet (Fig. 6b), and that the excited state manifold will consist of doublet, quartet, and hextet excited states (limiting the number of unpaired electrons to five for simplicity). With a hole in the (t_{2g}) levels, the lowest $Fe^{III}(btz)_3$ excitation was characterized as a doublet ligand-to-metal charge transfer state.[38] Metal centered quartet and hextet states could also be identified computationally, and involve population of e_g-levels on the iron center. Such intermediate- and high-spin MC states are naturally stabilized by elongated metal–ligand bonds, similar to the MC states for d^6 complexes.[1] Calculated spin density plots of such ^{4}MC and ^{6}MC states for $Fe^{III}(btz)_3$ are shown in Fig. 6c and d panels, respectively.

It is thus clear that the Fe^{II} and Fe^{III} forms of iron carbene complexes like $Fe(btz)_3$ differ fundamentally in their basic electronic properties, excitations, and excited state manifolds. In particular, the altered electronic structure effectively yields a reversal of the fundamental direction of the lowest charge-transfer excitations, and while there are metal centered scavenger states that facilitate excited state deactivation for both oxidation states, there are fundamental differences in the spin properties of the states and state crossings involved in the excited state cascade. It is therefore interesting to compare the excited state dynamics for the same complex in the two different oxidation states.

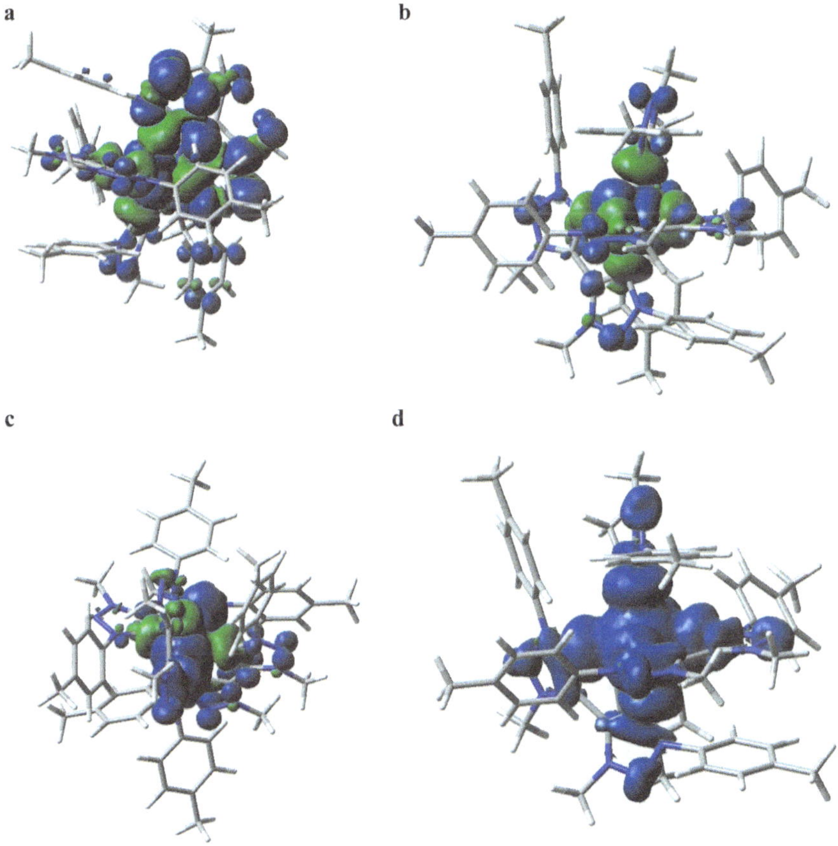

Fig. 6 Spin density plots (isovalue 0.0004) of selected $Fe^{II}(btz)_3$ and $Fe^{III}(btz)_3$ states from B3LYP*/6-311G(d)|MeCN calculations. Specific oxidation and electronic state assignments: (a) Fe^{II}:^{3}MLCT, (b) Fe^{III}:^{2}GS, (c) Fe^{III}:^{4}MC, (d) Fe^{III}:^{6}MC.

One potentially promising aspect of the photophysics of the $Fe^{III}(btz)_3$ complex for photosensitizer applications is that the excited state lifetime of the ^{2}LMCT state was characterized experimentally to be largely unaffected by the excitation energy over a significant part of the visible absorption band (probed between *ca* 400 and 550 nm).[39] This facilitates for the complex to perform well as a photosensitizer for a broad range of solar illumination. Here, the dynamics of the $Fe^{II}(btz)_3$ complex is also further explored experimentally *via* TA to enable a comparative discussion of the photophysics of the two oxidation states. It is in this context relevant to consider the excited state manifold that can be accessed at different excitation energies also for this Fe^{II} complex. Fig. 7 provides a graphical illustration of singlet and triplet excited states in the accessible energy region obtained from singlet–singlet and singlet–triplet TD-DFT calculations, respectively.

Discussion

Conventionally, the main electronic states considered for excited state evolution in d^6 TMCs include the singlet ground state (^{1}GS), ^{1}MLCT, ^{3}MLCT, ^{3}MC and, for

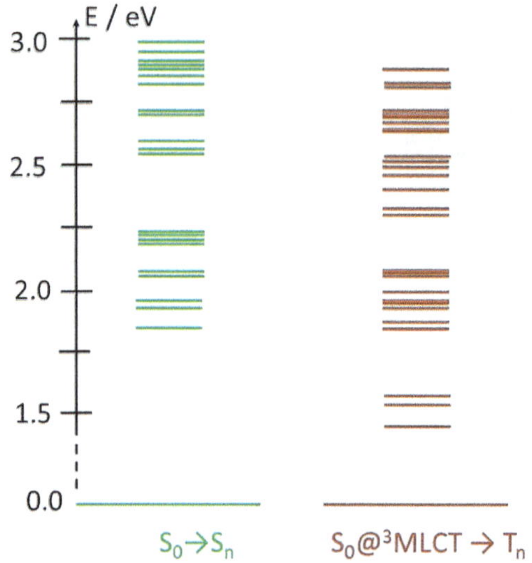

Fig. 7 TD-DFT excited state calculations for $Fe^{II}(btz)_3$. Shown excitations include singlet excitations at S0 geometry (left, green), and triplet excitations from the S0 state at the ^{3}MLCT geometry (right, red). Energies and assignments from Chábera *et al.*[39]

complexes with weak LFS, ^{5}MC, and in many cases modelling excited state dynamics as an interplay of these states has provided a satisfactory picture of the overall spectral changes associated with the excited state evolution including depopulation of the initial charge transfer state (loss of ESA) and ground state recovery (loss of GSB).

Our results for $Fe(btz)_3$ provide evidence that a more sophisticated description may generally be needed for a comprehensive understanding of the photoinduced dynamics of iron carbenes. A strong indication of the rather special features of $Fe(btz)_3$ is apparent already from analysis of the absorption spectra. In contrast to MLCT absorption of many typical TMCs, we record two well-resolved CT transitions centered at 450 and 700 nm, assigned to population of two different CT bands in $Fe^{II}(btz)_3$. For Fe^{II}, one can furthermore identify two high-energy shoulders at ~520 and ~825 nm for the two ^{1}MLCT absorption bands, respectively. We can attribute this rich CT absorption appearance as an effect of the strong σ-donating effect of the carbene ligands which causes an up-shift in the energy of the highest occupied metal-based t_{2g} levels of the ground state (correlated also to a shift towards more negative ground state reduction potentials).[1] This shifts the CT transition energies towards red to the extent that more of the CT manifold appears clearly below the more intense intra-ligand transitions. This observation is in very good agreement with the predictions of our TD-DFT calculations which, as outlined above, indicate that there is a broad range of charge-transfer excitations that cover the visible spectral region in the experimental linear absorption spectrum.

Having clearly distinct CT bands that are well separated in energy and furthermore have very different transition strength, provides an opportunity to investigate to what extent the dynamics following excitation to different CT states are similar or different. In the following discussion, we will first consider separately Fe^{II} and Fe^{III} cases before our final conclusions relating to the emergence of a more comprehensive view of the photoinduced dynamics of the iron carbenes.

As described in the results section, we observe differences in the excited state dynamics in $Fe^{II}(btz)_3$ after excitation into the two ^{1}MLCT bands. In particular, we observe some clear changes of the TA spectral shapes for different excitations, as exemplified by the more pronounced ~408 band in the ESA from excitations at low excitation energy at early timescales. Following the established picture of early Fe^{II} excited state evolution, we assume very efficient depopulation of the ^{1}MLCT due to fast ISC on a sub-picosecond timescale (not resolved in this study).[46] Our observations suggest that excitation into different parts of the ^{1}MLCT bands leads to different population dynamics and branching ratio between the ^{3}MLCT and ^{3}MC excited states. Initially hot conversion of ^{3}MLCT into ^{3}MC probably does not occur for delay times longer than ~1 ps. In agreement with our previous assignments for excitation at ~800 nm, we tentatively associate some changes to the blue part of the ESA at early times (see Fig. 3–5) mainly with intramolecular vibration relaxation (IVR).[39] Hot population of the energetically accessible ^{3}MC is likely to be limited to the early times prior to such IVR. It is tempting to conclude that population of the ^{3}MC occurs from ^{1}MLCT or "hot" ^{3}MLCT whereas the relaxed ^{3}MLCT has a lower probability of converting into ^{3}MC. This interpretation of the very fast excited state processes is visualized schematically in Fig. 8. This suggests that the efficiency of the ^{3}MC population is very sensitive to the exact crossings of the ^{1}MLCT and ^{3}MLCT with the ^{3}MC state in

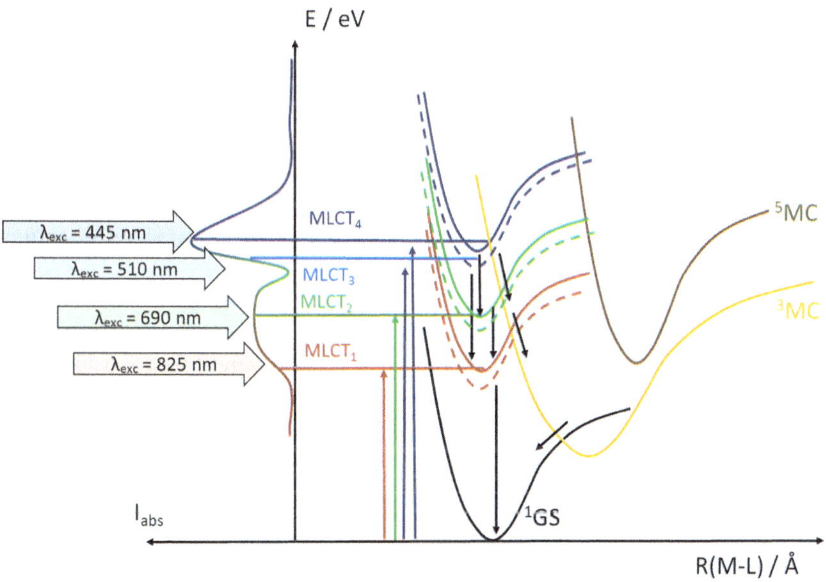

Fig. 8 Excited state manifold in the charge-transfer excitation energy region for $Fe^{II}(btz)_3$, indicating the different excitation energies/wavelengths employed experimentally to probe the excited state dynamics (red, green, light blue and blue arrows), as well as selected potential decay paths (black arrows). Singlet and triplet forms of representative MLCT states in the charge-transfer manifold energy region are distinguished as solid and dashed lines, respectively.

terms of the activation energy and structural rearrangement associated with such a transition.

Finally, irrespective of the rate and amplitude of this few tens of ps component, the overall relaxation of the excited complexes to the ground state occurs with the same rate within our signal-to-noise (S/N). These observations agree with dynamics on a manifold of ^{3}MLCT excited states, including energy relaxation within this manifold that occurs on timescales up to a few tens of ps. Based on the combined information from ground state absorption, first principles calculations, and time-resolved excited state dynamics, we infer that a consistent and complete interpretation of the results require a manifold of ^{1}MLCT and ^{3}MLCT excited states to be considered.

Taken together, the simultaneous excitation and probing of different parts of the CT manifold presented in this investigation provides a first glimpse of rich, and in some aspects band-selective, dynamics in these iron carbene complexes with remarkably long-lived (>100 ps) excited states. From the current understanding of the excited state manifold of these unconventional iron complexes the full dynamics potentially involves a combination of several excited state processes including ultrafast electronic processes such as IVR, IC, ISC within the manifold of excited states, together with molecular and solvent structural dynamics. The current evidence from the TA dynamics suggests that it will be important to carry out further work to fully elucidate key aspects of the ultrafast mechanisms responsible for the dynamics observations, including unambiguous assignments of the ultrafast dynamics and detailed interplay between initial singlet–triplet ISC as well as the deactivation cascade from CT to MC states for a full comparison with the ultrafast LIESST dynamics in conventional Fe complexes typically characterized by ultrafast ISC and CT → MC conversion. A range of further experimental and computational investigative steps appear promising to facilitate a better understanding, including complementary optical characterization *e.g.* including low-temperature characterization, time-resolved optical measurements with better time-resolution in the ultrafast domain, and studies of dynamics in different solvent environments, as well as complimentary experimental techniques in different energy regimes including *e.g.* time-resolved X-ray dynamics, as well as more sophisticated theoretical treatments such as high-level quantum dynamics simulations.

It is finally interesting to also compare these new results for Fe^{II} to those recently reported for the Fe^{III} form of the $Fe(btz)_3$ complex. Similar to Fe^{II}, two clearly separated transitions to CT states in the ground state absorption spectrum can be identified: a high energy band centered at ∼400 and a low energy band with two peaks at 525 and 560 nm. No significant excitation wavelength dependency in the TA spectra and kinetics was, however, noticed. It is important to note here that the excited state energy evolution does not exhibit fast and efficient ISC from the ^{2}LMCT state. On the contrary, after optical excitation from the doublet ground state (^{2}GS) to the ^{2}LMCT state, the primary energy relaxation occurs within the same excited state. This is evident as the excited state is emissive and exhibits a rather small Stokes shift. After reaching the bottom of the ^{2}LMCT manifold, the excited state population undergoes either radiative or non-radiative transition to the ground state in competition with activated ISC transition to ^{4}MC scavenger states. The relative efficiencies of the above listed transitions determine the excited state lifetime (identical to the ground state recovery time), as well as the yield of the photoluminescence.

Conclusions

Efforts to develop iron carbene photosensitizer complexes have recently yielded very significant improvements in excited state lifetimes for both conventional $3d^6$ Fe^{II} carbene complexes, and more unconventional low-spin $3d^5$ Fe^{III} complexes. These results were obtained in rapid succession for structurally similar Fe^{II} and Fe^{III} complexes and appear to rely on the same ligand design strategy to maximize the strong σ-donation by hexa-carbene environments around the metal in the two different formal oxidation states of the iron center. Our results on the excited state properties and dynamics of the $Fe(btz)_3$ molecule with iron in both oxidation state II and III provide an excellent opportunity to compare promising photophysics and photochemistry of iron carbene photosensitizers in the two different electronic configurations $3d^5$ and $3d^6$. Until recently, only a relatively limited number of d^5 ^{2}LMCT complexes, *e.g.* Re(II) and Tc(II), had received significant attention as potentially interesting for driving photocatalytic processes.[48–50] Our combined results for the $Fe(btz)_3$ complex suggest that the photophysics and photochemistry of these complexes provide interesting information regarding ultrafast dynamics following excitations into a manifold of charge-transfer excited states that may be relevant more broadly.

One key finding for the time-resolved TA dynamics of $Fe^{II}(btz)_3$ following higher energy excitations (from 825 nm to 445 nm excitation wavelengths) is that there remains a long-lived ground state recovery with a time constant of *ca* 500 ps independent of excitation wavelength accounting for a major part of the decay dynamics. This photophysical consistency in terms of a stable long-term excited state feature following excitation into the charge-transfer manifold of excited states/bands is similar to what was recently shown in our investigation of the $Fe^{III}(btz)_3$ form (in the range of 550–400 nm excitation wavelengths) where there was a *ca* 100 ps long component dominating the deactivation and ground state recovery regardless of excitation energy. This common feature of the two oxidation states is potentially quite important for the photofunctionality of the iron carbenes *e.g.* for light-harvesting in solar energy conversion applications profiting from exposure to the full solar spectrum, and not just limited to absorption in a small spectral region.

The detailed investigation of the ultrafast dynamics of the $Fe^{II}(btz)_3$ complex through multi-exponential fitting of selected TA kinetics, however, also reveals some interesting additional, smaller but clear, contributions to the dynamics on timescales of a few tens of ps that was not previously discussed. The quantum chemical calculations provide general support for the presence of a manifold of low excited-state energy metal-to-ligand charge transfer states, *e.g.* involving electronic rearrangements either among the $(t_{2g})^5$ electrons or low-energy reduced $(\pi^*_{btz})^1$ ligand orbitals in the MLCT configuration, at energies that become accessible for the higher energy excitations. It is therefore in itself not surprising that the complex excited state landscape leads to some additional fast dynamics, *e.g.* involving internal conversion or intersystem crossing. With the present TA and computational results, a detailed interpretation of the short-lived intermediate dynamics was not attempted. Instead, we note that given the current interest in hot dynamics in transition metal complexes in general, coupled with the current high interest in developing Earth-abundant photosensitizers based on iron, our new findings point to the importance of further investigations to characterize the ultrafast dynamics in this type of iron complexes with rich dynamics and extended excited state lifetimes in greater detail.

A comparison of the ultrafast charge-transfer dynamics and excited state cascade of Fe^{II} and Fe^{III} species finally provides a rare opportunity to compare largely complementary excited state dynamics including issues, *e.g.* about how internal de-excitation differ in charge-transfer excited manifolds that differ in the fundamental charge-transfer direction (MLCT *versus* LMCT in Fe^{II} and Fe^{III} species, respectively). The comparative analysis also points to fundamental differences in energy relaxation paths and state-crossings for species with fundamentally different sets of spin states (even *versus* odd electronic state multiplicities) that has significant impact on light-harvesting and light-emitting photofunctionality. Further studies of ultrafast dynamics of iron complexes with long-lived charge transfer excited states thus look important for providing further in-depth understanding relevant to the continued development of photofunctional iron complexes. More generally, it remains interesting to conduct investigations of the ultrafast photoinduced dynamics also for other transition metal complexes with unconventional excited state properties and novel ultrafast dynamics.

Conflicts of interest

The authors declare no conflicts of interest.

Acknowledgements

The Swedish Research Council (VR), the Swedish Energy Agency (Energimyndigheten), the Swedish Foundation for Strategic Research (SSF), and the Knut and Alice Wallenberg (KAW) Foundation are all acknowledged for financial support. The Swedish supercomputing centers LUNARC and NSC are gratefully acknowledged for providing computing facilities through the SNIC allocation system.

References

1 C. S. Ponseca, P. Chábera, J. Uhlig, P. Persson and V. Sundström, Ultrafast Electron Dynamics in Solar Energy Conversion, *Chem. Rev.*, 2017, **117**(16), 10940–11024.

2 V. Balzani, G. Bergamini, S. Campagna and F. Puntoriero, Photochemistry and Photophysics of Coordination Compounds: Overview and General Concepts, *Top. Curr. Chem.*, 2007, **280**, 1–36.

3 N. S. Lewis and D. G. Nocera, Powering the Planet: Chemical Challenges in Solar Energy Utilization, *Proc. Natl. Acad. Sci. U. S. A.*, 2006, **103**(43), 15729–15735.

4 G. J. Meyer, Molecular Approaches to Solar Energy Conversion with Coordination Compounds Anchored to Semiconductor Surfaces, *Inorg. Chem.*, 2005, **44**(20), 6852–6864.

5 A. Hagfeldt and M. Graetzel, Light-Induced Redox Reactions in Nanocrystalline Systems, *Chem. Rev.*, 1995, **95**(1), 49–68.

6 V. W.-W. Yam and K. M.-C. Wong, Luminescent Metal Complexes of D6, D8 and D10 Transition Metal Centres, *Chem. Commun.*, 2011, **47**(42), 11579.

7 L. A. Büldt and O. S. Wenger, Luminescent Complexes Made from Chelating Isocyanide Ligands and Earth-Abundant Metals, *Dalton Trans.*, 2017, **46**, 15175–15177.

8 B. Bozic-Weber, E. C. Constable and C. E. Housecroft, Light Harvesting with Earth Abundant D-Block Metals: Development of Sensitizers in Dye-Sensitized Solar Cells (DSCs), *Coord. Chem. Rev.*, 2013, **257**(21–22), 3089–3106.

9 O. S. Wenger, Photoactive Complexes with Earth-Abundant Metals, *J. Am. Chem. Soc.*, 2018, **140**(42), 13522–13533.

10 P. Sutra and A. Igau, Emerging Earth-Abundant (Fe, Co, Ni, Cu) Molecular Complexes for Solar Fuel Catalysis, *Curr. Opin. Green Sustain. Chem.*, 2018, **10**, 60–67.

11 B. M. Hunter, H. B. Gray and A. M. Müller, Earth-Abundant Heterogeneous Water Oxidation Catalysts, *Chem. Rev.*, 2016, **116**(22), 14120–14136.

12 T. Bessho, E. C. Constable, M. Graetzel, R. A. Hernandez, C. E. Housecroft, W. Kylberg, M. K. Nazeeruddin, M. Neuburger and S. Schaffner, An Element of Surprise-Efficient Copper-Functionalized Dye-Sensitized Solar Cells, *Chem. Commun.*, 2008, **32**, 3717–3719.

13 S. Otto, M. Grabolle, C. Förster, C. Kreitner, U. Resch-Genger and K. Heinze, $[Cr(Ddpd)_2]^{3+}$: A Molecular, Water-Soluble, Highly NIR-Emissive Ruby Analogue, *Angew. Chem., Int. Ed.*, 2015, **54**(39), 11572–11576.

14 Y. Liu, P. Persson, V. Sundström and K. Wärnmark, Fe N-Heterocyclic Carbene Complexes as Promising Photosensitizers, *Acc. Chem. Res.*, 2016, **49**, 1477–1485.

15 M. Abrahamsson, M. J. Lundqvist, H. Wolpher, O. Johansson, L. Eriksson, J. Bergquist, T. Rasmussen, H. C. Becker, L. Hammarström, P. O. Norrby, *et al.*, Steric Influence on the Excited-State Lifetimes of Ruthenium Complexes with Bipyridyl-Alkanylene-Pyridyl Ligands, *Inorg. Chem.*, 2008, **47**(9), 3540–3548.

16 M. Abrahamsson, Solar Energy Conversion Using Iron Polypyridyl Type Photosensitizers – a Viable Route for the Future?, in *Photochemistry*, ed. A. Albini and E. Fasani, RSC, Cambridge, 2017, vol. 44, pp. 285–295, ISBN: 978-1-78262-543-8, DOI: 10.1039/9781782626954.

17 J. E. Monat and J. K. McCusker, Femtosecond Excited-State Dynamics of an Iron(II) Polypyridyl Solar Cell Sensitizer Model, *J. Am. Chem. Soc.*, 2000, **122**(17), 4092–4097.

18 S. M. Fatur, S. G. Shepard, R. F. Higgins, M. P. Shores and N. H. Damrauer , A Synthetically Tunable System To Control MLCT Excited-State Lifetimes and Spin States in Iron(II) Polypyridines, *J. Am. Chem. Soc.*, 2017, **139**, 4493–4505.

19 W. Zhang, R. Alonso-Mori, U. Bergmann, C. Bressler, M. Chollet, A. Galler, W. Gawelda, R. G. Hadt, R. W. Hartsock, T. Kroll, *et al.*, Tracking Excited-State Charge and Spin Dynamics in Iron Coordination Complexes, *Nature*, 2014, **509**(7500), 345–348.

20 A. Cannizzo, C. J. Milne, C. Consani, W. Gawelda, C. Bressler, F. Van Mourik and M. Chergui, Light-Induced Spin Crossover in Fe(II)-Based Complexes: The Full Photocycle Unraveled by Ultrafast Optical and X-Ray Spectroscopies, *Coord. Chem. Rev.*, 2009, **254**, 2677–2686.

21 W. Zhang, R. Alonso-Mori, U. Bergmann, C. Bressler, M. Chollet, A. Galler, W. Gawelda, R. G. Hadt, R. W. Hartsock, T. Kroll, *et al.*, Tracking Excited-State Charge and Spin Dynamics in Iron Coordination Complexes, *Nature*, 2014, **509**(7500), 345–348.

22 J. K. Mccusker and A. Vlcek, Ultrafast Excited-State Processes in Inorganic Systems, *Acc. Chem. Res.*, 2015, **48**, 1207–1208.

23 Q. Sun, S. Mosquera-Vasquez, L. M. Daku, L. Guénée, H. A. Goodwin, E. Vauthey and A. Hauser, Experimental Evidence of Ultrafast Quenching of the 3 MLCT Luminescence in Ruthenium(II) Tris-Bipyridyl Complexes *via* a 3dd State, *J. Am. Chem. Soc.*, 2013, **135**, 13660–13663.

24 M. Chergui and E. Collet, Photoinduced Structural Dynamics of Molecular Systems Mapped by Time-Resolved X-ray Methods, *Chem. Rev.*, 2017, **117**(16), 11025–11065.

25 T. J. Penfold, M. Pápai, T. Rozgonyi, K. B. Møller and G. Vankó, Probing Spin–vibronic Dynamics Using Femtosecond X-Ray Spectroscopy, *Faraday Discuss.*, 2016, **194**(0), 731–746.

26 D. Leshchev, T. C. B. Harlang, L. A. Fredin, D. Khakhulin, Y. Liu, E. Biasin, M. G. Laursen, G. E. Newby, K. Haldrup, M. M. Nielsen, *et al.*, Tracking the Picosecond Deactivation Dynamics of a Photoexcited Iron Carbene Complex by Time-Resolved X-Ray Scattering, *Chem. Sci.*, 2018, **9**, 405–414.

27 W. Zhang and K. J. Gaffney, Mechanistic Studies of Photoinduced Spin Crossover and Electron Transfer in Inorganic Complexes, *Acc. Chem. Res.*, 2015, **48**, 1140–1148.

28 T. J. Penfold, E. Gindensperger, C. Daniel and C. M. Marian, Spin-Vibronic Mechanism for Intersystem Crossing, *Chem. Rev.*, 2018, **118**, 6975–7025.

29 C. Sousa, M. Alías, A. Domingo and C. de Graaf, Deactivation of Excited States in Transition-Metal Complexes: Insight from Computational Chemistry, *Chem.–Eur. J.*, 2019, **25**(5), 1152–1164.

30 J. Nance, D. N. Bowman, S. Mukherjee, C. T. Kelley and E. Jakubikova, Insights into the Spin-State Transitions in $[Fe(Tpy)_2]^{2+}$: Importance of the Terpyridine Rocking Motion, *Inorg. Chem.*, 2015, **54**, 11259–11268.

31 I. M. Dixon, F. Alary, M. Boggio-Pasqua and J.-L. Heully, The $(N_4C_2)^{2-}$ Donor Set as Promising Motif for Bis(Tridentate) Iron(II) Photoactive Compounds, *Inorg. Chem.*, 2013, **52**(23), 13369–13374.

32 Y. Liu, P. Persson, V. Sundström and K. Wärnmark, Fe N-Heterocyclic Carbene Complexes as Promising Photosensitizers, *Acc. Chem. Res.*, 2016, **49**(8), 1477–1485.

33 L. Liu, T. Duchanois, T. Etienne, A. Monari, M. Beley, X. Assfeld, S. Haacke and P. C. Gros, A New Record Excited State 3 MLCT Lifetime for Metalorganic Iron(II) Complexes, *Phys. Chem. Chem. Phys.*, 2016, **18**(18), 12550–12556.

34 Y. Liu, T. Harlang, S. E. Canton, P. Chábera, K. Suárez-Alcántara, A. Fleckhaus, D. A. Vithanage, E. Göransson, A. Corani, R. Lomoth, *et al.*, Towards Longer-Lived Metal-to-Ligand Charge Transfer States of Iron(II) Complexes: An N-Heterocyclic Carbene Approach, *Chem. Commun.*, 2013, **49**(49), 6412–6414.

35 Y. Liu, K. S. Kjaer, L. A. Fredin, P. Chabera, T. Harlang, S. E. Canton, S. Lidin, J. Zhang, R. Lomoth, K. E. Bergquist, *et al.*, A Heteroleptic Ferrous Complex with Mesoionic Bis(1,2,3-Triazol-5-Ylidene) Ligands: Taming the MLCT Excited State of Iron(II), *Chem.–Eur. J.*, 2015, **21**(9), 3628–3639.

36 L. A. Fredin, M. Papai, E. Rozsalyi, G. Vanko, K. Warnmark, V. Sundstrom and P. Persson, Exceptional Excited-State Lifetime of an Iron(II)-N-Heterocyclic Carbene Complex Explained, *J. Phys. Chem. Lett.*, 2014, **5**(12), 2066–2071.

37 T. C. B. Harlang, Y. Liu, O. Gordivska, L. A. Fredin, C. S. Ponseca, P. Huang, P. Chábera, K. S. Kjaer, H. Mateos, J. Uhlig, *et al.*, Iron Sensitizer Converts Light to Electrons with 92% Yield, *Nat. Chem.*, 2015, **7**(11), 883–889.

38 P. Chabera, Y. Liu, O. Prakash, E. Thyraug, A. El Nahhas, A. Honarfar, S. Essen, L. A. Fredin, T. C. B. Harlang, K. S. Kjaer, *et al.*, A Low-Spin Fe(III) Complex with 100-Ps Ligand-to-Metal Charge Transfer Photoluminescence, *Nature*, 2017, **543**, 695.

39 P. Chábera, K. S. Kjaer, O. Prakash, A. Honarfar, Y. Liu, L. A. Fredin, T. C. B. Harlang, S. Lidin, J. Uhlig, V. Sundström, *et al.*, Fe^{II}Hexa N-Heterocyclic Carbene Complex with a 528 Ps Metal-To-Ligand Charge-Transfer Excited-State Lifetime, *J. Phys. Chem. Lett.*, 2018, **9**(3), 459–463.

40 K. S. Kjær, N. Kaul, O. Prakash, P. Chábera, N. W. Rosemann, A. Honarfar, O. Gordivska, L. A. Fredin, K.-E. Bergquist, L. Häggström, *et al.*, Luminescence and Reactivity of a Charge-Transfer Excited Iron Complex with Nanosecond Lifetime, *Science*, 2019, **363**(6424), 249–253.

41 M. Reiher, O. Salomon and B. A. Hess, Reparameterization of Hybrid Functionals Based on Energy Differences of States of Different Multiplicity, *Theor. Chem. Acc.*, 2001, **107**, 48–55.

42 A. D. Becke, Density functional Thermochemistry III. The Role of Exact Exchange, *J. Chem. Phys.*, 1993, **98**(7), 5648–5652.

43 T. H. Dunning and P. J. Hay, Gaussian Basis Sets for Molecular Calculations, in *Methods of Electronic Structure Theory*, Springer US, Boston, MA, 1977, pp. 1–27.

44 J. Tomasi, B. Mennucci and R. Cammi, Quantum Mechanical Solvation Models, *Chem. Rev.*, 2005, **105**, 2999–3094.

45 M. J. Frisch, G. W. Trucks, H. B. Schlegel, G. E. Scuseria, M. A. Robb, J. R. Cheeseman, G. Scalmani, V. Barone, G. A. Petersson, H. Nakatsuji, X. Li, M. Caricato, A. Marenich, J. Bloino, B. G. Janesko, R. Gomperts, B. Mennucci, H. P. Hratchian, J. V. Ortiz, A. F. Izmaylov, J. L. Sonnenberg, D. Williams-Young, F. Ding, F. Lipparini, F. Egidi, J. Goings, B. Peng, A. Petrone, T. Henderson, D. Ranasinghe, V. G. Zakrzewski, J. Gao, N. Rega, G. Zheng, W. Liang, M. Hada, M. Ehara, K. Toyota, R. Fukuda, J. Hasegawa, M. Ishida, T. Nakajima, Y. Honda, O. Kitao, H. Nakai, T. Vreven, K. Throssell, J. A. Montgomery Jr, J. E. Peralta, F. Ogliaro, M. Bearpark, J. J. Heyd, E. Brothers, K. N. Kudin, V. N. Staroverov, T. Keith, R. Kobayashi, J. Normand, K. Raghavachari, A. Rendell, J. C. Burant, S. S. Iyengar, J. Tomasi, M. Cossi, J. M. Millam, M. Klene, C. Adamo, R. Cammi, J. W. Ochterski, R. L. Martin, K. Morokuma, O. Farkas, J. B. Foresman and D. J. Fox, *Gaussian 09, Revision A.02*, Gaussian, Inc., Wallingford CT, 2016.

46 G. Auböck and M. Chergui, Sub-50-Fs Photoinduced Spin Crossover in $[Fe(Bpy)_3]^{2+}$, *Nat. Chem.*, 2015, **7**(8), 629–633.

47 W. Zhang and K. J. Gaffney, Mechanistic Studies of Photoinduced Spin Crossover and Electron Transfer in Inorganic Complexes, *Acc. Chem. Res.*, 2015, **48**, 22.

48 A. S. Del Negro, C. J. Seliskar, W. R. Heineman, S. E. Hightower, S. A. Bryan and B. P. Sullivan, Highly Oxidizing Excited States of Re and Tc Complexes, *J. Am. Chem. Soc.*, 2006, **128**(51), 16494–16495.

49 S. Paulson, B. P. Sullivan and J. V. Caspar, Luminescent Ligand-to-Metal Charge-Transfer Excited States Based on Pentamethylcyclopentadienyl Complexes of Tantalum, *J. Am. Chem. Soc.*, 1992, 6905–6906.

50 Y. F. Lee and J. R. Kirchhoff, Absorption and Luminescence Spectroelectrochemical Characterization of a Highly Luminescent Rhenium(I1) Complex, *J. Am. Chem. Soc.*, 1994, **116**(14), 3599–3600.

Faraday Discussions

PAPER

Measuring local conformations and conformational disorder of $(Cy3)_2$ dimer labeled DNA fork junctions using absorbance, circular dichroism and two-dimensional fluorescence spectroscopy†

Dylan Heussman,‡[a] Justin Kittell,‡[b] Loni Kringle, [a] Amr Tamimi, [a] Peter H. von Hippel[c] and Andrew H. Marcus *[ac]

Received 21st December 2018, Accepted 25th January 2019

DOI: 10.1039/c8fd00245b

The sugar-phosphate backbone of DNA near single-stranded (ss)–double-stranded (ds) junctions likely fluctuates within a broad distribution of conformations to permit the proper binding of genome regulatory proteins that function at these sites. In this work we use absorbance, circular dichroism (CD), and two-dimensional fluorescence spectroscopy (2DFS) to study the local conformations and conformational disorder within chromophore-labeled DNA constructs. These constructs employ dimers of the fluorescent chromophore Cy3 that are site-specifically incorporated into the sugar-phosphate backbones of DNA strands at ss–ds DNA fork junctions. We show that these data can be analyzed to determine the local conformations of the $(Cy3)_2$ dimer, and the degree of conformational disorder. Our analysis employs an essential-state Holstein–Frenkel Hamiltonian model, which takes into account the internal electronic-vibrational motions within each Cy3 chromophore, and the resonant electronic interaction that couples the two chromophores together. Our results suggest that this approach may be applied generally to understand local backbone conformation and conformational disorder at ss–ds DNA fork junctions.

[a]*Department of Chemistry and Biochemistry, Center for Optical, Molecular and Quantum Science, University of Oregon, Eugene, OR 97403, USA. E-mail: ahmarcus@uoregon.edu*

[b]*Department of Physics, Center for Optical, Molecular and Quantum Science, University of Oregon, Eugene, OR 97403, USA*

[c]*Department of Chemistry and Biochemistry, Institute of Molecular Biology, University of Oregon, Eugene, OR 97403, USA*

† Electronic supplementary information (ESI) available: Temperature-dependent optimized parameters of model Hamiltonian fits to the absorbance spectra of Cy3 monomer duplex and fork DNA constructs. A comparison is shown between optimized fits to absorption and CD spectra using a point-dipole *versus* an extended-dipole model of the resonant coupling. Two-dimensional fluorescence spectroscopy (2DFS) measurements are presented for the Cy3 chromophore in methanol. See DOI: 10.1039/c8fd00245b

‡ These authors contributed equally to this work.

1. Introduction

The structure of genomic DNA is generally pictured as the static B-form conformation of the Watson–Crick (W–C) duplex.[1] However, many centrally important biological processes involve dynamic molecular events that require local regions of DNA to open spontaneously, thus allowing proteins to gain access to either single-stranded DNA base sequences, or to secondary structure motifs that depart from the stable W–C structure.[2] For example, the elongation events of DNA replication involve the binding of proteins to non-sequence-specific positions at or near single-stranded (ss)–double stranded (ds) DNA junctions. To facilitate elementary biochemical steps of the elongation process, the sugar-phosphate backbone of DNA near ss–ds junctions likely fluctuates into a broad distribution of functionally relevant conformations to permit the proper binding of replication proteins. Thus, the nature and extent of conformational disorder at DNA junctions may be central to the molecular mechanisms of the binding and the subsequent function of the protein–DNA complexes involved.

The carbocyanine dye Cy3 is often used as a chromophore label for proteins and nucleic acids. Such fluorescently labeled biomolecular constructs are employed in diverse applications, from rapid screening of single-molecules to imaging of subcellular components.[3–10] The Cy3 chromophore is comprised of an electronically conjugated trimethine group, which bridges two indole-like substituents (see Fig. 1A). The lowest energy $\pi \rightarrow \pi^*$ electronic transition occurs when the molecule is in its all-*trans* ground state configuration, with an electric dipole transition moment (EDTM) polarized parallel to the long axis of the trimethine bridge. The absorbance spectra of Cy3 labeled constructs exhibit a pronounced vibronic progression with energy spacing $\hbar\omega_0 \sim 1100\ \text{cm}^{-1}$.[3] The presence of the vibronic progression indicates that the electronic transition is effectively coupled (as characterized by the Huang–Rhys parameter, $\lambda^2 \approx 0.55$) to at least one local vibrational mode of relatively high frequency,[11–13] such as the Raman-active symmetric stretch of the trimethine bridge at $\sim$1200 cm^{-1}.[14]

Cyanine chromophores can be chemically attached to a nucleic acid base or to an amino acid side chain *via* a flexible linker.[3,9] Cy3 and Cy5 are often used as a double-label pair to perform Förster resonance energy transfer (FRET) experiments.[15] Such studies can provide information about inter-chromophore separation based on the resonant electronic coupling J between the optically excited Cy3 'donor' and the unexcited Cy5 'acceptor.' The relative fluorescence intensities of the Cy3/Cy5 donor–acceptor pair depends on the magnitude of J, which is most sensitive to changes in inter-chromophore separation on the scale of the Förster radius $R_0 \sim 50$ Å. FRET experiments are performed in the 'weak-coupling regime',[16] where the magnitude of J is small in comparison to the interactions between the chromophore and its local environment at room temperature ($\sim k_BT \approx 210\ \text{cm}^{-1}$).[11]

An alternative labeling scheme is to site-specifically incorporate the Cy3 chromophores 'internally' within the sugar-phosphate backbone of a single strand of DNA (see Fig. 1).[3,8,9] By annealing two complementary DNA strands, each containing a single internally-labeled Cy3 chromophore, a DNA construct can be formed with a $(Cy3)_2$ dimer (or a Cy3 monomer) at a predetermined position relative to a single-stranded (ss)–double-stranded (ds) DNA fork junction. The resulting chiral

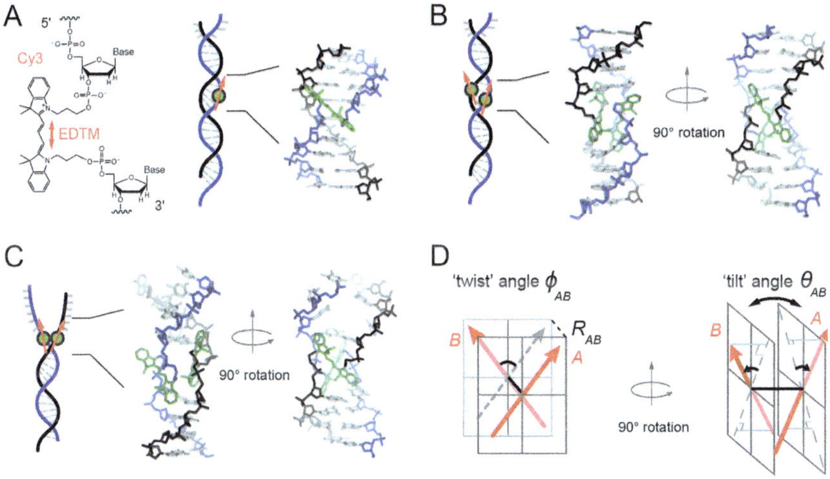

Fig. 1 Model structure of the internally labeled Cy3 monomer and $(Cy3)_2$ dimer in dsDNA. (A) The structural formula of the internally labeled Cy3 chromophore is shown with its insertion linkages to the 3′ and 5′ segments of the sugar-phosphate backbone of ssDNA. A red double-headed arrow indicates the orientation of the electric dipole transition moment (EDTM), which lies parallel to the axis of the trimethine bridge in the all-*trans* configuration. (B) A $(Cy3)_2$ dimer-dsDNA construct is formed by annealing two complementary DNA strands, which each contain a site-specifically positioned Cy3 chromophore. Space-filling structural models performed using the Spartan program (Wavefunction, Inc.) suggest that the dimer exhibits the same approximate D_2 symmetry as right-handed (B-form) helical dsDNA.[17] (C) A $(Cy3)_2$ dimer-fork DNA construct contains the dimer probe near the ss–ds DNA fork junction. The local conformation of the $(Cy3)_2$ dimer probe is expected to reflect the disruption of base-stacking interactions that occurs near the fork junction. In panels (A–C), the sugar-phosphate backbones of the conjugate strands are shown in black and blue, the bases in gray, and the Cy3 chromophores in green. (D) The structural parameters that define the local conformation of the $(Cy3)_2$ dimer are the inter-chromophore separation vector R_{AB}, the twist angle ϕ_{AB}, and the tilt angle θ_{AB}.

conformation of the $(Cy3)_2$ dimer is sterically constrained by the stabilizing interactions of the adjacently stacked bases within the DNA duplex. Conversely, the conformation of the $(Cy3)_2$ dimer can be destabilized by increasing the sample temperature or by positioning the probes close to the DNA fork junction where base stacking interactions are disrupted. Such internally-labeled Cy3 probes are thought to have only a minimal effect on the local DNA structure, as their presence in relatively short oligomeric dsDNA constructs (~30 base pairs) does not significantly alter the denaturation temperature ($T_m = 65$ °C).[9]

Due to the relatively small inter-chromophore separations within internally-labeled $(Cy3)_2$ dimer-DNA constructs ($R_{AB} \sim 5$ Å), the magnitude of the resonant coupling can be significantly greater than k_BT, approaching values comparable to the vibrational relaxation energy ($\lambda^2\hbar\omega_0 \sim 600$ cm^{-1}).[17] In this 'intermediate-to-strong coupling regime',[16] the $(Cy3)_2$ dimer forms delocalized excitons comprised of symmetric and anti-symmetric superpositions of the electronic-vibrational product states of the component Cy3 monomers.[18,19] While FRET-based experiments can provide limited information about inter-chromophore separation, experiments that can resolve the relative intensity contributions from the symmetric and anti-symmetric excitons of internally-labeled $(Cy3)_2$ dimer-DNA constructs can reveal detailed information about local dimer conformation and the ways in which fluctuations of the local environment broaden spectra and perturb the coupling. As we discuss further below, this information can be sensitively probed using absorbance, circular dichroism (CD), and two-dimensional fluorescence spectroscopy (2DFS).[17]

In previous work, we used absorbance and CD spectroscopy to study the monomer Cy3 and the dimer $(Cy3)_2$ duplex DNA constructs depicted in Fig. 1A and B, respectively, over the temperature range 15–60 °C below the dsDNA denaturation transition ($T_m = 65$ °C).[17] We analyzed our results with the aid of the Holstein–Frenkel Hamiltonian model for an electronically coupled dimer of two-electronic-level molecules, each with their electronic transition coupled to a single vibrational (harmonic) mode.[19] We found that the conformation of the $(Cy3)_2$-dsDNA construct undergoes systematic variation as a function of temperature, in terms of the inter-chromophore separation R_{AB}, the twist angle ϕ_{AB} (defined in Fig. 1D) and the degree of structural disorder. In the current work, we extend our approach to examine the local conformations of the $(Cy3)_2$ dimer-labeled fork DNA construct depicted in Fig. 1C. DNA forks and other ss–ds DNA junctions are sites of interaction for proteins that participate in DNA replication, recombination and repair.[20,21] The local conformation of the sugar-phosphate backbone and conformational disorder at the fork junction is a consequence of DNA 'breathing' – *i.e.*, thermally activated fluctuations of the folded secondary structure – that is a key factor affecting the assembly and stability of protein components.[2] To characterize the local conformations of the $(Cy3)_2$ dimer-labeled fork DNA construct, we introduce the structural parameter of the inter-chromophore tilt angle θ_{AB} (defined in Fig. 1D) to account for the disruption of hydrogen bonds and stacking of bases within the ss region adjacent to the $(Cy3)_2$ dimer. With the inclusion of this additional orientational parameter, we use the $(Cy3)_2$ dimer-labeled fork DNA construct as a test system to examine the reliability of the point-dipole approximation in estimating the resonant coupling strength J.

In general, the resonant electronic coupling J is determined by integrating the Coulomb interactions between electric transition charge densities of the component monomers.[11] Direct calculation of the atomic transition charge density by quantum chemical methods would provide the most accurate determination of the coupling strength, albeit at a significant computational cost.[22] For expediency, the point-dipole approximation is often invoked, where the transition charge density is focused at a single point, and J is described as the interaction between two point-dipole moments. Of course, for inter-chromophore separations comparable to the monomer size, the point-dipole approximation cannot reflect the details of the transition charge density, which is extended across the length of the trimethine bridge and indole substituents of the Cy3 molecule. To examine the reliability of the point-dipole model we here implement an extended-dipole model,[23] which includes a displacement parameter $\vec{l}$ to account for the length over which the transition charge q is distributed. In the extended dipole model, the displacement parameter is oriented parallel to the EDTM ($\vec{\mu}$), and the magnitudes of the transition charge and displacement satisfy $ql = \mu$.[22–24] We have performed our analyses for both the $(Cy3)_2$-ds DNA and $(Cy3)_2$-fork DNA constructs using both the point-dipole and extended dipole models. Our results indicate that both models yield very similar values for the structural parameters of these DNA constructs over the full range of temperatures studied.

In addition to determining the average local conformations adopted by the backbone labels within the $(Cy3)_2$-dsDNA and $(Cy3)_2$-fork DNA constructs, we further examine structural disorder by interpreting inhomogeneous spectral line broadening as an indicator of this effect. Molecular dynamics simulations suggest that the Cy3 chromophore(s), when incorporated internally within the DNA duplex, may exist in a variety of conformations.[25] This dispersion is due to the coupling between the Cy3 probe chromophores and the fluctuating local DNA environment, which induces the transition energies of the Cy3 probes to occur over a broad spectral range. We show that the results of 2DFS experiments performed on these systems – taken in combination with absorbance and CD spectroscopy – can be used to determine the homogeneous and inhomogeneous line broadening and the corresponding degree of structural disorder.

2. Experimental methods

2.1 Sample preparation

We show in Table 1 the sequences and nomenclature of the internally labeled Cy3 monomer and $(Cy3)_2$ dimer DNA constructs used in this work. Oligonucleotide samples were purchased from Integrated DNA Technologies (IDT, Coralville, IA) and used as received. For our absorbance, circular dichroism (CD), and two-dimensional fluorescence spectroscopy (2DFS) measurements, we prepared solutions with sample concentrations of 1 μM and a standard aqueous buffer of 10 mM Tris, 100 mM NaCl, and 6 mM $MgCl_2$. We combined complementary oligonucleotide strands to form the Cy3 monomer- and $(Cy3)_2$ dimer-labeled DNA constructs, which contain both ds and ss regions. The Cy3 monomer-labeled constructs contained a thymine base (T) in the complementary strand position directly opposite to the Cy3 probe chromophore. For the 'duplex' constructs, the Cy3 probes are positioned deep within the double-stranded region of the DNA. For the 'fork' constructs, the Cy3 probes are positioned near the ss–ds DNA fork junction. Prior to the experiments, the sample solutions were annealed by heating to 95 °C for 3 minutes before they were allowed to slowly cool to room temperature. For our control 2DFS experiments performed on the Cy3 chromophore in methanol (see ESI† section), we purchased 1,1′-diethyl-3,3,3′,3′-tetramethylindocarbocyanine iodide (Cy3) from Toronto Research Chemicals, Inc., which we used without further purification. Spectroscopic grade methanol was used to prepare solutions of concentration 1×10^{-3} M.

2.2 Absorbance and CD measurements

We performed absorbance measurements using a Cary 3E UV-Vis spectrophotometer, and CD using a Jasco model J-720 CD spectrophotometer. Temperature-dependent measurements were performed over the range 15–75 °C. Samples were held in a 1 cm quartz cuvette, and the temperature was controlled to within 0.1 °C using a Peltier thermoelectric heating block. We determined all spectra over the range 200–700 nm to examine the spectral region of the native bases (~275 nm), in addition to that of the Cy3 probe(s) (~540 nm). For all of the samples, we confirmed

Table 1 Base sequences and nomenclature for the Cy3 monomer and $(Cy3)_2$ dimer labeled DNA constructs used in these studies. The horizontal line indicates the regions of complementary base pairing

DNA construct	Nucleotide base sequences
Cy3 monomer duplex DNA	3′-GTC AGT ATT ATA CGC T**Cy3**C GCT AAT ATA TAC GTT TTT TTT TTT TTT TTT TTT TTT TTT TTT T-5′ 5′-CAG TCA TAA TAT GCG A T G CGA TTA TAT ATG CTT TTA CCA CTT TCA CTC ACG TGC TTA C-3′
$(Cy3)_2$ dimer duplex DNA	3′-GTC AGT ATT ATA CGC T**Cy3**C GCT AAT ATA TAC GTT TTT TTT TTT TTT TTT TTT TTT TTT TTT T-5′ 5′-CAG TCA TAA TAT GCGA**Cy3**G CGA TTA TAT ATG CTT TTA CCA CTT TCA CTC ACG TGC TTA C-3′
Cy3 monomer fork DNA	3′-GAG GGA GCA CAG CAG AGG TCA GTA TTA TAC GCT **Cy3**CG CTG GTA TAC CAC GTT T (x29)-5′ 5′-CTC CCT CGT GTC GTC TCC AGT CAT AAT ATG CGA T AT GCT TTT ACC ACT TTC ACT CAG GTG CTT A-3′
$(Cy3)_2$ dimer fork DNA	3′-GAG GGA GCA CAG CAG AGG TCA GTA TTA TAC GCT **Cy3**CG CTG GTA TAC CAC GTT T (x29)-5′ 5′-CTC CCT CGT GTC GTC TCC AGT CAT AAT ATG CGA **Cy3**AT GCT TTT ACC ACT TTC ACT CAG GTG CTT A-3′

that the ds region adopted the anticipated Watson–Crick B-form conformation by examination of the ultraviolet absorbance and CD of the native bases.[26]

In Fig. 2, we show absorbance and CD spectra at 25 °C for the Cy3 monomer- and dimer-labeled duplex DNA constructs, and the $(Cy3)_2$ dimer-labeled fork DNA construct. We note that the spectra of the Cy3 monomer fork DNA construct (not shown) are qualitatively similar to those of the corresponding duplex construct. The absorbance spectra of the Cy3 monomer duplex and fork DNA constructs exhibit a progression of vibronic features with the first (0–0) peak centered at 549 nm (18 280 cm^{-1}). The vibronic progression is still present in the spectrum of both the $(Cy3)_2$ dimer-labeled duplex DNA construct and the dimer-labeled fork DNA construct. However, individual vibronic features of the dimer are broadened relative to those of the monomer, and the ratio of the 0–0 to 1–0 vibronic peak intensities has decreased relative to that of the monomer $[I_{\text{mon}}^{(0-0)}/I_{\text{mon}}^{(1-0)} = 1.60]$. While the monomer CD signal is very weak (as expected), the CD of both the dimer duplex and fork constructs exhibit a progression of bisignate lineshapes (*i.e.* a change of sign within a given vibronic band), which is a signature of vibronic excitons in a chiral dimer.[19,27,28] Furthermore, the CD spectra of the $(Cy3)_2$ dimer-duplex and fork DNA constructs have opposite signs, indicating that the two systems have opposite chiral symmetries.

2.3 Phase modulated two-dimensional fluorescence spectroscopy (2DFS)

To obtain estimates of the homogeneous line widths for the samples listed in Table 1, we performed phase-modulated 2DFS experiments. These measurements were carried out as previously described.[29–32] 2DFS utilizes a train of four collinear laser pulses. The relative phase of pulses 1 and 2, and that of pulses 3 and 4, are continuously swept at the frequencies $\Omega_{21} = 5$ kHz and $\Omega_{43} = 8$ kHz, respectively. Fluorescence from the optically excited sample is detected as a function of the time delay between the first two pulses (t_{21}) and that of the second two pulses (t_{43}). For the experiments reported here, the delay between pulses two and three (t_{32}) is set to zero. The phase-selective fluorescence detection method[29,33] is used to extract the fourth-order population signals,[30] which vary in time at the sum $[(\Omega_{43} + \Omega_{21}) = 13$ kHz$]$ and difference $[(\Omega_{43}$ –

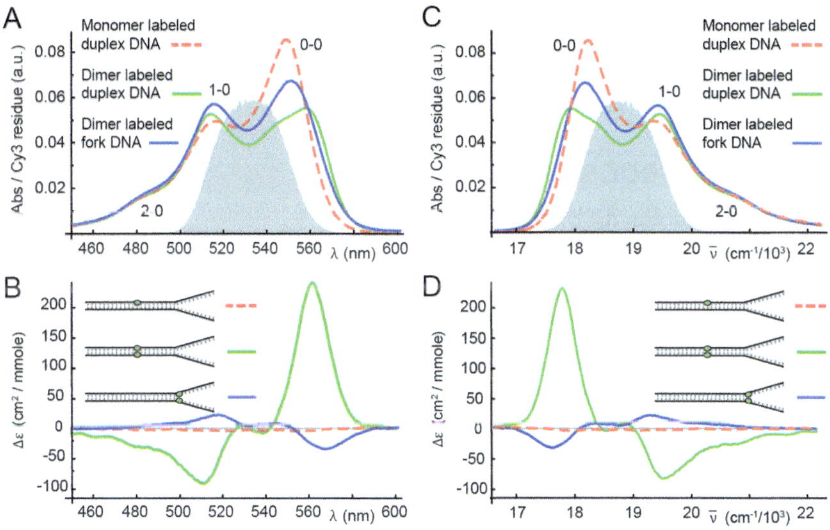

Fig. 2 Room temperature (25 °C) absorbance (A & C) and CD (B & D) spectra for Cy3 monomer duplex (dashed red), $(Cy3)_2$ dimer duplex (solid green), and $(Cy3)_2$ dimer fork (solid blue) DNA constructs. The $\Delta\varepsilon$ here is the differential absorbance of left and right circular polarized light. Nucleotide sequences and placement of the chromophore probes are indicated in Table 1. The spectra are shown as a function of optical wavelength (A & B) and as a function of wavenumber (C & D). The vibronic features of the monomer absorption spectra are labeled n_e–0, where n_e (=0, 1, 2) indicates the vibrational occupancy of the electronically excited monomer. An example laser spectrum used for 2DFS experiments is shown in gray. The laser spectrum spans a region containing both the 0–0 and 1–0 vibronic sub-bands.

$\Omega_{21}) = 3$ kHz] modulation frequencies associated with the non-rephasing (NRP) and rephasing (RP) signals, respectively.

The four laser pulses were generated using a single high-repetition-rate non-collinear optical parametric amplifier (NOPA) with the excitation centered at 532 nm (18 796 cm^{-1}) and a FWHM bandwidth of 31 nm (1100 cm^{-1}) for experiments on all samples performed at room temperature, and a bandwidth of 37 nm (1300 cm^{-1}) for control 2DFS measurements performed on the Cy3 chromophore in methanol (see ESI† section). Fluorescence was detected using a 615 nm long-pass filter (Chroma, HQ615LP), which served to reject scattered excitation light. To eliminate optical saturation effects, solutions were continuously circulated through the cuvette using a peristaltic pump. Pulses were compressed using a quadruple-pass fused-silica prism pair to compensate for dispersive media in the optical path preceding the sample. Pulse widths were characterized by placing a beta-barium borate (BBO) frequency doubling crystal at the sample position, where a phase-modulated train of pulse-pairs was incident. The frequency-doubled signal output was detected using a lock-in amplifier, which was referenced to the ac carrier signal used to modulate the relative phase of the pulses, as previously described.[29] We thus minimized the pulse width $\Delta\tau_L = \sim$14 fs by performing a pulse–pulse autocorrelation. We measured the laser bandwidth $\Delta\lambda_L$ centered at $\lambda_L = 532$ nm using an Ocean Optics mini-spectrometer. The measured time-bandwidth product was thus ~

$\Delta\tau_L(\Delta\lambda_L c/\lambda_L^2) \sim 0.55$, which is within 25% of the optimal value (0.44) for Fourier-transform-limited Gaussian pulses. Results from these measurements are presented below.

3. Theoretical modeling

3.1 Absorbance and CD spectra

Building upon previous work,[17] we implemented the Holstein–Frenkel (H–F) model to describe the electronic-vibrational structure of the $(Cy3)_2$ dimer probe, similar to the approach taken by others.[19,34–38] The H–F model accounts for the interactions between electronic and vibrational degrees of freedom internal to each chromophore, and the conformation-dependent resonant electronic coupling J between chromophores. The presence of both types of interactions (excitonic and vibronic) gives rise to delocalized symmetric and anti-symmetric transitions of well-defined energies and polarizations, which feature prominently in absorbance and CD spectra. Because the positions and amplitudes of the spectroscopic features depend sensitively on the conformation of the $(Cy3)_2$ dimer, an optimization procedure that compares simulated to experimental spectra[17] allows us to extract structural information about the $(Cy3)_2$ dimer.

In the H–F model, each Cy3 monomer is treated as a two-electronic-level molecule with ground state $|g\rangle$, excited state $|e\rangle$, and transition energy ε_{eg}. The monomers are labeled A and B, with electric dipole transition moment (EDTM) $\vec{\mu}_{eg}^{A(B)}$. The electronic transition of each Cy3 monomer is coupled to a single harmonic mode with frequency ω_0. The monomer Hamiltonian is given by

$$\hat{H}_{A(B)} = \varepsilon_{eg}|e\rangle_{A(B)}\langle e| + \hbar\omega_0\hat{b}^\dagger_{A(B)}\hat{b}_{A(B)} + \hbar\omega_0\{\lambda(\hat{b}^\dagger_{A(B)} + \hat{b}_{A(B)}) + \lambda^2\}|e\rangle_{A(B)}\langle e| \quad (1)$$

In eqn (1), the electronic-vibrational coupling is characterized by the Huang–Rhys parameter $\lambda^2 = d^2\omega_0/2\hbar$, where d is the vibrational coordinate displacement of the electronic excited state potential minimum relative to the electronic ground state. The value of λ^2 is the number of vibrational quanta absorbed by the system upon optical excitation. The operator $\hat{b}^\dagger_{A(B)}(\hat{b}_{A(B)})$ creates (annihilates) a vibrational excitation in the un-shifted electronic potential surface. The dimer Hamiltonian is

$$\hat{H}_{dim} = \hat{H}_A + \hat{H}_B + J\{|eg\rangle\langle ge| + |ge\rangle\langle eg|\} \quad (2)$$

where the final term couples the singly excited electronic states of the monomers. Here $|eg\rangle$ is the state in which monomer A is electronically excited and monomer B is unexcited, and $|ge\rangle$ is the state in which the A and B labels are interchanged. The resonant electronic coupling J depends on dimer conformation in terms of the Coulomb interaction between monomer site transition charge densities[11]

$$J = (4\pi\varepsilon\varepsilon_0)^{-1}\int_{-\infty}^{\infty} d\vec{r}_A \int_{-\infty}^{\infty} d\vec{r}_B \rho_A^{ge}(\vec{r}_A)\rho_B^{eg}(\vec{r}_B)/|\vec{R}_{AB}| \quad (3)$$

where $\rho_{A(B)}^{ge}(\vec{r}_{A(B)}) = \langle g|_{A(B)}\hat{\rho}(\vec{r}_{A(B)})|e\rangle_{A(B)}$.

The delocalized excitons and corresponding energies of the coupled AB dimer are obtained by diagonalization of the Hamiltonian given by eqn (2).[17] The singly excited states are symmetric (+) and anti-symmetric (−) superpositions of electronic-vibrational products $\left|e_{\pm}^{(\alpha)}\right\rangle = \sum_{n_e,n_g} c_{\pm}^{(\alpha)}[u_{n_e,n_g}|\mathrm{eg}\rangle \pm u_{n_g,n_e}|\mathrm{ge}\rangle]$. Here, the coefficients $c_{\pm}^{(\alpha)}u_{n_e,n_g}$ depend on the vibrational coordinates of monomers A and B, $n_{g(e)}$ (=0, 1,...) is the number of vibrational excitations in the un-shifted (shifted) potential of an unexcited (excited) monomer, and α (=0, 1,...) is an index in order of increasing state energy.[19]

The absorbance spectrum is the sum of symmetric (+) and anti-symmetric (−) exciton features

$$\sigma_{\mathrm{H}}(\varepsilon) = \sigma_{\mathrm{H}+}(\varepsilon) + \sigma_{\mathrm{H}-}(\varepsilon) \tag{4}$$

where $\sigma_{\mathrm{H}\pm}(\varepsilon) = \sum_{\alpha}\left|\left\langle 0\left|\vec{\mu}^{\mathrm{tot}}\right|e_{\pm}^{(\alpha)}\right\rangle\right|^2 L_{\mathrm{H}}(\varepsilon - \varepsilon_{\pm,\alpha})$, $\vec{\mu}^{\mathrm{tot}} = \vec{\mu}_{\mathrm{eg}}^{\mathrm{A}} + \vec{\mu}_{\mathrm{eg}}^{\mathrm{B}}$ is the collective EDTM, and $L_{\mathrm{H}}(\varepsilon) = \frac{1}{2}\Gamma_{\mathrm{H}}\Big/\left[\varepsilon^2 + \left(\frac{1}{2}\Gamma_{\mathrm{H}}\right)^2\right]$ is a Lorentzian function that represents the homogeneous lineshape of the transition with eigen-energy $\varepsilon_{\pm,\alpha}$ and FWHM line width Γ_{H}. Similarly, the CD spectrum is the sum of symmetric and anti-symmetric rotational strengths

$$\mathrm{CD}_{\mathrm{H}}(\varepsilon) = \sum_{\alpha}\left[\mathrm{RS}_{\mathrm{H}+}^{(\alpha)} L_{\mathrm{H}}(\varepsilon - \varepsilon_{+,\alpha}) + \mathrm{RS}_{\mathrm{H}-}^{(\alpha)} L_{\mathrm{H}}(\varepsilon - \varepsilon_{-,\alpha})\right] \tag{5}$$

where $\mathrm{RS}_{+}^{(\alpha)} = \frac{\varepsilon_{\mathrm{eg}}}{4\hbar c\left|\vec{\mu}_{\mathrm{eg}}\right|^2}\langle 0\left|\vec{\mu}_{\mathrm{eg}}^{\mathrm{A}}\right|e_{\pm}^{(\alpha)}\rangle\langle e_{\pm}^{(\alpha)}\left|\vec{\mu}_{\mathrm{eg}}^{\mathrm{B}}\right|0\rangle\vec{R}_{\mathrm{AB}}$. In the above expressions, we have defined the ground electronic-vibrational state of the AB dimer $|0\rangle = |\mathrm{gg}\rangle$. Because the dimer conformation is subject to disorder due to DNA breathing, the homogeneous absorbance and CD line shapes are convolved with an inhomogeneous distribution function $G_{\mathrm{I}}(\varepsilon_{\pm,\alpha}) = \exp[-(\varepsilon_{\pm,\alpha} - \bar{\varepsilon}_{\pm,\alpha})^2/2\sigma_{\mathrm{I}}^2]$, which is centered at the average transition energy $\bar{\varepsilon}_{\pm,\alpha}$ and has standard deviation σ_{I}. The final expressions for the absorbance and CD spectra are given by the Voigt profiles $\sigma(\varepsilon) = \int_{-\infty}^{\infty} \mathrm{d}\varepsilon'\sigma_{\mathrm{H}}(\varepsilon - \varepsilon')G_{\mathrm{I}}(\varepsilon')$ and $\mathrm{CD}(\varepsilon) = \int_{-\infty}^{\infty} \mathrm{d}\varepsilon'\mathrm{CD}_{\mathrm{H}}(\varepsilon - \varepsilon')G_{\mathrm{I}}(\varepsilon')$, respectively.

3.2 Two-dimensional fluorescence spectroscopy

We simulate 2DFS spectra based on the H–F Hamiltonian using previously developed methods.[30,39] We use $|\mathrm{v}\rangle$ to represent the electronic ground state with vibrational quantum number $\mathrm{v} = \mathrm{v_A} + \mathrm{v_B}$. For example, the state $|0\rangle$ is the electronic ground state with zero vibrational occupancy. The states $|\mathrm{e}\rangle$ and $|\mathrm{e}'\rangle$ represent any two of the symmetric and anti-symmetric excitons $|e_{\pm}^{(\alpha)}\rangle$ within the singly excited manifold, and the state $|\mathrm{f}\rangle$ represents any one state within the doubly excited manifold. Rephasing (RP) and non-rephasing (NRP) response functions are written[30]

$$S_{\mathrm{RP}}(t_{21}, t_{32} = 0, t_{43}) \propto -\left(Q_{4\mathrm{a}} + Q_{3\mathrm{a}} + Q_{2\mathrm{b}}^{*} - \Gamma_{2\mathrm{D}}Q_{8\mathrm{b}}^{*}\right) \tag{6}$$

and

$$S_{\mathrm{NRP}}(t_{21}, t_{32} = 0, t_{43}) \propto -\left(Q_{5\mathrm{a}}^{*} + Q_{2\mathrm{a}} + Q_{3\mathrm{b}}^{*} - \Gamma_{2\mathrm{D}}Q_{7\mathrm{b}}\right) \tag{7}$$

In eqn (6) and (7), the first two terms on the right hand side of the proportionalities represent, respectively, ground state bleach and stimulated emission. The final two terms are excited state absorption for the singly and doubly excited state manifolds, respectively. The parameter Γ_{2D}, which may assume values between zero and two, is the fluorescence quantum yield of the doubly excited state manifold relative to that of the singly excited state manifold. When the effects of inhomogeneous broadening are included in the response function,[40] the individual terms are written for RP:

$$Q_{4a} = \sum_{v,e,e'} \left[\mu_{0e}\mu_{ev}\mu_{0e'}\mu_{ve'}\right]_{e_1e_2e_3e_4} e^{-\Gamma_H(t_{21}+t_{43}) - \frac{1}{2}\sigma_I^2(t_{21}-t_{43})^2 + i\left(\omega_{e'v}t_{43} - \omega_{e0}t_{21}\right)} \tag{8}$$

$$Q_{3a} = \sum_{v,e,e'} \left[\mu_{0e}\mu_{e'0}\mu_{ev}\mu_{ve'}\right]_{e_1e_2e_3e_4} e^{-\Gamma_H(t_{21}+t_{43}) - \frac{1}{2}\sigma_I^2(t_{21}-t_{43})^2 + i\left(\omega_{e'v}t_{43} - \omega_{e0}t_{21}\right)} \tag{9}$$

$$Q_{2b}^* = Q_{8b}^* = \sum_{e,e',f} \left[\mu_{0e}\mu_{e'0}\mu_{fe'}\mu_{ef}\right]_{e_1e_2e_3e_4} e^{-\Gamma_H(t_{21}+t_{43}) - \frac{1}{2}\sigma_I^2(t_{21}-t_{43})^2 + i\left(\omega_{fe}t_{43} - \omega_{e0}t_{21}\right)} \tag{10}$$

and for NRP:

$$Q_{5a}^* = \sum_{v,e,e'} \left[\mu_{e0}\mu_{ve}\mu_{e'v}\mu_{0e'}\right]_{e_1e_2e_3e_4} e^{-\Gamma_H(t_{21}+t_{43}) - \frac{1}{2}\sigma_I^2(t_{21}+t_{43})^2 + i\left(\omega_{e'0}t_{43} + \omega_{e0}t_{21}\right)} \tag{11}$$

$$Q_{2a} = \sum_{v,e,e'} \left[\mu_{e0}\mu_{0e'}\mu_{e'v}\mu_{ve}\right]_{e_1e_2e_3e_4} e^{-\Gamma_H(t_{21}+t_{43}) - \frac{1}{2}\sigma_I^2(t_{21}+t_{43})^2 + i\left(\omega_{e'v}t_{43} + \omega_{e0}t_{21}\right)} \tag{12}$$

$$Q_{3b}^* = Q_{7b} = \sum_{e,e',f} \left[\mu_{e0}\mu_{0e'}\mu_{fe'}\mu_{e'f}\right]_{e_1e_2e_3e_4} e^{-\Gamma_H(t_{21}+t_{43}) - \frac{1}{2}\sigma_I^2(t_{21}+t_{43})^2 + i\left(\omega_{fe}t_{43} + \omega_{e0}t_{21}\right)} \tag{13}$$

In eqn (8)–(13), the factors $[\mu_{ab}\mu_{cd}\mu_{jk}\mu_{lm}]_{e_1e_2e_3e_4}$ denote the four-point product $\langle(\mu_{ab}\cdot e_1)(\mu_{cd}\cdot e_2)(\mu_{jk}\cdot e_3)(\mu_{lm}\cdot e_4)\rangle$ that takes into account the projections of the transition dipole moments onto the (parallel) plane polarizations of the four laser pulses, which are averaged over an isotropic distribution of dimer orientations. The complex-valued 2DFS spectra are obtained by Fourier transformation of the RP and NRP response functions [eqn (6) and (7)] with respect to the delay variables t_{21} and t_{43}.

3.3 Estimation of the resonant electronic coupling

In the point-dipole approximation, the electronic coupling is given by[30]

$$J = |\mu_{eg}|^2(4\pi\varepsilon\varepsilon_0)^{-1}|\vec{R}_{AB}|^{-3}[\hat{d}_A\cdot\hat{d}_B - 3(\hat{d}_A\cdot\hat{R}_{AB})(\hat{R}_{AB}\cdot\hat{d}_B)] \tag{14}$$

where $\hat{d}_A$ and $\hat{d}_B$ are unit vectors that specify the monomer EDTM directions (see Fig. 1D). The point-dipole approximation provides a reasonable value for J when the inter-chromophore separation is much greater than the molecular size.[11] In our previous studies, we estimated the long-axis dimension of the Cy3 chromophore to be ~14 Å and the inter-chromophore separation to be $R_{AB} \sim 6$ Å.[17] In our current work, we have carried out additional analyses using the extended-dipole model,[23,24] which includes a one-dimensional displacement parameter $\vec{l}$ to account for the finite size of the chromophore. Each dipole is represented as two point charges of equal magnitude and opposite sign ($\pm q$) separated by the distance l. The direction of $\vec{l}$ is the same as that of the monomer EDTM. The resonant coupling between monomers A and B in the extended-dipole model is

$$J = \left|\mu_{\mathrm{eg}}\right|^{2}(4\pi\varepsilon\varepsilon_{0})^{-1}l^{-2}\left[\frac{1}{R_{\mathrm{AB}}^{++}} - \frac{1}{R_{\mathrm{AB}}^{-+}} - \frac{1}{R_{\mathrm{AB}}^{+-}} + \frac{1}{R_{\mathrm{AB}}^{++}}\right] \tag{15}$$

where $ql = \mu_{\mathrm{eg}}$ and the distances between point charges are $R_{\mathrm{AB}}^{\pm\pm} = [R_{\mathrm{AB}} \pm l(\hat{d}_{\mathrm{A}} - \hat{d}_{\mathrm{B}})/2]$, $R_{\mathrm{AB}}^{-+} = [R_{\mathrm{AB}} - l(\hat{d}_{\mathrm{A}} + \hat{d}_{\mathrm{B}})/2]$, and $R_{\mathrm{AB}}^{+-} = [R_{\mathrm{AB}} + l(\hat{d}_{\mathrm{A}} + \hat{d}_{\mathrm{B}})/2]$.[24] For all of the calculations that follow, we used the measured EDTM value $\mu_{\mathrm{eg}} = 12.8D$, which we determined by integration of the absorbance lineshapes of the Cy3 monomer duplex and fork DNA constructs.[17] We estimated $l = 7$ Å by assuming the same value that was used by Knoester and co-workers to model the coupling between carbocyanine dyes of similar structure to Cy3.[24] The above values imply $q = 0.38e$ (where e is the charge of an electron). In both extended- and point-dipole models, the value of J depends on structural parameters that specify the Cy3 dimer conformation – *i.e.*, the inter-chromophore separation R_{AB}, the twist angle ϕ_{AB}, and the tilt angle θ_{AB} (see Fig. 1D).

In Fig. 3, we compare the results of calculations for the resonant electronic coupling based on the point-dipole model (shown in green) *versus* the extended-dipole model (shown in blue). In these calculations, we have set the inter-chromophore separation $R_{\mathrm{AB}} \sim 5$ Å, which is close to the value obtained from our previous analyses for the $(\mathrm{Cy3})_2$ dimer duplex DNA construct.[17] In Fig. 3A–C, the twist angle ϕ_{AB} is varied for fixed values of the tilt angle $\theta_{\mathrm{AB}} = 0$, 60 and 100°, respectively. For small θ_{AB}, the point-dipole and extended-dipole models exhibit qualitatively similar sinusoidal dependencies on ϕ_{AB}. However, the point-dipole approximation generally overestimates the magnitude of the coupling strength for arbitrary angles. The effect of the finite size of the chromophores, which is qualitatively captured by the extended-dipole model, is revealed for the case of a side-by-side geometry with $\phi_{\mathrm{AB}} = 0°$, and subsequently varying the tilt angle (see Fig. 3D). Here the extended-dipole model exhibits divergences at $\theta_{\mathrm{AB}} = 90°$ and 270°, which are due to the repulsive close-encounters between like charges, and a local minimum at the head-to-tail geometry of $\theta_{\mathrm{AB}} = 180°$, which is dominated by the attractive interaction between opposite charges. This 'finite-size effect' is entirely neglected by the point-dipole model, which varies smoothly with tilt angle, and reaches a local maximum at $\theta_{\mathrm{AB}} = 180°$. The stability of the head-to-tail conformation is most significant for increasing values of ϕ_{AB} (see Fig. 3E and F), in which case the repulsive barriers that are present for small values of ϕ_{AB} vanish. The above comparison indicates that, in principle, very different results might be obtained from analyses that employ the point-dipole model *versus* the physically more realistic extended-dipole model. Nevertheless, as we discuss further below, the results of our optimization analysis for the $(\mathrm{Cy3})_2$ dimer DNA constructs reveal that the two models produce very similar values for the conformational parameters, thus suggesting that the states accessible to the dimer are those for which the two models agree.

3.4 Multi-parameter optimization procedure

To efficiently explore the space of structural parameters needed to model the absorbance, CD and 2DFS of the $(\mathrm{Cy3})_2$ dimer DNA constructs, we implemented an automated multi-variable regression analysis. The procedure is similar to one

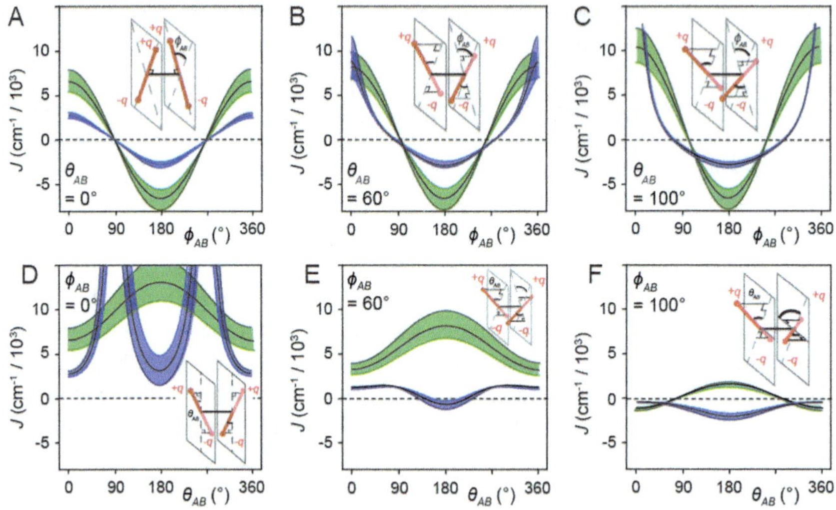

Fig. 3 Calculations of the resonant electronic coupling J based on the point-dipole (shown in green) and extended-dipole (shown in blue) models. In the top row, J is plotted as a function of the twist angle ϕ_{AB} for tilt $\theta_{AB} = 0°$ (A), 60° (B), and 100° (C), and in the bottom row as a function of θ_{AB} for $\phi_{AB} = 0°$ (D), 60° (E), and 100° (F). In these calculations, we have used the following values for the transition dipole moment $\mu_{eg} = 12.8D$, the extension length $l = 7$ Å, the transition charge $q = 0.38e$, and the inter-chromophore separation $R_{AB} = 5 \pm 0.3$ Å. The error bars in J correspond to the upper and lower values adopted for $R_{AB} = 5 \pm 0.3$ Å.

we have used in the past,[17,30,39,41] in which a random search algorithm generates an initial set of input parameters, and commercial software (KNITRO)[42] is used to refine the corresponding solutions. For each set of input trial parameters, we calculate a linear least-squares target function χ^2, which guides the selection of parameter values for subsequent iterations. The optimized solutions correspond to minimization of the target function. Error bars associated with the optimized parameters were determined by a 1% deviation of the target function from its minimized value. The results of our optimization analysis of the dimer spectra are presented in Table 2 and 3, and discussed further below.

3.5 Comparison between point-dipole and extended-dipole models

We performed optimization calculations to simulate the absorbance and CD spectra of the Cy3 labeled DNA constructs (see Table 1). For these calculations, we used values that were previously established for the monomer EDTM, $\mu_{eg} = 12.8D$, and the homogeneous line width, $\Gamma_H = 186$ cm^{-1}.[17] For calculations of the Cy3 monomer duplex and fork DNA constructs, we applied the monomer Hamiltonian given by eqn (1) to our temperature-dependent data, and thereby determined optimized values for the monomer electronic transition energy ε_{eg}, the Huang–Rhys parameter λ^2, the vibrational frequency ω_0, and the spectral inhomogeneity for the monomer specified by the Gaussian standard deviation $\sigma_{I,mon}$. The results of this analysis for the Cy3 monomer duplex and Cy3 monomer fork DNA constructs are listed in Tables S1 and S2 of the ESI† section, respectively. For both duplex and fork DNA constructs, the Cy3 monomer (intra-chromophore)

Table 2 Structural parameters of the $(Cy3)_2$ dimer DNA duplex construct at various temperatures using, alternatively, the point-dipole and extended-dipole models for the resonant coupling J. The reported values are based on optimized fits of the Holstein–Frenkel Hamiltonian model to absorbance and circular dichroism spectra. The calculations use as inputs the EDTM $\mu_{eg} = 12.8D$, the homogeneous line width $\Gamma_H = 186\ \text{cm}^{-1}$,[17] and for each temperature, the electronic transition energy ε_{eg}, the vibrational mode frequency ω_0, and the Huang–Rhys parameter λ^2 obtained from analyses of the Cy3 monomer DNA construct absorbance spectra (see Tables S1 and S2 of the ESI section). For the extended dipole model calculations, the extension length $l = 7.1$ Å and the effective charge $q = 0.38e$ were used. The parameters listed are the resonant coupling strength J, the inter-chromophore twist angle ϕ_{AB}, the inter-chromophore tilt angle θ_{AB}, the inter-chromophore separation R_{AB}, and the standard deviation of the Gaussian inhomogeneous disorder function $\sigma_{I,dim}$. Structural parameters are presented at temperatures below the melting transition at 65 °C, for which the dimer model may be reasonably applied. Error bars were calculated based on a 1% deviation of the target function from its optimized value

T (°C)	J (cm^{-1})	ϕ_{AB} (°)	θ_{AB} (°)	R_{AB} (Å)	$\sigma_{I,dim}$ (cm^{-1})
Point dipole model					
15	527 + 32/−11	82.9 ± 0.14	1.52 + 6.2/−9.2	5.8 ± 0.04	292 ± 5.3
25	512 + 33/−11	79.9 ± 0.22	7.02 + 4.4/−18	6.6 ± 0.05	302 ± 5.1
35	495 + 34/−12	76.0 ± 0.32	0.00 ± 10.5	7.4 ± 0.06	313 ± 5.2
45	482 + 35/−13	72.0 ± 0.44	6.90 + 6.7/−20	8.1 ± 0.07	325 ± 5.2
55	465 + 33/−13	70.2 ± 0.48	9.28 + 5.8/−24	8.4 ± 0.07	336 ± 5.0
65	361 + 40/−10	73.5 ± 0.5	0.71 + 14/−16	8.6 ± 0.10	336 ± 8.2
Extended dipole model					
15	532 + 24/38	80.7 + 0.2/−0.1	18.1 + 2.9/−2.1	3.7 ± 0.1	289 ± 5.2
25	514 + 14/−29	79.6 ± 0.21	10.2 + 6.4/−27	4.4 ± 0.07	302 ± 5.1
35	496 + 26/−19	76.0 ± 0.32	3.0 + 13/−19	5.5 ± 0.08	315 ± 5.5
45	483 + 32/−16	71.9 ± 0.43	7.7 + 8.5/−24	6.4 ± 0.08	325 ± 5.2
55	467 + 31/−15	70.4 ± 0.46	7.0 + 8.4/−22	6.8 ± 0.09	336 ± 5.0
65	354 + 34/−17	73.7 ± 0.6	5.4 + 11/−22	7.1 ± 0.13	353 ± 5.1

parameters ε_{eg}, λ^2, and ω_0 do not exhibit significant temperature dependences, while the monomer inhomogeneity parameter $\sigma_{I,mon}$ increases with temperature.

We next performed calculations for the $(Cy3)_2$ dimer duplex and fork DNA constructs based on the dimer Hamiltonian given by eqn (2). For each temperature, we used as inputs the corresponding monomer parameters listed in Tables S1 and S2 of the ESI† section. We thus determined optimized values for the $(Cy3)_2$ inter-chromophore separation R_{AB}, the inter-chromophore twist angle ϕ_{AB}, the inter-chromophore tilt angle θ_{AB}, and the spectral inhomogeneity parameter for the dimer, $\sigma_{I,dim}$. As described above, we performed these calculations using, alternatively, the point-dipole [eqn (14)] and the extended-dipole [eqn (15)] model for the resonant coupling. In Fig. S1 of the ESI† section, we directly compare the results of our optimizations using either method to experimental spectra taken at 15 °C, for both the $(Cy3)_2$ dimer duplex and fork DNA constructs. We find that equally favorable agreement between experimental and simulated spectra can be achieved using either the point-dipole or the extended-dipole models. Moreover, both models produce similar values for the optimized parameters J, ϕ_{AB}, θ_{AB}, and $\sigma_{I,dim}$, while the extended-dipole model generally produces smaller values for inter-chromophore separation R_{AB} than does the point-dipole model. This behavior is consistent with the relatively small values of the tilt angle θ_{AB} for both

Table 3 Structural parameters of the $(Cy3)_2$ dimer DNA fork construct at various temperatures using, alternatively, the point-dipole and extended-dipole models to determine the resonant coupling *J*. The parameters and conditions for performing these calculations are the same as those described in Table 2

T (°C)	J (cm^{-1})	ϕ_{AB} (°)	θ_{AB} (°)	R_{AB} (Å)	$\sigma_{I,dim}$ (cm^{-1})
Point dipole model					
15	−538 + 74/−65	103 ± 0.48	44.1 + 2.7/−2.5	6.4 ± 0.10	356 ± 12
25	−488 + 86/−91	101 ± 0.45	44.2 + 2.9/−3.0	6.1 ± 0.13	364 ± 13
35	−390 + 64/−68	101 ± 0.41	44.7 + 2.6/−2.5	6.5 ± 0.13	362 ± 10
45	−351 + 58/−62	98 ± 0.16	54.0 + 1.0/−1.0	5.0 ± 0.10	364 ± 7.0
55	366 + 29/−28	−11 ± 7.9	73.0 + 2.3/−2.4	14.4 ± 0.23	347 ± 7.0
65	380 + 34/−34	−23 ± 3.4	69.5 + 2.5/−2.6	13.9 ± 0.25	414 ± 8.1
Extended dipole model					
15	−537 + 68/−79	101 ± 0.67	47.3 + 4.2/−4.5	5.3 ± 0.1	355 ± 12
25	−489 + 84/−91	101 ± 0.82	44.2 + 5.4/−6.5	5.4 ± 0.07	364 ± 13
35	−390 + 62/−69	101 ± 0.74	44.4 + 6.1/−7.4	6.1 ± 0.08	364 ± 10
45	−352 + 57/−61	96 ± 0.52	58.2 + 3.2/−3.5	5.6 ± 0.08	365 ± 9.5
55	366 + 33/−31	−11 ± 7.3	73.0 + 2.2/−2.3	14.4 ± 0.09	349 ± 7.5
65	384 + 40/−38	−23 ± 3.2	69.6 + 2.4/−2.5	13.7 ± 0.13	407 ± 8.8

the duplex and fork DNA constructs, where the primary difference between the point-dipole and extended-dipole models is the overestimation of the resonant coupling by the point-dipole model (see Fig. 3). In Tables 2 and 3, we list the results of our optimization analyses for the $(Cy3)_2$ dimer duplex and fork DNA constructs, respectively. In the discussion that follows, we focus on the results of the extended-dipole model, which are qualitatively the same as those of the point-dipole model.

4. Discussion of results

4.1 Determination of $(Cy3)_2$ dimer conformation and conformational disorder using absorbance and CD spectroscopy

We studied the temperature-dependence of the absorbance and CD spectra of the $(Cy3)_2$ dimer-labeled duplex and fork DNA constructs. In Fig. 4, we present experimental absorbance and CD spectra of these samples at representative temperatures. Overlaid with the experimental data are simulated symmetric (+, shown in blue) and anti-symmetric (−, shown in red) exciton components resulting from our optimization procedure, and which are based on the extended-dipole model. The agreement between experiment and theory is very good over the full range of temperatures that we studied. In Tables 2 and 3, we list as a function of temperature the optimized values of the inter-chromophore structural parameters (*i.e.*, J, ϕ_{AB}, θ_{AB}, R_{AB}, and $\sigma_{I,dim}$) for the $(Cy3)_2$ dimer-labeled duplex and fork DNA constructs, respectively. We note that for both of these dimer-labeled DNA constructs at temperatures above the melting transition ($T_m \approx 65$ °C), the spectra appear indistinguishable from those of the corresponding Cy3 monomer DNA constructs, signifying the complete separation between the conjugated single DNA strands.

Our temperature-dependent analyses of the absorbance and CD spectra shown in Fig. 4 reveal that for both $(Cy3)_2$ dimer-labeled duplex and fork DNA constructs,

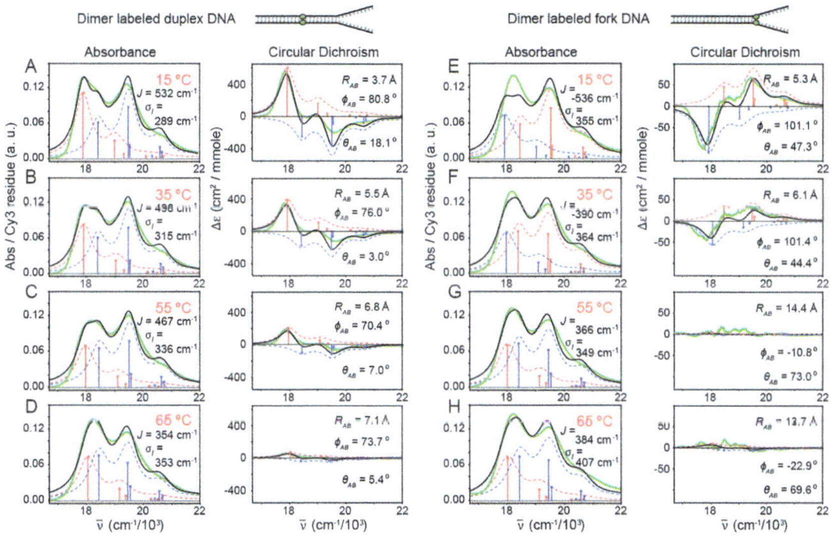

Fig. 4 Temperature-dependent absorbance and CD spectra for $(Cy3)_2$ dimer labeled duplex DNA construct (A–D), and for the $(Cy3)_2$ dimer labeled fork DNA construct (E–H). Experimental spectra are shown in solid green, and the simulated total lineshapes (inhomogeneous-plus-homogeneous) are shown in solid black. The optimized parameters shown in the insets were obtained using the extended-dipole model for the resonant coupling. Symmetric and anti-symmetric transitions determined from the model are shown as blue and red sticks, respectively. Symmetric and anti-symmetric contributions to the inhomogeneous lineshapes are shown as dashed blue and red curves, respectively.

the individual vibronic bands are well separated into symmetric and anti-symmetric exciton components with energy spacings that depend on the coupling strength J. For both types of DNA constructs, each exciton component contributes with comparable intensities to the absorbance spectra [as described by eqn (4)], which – as we discuss further below – is due to the occurrence in both species of the nearly orthogonal twist angles ϕ_{AB}. Moreover, we see from our analyses that the pronounced bisignate line shapes within individual vibronic bands of the CD spectra are due to the opposite sign contributions of the symmetric and anti-symmetric excitons [see eqn (5)].

We first discuss our results for the $(Cy3)_2$ dimer-labeled duplex DNA construct (see Fig. 4A) at 15 °C. The optimized values for the structural parameters are $J = 532\ cm^{-1}$, $\phi_{AB} = 80.7°$, $\theta_{AB} = 18.1°$, $R_{AB} = 3.7$ Å, and $\sigma_{I,dim} = 289\ cm^{-1}$. These values for R_{AB}, ϕ_{AB}, and θ_{AB} are consistent with space-filling models for the local conformation of the $(Cy3)_2$ dimer depicted in Fig. 1B, which shows the two Cy3 monomers positioned closely within the DNA duplex with an approximately coplanar and orthogonal relative orientation. Furthermore, this dimer conformation reflects the anticipated structure of the opposing segments of the sugar-phosphate backbone deep within the duplex region of the DNA construct. The value of the resonant coupling strength J is positive, indicating that the symmetric (anti-symmetric) exciton manifold is blue- (red-)shifted relative to the transition energy of the uncoupled monomers. The positive sign of J corresponds to the right-handed chirality of the dimer (with $\phi_{AB} < 90°$), which manifests as a positive

Cotton effect in the CD spectrum.[43] For both duplex and fork constructs, the magnitude of J is greater than that of the spectral inhomogeneity $\sigma_{\rm I,dim}$, which is a necessary condition for the dimer to support delocalized excitons.[11] Because the coupling strength is comparable to the intramolecular vibrational relaxation energy (*i.e.* $J \sim \lambda^2\hbar\omega_0 = 602\ \text{cm}^{-1}$, where we have used $\lambda^2 = 0.54$ and $\hbar\omega_0 = 1116\ \text{cm}^{-1}$), the dimer exists in the intermediate-to-strong exciton-coupling regime.

The results for the $(\text{Cy3})_2$ dimer-labeled duplex DNA construct discussed above serve to confirm our general expectations for the right-handed helical structure of B-form DNA. In contrast, much less is known about local conformations of the sugar-phosphate backbone near DNA fork junctions. It is therefore interesting to compare our results for the two different $(\text{Cy3})_2$ dimer-labeled DNA constructs. The optimized structural parameters for the dimer-labeled DNA fork construct at 15 °C (see Fig. 4E) are $J = -537\ \text{cm}^{-1}$, $\phi_{\rm AB} = 101°$, $\theta_{\rm AB} = 47.3°$, $R_{\rm AB} = 5.3$ Å, and $\sigma_{\rm I,dim} = 355\ \text{cm}^{-1}$. Here the values for $R_{\rm AB}$, $\phi_{\rm AB}$, and $\theta_{\rm AB}$ are consistent with the local conformation of the $(\text{Cy3})_2$ dimer depicted in Fig. 1C, for which the two Cy3 monomers are positioned closely at the DNA fork junction with a non-coplanar and nearly orthogonal orientation. The increased value of the tilt angle $\theta_{\rm AB}$ reflects the loss of cylindrical symmetry at the fork junction. While the magnitude of the resonant coupling J is nearly the same for the fork and duplex constructs, the sign of the coupling is negative for the DNA fork construct. This indicates that the symmetric (anti-symmetric) exciton manifold is red- (blue-)shifted relative to the transition energy of the uncoupled monomers. Thus, near the DNA fork junction the local backbone segments labeled by the $(\text{Cy3})_2$ dimer have adopted a left-handed chiral symmetry (*i.e.* $\phi_{\rm AB} > 90°$) such that the CD exhibits a negative Cotton effect. The spectral inhomogeneity for the dimer-labeled fork DNA construct is greater than that of the duplex. This latter finding is consistent with the notion that the distribution of backbone conformations near the DNA fork junction may be broadened due to biologically significant 'breathing' fluctuations, which uniquely occur at this position.[2]

As the temperature was increased over the range 15–65 °C, (see Fig. 4) the splittings between the symmetric and anti-symmetric exciton components decreased continuously, as did the finite amplitudes of the CD signal. The CD signal of the DNA fork construct decreased much more rapidly with increasing temperature than that of the DNA duplex, which is due to a loss of chiral symmetry at ~55 °C, well below the denaturation transition. For both the $(\text{Cy3})_2$ dimer-labeled DNA duplex and fork constructs, the temperature-dependent properties are correlated to systematic changes in the resonant coupling strength J. This is due to the temperature sensitivity of cooperative interactions between constituent nucleic acid bases (*i.e.*, base stacking interactions, Watson–Crick hydrogen bonding, *etc.*), which stabilize the right-handed helical structure of the duplex DNA construct, and only partially stabilize the left-handed conformation that we observe at the DNA fork junction. The temperature-dependent disruption of local DNA secondary structure is reflected by systematic changes in the conformation of the $(\text{Cy3})_2$ dimer DNA duplex and DNA fork constructs characterized by the structural parameters listed in Tables 2 and 3, respectively, and are plotted in Fig. 5.

As shown in Fig. 5, the majority of structural parameters of the $(\text{Cy3})_2$ dimer-labeled DNA duplex and fork constructs vary continuously over the range of

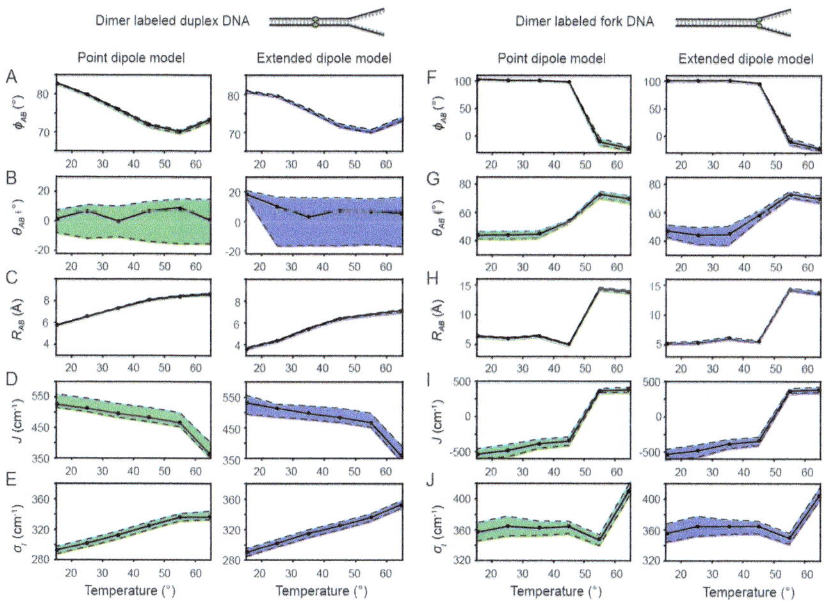

Fig. 5 Temperature-dependent optimization parameters from $(Cy3)_2$ dimer labeled DNA duplex (A–E) and fork (F–J) absorbance and CD spectra. Error bars were calculated based on a 1% deviation of the target function from its optimized value. The melting transition temperature T_m of the DNA constructs is 65 °C. (A and F) Inter-chromophore twist angle; (B and G) inter-chromophore tilt angle; (C and H) inter-chromophore separation; (D and I) resonant electronic coupling; and (E and J) spectral inhomogeneity parameter.

temperatures 15–55 °C. As discussed above, our results are nearly indistinguishable using either the point-dipole (shown in green) or the extended-dipole (shown in blue) models for the resonant coupling strength J. For our remaining discussion, we focus on the values given by the extended-dipole model.

For the $(Cy3)_2$ dimer-labeled duplex DNA construct, the inter-chromophore separation R_{AB} increases from 3.7–6.8 Å, the inter-chromophore twist angle ϕ_{AB} decreases from 80.7–70.4°, the tilt angle θ_{AB} decreases relatively quickly from 18.1–7.0°, the resonant coupling J decreases from 532–467 cm^{-1}, and the spectral inhomogeneity parameter $\sigma_{I,dim}$ increases from 289–336 cm^{-1}. The spectral inhomogeneity is a measure of the disorder of the local DNA environment experienced by the chromophores. In Table S1 of the ESI,† we list the spectral inhomogeneity for the Cy3 monomer-labeled duplex DNA construct as a function of temperature. The values of both parameters $\sigma_{I,mon}$ and $\sigma_{I,dim}$ increase with temperature, which suggests that for both the monomer and dimer labeled duplex DNA constructs, there exists a distribution of thermally populated conformational sub-states separated by thermal barriers, which are overcome with increasing temperature. The level of disorder appears to be greater in the monomer *versus* the dimer duplex DNA construct. The disorder parameter of the monomer $\sigma_{I,mon}$ increases monotonically with temperature over the range 15–45 °C, and then undergoes a gradual decrease over the range 45–65 °C to nearly the same value as that of the dimer $\sigma_{I,dim}$ at the melting transition. This is likely a reflection of the less favorable packing conditions of the Cy3 monomer within the duplex DNA construct, for which a single thymine base is positioned across from the Cy3 chromophore on the opposing DNA single-strand (see Table 1).

For the $(Cy3)_2$ dimer-labeled fork DNA construct, the inter-chromophore separation R_{AB} increases from 5.3–14.4 Å, the inter-chromophore twist angle ϕ_{AB} decreases from 101 to −11°, the tilt angle θ_{AB} increases from 47.3–73.0°, and the resonant coupling J increases from −537 to 366 cm^{-1}. An interesting property of the dimer-labeled fork DNA construct is that the spectral inhomogeneity parameter $\sigma_{I,dim}$ remains essentially constant (~365 cm^{-1}) over the temperature range 15–55 °C, before abruptly increasing to 407 cm^{-1} at the denaturation temperature 65 °C. This weak temperature dependence of the inhomogeneity parameter suggests the absence of thermal barriers separating the broad distribution of conformational sub-states associated with the probe labels at the fork junction. It is interesting to compare the temperature-dependent inhomogeneity parameter of the Cy3 monomer-labeled DNA fork construct (see Table S2 of the ESI†) to that of the $(Cy3)_2$ dimer fork DNA construct. At 15 °C, we see that the value of $\sigma_{I,mon}$ is approximately equal to the value of $\sigma_{I,dim}$. However, over the temperature range that the dimer disorder parameter remains constant, the monomer disorder increases monotonically with temperature, approaching a value of 405 cm^{-1}, which is nearly equal to that of the dimer $\sigma_{I,dim}$ at the denaturation temperature 65 °C.

4.2 Determination of inhomogeneous line widths using two-dimensional fluorescence spectroscopy

From our previous analyses of the linear absorbance and CD spectra of the $(Cy3)_2$ dimer-labeled DNA constructs, we see that it is possible to obtain significant information about the average conformation and conformational disorder of the chromophore probes within the local DNA framework. These analyses made use of a separate determination of the room temperature homogeneous line width parameter $\Gamma_H = 186$ cm^{-1}.[17] To achieve a more accurate characterization of local conformation and conformational disorder, we extend our analysis to include the results of 2DFS experiments, as described in Section 2. When combined with absorbance and CD, these 2DFS measurements can separately measure homogeneous and inhomogeneous contributions to the spectral lineshapes.

The rephasing (RP) and non-rephasing (NRP) 2DFS signals are complex-valued response functions described by eqn (6) and (7), respectively.[29,30] These signals are Fourier transformed with respect to the inter-pulse delays t_{21} and t_{43} to produce the complex-valued RP and NRP 2DFS spectra as a function of the frequency variables $\bar{\nu}_{21}$ and $\bar{\nu}_{43}$, respectively. The RP and NRP spectra may be combined to produce the total correlation (TC) spectra (*i.e.*, RP + NRP = TC). From eqn (8)–(10), we see that the RP spectra have the property that inhomogeneous line broadening does not contribute to the 2D lineshapes along the anti-diagonal direction ($\bar{\nu}_{43} - \bar{\nu}_{21}$), while homogeneous line broadening does contribute along both the diagonal ($\bar{\nu}_{21} + \bar{\nu}_{43}$) and the anti-diagonal directions. From eqn (11)–(13), we see that the NRP spectra contain both inhomogeneous and homogeneous contributions along the diagonal and the anti-diagonal directions. These properties of 2DFS spectral lineshapes allow us to separately resolve homogeneous and inhomogeneous parameters.

As a control study, we performed 2DFS measurements of the Cy3 chromophore in methanol solution at 25 °C (see Fig. S2 of the ESI† section). For these measurements, the laser center wavelength was set to 532 nm (18 796 cm^{-1}) with bandwidth 37 nm (1300 cm^{-1}). This laser spectrum spans the 0–0 and 1–0 vibronic sub-bands of the Cy3 absorbance spectrum, which occurs over the wavelength range 515–550 nm (18 200–19 400 cm^{-1}). In Fig. S2C,† we show the real and imaginary parts of the NRP (left column), RP (middle column), and TC (right column) spectra. The 2DFS spectra are dominated by the 0–0 vibronic transition, with a minor contribution from the 1–0 transition. The effect of these two overlapping vibronic features on the 2DFS spectra is to produce quasi-elliptical 2D lineshapes with long (short) axes oriented along the anti-diagonal (diagonal) directions of the NRP (RP) spectra. The combination of RP and NRP spectra produces a TC spectrum that is relatively symmetric with respect to diagonal and anti-diagonal axes. We note that there are only minor cross-peak intensities that appear to couple the 0–0 and 1–0 transitions. From an analysis of the diagonal and anti-diagonal cross-sectional line widths,[17] we obtain values for the homogeneous Lorentzian FWHM parameter $\Gamma_H = 400$ cm^{-1}, and the inhomogeneous Gaussian standard deviation $\sigma_I = 250$ cm^{-1}.

We next discuss the results of our 2DFS measurements performed on the Cy3-labeled DNA constructs at 25 °C. In Fig. 6, we present experimental absorbance spectra (top row), experimental RP 2DFS spectra (middle row) and simulated RP 2DFS spectra (bottom row) corresponding to the Cy3 monomer-labeled duplex DNA construct (Fig. 6A), the $(Cy3)_2$ dimer-labeled duplex DNA construct (Fig. 6B), and the $(Cy3)_2$ dimer-labeled fork DNA construct (Fig. 6C). Overlaid with the experimental absorbance spectra are the exciton features resulting from our analyses of the absorbance and CD spectra, which we discussed in previous sections. Also shown is a representative laser spectrum with center wavelength 532 nm and bandwidth 31 nm (1100 cm^{-1}). In all cases, the laser spectrum spans the 0–0 and 1–0 vibronic sub-bands. Similar to the RP 2D spectra obtained from the Cy3 chromophore in methanol (see Fig. S2C†), the RP 2D spectra of the Cy3 DNA constructs exhibit a dominant feature associated with the 0–0 transition, and a minor feature associated with the 1–0 transition. Yet the most striking observation for all three of the DNA constructs is the overall asymmetry of the 2D spectra, which indicates the presence of overlapping diagonal peaks and off-diagonal cross-peaks. Furthermore, the 2D lineshapes of individual spectral features are disproportionately broadened along the diagonal direction of the 2D spectrum due to the influence of conformational disorder, which uniquely affects the RP spectrum for each of the DNA constructs.

In Fig. 6 (bottom row), we present simulated RP spectra that assume as input structural parameters the same values that we obtained from our optimization analyses for the corresponding Cy3-labeled DNA construct. In these calculations, we have used the homogeneous line width parameter $\Gamma_H = 186$ cm^{-1}, in accordance with our previous analyses. We also used the value $\Gamma_{2D} = 0.5$ (where Γ_{2D} can take on values between zero and 2), which characterizes the relative fluorescence quantum yield of the doubly excited state relative to the singly excited state manifolds [see eqn (6) and (7)]. We note that our simulated 2D spectra for the $(Cy3)_2$ dimer system are relatively insensitive to the value of Γ_{2D}. For each of the simulated RP spectra, we adjusted the inhomogeneous line broadening parameter to obtain the most favorable comparison to the experimental data. For both the Cy3 monomer-labeled duplex DNA construct (Fig. 6A) and the $(Cy3)_2$ dimer-labeled duplex DNA construct (Fig. 6B), we obtain the value $\sigma_I = 125$ cm^{-1}. For

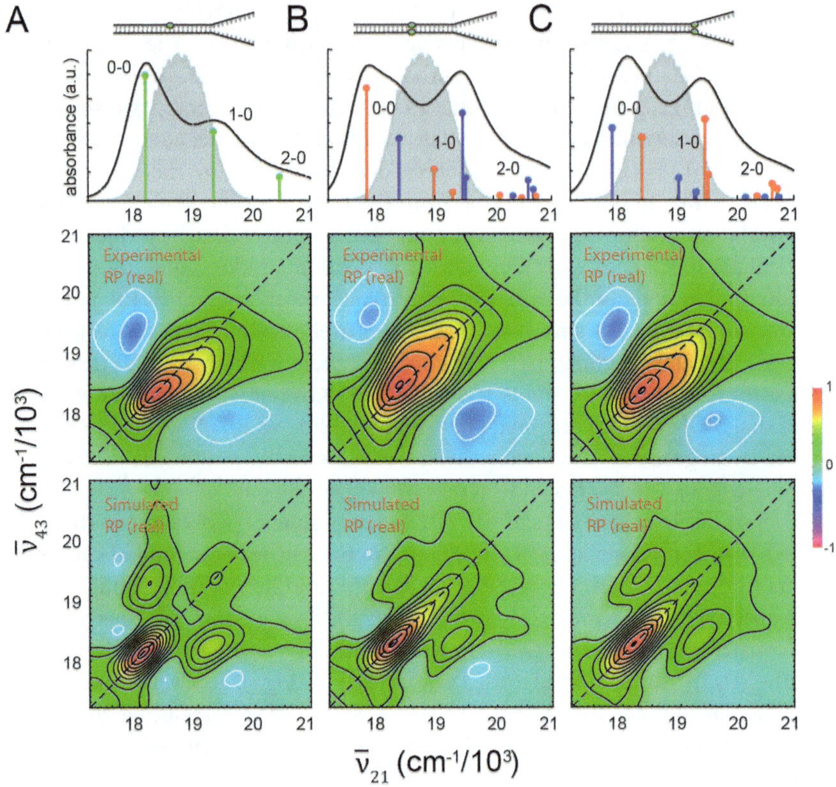

Fig. 6 Room temperature (25 °C) experimental and simulated 2DFS measurements performed on (A) the Cy3 monomer duplex DNA construct, (B) the $(Cy3)_2$ dimer duplex DNA construct, and (C) the $(Cy3)_2$ dimer fork DNA construct. Absorbance spectra are shown overlaid with the vibronic spectral features obtained from the H–F Hamiltonian given by eqn (1) and (2). The associated parameters are listed in Table S1† for the monomer duplex DNA construct, in Table 2 for the dimer duplex DNA construct, and in Table 3 for the dimer fork DNA construct. Also shown is the laser spectrum (in gray) with center frequency $\bar{\nu}_L = 18\,796$ cm^{-1} ($\lambda_L = 532$ nm) and FWHM bandwidth $\Delta\bar{\nu}_L = 1100$ cm^{-1} ($\Delta\lambda_L = 31$ nm). The laser spectrum spans a region containing both the 0–0 and 1–0 vibronic sub-bands. Experimental RP spectra (middle row) are compared to simulated RP spectra (bottom row). Simulated spectra are based on the homogeneous line width parameter $\Gamma_H = 186$ cm^{-1}, the fluorescence quantum yield parameter $\Gamma_{2D} = 0.5$ [eqn (6)], the structural parameters obtained from our analyses of absorbance and CD spectra, and inhomogeneous lineshape parameters (A) $\sigma_I = 125$ cm^{-1}, (B) $\sigma_I = 125$ cm^{-1} and (C) $\sigma_I = 250$ cm^{-1}.

the $(Cy3)_2$ dimer-labeled fork DNA construct, we obtain the value $\sigma_I = 250$ cm^{-1}. We note that the agreement between experimental and simulated RP spectra is good. The simulated spectra exhibit diagonal peaks and off-diagonal cross-peaks that give rise to the spectral features seen in the experiment. The effects of disorder can be understood in terms of their disproportionate contribution along the diagonal to the various 2D spectral line shapes. These results are in agreement with those of our previous analyses, which indicate that there is significantly greater conformational disorder associated with the $(Cy3)_2$ dimer-labeled fork DNA construct than that exhibited by the Cy3 monomer- and $(Cy3)_2$ dimer-labeled duplex DNA constructs.

5. Conclusions

We studied the absorbance, circular dichroism (CD) and two-dimensional fluorescence (2DFS) spectra of a $(Cy3)_2$ dimer that was site-specifically positioned within the sugar-phosphate backbone at the single-stranded (ss)–double-stranded (ds) DNA fork junction. We compared our results to those obtained from a $(Cy3)_2$ dimer that was positioned deep within the DNA duplex. We adopted the Holstein–Frenkel (H–F) Hamiltonian model to characterize the symmetric and anti-symmetric excitons supported by the dimer over a range of temperatures below the denaturation transition of the dsDNA. From this analysis, we obtained a temperature-dependent parameterization of the average dimer conformation and the degree of conformational disorder (see Table 3). We compared the results of our analyses using alternatively the point-dipole and the extended-dipole models, and we found that both models yield essentially the same temperature-dependent values for the conformational parameters.

At the lowest temperature we studied (15 °C), the $(Cy3)_2$ dimer-labeled DNA fork construct exhibits intermediate-to-strong resonant coupling ($J \sim -537$ cm^{-1}), comparable in magnitude to the vibrational relaxation energy of the constituent monomers ($\lambda^2\hbar\omega_0 \sim 600$ cm^{-1}). Under these conditions, the dimer can support delocalized excitons composed of symmetric and anti-symmetric superpositions of electronic-vibrational product states. The delocalized electronic structure is a consequence of the $(Cy3)_2$ dimer being held within the sugar-phosphate backbone at a very small inter-chromophore separation (R_{AB} = 5.3 Å) and a nearly orthogonal inter-chromophore twist angle. The conformation of the $(Cy3)_2$ dimer at the DNA fork junction lacks cylindrical symmetry with tilt angle $\theta_{AB} = 47.3°$ and inter-chromophore twist angle $\phi_{AB} = 101°$, which markedly deviates from the Watson–Crick B-form structure that we observe in the DNA duplex.

As the temperature is increased towards the ds–ss DNA denaturation temperature (T_m = 65 °C), the resonant electronic coupling strength of the $(Cy3)_2$ dimer-labeled DNA fork construct gradually decreases in magnitude over a ~200 cm^{-1} range, and undergoes a sign inversion at ~55 °C. The Hamiltonian parameters characteristic of the Cy3 monomer-labeled DNA fork construct (*i.e.* the transition energy ε_{eg}, the Huang–Rhys electronic-vibrational coupling parameter λ^2, and the vibrational frequency ω_0) remain approximately independent of temperature (see Table S1 of the ESI†). This behavior is due to the sensitivity of the local secondary structure of the $(Cy3)_2$ dimer-labeled fork (and duplex) DNA construct(s) to temperature, which affects the inter-chromophore separation, twist angle, and tilt angle, but not the electronic-vibrational properties internal to each Cy3 monomer.

While the magnitude of spectral inhomogeneity (~365 cm^{-1}) is significant across the 15–65 °C temperature range, the effects of exciton delocalization within the $(Cy3)_2$ dimer are not dominated by spectral inhomogeneity. The spectral inhomogeneity parameter of the Cy3 monomer-labeled fork DNA construct increases systematically over the 15–65 °C temperature range ($\sigma_{I,mon}$ = 355–398 cm^{-1}), signifying that the monomer probe experiences locally disordered, thermally activated regions of the DNA fork, well below the melting transition. However, the inhomogeneity parameter of the $(Cy3)_2$ dimer-labeled fork DNA construct is relatively constant over the 15–55 °C temperature range ($\sigma_{I,dim} \sim$

365 cm^{-1}) before abruptly increasing to 407 cm^{-1} at the denaturation temperature 65 °C. Our observation of a weak temperature-dependence for the inhomogeneity parameter $\sigma_{I,dim}$ suggests that thermal activation does not play a significant role in populating a broad distribution of $(Cy3)_2$ dimer conformational sub-states at the ss–ds DNA fork junction. This finding is consistent with the hypothesis that the sugar-phosphate backbone of DNA near ss–ds junctions fluctuates into a broad distribution of conformations to permit the proper binding of genome regulatory proteins.

Our accompanying two-dimensional fluorescence spectroscopy (2DFS) measurements allow us to test our H–F Hamiltonian model based on the analysis of absorbance and CD data, and to potentially examine the effects of conformational disorder in more detail. We find that our room temperature 2DFS data are consistent with the H–F Hamiltonian parameters that we determined from our analysis of absorbance and CD, in addition to our previously determined value for the homogeneous spectral line width (FWHM Γ_H = 186 cm^{-1}). Furthermore, rephasing (RP) spectra are extremely sensitive to the effects of disorder, which contribute disproportionately to the 2D lineshapes along the diagonal ($\bar{\nu}_{21} + \bar{\nu}_{43}$) and anti-diagonal ($\bar{\nu}_{43} - \bar{\nu}_{21}$) axis directions. Although our current analysis confirms that disorder at the DNA fork junction exceeds that within the DNA duplex, the magnitude of the disorder determined by 2DFS appears to be less (~100 cm^{-1}) than that determined from our analyses of linear spectra. Additional temperature-dependent analyses of RP and NRP spectral lineshapes should provide a comprehensive description of disorder at the ss–ds DNA fork junction.

Although the H–F model for the exciton-coupled $(Cy3)_2$ dimer is relatively simple as it assumes a single internal vibrational mode for each Cy3 monomer, the model appears to capture the essential features of the experimental absorbance, CD and 2DFS spectra. Temperature variation corresponds to systematic changes in local dimer conformation that allows the resonant coupling strength to be 'tuned' across the intermediate-to-strong exciton-coupling regime. The significant spectral inhomogeneity is due to the presence of local structural fluctuations of the DNA backbone and base stacking that influence the packing of the chromophore probes. Such local fluctuations of DNA are termed DNA 'breathing,' which likely play a functional role in the binding and assembly of gene regulatory proteins.[2]

Conflicts of interest

There are no conflicts of interest to declare.

Acknowledgements

We thank members of the von Hippel group for useful discussions pertaining to DNA breathing and other factors that affect DNA stability. We also thank Prof. Michael Raymer and Prof. Brian Smith for useful discussions about exciton coupled molecular networks, and Prof. Frank Spano and Prof. Jasper Knoester for discussions about extended dipole models for the resonant electronic coupling. This work was supported by the John Templeton Foundation (RQ-35859 to A. H. M. as a co-PI), by the National Science Foundation Chemistry of Life Processes

Program (CHE-1608915 to A. H. M.), and by the National Institutes of General Medical Sciences (NIGMS Grant GM-15792 to A. H. M. and P. H. vH. as co-PIs).

References

1 B. Alberts, A. Johnson, J. Lewis, M. Raff, K. Roberts and P. Walter, *Molecular Biology of The Cell*, Garland Science, New York, 5th edn, 2008.

2 P. H. von Hippel, N. P. Johnson and A. H. Marcus, 50 Years of DNA 'Breathing': Reflections on Old and New Approaches, *Biopolymers*, 2013, **99**, 923–954.

3 M. Levitus and S. Ranjit, Cyanine Dyes in Biophysical Research: The Photophysics of Polymethine Fluorescent Dyes in Biomolecular Environments, *Quat. Rev. Biophys.*, 2011, **44**, 123–151.

4 A. Mishra, R. K. Behera, P. K. Behera, B. K. Mishra and G. B. Behera, Cyanines During the 1990s: A Review, *Chem. Rev.*, 2000, **100**, 1973–2011.

5 F. Würthner, T. E. Kaiser and C. R. Saha-Möller, 75 Years of J-aggregates, *Angew. Chem., Int. Ed.*, 2000, **50**, 3376–3410.

6 U. Resch-Genger, M. Grabolle, S. Cavaliere-Jaricot, R. Nitschke and T. Nann, Quantum Dots Versus Organic Dyes as Fluorescent Labels, *Nat. Methods*, 2008, **5**, 763–775.

7 M. E. Sanborn, B. K. Connolly, K. Gurunathan and M. Levitus, Fluorescence Properties and Photophysics of the Sufoindocyanine Cy3 Linked Covalently to DNA, *J. Phys. Chem. B*, 2007, **111**, 11064–11074.

8 W. Lee, D. Jose, C. Phelps, A. H. Marcus and P. H. von Hippel, A Single-Molecule View of the Assembly Pathway, Subunit Stoichiometry and Unwinding Activity of the Bacteriophage T4 Primosome (Helicase-Primase) Complex, *Biochemistry*, 2013, **52**, 3157–3170.

9 W. Lee, P. H. von Hippel and A. H. Marcus, Internally Labeled Cy3/Cy5 DNA Constructs Show Greatly Enhanced Photostability in Single-Molecule FRET Experiments, *Nucleic Acids Res.*, 2014, **42**, 5967–5977.

10 C. Phelps, W. Lee, D. Jose, P. H. von Hippel and A. H. Marcus, Single-Molecule FRET and Linear Dichroism Studies of DNA 'Breathing' and Helicase Binding at Replication Fork Junctions, *Proc. Natl. Acad. Sci. U. S. A.*, 2013, **110**, 17320–17325.

11 L. Valkunas, D. Abramavicius and T. Mančal, *Molecular Excitation Dynamics and Relaxation: Quantum Theory and Spectroscopy*, Wiley-VCH, 2013.

12 T. Azumi and K. Matsuzaki, What Does the Term "Vibronic Coupling" Mean?, *Photochem. Photobiol.*, 1977, **25**, 315–326.

13 G. Fischer, *Vibronic Coupling: The Interaction Between the Electronic and Nuclear Motions*, Academic Press, London, 1984.

14 M. Aydin, Ö. Dede and D. L. Akins, Density Functional Theory and Raman Spectroscopy Applied to Structure and Vibrational Mode Analysis of 1,1′,3,3′-Tetraethyl-5,5′,6,6′-Tetrachloro-Benzimidazolocarbocyanine Iodide and its Aggregate, *J. Chem. Phys.*, 2011, **134**, 064325.

15 R. Roy, S. Hohng and T. Ha, A Practical Guide to Single-Molecule FRET, *Nat. Methods*, 2008, **5**, 507–516.

16 T. Förster, *Delocalized Excitation and Excitation Transfer*, Florida State University, Tallahassee, Florida, 1965, vol. Bulletin no. 18, Division of Biology and Medicine, U.S. Atomic Energy Commission.

17 L. Kringle, N. P. D. Sawaya, J. Widom, C. Adams, M. G. Raymer, A. Aspuru-Guzik and A. H. Marcus, Temperature-Dependent Conformations of Exciton-Coupled Cy3 Dimers in Double-Stranded DNA, *J. Chem. Phys.*, 2018, **148**, 085101.

18 F. C. Spano, The Spectral Signatures of Frenkel Polarons in H- and J-Aggregates, *Acc. Chem. Res.*, 2010, **43**, 429–439.

19 K. A. Kistler, C. M. Pochas, H. Yamagata, S. Matsika and F. C. Spano, Absorption, Circular Dichroism, and Photoluminescence in Perylene Diimide Bichromophores: Polarization-Dependent H- and J-Aggregate Behavior, *J. Phys. Chem. B*, 2012, **116**, 77–86.

20 P. H. von Hippel and J. D. McGhee, DNA-Protein Interactions, *Annu. Rev. Biochem.*, 1972, **41**, 231–300.

21 P. H. von Hippel, From "Simple" DNA-Protein Interactions to the Macromolecular Machines of Gene Expression, *Annu. Rev. Biophys. Biomol. Struct.*, 2007, **36**, 79–105.

22 I. A. Howard, F. Zutterman, G. Deroover, D. Lamoen and C. Van Alsenoy, Approaches to Calculation of Exciton Interaction Energies for a Molecular Dimer, *J. Phys. Chem. B*, 2004, **108**, 19155–19162.

23 V. H. D. F. Czikklely and H. Kuhn, Extended Dipole Model for Aggregates of Dye Molecules, *Chem. Phys. Lett.*, 1970, **6**, 207–210.

24 C. Didraga, A. Pugžlys, P. R. Hania, H. von Berlepsch, K. Duppen and J. Knoester, Structure, Spectroscopy, and Microscopic Model of Tubular Carbocyanine Dye Aggregates, *J. Phys. Chem. B*, 2004, **108**, 14976–14985.

25 E. M. S. Stennett, N. Ma, A. van der Vaart and M. Levitus, Photophysical and Dynamical Properties of Doubly Linked Cy3-DNA Constructs, *J. Phys. Chem. B*, 2014, **118**, 152–163.

26 C. R. Cantor and P. R. Schimmel, *Biophysical Chemistry Part II: Techniques for the Study of Biological Structure and Function*, Freeman, New York, 1980, vol. 2, Biophysical Chemistry.

27 E. Charney, *The Molecular Basis of Optical Activity: Optical Rotatory Dispersion and Circular Dichroism*, John Wiley & Sons, New York, 1979.

28 N. Berova and K. Nakanishi, in *Circular Dichroism: Principles and Applications*, ed. N. Berova, K. Nakanishi and R. W. Woody, John Wiley & Sons, Inc., New York, 2000, pp. 337–382.

29 P. F. Tekavec, G. A. Lott and A. H. Marcus, Fluorescence-Detected Two-Dimensional Electronic Coherence Spectroscopy by Acousto-Optic Phase Modulation, *J. Chem. Phys.*, 2007, **127**, 214307.

30 A. Perdomo, J. R. Widom, G. A. Lott, A. Aspuru-Guzik and A. H. Marcus, Conformation and Electronic Population Transfer in Membrane Supported Self-Assembled Porphyrin Dimers by Two-Dimensional Fluorescence Spectroscopy, *J. Phys. Chem. B*, 2012, **116**, 10757–10770.

31 J. R. Widom, N. P. Johnson, P. H. von Hippel and A. H. Marcus, Solution Conformation of 2-Aminopurine (2-AP) Dinucleotide by Ultraviolet 2D Fluorescence Spectroscopy (UV-2D FS), *New J. Phys.*, 2013, **15**, 025028.

32 K. J. Karki, J. R. Widom, J. Seibt, I. Moody, M. C. Lonergan, T. Pullerits and A. H. Marcus, Coherent Two-Dimensional Photocurrent Spectroscopy in a PbS Quantum Dot Photocell, *Nat. Commun.*, 2014, **5**, 5869.

33 P. F. Tekavec, T. R. Dyke and A. H. Marcus, Wave Packet Interferometry and Quantum State Reconstruction by Acousto-Optic Phase Modulation, *J. Chem. Phys.*, 2006, **125**, 194303.

34 S. Polyutov, O. Kühn and T. Pullerits, Exciton-Vibrational Coupling in Molecular Aggregates: Electronic Versus Vibronic Dimer, *Chem. Phys.*, 2012, **394**, 21–28.

35 A. Eisfeld, L. Braun, W. T. Stunz, J. S. Briggs, J. Beck and V. Engel, Vibronic Energies and Spectra of Molecular Dimers, *J. Chem. Phys.*, 2005, **122**, 134103.

36 V. Tiwari, W. K. Peters and D. M. Jonas, Electronic Resonance with Anticorrelated Pigment Vibrations Drives Photosynthetic Energy Transfer Outside the Adiabatic Framework, *Proc. Natl. Acad. Sci. U. S. A.*, 2013, **110**, 1203–1208.

37 A. J. Halpin, P. J. M. Johnson, R. Tempelaar, R. S. Murphy, J. Knoester, T. L. C. Jansen and R. J. D. Miller, Two-Dimensional Spectroscopy of a Molecular Dimer Unviels the Effects of Vibronic Coupling on Exciton Coherences, *Nat. Chem.*, 2014, **6**, 196–201.

38 H.-G. Duan, P. Nalbach, V. I. Prokhorenko, S. Mukamel and M. Thorwart, On the Origin of Oscillations in Two-Dimensional Spectra of Excitonically-Coupled Molecular Systems, *New J. Phys.*, 2015, **17**, 072002.

39 G. A. Lott, A. Perdomo-Ortiz, J. K. Utterback, J. R. Widom, A. Aspuru-Guzik and A. H. Marcus, Conformation of Self-Assembled Porphyrin Dimers in Liposome Vesicles by Phase-Modulation 2D Fluorescence Spectroscopy, *Proc. Natl. Acad. Sci. U. S. A.*, 2011, **108**, 16521–16526.

40 P. Grégoire, A. R. S. Kandada, E. Vella, C. Tao, R. Leonelli and C. Silva, Incoherent Population Mixing Contributions to Phase-Modulation Two-Dimensional Coherent Excitation Spectra, *J. Chem. Phys.*, 2017, **147**, 114201.

41 C. Phelps, B. Israels, D. Jose, M. Marsh, P. H. von Hippel and A. H. Marcus, Using Microsecond Single-Molecule FRET to Determine the Assembly Pathways of T4 ssDNA Binding Protein onto Model DNA Replication Forks, *Proc. Natl. Acad. Sci. U. S. A.*, 2017, **114**, E3612–E3621.

42 R. H. Byrd, J. Nocedal and R. A. Waltz, *KNITRO: An Integrated Package for Nonlinear Optimization*, Springer-Verlag, Berlin, Germany, 2006, Large-Scale Nonlinear Optimization, pp. 35–59.

43 B. Nordén, A. Rodger and T. Dafforn, *Linear Dichroism and Circular Dichroism: A Textbook on Polarized-Light Spectroscopy*, RSC Publishing, Cambridge, UK, 2010.

Faraday Discussions

PAPER

How charges separate: correlating disorder, free energy, and open-circuit voltage in organic photovoltaics

Débora P. Mroczek,[a] Vladimir Lankevich†[b] and Eric R. Bittner *[b]

Received 19th November 2018, Accepted 22nd January 2019
DOI: 10.1039/c8fd00182k

In order for a photovoltaic cell to function, charge carriers produced by photoexcitation must fully dissociate and overcome their mutual Coulomb attraction to form free polarons. This becomes problematic in organic systems in which the low dielectric constant of the material portends a long separation distance between independent polaron pairs. In this paper, we discuss our recent efforts to correlate the role of density of states, entropy, and configurational and energetic disorder to the open-circuit voltage, V_{OC}, of model type-II organic polymer photovoltaics. By comparing the results of a fully interacting lattice model to those predicted by a Wigner–Weisskopf type model we find that energetic disorder does play a significant role in determining the V_{OC}; however, mobility perpendicular to the interface plays the deciding role in the eventual fate of a charge-separated pair.

1 Introduction

The performance of an organic photovoltaic devices depends on successful charge-separation following photoexcitation, which in turn hinges upon whether or not the initial excitonic state has sufficient electronic or vibronic energy to overcome the Coulomb attraction between an electron and a hole.[1–4] Whether this happens rapidly through "hot" excitonic states or through thermalized "cold" states has been a matter of considerable recent debate, and the answer to this question contains information crucial for the design of highly efficient organic solar cells. "Hot excitons" are attributed to the prompt formation of mobile polarons observed in fullerene/polymer blends. While many groups argue that "hot excitons" are the main source of photocurrent, experiments conducted by Vandewal *et al.* indicate that it is "cold excitons", *i.e.* thermalized excited states are the ones that give the predominant contribution to the photocurrent.

[a]*University of Houston, Department of Physics, Houston, TX, USA. E-mail: dpmroczek@Central.UH.EDU*
[b]*University of Houston, Department of Chemistry, Department of Physics, Houston, TX, USA. E-mail: ebittner@central.uh.edu*
† Present address: Rice University, Department of Chemistry.

The dissociation of electron–hole pairs into free charge carriers is especially puzzling since it requires an electron and a hole to overcome the strong electrostatic attractions under seemingly unfavorable conditions. Heuristically, charge carriers are *free* when thermal energy is comparable to the coulombic attraction,

$$\frac{e^2}{4\pi\varepsilon_0\varepsilon r} \leq k_B T, \tag{1}$$

where e is the electron charge, ε_0 is the vacuum permittivity, ε is a dielectric constant of the material and r is the distance of separation between the charges. Due to the low dielectric constant ($\varepsilon = 2$–4) of organic materials typically used in OPVs, an electron finds itself with a 0.5 eV barrier to surmount, corresponding to the Coulomb capture radius of 15 to 28 nm. It is highly improbable that charges would be able move this distance before recombining; however, it has been observed that free charge carriers can be formed at separations of 4 nm on the femtosecond timescale.[5,6] Thus, a goal of our recent work has been to rectify both "hot" and "cold" exciton dissociation mechanisms within the context of a common theoretical model.

From thermodynamics, we can consider the dissociation in terms of the reversible work (*i.e.* Helmholtz energy) required to separate an electron from a hole to a given radius.

$$F(r) = U(r) - TS(r) = U(r) - k_B T \ln \Omega(r) \tag{2}$$

where $\Omega(r)$ is the number of equivalent electron/hole states with separation r, T is the absolute temperature, k_B is the Boltzmann constant, $U(r)$ is the electron–hole interaction potential, and S is the entropy of the electronic degrees of freedom.[7] For a single polymer chain, a π-electron that is confined to move along a single quasi-one-dimensional polymer chain has only one defined path and consequently the electronic entropy contribution. For thin-films (2-D) and fullerene-based acceptors (3-D) the number of electron/hole configurations available to the system with a given electron/hole separation radius scales with the surface area

$$\Omega \propto (r/r_o)^{d-1}. \tag{3}$$

where r_o is the unit length and $d = 1, 2, 3$ is the dimensionality of the system. Consequently the entropy

$$S = (d - 1)\ln(r/r_o) \tag{4}$$

can become energetically comparable to the coulombic energy of the electron–hole pair in two and three dimensions.[7,8] This has the desired result of providing a zero-work pathway for the dissociation of an exciton to a free electron/hole pair. This estimate, however, is only valid for the scenarios with an immobile hole.

Allowing the hole to move adds additional degrees of freedom and increases the number of available electronic states further emphasizing the importance of the entropic contribution. The reversible work theorem also implies an equality between chemical potential and the open-circuit voltage, qV_{OC}, of the photovoltaic device[9] and establishes a crucial connection between theoretical and laboratory investigations of the current generation in OPVs. From such considerations, Burke *et al.* arrive at the following expression for the V_{OC} from the canonical ensemble:

$$qV_{OC} = E_{CT} - \frac{\sigma_{CT}{}^2}{2k_B T} - k_B T \ln\left(\frac{qfN_0 L}{\tau_{CT} J_{SC}}\right) \tag{5}$$

where f is the volume fraction of the device that is mixed or interfacial, L is the thickness of the solar cell, J_{SC} is the short-circuit current of the cell, q is the electronic charge, and N_0 is the density of the electronic states in the device. Eqn (5) includes the necessary dependence of qV_{OC}, and therefore of F, on the average energy of the CT state, the disorder in the CT energies, E_{CT}, expressed through standard deviation σ_{CT}, and the life-time of the CT-state, τ_{CT}.[9] Nonetheless, this expression is composed of variables that refer to the entire device making eqn (5) difficult to connect to a microscopic model.

More recently, Hood and Kassal have shown that the free energy better reflects the energy landscape that an electron and a hole traverse, since the entropic term effectively lowers the energy barrier needed for the transferred charge to become a free charge carrier. Additional dissociation paths and energetic disorder make the coulombic interaction comparable to the thermal fluctuations. Consequently, the Coulomb attraction no longer defines how far an electron and a hole can separate as suggested by eqn (1). Within the Hood and Kassal model, a bulk-heterojunction is an ensemble of energetically disordered hexagonal lattice sites. Under an assumption of electronic equilibrium, they then compute the free energy as[8]

$$F = -\langle k_B T \ln Z \rangle \tag{6}$$

where Z is the partition function that describes specific energy states and the bracket $\langle \cdots \rangle$ denotes a statistical average over realizations of the disordered lattice. Both eqn (5) and (6) are derived in the canonical ensemble and carry the same information. However, the latter approach is far more suitable for connecting to microscopic details such as energetic and structural disorder. There are, however, a few drawbacks of the Hood–Kassal model. The electrons and holes are not permitted to cross from one domain to the other, and a hole is restricted to move only perpendicular to the interface under the assumption of translational symmetry. The model includes the electrostatic potential, but does not take into account important quantum effects such as delocalization, mixing between excitonic and charge-transfer configurations, and electronic exchange effects.

We recently examined this approach using a fully interacting 2D electron/hole lattice model that our group developed for studying exciton and charge transfer dynamics at organic heterojunction interfaces[10–16] and compared it to experimental values for the open-circuit voltage by including site-energy disorder and thermal fluctuations within the lattice itself. We show that this generally facilitates charge separation; however, due to the excess energy supplied by the initial photoexcitation, highly energetic electron–hole pairs can dissociate in unfavorable directions, potentially never contributing to the photocurrent while "cold" thermalized states follow the free energy curve defined at the operating temperature of the device.[17] In this discussion, we pursue an heuristic/thermodynamical model that can be solved analytically for computing the reversible work to separate charges at a given temperature. We find that the analytical model reproduces many of the features of a more detailed model in the limit of very low mobility of the transferred charge. We also explore the use of singular value decomposition of the electron/hole wave functions as a means to quantify when the separated carriers are truly independent.

2 Theoretical models

2.1 Simple lattice model

We begin by considering a model donor/acceptor heterojunction system as sketched in Fig. 1 whereby the physical system is represented as a series of lattice points in three physical dimensions. The system is partitioned such that all points

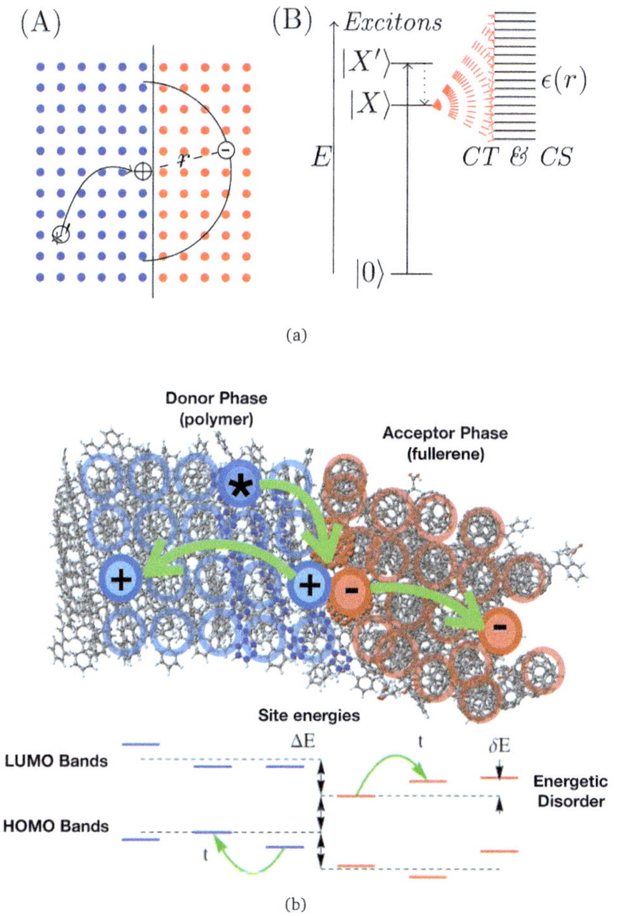

Fig. 1 (a) Sketch of the model. Our analytical model assumes that the hole remains pinned to the interface and that the donated charge tunnels some distance r into the acceptor phase. While the sketch is in a two-dimensional plane, the full physical system and our calculations is in three-dimensions. (b) Atomistic representation of the interface and its mapping onto a site model that is represented as a set of two-level systems with energy gaps modulated about their average HOMO–LUMO gap.

with $x < 0$ are considered to be in the donor-phase while all points $x > 0$ are in the acceptor phase. Charge generation occurs when a neutral exciton, denoted as *′ in our sketch and created by photoexcitation, finds itself close to the phase boundary where there is sufficient driving force to separate the exciton into an electron/hole state. Within the context of our sketch, we will assume for purposes of this model that the vacated hole is either located at the origin or at some point $\vec{r}_o$ within the donor phase and the transferred electron is located at some lattice point $\vec{r}$ away from this site. This allows us to describe the resulting electron/hole configuration energy as $\varepsilon(r)$ which we will take to be

$$\varepsilon(r) = \Delta E - E_{\text{off}} + J(r) \tag{7}$$

where ε is the dielectric constant and ΔE is the energy gap between local LUMO and HOMO orbitals and E_{off} is the offset between the HOMO orbitals in the acceptor phase and the HOMO orbitals in the donor phase. For purposes of constructing a model we can analyze analytically, we assume that the local site energy for an electron/hole configuration is coulombic ($J = e^2/\varepsilon r$) at long range and equal to U when the electron and hole are localized on the same site as in an excitonic configuration. Hence, $\varepsilon(0)$ corresponds to the energy of a non-dissociated exciton at the interface in our model. The interpolation between the local interaction U and the long-range Coulomb interaction is given by the Mataga–Nishimoto potential

$$\varepsilon(r) = \frac{U}{1 + U\varepsilon_r r/r_o}, \tag{8}$$

where ε_r is the relative permittivity between the donor and acceptor materials (and can be set to 1 since we assume the two materials to be conjugated organics) and $r_o = 14.397$ when distances are in Å and energies are in eV. Finally, we take the coupling between the exciton and the charge-separated states to be due to tunneling and of the form

$$v(r) = V_o e^{-\beta r} \tag{9}$$

where β^{-1} defines a tunnelling length. Within this model, the idempotent operator is given by

$$1 = |X\rangle\langle X| + \sum_r |r\rangle\langle r| \tag{10}$$

where $|X\rangle$ denotes a local exciton state and $|r\rangle$ denotes a charge-separated state with electron/hole radius r. The energy eigenstates can be determined by expanding in this basis

$$|\mu\rangle = \langle\mu|X\rangle|\mu\rangle + \sum_r |r\rangle\langle r|\mu\rangle \tag{11}$$

and then writing out the Hamiltonian matrix elements as

$$\varepsilon(r)\langle r|\mu\rangle + v(r)\langle X|\mu\rangle = E_\mu\langle r|\mu\rangle \tag{12}$$

$$\sum_r v(r)\langle r|\mu\rangle = E_\mu\langle X|\mu\rangle. \tag{13}$$

This leads to a transcendental equation for the energy eigenvalues,

$$E_\mu = \sum_r \frac{v^2(r)}{E_\mu - \varepsilon(r)}. \tag{14}$$

Similarly, one finds the projection of the exciton onto the $|\mu\rangle$ eigenstate as

$$|\langle X|\mu\rangle|^2 = \left(1 + \sum_r \frac{v^2(r)}{\left(E_\mu - \varepsilon(r)^2\right)}\right)^{-1}. \tag{15}$$

In the Wigner–Weisskopf model, one sets $\varepsilon(r) = r$ corresponding to a uniform density of states and assumes the coupling is constant. This allows both summations to be performed analytically. For the case at hand, we assume the sums can be transformed into radial integrations and then use the Laplace method to evaluate.‡

For comparison, we consider the case of a 2D and a 3D lattice with similar parameters: $\{E_x \rightarrow 2.5$ eV, $E_o \rightarrow 5$ eV, $V_o \rightarrow -1$ eV, $U \rightarrow 3$ eV, and $\beta \rightarrow 0.75^{-1}\}$ where $E_x = E_{D,lumo} - E_{D,homo} - U$ is the local exciton energy taken as the energy difference between the donor's LUMO and HOMO orbitals plus the local electron/hole attraction U, and $E_o = E_{A,lumo} - E_{D,homo}$ is the energy for a non-interacting electron/hole pair. Finally, we set the local electron/hole site energy to be

$$E(\vec{r}) = E_o + \varepsilon(r) + \delta E \tag{16}$$

with $\langle\delta E\rangle = 0$ and $\sqrt{\langle\delta E^2\rangle} = 0.1$ eV to account for the fact that the local electron/hole site energies are distributed about E_o to reflect inhomogeneous local environments. While the model is very simple and ignores interactions between charge-separated configurations, the resulting energies and density of states are similar to what we obtain with our more elaborate, fully interacting model, supporting that this heuristic model captures the salient physics of a more sophisticated model. Parametrically, the model presented here is identical to the more realistic, fully interacting heterojunction model we have published extensively upon over the past 15 years. We shall comment further on this model later in this report.

It is interesting to note that in the 2D case, there are far *more* states below the exciton energy (at $E = 0$) than above, which suggests that dimensionality plays a central role in the dissociation of an initially prepared exciton state. This notion is further enforced when we do a head-to-head comparison of the normalized density of states for the 2D and 3D lattices as shown in Fig. 2. First, we note that the DOS in the 3D lattice is skewed towards higher energies as compared to the 2D lattice. Furthermore, in 3D the number of local charge-transfer states with energy lower than the exciton's energy is lower than in the 2D case. From this simple model, we are drawn to the conclusion that dimensionality and density of states are every bit as important as energetics when building models for exciton dissociation.

‡ While the model can be solved fully analytically, it is computationally far faster to simply diagonalize the Hamiltonian matrix given by eqn (12) and (13).

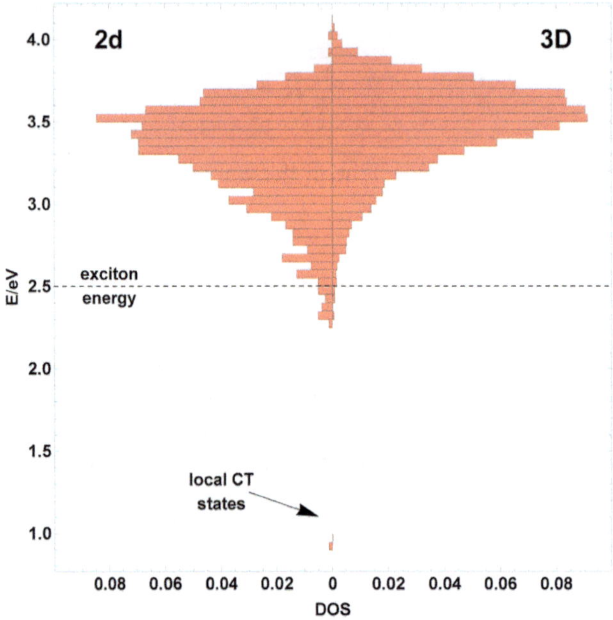

Fig. 2 Normalized DOS for 2D and 3D model lattices.

Having determined the eigenstates of the model, we next construct a function describing the free energy required to separate charges to a given radius, r_{sep}. From statistical mechanics, we know that

$$F = -k_{\mathrm{B}}T \ln Z \tag{17}$$

where Z is the canonical partition function given in the usual way *via* summing over states

$$Z = \sum_{\mu} \exp(-E_{\mu}/k_{\mathrm{B}}T). \tag{18}$$

We write this as an integral over r and define $E_{\mu}(r)$ as the energy of a charge-separated state whose mean radius is equal to $r = |\langle\mu|\vec{r}|\mu\rangle|$. Thus

$$Z = \int_0^{\infty} g(r)\exp(-E_{\mu}(r)/k_{\mathrm{B}}T)\mathrm{d}r \tag{19}$$

where $g(r)$ is the density of eigenstates with mean radius r. In Fig. 3 we show that $E_{\mu}(r)$ can be described as a nearly single-valued function of r except in cases where the exciton is resonant with the separation energy. Moreover, to a good approximation, $E_{\mu}(r) \approx \varepsilon(r)$ when we shift the spectral range to place the lowest energy eigenvalue at the origin with $r = 1$ and shift the interaction energy $\varepsilon(1) = 0$. This shift in the energy origin is typically done when constructing the partition function so that all energies are considered to be excitations from the lowest energy state. Thus, we define the "radial" partition function and free energy as simply

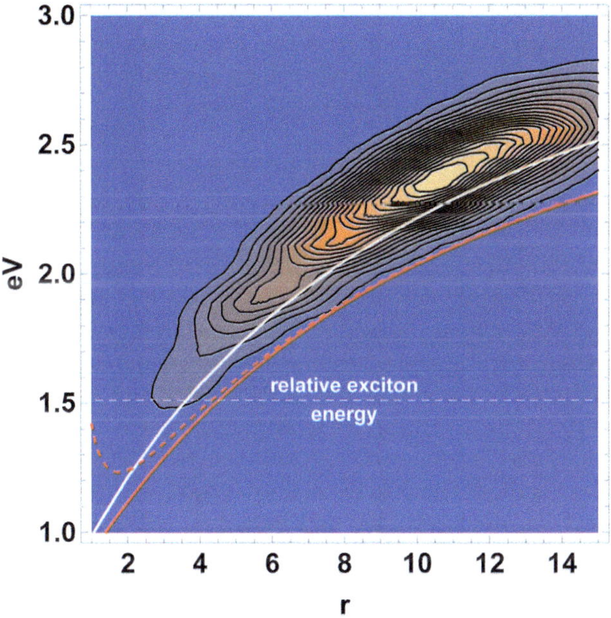

Fig. 3 Energetics and approximate free energy from the Statistical Hopping model for exciton dissociation. The contours give the density of eigenstates with a given average electron/hole separation radius and eigenvalue energy. The white, red, and red-dashed curves are theoretical models for the energy and free energy as described in the text.

$$Z_{ct}(r) = \exp(-(\varepsilon(r) - k_B T \ln(g(r)))/k_B T) \tag{20}$$

and

$$F_{ct}(r) = (\varepsilon(r) - k_B T \ln(g(r))). \tag{21}$$

Taking $g(r) = 4\pi r^2$ as the 3D radial volume element we have an analytical expression for the free energy of a charge-separated state with radius r. Finally, the initial exciton state has an entropy of zero since it is a local state within our initial basis, and we write the reversible work as

$$\Delta F(r) = F_{ct}(r) - E_{ex}. \tag{22}$$

Finally, we introduce a correction for the relative kinetic energy of the charge-transfer state as function of the electron/hole separation radius. This can be justified from simple argument based on the uncertainty principle $\Delta x \Delta p \geq \hbar/2$ and $\langle K \rangle = \langle \Delta p^2 \rangle / 2m$ which immediately yields that $\langle K \rangle \approx \hbar^2/(2mr^2)$ upon taking $r^2 = \langle \Delta x^2 \rangle$.

In Fig. 3 we compare the distribution of free energies compared to the electron/hole site energy, $\varepsilon(r)$ and the F_{ct} as computed by eqn (21), superimposed in white and red respectively. The dashed red-line includes the short-ranged kinetic energy correction which does appear to limit the closest point of contact between the electron and hole, but dies off rapidly at long distances. Further, to a good approximation, the free energy is well approximated by simply the charge-transfer site energy $\varepsilon(r)$ and shows no indication of rolling over at ranges relevant for exciton dissociation. We conclude from this that models that do not allow the electron and hole to move independently and quantum mechanically or only include intermediate hopping or tunneling effects are poor representations of semiconductor interfaces.

2.2 A fully interacting model

Missing from this simple model is the fact that both the electron and the hole are mobile and consequently are delocalized over multiple sites. Also, different electron/hole configurations interact. With these in mind, we next describe a fully interacting electron/hole model. Adding the additional degrees of freedom dramatically increases the complexity of the model, and hence the model is currently restricted to a 2D lattice. We recently used this model to assess the free energy of a 2D model heterojunction system[17] and we shall recapitulate some of the results here as they pertain to the heuristic model we presented above.

The model is described by the following electron/hole- + -phonon Hamiltonian,[10,15]

$$\begin{aligned}\hat{H} &= \hat{H}_{\text{el}} + \hat{H}_{\text{el-ph}} + \hat{H}_{\text{ph}} \\ &= \sum_{\mathbf{mn}} (F_{\mathbf{mn}} + V_{\mathbf{mn}})|\mathbf{m}\rangle\langle\mathbf{n}| \\ &+ \sum_{\mathbf{mn},a,\mu} \left(\frac{\partial F_{\mathbf{mn}}}{\partial q_{a,\mu}}\right) q_{a,\mu}|\mathbf{m}\rangle\langle\mathbf{n}| \\ &+ \frac{1}{2}\sum_{\mu,a} \omega_a{}^2\left(q_{a,\mu}{}^2 + \lambda q_{a,\mu,\mu+1}\right) + q_{a,\mu}{}^2,\end{aligned} \tag{23}$$

where electron–hole configurations $|\mathbf{n}\rangle = |\mathbf{h}_i\mathbf{e}_j\rangle$, whereby a hole in a valence orbital on site i and an electron in a conduction orbital on site j form the basis for the Hamiltonian. We have published the details and parameterization of the model previously and we briefly review its salient features. The term $F_{\mathbf{mn}}$ describes the single-particle motion for a non-interacting electron/hole pair. In its simplest form for configurations $|\mathbf{n}\rangle = |\mathbf{h}_i\mathbf{e}_j\rangle$ and $|\mathbf{m}\rangle = |\mathbf{h}_k\mathbf{e}_l\rangle$ it is given by

$$F_{\mathbf{mn}} = f^{\mathrm{e}}_{jl}\,\delta_{ik} + f^{\mathrm{h}}_{ik}\,\delta_{jl} \tag{24}$$

where f^{e} and f^{h} have the meaning of localized energy levels and transfer integrals for conduction band electrons and valence band holes. In the absence of disorder, these quantities obey charge-conjugation symmetry with $f^{\mathrm{h}}_{mn} = -f^{\mathrm{e}}_{mn}$, and we assume that only nearest neighbors on the 2D lattice are coupled by these terms $(t = f_{i,i+1})$. The donor and acceptor domains are differentiated by imposing an energetic off-set, ΔE, in the site-energies at the domain boundary as sketched in Fig. 1b. $V_{\mathbf{mn}}$ describes spin-dependent two-particle coulombic and exchange interactions for each configuration, as well as interactions between different singlet geminate electron–hole pairs. For the case of singlet excitations,

$$V_{\mathbf{mn}} = -\langle h_i e_j|V|h_i e_j\rangle + 2\langle h_j e_i|V|h_i e_j\rangle. \tag{25}$$

We assume that the inter-unit overlap of the primitive site-basis functions is small and that three types of integrals contribute to this interaction, each with well known meaning and long-range behavior.[10] First, an electron and a hole on sites separated by a distance r will experience a long-range Coulomb attraction of the Mataga-type as used above

$$\langle h_i e_j | v | h_i e_j \rangle = \frac{J_o}{1 + r_{ij}/r_o}, \tag{26}$$

a short range exchange term which decays exponentially with electron/hole separation

$$\langle h_j e_i | v | h_i e_j \rangle = K_o \exp(-r_{ij}/r_o). \tag{27}$$

Secondly, because differential overlap between an electron and a hole sitting on the same site is substantial, we include the transition dipole–dipole coupling between singlet geminate electron–hole pairs,

$$\langle h_i e_i | v | h_j e_j \rangle = \frac{D_o}{\left(r_{ij}/r_o\right)^3}. \tag{28}$$

where J_o and K_o indicate coulombic and exchange interactions between an electron and a hole on the same site. D_o is a dipole moment of an electron–hole pair located on site i. It is important to note that these terms carry information pertaining to the spatial disorder in our model. While the model can treat singlet or triplet states, we are analyzing post-photoexcitation charge-transfer states and will only focus on singlets in this study.

The phonon term, $\hat{H}_{\text{ph}}$, assigns to each lattice site a high and a low frequency vibrational mode described by a local harmonic oscillator with weak nearest-neighbor coupling. These terms are determined from spectroscopic Huang–Rhys parameters typifying organic conjugated polymer systems. Such contributions modulate *both* the on-site band-gap as well as the site-to-site hopping integrals.

For a given lattice configuration, the eigenstates of the Hamiltonian in eqn (23) are linear combinations of all configurations allowed on a particular lattice

$$(\hat{H}_{\text{el}} + \hat{H}_{\text{el-ph}}(\{\mathbf{q}\}))|\psi^k\rangle = E_k(\{\mathbf{q}\})|\psi^k\rangle \tag{29}$$

with the k-th eigenstate defined as $|\psi^k\rangle = \sum_{ij} c_{ij}{}^k |h_i e_j\rangle$. We construct a 10 site × 10 site lattice with a donor–acceptor energy offset of $\Delta E = 0.5$ eV, the transfer energy between nearest sites set to $t = 0.536$ eV and lattice constant $a = 1$ nm in all directions. We deliberately choose a higher value for the hopping parameter (previously calculated to represent polymer intra-chain transfer energy[15]) to allow an electron and a hole easy passage throughout the lattice and to ensure that nothing but the presence of disorder inhibits or facilitates electron/hole dissociation.

Since the fully interacting model allows valance holes in the acceptor region and conduction electrons in the donor domain, we need to be careful when defining the free energy as a function of electron/hole separation distance since we only want to include states in which the conduction electron is in the acceptor domain and the hole is in the donor domain. With this in mind, we define this as

$$\langle R_D{}^k \rangle = \sum_{ij} |\rho_{ij}{}^k| \vec{r}_{ij} \cdot \vec{n} \tag{30}$$

where $\rho_{ij}{}^k$ is the eigenstate amplitude at the electron/hole configuration and $\vec{n}$ is a unit vector normal to the donor/acceptor interface. The dotted curves in Fig. 4 show the distribution of charge-transfer energies *vs.* mean electron/hole separation radius R_D as averaged over 1000 realizations of the site energies with a $\delta E =$ 0.01 eV − 0.20 eV fluctuation in the local site energy. The solid curves are the free energies which include the density of states contribution at 300 K. Lastly, the shaded region indicates a range of literature values for qV_{OC} for a variety of polymer/fullerene based heterojunction systems. This figure provides a connection between the "hot" and "cold" dissociation mechanisms for exciton dissociation.

One can think of a "hot" exciton process as occurring on a potential energy surface described by the average CT energy (*i.e.* the dotted curve in Fig. 4). In this case, an exciton is injected into the system at an energy above ≈2.0 eV. While some energetic relaxation may occur, the system encounters a small barrier and can efficiently dissociate into separate charge-carriers. In this case, energy fluctuations appear to increase the effective energy barrier between the initial exciton at $R_D = 0$ and the fully separated charge carriers. Conversely, the "cold" exciton pathway can be understood as following a potential of mean force path along the free-energy minimum plotted as the solid curves. While the energy landscape is somewhat bumpy compared to the "hot" dissociation path, free energy

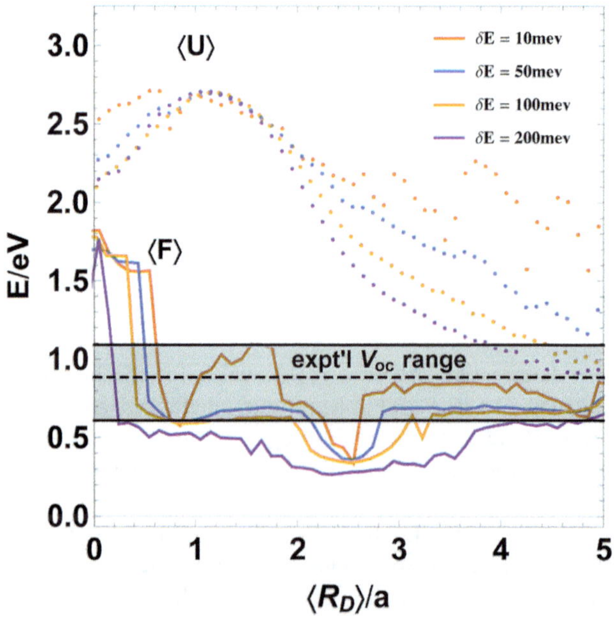

Fig. 4 Average (points) and free energy (solid) curves for the fully-interacting model at different values of site-energy disorder δE. The shaded region represents the range of experimental values for open-circuit voltage with the average indicated by the dashed black line.

fluctuations tend to be smoothed out by increasing the fluctuations in the site energies. In both cases the energetics converge to the experimental qV_{OC} range. This leads to the conclusion that both "hot" and "cold" exciton mechanisms contribute to the final production of charge-carriers.

In Fig. 5(a–d) we present what happens when we modify the lattice model such that the transferred electron is far less mobile in the direction perpendicular to the interface than in the direction parallel to the interface. In this, we model more closely the case of π-stacked oligomers in the acceptor phase. The total energy and free energy curves are superimposed over the density of eigenstates in which the local site HOMO/LUMO gap included a random (static) fluctuation about a mean gap. The computed results reproduce many of the features of the exact model presented in Sec. 2.1 including the short-range kinetic energy contribution and the long-range tail of the average energy and free energy. This would imply that lower mobility in one of the phases has a profound impact on the ability of charges to escape from the interfacial region. Only in the instances of higher energetic disorder do we begin to see the free energy contribution begin to roll-over and plateau suggesting that local trapped states may facilitate the separation process.

2.3 Transition rates between states

In computing the eigenstates of eqn (23) we took $q_{a\mu} = 0$ and considered only the vertical excitations, assuming that fluctuations in the energy are due to the local electrostatics and not due to fluctuations in the phonons from their equilibrium positions. The remaining two terms in eqn (23) allow for transitions between eigenstates. For this, we allow for both high and low-frequency phonon branches and assume that the coupling term $(\partial F_{nm}/\partial q_{a,\mu})$ can be determined from spectroscopic Huang–Rhys factors. It is then a simple matter to derive the golden-rule rate constant between eigenstates of $H_{\rm el}|k\rangle = \varepsilon_k|k\rangle$. For this, we first perform a unitary (polaron/shift) transform on the entire H.[18]

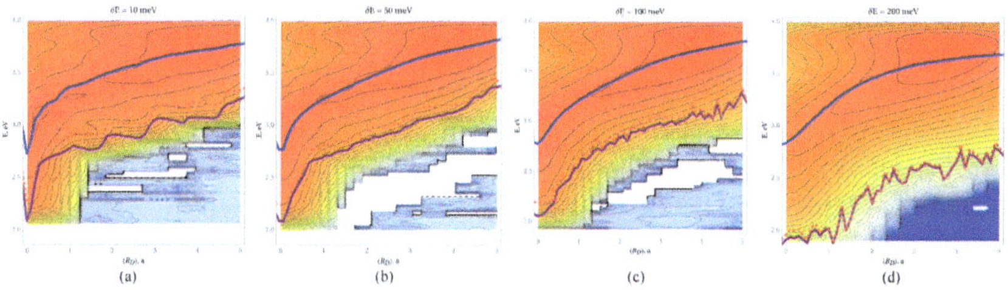

Fig. 5 Radial energy distributions for lattice model with decreased mobility perpendicular to interface. The blue curve represents thermodynamic energy in the high temperature regime as a function of electron–hole separation; the dashed red curve is the thermodynamic energy at 298 K; the purple curve is the free energy at 298 K. Contours and background color indicate the normalized probability to find an eigenstate with a given energy and with a given electron/hole separation as sampled over fluctuations in the local site energies. In these cases we model the donor phases as a series of π-stacked oligomers with high electron/hole mobility parallel to the interface and much lower perpendicular to the interface. Both the total energy and free energies at $T = 300$ K are similar to the model of section 2.1 where the transferred charge is effectively immobile.

$$\tilde{H} = e^{-S} H e^{S} \tag{31}$$

where S is an anti-hermitian operator

$$S = -\sum_{ka} (g_{kkq}/\omega_q)|k\rangle\langle k|\left(a_q^{\dagger} - a_q\right) \tag{32}$$

which gives

$$e^{S} = \sum_{k} |k\rangle\langle k| \exp\left[-\sum_{q} \frac{g_{kkq}}{\omega_q}\left(a_q^{\dagger} - a_q\right)\right] \tag{33}$$

where the sum over q is over all normal modes, g_{kkq} is the force term in eqn (23) transformed into the normal mode and eigenstate basis, and the sum over k is over eigenstates of H_{el}. The polaron transform renormalizes the electronic energies

$$\tilde{\varepsilon}_k = \varepsilon_k - \sum_{q} \frac{{g_{kkq}}^2}{\omega_q} \tag{34}$$

and shifts the electronic coupling

$$\begin{aligned} \tilde{V}_{kk'} &= \left(e^{-S} H_{\mathrm{el\text{-}ph}} e^{S}\right)_{kk'} \\ &= \sum_{k \neq k'} |k\rangle\langle k'| M_{kk'} \end{aligned} \tag{35}$$

where

$$M_{kk'} = \sum_{q} g_{kk'q}\left(a_q^{\dagger} + a_q - \frac{2g_{kkq}}{\omega_q}\right) \mathrm{e}^{\sum_{q'}\left(\frac{g_{kkq'} - g_{k'k'q'}}{\omega_j}\left(a_j^{\dagger} - a_j\right)\right)} \tag{36}$$

is the dressed off-diagonal coupling. Transforming to the Heisenberg representation and integrating over the phonon degrees of freedom leads to an expression for the golden-rule rate in terms of the autocorrelation of these dressed operators

$$W_{kk'} = 2\mathrm{Re}\int_0^{\infty} \sum_{qq'} \mathrm{Tr}\left[M_{kk'q} M_{k'kq'}(t) \rho_{\mathrm{eq}}^{\mathrm{ph}}\right] \mathrm{e}^{-i(\tilde{\varepsilon}_k - \tilde{\varepsilon}_{k'})t} \mathrm{d}t \tag{37}$$

In general, the time-correlation function has recurrences (since it is a sum of periodic functions) and in practice we perform the time iteration only over a sufficiently long enough interval for the correlation function to decay and truncate the integration before the first recurrence.

While this approach is accurate and we have used it extensively for computing rates, the drawback is that it is time-consuming to compute this between each and every possible eigenstate of H_{el} and then average over energetic fluctuations. Since we are primarily interested in transitions from a local exciton to charge-separated states, we developed a machine-learning approach to rapidly parse through the H_{el} eigenstates and select *only* excitonic and charge-separated eigenstates using a training set of pre-determined exciton and charge-separated states.

Secondly, we wanted to understand whether or not the charge-separated/polaron states are entangled electron/hole polarons or if the state is separable into a product of polaron states. In other words, can we write a given charge-separated state as a linear combination of site-local wavefunctions

$$\psi_k(r_e, r_h) = \sum_{ij} c_{ij}{}^k \phi_i(r_e)\bar{\phi}_j(r_h) \tag{38}$$

as a sum over a single index

$$\psi_k(r_e, r_h) = \sum_n \lambda_n{}^k f_n{}^k(r_e) g_n{}^k(r_h) \tag{39}$$

where $f_n{}^k(r_e)$ and $g_n{}^k(r_h)$ are orthogonal polynomials given by Schmidt decomposition of $\psi^k(r_e,r_h)$. These are the so-called "Schmidt modes" from quantum information theory. Using the normalized Schmidt eigenvalues $\tilde{\lambda}_n{}^k$ for a given eigenstate, we can define the Shannon entropy as

$$S_k = -\sum(\tilde{\lambda}_n{}^k)^2 \ln((\tilde{\lambda}_n{}^k)^2). \tag{40}$$

When $S_k = 0$, only one term contributes to the sum (with only one of the $\tilde{\lambda}_k = 1$). Furthermore $S = \ln N$ where N is the dimensionality of the information space needed to describe a given eigenstate. In essence, $N = \exp(S)$ gives the minimum number of single-particle reducible states needed to describe a given many-body wavefunction. For an entangled two particle wavefunction as given in eqn (38), that decomposition is expressed in eqn (39). If $S = 0$, only one term will contribute to this sum, and the CI eigenstate would be a simple product of a single valance-band electron orbital times a single conduction-band hole orbital. Consequently, one anticipates that localized polaron states would be states with minimal entanglement and hence describable as separable electron/hole states and excitonic states to be highly entangled and generally non-separable.

In Fig. 6 we compare the state energy and its associated entropy for a given lattice realization. Here we determined whether a given eigenstate is charge-

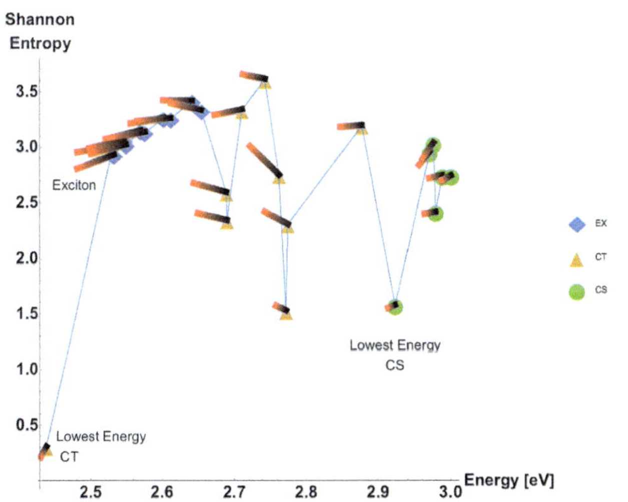

Fig. 6 Correlation between state energy, Shannon entropy, and net flux from a given state for a model heterojunction system.

separated (green dots), charge-transfer (yellow triangles), or excitonic (blue diamonds) using a machine learning method that compared computed electron/hole eigenstates to "idealized" CS, CT, or excitonic states. For clarity we show the results from only a single lattice realization. First, we note that excitonic states tend to cluster with high entanglement entropies which one anticipates since the electron and hole are occupying the same set of sites on the lattice. Similarly, the lowest energy CT state, which is best described as an electron/hole pair separated across the interface, but pinned to each other *via* coulombic interactions. Interestingly, even though their center of mass motions are coupled, the state itself is separable into a product of polaron states. Energetically above the excitons lie another band of charge-transfer states. While these states are CT in character, the electron and hole remain highly entangled and in general these states contain some excitonic configurations. Finally, energetically above these are true charge-separated states; however, Schmidt decomposition reveals that these are not strictly separable into electron and hole single-particle polaron states.

Fig. 6 also shows the net rate that population leaves a given energy state

$$\eta_n = \sum_i (W_{in} - W_{ni}) \tag{41}$$

as indicated by a flag whose magnitude indicates the magnitude of η_n and whose direction indicates the flow on the (E,S) plane. Not surprisingly, the net "flow" is towards the lowest energy CT state. However, it is interesting to note that the lowest energy CS state is a kinetic trap.

3 Conclusions

In this work, we have discussed various approaches we have followed to connect the interfacial energetics to the observed open-circuit voltage for organic polymer-based photovoltaic cells. Such coarse-grained models provide considerable insight into to the energetics and kinetics of charge-transfer processes in meso-scale systems that are currently too large to be considered using more atomistic/*ab initio* based models. In particular, this discussion has focused upon the role of the density of states and entropic contributions to the charge-separation process following photoexcitation. One of the principle results of this analysis is a comparison between an analytical model based upon the Wigner–Weisskopf model for the decay of a quantum state into a broad continuum to a more detailed lattice model our group has developed. Surprisingly, the two models give remarkably similar results in the limit that the mobility of the transferred charge is sharply reduced in the direction perpendicular to the interface.

Secondly, we test the hypothesis that the charge-separated state is a simple product of an electron and a hole state of the form

$$\psi_{\rm cs}(r_{\rm e},r_{\rm h}) = \phi_{\rm e}(r_{\rm e})\phi_{\rm h}(r_{\rm h}).$$

In other words, that the singular value decomposition of the full electron/hole wave function of eqn (38) should contain exactly one and only one non-trivial term in eqn (39) resulting in a Shannon entropy (eqn (40)) of exactly 0. Rather, we find that only the lowest energy CT state is the least entangled electron/hole state. We shall continue to explore the connection between dissociated CS states and their Shannon entropy since it pertains as to the range at which electrons and holes are truly separable species.

Conflicts of interest

There are no conflicts of interest.

Acknowledgements

This work was funded in part by the National Science Foundation (CHE-1664971 and MRI-1531814) and the Robert A. Welch Foundation (E-1337). DPM acknowledges the support from a UH Summer Undergraduate Research Fellowship (SURF).

References

1 M. D. Chatzisideris, A. Laurent, G. C. Christoforidis and F. C. Krebs, *Appl. Energy*, 2017, **208**, 471–479.
2 S. H. Park, A. Roy, S. Beaupre, S. Cho, N. Coates, J. S. Moon, D. Moses, M. Leclerc, K. Lee and A. J. Heeger, *Nat. Photonics*, 2009, **3**, 297–302.
3 M. Scharber and N. Sariciftci, *Prog. Polym. Sci.*, 2013, **38**, 1929–1940.
4 Z. He, B. Xiao, F. Liu, H. Wu, Y. Yang, S. Xiao, C. Wang, T. P. Russell and Y. Cao, *Nat. Photonics*, 2015, **9**, 174–179.
5 S. Gélinas, A. Rao, A. Kumar, S. L. Smith, A. W. Chin, J. Clark, T. S. van der Poll, G. C. Bazan and R. H. Friend, *Science*, 2014, **343**, 512–516.
6 A. C. Jakowetz, M. L. Böhm, J. Zhang, A. Sadhanala, S. Huettner, A. A. Bakulin, A. Rao and R. H. Friend, *J. Am. Chem. Soc.*, 2016, **138**, 11672–11679.
7 B. A. Gregg, *J. Phys. Chem. Lett.*, 2011, **2**, 3013–3015.
8 S. N. Hood and I. Kassal, *J. Phys. Chem. Lett.*, 2016, **7**, 4495–4500.
9 T. M. Burke, S. Sweetnam, K. Vandewal and M. D. McGehee, *Adv. Energy Mater.*, 2015, **5**, 1500123.
10 S. Karabunarliev and E. R. Bittner, *J. Chem. Phys.*, 2003, **119**, 3988–3995.
11 S. Karabunarliev and E. R. Bittner, *J. Phys. Chem. B*, 2004, **108**, 10219–10225.
12 E. R. Bittner, J. G. S. Ramon and S. Karabunarliev, *J. Chem. Phys.*, 2005, **122**, 214719.
13 E. R. Bittner, J. G. S. Ramon and S. Karabunarliev, *J. Chem. Phys.*, 2005, **122**, 214719.
14 E. R. Bittner, S. Karabunarliev and A. Ye, *J. Chem. Phys.*, 2005, **122**, 034707.
15 S. Karabunarliev and E. R. Bittner, *J. Chem. Phys.*, 2003, **118**, 4291–4296.
16 S. Karabunarliev and E. R. Bittner, *Phys. Rev. Lett.*, 2003, **90**, 057402.
17 V. Lankevich and E. R. Bittner, *J. Chem. Phys.*, 2018, **149**, 244123.
18 A. Pereverzev and E. R. Bittner, *J. Chem. Phys.*, 2006, **125**, 104906.

Faraday Discussions

PAPER

Donor–acceptor preassociation, excited state solvation threshold, and optical energy cost as challenges in chemical applications of photobases†

Jonathan Ryan Hunt, Cindy Tseng and Jahan M. Dawlaty *

Received 30th November 2018, Accepted 11th December 2018
DOI: 10.1039/c8fd00215k

Photobases are molecules with increased pK_a in the excited state that can serve to transduce light energy into proton removal capability. They can be used to control chemical reactions using light, such as removing protons from a catalytic site in reactions that are rate-limited by proton transfer. We identify and explore several major challenges toward their practical applications. Two important challenges are the need for pre-association (or ground state hydrogen bonding) between the proton donor and the photobase, and the need for excited state solvation of the photogenerated products. We investigate these two challenges with the photobase 5-methoxyquinoline as the proton acceptor and a low-pK_a alcohol, 2,2,2-trifluoroethanol, as the proton donor. We vary the concentration of the donor in a background non-hydrogen-bonding solvent. Using absorption spectroscopy, we have identified that the donor–acceptor concentration ratio must exceed 100 : 1 to achieve appreciable ground state hydrogen bonding. Interestingly, emission spectroscopy reveals that the onset of ground state hydrogen bonding does not guarantee successful excited state proton transfer. It takes an additional order of magnitude increase in donor–acceptor ratio to achieve that goal, revealing that it is necessary to have excess donor molecules to reach the solvation threshold for the photogenerated products. The next challenge is reducing the large ground-excited state energy gap, which often requires UV photons to drive proton transfer. We show experimental and computational data comparing the photobasicity and optical energy gap for a few N-aromatic heterocyclic photobases. In general, we find that reducing the energy gap by increasing the conjugation size necessarily reduces photobasicity, while adding substituents of varying electron-withdrawing strength allows some fine-tuning of this effect. The combination of these two factors provide a preliminary design space for creating new photobasic molecules.

University of Southern California, 920 Bloom Walk, Los Angeles, USA. E-mail: dawlaty@usc.edu; Tel: +1-213-740-9337

† Electronic supplementary information (ESI) available. See DOI: 10.1039/c8fd00215k

Proton transfer is important in a wide range of chemical and biological phenomena. Often redox reactions, especially those that are relevant for energy storage, are coupled with proton transfer and cannot be explained as electron transfer in isolation. Proton gradients are the driving force for a range of processes in living organisms, most notably synthesis of ATP. Most proposals on storing light energy in chemical bonds require driving multi-electron and multi-proton reactions. Given the importance of proton transfer in such diverse areas, understanding optical control over proton transfer is not only of fundamental importance, but it also opens opportunities for applications. Understanding excited state proton transfer in photoacids and photobases is a fundamental first step towards these goals.

Photoacids, which are molecules that become acidic in the electronically excited state, have been well-known for several decades.[1–6] Prime examples are conjugated systems such as pyrenols[7–9] and naphthols,[10–12] in which the OH functionality becomes more acidic in the excited state, often by more than 6 pK_a units. Photoacids have been used in a wide range of applications. Some examples are triggering acid initiated protein folding,[13,14] acid-catalyzed reactions,[15] pH-dependent enzymatic reactions,[16,17] synthetic organic reactions,[18,19] generation of a protonic potential difference with light,[20] and photo-control of protonic conductivity of electrolytes with light.[21]

Photobases are molecules that achieve the reverse effect and become more basic in the excited state. Some examples are quinolines,[22–27] Schiff bases,[28] acridines,[29–32] 3-styrylpyridine,[33] xanthone,[34] and curcumin.[35] Both fundamental studies and applications of photobases to chemical problems are far more limited compared to those of their photoacid counterparts. They offer rich prospects for applications, for example in catalytic reactions that are rate-limited by sluggish proton removal and require the use of strong bases. One may use photobases to create the proton removal drive using light in the vicinity of such catalysts. In a recent work, we have installed pendant photobases on an Ir-core organometallic catalyst with this goal in mind.[36] In another work, we have used photobases to deprotonate a family of alcohols, with the broader goal of generating reactive alkoxides using light.[37]

Despite promises, several challenges exist in the way of applying photobases as optically driven proton removal agents in catalysis or for inducing large scale pH change. We note four of these challenges. The first challenge is pre-association of the proton donor and photobase. Due to the often limited excited state lifetime of the photobase, it is desirable for it to be pre-hydrogen bonded with the target proton donor in the ground state. Upon optical excitation, the proton is then removed and the photobase is not required to search for and find its target during its limited excited state lifetime.

The second challenge is solvation of the photogenerated products or dissociation of ion-pairs after successful proton transfer. In order for the initial proton transfer step to occur, certain solvation conditions must be met. Often for downstream chemistry to proceed on a deprotonated target molecule, that molecule must then dissociate from the conjugate acid form of the photobase. Such dissociation must beat the excited state lifetime of the photobase. It is possible to tune this dissociation by the choice of the medium that would allow faster diffusion and better screening of the Coulomb attraction between ion pairs. Previously, in a work on photoinduced proton conductivity using photoacids we have shown that the separation of ions within the excited state lifetime is possible.[38] Such considerations are needed when photobases are required to induce bulk pH changes.

The third challenge is the energetic cost of proton removal. Many photobases operate with photon energies higher than 3 eV (<400 nm). While for certain applications using such high energies may not pose a problem, it is desirable to lower this energetic cost. This necessitates finding out how the excited state pK_a scales with the optical gap in various families of photobases.

The fourth challenge is the limited excited state lifetime of photobases, and is intimately tied to the challenges described above. A photobase, after successful deprotonation of a target molecule, will only retain the proton for as long as it remains in the excited state, which could be several nanoseconds for singlet states. Upon returning to the ground state, it releases the proton back. To give the often sluggish chemical steps a chance to proceed in the aftermath of a proton transfer, it is desirable to hold on to the removed proton for longer. An immediate strategy is to extend the excited state lifetime by intersystem crossing and creating triplets. However, the pK_a of triplet states is often lower than that of singlets, and one must find strategies to optimize the balance between pK_a^* and excited state lifetime.

In this work, we will take steps towards understanding the first three of the above challenges – the pre-association of the photobase and proton donors, the solvation threshold for photogenerated products, and the energetic cost of optical excitation. We will present experimental and theoretical results for better understanding these problems and propose strategies to mitigate them.

In Section 1, we will review the thermodynamics and kinetics of photobasicity in quinolines. In Section 2 we will show our experimental work that quantifies pre-association and the threshold for excited state proton transfer. We will briefly discuss proposed strategies to overcome the limitations posed by these problems. In Section 3 we will discuss that while increasing the conjugation size of the chromophore may decrease the optical gap, it will also reduce the thermodynamic drive for proton removal. We will propose strategies for finding a balance between these opposing effects.

1 Brief overview of thermodynamics and kinetics of photobases

To set the stage for the rest of this work, in this section we review the fundamental ideas of excited state proton transfer, with emphasis on photobasicity.

The excited state acidity and basicity is commonly inferred from a thermodynamic cycle originally reported by Förster,[39] depicted in Fig. 1. The approximations and conditions for the validity of the Förster cycle are discussed in several references.[2,40] In brief, the Förster cycle for a photobase starts with the unprotonated form that absorbs light, undergoes proton transfer with a forward thermodynamic drive ΔG^*, emits from the protonated form, and finally releases the proton and returns to the initial state. The energy difference ΔG between the protonated and unprotonated forms is determined by measuring the pK_a of the molecule by conventional ground state titration and using $pK_a = \Delta G/(2.3RT)$. Absorption and emission spectroscopies are used to determine the energy gaps

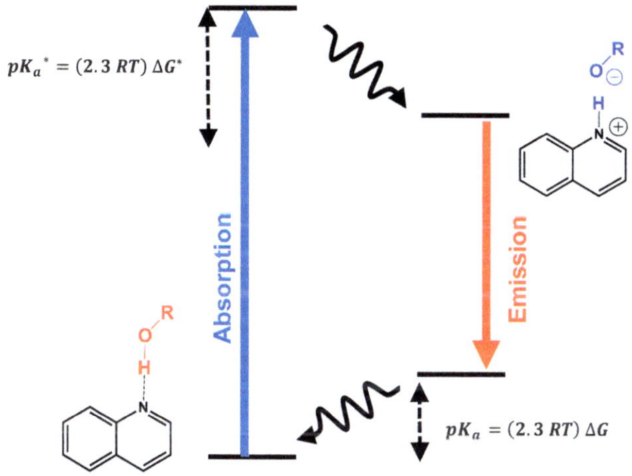

Fig. 1 The Förster thermodynamic cycle for proton transfer in a quinoline photobase. The excited state pK_a^* can be inferred from the energy gaps of the protonated and unprotonated forms and the ground state pK_a.

between the ground and excited states of the protonated and unprotonated forms. These numbers allow for estimating the gap ΔG^*, which is related to the excited state pK_a by $pK_a^* = \Delta G^*/(2.3RT)$. As a convenient reference number, an energy gap of 477 cm^{-1} (59 meV or nearly 10 nm in the visible range) corresponds to one unit of pK_a at $T = 298$ K. Various approximations exist that help with determination of 0–0 vibronic gaps for the unprotonated and protonated forms in the presence of broad absorption and emission spectra. Often a large separation of time-scales exist between proton transfer times (few tens of picoseconds) and relaxation to ground state times (few nanoseconds). This justifies the conventional use of equilibrium thermodynamic language of pK_a^* for expressing the drive for proton transfer in the excited state.

Even though the Förster cycle described above is quite useful, it only informs us about the existence of a thermodynamic drive for proton transfer. In fact, the energy gap for the protonated form in the cycle need not be determined by an emission spectrum, and instead can be determined by measuring the absorption gap of a molecule that is protonated in the ground state by lowering the pH. Therefore, a Förster cycle can be constructed even if excited state proton transfer does not occur due to kinetic limitations.

Several processes can compete against proton transfer in the excited state. It seems reasonable to think a free energy relation (FER) must exist between the speed of protonation and the thermodynamic drive or pK_a^* of the molecule. However, experiments suggest that the excited state dynamics are often more complicated due to a variety of competing factors. In our previous work, we followed two strategies to investigate kinetic trends for proton transfer in the quinoline family. First, we tuned the pK_a^* of a quinoline photobase family by varying the electron withdrawing strength of a substituent (as described by its Hammett parameter) attached to the conjugated rings.[26] According to the Förster analysis for this series, the quinoline pK_a^* linearly decreased with the increasing Hammett parameter. However, there was no clear trend in the kinetics of the proton transfer, and in some members signatures of proton transfer were convoluted with triplet dynamics in the excited state.[27] Second, we varied the pK_a of the proton donor by choosing a series of alcohols, while keeping the photobase molecule the same.[37] In this case, a clearer trend in kinetics was observed, with the more acidic donors undergoing faster proton transfer. Apart from the intramolecular factors, solvation of the conjugate base and conjugate acid likely influences the kinetics. In brief, Förster analysis is an important and powerful tool for assessing the thermodynamic drive of photobases. However, it is unable to predict the competing kinetic pathways, and barriers to proton transfer.

2 Pre-association in the ground state and solvation threshold for excited state proton transfer

Successful proton transfer in the excited state often requires the proton donor and acceptor to be hydrogen bonded or pre-associated in the ground state. Achieving such ground state pre-association is often difficult, especially when the proton donor exists in small concentrations in a background solvent. Pre-association is necessary due to the typically short (several ns) lifetimes of the singlet excited states of the photobase. The excited states of photobases likely have larger drives for complexation than their ground states. However, if the diffusion time necessary to find a target proton donor partner is longer than the excited state lifetime, the photobase will return to the ground state without successful deprotonation.

To understand the problem of pre-association, we consider a proton donor and a photobase in a background non-hydrogen-bonding solvent. Although the enthalpy of hydrogen bonding favors complexation of the photobase with the donor, the Gibbs energy of solvation of the two partners likely works against it. Even if the enthalpy of solvation of separated species compared to the solvation of the complex is negligible, the entropy of the dissociated species is vastly larger compared to the complex. Therefore, the entropic drive for decomplexation can overwhelm the enthalpic drive for hydrogen bonding. To enforce a significant population of hydrogen bonded complexes against this entropic drive, one may increase the concentration of the donor.

We have two purposes in this section. First, we will demonstrate the threshold concentration ratio between proton donor and photobase that is necessary to create a significant ground state hydrogen bonded population countering the entropy of dissociation. Second, we will demonstrate that slightly above this threshold concentration of the donor, even though ground state hydrogen bonded complexes are formed, excited state deprotonation still fails. We will show that to achieve successful excited state deprotonation, the concentration of donor must go even higher. The excess donors are necessary to solvate the conjugate base of the proton donor after proton transfer. We will demonstrate the threshold concentration of donors that makes this solvation possible.

Our photobase in this study is 5-methoxyquinoline (MeOQ). It has a pK_a^* of approximately 15.5 and has been studied previously by us.[26,27,37] The proton donor is 2,2,2-trifluoroethanol (TFE), which is a low pK_a alcohol (pK_a = 12.4) that is known in the literature as a good hydrogen-bond donor.[41] We have shown in a previous work that MeOQ deprotonates TFE rather rapidly (~2 ps) due to the large thermodynamic drive for deprotonation ($pK_a^* - pK_a = 3.1$). The background non-hydrogen-bonding solvent is dichloromethane (DCM). The concentration of the photobase MeOQ is kept at 5×10^{-5} M, which is a convenient value for absorption and emission measurements. The concentration of the donor TFE relative to the photobase MeOQ is scanned over several orders of magnitude (1 : 1 to 200 000 : 1 with respect to MeOQ) in the background DCM solvent. For each solution of a given concentration ratio, absorption spectra were measured for signs of ground state hydrogen bonding, while emission spectra were monitored for identifying the threshold of excited state deprotonation.

We studied the growth of the ground state hydrogen-bonded complex using absorption spectroscopy as shown in Fig. 2. As the ratio of donor TFE to photobase MeOQ increases, the spectrum of the photobase red-shifts, indicating hydrogen bond formation with TFE. Most of the shift is seen before TFE constitutes a significant portion of the total solvent volume, showing that this shift is not due to average modification of the dielectric constant. Furthermore, since TFE and DCM have roughly the same dielectric constant[42,43] (8.55 and 8.93, respectively), we can be certain that the large shifts seen in these experiments (>10 nm) are due to hydrogen-bonding.

As discussed in the Experimental methods section, this continuous red shift of the absorption spectrum can be used to identify the relative populations of the hydrogen-bonded and non-hydrogen-bonded photobases. The blue trace in Fig. 3 shows the fraction of the hydrogen bonded photobases as a function of increasing TFE concentration.

The figure shows that hydrogen bonding does not begin until there are approximately 200 TFE molecules for every photobase MeOQ molecule. It takes a TFE to photobase ratio of 2000 : 1 for approximately half of the photobase molecules to hydrogen bond, and it takes a ratio of 50 000 : 1 to see hydrogen bonding reach the same level as in bulk TFE. These observations indicate that, even though there is an enthalpic drive for pre-association by hydrogen bonding, dissociation is far more favored and a great excess of hydrogen bond donor is necessary to overcome the dissociation entropy. Since pre-association is necessary

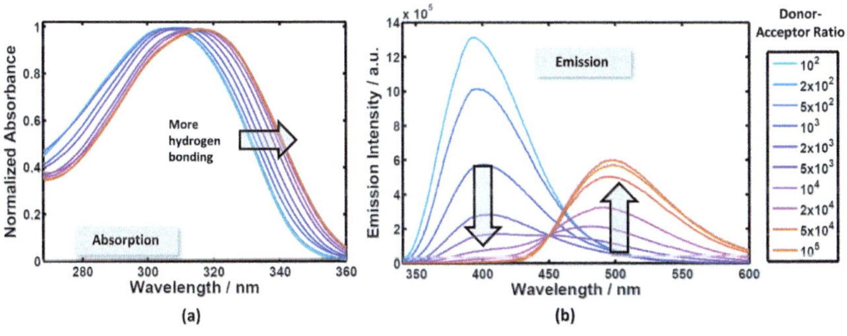

Fig. 2 (a) Absorption spectra of the photobase MeOQ as a function of TFE concentration. The hydrogen bonded photobase has a red-shifted spectrum. Therefore, with the increasing ratio of proton donor TFE the spectrum red-shifts. (b) Emission spectra of the photobase MeOQ. The protonated photobase in the excited state has a distinctly red shifted emission, which increases with increasing donor TFE concentration.

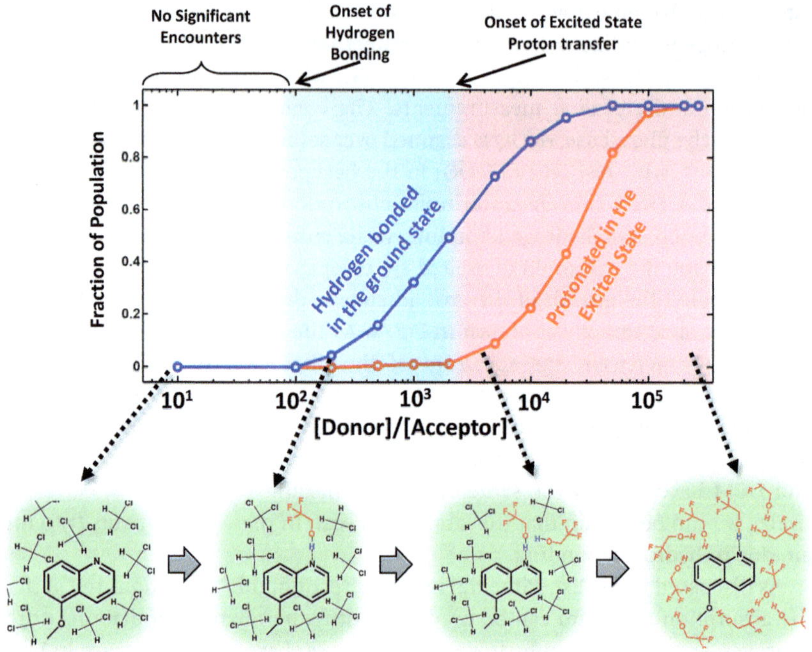

Fig. 3 Data highlighting the importance of pre-association in the ground state and the solvation threshold for successful excited state deprotonation. The blue line is extracted from the red shift of the hydrogen bonded absorption spectra and indicates the rise of the relative fraction of ground state hydrogen bonded complexes. The red line is extracted from the rise of the emission of the protonated form and indicates the fraction of excited state molecules that can successfully deprotonate the donor. To begin hydrogen bonding in the ground state, a large excess of nearly 100 : 1 of proton donor to photobase is needed. A lag of nearly one order of magnitude between the onset of hydrogen bonding and the onset of emission from protonated form indicates that more than one molecule of donor in the solvation shell of a donor–acceptor complex are needed to solvate the resulting conjugate base and therefore allowing successful deprotonation. The cartoon in the bottom indicates the progression of increasing the donor concentration.

for excited state proton transfer, this difficulty in making hydrogen bonded complexes has an impact on the application of photobases for driving deprotonation reactions where the donor is dilute. Such problems are likely to be even greater for weaker hydrogen-bond donors, compared to the strong hydrogen bond donor in our work.

To identify the threshold concentration for proton transfer, we have measured emission spectra as a function of relative ratio of donor to photobase (Fig. 2b). Fortunately, the protonated and unprotonated forms of the photobase have distinctly different emission spectra, with the protonated form red shifted by approximately 100 nm, which makes analysis of the relative populations of the protonated and unprotonated forms relatively convenient as described in the methods section. Since in the ground state there is no proton transfer, emission from the protonated (unprotonated) form indicates the degree of successful (unsuccessful) proton transfer in the excited state. We were particularly interested in whether emission from protonated MeOQ would track the ground state hydrogen bonding pattern. In other words, we aim to answer whether excited state proton transfer is possible for all ground state hydrogen-bonded complexes, even those at the onset of the hydrogen bonding threshold.

The red trace in Fig. 3, shows the rise of the protonated MeOQ emission as a function of TFE concentration. It is clearly seen that there is a distinct lag between the onset of ground state hydrogen bonding and successful excited state proton transfer. To achieve successful deprotonation in the excited state, the concentration of the donor must be much higher than that required to achieve ground state hydrogen bonding.

At the onset of H-bonding, noticeable small red-shifts and a decrease in the intensity of the emission spectrum of MeOQ were observed. However, no rise in the emission of the protonated form of MeOQ was seen. We therefore assign these early spectral changes exclusively to hydrogen bond formation. Although hydrogen-bonding was first observed in the absorption spectra at a ratio of 200 : 1, emission from the protonated form of MeOQ was not observed until a ratio of 5000 : 1, a ratio where approximately 75% of the MeOQ molecules were hydrogen bonded in the ground state. The lag seen between the hydrogen-bond complex formation threshold and proton transfer threshold is more than an order of magnitude in TFE concentration.

This lag between ground state hydrogen bond formation and successful excited state proton transfer points towards the importance of solvation of the photogenerated products (conjugate base of the donor and conjugate acid of the acceptor) after proton transfer. If only a single donor is present in a background non-hydrogen-bonding solvent, solvation of the conjugate base is difficult. However, if a small cluster of several donor molecules exists in the vicinity of the donor–acceptor complex, it becomes much easier to solvate the resulting conjugate base due to the formation of a small hydrogen bond network. This scenario is shown in Fig. 4. The ability of small alcohol clusters to solvate charges has been reported in the photoacid literature previously.[44]

Emission from the protonated form of MeOQ begins while there is still an appreciable population of non-hydrogen bonded MeOQ in the ground state and

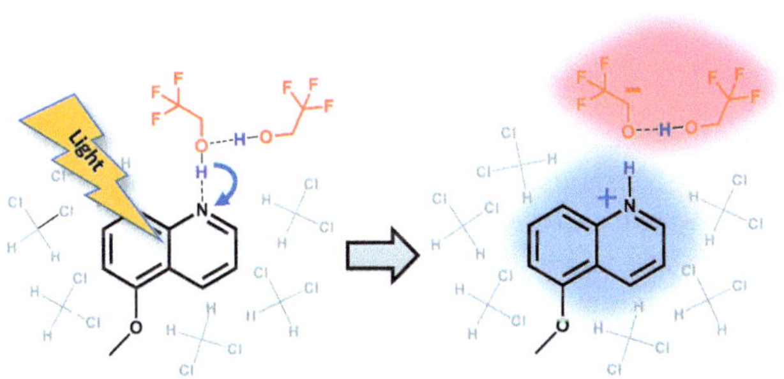

Fig. 4 The data supports a model in which at least two proton donors form a hydrogen bond complex with a photobase. Upon optical excitation and proton capture from one of the donors, the resulting anion is stabilized by a second donor. A photobase hydrogen bonded with only a single donor is unable to achieve this. For that reason, proton transfer becomes possible at a higher threshold concentration of donors than the threshold to create singly hydrogen bonded complexes.

while the contribution of TFE to the solution volume is still small (approximately 1%). Therefore, any alcohol clusters that form near MeOQ are likely to be small as well, especially since the enthalpic drive for cluster formation is likely smaller than the enthalpic drive for hydrogen bonding with the photobase. Previous literature has studied the kinetics of photoacid dissociation as a function of alcohol concentration and determined that two alcohol molecules are capable of solvating the resulting protonic charge.[44] Therefore, it is reasonable and logical to estimate that two TFE molecules participate in the clusters required to solvate the excited state photobasic products studied here. A pictorial representation of the proposed two-molecule TFE cluster can be seen in Fig. 4.

We reiterate that the dielectric constants of the background solvent and TFE are very close to each other[42,43] (8.55 and 8.93 respectively). It is therefore unlikely that the solvation threshold can be attributed to the interpolated dielectric constant between the two solvents. We also point out that hydrogen bonding to the methoxy terminal of our photobase is possible. The contribution of that hydrogen bond to stabilizing the protonated excited state of the photobase is likely minimal and, if non-negligible, is probably slightly unfavorable due to the resulting dislocation of electron density away from the aromatic rings and therefore away from the photobasic nitrogen atom.

These results reveal the challenge of pre-association and solvation of photogenerated products in practical applications of photobases. Even if there is a large thermodynamic drive for deprotonation of a proton donor (as there is in this scenario), dissociation entropy must be overcome with a large excess of the donor before the necessary pre-association is achieved. Furthermore, solvation of the photogenerated products must be ensured for excited state proton transfer.

Based on the above, two strategies are proposed for overcoming these problems. The first is directly tethering the photobase to a location where it is needed, thereby avoiding the necessity for pre-association. For example, a photobase can be covalently bound to the ligand periphery of an organometallic catalyst as a pendant moiety. The center of action for deprotonation is the catalytic metal. When a substrate binds to the metal and the reaction progresses to the stage that it becomes limited by proton transfer, excitation of the photobase can facilitate proton removal from the substrate. Recently, work by us and our collaborators has resulted in the synthesis of a model iridium complex with a pendant photobase.[36] Even though the catalytic properties of the created complex is still unclear, it was shown that the pendant quinoline maintains its photobasicity in the vicinity of the metal center. This is a first step toward rationally incorporating photobases in catalytic centers. The second strategy is the use of non-hydrogen bonding solvents with high dielectric constants, since they will not compete for hydrogen bonding with the donor and will assist with charge separation and screening of the photogenerated conjugate acid and base. In addition to these restrictions, optical transparency of the solvent also needs to be accounted for to ensure that light reaches the photobase.

3 Energy cost for photodriven deprotonation

The second challenge in using photobases in applications is the energy cost of the optical excitation. Typical unprotonated quinoline photobases have their first absorption in the range of 290–350 nm (3.5–4.2 eV). In some applications using such high energy photons may be justified, particularly when alternate photochemical processes do not interfere. However, it is desirable to reduce the necessary excitation energy. In this section we address this problem using two approaches. The first is reducing the ground-excited state energy gap by increasing the size of the N-heterocyclic conjugated systems. The second approach is addition of substituents to the aromatic system. We will show that while both of these handles can tune the excitation energy, they also have a large effect on photobasicity. The combination of these two factors provide a preliminary design space for creating new photobasic molecules.

To begin with, we demonstrate a correlation between experimental $\Delta pK_a = pK_a^* - pK_a$ and computed electron density build up on nitrogen upon photoexcitation $\Delta q_N = q_N^* - q_N$. As shown in Fig. 5 through visualization of electron density differences, there is an increase in electron density on the heteroatomic nitrogen of N-heterocycles upon excitation to the L_a state. In Fig. 6, we show the experimentally measured ΔpK_a values determined from Förster cycle analysis for various 5-substituted quinolines plotted *versus* the computationally calculated change in electron density Δq_N of the L_a state upon excitation, determined using Lowdin analysis. The L_a state was chosen because it is known to be the photobasic state in N-heterocyclic photobase molecules, as discussed in previous work.[26] A clear linear relation is observed between the excess charge build up on nitrogen and the ΔpK_a. This correlation is mechanistically justified since the excess negative charge on nitrogen in the excited state can be reasonably interpreted as the initial drive for attracting the proton. We also include the experimentally measured Förster cycle ΔpK_a for acridine and its corresponding computed Δq_N in the plot. Even though it is an entirely different molecule with a larger conjugation size, it roughly falls within the trend for the substituted quinoline family. This correlation shows that excess charge density Δq_N can reasonably be used as a computational proxy for photobasicity when exploring the wide space of N-heterocyclic aromatic photobases.

Armed with the above knowledge, we proceed to use Δq_N as a convenient computational proxy for ΔpK_a to explore the conjugation size and substituents parameter space of N-heterocyclic aromatics. We have chosen quinolines, acridines, and α-napthoquinoline (benzacridine), which have two, three, and four conjugated rings, respectively (Fig. 7). We performed calculations on these molecules with a range of substituents of varying electron-withdrawing strengths. The substituent locations on the conjugated systems were chosen as shown in

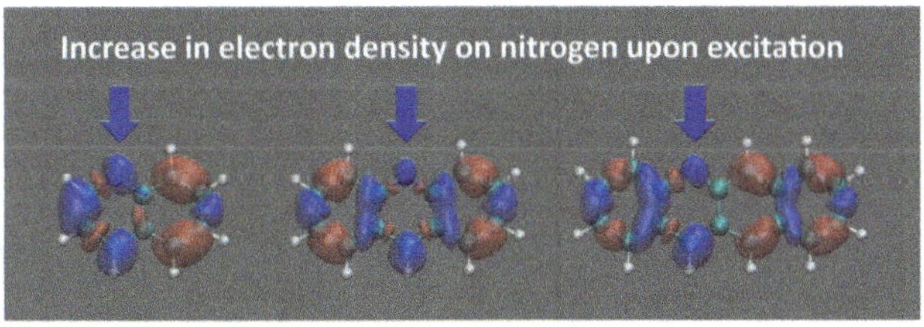

Fig. 5 Electron density difference maps of three N-heterocycles (quinoline, acridine and benzacridine) for the L_a excited electronic state. Blue surfaces indicate an increase in electron density and red surfaces indicate a decrease in electron density. For all molecules, an increase in the electron density of the heteroatom is observed.

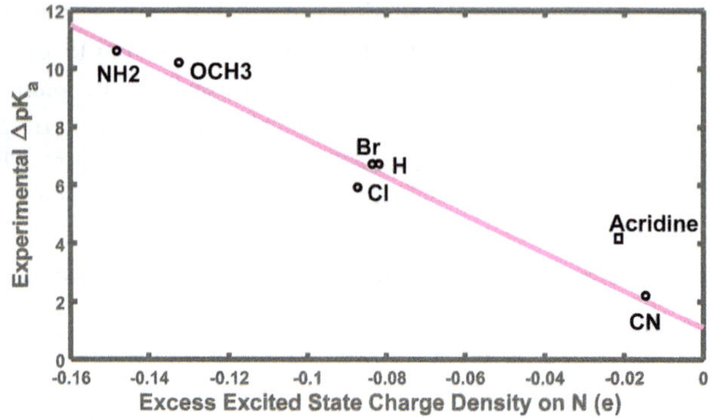

Fig. 6 Experimental ΔpK_a of substituted quinolines determined from Förster cycle analysis, *versus* their calculated excess charge density build up on nitrogen upon photoexcitation Δq_N using Lowdin analysis. The experimental ΔpK_a and computed Δq_N for acridine is also included, which, even though is a different conjugation size, falls reasonably near the trend line.

Fig. 7. We chose the 5 position for the quinoline substituents because of the existence of both experimental Förster analysis and ultrafast kinetic data for proton transfer for the 5-*R*-quinoline family reported by us previously.[37,45] The

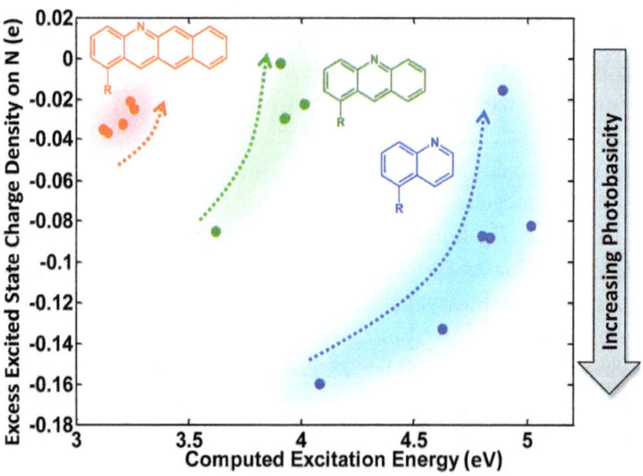

Fig. 7 The influence of conjugation size and substitution on photobasicity, serving as a guide for designing photobasic N-heterocyclic molecules. Excess electronic charge build up on nitrogen upon photoexcitation is plotted *versus* the electronic energy gap. The excess negative charge correlates well with the experimental Förster cycle ΔpK_a values (Fig. 6) and therefore can serve as a proxy for photobasicity. As the size of the conjugated system is increased, the excitation energy goes down as expected. However, the excess electron density on nitrogen and correspondingly the photobasicity decreases. The dotted arrows for each molecule indicate the direction of increasing electron withdrawing power of the substituent R. Generally, more electron withdrawing substituents polarize the electron density away from nitrogen and result in lower photobasicity. The larger the conjugated system, the less sensitive its nitrogen is to varying substituents. Finally, we note that the computed energies are systematically overestimated by nearly 0.7 eV compared to experimental values. Here the relative trends, rather than the absolute values, are meaningful.

positions on acridine and benzacridine are the analogous positions with respect to the heteroatom. The substituents chosen in the order of increasing electron withdrawing power (Hammett parameter) are $-NH_2$, $-OCH_3$, –Cl, –Br, and –CN. Unsubstituted molecules (Hammett parameter of zero) are also included.

Several important lessons may be learned from Fig. 7. First, as expected, the excitation energy gap is reduced as we move to larger conjugated systems. However, this reduction in excitation energy correlates with a reduction of the excess electron density on nitrogen Δq_N upon excitation, which can be interpreted as a smaller ΔpK_a. This observation is logically justifiable, because in a large N-heterocyclic system a single electronic excitation is spread over a larger space and the share of electron density moved to nitrogen from that excitation is smaller. The lesson here is that the size of the conjugated system is intimately coupled with photobasicity and cannot be independently tuned.

The next observation in the figure is the influence of the substituents on the energy gap and photobasicity. First, it can be clearly observed that the substituents make a larger difference for the smaller conjugated systems. For the 4-ring system (α-napthoquinoline), the substituents barely change the excitation gap and the excess charge on nitrogen, while for quinoline a much wider spread in both of these quantities is observed. This is justifiable since for the larger systems, the substituent is a smaller perturbation to both the ground and excited states of the molecule. The lesson from this observation is that while one may use the substituent effect to tune photobasicity and the electronic energy gap, the tunability range is wider for smaller systems.

Next, we note that as the electron-withdrawing strength (Hammett parameter) of the substituent is increased (the direction of the dotted arrows in Fig. 7), the photoexcited excess charge on nitrogen $\Delta q_N = q_N^* - q_N$ is diminished. It is expected that increasing the Hammett parameter should reduce both q_N and q_N^*. The figure shows that their differences are also reduced with the increasing Hammett parameter. This can only be true if q_N^* is more sensitive to the Hammett parameter and reduces more rapidly compared to q_N. This is justifiable because the excited states of molecules are more polarizable and therefore more sensitive to electron withdrawing effects of the substituents. The lower Hammett values (*i.e.* electron releasing groups) make the nitrogen more basic in the ground state. The figure shows that such groups also have larger photobasicity. The lesson learned from this observation is that within the N-heterocyclic aromatic family, stronger bases are in general also stronger photobases.

Finally, while Fig. 7 is useful as a design guide, a cautionary point must be made about its interpretation. The first point is that the diagram is only constructed for optically active singlet states, which also happen to be mostly the lowest lying excited states in these molecules. The photoexcited molecules may be capable of intersystem crossing and going to triplet states. We have previously explored the importance of triplet states in the excited state kinetics of substituted quinolines.[27] Acridines are also known to undergo singlet to triplet transitions.[46] When transition to triplet states is faster than proton transfer, proton transfer is out-competed by intersystem crossing even if the proton transfer is thermodynamically favorable. It is possible for a triplet state to be photobasic as well (*i.e.* proton transfer after intersystem crossing). However, since the accessible triplet states are lower in energy than the photoexcited singlet state, part of the excited state drive may already be consumed in intersystem crossing, rendering the triplet states less photobasic. The lesson, therefore, is to interpret Fig. 7 as a thermodynamic guide that depicts necessary conditions for proton transfer, but on its own is not sufficient to predict its kinetic feasibility. Computing the kinetics of proton transfer is more complicated and beyond the scope of this work.

4 Conclusions

We have highlighted the challenges towards chemical application of photobases, with particular emphasis and experimental results on donor–acceptor pre-association, the excited state solvation threshold, and engineering the optical gap of photobases while retaining their properties. Our results show that driving bimolecular reactions such as proton capture from a molecule by the photobase requires overcoming the entropic cost of ground state pre-association. In the case studied here it was only possible by increasing the concentration of the donor relative to acceptor by a factor of more than 100. An interesting result of this work is that not all ground state pre-associated molecules undergo proton transfer, even in the presence of a large pK_a difference. It is necessary to have a desirable solvation environment to stabilize the photoproducts (conjugate acid and base). We have found that to achieve such an environment, the donor concentration needed to be increased even further to allow solvation of the deprotonated donor amongst a cluster of molecules of its own kind. Covalently tethering photobases near catalytic sites, as has been explored by us previously,[36] and exploring solvents that would facilitate hydrogen bonding in the ground state and charge separation in the excited state are possible directions of further investigation.

We have also found some design trends for tuning the optical gap for photobases, with the intention of finding molecules that will achieve photobasicity at lower excitation energies. We have realized that this design space, at least in the dimensions studied here, is rather constrained. Increasing the conjugation size lowers the optical gap, but also adversely affects photobasicity. Tuning the gap *via* substituents is more efficient for smaller conjugated systems. Our study lays the foundation for exploring this design space further. For example, multiple substituents, multiple heterocyclic nitrogens or other heteroatoms in the conjugated system, exploring non-benzene rings, and identifying the influence of triplet states on the photobase kinetics are all potentially fruitful directions of future work.

5 Experimental and computational methods

We took the absorption and emission spectra for a series of eleven solutions, each with varying molar concentration ratios of 5-methoxyquinoline (MeOQ) to trifluoroethanol (TFE) in dichloromethane (DCM). These ratios were 10^2, 2×10^2, 5×10^2, 10^3, 2×10^3, 5×10^3, 10^4, 2×10^4, 5×10^4, 10^5, and 2×10^5. We also performed these experiments in bulk DCM and TFE. In all cases, the concentration of MeOQ was 5×10^{-5} M. All chemicals were purchased from Sigma-Aldrich and used without further purification.

Absorption spectra were obtained using a Cary 50 UV/Vis spectrophotometer. Measurements were made in a 1 cm fused quartz cuvette. Absorption spectra were background subtracted and normalized for analysis. In order to analyze the absorption data, we fit the spectra at the concentration ratio extremes (10^2 and 5×10^4) to three Gaussians using MATLAB.[47] We did so to obtain robust analytical forms of these spectra, not to extract physical significance. The absorption spectrum at each TFE concentration ratio was then fit as a linear combination of these two basis spectra. The contribution of the hydrogen-bonded basis spectrum was used as a measure of the relative hydrogen-bonded MeOQ population. These values are shown in the blue curve in Fig. 3. We chose the spectrum with the donor to acceptor ratio of 10^2 to represent the initial non-hydrogen bonded species, and we chose the spectrum with the donor to acceptor ratio of 5×10^4 to represent the final state where the photobase was completely hydrogen bonded. We chose to fit the concentration ratio of 5×10^4, rather than the higher ratios of 10^5 or 2×10^5, in order to ignore any bulk dielectric effects accompanying the introduction of TFE as a significant volumetric component of the solvent. At these higher ratios a blue shift in the absorption spectrum of almost 1 nm was observed, contrary to the red shifting pattern seen at lower concentrations. The red shift indicative of an increase in hydrogen bonded population had completed by a ratio of 5×10^4, so we are confident that this choice in basis spectrum is appropriate.

Emission spectra were collected on a Jobin-Yvon Fluoromax 3 fluorometer. Measurements were made in a 1 cm fused quartz cuvette. All samples were excited at 310 nm. For the emission data, there was no straightforward choice of basis spectra due to the convolution of ground-state hydrogen bonding effects and excited state proton transfer. Instead, we chose to plot the emission intensities at 575 nm as a function of TFE concentration, normalized with respect to minimum and maximum intensities. These values are shown in the red curve in Fig. 3. Because there is virtually no emission intensity from the unprotonated form at 575 nm, we can be confident that the rise in intensity we see is due entirely to the rise of the protonated form of MeOQ. After the onset of emission from the protonated form of MeOQ, a clear trend with respect to growth of emission from the protonated form was observed and most of the emission spectra shared an isosbestic point at approximately 450 nm. Two of the collected spectra were offset in the overall emission intensity, but not in the relative peak ratios, likely due to scattering at the sample. To correct for this deviation, we multiplied each spectrum by a constant in order to enforce the isosbestic point at 450 nm.

Electronic structure calculations of quinolines, acridines, and benzacridines were carried out using the Q-Chem software.[48] Ground state geometries were optimized using the ωB97x-D/6-31+G* level of theory. Singlet excited states were then calculated with the TDDFT/ωB97x-D/6-31+G* level of theory. Lowdin charge densities were used for population analysis. The excited states chosen for analysis in all cases were the singlet L_a states, identified *via* transition dipole moment. Electron density difference maps shown in Fig. 5 were visualized with VMD.[49]

Details of the determination of the experimental ΔpK_a of the 5-substituted quinolines seen in Fig. 6 can be found in a previous publication.[26]

Conflicts of interest

The authors declare no conflict of interest.

Acknowledgements

The authors gratefully acknowledge support from the NSF CAREER Award (1454467), the NSF Graduate Research Fellowship, and the Cottrell Scholar Award from the Research Corporation.

Notes and references

1 T. Förster, *Z. Elektrochem. Angew. Phys. Chem.*, 1950, **54**, 42–46.
2 A. Weller, *Prog. React. Kinet. Mech.*, 1961, **1**, 187–214.
3 L. M. Tolbert and K. M. Solntsev, *Acc. Chem. Res.*, 2002, **35**, 19–27.
4 R. Simkovitch, K. Akulov, S. Shomer, M. E. Roth, D. Shabat, T. Schwartz and D. Huppert, *J. Phys. Chem. A*, 2014, **118**, 4425–4443.
5 R. Simkovitch, S. Shomer, R. Gepshtein and D. Huppert, *J. Phys. Chem. B*, 2015, **119**, 2253–2262.
6 D. Pines, E. T. J. Nibbering and E. Pines, *Isr. J. Chem.*, 2015, **55**, 1240–1251.
7 E. Pines and D. Huppert, *J. Phys. Chem.*, 1983, **87**, 4471–4478.
8 D. Spry, A. Goun and M. Fayer, *J. Phys. Chem. A*, 2007, **111**, 230–237.
9 D. B. Spry and M. D. Fayer, *J. Chem. Phys.*, 2008, **128**, 084508.
10 M. Gutman, D. Huppert and E. Pines, *J. Am. Chem. Soc.*, 1981, **103**, 3709–3713.
11 N. Agmon, *J. Phys. Chem. A*, 2005, **1019**, 13–35.
12 M. Prémont-Schwarz, T. Barak, D. Pines, E. T. J. Nibbering and E. Pines, *J. Phys. Chem. B*, 2013, **117**, 4594–4603.
13 S. Abbruzzetti, E. Crema, L. Masino, A. Vecli, C. Viappiani, J. R. Small, L. J. Libertini and E. W. Small, *Biophys. J.*, 2000, **78**, 405–415.
14 T. P. Causgrove and R. B. Dyer, *J. Chem. Phys.*, 2006, **323**, 2–10.
15 J. L. Dempsey, J. R. Winkler and H. B. Gray, *J. Am. Chem. Soc.*, 2010, **132**, 16774–16776.
16 S. Kohse, A. Neubauer, A. Pazidis, S. Lochbrunner and U. Kragl, *J. Am. Chem. Soc.*, 2013, **135**, 9407–9411.
17 H. Peretz-Soroka, A. Pevzner, G. Davidi, V. Naddaka, M. Kwiat, D. Huppert and F. Patolsky, *Nano Lett.*, 2015, **15**, 4758–4768.
18 B. K. Keitz and R. H. Grubbs, *J. Am. Chem. Soc.*, 2009, **131**, 2038–2039.
19 A. Das, S. Ayad and K. Hanson, *Org. Lett.*, 2016, **18**, 5416–5419.
20 W. White, C. D. Sanborn, R. S. Reiter, D. M. Fabian and S. Ardo, *J. Am. Chem. Soc.*, 2017, **139**, 11726–11733.
21 S. Haghighat, S. Ostresh and J. M. Dawlaty, *J. Phys. Chem. B*, 2016, **120**, 1002–1007.
22 N. Mataga, Y. Kaifu and M. Koizumi, *Bull. Chem. Soc. Jpn.*, 1956, **29**, 373–379.
23 E. Pines, D. Huppert, M. Gutman, N. Nachliel and M. Fishman, *J. Phys. Chem.*, 1986, **90**, 6366–6370.
24 O. Poizat, E. Bardez, G. Buntinx and V. Alain, *J. Phys. Chem. A*, 2004, **108**, 1874–1880.
25 N. Munitz, Y. Avital, D. Pines, E. T. J. Nibbering and E. Pines, *Isr. J. Chem.*, 2009, **49**, 261–272.
26 E. W. Driscoll, J. R. Hunt and J. M. Dawlaty, *J. Phys. Chem. Lett.*, 2016, **7**, 2093–2099.
27 E. W. Driscoll, J. R. Hunt and J. M. Dawlaty, *J. Phys. Chem. A*, 2017, **121**, 7099–7107.

28 W. Sheng, M. Nairat, P. D. Pawlaczyk, E. Mroczka, B. Farris, E. Pines, J. H. Geiger, B. Borhan and M. Dantus, *Angew. Chem., Int. Ed.*, 2018, 14742–14746.
29 A. Kellmann and Y. Lion, *Photochem. Photobiol.*, 1979, **29**, 217–222.
30 E. Nachliel, Z. Ophir and M. Gutman, *J. Am. Chem. Soc.*, 1987, **109**, 1342–1345.
31 E. T. Ryan, T. Xiang, K. P. Johnston and M. A. Fox, *J. Phys. Chem. A*, 1997, **101**, 1827–1835.
32 T. T. Eisenhart and J. L. Dempsey, *J. Am. Chem. Soc.*, 2014, **136**, 12221–12224.
33 G. Favaro, U. Mazzucato and F. Masetti, *J. Phys. Chem.*, 1973, **77**, 601–604.
34 S. Vogt and S. G. Schulman, *Chem. Phys. Lett.*, 1983, **97**, 450–453.
35 K. Akulov, R. Simkovitch, Y. Erez, R. Gepshtein, T. Schwartz and D. Huppert, *J. Phys. Chem. A*, 2014, **118**, 2470–2479.
36 I. Demianets, J. R. Hunt, J. M. Dawlaty and T. Williams, *Organometallics*, 2019, **38**, 200–204.
37 J. R. Hunt and J. M. Dawlaty, *J. Phys. Chem. A*, 2018, **122**, 7931–7940.
38 S. Haghighat, S. Ostresh and J. M. Dawlaty, *J. Phys. Chem. B*, 2016, **120**, 1002–1007.
39 T. Förster, *Z. Elektrochem. Angew. Phys. Chem.*, 1950, **54**, 42–46.
40 J. F. Ireland and P. A. H. Wyatt, *Adv. Phys. Org. Chem.*, 1976, **12**, 131–221.
41 D. Roccatano, G. Colombo, M. Fioroni and A. E. Mark, *Proc. Natl. Acad. Sci. U. S. A.*, 2002, **99**, 12179–12184.
42 R. Chitra and P. E. Smith, *J. Chem. Phys.*, 2001, **114**, 426–435.
43 A. Vogel and B. Furniss, *Vogel's Textbook of Practical Organic Chemistry Including Quantitative Organic Analysis*, 1978.
44 S.-Y. Park, Y. M. Lee, K. Kwac, Y. Jung and O.-H. Kwon, *Chem.–Eur. J.*, 2016, **22**, 4340–4344.
45 E. W. Driscoll, J. R. Hunt and J. M. Dawlaty, *J. Phys. Chem. Lett.*, 2016, 7, 2093–2099.
46 I. Y. Goryacheva, G. Mel'nikov and S. Shtykov, *J. Anal. Chem.*, 2000, **55**, 874–878.
47 *MATLAB, version 7.10.0 (R2010a)*, The MathWorks Inc, Natick, Massachusetts, 2010.
48 Y. Shao, Z. Gan, E. Epifanovsky, A. T. Gilbert, M. Wormit, J. Kussmann, A. W. Lange, A. Behn, J. Deng, X. Feng, D. Ghosh, M. Goldey, P. R. Horn, L. D. Jacobson, I. Kaliman, R. Z. Khaliullin, T. KuÅŻ, A. Landau, J. Liu, E. I. Proynov, Y. M. Rhee, R. M. Richard, M. A. Rohrdanz, R. P. Steele, E. J. Sundstrom, H. L. Woodcock III, P. M. Zimmerman, D. Zuev, B. Albrecht, E. Alguire, B. Austin, G. J. O. Beran, Y. A. Bernard, E. Berquist, K. Brandhorst, K. B. Bravaya, S. T. Brown, D. Casanova, C.-M. Chang, Y. Chen, S. H. Chien, K. D. Closser, D. L. Crittenden, M. Diedenhofen, R. A. DiStasio Jr, H. Do, A. D. Dutoi, R. G. Edgar, S. Fatehi, L. Fusti-Molnar, A. Ghysels, A. Golubeva-Zadorozhnaya, J. Gomes, M. W. Hanson-Heine, P. H. Harbach, A. W. Hauser, E. G. Hohenstein, Z. C. Holden, T.-C. Jagau, H. Ji, B. Kaduk, K. Khistyaev, J. Kim, J. Kim, R. A. King, P. Klunzinger, D. Kosenkov, T. Kowalczyk, C. M. Krauter, K. U. Lao, A. D. Laurent, K. V. Lawler, S. V. Levchenko, C. Y. Lin, F. Liu, E. Livshits, R. C. Lochan, A. Luenser, P. Manohar, S. F. Manzer, S.-P. Mao, N. Mardirossian, A. V. Marenich, S. A. Maurer, N. J. Mayhall, E. Neuscamman, C. M. Oana, R. Olivares-Amaya, D. P. O'Neill, J. A. Parkhill, T. M. Perrine, R. Peverati,

A. Prociuk, D. R. Rehn, E. Rosta, N. J. Russ, S. M. Sharada, S. Sharma, D. W. Small, A. Sodt, T. Stein, D. StÃijck, Y.-C. Su, A. J. Thom, T. Tsuchimochi, V. Vanovschi, L. Vogt, O. Vydrov, T. Wang, M. A. Watson, J. Wenzel, A. White, C. F. Williams, J. Yang, S. Yeganeh, S. R. Yost, Z.-Q. You, I. Y. Zhang, X. Zhang, Y. Zhao, B. R. Brooks, G. K. Chan, D. M. Chipman, C. J. Cramer, W. A. Goddard III, M. S. Gordon, W. J. Hehre, A. Klamt, H. F. Schaefer III, M. W. Schmidt, C. D. Sherrill, D. G. Truhlar, A. Warshel, X. Xu, A. Aspuru-Guzik, R. Baer, A. T. Bell, N. A. Besley, J.-D. Chai, A. Dreuw, B. D. Dunietz, T. R. Furlani, S. R. Gwaltney, C.-P. Hsu, Y. Jung, J. Kong, D. S. Lambrecht, W. Liang, C. Ochsenfeld, V. A. Rassolov, L. V. Slipchenko, J. E. Subotnik, T. V. Voorhis, J. M. Herbert, A. I. Krylov, P. M. Gill and M. Head-Gordon, *Mol. Phys.*, 2015, **113**, 184–215.

49 W. Humphrey, A. Dalke and K. Schulten, *J. Mol. Graphics*, 1996, **14**, 33–38.

Faraday Discussions

DISCUSSIONS

Photovoltaics and bio-inspired light harvesting: general discussion

Michael Ashfold, Jon Bender, Eric Bittner, Jeffrey Cina, Carlos E. Crespo-Hernández, Natércia das Neves Rodrigues, Jahan Dawlaty, Ryan Dill, Amro Dodin, Marta Duchi, Laura Estergreen, Gregory Gate, James Gaynor, Naomi Ginsberg, Christopher Grieco, Sharon Hammes-Schiffer, Vanessa Huxter, Shawn Irgen-Gioro, Jimmy Joy, Bern Kohler, Nadia Korovina, Johannes Mahl, Andrew Marcus, Todd Martinez, Karen Morenz, Amanda Morris, Jennifer Ogilvie, Thomas A.A. Oliver, Tom Penfold, Petter Persson, Trevor Roberts, Benjamin Schwartz, Yin Song, Mark Thompson, Michael Wasielewski, Emily Weiss and Kristopher Williams

DOI: 10.1039/C9FD90028D

Nadia Korovina opened discussion of the paper by Emily Weiss: Can you rule out the possibility of singlet energy transfer from acene to nanoparticle, followed by triplet energy transfer from the nanoparticle to the acene as a possible explanation for the faster growth of the triplet population in the presence of nanoparticles? This mechanism was recently presented in a paper by J. C. Johnson *et al.*[1]

1 D. M. Kroupa, D. H. Arias, J. L. Blackburn, G. M. Carroll, D. B. Granger, J. E. Anthony, M. C. Beard and J. C. Johnson, *Nano Lett.*, 2018, **18**, 865–873.

Emily Weiss replied: Yes, I think so. We do not pump enough energy into the system to create the singlet excited state of the acene.

Nadia Korovina followed: What is the band gap of the nanoparticle which results in maximum triplet yield when doped into the acene?

Emily Weiss answered: The band gap is ~1 eV.

Tom Penfold said: Can you rule out direct intersystem crossing, driven by an external heavy atom effect, contributing to the enhanced rate of triplet formation in the presence of the PbS quantum dots?

Emily Weiss responded: We cannot rule it out; we did not calculate a quantum yield of triplet formation. The 2-15 ps component we attribute to singlet fission is most likely too fast to be assigned to ISC within diphenylpentacene (DPP), but we do observe a longer-time growth of the DPP triplet that could have a contribution from ISC.

Vanessa Huxter asked: Could you comment on the possibility of using different nanocrystal shapes to tune the surface energy and influence singlet fission rates in addition to the effect of the ligands?

Emily Weiss responded: Yes, tuning the shape should be a powerful way to tune the singlet fission rate because the shape controls the facet distribution, which controls not only surface energy but ligand coverage. The ligand coverage, in turn, controls affinity for diphenylpentacene, and (presumably) the affinity for different types of aggregates of this molecule.

Gregory Gate said: It is clear that DPP having a perpendicular alignment affects the rate of singlet fission. Have you considered any ways to artificially change the alignment of the system?

Emily Weiss answered: There probably are ways—changing substituent patterns on the pentacene and using CVD to deposit on a templating substrate—but we haven't pursued them yet.

Kristopher Williams asked: According to this work (DOI: 10.1039/c8fd00157j), the change in the rate of singlet fission was dependent on the ligand (PEG) density and dielectric environment. That being the case, could I achieve the same increase in singlet fission rate without using quantum dots? *i.e.* Could I use any macroscopic surface with the correct PEG density seen here?

Emily Weiss replied: Yes, I think as long as you modify the morphology of the aggregates and/or the local dielectric, you should see a change in fission rate. The reason that using QDs is convenient is because you can do the measurements in solution – *i.e.*, you will have a large concentration to look at spectroscopically. A bulk or 2D surface limits the amount of pentacene to that which will fit in a monolayer (or so) at the interface.

Christopher Grieco remarked: A necessary question for any so-called "singlet fission material" is whether ultrafast intersystem crossing or singlet fission is the triplet generation mechanism. In the former case, the maximum possible triplet quantum yield is 1, while in the latter case it is 2 (exciton multiplication). One way to prove if singlet fission is operative is to measure the triplet quantum yield of the material; if the quantum yield is greater than 1, then singlet fission is likely to be the mechanism of triplet generation.

In addition, spatial separation of a correlated triplet pair produced by singlet fission is a characteristic process that does not occur for triplets formed *via* intersystem crossing. Such a process was identified in this paper (DOI: 10.1039/c8fd00157j), and so this provides additional evidence for singlet fission over ultrafast intersystem crossing. Regarding this, a more compelling argument could be made using the data presented in this paper (DOI: 10.1039/c8fd00157j). Pensack *et al.* demonstrated that the excited state absorption (ESA) spectrum of the correlated triplet pair state produced *via* singlet fission is sensitive to the spatial separation of the triplet excitons.[1] Before the triplet pair has spatially separated, the triplets are close enough to interact with each other, causing a perturbation to their ESA spectrum. When the triplets spatially separate, the triplet ESA spectrum changes in lineshape (*e.g.* changes in vibronic peak intensity ratios). I believe such a phenomenon was present in the data presented in this paper (DOI: 10.1039/c8fd00157j), which could be used to further corroborate the claim that singlet fission is occurring.

1 R. D. Pensack, E. E. Ostroumov, A. J. Tilley, S. Mazza, C. Grieco, K. J. Thorley, J. B. Asbury, D. S. Seferos, J. E. Anthony and G. D. Scholes, *J. Phys. Chem. Lett.*, 2016, 7, 2370–2375.

Emily Weiss responded: I agree with the statements here. We did not measure a quantum yield of triplets produced, but we probably could estimate it given the well-characterized excitation source of the transient absorption experiment.

We do see spectral diffusion of the triplet signal, and it is the dynamics of this spectral diffusion that we assigned to migration of the bound triplet pair into spatially separated triplets.

Benjamin Schwartz commented: The dielectric environment that stabilizes the intermediate responsible for singlet fission is quite complex, with layers consisting of the nanocrystal, the ligands and the solvent. Is it possible to model this environment and the way it would stabilize a charge-separated state? Or at least put a standard solvatochromic probe in the same location to learn more about the dielectric environment and thus the nature of the singlet fission intermediate? Conversely, if the singlet fission rate is sensitive enough to the surroundings, could you use the DPP molecule as a solvatochromic probe of other local environments?

Emily Weiss answered: Yes, I believe modeling this environment would have a significant amount of uncertainty considering the degree of disorder in both the ligand arrangements and structure of the DPP aggregate, but placing a dielectric probe or "ruler" molecule at the surfaces of these particles would be a straightforward way to correlate dielectric constant with SF rate.

Naomi Ginsberg asked: Would one way to answer Prof. Schwartz's question about systematically varying the dielectric environment be to employ different lengths of PEG ligands on the nanoparticles?

Emily Weiss replied: I believe changing the length of the PEG ligands would alter the affinity of the surface for the DPP aggregate more than it would alter the dielectric environment. Perhaps systematically introducing lower or higher dielectric ligands (as a mixed monolayer with the PEG) would be a way to control this dielectric.

Jimmy Joy queried: Does DPP intercalate into the PEG layer on the nanoparticle surface? Or does it remain in solution?

Emily Weiss answered: It definitely intercalates. That is why the QDs influence its singlet fission rate.

Laura Estergreen remarked: The paper (DOI: 10.1039/c8fd00157j) indicates that the PbS nanoparticles may disturb the perpendicular stacking allowing for increased coupling between DPP molecules. Was there any evidence in the blue part of the steady-state absorption spectrum that showed H-aggregate formation of the S0-S2 transition in pentacene, whose transition dipole is oriented along the length of the pentacene chromophore? This transition is about an order of magnitude brighter than the S0-S1 transition, so it may be more visible (as shown by the Damrauer group in their norbornyl-bridged pentacene dimers).[1] Or is it accurate to assume that the DPP aggregates will stack in their crystalline geometry?

1 J. D. Cook, T. J. Carey, D. H. Arias, J. C. Johnson and N. H. Damrauer, *J. Phys. Chem. A.*, 2017, **12**, 9229–9242; A. T. Gilligan, E. G. Miller, T. Sammakia and N. H. Damrauer and *J. Am. Chem. Soc.*, 2019, **141**(14), 5961–5971.

Emily Weiss answered: We would very much like to investigate this, but the pentacene spectrum in this region is covered up by the spectrum of the PbS QDs, which absorb into a quasi-continuum at energies far above the band-edge.

Jon Bender asked: I'm very interested in the organization of these molecular aggregates, at both the level of molecular organization, *i.e.* molecular polymorphs and disorder in the aggregates, and the organization between PbS NCs and DPT nanocrystal agglomerations you observe in the TEM images. As you know, singlet exciton fission and ensuing triplet exciton dynamics are highly sensitive to small changes in intermolecular organization, such as deviations from bulk crystal structures that may arise from "thin-film phases" that, in this case, may manifest themselves on a smaller scale as you grow the molecular aggregates on PEG surfaces with different curvatures, etc. Additionally, disorder may be playing a role in the differences in singlet exciton fission dynamics that you observe. Have you performed Raman or X-ray diffraction studies on these molecular aggregates to verify their structure or monitor changes in their structure? Also, what role do you think disorder or polymorphic changes may be playing in the data you present?

Emily Weiss replied: We believe that changes in intermolecular disorder is one of the two (along with changes in local dielectric constant) primary factors in the change in SF rates we observe as a function of QD size. We did not do any direct structural measurements on these systems. Raman would probably be more useful than XRD given the high curvature of the QD substrates.

Jon Bender followed: What evidence do you have to discount or verify the existence of agglomerated ensembles of molecular NCs and PbS NCs in solution similar to those you observe in the TEM data? If these agglomerated ensembles exist in solution for these intermediate sized PbS NCs, what role do you think singlet exciton or triplet exciton energy transfer between distinct molecular aggregates plays in the data you observe?

Emily Weiss replied: The NCs are likely somewhat agglomerated in solution, and there is certainly contact between DPP aggregates on different NCs. It is unlikely that energy transfer is competitive with SF, however, so we don't believe it greatly influences the trend we observe.

Karen Morenz queried: I have two questions. First, I noticed that you used the Hines prep to make your lead sulfide quantum dots. One of my labmates, Philippe Green, is really interested in different lead sulfide dot preps, so I'm asking this question on his behalf. He has seen that if you use lead chloride as a precursor there's some evidence that you get an inherent shell on the dot and I'm wondering how that would effect the results you see here and if you've considered using alternative dot preps. Second, there's a related paper by Davis *et al.* from Akshay Rao's group where they put tetracene ligands on lead sulfide quantum dots and see triplet transfer into the dot, so can you comment on how your results connect up with theirs?

1 N. J. L. K. Davis, J. R. Allardice, J. Xiao, A. J. Petty, N. C. Greenham, J. E. Anthony and A. Rao, *J. Phys. Chem. Lett.*, 2018, **9**, 1454–1460.

Emily Weiss replied: There are several preps for PbS quantum dots (QDs) depending on your application. Some use intermediate-strength ligands, like oleate. Some produce QDs with weaker ligands, like amines, as the final surfactant. Chloride is an option for very polar solvents or water, but it is an extremely weak ligand (and, in fact, readily replaceable by OH- in even mildly basic water), so it is probably better for film applications. For a hydrophobic adsorbate like substituted pentacene, a more hydrophobic ligand creates greater affinity for the QD surface. There are several papers that document triplet energy transfer between QDs and molecules. It appears to be a facile process if the energy levels are right. We do not observe it with diphenylpentacene (DPP), but that might be because a redox process (photoinduced hole transfer from the QD to the DPP) is faster than energy transfer.

Kristopher Williams opened discussion of the paper by Amanda J. Morris: Both singlet and triplet states were discussed in the paper (DOI: 10.1039/c8fd00194d), but inter-system crossing (ISC) was not mentioned explicitly. ISC can also significantly quench singlets in porphyrins – which can be even more efficient in the MOF environment due to the heavy atoms – could you speak to how this factored into your study on the quenching of singlets in the porphyrin-based MOF environment?

Amanda Morris answered: A heavy atom effect imposed by the Zr-nodes of the MOF should increase both the rate of intersystem-crossing and triplet decay. In similar Zr-containing MOFs functionalized with ruthenium chromophores, we have observed an increase in the triplet lifetime in comparison to solution due to rigidification in the MOF structure.[1] Therefore, we do not feel that a heavy atom effect from the presence of Zr is significant. However, we do not directly measure the ISC rate in this work, due to complications in triplet quantum yield determination. For PCN-223, we do see a decrease in the triplet lifetime between homogeneous solution and the MOF, which we hypothesize is due to energy transfer between neutral porphyrins to *N*-protonated trap states. It is possible this could be due to a heavy atom effect and more studies are needed to elucidate the true nature of the lifetime decrease.

1 W. A. Maza and A. J. Morris, *J. Phys. Chem. C.*, 2014, **118**, 8803–8817.

Kristopher Williams followed: To clarify the crystal structure, do the porphyrin molecules in this MOF structure stack face-to-face (pi-stack) as that could increase energy transfer rates/lengths in the system?

Amanda Morris replied: The porphyrin linkers are not stacked in a face-to-face manner. The porphyrin-porphyrin torsional angle is ~55 degrees. Orienting the porphyrin linkers in a face-to-face manner would align their transition dipole moments, leading to increased energy transfer rates and longer hopping distances.[1]

1 H.-J. Son, Shengye Jin, S. Patwardhan, S. J. Wezenberg, N. C. Jeong, M. So, C. E. Wilmer, A. A. Sarjeant, G. C. Schatz, R. Q. Snurr, O. K. Farha, G. P. Wiederrecht, and J. T. Hupp, *J. Am. Chem. Soc.*, 2013, **135**, 862–869.

Bern Kohler commented: You assign the time constants of 57 ± 22 ps, measured for the free base porphyrin in 1 : 1 water-ethanol solvent, and 115 ± 22 ps, measured for the porphyrin-containing MOF in water, to vibrational relaxation (cooling) in the S_1 state. However, vibrational cooling in a polar, hydrogen-bonding solvent usually occurs with a time constant on the order of 10 ps or even shorter. F. Arcudi *et al.* recently studied the same porphyrin in aqueous solution at pH 7.2.[1] They observed a decay component of 84.9 ps, which they assigned to internal conversion from the S_1 state to the ground state in aggregates. Do you think that the porphyrin TCPP could be aggregating under your conditions? As far as the MOF is concerned, it seems likely that new decay pathways will appear that are not observed for TCPP in solution. You describe an exciton hopping time of 0.53 ps. Could energy transfer to a trap site contribute to the subnanosecond dynamics you observe?

1 F. Arcudi, V. Strauss, L. Đorđević, A. Cadranel, D. M. Guldi and M. Prato, *Angew. Chem. Int. Ed.*, 2017, **56**, 12097.

Amanda Morris responded: Porphyrin-based derivatives, such as TCPP, are known to aggregate under acidic conditions. However, we have observed some TCPP aggregation in water at pH = 8 also. To improve the solubility of TCPP and avoid aggregation, we used a 1 : 1 water-ethanol mixture as the solvent for fsTA measurements. Typically, the UV-vis absorption spectrum of porphyrin aggregates shows a split Soret band. We did not observe Soret band splitting in our TCPP samples and concluded that TCPP existed in its monomeric form. However, we cannot completely rule out aggregation based on this observation and it is possible that the time constant of 57 ± 22 ps corresponds to aggregated species.

Jiang and co-workers have reported time constants of 18 ± 1 ps and 190 ± 11 ps for PCN-222, which is also a zirconium-MOF based on *meso*-tetracarboxyphenylporphyrin (TCPP).[1] They proposed that these time constants may reflect the electron dynamics associated with the electron trap states that are energetically located within the bandgap of PCN-222. We are not fully convinced with the idea of treating a zirconium-MOF as a semiconductor and, therefore, assigned the time constants to solvent reorganization and vibrational cooling processes. It is possible that energy transfer to trap states (*N*-protonated porphyrins) contributes to the sub-nanosecond dynamics that we observed and further investigation is required to provide evidence in favor of this argument.

1 H.-Q. Xu, J. Hu, D. Wang, Z. Li, Q. Zhang, Y. Luo, S.-H. Yu and H.-L. Jiang, *J. Am. Chem. Soc.*, 2015, **137**, 13440–13443.

Naomi Ginsberg said: How does the MOF geometry impact the ultrafast decay times that you observe that Prof. Kohler referred to as being surprisingly slow? Does the connectivity of monomers rigidify the individual monomers to lengthen these time scales compared to in solution?

Amanda Morris answered: The restriction of the molecules both by rigidification and confinement would slow the observed decay rates in comparison to solution. We think of the rigidification similar to examples of ruthenium trisbipyridine in poly-methylmethacrylate films, in which the lifetimes of the excited state are greatly increased. In terms of confinement effects, our observations are consistent with those of Fayer *et al.* for MOFs and micelles.[1]

1 J. Nishida, A. Tamimi, H. Fei, S. Pullen, S. Ott, S. M. Cohen and M. D. Fayer, *Proc. Natl. Acad. Sci. U. S. A.*, 2014, **111**, 18442–18447.

Naomi Ginsberg followed: Has anyone measured the ultrafast dynamics of the solvent directly to determine to what extent it is constrained?

Amanda Morris replied: We are aware of one such study by Sascha Ott, Seth Cohen, Michael Fayer *et al.* looking at the structural deformations and solvent motions in UiO-66. They found that "the dynamics of the DMF molecules [were] greatly restricted."[1] This was inferred from the difference in homogeneous line widths of a probe in the MOF *vs.* bulk solution.

1 J. Nishida, A. Tamimi, H. Fei, S. Pullen, S. Ott, S. M. Cohen and M. D. Fayer, *Proc. Natl. Acad. Sci. U. S. A.*, 2014, **111**, 18442–18447.

Jahan Dawlaty asked: Is it possible to metalate the resulting prophyrin structure after it is constructed?

Also, is it possible to start from metalated porphyrins and construct the MOFs?

Amanda Morris responded: It is possible to metallate porphyrin-based MOFs after formation. Ultimately, the efficiency of that metallation depends on the ability for the metallation salt to diffusion through the MOFs pores. In general, metallation is not 100% effective.

You can start with metallated porphyrins in MOF synthesis. However, using metallated porphyrins often changes the 3D structure that results. Hong-Cai Zhou is a leader in this field. For an example publication relevant to the two questions posed, I point the readers to a paper from his group.[1]

1 D. Feng, W.-C. Chung, Z. Wei, Z.-Y. Gu, H.-L. Jiang, Y.-P. Chen, D. J. Darensbourg and H.-C. Zhou, *J. Am. Chem. Soc.*, 2013, **135**, 17105–17110.

Michael Wasielewski questioned: Have you thought about examining the corresponding metalloporphyrins?

Amanda Morris replied: Yes, this is a direction we will probe in the near future.

Jon Bender commented: Did you observe any changes in polymorph with temperature or pH? If so, how do these structural changes manifest themselves in your data?

Amanda Morris replied: No, we have not observed polymorph changes with either pH or temperature.

Sharon Hammes-Schiffer commented: Do you know where the protons are on the porphyrins? We've seen cases where the carbon atoms of metalloporphyrins can become protonated, forming phlorins or possibly chlorins or bacteriochlorins (although these pathways may not be possible for these systems). Also, are the protons moving during this process? Have you computed a hydrogen/deuterium isotope effect to investigate this issue?

Amanda Morris responded: UV-visible absorption spectroscopy is a useful tool for structural characterization of porphyrins, phlorins, chlorins and bacteriochlorins. The absorption spectra of phlorin are shown in Fig. 1, and those of chlorins and bacteriochlorins are shown in Fig. 2.[1,2] The UV-visible absorption spectrum of *meso*-tetrakis(4-carboxyphenyl)porphyrin (TCPP), the ligand used for the synthesis of PCN-223(fb) MOF, has been studied in neutral and acidic conditions. Under neutral conditions, the absorption spectrum of TCPP shows an intense Soret band at ~420 nm and four Q bands at 520 nm, 560 nm, 595 nm and 655 nm (Fig. 3).[3] Under acidic conditions, the Soret band position is red-shifted to 434 nm and the Q-band appears at 660 nm with a shoulder at 603 nm.[4,5] The absorption spectrum of TCPP under acidic conditions does not show the characteristic peaks shown by phlorins, chlorins and bacteriochlorin (phlorins show a shoulder peak at 350 nm, chlorins show a peak at 500 nm, and bacteriochlorins show a peak at 780 nm). Thus, the protons in TCPP solution may not be reducing the pyrrolic double bonds and the double bonds that link the pyrrole rings. The protons are most likely located on the pyrrolic nitrogen atoms.

We have not studied the hydrogen/deuterium isotope effect to investigate the movement of protons in the macrocyclic ring.

1 A. J. Pistner, G. P. A. Yap and J. Rosenthal, *J. Phys. Chem. C*, 2012, **116**, 16918–16924.
2 N. Kashef, Y.-Y. Huang and M. R. Hamblin, *Nanophotonics*, 2017, **6**, 853–879.
3 M. Aydin, *Molecules*, 2014, **19**, 20988–21021.
4 S. E. Clarke, C. C. Wamser and H. E. Bell, *J. Phys. Chem. A*, 2002, **106**, 3235–3242.
5 T.Bruhn and C. Brückner, *Phys. Chem. Chem. Phys.*, 2015, **17**, 3560–3569.

Mark Thompson asked: I was assuming that the compounds are really porous, based on the crystal structures, but are they really? Is the full pore volume accessible? Is this related to why you can only metallate to the level of 85%?

Amanda Morris responded: These structures are highly porous and the internal surface area is accessible, at least as measured in the gas phase – 1600 m^2 g^{-1}. The internal volume is harder to access when in solvent, as required for metallation. This is mainly due to the much-restricted diffusion of a species through a constrained environment – collisions with and rearrangement of solvent molecules in the pores.

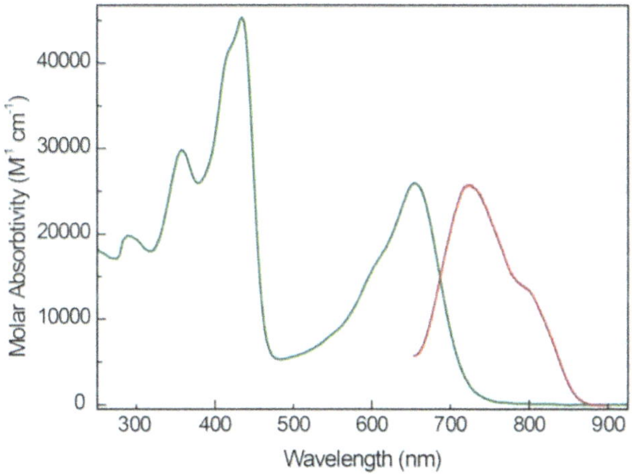

Fig. 1 UV-vis absorption (green) and normalized emission (red) spectra recorded for a phlorin solution (10 μM) in THF.[1]

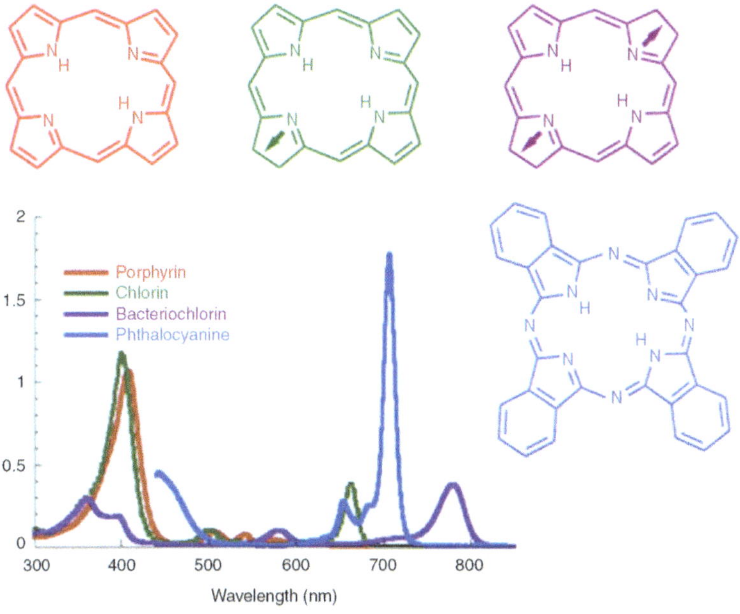

Fig. 2 UV-Vis absorption spectra of porphyrins, chlorins, bacteriochlorins, and phthalocyanines.[2]

Naomi Ginsberg said: Do you mean that at the molecular scale the solvent molecules are able to move fairly freely to reorganise around a porphyrin upon photoexcitation, but that bulk flow of the same solvent molecules into and out of the entire MOF is more difficult?

Amanda Morris answered: Diffusion through metal–organic framework structures is known to be on the order of 10^{-9} to 10^{-12} cm^2 s^{-1}.[1,2] In comparison

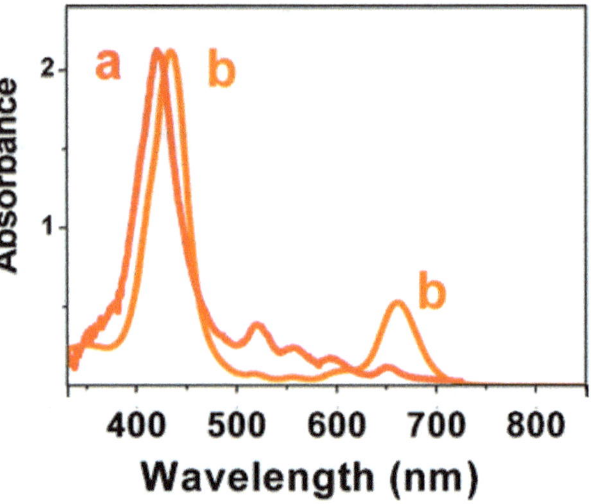

Fig. 3 Steady state absorption spectra of TCPP 10 μM solution in: (a) neutral conditions (1:1 water/ethanol), and (b) acidic conditions (HCl).[3]

to solution, the diffusion in the pore space is therefore 3-7 orders of magnitude slower.

1 T. G. Grissom, C. H. Sharp, P. M. Usov, D. Troya, A. J. Morris and J. R. Morris, *J. Phys. Chem. C.*, 2018, **122**, 16060–16069.
2 S. R. Ahrenholtz, C. C. Epley and A. J. Morris, *J. Am. Chem. Soc.*, 2014, **136**, 2464–2472.

Emily Weiss queried: How does a MOF substrate affect the surface tension of a liquid? Are there ways to think about lowering the surface tension of a solvent so that we may more easily load a solute into the pores of a MOF?

Amanda Morris replied: The chemical composition and texture (roughness) of a solid surface can influence its wetting properties. We hypothesize that surface functionalities and the morphology of the MOF particles can control the nature of intermolecular interactions between the solvent and MOF substrate, and, therefore, affect the surface tension of solvents.

We are not aware of reports in the literature that describe ways to lower the surface tension of solvents *via* interactions with the MOF substrate. Matzger and co-workers have shown the soaking MOFs in ultralow surface tension solvents, such as *n*-hexane and perfluoropentane, can maximize their surface area by efficiently removing the residual reaction solvent and modulator molecules trapped in the pores.[1]

1 J. Ma, A. P. Kalenak, A. G. Wong-Foy and A. J. Matzger, *Angew. Chemie Int. Ed.*, 2017, **56**, 14618–14621

Petter Persson asked: The use of porphyrin-based metal organic frameworks (MOFs) that you present appears to be a very appealing strategy to achieve a high degree of structural control on the atomistic level for energy and charge transfer processes between chromophores. In previous studies of molecular and interfacial systems for electron transfer employing molecular rigid rod linker systems developed synthetically by the group of Elena Galoppini, we were often concerned about the influence of remaining low-energy torsional flexibility (*i.e.* relating to the relative orientation of planar ring structures along the rigid rods) on donor–acceptor couplings (see *e.g.* Lundqvist *et al.*[1] and Thyagarajan *et al.*[2]). To what extent do you believe that similar effects still play a role in MOFs, and to what extent do you think that careful design of new MOFs, along the lines that you discuss in your paper (DOI: 10.1039/c8fd00194d), provides new opportunities to achieve further improved structural control of the linker parts to effectively overcome all such issues relating to residual structural variations at least in this particular context?

1 M. J. Lundqvist, E. Galoppini, G. J. Meyer and P. Persson, *J. Phys. Chem. A*, 2007, **111**, 1487.
2 S. Thyagarajan, E. Galoppini, P. Persson, J. M. Giaimuccio and G. J. Meyer, *Langmuir*, 2009, **25**, 9219.

Amanda Morris The conformation of the peripheral phenyl rings with respect to the porphyrin ring is are are of active research. Structures that exhibit coplanarity between these features have been hypothesized. However, the steric hindrance in this conformation induces quite a energy penalty. When coupled to the conjugation that occurs between the carboxylate and the phenyl rings and rotation is highly energetically non-preferred. Therefore, I do not think the phenyl rings rotate as freely as one would imagine in the MOF structure - as opposed to linear molecular structures like those referenced in the question. For more on this topic I point the readers to *J. Am. Chem. Soc.*, 2015, **137**, 413–419, which provides an in depth discussion of phenyl ring rotation and methods to force this conformational change in MOFs.

Jennifer Ogilvie remarked: These materials seem very exciting. Can you please comment on the biggest challenges of achieving directional energy transfer and coupling to acceptor species?

Amanda Morris replied: One of the key challenges is to identify framework topologies that are conducive for directional exciton transport. For instance, MOF structures in which porphyrin struts are stacked in a face-to-face manner (pi-stack) are highly desirable because transition dipole moments of the porphyrin units would be aligned, leading to increased energy transfer rates and longer hopping distances.[1] Ultimately, you would want this alignment in a framework that exhibits stability in a variety of environments - Zr-based, Fe^{3+}-based, *etc.* The structure of the chromophores also needs careful attention to fine-tune their photophysical properties, such as fluorescence quantum yield and molar absorption coefficient. Long-range crystalline order is also a key factor for ensuring directional exciton transport. Thus, synthesis strategies the promote high crystallinity and less defects need to be explored.[1]

1 M. C. So, G. P. Wiederrecht, J. E. Mondloch, J. T. Hupp and O. K. Farha, *Chem. Commun.*, 2015, **51**, 3501–3510.

Carlos E. Crespo-Hernández commented: My question concerns the assignments of τ_1 and τ_2 lifetimes presented in Tables 3 and 4 in your paper (DOI: 10.1039/c8fd00194d): have you considered performing experiments at different excitation wavelengths to probe the proposed role of excess vibrational energy in the dynamics of the S_1 state in the TCPP and PCN-223 MOF samples?

Amanda Morris responded: We agree that fsTA measurements at different excitation wavelengths would give us more information about the dynamics of the S_1 state in the PCN-223 MOF and would like to perform these measurements in the future. Zewail and co-workers have looked into the relaxation dynamics of the free base tetraphenylporphyrin (TPP) in solution following Soret (400 nm), Qy (550 nm), and Qx band (585 nm) excitation.[1] Excitation of the Qx band revealed the presence of a 100-200 fs component, which is assigned to intramolecular vibrational dephasing. It would be interesting to know whether we are able to observe this component in the MOF as well.

1 J. S. Baskin, H.-Z. Yu and A. H. Zewail, *J. Phys. Chem. A*, 2002, **106**, 9837–9844.

Ryan Dill opened discussion of Petter Persson's paper: As a clarification, the reason you believe the MLCT state in your iron(II) btz complex is longer lived than in many similar iron(II) complexes is because the strong ligand field pushes the metal-centered states higher in energy, making them harder to access? If you look at the Tanabe-Sugano diagrams, the energies of states depend on the ligand-field strength, but also on the Racah B parameter/nephelauxetic effect. Making changes to the Racah parameter is of course more conceptually challenging, but have you considered such an approach?

Petter Persson answered: Yes, many traditional Fe(II) complexes, such as polypyridyl complexes, are known to suffer from quick deactivation of photo-excited metal-to-ligand charge-transfer (MLCT) excited states on a sub-ps timescale to low-energy metal centered (MC) scavenger states in a way that does not happen *e.g.* for corresponding Ru(II) complexes where the MC states are less accessible. This view is corroborated by our calculations on several iron carbene complexes that show that the introduction of strongly σ-donating carbene ligands does indeed significantly destabilize the MC scavenger states. An interesting question in this respect concerns the apparent similarity in terms of improved excited state lifetimes in both Fe(II) and Fe(III) carbenes (see Chábera *et al.*[1] and Chábera *et al.*,[2] respectively) where the latter case of less explored $3d^5$ complexes seems to suggest that the destabilization of the MC states could be an important consideration also for achieving long-lived doublet Ligand-to-Metal Charge Transfer (^{2}LMCT) excited states. This will be interesting to consider in further detail, hopefully including also a wider range of LMCT complexes, in the near future.

Regarding the ligand field considerations, our primary goal in this early stage of developing iron-based photosensitizers has been to reach a new general level of photo-functionality with stronger ligand field splitting in the iron carbenes using a combination of strategies to modify the metal–ligand σ- and π-bonding properties, as well as to consider different coordination geometries, and different types of auxiliary ligands with electron donating group (EDG) and electron withdrawing group (EWG) properties as largely described in the paper by Liu *et al.*[3] While we have not explicitly considered the particular ligand field aspects that you outline, the recent general progress in the challenging area to develop photo-functional first-row transition metal complexes of iron and other metals is clearly intimately connected with finding good strategies to influence the manifold of excited states in new ways where ligand field effects play a central role, and I believe that it will be exciting to see if new strategies and ideas can be explored synthetically to make further improvements in the photophysical properties.

1 P. Chábera, K. S. Kjaer, O. Prakash, A. Honarfar, Y. Liu, L. A. Fredin, T. C. B. Harlang, S. Lidin, J. Uhlig, V. Sundström, R. Lomoth, P. Persson and K. Wärnmark, *J. Phys. Chem. Lett.*, 2018, **9**, 459.
2 P. Chábera, Y. Liu, O. Prakash, E. Thyrhaug, A. El Nahhas, A. Honarfar, S. Essén, L. A. Fredin, T. C. B. Harlang, K. S. Kjær, K. Handrup, F. Ericson, H. Tatsuno, K. Morgan, J. Schnadt, L. Häggström, T. Ericsson, A. Sobkowiak, S. Lidin, P. Huang, S. Styring, J. Uhlig, J. Bendix, R. Lomoth, V. Sundström, P. Persson and K. Wärnmark, *Nature*, 2017, **543**, 695.
3 Y. Liu, P. Persson, V. Sundström and K. Wärnmark, *Acc. Chem. Res.*, 2016, **49**, 1477.

Johannes Mahl asked: Could you provide some more details regarding the photostability of the system? What average laser powers (mW) and laser pulse fluences (mJ cm^{-2}) may be used in a pump–probe experiment? How does the system's photostability compare to ruthenium- or iron-tris(bipyridine)?

Petter Persson replied: The iron carbenes generally have very good photostability in our experience, significantly adding to their appeal for photochemical applications and typically allowing us to work in normal ambient conditions, with no low-light, glove-box, or inert gas storage necessary. However, as a precaution we typically purge the solvent with N_2 prior to experiments. In one of our recent publications we compared the photostability of a related iron carbene complex with ruthenium-tris(bipyridine) through continuous illumination of aerated acetonitrile solutions over a period of one week, and found the iron carbene complex to display significantly better photostability compared to the ruthenium-tris(bipyridine) complex.[1] Regarding the technical aspects of the measurements, the excitation power was kept as low as possible in order to get reasonable signal/noise ratio for the Transient Absorption (TA) experiments presented here. The highest excitation fluency used was 3×10^{15} photons per pulse per cm^2 which in our case corresponded to an excitation power of $\sim$1 mW or excitation energy of $\sim$2 μJ (rep. rate is 500 Hz due to a chopper in the pump beam). Moreover, the power dependency was measured, where the amplitude of the TA signal was compared over a wide range of excitation powers, varying from 100 to 2500 μW, with no deviation from linear dependence.

1 K. S. Kjær, N. Kaul, O. Prakash, P. Chábera, N. W. Rosemann, A. Honarfar, O. Gordivska, L. A. Fredin, K.-E. Bergquist, L. Häggström, T. Ericsson, L. Lindh, A. Yartsev, S. Styring, P. Huang, J. Uhlig, J. Bendix, D. Strand, V. Sundström, P. Persson, R. Lomoth and K. Wärnmark, *Science*, 2019, **363**, 249.

Todd Martinez remarked: In Fig. 8 of your paper (DOI: 10.1039/c8fd00232k), you plot the x axis as the M-L distance. Much of inorganic photochemistry is discussed in one-dimensional terms with reference to this particular coordinate. To what extent do other coordinates participate? Have you searched for minimal energy crossing points and/or minimal energy conical intersections connecting the relevant states? What coordinates are displaced (relative to the Franck–Condon point) in these structures?

Petter Persson replied: This is a very good point that while it is often convenient for illustration purposes to use a single Metal–Ligand (M-L) distance as a reaction coordinate to represent the structural difference between key states (typically including the ground state, charge-transfer states, and metal centered states), it is actually very important to consider the intrinsically multi-dimensional nature of the excited state landscapes to understand many photo-physical and photochemical processes. We typically find it useful to first perform full optimizations without any symmetry constraints of all different excited states of the transition metal complexes we are interested in to identify the most important structural coordinates from a multidimensional perspective that can then be used as a basis for *e.g.* more comprehensive calculations of excited state potential energy surfaces (see *e.g.* Österman *et al.*).[1] In our experience, the M-L distances are often key to understand much of the photoexcited dynamics, but it is important to simultaneously consider the effects of varying several M-L bonds, *e.g.* in terms of symmetry-lowering Jahn–Teller type distortions associated with different populations of the metal d-levels in different states such as the open-shell $(t_{2g})^5(e_g)^1$ orbital population in ^{3}MC states often involved in deactivation cascades in transition metal complexes of iron and other metals studied by us and many others. We have often found that the simple one-dimensional schematic drawings of potential energy curves can nevertheless be understood as practically useful for displaying the information of the multi-dimensional M-L landscape on an effective coordinate that can contain much of the most relevant information in the sense of it effectively being a projection of key points of the multidimensional M-L landscape on a one-dimensional representation (see *e.g.* Fredin *et al.*).[2] One of the exciting current developments in this field is the emerging possibility to use a range of experimental time-resolved X-ray techniques to probe transient structural dynamics and short-lived intermediate structures that makes for very useful direct comparisons between experiments and calculations as illustrated *e.g.* in the opening presentation of this Faraday Discussion by Prof. Chergui (DOI: 10.1039/c9fd00036d), and also something that we have found very rewarding ourselves for the iron-carbene molecules (see *e.g.* Leshchev *et al.*).[3] In terms of locating key points and pathways such as transition states, minimum energy pathways, and conical intersections, we have often been limited by difficulties to locate points such as true transition states on extremely flat excited state surfaces and other computational complications relating to limited computational accuracy. Our approach has therefore mainly been to scan relevant potential energy surface regions in systematic fashion. This is, however, an area where there is currently a lot of interesting progress and development by several groups to use better methods for excited state calculations, find low energy transition pathways by *e.g.* nudged-elastic-band (NEB) methods, and also in terms of more advanced quantum dynamics simulations of interesting prototype complexes that promises to make this an exciting and rapidly developing area for advanced computational work in the near future.

1 T. Österman, M. Abrahamsson, H. C. Becker, L. Hammarström and P. Persson, *J. Phys. Chem. A*, 2012, **116**, 1041.
2 L. A. Fredin, M. Papai, E. Rozsalyi, G. Vanko, K. Wärnmark, V. Sundström and P. Persson, *J. Phys. Chem. Lett.*, 2014, **5**, 2066.
3 D. Leshchev, T. C. B. Harlang, L. A. Fredin, D. Khakhulin, Y. Liu, E. Biasin, M. G. Laursen, G. E. Newby, K. Haldrup, M. Nielsen, K. Wärnmark, V. Sundström, P. Persson, K. S. Kjær and M. Wulff, *Chem. Sci.*, 2018, **9**, 405.

Michael Ashfold remarked: The previous question addressed aspects of the dynamics of the iron carbene complexes following photoexcitation. This question focuses more on the initial photoexcitation itself. Given the very fast intersystem crossing rate (sub 100 fs timescale, assumed but not resolved in the present study), I wonder to what extent labels like singlet, triplet, *etc.* are helpful in describing either the excitation or the early time evolution of the excited state in such molecules?

Petter Persson replied: We retain these spin labels here as used conventionally for a long time and as a useful first order approximation in descriptions neglecting explicit spin–orbit coupling, but in terms of a more detailed description of ultrafast dynamics this question about the nature of the spin states seems both interesting and challenging for current research including both ultrafast experiments and high-level calculations. Iron (atomic number 26) is among the lighter first row transition metal elements so that spin–orbit coupling should be considerably less than in many well-known heavier second and third row transition metal complexes of *e.g.* ruthenium and iridium *etc.*, and therefore it looks reasonable to us to initially retain the spin assignments mainly for the initial excitation and for thermally relaxed states after the ultrafast initial processes on the sub-ps timescales. The initial ultrafast femtosecond dynamics involving the mixing of many states close in energy (typical for many transition metal complexes) following initial photoexcitation, however, makes this a rich area for current research where several groups have made contributions in the last few years to present convincing evidence for a much more elaborate view including considerable mixing of electronic and spin states. This looks particularly relevant for the cases where there is rich photoinduced dynamics involving a manifold of accessible excited states as we also see some evidence for in the data for the iron carbene molecule we present here (DOI: 10.1039/c8fd00232k), so this certainly seems like an interesting area for further investigation, especially for also trying to provide a better understanding of the fastest photoinduced dynamics on the fs timescale.

Carlos E. Crespo-Hernández asked: The B3LYP functional is known to incorrectly describe charge transfer excitations in molecules, and basis sets that incorporate relativistic effects are usually preferred for describing the electronic structure of molecules with heavy-atoms or transition metal atoms. Could you comment on the rationale behind the selection of the B3LYP* functional and the 6-311G(d) basis set used for the calculations? Concerning the excited-state model presented in Fig. 8 in your paper (DOI: 10.1039/c8fd00232k), could you comment on whether a description of the electronic excited states as exclusively singlet or triplet states is justified given the expected significant spin–orbit coupling interactions and the high density of states in the $Fe^{II}(btz)_3$ and $Fe^{III}(btz)_3$ complexes?

Petter Persson answered: These are all very legitimate methodological concerns relating to the functional, basis set, and spin–orbit coupling where we face many of the same concerns as many other groups in trying to balance accuracy against efficiency and broad applicability for excited state calculations of transition metal complexes that are demanding to treat with high accuracy. The choice of the B3LYP* functional has been guided by findings of others as well as our own impression (see *e.g.* Leshchev *et al.*[1] and references therein) that suggests that this offers a fair description *e.g.* in regards to the important and often sensitive energetic balance between charge-transfer (CT) and metal centered (MC) states. We have also generally found the here adopted level of theory (the combination of functional and basis set) to offer a useful level of theory that allows us to maintain consistency with other work, and that makes it useful for comparative purposes. In our view this offers a useful guide to many of the experimental trends that we observe, and we regularly make validation tests where we *e.g.* compare different functionals and basis sets. For example, this level of theory allows us to both screen excited state energy surfaces relevant for many of the photochemical properties with reasonable efficiency at the same time as it has offered a good balance between accuracy and affordability, also for more demanding interfacial investigations of molecularly sensitized nanomaterials of interest for photosensitization applications (see *e.g.* Fredin *et al.*).[2] While we thus feel that we have good support for believing that our calculations are valuable as a guide for understanding many of the phenomena that we are exploring experimentally, we at the same time fully recognize that higher-level calculations are desirable for achieving better and quantitatively more reliable results of the complex excited state properties that feature both many electronic states close in energy, with significant spin–orbit mixing effects that we are not accounting for here (as discussed also in the response to the previous question), and dynamical processes that are very interesting from an ultrafast perspective. Thus, we continuously try to be careful not to overinterpret the results obtained at the current level of standard DFT/TD-DFT theory, and in fact we in many respects regard our present calculations as exploratory in nature where hopefully more accurate calculations and advanced quantum dynamics simulations can subsequently be applied on the most interesting and/or challenging cases to describe the ultrafast dynamics in a more accurate and complete manner.

1 D. Leshchev, T. C. B. Harlang, L. A. Fredin, D. Khakhulin, Y. Liu, E. Biasin, M. G. Laursen, G. E. Newby, K. Haldrup, M. Nielsen, K. Wärnmark, V. Sundström, P. Persson, K. S. Kjær and M. Wulff, *Chem. Sci.*, 2018, **9**, 405.
2 L. A. Fredin, K. Wärnmark, V. Sundström and P. Persson, *ChemSusChem*, 2016, **9**, 667.

Naomi Ginsberg addressed Petter Persson and Mark Thompson: Why was it initially so difficult to make these complexes with iron? And what was the advance that allowed the iron complexes to finally be formed?

Petter Persson responded: Making useful iron-based light-harvesting complexes with sufficiently long excited state lifetimes for photochemical applications has remained a significant challenge for several decades. This problem is intimately connected with the difficulty to achieve sufficiently strong ligand-field splitting in first-row transition metal complexes that in turn can be traced back to the radial distribution of the d-levels on the metal (see *e.g.* a nice recent exposition by McCusker[1]) and how this influences the interaction with the ligands. The successful introduction of *N*-Heterocyclic Carbene (NHC) ligands for photofunctional transition metal complexes of iron and other transition metals makes use of recent advances in ligand design from the field of homogeneous catalysis and has offered significant new opportunities to synthesize new types of transition metal complexes that are both stable and have long-sought electronic properties in terms of promoting strong ligand field splitting (LFS) through strong σ-donation *etc.*, making them very interesting as light-harvesting complexes (for a more detailed discussion of the iron carbenes see *e.g.* Liu *et al.*).[2] Hopefully, further improvements in the photo-functionality of Earth-abundant transition metal complexes can now be achieved using a further influx of new synthetic ideas and improvements in design strategies.

1 J. K. McCusker, *Science*, 2019, **363**, 484.
2 Y. Liu, P. Persson, V. Sundström and K. Wärnmark, *Acc. Chem. Res.*, 2016, **49**, 1477.

Mark Thompson answered: This was primarily for Petter, but I can give you some background. The problem is not making iron compounds, but it is making iron complexes with high enough ligand fields that the d-d transitions are not the lowest energy excited states. d-d transitions lead to very rapid deactivation of the excited state. The key here was to switch to a ligand that has the highest ligand field known, *i.e.* carbenes. The carbene was first used as a ligand for metal complexes in the 1980s, on homogeneous catalysts. About 10 years ago people started looking at the photophysics of carbene ligands. There is a good deal of iridium, platinum and copper carbene photophysics out there. In Petter's paper (DOI: 10.1039/c8fd00232k) they show that for both d5 and d6 systems they can push the ligand field states above the MLCT and get complexes with reasonable lifetimes.

Naomi Ginsberg asked: How do you establish what is the reaction coordinate/potential energy surfaces in this case?

Petter Persson responded: In our experience this is a very important consideration to get a good description of the excited state evolution that is often far from trivial. This is especially true for the excited state dynamics of transition metal complexes as there is more often than not significant simultaneous variations of several metal–ligand bonds, *e.g.* in going from charge-transfer to metal centered states, that typically makes this a multi-dimensional problem that is more complicated than say if a single bond is broken in an organic photo-dissociation reaction.

There is often very limited experimental information about the structural dynamics to guide this understanding, although there are very exciting developments in the field of time-resolved X-ray spectroscopies that provide increasing opportunities to characterize also photoinduced structural dynamics and short-lived excited states also for metal complexes.

From a theoretical/computational perspective, we have found on several occasions that a good starting point is to obtain fully optimized geometries for all states involved in the excited state cascade, and then map out paths (if necessary using multi-dimensional potential energy surface scans) between these states. This effort can often be guided by fundamental considerations of the electronic structure properties involved, *e.g.* population of specific anti-bonding orbitals naturally couple with weakening of those bonds *etc.* Scanning extensive potential energy surfaces in many dimensions quickly becomes quite laborious, but better computational capabilities look promising to alleviate this problem by offering automated scans or on-the-fly dynamics/energy evaluations capable of identifying the most relevant regions to map of the excited state landscape.

Marta Duchi opened a discussion of the paper by Andrew H. Marcus: In your study (DOI: 10.1039/c8fd00245b), the (Cy3)2 fluorophore is used as a local probe of DNA structure in several constructs. On the 3′ chain Cy3 is always inserted between C and T bases, is it possible to move the fluorophore into a different position or change the order of the adjacent nucleo bases? Are your observations dependent on the position of the fluorophore within the duplex or fork DNA? Further, is the DNA conformational disorder dependent on the specific base sequence?

Andrew Marcus replied: I agree these are very interesting and important questions, which we are working to address in follow-up studies. Yes, the studies we have completed thus far suggest that it is possible to change the nucleotide sequence to investigate the effects of base sequence on local DNA backbone conformation and disorder. The differences we have observed between local conformations deep within the duplex in comparison to those at the fork junction suggest that we might determine the position dependence of the local backbone conformation and disorder relative to the fork junction in this way. This would be useful to understand how protein binding near a junction might disrupt the secondary structure further into the duplex region.

Shawn Irgen-Gioro asked: In your work (DOI: 10.1039/c8fd00245b), you look at how the displacement of this high-frequency mode changes when the dye aggregates. Do you expect that the electronic coupling affects all vibrations equally? I ask this because you are trying to extend your model beyond a dipole moment. If we were to see non-uniform changes in displacement, maybe we can map out the electronic moment in detail

Andrew Marcus answered: In the actual system, there is no reason to expect the electronic transition to couple to all vibrations equally. However, for the Holstein model that we used in our work, only one effective high-frequency vibrational mode was required to successfully simulate our data over a wide range of conditions. It might be possible to experimentally distinguish between different models that use successively higher numbers of vibrational modes, but for the cyanine dimer DNA system we studied, the single mode Holstein model appears to capture the essential features of exciton delocalization.

Naomi Ginsberg remarked: Why do you think it is possible to capture the essence of the experimental data using a model with a single vibrational mode?

Andrew Marcus replied: This is an interesting question. Of course, the appeal of using toy models to describe essential features of complex molecular systems is that they may provide insight to the underlying physics of said features. The relatively simple Holstein model appears to account for many spectroscopic observables of the Cy3 dimer in the normal duplex and forked DNA environments. Even the Cy3 monomer in DNA is well described as a two-electronic-level system coupled to a single high-frequency vibrational mode. I think of the Holstein model as a course-grained approximation to a more realistic (and computationally expensive) model, which would include, in principle, all of the vibrational modes and their electronic-vibrational couplings. The presence of the additional (mostly lower frequency) vibrational modes increases the spectral density of the vibronic states, but their presence does not heavily influence the overall shapes of the absorbance and CD spectra, which are dominated by the single high-frequency mode at ~1,100 cm^{-1}. In the presence of homogeneous and inhomogeneous line-broadening mechanisms, the Holstein model, with one "effective" vibrational mode, seems to be sufficient to capture the overall shapes of these spectra, and how these spectra vary as the resonant coupling strength is adjusted by temperature. See Roden *et al.* for an interesting paper that looks at this issue in the context of high-resolution vibronic spectroscopy of oligomers in helium nanodroplets.[1]

1 J. Roden, A. Eisfeld, M. Dvořák, O. Bünermann and F. Stienkemeier, *J. Chem. Phys.*, 2011, **134**, 054907.

Jennifer Ogilvie remarked: Can you please comment on some of the differences between how homogeneous broadening is extracted in the fluorescence-detected 2D experiment compared to the more conventional coherent-detected 2D approach? For example, in our recent fluorescence-detected 2D studies of LH2,[1] we find that the two oppositely-signed excited-state absorption signals completely cancel, leading to a considerably broader anti diagonal width than in the conventional 2D spectra which contain a single excited-state absorption contribution that is opposite in sign to the other signal components.

1 T. Kunsel, V. Tiwari, Y. A. Matutes, A. T. Gardiner, R. J. Cogdell, J. P. Ogilvie and T. L. C. Jansen, *J. Phys. Chem. B*, 2019, **123**, 394–406.

Andrew Marcus replied: The theoretical relationship between 2D electronic spectroscopy (2DES) and 2D fluorescence spectroscopy (2DFS) is discussed in an earlier paper.[1] In both 2DES and 2DFS, the spectral lineshapes are sums of the three coherent (excitation pathway) terms: ground state bleach (GSB), stimulated emission (SE) and excited-state absorption (ESA). In both 2DES and 2DFS, the response functions depend on the system's transition frequencies and homogeneous dephasing times in much the same ways.

In 2DES the signal is given by the usual sum: GSB + SE − ESA. In 2DFS there are two separate ESA contributions to the signal; a one-photon absorbing ESA term that terminates with population on the singly-excited state manifold, and a two-photon absorbing term that terminates with population on the doubly-excited state manifold. The 2DFS signal is given by the sum: GSB + SE + ESA * (1 − Gamma), where Gamma (0 ≤ Gamma ≤ 2) is a constant that describes the quantum yield contribution of the two-photon absorbing ESA term relative to that of the one-photon absorbing ESA term. This means that in 2DFS, the sign and magnitude of the ESA term can vary relative to the GSB and SE terms, depending on the value of Gamma. For example, if Gamma = 2, all of the population excited in the doubly-excited state is converted to population in the singly-excited state (which contributes to the detected fluorescence), in which case the 2DFS and 2DES signals are identical. If Gamma = 0, no doubly-excited state population is converted to fluorescence so that the three terms add equally with the same sign. There is the special case Gamma = 1 (*i.e.*, exactly 50% conversion of the doubly-excited state population to singly-excited state population), where the ESA terms exactly cancel and do not contribute to the 2DFS signal. The value of Gamma is system-specific, as it depends on the deactivation pathways (radiative and non-radiative) of the doubly-excited state population.

Sometimes the peaks of spectral features arising from ESA terms are shifted relative to those arising from GSB and SE terms. In this case, the 2D spectral lineshapes for 2DES can appear to be "effectively narrowed" because they result from adding a negative ESA peak to positive GSB and/or SE peak(s) that are spectrally displaced from one another. In a 2DFS experiment on the same system, if Gamma takes a value close to zero, the spectrally displaced ESA and GSB/SE peaks add with the same sign such that the lineshape can appear to be "effectively broadened". As stated above, values of Gamma approaching 1 will lead to diminishing contributions from the ESA term so that these effects of peak overlap will become less pronounced. In both 2DES and 2DFS experiments, it is important to compare the experimentally derived lineshapes to theoretical models in order to accurately extract the homogeneous lineshapes of individual spectral features.

1 A. Perdomo-Ortiz, J. R. Widom, G. A. Lott, A. Aspuru-Guzik and A. H. Marcus, *J. Phys. Chem. B*, 2012, **116**, 10757–10770.

Nadia Korovina asked: The cyanine dyes used in this study as fluorescent labels have substantial mass and size. Do they affect the dynamics and do they alter the backbone of the DNA in a way that affects its conformations?

Have you tried other cyanine dyes to ensure that the conformational dynamics of DNA are independent of the molecular structure of the dye?

Andrew Marcus replied: In our current paper (DOI: 10.1039/c8fd00245b), we studied internally labeled cyanine dyes that were inserted into the sugar-phosphate backbone of duplex and fork DNA constructs. We studied the ground-state-accessible excitons of the Cy3 homodimer in DNA to see if the local conformations of the dyes would resemble those of the unlabeled sugar-phosphate backbone. Of course, the presence of the Cy3 labels is expected to have some affect on the local chemical environment of the sugar-phosphate backbone since they replace native nucleotides in the probe-labeled DNA construct. The question is whether the presence of the dyes dominates the experimentally observed behavior, or if their presence introduces relatively small perturbations. The results of our current studies suggest that the local conformations determined for the Cy3 labels do indeed reflect reasonable temperature-dependent behavior of the sugar-phosphate backbone in both the duplex and the fork DNA constructs. This is consistent with our observations from prior work that show that the presence of the cyanine dyes in similarly labeled duplex DNA lowers the denaturation temperature by only $\sim$2 °C, which is a much smaller affect then the $\sim$6 °C destabilization observed in cyanine labeled DNA constructs that attach the dyes using flexible linkers.[1] We've also found that the presence of the internally labeled Cy3 dyes do not appear to disrupt specific biological functions such as the correct assembly of replication nucleoprotein complexes, "breathing" at model DNA fork junctions, and the functional unwinding of double-stranded DNA by a replication helicase-primase complex.[2,3]

We have performed studies on Cy5 dimer labeled DNA constructs, analogous to those reported in the current paper (DOI: 10.1039/c8fd00245b) for the Cy3 dimer. This work is in progress, but our preliminary results suggest that Cy5 labeled DNA constructs exhibit very similar behavior to those for Cy3.

1 W. Lee, P. H. von Hippel and A. H. Marcus, *Nucl. Acids Res.*, 2014, **42**, 5967–5977.
2 C. Phelps, W. Lee, D. Jose, P. H. von Hippel and A. H. Marcus, *Proc. Natl. Acad. Sci. U. S. A.*, 2013, **110**, 17320–17325.
3 W. Lee, D. Jose, C. Phelps, A. H. Marcus and P. H. von Hippel, *Biochemistry*, 2013, **52**, 3157–3170.

Yin Song queried: Spano reported that two-particle state (with electronic excitation on molecule a and vibrational excitation on molecule b) can modulate the fluorescence spectrum of the dimer and aggregates.[1] Will these two-particle states also contribute to 2DFS? Can 2DFS be a better tool to resolve these states than steady-state fluorescence spectroscopy since it has one additional dimension to separate signals from different pathways?

1 F. C. Spano, *Acc. Chem. Res.*, 2010, **43**, 429–439.

Andrew Marcus replied: I do think that 2DFS could be used to spectrally filter the photoluminescence signal to separate contributions from the H and J aggregate. My group has performed experiments along these lines in the context of a Zn-porphyrin dimer embedded in a biological membrane, in which we used spectral filters to separately measure 2DFS signals derived from components of the sample configured as an H-dimer and components that were configured as a J-dimer.[1] Although we did not make the formal connection to one- and two-particle states as done by Spano, I think this should be possible using 2DFS.

1 J. R. Widom, W. Lee, A. Perdomo-Ortiz, D. Rappoport, T. F. Molinski, A. Aspuru-Guzik and A. H. Marcus, *J. Phys. Chem. A*, 2013, **117**, 6171–6184.

Jeffrey Cina remarked: I have a comment concerning the interpretation of the two contributions to the 2DFS signal for an energy-transfer dimer involving the doubly-excited electronic state, which are sometimes referred to as ESA1 and ESA2: That generating a quadrilinear population in the doubly excited state can be regarded as excited state absorption proper. The one accessing the doubly-excited state, but generating population in the singly-excited manifold represents excited-state bleach – it accounts for population loss from the singly-excited manifold due to its transfer to the doubly excited state. This description is consistent with the equal-and-opposite nature of these two quadrilinear population increments. In the particular case of equal relative fluorescence quantum yields from single and double excitations (the Kasha's-rule case), this population transfer has no effect on the signal, giving rise to a cancelation between excited-state absorption and excited-state bleach; but of course they do not cancel in general.

My question is to ask for the speaker's thoughts on the prospect of achieving partial spatial orientation of the intercalated dimer by flowing or stretching its DNA host. Complemented with polarization spectroscopy, this sort of spatial orientation might approach the ideal situation in which different laser pulses could selectively address either the energy-transfer donor or the acceptor.

Andrew Marcus answered: I agree with the essence of the above comment. However, I think it may be inaccurate to refer to the case of exact cancellation between the two excited state absorption terms as "the Kasha's-rule case." The 2DFS signal is the sum of terms: GSB + SE + ESA * (1 − Gamma), where Gamma is a constant that describes the quantum yield contribution of the two-photon absorbing ESA term relative to that of the one-photon absorbing ESA term. Depending on the branching ratio between radiative and non-radiative relaxation pathways of the doubly-excited state (which are specific to the system), Gamma can take on any value between 0–2. The case of exact cancellation of the two ESA terms corresponds to Gamma = 1 (*i.e.*, the case in which exactly 50% of the doubly-excited state population is converted to singly-excited state population). The value of Gamma = 1 has no deeper connection to Kasha's rule than does any of its other possible values. Kasha's rule states that "the emitting level of a given multiplicity is the lowest (energy) excited level of that multiplicity".[1] I don't think that Kasha's rule makes a general statement about the fluorescence quantum yield resulting from relaxation of population in the doubly-excited state.

It is an excellent suggestion to use solvent flow to control the orientation of DNA segments in optical experiments. We've had some success in implementing this idea in the context of single-molecule polarized light experiments on the Cy3 dimer DNA systems discussed in our paper. Because the symmetric and anti-symmetric excitons of the dimer have transition dipole moments oriented in orthogonal directions, single-molecule experiments using modulated light polarization can, in principle, be sensitive to local conformation fluctuations of the DNA backbone.

1 M. Kasha, *Discuss. Faraday Soc.*, 1950, **9**, 14–19.

Naomi Ginsberg commented: In my experience working out the maths, generating orientational anisotropy in the sample (by alignment) for a nonlinear optical experiment, in the way that one might for linear dichroism, does not allow one to obtain more information than employing different light polarisation combinations.

Amro Dodin asked: Does your group have any insight (*e.g.* from single molecule experiments) if the dye molecules embedded in the DNA helix pack similarly across different complexes in the ensemble or if there is heterogeneity in their relative configurations?

Andrew Marcus responded: In our current paper, which uses linear spectroscopy and 2DFS on ensemble cyanine labeled DNA samples, we observe significant spectral inhomogeneity. We interpreted this spectral inhomogeneity in terms of structural heterogeneity of the cyanine dimer, which depends on DNA backbone conformation. My group is currently performing polarization-selective single-molecule experiments on these same systems, with the goal to observe transitions between the microstates that underlie the structural heterogeneity.

Mark Thompson opened a discussion of the paper by Eric Bittner: In your paper (DOI: 10.1039/c8fd00182k) you state that the most efficient charge separation will take place in systems where the mobilities of the two carriers (holes in the donor and electrons in the acceptor) are disparate. This seems counter intuitive. What is the explanation for this? It seems that this situation would favor charge recombination.

Eric Bittner responded: I believe the statement you are referring to is "*This would imply that lower mobility in one of the phases has a profound impact on the ability of charges to escape from the interfacial region*". Here, we examine what happens if you have a higher degree of charge mobility along the interface compared to perpendicular to the interface (as would be the case if you had stacked polymers). Within the model, this is set by setting the hopping integrals between sites along the interface to be much larger than those perpendicular to the interface. In this case, adding energetic site disorder more or less equalizes the charge's ability to move in one direction *vs.* the other.

Tom Oliver asked: Two-dimensional electronic spectroscopic studies of bulk heterojunction P3HT/PCBM polymer thin films by the groups of Scholes[1] and Cerullo[2] have found that coherent vibronic coupling induces ultrafast charge separation and polaron formation circumventing the inherent disorder in the system. Your study implies that some disorder is imperative for the efficiency of charge separation. So what amount of disorder is optimal?

1 Y. Song, S. N. Clafton, R. D. Pensack, T. W. Kee and G. D. Scholes, *Nat. Commun.*, 2014, **5**, 4933.
2 A. De Sio, F. Troiani, M. Maiuri, J. Réhault, E. Sommer, J. Lim, S. F. Huelga, M. B. Plenio, C. A. Rozzi, G. Cerullo, E. Molinari and Christoph Lienau, *Nat. Commun.*, 2016, 7, 13742.

Eric Bittner replied: Our model introduces disorder *via* a random energetic landscape in the local site energies with fluctuations $\delta\varepsilon$ as proportional to the thermal fluctuations in the phonons and we find that for this model, a range of disorder is optimal. More coherent models, such as those you reference, as well as the "Noise-Induced Coherence" model that Carlos Silva and I developed,[1] are able to "circumvent" the local disorder by providing an ultrafast tunnelling escape route from the interfacial states to long-ranged charge-separated states. Here we found that indeed the exciton can access these long-ranged charge-separated state on the 40fs time-scale – which jives with the coherent coupling model. We also in a separate paper connect this tunnelling with the observed 2D photocurrent.[2] The model Hamiltonian we used in this paper (DOI: 10.1039/c8fd00182k) is the same Hamiltonian as we used in the previous work.[1,2] Consequently, I believe both the coherent coupling and the incoherent hopping models are really two sides of the same coin.

1 E. R. Bittner and C. Silva, *Nat. Commun.*, 2014, 5, 3119.
2 E. Vella, H. Li, P. Gregoire, S. M. Tuladhar, M. S. Vezie, S. Few, C. M. Bazan, J. Nelson, C. Silva-Acuna and E. R.Bittner, *Sci. Rep.*, 2016, **6**, 29437.

Benjamin Schwartz remarked: There's been a lot of recent work on the idea that in organic photovoltaics, the electron rapidly separates from the hole on the polymer by delocalizing over a number of fullerenes. For example, Friend and co-workers found direct evidence that the electron is able to coherently move ~4 nm from the hole in a time ≤ 100 fs because of electron delocalization.[1] The idea of electron delocalization also fits with work from Barry Thompson's group, which showed that the open circuit voltage of OPVs containing fullerene mixtures is an average of the two individual components rather than being pinned by the lower-energy component, also providing strong evidence that a single electron experiences multiple fullerenes following charge separation.[2] Is this type of effect accounted for in your model, and if not, how would it affect your conclusions?

1 S. Gélinas, A. Rao, A. Kumar, S. L. Smith, A. W. Chin, J. Clark, T. S. van der Poll, G. C. Bazan and R. H. Friend, *Science*, 2014, **343**, 512.
2 R. A. Street, D. Davies, P. P. Khlyabich, B. Burkhart and B. C. Thompson, *J. Am. Chem. Soc.*, 2013, **135**, 986.

Eric Bittner responded: To some extent I addressed this in my reply to Prof. Oliver. I think *both* the ultrafast "ballistic" limit and the kinetic limit are encapsulated in this model. I'd add to the Friend paper you mention, the work by Silva's group which reports the prompt formation of polarons at a polymer : fullerene interface within 300fs.[1] Subsequent to this we proposed that energy level fluctuations are sufficient to induce tunneling from the local excitonic state to long-ranged charge-separated states giving rise to tunneling times-scales of about 50 fs (comparable to Friend's results).[2] Also, our 2016 paper on 2d-photocurrent spectroscopy also adds evidence to the fact/coherent mechanism.[3] However, I think one has to conclude that while it is physically reasonable to observe prompt formation of polarons on the ultrafast times-scale, those processes are in the minority compared to slower relaxation processes to lower CT states. This doesn't mean that the fast processes are uninteresting from the view of fundamental photophysics.

1 F. Provencher, N. Bérubé, A. W. Parker, G. M. Greetham, M. Towrie, C. Hellmann, M. Côté, N. Stingelin, C. Silva and S. C. Hayes, *Nat. Commun.*, 2014, **5**, 4288.
2 E. R. Bittner and C. Silva, *Nat. Commun.*, 2014, **5**, 3119.
3 E. Vella, H. Li, P. Grégoire, S. M. Tuladhar, M. S. Vezie, S. Few, C. M. Bazán, J. Nelson, C. Silva-Acuña and E. R. Bittner, *Sci. Rep.*, 2016, **6**, 29437.

Amro Dodin asked: An earlier study using a related semiclassical model for charge transfer excitons on a lattice found that there was an optimal level of disorder for charge separation.[1] This arose when the energetic disorder "lined up" to create pathways with decreasing energy, mitigating the Coulombic cost of separating the electron and hole, but when the disorder was small enough that the formation of two energy traps was not too likely. Does the quantum model show the same optimal level of disorder and do quantum phenomena modify the competition or nature between the two (mitigation and trapping) mechanisms?

1 L. Shi, C. K. Lee and A. P. Willard, *ACS Cent. Sci.*, 2017, **3**, 1262–1270.

Eric Bittner answered: Yes. We discussed this in earlier work.[1,2]

1 E. R. Bittner, V. Lankevich, S. Gelinas, A. Rao, D. A. Ginger and R. H. Friend, *Phys. Chem. Chem. Phys.*, 2014, **16**, 20321–20328.
2 E. R. Bittner and C. Silva, *Nat. Commun.*, 2014, **5**, 3119.

Michael Wasielewski remarked: What is the magnitude of the electronic coupling in this regime? Time-resolved EPR spectroscopy indicates that this coupling is relatively small.

Eric Bittner responded: The coupling constants are given in the manuscript (DOI: 10.1039/c8fd00182k) and are fairly short-ranged since there's a screening length on the order of the lattice constant.

Petter Persson remarked: To what extent can the kind of coarse-grained/lattice models for interfacial charge separation in organic photovoltaics (OPVs) presented in your paper provide information about different types of materials and processes of current interest for OPV devices? In particular, the model that you present is formulated in a quite general way but the presented discussion, *e.g.* in Fig. 1b of the paper (DOI: 10.1039/c8fd00182k), seems to relate primarily to a polymer : fullerene blend. Fullerenes have been discussed to have special properties *e.g.* in terms of forming unique domain structures in the OPV blends, while there are significant current efforts to instead develop *e.g.* polymer-: polymer blend OPVs. Can such differences in the constituent materials be accounted for in your models, *e.g.* in terms of different parameter values or site distributions, and do you think that the models can be used to say something about what similarities or differences in OPV performance can be expected from such material variations? Secondly, in terms of fundamental processes, charge separation and subsequent charge extraction seems to depend significantly on electron and hole mobilities, *e.g.* as discussed for the fully interacting model in section 2.2 of your paper (DOI: 10.1039/c8fd00182k). In our experience, electron and hole mobilities often depend strongly on subtle conformational effects so that they can vary by several orders of magnitude depending *e.g.* on annealing conditions, and where the mobilities for the electrons and holes furthermore respond differently to such annealing (see *e.g.* L. M. Andersson *et al.*[1]). To what extent can the influence of such effects on OPV performance be rationalized, or even predicted, using the kind of models discussed in your paper?

1 L. M. Andersson, S. Hedström and P. Persson, *Appl. Phys. Lett.*, 2013, **103**, 213303.

Eric Bittner answered: We believe that the basic lattice model can be extended to a more general context *via* coarse graining and there are a number of other works looking at excitons (as opposed to charge-transfer processes) in coarse grained models. So, it's entirely possible to extend the model and we've made some forays towards that. However, one rapidly encounters a formidable computational barrier since each coarse grained site (in general) can accommodate either a conduction electron or a valance band hole, so the size of the state space scales as N^2. Also, in the past, we've considered stacked polymer : fullerene blends as well as various lattice geometries and connectivities as a way to address the varieties of materials as you mention. In so far as OPV performance, I think we have a good handle on the V_{OC}. J_{SC} is another matter. However, in principle, given that we can compute rate constants, we should be able to formulate or provide input into semi-empirical models for J–V relations. We have not done this, thus far.

Naomi Ginsberg questioned: Might there not be different qualitative characteristics of disorder that either enhance or suppress charge separation at the interface? In your paper (DOI: 10.1039/c8fd00182k) you also address the role of dimensionality and of entropy. Can you comment on those as well?

Eric Bittner responded: Yes. Defects, asperities, and incursion of one phase into the other, just to name a few. Add to this local ordering and micro-crystalinity at the interface; we've looked at some of these over the years. However, in regard to the dimensionality, extending our model to 3D is straightforward, but computationally costly, so we have not examined 3D lattices. However, we have examined 1D junctions, in particular triplet recombination in fullerene/polymer systems from interfacial CT states to localized excitons.[1] In one-dimensional co-linear and co-facial models of the interface, increasing local energetic disorder in one phase localizes the interfacial triplet charge-transfer (3CT) states and increases the rate at which these states relax to form lower-energy triplet excitons. In the 2D-BHJ model, increasing the energetic disorder within the fullerene phase has a small effect on the average localization of the initial 3CT wave function and the most-likely rate (*i.e.* the mode) shows very little variation with increasing energetic disorder.

1 E. R. Bittner, V. Lankevich, S. Gelinas, A. Rao, D. A. Ginger and R. H. Friend, *Phys. Chem. Chem. Phys.*, 2014, **16**, 20321–20328.

Carlos E. Crespo-Hernández queried: The morphology of bulk heterojunction organic photovoltaic cells is thought to control many of the performance characteristics of their devices. Could you comment on whether a change of the morphology of the donor and acceptor interface may affect the results and conclusions of the exciton dissociation dynamics presented in your contribution?

Eric Bittner responded: The specifics of the interfacial morphology (in our model) has some impact on the results we obtain. The free energy and averaged energies we find tend to agree with the V_{OC} measurements of a wide range of organic-based photovoltaic devices. However, in my opinion, the scale of our models is too small to reflect the morphology of a real physical device. Hence, a flat interface (as per our model) is probably more realistic than something rough. A bit of data mining by Ratner's group correlated power conversion efficiency (PCE) with device performance over about 150 different devices.[1] They conclude that there is a strong correlation between power conversion efficiency and morphology-dominated properties and a much weaker correlation between PCE and energy level positioning and interfacial properties. In our model, the V_{OC} is more or less predetermined by the energy level alignment and off-set, so it is not too surprising that we fall into the observed range. However, device performance also depends upon current, which is going to be highly sensitive to morphology once the charges separate. In fact, I think that the free energy barrier we see in our results is consistent with the notions suggested that once the charges are able to escape the interface, no additional work is required to separate further and that this also provides a barrier for recombination.

1 N. E. Jackson, B. M. Savoie, T. J. Marks, L. X. Chen and M. A. Ratner, *J. Phys. Chem. Lett.*, 2015, **6**, 77.

Jahan Dawlaty asked: Do effects such as Anderson localization manifest themselves in these situations? One would expect that beyond a certain threshold of disorder qualitatively different behavior would appear.

Eric Bittner answered: In any disordered energy landscape, Anderson (or Anderson-like) localization will always manifest. In the paper, we focused on just site energy disorder and indeed one does see qualitatively different behaviour with increasing energetic disorder. However, we do not see evidence of a sudden transition from band-like transport to localized hopping. My suspicion is that the sample domain is too small.

Bern Kohler remarked: You write that localized electron and hole polaron states are expected to show reduced entanglement. Can you say more about the relation between polaron localization and entanglement? In your paper (DOI: 10.1039/c8fd00182k), you write that high-lying, true charge separated states still exhibit entanglement. Can you comment on this and on just how localized the polaronic states are in this case?

Eric Bittner replied: This result surprized us. Formally, I would expect that for a properly separated electron/hole pair, that the wavefunction could be written as $\Psi = \varphi_e(r_e)\varphi_h(r_h)$, which is perfectly separable and would have a single non-trivial SVD coefficient (of 1).

It's more likely that the states we obtain are not as asymptotically separated as we'd like since the spatial domain is still pretty small. This indicates that even though the energetics (and free energy) appear to be hitting the asymptotic limit, there is still spatial overlap between the conduction band electron and valance band hole's wavefunction.

Naomi Ginsberg commented: Since your simulated interfaces are idealised in their tidiness, and since it is more straightforward to work in 2D rather than 3D, the model appears to be ideally suited to treat charge separation in 2D transition metal dichalcogenide heterostructures. There are similar questions to those that exist about the nature of charge separation in bulk heterojunction organic photovoltaics, except that the structure has far less disorder and is far more 'model'-like.

Natércia das Neves Rodrigues opened a discussion on the paper by Jahan Dawlaty: You mentioned in your talk that, in some cases, proton transfer did not take place along an intermolecular hydrogen bond, despite a great 'drive' towards this process. I was wondering if this 'drive' you referred to was an actual, calculated/measured energetic drive for this process or if you were just referring to the fact that proton transfer may intuitively be expected to be occur in these cases due to the intermolecular hydrogen bond? I have found in some of my studies that, despite (intramolecular) hydrogen bonding, a (intramolecular) proton transfer coordinate may not be energetically favorable, according to computational chemistry simulations.

Jahan Dawlaty responded: This is a good question. The drive for proton transfer can be estimated from a Förster cycle, as explained in our manuscript (DOI: 10.1039/c8fd00215k) and based on our previous work. The comment about a large drive for deprotonation is based on the numbers from the Förster cycle. The drive for proton transfer in the excited state of 5-methoxy quinoline, for example, is nearly 15.5 units of pK_a. The pK_a of our proton donor is nearly 12. Therefore one may estimate that the equilibrium constant should be tipped towards deprotonating the donor by ~3 units in pK_a (*i.e.* ratio of protonated to unprotonated excited quinolines in a pure solution of the donor should be ~1000).

In this work, we have shown that for the above to be true, one must have more than one donor clustered near the photobase. If just one donor is available, the above is not true, even if the donor is pre-hydrogen bonded. We propose that the second donor serves to stabilize the photogenerated anion. Therefore, the 'drive' for deprotonation is not entirely due to the excited state properties of quinoline and necessarily includes the solvation energy of the photogenerated products.

As for the intramolecular proton transfer case, the comparison to quinoline intermolecular proton transfer may only be partially valid. For intramolecular proton transfers, the geometry is often much more well-defined than this problem. Certain low frequency skeletal modes are often at play, especially those that are Franck–Condon active. Finally, the electronic structure of the molecule changes in very subtle ways that are different from this problem.

Natércia das Neves Rodrigues asked: Following from my last question, and given the observation that hydrogen bonding does not necessarily result in proton transfer, as may be intuitively expected, I was wondering if you had found some factors that generally tend to influence (hinder or enable) proton transfer? And can parallels be drawn between the intra- and inter-molecular cases?

Jahan Dawlaty responded: Some general and very rough guidelines that influence proton transfer in the excited state would be the following: (1) Does the excitation result into shifting of charge densities to favor proton transfer? This is discussed in the manuscript to some extent (DOI: 10.1039/c8fd00215k), and remains somewhat reasonable for intramolecular proton transfer case as well; (2) are there means of solvation for the photogenerated products? This may include hydrogen bonding by the neighboring molecules, as proposed by us; (3) for intramolecular proton transfer it is probably quite important to consider donor–acceptor distances and how Franck–Condon active modes affect these distances.

James Gaynor asked: Could you please comment on the connection between the energy difference of the donor and acceptor vibrational mode frequencies and the thermodynamic driving force for excited state proton transfer? Specifically, the OH stretch frequency of the alcohol donor and the formed NH stretch frequency of the photobase following proton transfer. As you've discussed, an intermolecular proton donor concentration sufficient for pre-association is important. This should also broaden and red-shift the OH stretch frequency of the proton donor due to intermolecular hydrogen bonding, which would presumably increase the difference in energy between the OH and NH frequencies.

Jahan Dawlaty responded: While the OH and NH vibrational frequencies are distinctly affected by hydrogen bonding both in the ground and excited states, it is hard to draw a quantitative link between the frequencies and the proton transfer drive. The reason is that the drive arises from a shift in the minima of the potential energy surfaces of protonated and unprotonated form, while vibrational frequencies reflect the curvatures of the potential energy surfaces at those minima. While the two are related at some level, I can not comment on a quantitative relationship between them.

Todd Martinez commented: In your previous responses, you have discussed both intermolecular and intramolecular proton transfer processes as though these involved the transfer of a proton. Yet there is computational evidence that excited state intramolecular proton transfer proceeds through a two step process where the skeletal bond lengths first rearrange (double bonds become single bonds and *vice versa*) and then a hydrogen atom transfers from the "proton donor" to the "proton acceptor".[1–3] Is there experimental evidence that these involve proton transfer as opposed to hydrogen atom transfer?

1 J. D. Coe and T. J. Martínez, *J. Phys. Chem. A*, 2006, **110**, 618–630.
2 J. D. Coe, B. G. Levine and T. J. Martínez, *J. Phys. Chem. A*, 2007, **111**, 11302–11310.
3 J. D. Coe and T. J. Martinez, *Mol. Phys.*, 2008, **106**, 537–545.

Jahan Dawlaty responded: This is a very fascinating and interesting question. In brief, going beyond the description of the excited state in terms of Lewis structures (*i.e.* double/single bonds), this interesting point may be rephrased as the influence of the symmetry of the orbitals in the excited state on proton/hydrogen atom transfer. That is, the symmetry of the orbitals in the excited state for ESIPT may necessitate that not a proton, but rather a hydrogen atom be transferred. This seems like a valid proposal that has computational support and could very well be true. However, I can not propose a spectroscopic experiment that will test this proposal directly. All ultrafast electronic spectroscopies will feel the changes in the electronic state of the entire molecule, and all ultrafast vibrational spectroscopies will not know whether a proton got transferred alone or along with a cloud of electrons (*i.e.* a hydrogen atom).

I may loosely anticipate that a direct test may be possible with an experiment that feels and measures the state of individual atoms in a molecule, rather than the entire molecule as a whole. Ultrafast core-level experiments may be an option. Scanning Tunneling Microscopy can do this, but does not have the time resolution.

In any event, with or without an experimental test, this is a very interesting idea, and actually calls into question whether we should call ESIPT processes by their name. Maybe they should be called Excited State Intramolecular Hydrogen Atom Transfer (ESIHAT).

Lastly, for the case of quinoline photobases presented in this study we are not dealing with intramolecular transfer. The process is intermolecular. The experimental signatures of quinolinium ion in the excited state is manifestly observed indicating that in this case we have an excited state proton transfer and not hydrogen atom transfer.

Todd Martinez followed: Can you compare and contrast excited state proton transfer in photoacids and photobases? Are there essential differences in the two cases?

Jahan Dawlaty replied: Some parallels can be drawn between them. For example, photoacidity in 5-substituted naphthol photoacids are affected by the 5-substituent Hammett parameter as previously studied by Nibering and others.[1] In direct parallel to that, photobasicity in 5-substituted quinolines (at least as inferred from the Förster analysis), also follows the same trend.[2] In the former, more electron withdrawing groups increase photoacidity, while in the latter, they decrease the photobasicity. This correlates with the charge density arguments that we have presented in the current work.

Another delegate in this meeting, Yuan-Chung Cheng from Nationa Taiwan University, has studied the electrostatic potential on oxygen atom in photoacids and has found that it is a good predictor of reported photoacidities (DOI: 10.1039/c8fd00205c).

A thorough computational and experimental comparison of photoacids and photobases in related families (*i.e.* naphthols and quinolines, or acridines and hydroxyanthracenes) may be necessary to reveal further similarities and differences.

1 M. Prémont-Schwarz, T. Barak, D. Pines, E. T. J. Nibbering and E. Pines, *J. Phys. Chem. B*, 2013, **117**, 4594–4603.
2 E. W. Driscoll, J. R. Hunt and J. M. Dawlaty, *J. Phys. Chem. Lett.*, 2016, 7, 2093–2099.

Trevor Roberts said: I appreciated how you mentioned one possible method for circumventing the preassociation concentration requirements (*i.e.* orders of magnitude more H-bond donors than photobase) is to covalently anchor the photobase to a system like an organometallic catalyst. I'm wondering, generally, if there are other strategies to overcome this large concentration difference between photobase and proton donors. Have photobases ever been incorporated in more rigid media, like polymers? I ask because controllably quenching the propagation of protons in lithographic photoresists (from photoacid generators) is still a very open and sought after question. Could incorporation of photobases in the backbone of the polymer allow for a "solid state" solvation of sorts to facilitate the photobasicity? Anyway, many thanks for the talk and very interesting insight.

Jahan Dawlaty answered: That is an excellent question. Addition of pendant acid/base moieties to organometallic catalysts is quite common. However, addition of optically driven pendant acids or bases with the explicit intention of converting photoexcitation to proton motion was not reported. Along with a collaborator, we have recently published a paper that shows tethering of a quinoline photobase to an iridium catalyst.[1] However, we have not been able to prove the reactivity of the pendant photobase for a catalytic reaction. The topic remains open for further research.

As for polymers, photoacids have been incorporated within polymer backbones with the intention of using them for generating optically induced proton populations. We have previously used unbound photoacids within a matrix to modulate proton conductivity of that matrix.[2] However, I do not know of any work in which photobases are tethered to a polymer backbone.

Control over both release and uptake of protons with light within a polymer is likely to be relevant for many applications. The one distinction that one must remember when contrasting this work with photolithography is that in the latter the reactions are often intentionally irreversible. In quinolines (as with traditional photoacids), the major chemistry that happens in the excited state is deprotonation/protonation which is reversed upon return to the ground state.

1 I. Demianets, J. R. Hunt, J. M. Dawlaty and T. Williams, *Organometallics*, 2019, **38**, 200–204.
2 S. Haghighat, S. Ostresh and J. M. Dawlaty, *J. Phys. Chem. B*, 2016, **120**, 1002–1007.

Sharon Hammes-Schiffer commented: These types of reactions could be viewed as proton-coupled electron transfer (PCET) because they involve significant electron rearrangement along with proton transfer. The gating that Todd Martinez mentioned, which involves bond framework rearrangement, is related to inner-sphere reorganization in Marcus theory and is also related to solvent reorganization in Marcus theory. In photoinduced PCET, we have found that solvent reorganization can gate proton transfer following photoexcitation, analogous to the gating involving bond framework rearrangement. To address the question of whether this reaction is proton transfer or hydrogen atom transfer, it is probably somewhere in between because the electron is unlikely to move exactly with the proton, which would be a fully electronically adiabatic process. We can define the initial and final states in terms of chemical bonding and electronic structure, but it is difficult to determine the path connecting these two states. The electron and proton could even follow different paths. It is difficult to know which moves first, the electron or the proton, or whether they move concertedly, or somewhere in between. This is a chicken and egg problem, and it is challenging to distinguish cause and effect. Calculations that treat the proton, as well as the electrons, quantum mechanically may provide insights into how electron and proton densities move along the reaction path. The main point to keep in mind is that one picture will not cover all cases of photoinduced proton transfer or PCET because these are complex processes exhibiting many subtle effects.

Jahan Dawlaty responded: I fully agree with this assessment. The reaction coordinate for these processes are quite subtle and as you mentioned may not be fully described by the adiabatic motion of electrons with protons. After further discussion with Todd Martinez, I understood that he is alluding to yet another subtlety in excited state intra-molecular proton transfer (ESIPT), and that is the symmetry of orbitals. Based on his proposal, the symmetry of orbitals in the excited state for ESIPT cases may necessitate that not a proton, but rather a hydrogen atom be transferred. This is quite an interesting proposal. However, I can not propose a spectroscopic experiment that will test this proposal directly. All ultrafast electronic spectroscopies will feel the changes in electronic state of the entire molecule, and all ultrafast vibrational spectroscopies will not know whether a proton got transferred alone or along with a cloud of electrons. I loosely conjecture that a direct test may be possible with an experiment that feels and measures the state of individual atoms in a molecule, rather than the entire molecule as a whole. Ultrafast core-level experiments may be an option.

Nadia Korovina asked: After the excited state proton transfer, when the quinone relaxes to the ground state, does it retain the proton that it acquired in the excited state?

Jahan Dawlaty replied: No. It will release the proton back at the earliest opportunity upon returning to the ground state. We do not have a convenient way to measure the time-scale of that event.

Tom Oliver said: In the excited state photobase-acid reactions you have investigated, are any of the immediate photoproducts produced in the electronically excited states? And if so, how stable are these species?

Jahan Dawlaty replied: The photoproducts in this process are a quinolinium cation and an alkoxide anion. Upon return to the ground state, the photoproducts reverse back to quinoline and alcohol. Solvation of the alkoxide by other alcohols, based on our analysis, seems to be necessary for proton transfer.

If you are referring to the irreversible products of photoexcitation, we have not investigated them. There is reason to believe that there are some irreversible products (from our perspective, photo-damage to quinoline) that arise from electron transfer to quinolinium and further modification of the rings. These are lower yield channels to our knowledge and have been studied by others.

Bern Kohler commented: I would like to comment that light-enhanced acidity and basicity do not always require a photoexcited molecule with a proton donating or accepting site. While this is the case for Brønsted photoacids and photobases, the alternative framework of Lewis in which an acid (base) accepts (donates) an electron pair can relax this constraint. For excited states, one might prefer the perspective of Mulliken that ascribes acid/base properties to partial charge transfers. A Lewis or Mulliken photoacid can promote the hydrolysis of a nearby water molecule and some examples involving strongly oxidizing excited states have been described.[1,2]

1 J. D. Henrich, S. Suchyta and B. Kohler, *J. Phys. Chem. B*, 2015, **119**, 2737–2748.
2 R. A. Danforth and B. Kohler, *Chem. Phys. Lett.*, 2017, **683**, 315–321.

Jahan Dawlaty replied: Yes, I agree with this assessment. As an additional comment, we have paired our quinoline photobases with a Lewis acid (tris-pentafluorophenyl borane). We have observed that the emission of quinoline red-shifts in the same manner as when protonated. We also have some very preliminary computational evidence that upon electronic excitation of a quinoline-Lewis acid adduct, the Lewis acidic boron is drawn closer towards the photobase in analogy to a proton.

Conflicts of interest

There are no conflicts to declare.

Faraday Discussions

PAPER

Quantum interferences among Dexter energy transfer pathways†

Shuming Bai,[a] Peng Zhang,[a] Panayiotis Antoniou,[b] Spiros S. Skourtis[b] and David N. Beratan[acd]

Received 14th January 2019, Accepted 21st January 2019
DOI: 10.1039/c9fd00007k

Dexter energy transfer in chemical systems moves an exciton (*i.e.*, an electron–hole pair) from a donor chromophore to an acceptor chromophore through a bridge by a combination of bonded and non-bonded interactions. The transition is enabled by both one-electron/one-particle and two-electron/two-particle interaction mechanisms. Assuming that there is no real intermediate state population of an electron, hole, or exciton in the bridge, the transport involves two states that are coupled non-adiabatically. As such, coherent quantum interferences arise among the Dexter energy coupling pathways. These interferences, while related to well understood interferences in single-electron transfer, are much richer because of their two particle nature: the transfer of a triplet exciton involves the net transfer of both an electron and a hole. Despite this additional complexity, simple rules can govern Dexter coupling pathway interferences in special cases. As in the case of single-electron transfer, identical parallel coupling pathways can be constructively interfering and may enhance the Dexter transfer rate. Because of the virtual particle combinatorics associated with two-particle superexchange, two parallel Dexter coupling routes may be expected to enhance Dexter couplings by more than a factor of two. We explore Dexter coupling pathway interferences in non-covalent assemblies, employing a method that enables the assessment of Dexter coupling pathway strengths and interferences, in the context of one-particle and two-particle coupling interactions.

1. Introduction

"Any other situation in quantum mechanics, it turns out, can always be explained by saying, 'You remember the case of the experiment with two holes? It's the same thing'." – R. P. Feynman, *The Character of Physical Law*, 1965, p. 130.

[a]Department of Chemistry, Duke University, Durham, NC 27708, USA
[b]Department of Physics, University of Cyprus, 1678 Nicosia, Cyprus
[c]Department of Biochemistry, Duke University, Durham, NC 27710, USA
[d]Department of Physics, Duke University, Durham, NC 27708, USA

† Electronic supplementary information (ESI) available. See DOI: 10.1039/c9fd00007k

The directed flow of electronic excitation energy through molecules and their assemblies is required for the efficient capture and conversion of solar energy in living and man-made structures.[1,2] Excitation energy transport to a photocatalytic site, or to an interface that generates free carriers, allows solar energy to be stored in forms that are less ephemeral than electronic excited states. The mechanism of electronic excitation energy transfer depends on the spin states of the chromophores. When the excited energy donor and ground state acceptor species have different spins, the process occurs by the so-called Dexter mechanism. The Dexter coupling originates in the exchange interaction between donor and acceptor and, as such, is mediated by through-bond and through-space couplings. This contrasts to the case of dipole–dipole coupling for the more familiar case of spin-allowed Förster energy transfer.[2,3]

Here, we explore the mechanism of bridge-mediated Dexter energy transfer. Triplet energy transfer is the most familiar example of a Dexter process, *i.e.*, $[^3D^{*1}A] \rightarrow [^1D^3A^*]$. Here, D and A represent the energy donor and acceptor species, respectively. Dexter transfer may appear to be spin forbidden, since both D and A undergo a spin change, but it is not. The *overall* spin is conserved, and the reaction is allowed. While Dexter energy transfer reactions are ubiquitous, the molecular features that control the DA coupling, and thus the energy transfer kinetics, are poorly understood. The changes in D and A local spin states cause Dexter transfer to be electric-dipole forbidden, disabling the Förster mechanism that underpins fluorescence resonant energy transfer (FRET), a workhorse mechanism used as a "molecular ruler" at the nanoscale. Strategies to control Dexter energy transfer rates are poorly developed in comparison with Förster mechanisms, despite the fact that Dexter mechanisms are potentially much richer in their molecular structure dependence. Fig. 1 summarizes some of the energy capture and conversion processes that rely on Dexter-mechanism energy transfer.

Here, we develop a framework to explore the strength and quantum mechanical interferences among Dexter coupling pathways. Dexter couplings are similar to the couplings that arise in bridge-mediated electron transfer.[4] However, the landscape for Dexter energy transfer is complicated by the transit of two particles between donor and acceptor.

The theoretical methods used here, and the molecular insights thus derived, may prove useful in the design and interpretation of light harvesting structures, as well as for developing phosphorescence resonant energy transfer probes of distance and structure at the nanoscale.

1.1 Dexter energy transfer

In a weak donor–acceptor electronic coupling regime, the initial and final (diabatic) states are well localized, and the anti-symmetrized approximate initial (D*A) and final (DA*) triplet states are indicated in eqn (1a) and (1b).[4] The operator $V = 1/r_{12}$ couples these states and enables energy transfer. The first term in eqn (2), through a multipole expansion, produces the Förster interaction (nonzero when D and D* have the same spin, and A and A* have the same spin).[5]

$$|\Psi_{\rm I}\rangle = \frac{1}{\sqrt{2}}[|{\rm D}^*(1){\rm A}(2)\rangle - |{\rm D}^*(2){\rm A}(1)\rangle] \tag{1a}$$

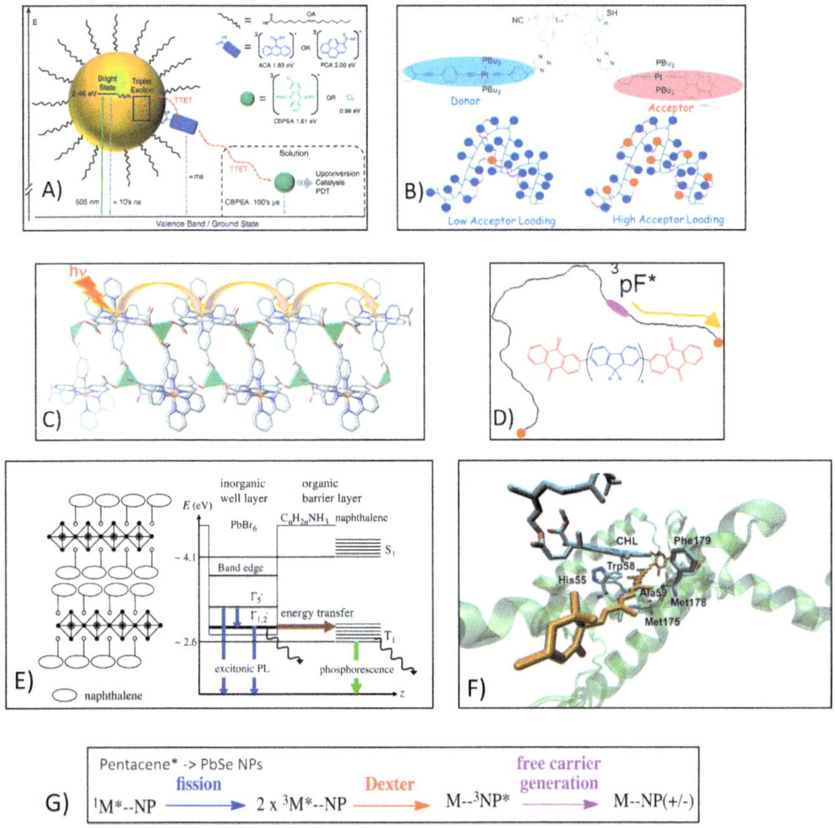

Fig. 1 Examples of Dexter energy transfer reactions in energy capture and conversion structures. (A) Triplet–triplet energy transfer between a nanoparticle (NP) and organic chromophores at the NP surface and with a second acceptor species in solution; (B) exciton migration and trapping in polymer networks consisting of Pt chromophores;[17] (C) multi-step exciton hopping among Ru chromophores in a metal–organic framework structure with lightly doped Os exciton traps;[13] (D) triplet exciton hopping over 100 nm along a *single* organic polymer chain, consisting of organic fluorene chromophores with traps at chain termini;[18] (E) Wannier to Frenkel exciton conversion in perovskites *via* Dexter energy transfer between high spin $PbBr_6$ donors and naphthalene-derivative acceptors;[19] (F) leutine 2 (conjugated structure, drawn in orange) binds to the light harvesting complex II (LHCII) and quenches triplet Chl excited states *via* Dexter energy transfer, serving a photo-protectant;[20] (G) Dexter energy transfer enables the injection of triplet excitons from small molecules to nanoparticles following exciton fission. Transport in the nanoparticle itself involves Dexter transfer, finally leading to free carrier generation.[12] Figure adapted from cited references.

$$|\Psi_F\rangle = \frac{1}{\sqrt{2}}[|D(1)A^*(2)\rangle - |D(2)A^*(1)\rangle] \tag{1b}$$

$$\langle\Psi_I|V|\Psi_F\rangle \approx \langle D^*(1)D(1)|V|A(2)A^*(2)\rangle - \langle D^*(1)A^*(1)|V|D(2)A(2)\rangle \tag{2}$$

Förster interactions decay with the inverse cube of the DA distance, while the Dexter (exchange) term has an approximately exponential distance dependence.[6]

Since Dexter's pioneering analysis of this problem in 1953, it was found that the second-order application of the one-electron interaction operator couples D*A and DA* states, also with an approximately exponential distance dependence.

When D* and A are linked chemically, the Dexter coupling is expected to be enhanced as the barrier for neutral and charge-transfer exciton propagation is lowered. Dexter transfer is mediated by both virtual exciton states of the bridge (both electron and hole on bridge) and by charge-transfer states (just one virtual particle on the bridge or acceptor). The virtual states can be created by the $1/r_{12}$ operator or by repeated action of the one-electron Hamiltonian operator. We will explore both sources of DA couplings, coupling pathways, and coupling pathway interferences.

1.2 Dexter energy transfer in energy science

Natural and synthetic light-harvesting structures employ Dexter energy transfer. Synthetic solar energy harvesting and concentrating schemes based on transition metals (*e.g.*, ruthenium and osmium),[7,8] lanthanides,[9] quantum dots (*e.g.*, PbSe and CdSe),[10–12] organic and organometallic polymers, molecule–nanoparticle junctions, and metal–organic framework materials (MOFs)[13,14] provide examples (see Fig. 1). In biology, Dexter energy transfer degrades potentially destructive triplet excited states that can otherwise produce oxidative damage.[15] Dexter energy transfer also enables electronic excitation energy up-conversion in some systems, a process that creates high energy electronic excited states from multiple lower-energy excitons.[16]

Coupling pathway analysis based on Green's function methods and the CIS formalism were used recently to explore the contributions of bridge exciton and charge-transfer virtual states to Dexter transfer.[3] Those studies found that the triplet–triplet Dexter coupling contains contributions from bridge exciton intermediates that had been ignored in previous theories (*vide infra*). It was found that virtual exciton states may provide most of the triplet–triplet coupling for long bridges or bridges with energetically low-lying excited states. The bridge excitons can contribute to one-electron and two-electron pathway components. Ref. 21 describes simple analytical models for bridge exciton and charge-transfer mediated pathways, allowing the study of quantum interferences among Dexter paths.

1.3 Dexter paths, electron transfer paths, and quantum interferences

Dexter transport is mediated by bridge virtual exciton states ($B^{+/-} = B_1{}^{+/-}B_2$, $B_1B_2{}^{+/-}$, $B_1{}^{+}B_2{}^{-}$ or $B_1{}^{-}B_2{}^{+}$ states in a two-unit bridge) *and* by charge-transfer (CT) exciton states (D^+B^-A, D^-B^+A, D^+BA^-, *etc.*). Fig. 2 and 3 show schematic representations of virtual states and coupling pathways in DBA structures. When B → B*, D → D* and A → A* transition energies are close in energy, virtual bridge excitons contribute strongly to the Dexter coupling. D^+A^- charge transfer excitons dominate when the B → B* transition energies are of much higher energy than the D → D* and A → A* excitation energies. However, there is a competition between the energy cost required to accomplish electron–hole separation (disfavoring the formation of CT states) and the energy cost to inject both the electron and the hole to the bridge (disfavoring the formation of B* ($B^{+/-}$) states).

An early model of Closs, Miller, Piotrowiak, Johnson, MacInnis, and Fleming described the Dexter coupling as a product of electron and hole superexchange

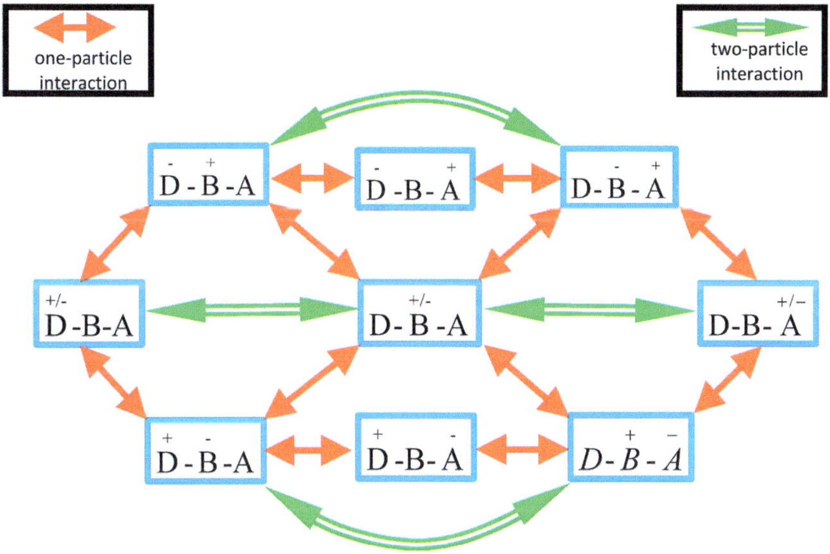

Fig. 2 Schematic summary of the kinds of interaction that cause Dexter energy transfer in a bridged system from D*BA to DBA*. The action of the one-particle Hamiltonian operator on an excitonic state can propagate either an electron or a hole but not both. The action of a two-particle interaction Hamiltonian operator on an excitonic state simultaneously propagates an electron and a hole. In our pathway analysis, these interactions are computed as couplings between initial and final diabatic states of the D, B, and A species, obtained using the constrained DFT method.

interactions, valid when: (1) the bridge excitation energy is much larger than the D and A excitation energies and (2) the one-electron coupling operator dominates the propagation mechanism.[22,23] These early theoretical and experimental studies, and later seminal studies of Scholes, Harriman, Albinsonn, Speiser, and Baldo, established a semi-quantitative linkage between bridge-mediated electron-transfer and Dexter energy-transfer reactions.[4,15,22,24–27]

2. Early links between bridge-mediated Dexter energy transfer and electron transfer

The Dexter coupling approximation of Closs *et al.* is:[22,23]

$$H_{\mathrm{DA}}^{\mathrm{Dexter}} \propto V^{\mathrm{e}} \times V^{\mathrm{h}}/\Delta E_{\mathrm{CT}} \tag{3}$$

The electron- and hole-mediated superexchange interactions (defined as V^{e} and V^{h}, respectively) are one-electron transfer couplings. This approximation captures charge-transfer virtual states, neglecting bridge exciton virtual states and two-electron (Coulomb) mediated interactions. The values of $V_{\mathrm{DA}}^{\mathrm{e}}$ and $V_{\mathrm{DA}}^{\mathrm{h}}$ are the empirical electron- and hole-mediated superexchange interactions, and ΔE_{CT} is the average energy gap between the D*BA excited state and the charge-transfer states, $D^{+}BA^{-}$ and $D^{-}BA^{+}$. This formulation was validated in experiments that compared electron and energy transfer rates in rigid donor–bridge–acceptor structures with large B $\rightarrow$ B* excitation energy gaps.

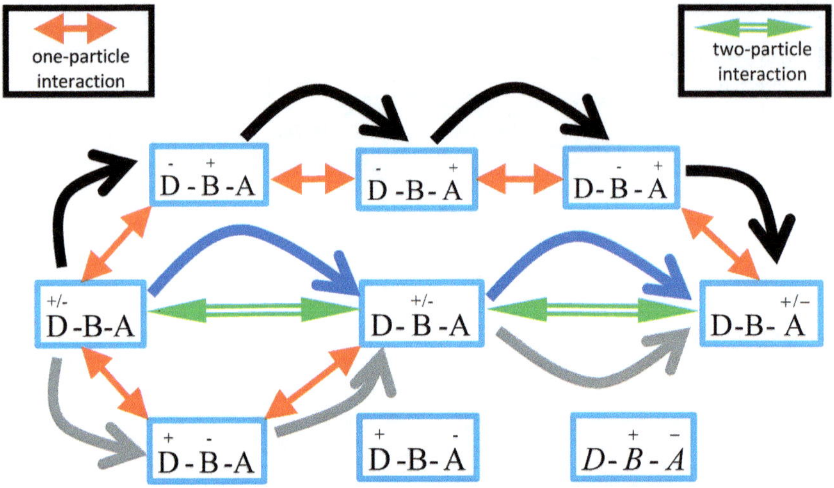

Fig. 3 Schematic representation of Dexter coupling pathways in a simple DBA system. One sided arrows represent single steps within a Dexter pathway from D*BA to DBA*. The black one-sided arrows demark a four-step pathway with a DA charge transfer exciton intermediate (D^-BA^+) mediated solely by one-particle interactions. The blue one sided arrows indicate a two-step pathway with a bridge exciton intermediate ($DB^{+/-}A$) mediated solely by two-particle interactions. The gray one sided arrows denote a three-step pathway with a bridge exciton intermediate ($DB^{+/-}A$ = DB*A) mediated by one particle interactions (first two steps) and a two-particle interaction (last step).

2.1 Coupling pathways for electron transfer

Before addressing Dexter couplings and their dependence on bridge structure in more detail, we revisit the case of electron or hole transfer though a single linear pathway and through two parallel paths.

2.1.1 Linear DBA pathways. Assuming only nearest-neighbor interactions between sites, the lowest-order contribution to the bridge-mediated electron (hole) transfer coupling in the linear chain indicated in Fig. 4 is $V_{DB}^{e(h)}V_{BA}^{e(h)}/\Delta E^{e(h)}$. $\Delta E^{e(h)}$ is the energy gap between the tunneling energy (for electron or hole transfer) and the mediating bridge state. The D and A state energies are assumed to be equal at the ET transition state.

2.1.2 Parallel DBA pathways. In the case of two parallel coupling routes with the same coupling interaction from D to the upper (U) and lower (L) bridge orbitals (Fig. 5), the coupling is twice as strong as the interaction for the linear

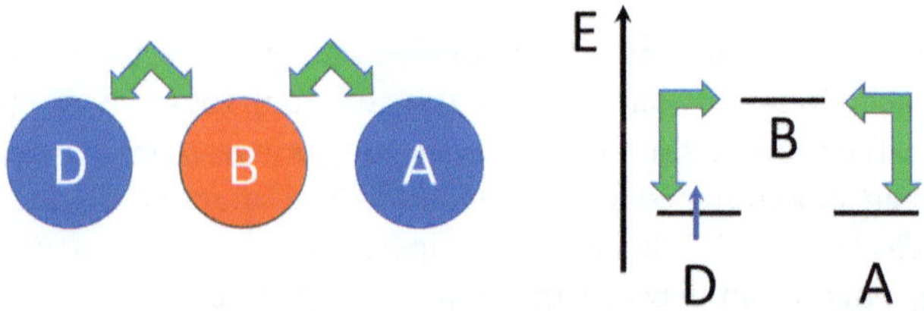

Fig. 4 Schematic representation of a bridge-mediated donor–acceptor interaction in a linear DBA chain. The lowest order superexchange coupling is $V_{DB}V_{BA}/\Delta E$ where V_{DB} and V_{BA} are the one-electron couplings between the D–B and B–A orbitals, respectively.

chain of Fig. 4.[28] This doubling is expected when direct through-space DA interactions are weaker than the through-bond interactions, and when the value of V_{DB} and V_{BA} are equal in the two systems.

Recent studies also probed (theoretically) the effective coupling for two-electron transfer from $D^{=}A$ to $DA^{=}$, indicated in Fig. 6. The lowest order superexchange pathway for two-electron transfer generated by the one electron operator has the strength $-\frac{2V^4}{\Delta(\varepsilon_B + \Delta)^2} - \frac{-4V^4}{2\varepsilon_B(\varepsilon_B + \Delta)^2}$ (see Fig. 6 for a definition of the symbols).[29] Two-electron superexchange interactions were not considered as they are assumed much weaker (proportional to squares of orbital overlaps rather than terms linear in overlaps).

2.2 Dexter coupling pathways

What are the structure–function relationships that govern the Dexter mechanism? Fig. 7 shows the qualitative energy landscape that governs bridge-mediated Dexter energy transfer. Here, d and d*, b and b*, and a and a* denote HOMOs and LUMOs of the donor, bridge and acceptor fragments, respectively.

The simplest orbital description of a triplet donor exciton places a hole in the $|d\rangle$ orbital and an electron in the $|d^*\rangle$ orbital ($D^{+/-}$ configuration). The exciton can shift to the acceptor, producing a hole in $|a\rangle$ and an electron in $|a^*\rangle$ ($A^{+/-}$ configuration). The donor–acceptor charge-transfer (DA/CT) exciton intermediate states (D^+BA^- and D^-BA^+) define the one-particle contribution that was emphasized in early studies of Closs *et al.*, and more recently in the analysis of Scholes and co-workers.[26,30] To the best of our knowledge, the bridge exciton (BE) term was not explored in detail prior to our recent study.[3,21] The BE term arises from bridge exciton virtual intermediates (DB*A), so the full Dexter coupling may be partitioned into three terms:

$$V_{\text{triplet}} \approx 2 \times \frac{V^e \times V^h}{\Delta E_{CT}} + \text{BE} - (\text{DA}|\text{D*A*}) \qquad (4)$$

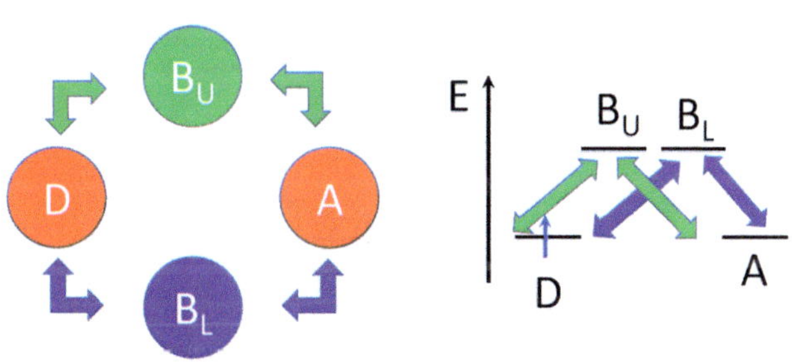

Fig. 5 Schematic representation of a bridge with two equivalent parallel coupling pathways. The lowest-order contribution to the bridge-mediated superexchange coupling is $2V_{DB}V_{BA}/\Delta E$, where V_{DB} is the donor-bridge one-electron coupling, V_{BA} is the bridge-acceptor one-electron coupling and ΔE is the tunneling energy gap at the transition state for electron transfer. This enhanced coupling produces an electron or hole transfer rate that is four-times the rate in the linear bridge, since the golden rule rate is proportional to the squared coupling.

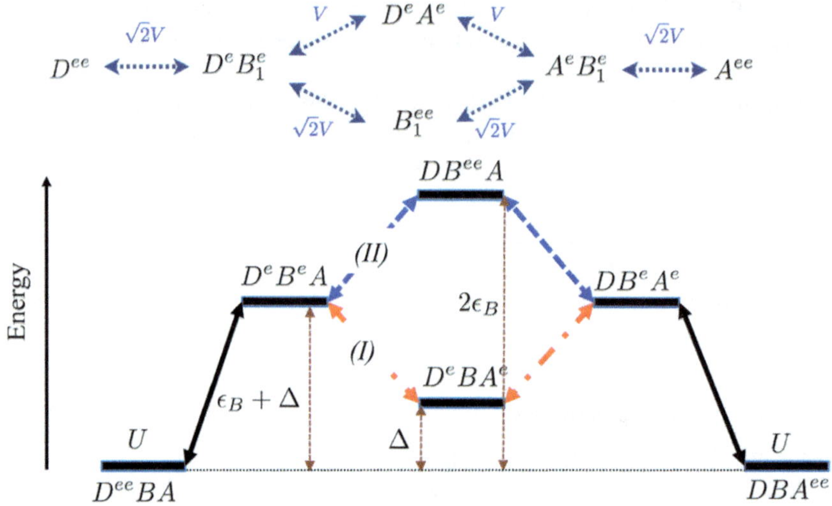

Fig. 6 Schematic representation of a linear two-electron transfer pathway. The two-electron nature of the transition creates interferences between pathways with a doubly reduced bridge virtual state and a D → A charge transfer intermediate state. In this figure, e represents the positions of the mobile electrons (D^{ee} represents the doubly reduce donor, D^e represents the singly reduced donor, and D represents the fully oxidized donor; analogous notation is used for B and A).

The third term on the right side of eqn (4) is the direct two-particle exchange coupling:

$$(\mathrm{DA}|\mathrm{D^*A^*}) = \iint d^3r_1 d^3r_2 \phi_\mathrm{D}(r_1)\phi_\mathrm{A}(r_1){r_{12}}^{-1}\phi_{\mathrm{D^*}}(r_2)\phi_{\mathrm{A^*}}(r_2) \tag{5}$$

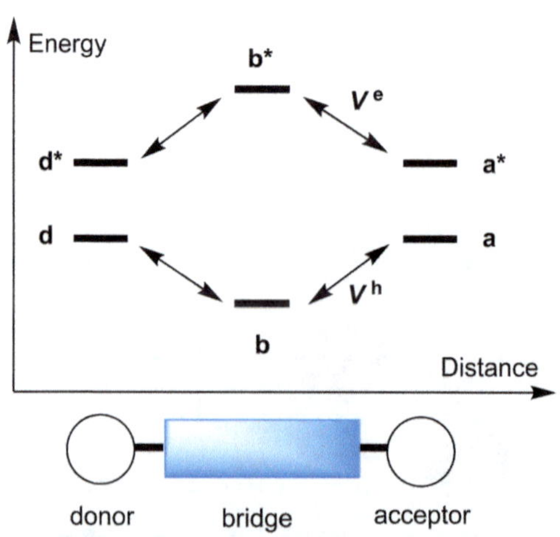

Fig. 7 Electronic orbital landscape for Dexter coupling through-bond. V_h and V_e denote one-particle (electron and hole, respectively) Hamiltonian matrix elements between the fragment HOMOs and LUMOs, respectively. These electronic couplings produce bridge-mixed HOMOs and LUMOs that are donor centered ($|\mathrm{D}\rangle \approx |\mathrm{d}\rangle + |\delta_\mathrm{D}^{\mathrm{br}}\rangle$ and $|\mathrm{D^*}\rangle \approx |\mathrm{d^*}\rangle + |\delta_\mathrm{D}^{\mathrm{*br}}\rangle$, respectively) and acceptor-centered ($|\mathrm{A}\rangle \approx |\mathrm{a}\rangle + |\delta_\mathrm{A}^{\mathrm{br}}\rangle$ and $|\mathrm{A^*}\rangle \approx |\mathrm{a^*}\rangle + |\delta_\mathrm{A}^{\mathrm{*br}}\rangle$, respectively).

Hartree–Fock single configuration interaction analysis and pathway analysis found that the BE contribution can be larger than the CT/DA contributions for long bridges or for bridges with low excitation energies.[3,21] The contributions that arise from bridge exciton interactions can soften the approximately exponential distance decay of the Dexter coupling.[21]

In summary, either the CT or BE contributions may dominate the Dexter coupling, depending on the molecular structure.[3,21] These findings were based on a CIS formalism, where the one and two-electron interactions (beyond Hartree–Fock) are separated in the Hamiltonian. In the analysis below, we will use constrained DFT (CDFT) methods to analyze bridge-mediated Dexter interactions as this simplifies the pathway analysis.

3. Assessing Dexter pathway interferences in non-covalent assemblies

Fig. 8 shows three virtual states that contribute to exciton migration through a DBA system: the bridge exciton (DB*A = $DB^{+/-}A$), the DB and BA electron/hole charge transfer states (*e.g.*, D^+B^-A/D^-B^+A), and the DA charge transfer (CT) exciton state (D^+BA^-/D^-BA^+).

We next explore Dexter coupling and transport in model van der Waals structures, using naphthalene donors and acceptors with single and double (parallel) ethylene bridges. The aim of these studies is to explore quantum interference among the Dexter coupling pathways.

The diabatic states related to Dexter energy transfer are constructed directly using constrained density functional theory (CDFT),[31–33] which is implemented in the QCHEM 4.4 quantum chemistry package.[34] In all of the CDFT calculations, charge and spin density constraints on each fragment (D/B/A) are enforced, and the lowest energy state subject to the density constraint is computed. With the Kohn–Sham wave function, the configuration interaction CDFT method (CDFT-CI) is used to calculate the diabatic couplings,[32] where the one- and two-electron matrix elements are reformed using the KS orbitals.[32] The DFT calculations were performed with the PBE functional[35] and the 6-31+G(d) basis set.[36] The Dexter coupling in stacked naphthalene–naphthalene dimers was computed and compared with results from the fragment spin difference (FSD) method,[37] and the excellent performance confirms the validity of the CDFT-CI approach (see Fig. S3 of the ESI†).

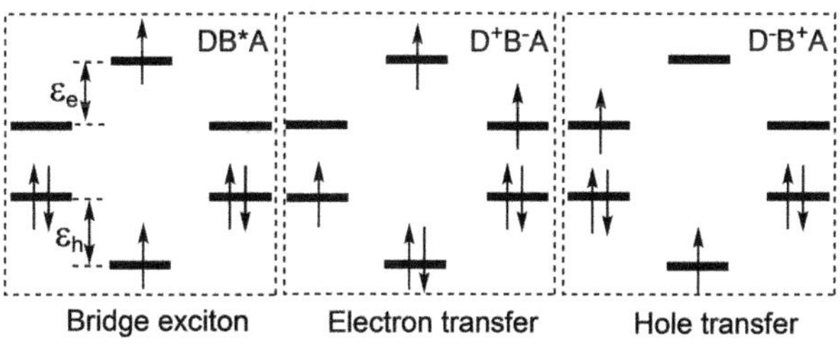

Fig. 8 Electronic configurations of bridge exciton and charge transfer (electron and hole) excitons.

Here, we examine Dexter energy transfer in the deep tunneling regime (*i.e.*, there is a very large energy gap between the naphthalene and the ethylene electronic excitation energies), and the donor–acceptor coupling is computed using a Green's function approach.[3,38]

We define the donor-bridge one-particle electron and hole interactions in the CDFT methodology, for example in a DBA system when the e or h transfers from D to B the couplings are defined as:

$$V^{e}_{DB} = \langle D^{+/-}BA|\hat{H}|D^{+}B^{-}A\rangle \tag{6a}$$

$$V^{h}_{DB} = \langle D^{+/-}BA|\hat{H}|D^{-}B^{+}A\rangle \tag{6b}$$

All other one-particle interactions are constructed by analogy with these terms.

We define the two-particle interactions among diabatic states in the CDFT methodology, for example the two-particle interaction between states $D^{+}B^{-}A$ and $DB^{+}A^{-}$ is:

$$V^{eh}_{D^{+}B^{-}A/DB^{+}A^{-}} = \left\langle D^{+}B^{-}A\middle|\hat{H}\middle|DB^{+}A^{-}\right\rangle \tag{7}$$

All other two-particle interactions are constructed by analogy with these terms. As a special case, we refer to the two-particle interaction between the initial and final states as the direct coupling ($D^{*} = D^{+/-}$, $A^{*} = A^{+/-}$):

$$V_{direct} = V^{eh}_{D^{*}BA/DBA^{*}} = \langle D^{*}BA|\hat{H}|DBA^{*}\rangle \tag{8}$$

The geometries of the molecular assemblies are shown in Fig. 9. We begin with three naphthalene dimer structures, two in a "sandwich" geometry, and one in a "zigzag" geometry (see Fig. 9a–c). Then, we introduce one ethylene molecule between the naphthalenes to produce D–B–A structures with a stacked or an offset architecture (Fig. 8d and e). Finally, we introduce two ethylene bridging molecules to produce the D–B_U/B_L–A structure with two parallel pathways (Fig. 8f).

3.1 Pathway interference in the D–A structure

In the D–A structure, Dexter coupling is limited to one-particle (1p) pathways involving the charge transfer (CT) exciton states ($D^{+}A^{-}/D^{-}A^{+}$) and the two-particle (2p) pathway involving the exchange coupling. We calculated all of the state energies and couplings among the diabatic states associated with the three pathways, and the results are shown in the ESI.† With these computed interaction elements, the effective donor–acceptor couplings were computed using a Green's function approach (see Table 1).

The Nap–Nap dimer at 3.5 Å is shown in Fig. 9a. Because of the short Nap–Nap distance and the mirror symmetry, both the one-electron and two-electron interactions are larger than 0.1 eV, although the one-electron coupling is larger. The one-electron terms in eqn. (4) gives:

$$V^{eff}_{1e} = -2\frac{V^{h}_{DA} \times V^{e}_{D^{*}A^{*}}}{\Delta E_{CT}} = -2\frac{0.67 \times 0.60}{2.3}\ eV = -0.35\ eV \tag{9}$$

The direct two-particle coupling interaction (V_{direct}) is −0.14 eV, which is somewhat less than the one-electron coupling.

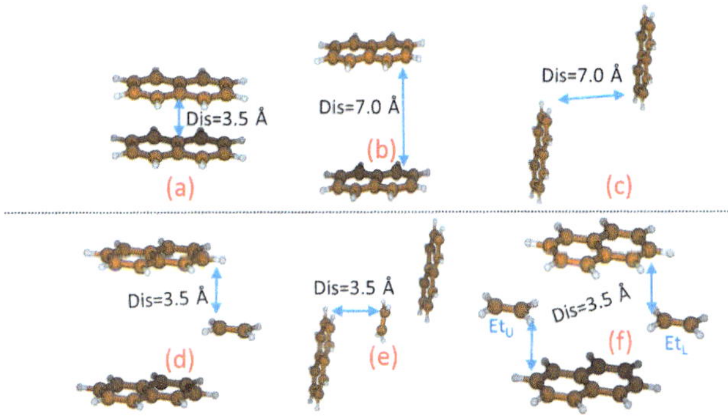

Fig. 9 The geometries of the structures containing naphthalene (Nap) and ethylene (Et): (a) the sandwich Nap–Nap dimer with a 3.5 Å distance; (b) the sandwich Nap–Nap dimer with a 7.0 Å distance between ring planes; (c) zigzag Nap–Nap dimer with 7.0 Å between ring planes; (d) sandwich Nap–Et–Nap complex; (e) zigzag Nap–Et–Nap complex; (f) sandwich-Nap–Et_U/Et_L–Nap complex.

Table 1 Calculated couplings of the molecular systems of Fig. 9: the direct (d) coupling for electron transfer (V_{ET_d}); the bridge mediated coupling for electron transfer (V_{ET_br}); the direct coupling for hole transfer (V_{HT_d}); the bridge mediated coupling for hole transfer (V_{HT_br}); the sum of couplings for Dexter transfer from one electron bridge mediated pathways (V_{1e}^{eff}); the sum of couplings for bridge-mediated Dexter transfer (V_{br}^{eff}); the direct coupling for two-particle Dexter coupling (V_{direct}); the total effective Dexter coupling (V_{total}^{eff}). Direct one-particle couplings are computed using eqn (6); direct two-particle couplings are computed using eqn (8). The bridge-mediated couplings are computed using Green's function methods described in ref. 3

Structure	Sandwich Nap–Nap Fig. 8a	Sandwich Nap–Nap Fig. 8b	Zigzag Nap–Nap Fig. 8c	Sandwich Nap–Et–Nap Fig. 8d	Zigzag Nap–Et–Nap Fig. 8e	Sandwich Nap-$Et_U Et_L$Nap Fig. 8f
V_{ET_d} eV	6.0×10^{-1}	4.6×10^{-3}	1.8×10^{-3}	3.4×10^{-3}	7.1×10^{-4}	-2.3×10^{-3}
V_{ET_br}/eV	-7.4×10^{-2}	1.8×10^{-9}	-5.8×10^{-10}	-2.1×10^{-3}	-1.9×10^{-3}	3.4×10^{-3}
V_{HT_d}/eV	6.7×10^{-1}	2.4×10^{-3}	3.5×10^{-4}	-4.5×10^{-3}	-3.4×10^{-4}	-7.0×10^{-3}
V_{HT_br}/eV	-6.8×10^{-2}	3.4×10^{-9}	-3.0×10^{-9}	-3.0×10^{-3}	-4.1×10^{-3}	-5.8×10^{-3}
V_{1e}^{eff}/eV	-3.5×10^{-1}	-6.9×10^{-6}	-3.4×10^{-7}	-6.0×10^{-6}	3.3×10^{-6}	-1.2×10^{-5}
V_{br}^{eff}/eV	-4.1×10^{-1}	-6.9×10^{-6}	-3.4×10^{-7}	-1.4×10^{-7}	3.1×10^{-5}	-6.0×10^{-5}
V_{direct}/eV	-1.4×10^{-1}	-1.1×10^{-5}	-5.6×10^{-8}	1.1×10^{-5}	5.4×10^{-6}	8.2×10^{-6}
V_{total}^{eff}/eV	-5.5×10^{-1}	-1.8×10^{-5}	-4.0×10^{-7}	1.1×10^{-5}	3.6×10^{-5}	-5.2×10^{-5}

The couplings from the one-electron pathways and direct two-particle interaction, V_{1e}^{eff} and V_{direct}, are the same order of magnitude in the other two Nap–Nap dimers at larger distances (Fig. 8b and c). V_{direct} can be larger than V_{1e}, so the direct coupling mechanism may not generally be ignored. Although each one-electron interaction is always larger than the two-electron interactions, the V_{1e} Dexter coupling is the product of two one-electron interactions divided by ΔE_{CT}. As a result, the calculated V_{1e}^{eff} Dexter coupling from eqn (9) is the same order of magnitude as V_{direct} in these systems.

At 7 Å (Fig. 8b and c), the couplings decrease substantially. The electron transfer coupling is reduced by a factor of 10^3, while V_{1e}^{eff} for Dexter transfer is decreased by 10^6. When we slide the Nap–Nap dimer into a zigzag geometry, the coupling decrease because of the weakened orbital overlap.

There is quantum interference between the two one-electron Dexter coupling pathways in the Nap–Nap structure ($D^*A \rightarrow D^+A^- \rightarrow DA^*$ and $D^*A \rightarrow D^-A^+ \rightarrow DA^*$). The difference between V_{1e}^{eff} and V_{br} arises from quantum interferences among the couplings (see eqn (6a), (6b) and (7)). The coupling between the two CT states enhances the bridge-mediated coupling by 0.06 eV. We can understand this constructive interference from two different perspectives.

When the coupling $V_{D^+A^-/D^-A^+}^{eh}$ is non-zero, two adiabatic mixed CT states are generated (from a mixture of D^-A^+ and D^+A^- states). The symmetric combination couples the donor and acceptor states effectively, while the antisymmetric combination gives zero coupling because of destructive interference. Whether the final coupling V_{br} is enhanced or decreased depends on the energy of the symmetric combination state, which is determined by the sign of the interstate coupling. The influence of $V_{D^-A^+/D^+A^-}$ on the coupling is:

$$V_{br}^{eff} = -2\frac{V_{DA}^{h} \times V_{D^*A^*}^{e}}{\Delta E_{CT} + V_{D^+A^-/D^-A^+}^{eh}} \tag{10}$$

for the symmetric dimer.

The interstate coupling also activates two more pathways: $D^*A \rightarrow D^-A^+ \rightarrow D^+A^- \rightarrow DA^*$ and $D^*A \rightarrow D^+A^- \rightarrow D^-A^+ \rightarrow DA^*$. Adding the coupling of the new pathways to V_{1e}, the full bridge-mediated coupling is:

$$\begin{aligned} V_{br}^{eff} &= -2\frac{V_{DA}^{h} \times V_{D^*A^*}^{e}}{\Delta E_{CT}} + 2\frac{V_{DA}^{h} \times V_{D^*A^*}^{e} \times V_{D^+A^-/D^-A^+}^{eh}}{\Delta E_{CT}\Delta E_{CT}} \\ &= -2\frac{V_{DA}^{h} \times V_{D^*A^*}^{e}}{\Delta E_{CT}{}^2}\left(\Delta E_{CT} - V_{D^+A^-/D^-A^+}^{eh}\right) \\ &= -2\frac{V_{DA}^{h} \times V_{D^*A^*}^{e}}{\Delta E_{CT} + V_{D^+A^-/D^-A^+}^{eh}}\left(1 - \frac{V_{D^+A^-/D^-A^+}^{eh}{}^2}{\Delta E_{CT}{}^2}\right) \end{aligned} \tag{11}$$

When $V_{D^-A^+/D^+A^-} \ll \Delta E_{CT}$, the effective couplings in eqn (7) and (8) will be nearly the same. In this case, the sign of the additional pathway contributions determines the nature of the interference (constructive/destructive).

Eqn (11) is exact but difficult to use for complicated systems. Eqn (10) is linked directly to the coupling pathways and is more convenient to assess interferences.

3.2 Dexter coupling pathways in bridged systems

For bridge-mediated systems, such as the D–B–A complexes explored here, the CDFT analysis allows us to compute the one- and two-particle interactions needed to quantify the couplings indicated in Fig. 2 and to assess the coupling pathways indicated in Fig. 3.

We used the sandwich-Nap–Et–Nap structure in Fig. 9d, and the contributing Dexter coupling interactions are shown in Fig. 10. Among the six one-electron bridge-mediated pathways shown in Fig. 10, four arise from the BE contribution, and the others arise from CT contributions. From the interactions shown in Fig. 10, we find that the effective coupling from the two CT pathways is -1.4×10^{-6} eV, while the effective coupling from the four BE pathways is -4.6×10^{-6} eV. The sum of the single-particle pathway couplings provides the total effective one-electron pathway coupling, V_{1e}^{eff}. The larger magnitude BE contribution arises because the BE states have lower energies than the CT states, and the number of BE pathways is twice as large as the number of CT pathways.

In the D–A systems, we found that the direct two-particle Dexter coupling is comparable in magnitude to the effective one-electron Dexter coupling. In D–B–A systems, the two-particle operator introduces more mediating states than does the pure one-particle operator, including mixed 1e/2e pathways. Although the pure 2e coupling term is usually smaller than the 1e coupling term, 2e pathways and mixed 1e/2e pathways can have fewer steps than pure 1e pathways, which may produce couplings comparable to or stronger than the pathways based only on 1e interactions. As an example, the pure 2e pathway (Fig. 10) linking states 1, 5, and 9 has a pathway coupling of -4.0×10^{-6} eV, which is very close to the pure V_{1e} coupling value.

The interactions represented by dark yellow lines in Fig. 10 are dominated by the through space interaction between the two naphthalene molecules while the bridging ethylene makes little contribution. If we remove all of these through-space coupling pathways denoted in the figure with dark yellow lines, then the effective bridge-mediated coupling grows to -1.2×10^{-5} eV. If we retain just the dark yellow terms, the bridge-mediated coupling is 9.3×10^{-6} eV. These two values are similar in magnitude to each other, but have opposite signs, so they interfere destructively. This interference explains the small overall bridge-mediated coupling V_{br}^{eff} in the structure. In the next structure studied, we weaken the naphthalene–naphthalene interaction through space.

In the zigzag-Nap–Et–Nap system with offset naphthalene units, the orbital overlap between the naphthalene molecules is much smaller, and the through-space interaction between them is therefore much weaker. The pathway map is shown in Fig. 11. In contrast to the sandwich-geometry Nap–Et–Nap structure, where the bridge-mediated ET/HT couplings and through-space couplings are

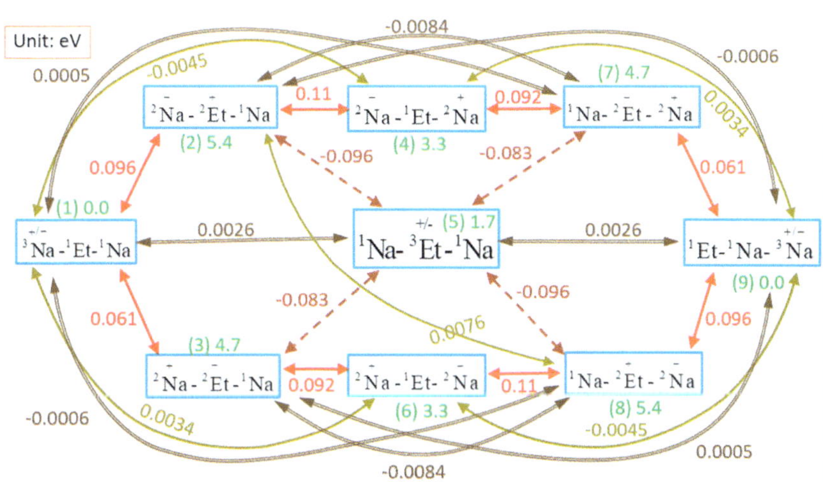

Fig. 10 The state energies and couplings (eV) between the diabatic states for the sandwich-Nap–Et–Nap structure. Numbers in green indicate state energies, numbers near single (red and dark yellow) lines indicate one-electron couplings, numbers near double (brown) lines indicate two-electron couplings.

similar in magnitude, the zigzag structure has much smaller Nap–Nap through-space couplings.

The effective 1e coupling (V_{1e}^{eff}) in the zigzag Nap–Et–Nap geometry is 3.3 × 10^{-6} eV, which is slightly smaller than the value in the sandwich Nap–Et–Nap structure. However, the overall bridge-mediated coupling (V_{br}) is 3.1 × 10^{-5} eV, much larger than the value in the sandwich-Nap–Et–Nap structure. The difference arises from the dark yellow through-space interaction pathways indicated in Fig. 11. The couplings along these pathways are smaller, and more importantly, have the same sign as the other bridge-mediated pathways. As a consequence, destructive interference of bridge-mediated and through-space interaction pathways in the Nap–Et–Nap sandwich structure changes to a constructive interference effect in the zigzag-Nap–Et–Nap structure, enhancing the bridge-mediated coupling, V_{br}^{eff}.

3.3 Dexter pathways in parallel-bridge-mediated systems

We now examine the case of the two parallel ethylene-mediated system shown in Fig. 9f. For single electron or hole transfer, the bridge-mediated superexchange interaction is expected to be doubled for two equivalent bridges compared to just one, which is confirmed by the V_{HT_br} and V_{HT_br} values in Table 1. For the Dexter couplings, the number of one-electron pathways grow by a factor of four, since each transiting electron and each hole may choose the upper (U) or the lower (L) superexchange pathway (2 × 2 = 4). Note, however, that the superexchange pathways have two different energies for the BE virtual states, depending upon whether or not the electron and hole reside on the same or different ethylene molecules. As a result, there are 24 unique pathways generated by the one-electron operator in the structures with two ethylene bridges (see Fig. S2 of ESI†), rather than just 6 one-electron operator generated pathways in the case of one ethylene bridge (see ESI† for further discussion). If the coupling mediated by

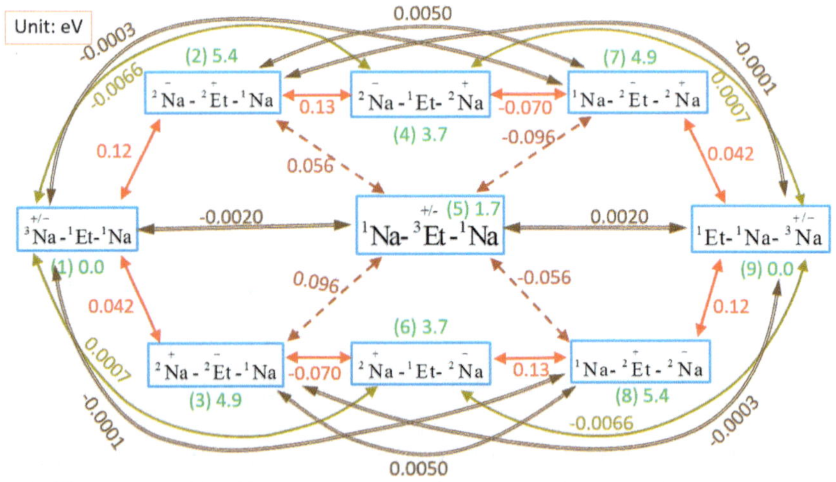

Fig. 11 The state energy and coupling between related diabatic states of the zigzag-Nap–Et–Nap, with the same labels as in Fig. 10.

each pathway were identical, then the sum of the coupling strengths V_{1e}^{eff} would be increased by a factor of 4.

A factor of 2 instead of a factor of 4 arises for V_{1e}^{eff} for the one *vs.* two ethylene sandwich bridge structures. This is because the 24 one-particle Dexter pathways have a wide range of excited state energies. The BE energies are also very different when the electron and hole reside on the same *vs.* different ethylene molecules of the bridge (*i.e.*, 1.7 eV *vs.* 7.9 eV, respectively). If we consider only the lower energy BE pathways, then the number of sandwich-Nap–Et–Nap pathways is 4, while the number of sandwich-Nap–Et_U/Et_L–Nap is 8, which explains the factor of 2 coupling enhancement for the double ethylene structure. It will be interesting to explore how the enhancement factor may be manipulated by further modifying the bridging structure.

For the sandwich-Nap–Et–Nap structure, coupling that arises from the through-space interaction between the two naphthalenes (represented by dark yellow lines) is close in magnitude to the effective bridge-mediated coupling (but with opposite sign), leading to the very small overall bridge-mediated coupling. This naphthalene–naphthalene through-space interaction will change little as the ethylene bridges are added, while the bridge-mediated coupling is expected to grow with the number of parallel coupling pathways. As a result, the sum of the through-space (one particle) and the effective bridge-mediated coupling increases from 1×10^{-7} eV to 6×10^{-5} eV (as we change from the one ethylene to the two ethylene bridge), and the total effective Dexter coupling grows from 1×10^{-5} to 5×10^{-5} eV as the second ethylene is added.

The effective 1e bridge-mediated pathway coupling (V_{1e}^{eff}) is just 20% of the overall bridge-mediated coupling value V_{br}^{eff}, so the one-particle, two-particle and mixed one-particle/two-particle pathways are all essential for calculating the effective coupling in this structure. As such, the nature of Dexter pathway interference is much richer and more varied than in the case of single-electron superexchange.

4. Summary and conclusions

We have studied the dependence of the bridge-mediated coupling in Dexter energy transfer systems on the molecular architectures: D–A, D–B–A and D–B_UB_L–A using constrained DFT methods. The molecular nature of the Dexter coupling pathway interferences were analyzed and contrasted with the results for single electron or hole transfer.

We find that the two particle coupling interactions establish effective coupling pathways for some of the Dexter energy transfer systems studied here. Although the two-electron couplings are smaller than the one-electron couplings, they can produce pathways with fewer steps, and can dominate. Mixed one- and two-electron transfer pathways also contribute significantly to the bridge-mediated couplings. For electron transfer, the strongest coupling pathways are usually defined by a sequence of one-electron or one-hole steps.

Simple pathway counting arguments are not useful for predicting Dexter interactions because the energies of the virtual states depend strongly on the physical position of the electron and the hole and both one and two particle interactions need to be included. In the case of the noncovalent ethylene bridged systems studied here, two parallel pathways produce a constructive single particle interference effect of a factor of two, similar to the case in single electron transfer. However, the net Dexter coupling grows by a factor of four when adding the second pathway, as a result of two-particle and mixed one and two particle interactions. Further studies of bridge-mediated coupling effects in covalent and non-covalent systems should reveal whether systems could be designed that capture further coupling pathway enhancements that may arise from multiple constructively interfering Dexter pathways. It is also interesting to reflect on the possibility of developing phosphorescence probes that could be sensitive to the molecular structure of the material intervening between the absorbing and phosphorescent chromophores.

Conflicts of interest

The authors declare not conflicts of interest.

Acknowledgements

This research was supported by the U.S. Department of Energy, Office of Science, Office of Basic Energy Sciences, under Award DE-SC0019400.

References

1 R. E. Blankenship, *Molecular Mechanisms of Photosynthesis*, Wiley-Blackwell, 2nd edn, 2014.
2 V. May and O. Kühn, *Charge and Energy Transfer Dynamics in Molecular Systems*, Wiley, Weinheim, 3rd edn, 2011.
3 S. S. Skourtis, C. R. Liu, P. Antoniou, A. M. Virshup and D. N. Beratan, *Proc. Natl. Acad. Sci. U. S. A.*, 2016, **113**, 8115–8120.
4 S. Speiser, *Chem. Rev.*, 1996, **96**, 1953–1976.
5 T. Förster, *Ann. Phys.*, 1948, **2**, 55–75.
6 D. L. Dexter, *J. Chem. Phys.*, 1953, **21**, 836–850.
7 A. Ito, D. J. Stewart, Z. Fang, M. K. Brennaman and T. J. Meyer, *Proc. Natl. Acad. Sci. U. S. A.*, 2012, **109**, 15132–15135.
8 F. Schmitt, J. Freudenreich, N. P. E. Barry, L. Juillerat-Jeanneret, G. Suss-Fink and B. Therrien, *J. Am. Chem. Soc.*, 2012, **134**, 754–757.
9 S. Omagari, T. Nakanishi, Y. Kitagawa, T. Seki, K. Fushimi, H. Ito, A. Meijerink and Y. Hasegawa, *Sci. Rep.*, 2016, **6**, 37008.
10 X. Li, Z. Y. Huang, R. Zavala and M. L. Tang, *J. Phys. Chem. Lett.*, 2016, **7**, 1955–1959.
11 C. Mongin, S. Garakyaraghi, N. Razgoniaeva, M. Zamkov and F. N. Castellano, *Science*, 2016, **351**, 369–372.
12 M. Tabachnyk, B. Ehrler, S. Gelinas, M. L. Bohm, B. J. Walker, K. P. Musselman, N. C. Greenham, R. H. Friend and A. Rao, *Nat. Mater.*, 2014, **13**, 1033–1038.
13 C. A. Kent, D. Liu, T. J. Meyer and W. Lin, *J. Am. Chem. Soc.*, 2012, **134**, 3991–3994.
14 M. C. So, G. P. Wiederrecht, J. E. Mondloch, J. T. Hupp and O. K. Farha, *Chem. Commun.*, 2015, **51**, 3501–3510.
15 R. J. Cogdell and H. A. Frank, *Biochim. Biophys. Acta*, 1987, **895**, 63–79.

16 Z. Y. Huang, P. Xia, N. Megerdich, D. A. Fishman, V. I. Vullev and M. L. Tang, *ACS Photonics*, 2018, **5**, 3089–3096.
17 Z. Chen, H. Y. Hsu, M. Arca and K. S. Schanze, *J. Phys. Chem. B*, 2015, **119**, 7198–7209.
18 X. Li, M. Bird, G. Mauro, S. Asaoka, A. R. Cook, H. C. Chen and J. R. Miller, *J. Phys. Chem. B*, 2015, **119**, 7210–7218.
19 K. Ema, M. Inomata, Y. Kato and H. Kunugita, *Phys. Rev. Lett.*, 2008, **100**, 257401.
20 J. M. Ho, E. Kish, D. D. Mendez-Hernandez, K. WongCarter, S. Pillai, G. Kodis, J. Niklas, O. G. Poluektov, D. Gust, T. A. Moore, A. L. Moore, V. S. Batista and B. Robert, *Proc. Natl. Acad. Sci. U. S. A.*, 2017, **114**, E5513–E5521.
21 P. Antoniou, Ph.D. thesis, University of Cyprus, 2017.
22 G. L. Closs, M. D. Johnson, J. R. Miller and P. Piotrowiak, *J. Am. Chem. Soc.*, 1989, **111**, 3751–3753.
23 G. L. Closs, P. Piotrowiak, J. M. Macinnis and G. R. Fleming, *J. Am. Chem. Soc.*, 1988, **110**, 2652–2653.
24 M. A. Baldo, D. F. O'Brien, Y. You, A. Shoustikov, S. Sibley, M. E. Thompson and S. R. Forrest, *Nature*, 1998, **395**, 151–154.
25 V. Grosshenny, A. Harriman, M. Hissler and R. Ziessel, *J. Chem. Soc., Faraday Trans.*, 1996, **92**, 2223–2238.
26 R. D. Harcourt, G. D. Scholes and K. P. Ghiggino, *J. Chem. Phys.*, 1994, **101**, 10521–10525.
27 G. D. Scholes, *ACS Nano*, 2008, **2**, 523–537.
28 D. N. Beratan and J. J. Hopfield, *J. Am. Chem. Soc.*, 1984, **106**, 1584–1594.
29 J. X. Lin, D. Balamurugan, P. Zhang, S. S. Skourtis and D. N. Beratan, *J. Phys. Chem. B*, 2015, **119**, 7589–7597.
30 G. D. Scholes and K. P. Ghiggino, *J. Chem. Phys.*, 1994, **101**, 1251–1261.
31 T. Van Voorhis, T. Kowalczyk, B. Kaduk, L. P. Wang, C. L. Cheng and Q. Wu, *Annu. Rev. Phys. Chem.*, 2010, **61**, 149–170.
32 Q. Wu and T. Van Voorhis, *J. Chem. Phys.*, 2006, **125**, 164105.
33 S. Yeganeh and T. Van Voorhis, *J. Phys. Chem. C*, 2010, **114**, 20756–20763.
34 Y. H. Shao, Z. T. Gan, E. Epifanovsky, A. T. B. Gilbert, M. Wormit, J. Kussmann, A. W. Lange, A. Behn, J. Deng, X. T. Feng, D. Ghosh, M. Goldey, P. R. Horn, L. D. Jacobson, I. Kaliman, R. Z. Khaliullin, T. Kus, A. Landau, J. Liu, E. I. Proynov, Y. M. Rhee, R. M. Richard, M. A. Rohrdanz, R. P. Steele, E. J. Sundstrom, H. L. Woodcock, P. M. Zimmerman, D. Zuev, B. Albrecht, E. Alguire, B. Austin, G. J. O. Beran, Y. A. Bernard, E. Berquist, K. Brandhorst, K. B. Bravaya, S. T. Brown, D. Casanova, C. M. Chang, Y. Q. Chen, S. H. Chien, K. D. Closser, D. L. Crittenden, M. Diedenhofen, R. A. DiStasio, H. Do, A. D. Dutoi, R. G. Edgar, S. Fatehi, L. Fusti-Molnar, A. Ghysels, A. Golubeva-Zadorozhnaya, J. Gomes, M. W. D. Hanson-Heine, P. H. P. Harbach, A. W. Hauser, E. G. Hohenstein, Z. C. Holden, T. C. Jagau, H. J. Ji, B. Kaduk, K. Khistyaev, J. Kim, J. Kim, R. A. King, P. Klunzinger, D. Kosenkov, T. Kowalczyk, C. M. Krauter, K. U. Lao, A. D. Laurent, K. V. Lawler, S. V. Levchenko, C. Y. Lin, F. Liu, E. Livshits, R. C. Lochan, A. Luenser, P. Manohar, S. F. Manzer, S. P. Mao, N. Mardirossian, A. V. Marenich, S. A. Maurer, N. J. Mayhall, E. Neuscamman, C. M. Oana, R. Olivares-Amaya, D. P. O'Neill, J. A. Parkhill, T. M. Perrine, R. Peverati, A. Prociuk, D. R. Rehn, E. Rosta, N. J. Russ, S. M. Sharada, S. Sharma,

D. W. Small, A. Sodt, T. Stein, D. Stuck, Y. C. Su, A. J. W. Thom, T. Tsuchimochi, V. Vanovschi, L. Vogt, O. Vydrov, T. Wang, M. A. Watson, J. Wenzel, A. White, C. F. Williams, J. Yang, S. Yeganeh, S. R. Yost, Z. Q. You, I. Y. Zhang, X. Zhang, Y. Zhao, B. R. Brooks, G. K. L. Chan, D. M. Chipman, C. J. Cramer, W. A. Goddard, M. S. Gordon, W. J. Hehre, A. Klamt, H. F. Schaefer, M. W. Schmidt, C. D. Sherrill, D. G. Truhlar, A. Warshel, X. Xu, A. Aspuru-Guzik, R. Baer, A. T. Bell, N. A. Besley, J. D. Chai, A. Dreuw, B. D. Dunietz, T. R. Furlani, S. R. Gwaltney, C. P. Hsu, Y. S. Jung, J. Kong, D. S. Lambrecht, W. Z. Liang, C. Ochsenfeld, V. A. Rassolov, L. V. Slipchenko, J. E. Subotnik, T. Van Voorhis, J. M. Herbert, A. I. Krylov, P. M. W. Gill and M. Head-Gordon, *Mol. Phys.*, 2015, **113**, 184–215.

35 J. P. Perdew, K. Burke and M. Ernzerhof, *Phys. Rev. Lett.*, 1996, **77**, 3865–3868.

36 W. J. Hehre, R. Ditchfield and J. A. Pople, *J. Chem. Phys.*, 1972, **56**, 2257–2261.

37 Z. Q. You and C. P. Hsu, *J. Chem. Phys.*, 2010, 133.

38 S. S. Skourtis and D. N. Beratan, *Adv. Chem. Phys.*, 1999, **106**, 377–452.

Faraday Discussions

PAPER

Quantum coherence in ultrafast photo-driven charge separation†

Brian T. Phelan, ‡ Jonathan D. Schultz, ‡ Jinyuan Zhang, Guan-Jhih Huang, Ryan M. Young * and Michael R. Wasielewski *

Received 30th November 2018, Accepted 8th January 2019
DOI: 10.1039/c8fd00218e

Coherent interactions are prevalent in photodriven processes, ranging from photosynthetic energy transfer to superexchange-mediated electron transfer, resulting in numerous studies aimed towards identifying and understanding these interactions. A key motivator of this interest is the non-statistical scaling laws that result from coherently traversing multiple pathways due to quantum interference. To that end, we employed ultrafast transient absorption spectroscopy to measure electron transfer in two donor–acceptor molecular systems comprising a *p*-(9-anthryl)-*N*,*N*-dimethylaniline chromophore/electron donor and either one or two equivalent naphthalene-1,8:4,5-bis(dicarboximide) electron acceptors at both ambient and cryogenic temperatures. The two-acceptor compound shows a statistical factor of 2.1 ± 0.2 rate enhancement at room temperature and a non-statistical factor of 2.6 ± 0.2 rate enhancement at cryogenic temperatures, suggesting correlated interactions between the two acceptors with the donor and with the bath modes. Comparing the charge recombination rates indicates that the electron is delocalized over both acceptors at low temperature but localized on a single acceptor at room temperature. These results highlight the importance of shielding the system from bath fluctuations to preserve and ultimately exploit the coherent interactions.

1. Introduction

Extensive work during the past several decades suggests that quantum coherence may play a functional role in various natural and artificial processes,[1] including energy transfer (EnT),[2–6] singlet exciton fission (SF),[7,8] and electron transfer (ET) reactions.[9–16] Much of this evidence manifests as damped oscillations superimposed on population dynamics, or extracted from spectral lineshape analysis, that result from coherent nuclear motion along the Franck–Condon active modes following excitation by a sufficiently short laser pulse.[17] These vibrational wave

Department of Chemistry, Institute for Sustainability and Energy at Northwestern, Northwestern University, Evanston, IL 60208-3113, USA. E-mail: m-wasielewski@northwestern.edu; ryan.young@northwestern.edu

† Electronic supplementary information (ESI) available. See DOI: 10.1039/c8fd00218e

‡ Brian T. Phelan and Jonathan D. Schultz contributed equally.

packets have lifetimes of 100s of femtoseconds, even in dynamic or disordered environments,[6,7,11,18–23] which can be similar to the timescales of fast excited-state reactions. In some cases, these wave packets have even been observed to survive reactions, *e.g.* in SF and ET.[7,11,13,21–23] The timescales for damping of the oscillations, *i.e.* dephasing of the wave packets, indicate that the relevant chromophores are not vibrationally relaxing before the reaction, therefore implicating coherence as being functionally relevant to the excited-state reaction.

While the coherences observed in EnT in photosynthetic light harvesting (PS) are believed to be primarily vibrational or vibronic in nature,[5,16,24,25] electronic coherence, in the context of donor–acceptor (DA) electronic coupling, plays a key role in ET reactions occurring *via* superexchange interactions in donor–bridge–acceptor (DBA) supramolecular architectures and manifests in a variety of ways. Increasing the bridge length has been shown to induce a transition from coherent superexchange to an incoherent hopping mechanism which results from the distance dependences of the bridge state-mediated ET coupling.[26–30] ET rates across norbornyl bridges were shown to be attenuated due to symmetry constraints, *i.e.* destructive interference, between the electron donor and acceptor orbitals.[31–34] Theoretical and experimental investigations of conductance[35,36] and ET rates[37–39] through conjugated and cross-conjugated bridges demonstrate marked decreases in coupling element magnitudes for the cross-conjugated systems resulting from destructive interference. More generally, quantum interference between multiple equivalent pathways results in rates that may scale non-statistically with the number of available pathways where constructive and destructive interference lead to ET rate enhancement and suppression, respectively.[40–49] One of the key results of these studies has been determining the role of dephasing, usually resulting from fluctuations in the donor, acceptor, or bridge site energies, with small fluctuations or weak system-bath coupling allowing for longer correlations between bridge sites and therefore preservation of interference effects.

The coherent effects arising from quantum interference between multiple equivalent pathways is particularly interesting for potential molecular electronics applications, and Nature has selected such a topology for the photosynthetic reaction centers (RCs). It is unlikely that the excited electron propagates coherently down both pathways in the RCs since (i) the two pathways are not strongly interacting and (ii) the electron is observed to populate the intermediate chlorophyll and phylloquinone acceptors. However, experimental observations of long-lived vibrational wave packet dynamics in PS[18,20,50,51] indicate that the protein matrix may work to alleviate sources of dephasing, which may also apply to site fluctuations in the RCs, and theoretical modeling suggests that quantum interference effects are possible.[44] In artificial systems, interest is driven primarily by the potential to enhance or control ET dynamics through use of supramolecular systems with multi-path topologies. Beratan and co-workers have described molecules and experiments where vibrational excitation provides a handle to turn quantum interference on or off,[48] while Rubtsov *et al.* and Weinstein *et al.* have used such experiments for related problems.[52,53] Demonstrating rate enhancement due to constructive interference experimentally is necessary to better understand when it can be utilized to enhance function. However, it is particularly challenging because the most straightforward method relies upon a comparison between two separate molecular systems, where structural differences can lead to larger enhancements than that from constructive interference.[54] Several groups, however, have demonstrated strong experimental evidence for constructive interference by measuring conductance *via* STM-break junction experiments[55] and the exchange coupling between two spin centers *via* electron paramagnetic resonance spectroscopy,[56] with the latter example operating over an impressive 3.9 nm. While these elegant experiments clearly demonstrate constructive interference, neither observes ET dynamics using time-resolved techniques. Though synthesis of full DBA systems featuring one or two equivalent pathways has yet to be realized, here we report on ultrafast charge separation (CS) and nanosecond charge recombination (CR) in two DA molecules containing either one or two equivalent acceptors. The DA architecture studied here also lends itself to closure into a full two-path system. Though the pair of acceptors is found to be highly susceptible to dephasing, the system still demonstrates coherent interactions suggesting it could be used for future studies on ET in closed two-pathway DBA architectures.

2. Experimental

2.1 Synthesis

The full synthetic routes as well as characterization of the intermediates and final products are provided in detail in the ESI Section 2 (ESI†).

2.2 Electrochemistry

Differential pulse voltammetry (DPV) experiments were conducted in argon-purged room temperature solutions in dichloromethane (DCM) using a CH Instruments, Inc. model 660 electrochemical workstation interfaced to a computer. The DPV experiments were performed using Pt as the working and counter electrodes, Ag/AgCl as the reference electrode, and ferrocene as an internal standard. The concentration of the supporting electrolyte, tetrabutylammonium hexafluorophosphate ($TBAPF_6$), was 0.1 M.

2.3 Optical spectroscopy

Steady-state UV-visible absorption spectra were collected using a Shimadzu UV-1800 spectrophotometer for **1NDI** and **2NDI** prepared as room temperature solutions (less than 1 mM) in 1,4-dioxane with optical densities less than 1.0 in quartz cuvettes with a 1 mm path length. Transient absorption (TA) spectroscopy experiments were performed using a 1 kHz regeneratively-amplified Ti:sapphire laser system (Tsunami oscillator/Spitfire Pro amplifier, Spectra-Physics Inc.) described previously.[57] Briefly, half of the about 80 fs 1 mJ per pulse fundamental was used to seed a commercial non-collinear optical parametric amplifier (TOPAS-White, Light Conversion, LLC) to generate the pump pulse centered at 435 nm, less than 50 fs (Fig. S1–S3†). The pump pulse was directed through an external prism compressor (LaK21 glass) with a 154 mm separation to pre-compensate for second-order dispersion accumulated in transmissive optics and the sample cells. The polarization of the pump pulse was set to magic angle (54.7°) with respect to the probe pulse and attenuated to 300–500 nJ per pulse at the sample. About 5% of the remaining fundamental was directed along a delay stage before about 2 μJ per pulse were focused into a translated CaF_2 crystal (uncoated, 2 mm, <001> cut, Newlight Photonics Inc) to generate the probe pulse. The probe pulse was split so that one portion interacted with the sample and the other provided a reference pulse. The signal and reference probe pulses were coupled into optical fibers and detected using a customized Helios spectrometer and Helios software (Ultrafast Systems, LLC).

TA experiments were performed at room temperature (295 K) and at cryogenic temperatures (90 and 5.5 K). The room temperature samples were prepared as solutions (less than 1 mM) in 1,4-dioxane in 1 mm quartz cuvettes and were stirred to avoid the effects of local heating and sample degradation. The cryogenic temperature samples were prepared as solutions in 2-methyltetrahydrofuran (Me-THF) in a glove box under nitrogen atmosphere (to exclude oxygen) and sealed between two quartz windows with an about 0.75 mm PTFE o-ring spacer. The sample holder was mounted between two copper plates in an STVP-100 cryostat (Janis Research Company, LLC) and the temperature was controlled with a Cryo-Con 32B temperature controller (Cryogenic Control Systems, Inc). The pump pulse was focused to a diameter of about 500 μm and the cryostat was translated normal to the probe pulse after each scan of the delay stage to minimize the effects of local heating and sample degradation and to average over inhomogeneities in the frozen solution. The instrument response duration of these experiments, τ_{IRF}, defined as the full-width at half-maximum of the Gaussian instrument response function (IRF), reached a minimum of about 90 fs at probe frequencies near that of the pump and increased significantly at probe frequencies further from the pump. Measurements were performed for each sample on three separate days with new samples prepared each day.

2.4 Density functional theory computation

Density functional theory (DFT) calculations were performed at the level of ωB97X-D/6-31G* *in vacuo* using Q-Chem 5.1 to obtain optimized geometries for the neutral, singly reduced, and singly oxidized **1NDI** and **2NDI** (see ESI section 5†).[58] The solubilizing octyl chains bonded to the NDI imide nitrogen were replaced with a methylene (–CH_3) group to simplify the calculation. Normal mode analyses were completed using the optimized structures at the same level of theory to confirm that no or only one imaginary frequencies were found (Table S1†). We then calculated the single-point energies of the singly oxidized and singly reduced **1NDI** and **2NDI** using the optimized neutral geometries to determine the internal reorganization energy for each sample (*vide infra*).

3. Results and discussion

3.1 Synthesis

The two DA complexes reported here employ a *p*-(9-anthryl)-*N*,*N*-dimethylaniline moiety (DMA–An) as the chromophore and initial electron donor state. We employ naphthalene-1,8:4,5-bis(dicarboximide) (NDI) as the electron acceptor, bonding either one NDI through the imide nitrogen to the 4-position of the anthracene (An) moiety in DMA–An or two NDIs to the 4- and 5-positions of the An moiety yielding **1NDI** and **2NDI**, respectively (Fig. 1A).

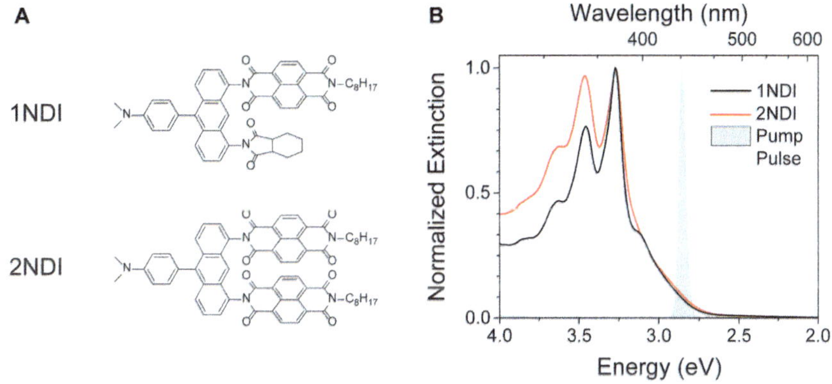

Fig. 1 (A) Molecular structures for the compounds in these studies and (B) normalized steady-state absorption spectra obtained in 1,4-dioxane at 295 K with the typical excitation pulse spectrum superimposed.

3.2. Steady-state electronic absorption spectroscopy

The normalized electronic absorption spectra of **1NDI** and **2NDI** dissolved in 1,4-dioxane at 295 K are shown in Fig. 1B. Both compounds display a prominent vibronic progression of about 1450 cm^{-1} primarily due to the NDI moieties with absorption maxima at 3.27, 3.46, and 3.63 eV (379, 358, and 342 nm).[59] The An moiety also contributes to the observed absorption maxima, though only weakly as a result of a charge-transfer (CT) interaction with DMA which is evidenced by the broad tail extending from the lowest energy vibronic feature to about 2.5 eV (496 nm) in both **1NDI** and **2NDI.**[60] The CT transition only results in partial ET between DMA and An to yield $DMA^{\delta\bullet+}$–$An^{\delta\bullet-}$, especially in low polarity environments where the TA spectra do not resemble a linear combination of the radical ion spectra.[60,61] In **2NDI** compared to **1NDI**, the 3.27 eV vibronic transition decreases in relative absorption with respect to the two higher energy vibronic transitions at 3.46 and 3.63 eV. This results from positive dipolar coupling between the two H-aggregated NDI moieties,[62] as has been observed previously for a similar arrangement of NDIs.[63]

3.3 Electrochemistry

DPV experiments were conducted on **1NDI** and **2NDI** to determine the reduction and oxidation (redox) potentials for calculation of the CS driving force ΔG_{CS} in each system. Fig. 2 shows voltammograms collected for both compounds and values for the measured redox potentials are shown in Fig. 2 as well as listed in Table 1. The oxidation potential for DMA was observed in both compounds at about 0.9 V *vs.* the saturated calomel electrode (SCE), though it shifted 40 mV more positive in **2NDI** compared to **1NDI**, possibly due to an additional electron withdrawing effect of the second NDI. In **1NDI**, two reduction events were observed at −0.54 V and −1.05 V *vs.* SCE, corresponding to one electron reductions generating $NDI^{\bullet-}$ and NDI^{2-}, respectively. In **2NDI**, however, the first reduction wave split by 0.18 V to yield two similar intensity reduction waves at −0.45 V and −0.63 V *vs.* SCE which indicates interaction between the lowest

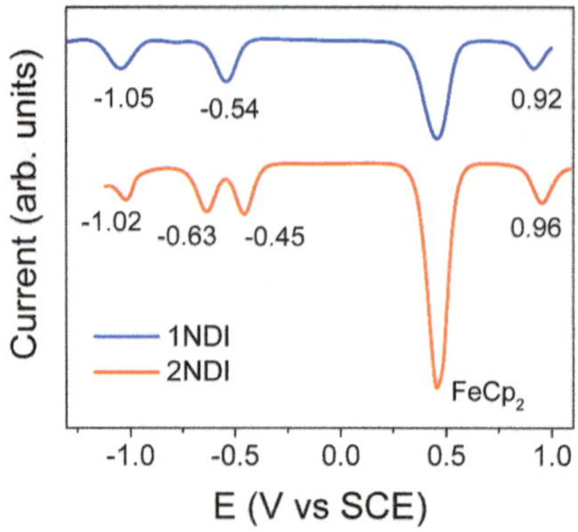

Fig. 2 Differential pulse voltammograms of **1NDI** and **2NDI** in DCM with 0.1 M $TBAPF_6$.

unoccupied molecular orbitals (LUMO) on the two NDI moieties. The more negative reduction wave shifted 30 mV positive to −1.02 V *vs.* SCE and would likely show a similar splitting as the less negative reduction wave if more negative potentials were scanned. As the splitting of the reduction waves in **2NDI** is driven by electronic interaction between the two NDI moieties, we employed the voltage of the lower energy reduction event for the calculation of ΔG_{CS} in **2NDI.**

3.4 Density functional theory computation and energetic parameters

DFT calculations were performed to obtain optimized geometries for **1NDI** and **2NDI** in the neutral, singly oxidized, and singly reduced states which were used to estimate ΔG_{CS} and the internal reorganization energy λ_v. We conducted normal mode analyses on each system to affirm the validity of the optimized structures as determined by the presence of no or only low frequency, less than 100 cm^{-1}, imaginary modes. While four of these normal mode analyses found either no or only one low frequency imaginary mode (Table S1†), two of the analyses were unable to converge under the tight convergence criteria specified (singly oxidized **1NDI** and singly reduced **2NDI**). Considering the frequencies of the one imaginary mode found for each of neutral **1NDI** and singly reduced **1NDI** are particularly low, less than 40 cm^{-1}, we assume their contribution to the determined value of

Table 1 Redox potentials in V *vs.* SCE along with donor–acceptor distance, internal reorganization energy, total reorganization energy in 1,4-dioxane, and driving force for charge separation in 1,4-dioxane and glassy Me-THF for **1NDI** and **2NDI**

	E_{ox} (V *vs.* SCE)	E_{red} (V *vs.* SCE)	R_{DA} (Å)	λ_v (eV)	λ^a (eV)	ΔG_{CS} (eV) 1,4-Dioxane[a]	ΔG_{CS} (eV) Me-THF[b]
1NDI	0.92	−0.54	6.65	0.73	0.92	−0.89	−1.03
2NDI	0.96	−0.45	6.51[c]	0.61	0.80	−0.93	−1.08

[a] For samples in 1,4-dioxane at 295 K; calculated λ_S using $n = 1.42$ and $\varepsilon_S = 2.22$.[68] [b] For samples at 90 and 5.5 K; calculated using $\varepsilon_S = 2.6$.[69] [c] Assumes the negative charge distribution is centered between both NDI moieties.

λ_v is negligible. Additionally, we note that the 1.0×10^{-8} tolerance for convergence used in these calculations, as compared to the more relaxed default 1.0×10^{-7} tolerance typically employed, is less likely to yield optimized structures with high frequency imaginary modes.

ΔG_{CS} is the free energy change between the initial donor state and the final radical ion pair (RP) state and can be calculated using eqn (1),[64]

$$\Delta G_{CS} = -E_{00} + e(E_{ox} - E_{red}) - \frac{e^2}{4\pi\varepsilon_0 R_{DA}\varepsilon_S} + \frac{e^2}{4\pi\varepsilon_0}\left(\frac{1}{2r_+} + \frac{1}{2r_-}\right)\left(\frac{1}{\varepsilon_S} - \frac{1}{\varepsilon_{sp}}\right) \quad (1)$$

where E_{00} is the energy of the initial excited state, E_{ox} and E_{red} are the donor (DMA) oxidation and acceptor (NDI) reduction potentials, respectively, e is the elementary charge, ε_0 is the permittivity of free space, ε_S is the static dielectric constant of the experimental medium, ε_{sp} is the static dielectric constant of the solvent in which the electrochemical measurements were performed, R_{DA} is the distance between the $DMA^{\bullet+}$ and $NDI^{\bullet-}$ moieties, and r_+ and r_- are the hard-sphere radii of $DMA^{\bullet+}$ and $NDI^{\bullet-}$, respectively. r_+ and r_- were assumed to each be equal to half of R_{DA}. ΔG_{CS} values calculated *via* eqn (1) are provided in Table 1. For these estimates, we assumed E_{00} was equal to the photon energy of the pump pulse since the $DMA^{\delta\bullet+}$–$An^{\delta\bullet-}$ excited state is known to have complex relaxation dynamics occurring on longer timescales than CS.[65] The estimated ΔG_{CS} values indicated that CS is more favorable for **2NDI** than for **1NDI**, primarily due to the less negative reduction potential.

The reorganization energy λ represents the energy associated with rearranging the system to accommodate the new charge distribution. λ is typically estimated by separately considering the reorganization energy of the solvent shell λ_S and λ_v which accounts for changes in the intramolecular vibrational modes of the donor and acceptor. λ_S was estimated for the room temperature solutions using the Born continuum model (eqn (2)) which treats the solvent as a dielectric continuum where n is the refractive index.

$$\lambda_S = \frac{e^2}{4\pi\varepsilon_0}\left(\frac{1}{2r_+} + \frac{1}{2r_-} - \frac{1}{R_{DA}}\right)\left(\frac{1}{n^2} - \frac{1}{\varepsilon_S}\right) \quad (2)$$

At cryogenic temperatures the solvent has frozen and is restricted to small translational motions, so λ_S approaches zero and becomes negligible with respect to λ_v.[66] Therefore, we estimated λ as $\lambda_S + \lambda_v$ at room temperature and λ_v at cryogenic temperatures. The energies determined for the optimized geometries along with single-point energies of the singly reduced and singly oxidized **1NDI** and **2NDI** at the optimized neutral geometries were used to calculate λ_v using eqn (3) and (4), values for which are given in Table 1.

$$\lambda_v = \lambda_v^{+} + \lambda_v^{-} \quad (3)$$

$$\lambda_v^{+/-} = E_{neutral}^{+/-} - E_{opt}^{+/-} \quad (4)$$

$\lambda_v^{+/-}$ is the internal reorganization energy for oxidation of the donor and reduction of the acceptor, $E_{neutral}^{+/-}$ is the energy of the oxidized donor or reduced acceptor at the neutral geometry, and $E_{opt}^{+/-}$ is the energy of the oxidized donor or reduced acceptor at the optimized geometry. λ_v^{+} contributes about 0.4–0.5 eV to the overall λ_v for **1NDI** and **2NDI**, nearly twice that of λ_v^- due to the planarization of the DMA moiety upon oxidation. That λ_v^- for **2NDI** is similar to that of **1NDI**, showing only a small increase, is expected regardless of how the electron interacts with the pair of NDI acceptors. If the electron is localized, then it interacts with a similar number of normal modes in both **1NDI** and **2NDI**, and, if the electron is delocalized, then while it interacts with significantly more modes in **2NDI** compared to **1NDI**, it leads to smaller changes in the modes since the electron is spread over the additional modes. Comparison of the estimated ΔG_{CS} and λ_v values for each compound indicate that CS is in the Marcus-inverted region at cryogenic temperatures, while at room temperature in 1,4-dioxane CS for **1NDI** is nearly activationless and CS for **2NDI** is slightly inverted.[67] Additionally, between **2NDI** and **1NDI**, the increased exoergicity combined with the decreased reorganization energy for **2NDI** indicates that CS is slightly further in the inverted region for **2NDI** compared to **1NDI**.

3.5 Transient electronic absorption spectroscopy

3.5.1 Charge separation dynamics. TA spectra were recorded following excitation of the DMA–An CT transition with a 435 nm, 50 fs pump pulse which, as discussed above, generates the polar singlet excited state with partial ET DMA$^{\delta\bullet+}$–An$^{\delta\bullet-}$. The TA spectra of **1NDI** following excitation at 435 nm are shown in Fig. 3A–C for solutions in 1,4-dioxane at 295 K and Me-THF at 90 and 5.5 K, respectively. At all the temperatures studied here, the transient spectrum of DMA$^{\delta\bullet+}$–An$^{\delta\bullet-}$ is either only briefly observed (295 K) or obscured due to sub-picosecond evolution of the TA spectra. Fig. 3A shows, at time delays immediately following the cross-phase modulation, negative absorption at 405 and 380 nm, assigned to ground state bleach (GSB) features of the CT transition and localized An transition, respectively, and positive absorption at 590 and 340 nm assigned to excited-state absorption (ESA) from DMA$^{\delta\bullet+}$–An$^{\delta\bullet-}$ to higher lying singlet excited states S_n.[61] These features, except for the 405 nm GSB, decay rapidly yielding new negative features at 378 and 357 nm, assigned as GSB of the NDI-localized absorption, and new positive ESA features at 475, 585, 607, 632, and 697 nm. The ESA at 475, 607, and 697 nm indicate formation of NDI$^{\bullet-}$.[59] The 585,

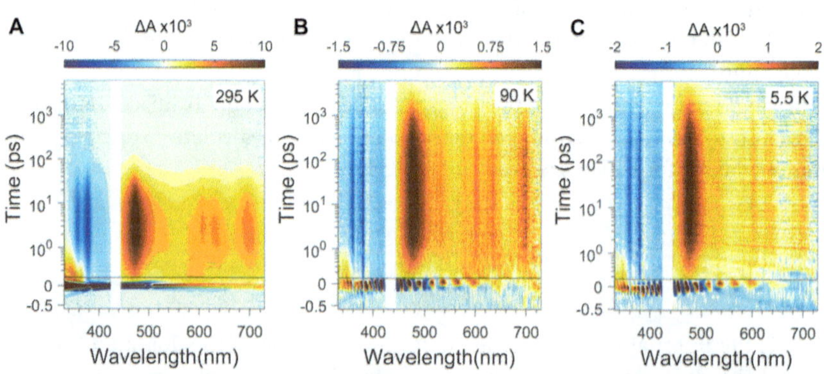

Fig. 3 TA spectra of **1NDI** in (A) 1,4-dioxane at 295 K, (B) glassy Me-THF at 90 K, and (C) glassy Me-THF at 5.5 K. Upper and lower sections of each plot are displayed with logarithmic and linear scales, respectively. Pump scatter is blocked for clarity.

632, and 697 nm ESA closely resemble the steady-state electronic absorption spectrum measured for DMA–An dissolved in 2-chlorobutane following photolysis with 355 nm light (Fig. S5†). We assign these transitions to ESA of $DMA^{•+}$. Notably, the 405 nm GSB does not decay rapidly with the other initial TA features, as is expected for the CT transition. Formation of the $NDI^{•-}$ and $DMA^{•+}$ ESA along with GSB of NDI and the CT transition provide clear evidence of rapid CS yielding $DMA^{•+}$–An–$NDI^{•-}$. Similar dynamics are observed for **1NDI** dissolved in Me-THF at 90 and 5.5 K with CS still occurring on the sub-picosecond timescale.

The TA spectra of **2NDI** following direct CT excitation under identical experimental conditions as **1NDI** are provided in Fig. 4. The signals observed for this compound parallel those described for **1NDI** apart from changes in the relative ESA band amplitudes. In contrast to the $NDI^{•-}$ ESA features, absorption by $DMA^{•+}$ in the **2NDI** compound is more intense compared to that in **1NDI** under all measured conditions. Moreover, comparison of Fig. 3 and 4 shows differences in the kinetics of species generated by CS in **1NDI** and **2NDI**. Namely, the anionic and cationic ESA features imply more rapid formation of $DMA^{•+}$–An–$NDI^{•-}$ in **2NDI** than **1NDI**.

We quantified the effect of the second NDI acceptor on the sub-picosecond CS dynamics seen in Fig. 3 and 4 by measuring the CS rate constants k_{CS} for both **1NDI** and **2NDI** and then evaluating the ratio k_{CS}(**2NDI**)/k_{CS}(**1NDI**). We conducted global fitting analyses of the most prominent anionic TA ESA feature at 475 nm for both **1NDI** and **2NDI** where group velocity mismatch between the pump and probe did not significantly stretch the time resolution τ_{IRF} of the experiment. CS was assumed to be a first-order kinetic process, so the kinetic traces were fit to a sum of exponential decays (see ESI section 1.4†). The CS rates extracted from these analyses are tabulated in Table 2; these values represent the average CS rate constant from measurements made on three days along with one standard deviation, $\langle k_{CS}\rangle \pm \sigma$. CS rate constants in these experiments were measured to range between $(200 \pm 20\ \text{fs})^{-1}$ and $(890 \pm 20\ \text{fs})^{-1}$, the most rapid of which are still resolvable with the approximately 90 fs τ_{IRF} of our experiments. As the temperature was decreased, we observed a corresponding decrease in the CS rate constant for both **1NDI** and **2NDI**.

To investigate the involvement of coherence in electron transfer within these single- and dual-acceptor systems, we evaluated the ratio of the **2NDI** to **1NDI** CS

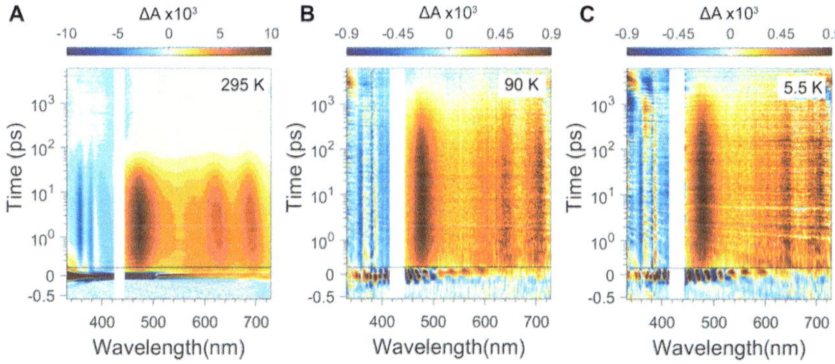

Fig. 4 TA spectra **2NDI** in (A) 1,4-dioxane at 295 K, (B) glassy Me-THF at 90 K, and (C) glassy Me-THF at 5.5 K. Upper and lower sections of each plot are displayed with logarithmic and linear scales, respectively. Pump scatter is blocked for clarity.

Table 2 Charge separation and charge recombination rate constants determined from global fitting of the short and long-time TA data, respectively, for **1NDI** and **2NDI** along with the ratio of **2NDI** rate constants to those of **1NDI**

	k_{CS} (fs^{-1})			k_{CR} (ps^{-1})		
	295 K, 1,4-dioxane	90 K, Me-THF	5.5 K, Me-THF	295 K, 1,4-dioxane	90 K, Me-THF	5.5 K, Me-THF
1NDI	$(410 \pm 20)^{-1}$	$(580 \pm 60)^{-1}$	$(890 \pm 20)^{-1}$	$(25 \pm 2)^{-1}$	$(1407 \pm 7)^{-1}$	$(1800 \pm 100)^{-1}$
2NDI	$(200 \pm 20)^{-1}$	$(240 \pm 20)^{-1}$	$(340 \pm 30)^{-1}$	$(32 \pm 2)^{-1}$	$(690 \pm 50)^{-1}$	$(1100 \pm 200)^{-1}$
Rate ratio	2.1 ± 0.2	2.4 ± 0.3	2.6 ± 0.2	0.8 ± 0.1	2.0 ± 0.2	1.7 ± 0.3

rate constants k_{CS}(**2NDI**)/k_{CS}(**1NDI**) at each temperature. These ratios along with one standard deviation are given in Table 2. At 295 K in 1,4-dioxane, k_{CS}(**2NDI**)/k_{CS}(**1NDI**) is within experimental error of two, whereas at cryogenic temperatures in glassy Me-THF k_{CS}(**2NDI**)/k_{CS}(**1NDI**) rises above two, specifically to 2.4 ± 0.3 at 90 K and of 2.6 ± 0.2 at 5.5 K. Fig. 5 shows example kinetic traces at a probe wavelength of 477 nm ($NDI^{\bullet-}$ peak maximum), produced from the average of three experimental trials, for both **1NDI** and **2NDI**. The solid lines and shading represent the average fit along with one standard deviation. Example fits at all experimental conditions are shown in Fig. S4†. Shifting and normalization of these data and fit traces illustrates the deviation of k_{CS}(**2NDI**) from twice that of **1NDI** (dashed line shown in Fig. 5).

To better understand the role of electronic coherence and dephasing in electron transfer, we examine our results within the framework of the spin-boson model.[70,71] There are three distinct physical scenarios depending on the nature of the

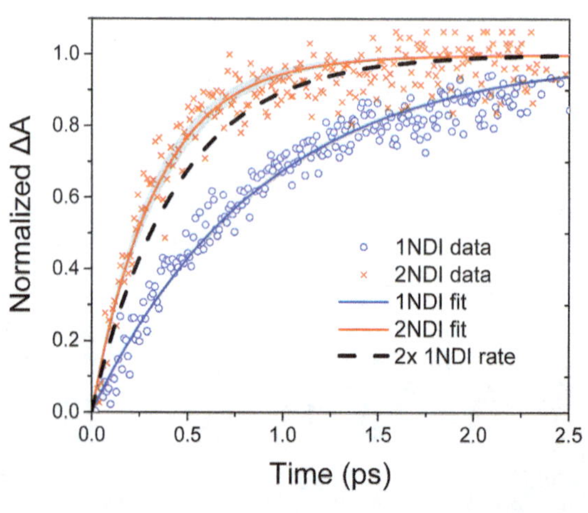

Fig. 5 Normalized kinetic traces and fits at the 477 nm probe wavelength for **1NDI** (blue) and **2NDI** (red) in glassy Me-THF at 5.5 K, along with a curve illustrating twice the CS rate (dashed black) found for the **1NDI** compound. Uncertainty in the fits shown here represents the standard deviation in the CS rate constants found for measurements made on three separate days. To aid direct comparison, we shifted both the data and the fits to intersect at the origin and subsequently normalized all traces from 0 to 1.

interaction between the DMA donor and two NDI acceptors and between the two NDI acceptors and the bath, assuming nonadiabatic ET, weak system-bath coupling, and fast relaxation of the bath modes (see ESI section 3†).[71] In scenario (I) molecular vibrations or interactions with the solvent disrupt the equivalency of the two NDI acceptors and k_{CS}(**2NDI**)/k_{CS}(**1NDI**) = 2; this is the incoherent scenario. In scenario (II) the two NDI acceptors remain equivalent but the bosonic bath modes are over-damped giving k_{CS}(**2NDI**)/k_{CS}(**1NDI**) < 2 with the high-temperature limit yielding k_{CS}(**2NDI**)/k_{CS}(**1NDI**) = 1; this is the coherent, over-damped scenario. In scenario (III) the two NDI acceptors are equivalent and the bosonic bath modes are under-damped giving k_{CS}(**2NDI**)/k_{CS}(**1NDI**) > 2 with the low-temperature limit yielding k_{CS}(**2NDI**)/k_{CS}(**1NDI**) = 4; this is the constructive coherent limit.

The value of k_{CS}(**2NDI**)/k_{CS}(**1NDI**) determined for CS in 1,4-dioxane at 295 K is indistinguishable from that described by the incoherent scenario (I) given the experimental error. This is expected since thermal motion at room temperature allows for dynamic solvent environments and molecular vibrations that can lead to uncorrelated fluctuations in the energies of the two NDI acceptors. As a result, the two NDIs couple as distinct, non-interacting sites to the donor and k_{CS} increases by a factor of 2. As the temperature is decreased, and specifically as the solvent freezes, the determined values of k_{CS}(**2NDI**)/k_{CS}(**1NDI**) increase to be statistically greater than 2, indicating a transition from the incoherent scenario (I) to the constructive coherent scenario (III). The observed values of k_{CS}(**2NDI**)/k_{CS}(**1NDI**) > 2 are unlikely to be caused by changes in ΔG_{CS} or λ. As discussed earlier, at low temperatures CS is in the Marcus-inverted region and, furthermore, λ for **2NDI** decreases compared to **1NDI** while ΔG_{CS} is slightly more favorable for **2NDI** than **1NDI**. Thus, based on ΔG_{CS} and λ alone, and excluding the statistical factor of 2 enhancement, one would expect k_{CS} to decrease, not increase. Instead, the frozen solvent likely shields the two acceptors from the rapid fluctuations present at ambient conditions resulting in longer correlations between the energies of the two NDIs so that they couple coherently to the donor state. In this scenario, it becomes more convenient to consider the superposition states $A_{+/-} = (|\mathrm{NDI}_1> \pm |\mathrm{NDI}_2>)/\sqrt{2}$ which couple to the donor state with $V_{+/-} = (V_{DA} \pm V_{DA})/\sqrt{2}$. Here $|\mathrm{NDI}_1>$ and $|\mathrm{NDI}_2>$ represent the acceptor states of the two equivalent NDI moieties as shown in Scheme 1, V_{DA} is the coupling between the donor and a single acceptor, and $V_{+/-}$ is the renormalized coupling between the donor and the $A_{+/-}$ superposition states. Scheme 1 indicates how coupling between the two NDI moieties in **2NDI** causes the A_+ state to be stabilized and the A_- state to be destabilized with respect to the original NDI energy levels. This treatment demonstrates how the antibonding state A_- is uncoupled from the donor while the bonding state A_+ has an increased coupling of $\sqrt{2}V_{DA}$. In the same manner that the coupling between A_+ and the donor is renormalized by a factor of $\sqrt{2}$, the coupling between A_+ and the bath V_{SB} should also be renormalized by a factor of $\sqrt{2}$. In the context of the spin-boson model, ET in a system that interacts only weakly with the bath, which has fast relaxation, can be approximated with eqn (5)

$$k_{ET} \approx ({V_{DA}}^2\kappa)/(\kappa^2 + {\varepsilon_{DA}}^2) \quad (5)$$

where V_{DA} is the donor–acceptor coupling, κ is a dephasing parameter that is proportional to ${V_{SB}}^2$, and ε_{DA} is the instantaneous DA energy gap (see ESI section

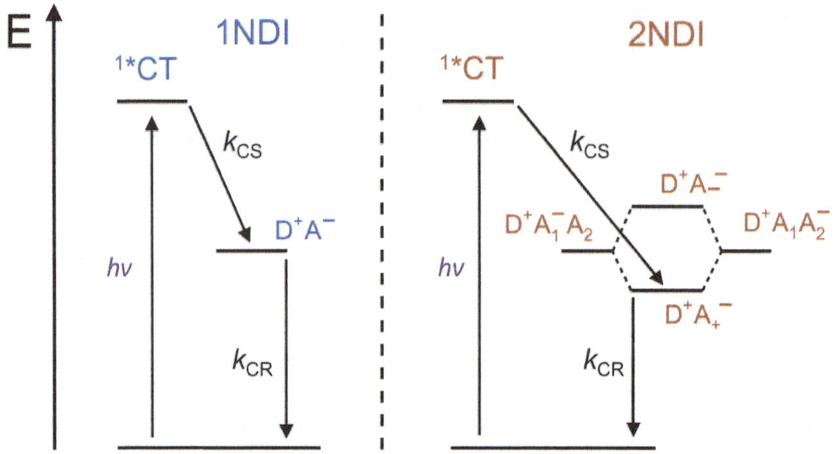

Scheme 1 Jablonski diagram showing the relevant states and processes (*i.e.* population flow) in **1NDI** (left) and **2NDI** (right) at cryogenic temperatures where the two acceptors A_1 and A_2 couple to form the bonding A_+ and antibonding A_- states. Local triplet levels are suppressed from this diagram for clarity.

3†).[71] At low enough temperatures $\kappa \ll \varepsilon_{DA}$ and eqn (5) reduces to $k_{ET} \approx (V_{DA}{}^2\kappa)/\varepsilon_{DA}{}^2$ indicating that, since V_{DA} and V_{SB} are both renormalized by a factor of $\sqrt{2}$, k_{ET} can be enhanced by up to a factor of 4 *via* the presence of the second acceptor in the constructive coherent limit, *i.e.* k_{CS}(**2NDI**)/k_{CS}(**1NDI**) = 4. The observed ratios at cryogenic temperatures clearly do not show a factor of four enhancement, which indicates that the assumption $\kappa \ll \varepsilon_{DA}$ is not completely valid even at 5.5 K. Physically, this suggests that the energy fluctuations of the two NDIs are still large enough to lead to dephasing on the CS timescale which manifests as a ratio statistically larger than 2 but not reaching the factor of 4 limit. The source of dephasing here could be due to low frequency modes of the acceptors or interaction between the bath and the relatively large aromatic surface or alkyl solubilizing tail. Regardless, these results highlight the importance of controlling the system-bath interactions in a manner that minimizes dephasing. While shielding the acceptors from dephasing is achieved here to a small degree by drastically decreasing the temperature, it is similar to the proposed role of the protein matrix in PS that enables relatively long lifetimes for vibrational coherences.[18,20,50]

3.5.2 Charge recombination dynamics. The $DMA^{•+}$–An–$NDI^{•-}$ ESA decays on the tens of picoseconds timescale for both **1NDI** and **2NDI** in 1,4-dioxane at 295 K as can be seen in Fig. 3A and 4A, respectively. Following the decay of the RP state, a weak absorption persists on the nanosecond timescale. At cryogenic temperatures, the $DMA^{•+}$–An–$NDI^{•-}$ ESA shows a marked increase in lifetime with respect to that at room temperature from the tens of picoseconds to single nanosecond timescale (Fig. 3 and 4). There is again a weak absorption that persists after the RP signal has decayed completely. At all temperatures, the absorption of this final state is too weak to interpret, though, given the energy of the RP state it is possible that it results from a fraction of the $DMA^{•+}$–An–$NDI^{•-}$ population undergoing CR to a local triplet state, *e.g.* ^{3*}NDI, *via* an enhanced intersystem crossing mechanism.[72] While the absorption of this state could not be interpreted, it was accounted for in the kinetic fits to ensure that it did not impact the extracted CR rate constants, k_{CR}. Additionally, the TA spectra shown in Fig. 3 and 4 indicate a clear enhancement in the charge recombination at cryogenic temperatures for **2NDI** compared to **1NDI**, but not at room temperature.

To quantify the apparent differences in k_{CR} between **1NDI** and **2NDI** we conducted global fitting analyses (i) on the TA data collected at 295 K using the same method as for k_{CS} and (ii) on the TA data collected at 90 and 5.5 K using a multi-wavelength global fitting methodology described in the ESI.† Briefly, the differential equations for the sequential A → B → C model were solved numerically and the kinetic parameters (*e.g.* rate constants) that best match the fits to the raw data were determined *via* least-squares fitting. These parameters were then used to extract population curves for each species which were then used to deconvolute the raw data and extract the evolution-associated spectrum (EAS) for each species. In these fits, we used an extra exponential decay to remove the cross-phase modulation contribution to the TA spectra. Species A, B, and C refer to the initial photoexcited CT state, the RP state, and the weak long-lived ESA discussed above, respectively. Furthermore, the rate constant for population transfer from A to B, *i.e.* k_{CS}, was fixed to the values listed in Table 2 and only the rate constant for population transfer from B to C, *i.e.* k_{CR}, was fit. The left panels of Fig. 6 and 7 show the EAS for **1NDI** and **2NDI** at 90 and 5.5 K, respectively, while the right

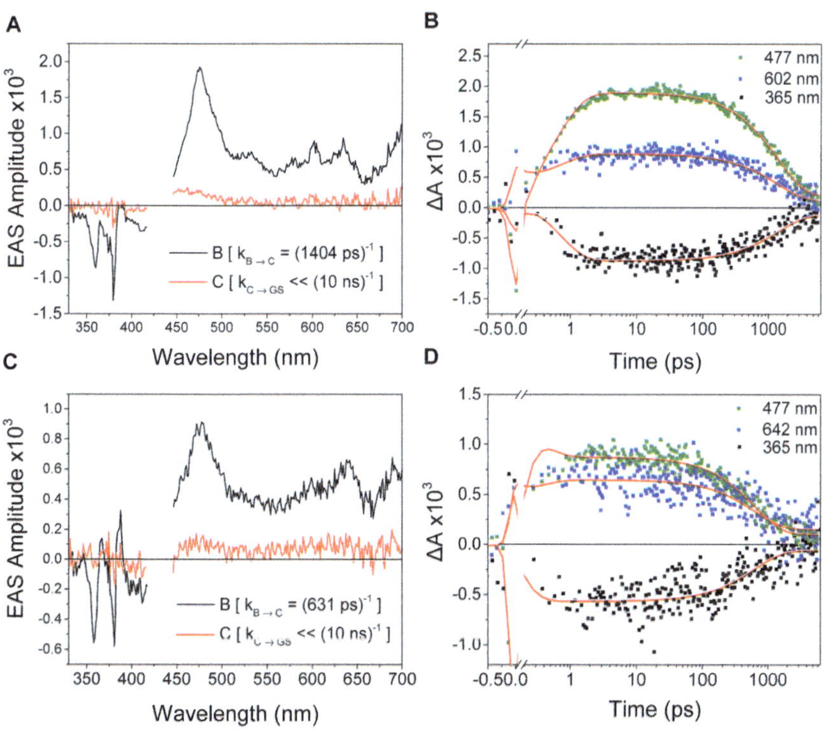

Fig. 6 Evolution-associated spectra (left) and selected single-wavelength kinetics (right) for (A and B) **1NDI** in Me-THF at 90 K and (C and D) **2NDI** in Me-THF at 90 K. Rate constants for the processes shown in (A) and (C) are given in parentheses, where GS signifies the terminal ground state of the system.

panels provide representative kinetics at probe wavelengths corresponding to key spectral features and the associated fits determined by this method.

Through comparison of panels (B) and (D) of both Fig. 6 and 7, it is evident that CR, which is predominantly to S_0, is accelerated when two NDI moieties are present in contrast to one. Values of k_{CR} determined by our model are given in Table 2 in the right column along with the determined ratio values comparing the CR rate constants for **2NDI** and **1NDI**, k_{CR}(**2NDI**)/k_{CR}(**1NDI**). At both 90 and 5.5 K in glassy Me-THF, we found the value of k_{CR}(**2NDI**)/k_{CR}(**1NDI**) to be 2 within the experimental uncertainty. In contrast, at 295 K in 1,4-dioxane, we found k_{CR}(**2NDI**)/k_{CR}(**1NDI**) to be slightly less than 1. Similar to CS, the changes in k_{CR} are unlikely to stem from shifts in the CR driving force ΔG_{CR} and the CR reorganization energy, λ_{CR}. The values for ΔG_{CR} equal the negative of the energies of the RP states (*i.e.* $\Delta G_{CR} < -1.77$ eV) and are much larger than the values for λ_{CR}, which are likely similar to those for CS (*i.e.* $\lambda_{CR} < 1.0$ eV). This demonstrates that CR is also far in the Marcus-inverted region. While ΔG_{CR} is less favorable for **2NDI** than for **1NDI**, λ_{CR} also decreases for **2NDI** and is likely to offset the decreased ΔG_{CR}. This conclusion is further supported by the large differences observed for k_{CR}(**2NDI**)/k_{CR}(**1NDI**) at room and cryogenic temperatures since the relative changes in ΔG_{CR} and λ_{CR} between **1NDI** and **2NDI** are similar in both temperature regimes. Instead, these k_{CR} ratio values are consistent with the CS observations and proposed models describing the interaction of the donor, two acceptors, and

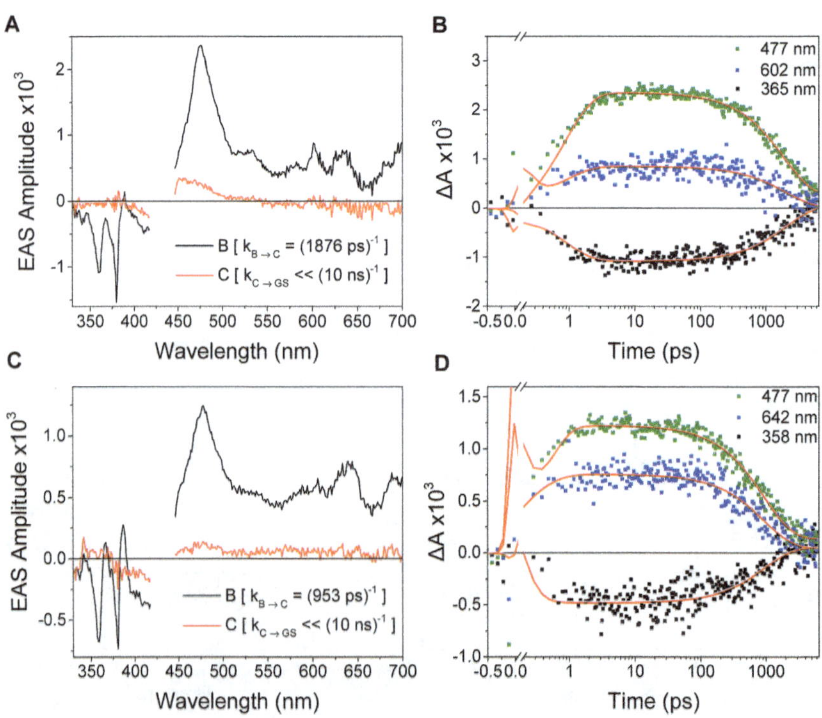

Fig. 7 Evolution-associated spectra (left) and selected single-wavelength kinetics (right) for (A and B) **1NDI** in Me-THF at 5.5 K and (C and D) **2NDI** in Me-THF at 5.5 K. Rate constants for the processes shown in (A) and (C) are given in parentheses, where GS signifies the terminal ground state of the system.

bath. At 295 K in 1,4-dioxane, we found k_{CS}(**2NDI**)/k_{CS}(**1NDI**) to be about 2 and attributed that to rapid fluctuations of the energies of the two NDI acceptors causing them to couple distinctly to the donor. Since, in this scenario, CS occurs to individual acceptor moieties, the electron likely remains localized on a single acceptor until CR occurs. If the electron only interacts with a single acceptor, then the presence of the second acceptor should have no effect and k_{CR}(**2NDI**)/k_{CR}(**1NDI**) ∼ 1. The deviation to a value slightly less than 1 could indicate that the electron is hopping between acceptors on a similar timescale than that of CR, a phenomenon we have previously found to decrease the CR rate.[73] At 90 and 5.5 K in Me-THF, we observed k_{CS}(**2NDI**)/k_{CS}(**1NDI**) > 2, providing evidence that CS occurs to a superposition of the two acceptors. If the electron remains in this delocalized superposition state for a timescale longer than that of CR, then the DA coupling for the CR reaction will again be renormalized by a factor of $\sqrt{2}$, contributing a factor of 2 increase to k_{CR}(**2NDI**) compared to k_{CR}(**1NDI**). The nature of the system-bath interactions is less clear here, though two plausible explanations are as follows: (i) the timescale of CR and the temperature place the system at an intermediate state between the under- and over-damped scenarios and (ii) CR is again predominantly driven by system-bath coupling *via* the acceptor, which in the CR reaction is $DMA^{•+}$. In this case there is no renormalization of the system-bath coupling upon addition of the second acceptor so CR in **2NDI** is limited to twice the rate of CR in **1NDI**.

4. Conclusion

The DA molecules reported here, **1NDI** and **2NDI**, were synthesized to probe the role of electronic coherence between two equivalent acceptor sites and the resultant impact on electron transfer rates. Following excitation to an intramolecular CT state, both compounds exhibit CS with sub-picosecond rate constants ranging between $(890\ \text{fs})^{-1}$ and $(200\ \text{fs})^{-1}$ followed by CR with rate constants ranging from $(25\ \text{ps})^{-1}$ in 1,4-dioxane at 295 K to $(1800\ \text{ps})^{-1}$ in Me-THF at 5.5 K. Comparison of the rate constants for **2NDI** and **1NDI** indicated that transitioning from a room temperature solution to a frozen matrix lead to non-statistical enhancement in the CS rate constant, with a small effect observed just below the glassing temperature and a larger effect observed near liquid helium temperatures. When system-bath interactions are considered in the context of the spin-boson model, the transition from a statistical CS rate enhancement at room temperature to greater-than-statistical rate enhancement at cryogenic temperatures can be attributed to a change in the magnitude of the fluctuations experienced by the two acceptor sites. At cryogenic temperatures, reduced fluctuations enable the two acceptors in **2NDI** to remain coherent for longer timescales where CS likely occurs to a superposition state of the two acceptors. Furthermore, coupling between this superposition state and the bath pushes the rate enhancement beyond the statistical limit. The observed CS and CR dynamics have been explained in the context of this model and demonstrate the importance of shielding the system from randomizing fluctuations if observing or manipulating coherences is desired, which is in concert with current explanations for long-lived coherences in protein matrices or the solid state. Though **2NDI** demonstrated the effects of coherent interactions, the fully coherent limit was not observed and further studies looking at other acceptors with different solubilizing groups or structures could provide additional insight into the precise source of the dephasing in these and related systems.

5. Funding

This work was supported by the National Science Foundation under grant number DMR-1710104. This material is based upon work supported by the National Science Foundation Graduate Research Fellowship Program under Grant No. DGE-1842165. Any opinions, findings, and conclusions or recommendations expressed in this material are those of the author(s) and do not necessarily reflect the views of the National Science Foundation.

Conflicts of interest

There are no conflicts to declare.

Acknowledgements

We thank Dr Mehdi Zarea for helpful discussions regarding the role of the bath, Dr Yilei Wu for performing the differential pulse voltammetry experiments, and Dr Noah Horwitz for performing the photolysis experiment.

References

1 G. D. Scholes, G. R. Fleming, L. X. Chen, A. Aspuru-Guzik, A. Buchleitner, D. F. Coker, *et al.*, Using coherence to enhance function in chemical and biophysical systems, *Nature*, 2017, **543**, 647.

2 S. Savikhin, D. R. Buck and W. S. Struve, Oscillating anisotropies in a bacteriochlorophyll protein: Evidence for quantum beating between exciton levels, *Chem. Phys.*, 1997, **223**(2), 303–312.

3 Y.-C. Cheng and G. R. Fleming, Dynamics of Light Harvesting in Photosynthesis, *Annu. Rev. Phys. Chem.*, 2009, **60**(1), 241–262.

4 G. S. Engel, T. R. Calhoun, E. L. Read, T.-K. Ahn, T. Mančal, Y.-C. Cheng, *et al.*, Evidence for wavelike energy transfer through quantum coherence in photosynthetic systems, *Nature*, 2007, **446**, 782.

5 A. Chenu and G. D. Scholes, Coherence in Energy Transfer and Photosynthesis, *Annu. Rev. Phys. Chem.*, 2015, **66**(1), 69–96.

6 E. Collini and G. D. Scholes, Coherent Intrachain Energy Migration in a Conjugated Polymer at Room Temperature, *Science*, 2009, **323**(5912), 369.

7 A. J. Musser, M. Liebel, C. Schnedermann, T. Wende, T. B. Kehoe, A. Rao, *et al.*, Evidence for conical intersection dynamics mediating ultrafast singlet exciton fission, *Nat. Phys.*, 2015, **11**, 352.

8 F. Mirjani, N. Renaud, N. Gorczak and F. C. Grozema, Theoretical Investigation of Singlet Fission in Molecular Dimers: The Role of Charge Transfer States and Quantum Interference, *J. Phys. Chem. C*, 2014, **118**(26), 14192–14199.

9 S. S. Skourtis, D. N. Beratan and D. H. Waldeck, Coherence in electron transfer pathways, *Procedia Chem.*, 2011, **3**(1), 99–104.

10 A. A. Bakulin, A. Rao, V. G. Pavelyev, P. H. M. van Loosdrecht, M. S. Pshenichnikov, D. Niedzialek, *et al.*, The Role of Driving Energy and

Delocalized States for Charge Separation in Organic Semiconductors, *Science*, 2012, **335**(6074), 1340.

11 S. M. Falke, C. A. Rozzi, D. Brida, M. Maiuri, M. Amato, E. Sommer, *et al.*, Coherent ultrafast charge transfer in an organic photovoltaic blend, *Science*, 2014, **344**(6187), 1001.

12 C. Andrea Rozzi, S. Maria Falke, N. Spallanzani, A. Rubio, E. Molinari, D. Brida, *et al.*, Quantum coherence controls the charge separation in a prototypical artificial light-harvesting system, *Nat. Commun.*, 2013, **4**, 1602.

13 Y. Song, S. N. Clafton, R. D. Pensack, T. W. Kee and G. D. Scholes, Vibrational coherence probes the mechanism of ultrafast electron transfer in polymer–fullerene blends, *Nat. Commun.*, 2014, **5**, 4933.

14 A. S. Rury, S. A. Sorenson and J. M. Dawlaty, Coherent Vibrational Probes of Hydrogen Bond Structure Following Ultrafast Electron Transfer, *J. Phys. Chem. C*, 2016, **120**(38), 21740–21750.

15 S. Rafiq and G. D. Scholes, From Fundamental Theories to Quantum Coherences in Electron Transfer, *J. Am. Chem. Soc.*, 2019, **141**(2), 708–722.

16 F. D. Fuller, J. Pan, A. Gelzinis, V. Butkus, S. S. Senlik, D. E. Wilcox, *et al.*, Vibronic coherence in oxygenic photosynthesis, *Nat. Chem.*, 2014, **6**, 706.

17 S. Ruhman, A. G. Joly and K. A. Nelson, Coherent molecular vibrational motion observed in the time domain through impulsive stimulated Raman scattering, *IEEE J. Quantum Electron.*, 1988, **24**(2), 460–469.

18 H. Lee, Y.-C. Cheng and G. R. Fleming, Coherence Dynamics in Photosynthesis: Protein Protection of Excitonic Coherence, *Science*, 2007, **316**(5830), 1462.

19 C. Fang, R. R. Frontiera, R. Tran and R. A. Mathies, Mapping GFP structure evolution during proton transfer with femtosecond Raman spectroscopy, *Nature*, 2009, **462**, 200.

20 G. Panitchayangkoon, D. Hayes, K. A. Fransted, J. R. Caram, E. Harel, J. Wen, *et al.*, Long-lived quantum coherence in photosynthetic complexes at physiological temperature, *Proc. Natl. Acad. Sci. U. S. A.*, 2010, **107**(29), 12766.

21 M. H. Vos, M. R. Jones, C. N. Hunter, J. Breton and J. L. Martin, Coherent nuclear dynamics at room temperature in bacterial reaction centers, *Proc. Natl. Acad. Sci. U. S. A.*, 1994, **91**(26), 12701.

22 M. H. Vos, M. R. Jones, C. N. Hunter, J. Breton, J.-C. Lambry and J.-L. Martin, Coherent Dynamics during the Primary Electron-Transfer Reaction in Membrane-Bound Reaction Centers of Rhodobacter sphaeroides, *Biochemistry*, 1994, **33**(22), 6750–6757.

23 M. H. Vos, F. Rappaport, J.-C. Lambry, J. Breton and J.-L. Martin, Visualization of coherent nuclear motion in a membrane protein by femtosecond spectroscopy, *Nature*, 1993, **363**, 320.

24 H.-G. Duan, V. I. Prokhorenko, R. J. Cogdell, K. Ashraf, A. L. Stevens, M. Thorwart, *et al.*, Nature does not rely on long-lived electronic quantum coherence for photosynthetic energy transfer, *Proc. Natl. Acad. Sci. U. S. A.*, 2017, **114**(32), 8493–8498.

25 F. Fassioli, R. Dinshaw, P. C. Arpin and G. D. Scholes, Photosynthetic light harvesting: excitons and coherence, *J. R. Soc., Interface*, 2014, **11**, 20130901.

26 F. D. Lewis, H. Zhu, P. Daublain, T. Fiebig, M. Raytchev, Q. Wang, *et al.*, Crossover from Superexchange to Hopping as the Mechanism for

Photoinduced Charge Transfer in DNA Hairpin Conjugates, *J. Am. Chem. Soc.*, 2006, **128**(3), 791–800.

27 R. H. Goldsmith, L. E. Sinks, R. F. Kelley, L. J. Betzen, W. Liu, E. A. Weiss, *et al.*, Wire-like charge transport at near constant bridge energy through fluorene oligomers, *Proc. Natl. Acad. Sci. U. S. A.*, 2005, **102**(10), 3540.

28 W. B. Davis, M. A. Ratner and M. R. Wasielewski, Conformational Gating of Long Distance Electron Transfer through Wire-like Bridges in Donor–Bridge–Acceptor Molecules, *J. Am. Chem. Soc.*, 2001, **123**(32), 7877–7886.

29 H. M. McConnell, Intramolecular Charge Transfer in Aromatic Free Radicals, *J. Chem. Phys.*, 1961, **35**(2), 508–515.

30 K. Kilså, J. Kajanus, A. N. Macpherson, J. Mårtensson and B. Albinsson, Bridge-Dependent Electron Transfer in Porphyrin-Based Donor–Bridge–Acceptor Systems, *J. Am. Chem. Soc.*, 2001, **123**(13), 3069–3080.

31 J. R. Reimers and N. S. Hush, Spectroscopic evidence for electronically forbidden but vibronically allowed long-range electron transfer in norbornylog-bridged naphthalene-diacyanoethylene systems, *Chem. Phys.*, 1990, **146**(1), 105–114.

32 J. R. Reimers, N. S. Hush, D. M. Sammeth and P. R. Callis, Two-photon fluorescence excitation spectrum of a naphthalene norbornylog: Implications for electron transfer, *Chem. Phys. Lett.*, 1990, **169**(6), 622–626.

33 Y. Zeng and M. B. Zimmt, Symmetry Effects in Photoinduced Electron-Transfer Reactions, *J. Am. Chem. Soc.*, 1991, **113**(13), 5107–5109.

34 Y. Zeng and M. B. Zimmt, Symmetry effects on electron-transfer reactions: temperature dependence as a diagnostic tool, *J. Phys. Chem.*, 1992, **96**(21), 8395–8403.

35 M. H. Garner, G. C. Solomon and M. Strange, Tuning Conductance in Aromatic Molecules: Constructive and Counteractive Substituent Effects, *J. Phys. Chem. C*, 2016, **120**(17), 9097–9103.

36 G. C. Solomon, D. Q. Andrews, R. H. Goldsmith, T. Hansen, M. R. Wasielewski, R. P. Van Duyne, *et al.*, Quantum Interference in Acyclic Systems: Conductance of Cross-Conjugated Molecules, *J. Am. Chem. Soc.*, 2008, **130**(51), 17301–17308.

37 E. Maggio, G. C. Solomon and A. Troisi, Exploiting Quantum Interference in Dye Sensitized Solar Cells, *ACS Nano*, 2014, **8**(1), 409–418.

38 G.-J. Huang, M. A. Harris, M. D. Krzyaniak, E. A. Margulies, S. M. Dyar, R. J. Lindquist, *et al.*, Photoinduced Charge and Energy Transfer within meta- and para-Linked Chlorophyll a-perylene-3,4:9,10-bis(dicarboximide) Donor–Acceptor Dyads, *J. Phys. Chem. B*, 2016, **120**(4), 756–765.

39 A. B. Ricks, G. C. Solomon, M. T. Colvin, A. M. Scott, K. Chen, M. A. Ratner, *et al.*, Controlling Electron Transfer in Donor–Bridge–Acceptor Molecules Using Cross-Conjugated Bridges, *J. Am. Chem. Soc.*, 2010, **132**(43), 15427–15434.

40 D. N. Beratan and J. J. Hopfield, Calculation of tunneling matrix elements in rigid systems: mixed-valence dithiaspirocyclobutane molecules, *J. Am. Chem. Soc.*, 1984, **106**(6), 1584–1594.

41 J. N. Onuchic and D. N. Beratan, Molecular bridge effects on distant charge tunneling, *J. Am. Chem. Soc.*, 1987, **109**(22), 6771–6778.

42 L. A. Curtiss, C. A. Naleway and J. R. Miller, Superexchange Pathway Calculation of Electronic Coupling through Cyclohexane Spacers, *J. Phys. Chem.*, 1995, **99**(4), 1182–1193.
43 R. H. Goldsmith, M. R. Wasielewski and M. A. Ratner, Scaling Laws for Charge Transfer in Multiply Bridged Donor/Acceptor Molecules in a Dissipative Environment, *J. Am. Chem. Soc.*, 2007, **129**(43), 13066–13071.
44 N. Renaud, D. Powell, M. Zarea, B. Movaghar, M. R. Wasielewski and M. A. Ratner, Quantum Interferences and Electron Transfer in Photosystem I, *J. Phys. Chem. A*, 2013, **117**(29), 5899–5908.
45 M. Zarea, D. Powell, N. Renaud, M. R. Wasielewski and M. A. Ratner, Decoherence and Quantum Interference in a Four-Site Model System: Mechanisms and Turnovers, *J. Phys. Chem. B*, 2013, **117**(4), 1010–1020.
46 S. S. Skourtis, D. H. Waldeck and D. N. Beratan, Inelastic Electron Tunneling Erases Coupling-Pathway Interferences, *J. Phys. Chem. B*, 2004, **108**(40), 15511–15518.
47 R. H. Goldsmith, M. R. Wasielewski and M. A. Ratner, Electron Transfer in Multiply Bridged Donor–Acceptor Molecules: Dephasing and Quantum Coherence, *J. Phys. Chem. B*, 2006, **110**(41), 20258–20262.
48 D. Xiao, S. S. Skourtis, I. V. Rubtsov and D. N. Beratan, Turning Charge Transfer On and Off in a Molecular Interferometer with Vibronic Pathways, *Nano Lett.*, 2009, **9**(5), 1818–1823.
49 K. Kumar, I. V. Kurnikov, D. N. Beratan, D. H. Waldeck and M. B. Zimmt, Use of Modern Electron Transfer Theories To Determine Electronic Coupling Matrix Elements in Intramolecular Systems, *J. Phys. Chem. A*, 1998, **102**(28), 5529–5541.
50 M. Mohseni, P. Rebentrost, S. Lloyd and A. Aspuru-Guzik, Environment-assisted quantum walks in photosynthetic energy transfer, *J. Chem. Phys.*, 2008, **129**(17), 174106.
51 S. S. Skourtis, D. H. Waldeck and D. N. Beratan, Fluctuations in Biological and Bioinspired Electron-Transfer Reactions, *Annu. Rev. Phys. Chem.*, 2010, **61**(1), 461–485.
52 Z. Lin, C. M. Lawrence, D. Xiao, V. V. Kireev, S. S. Skourtis, J. L. Sessler, *et al.*, Modulating Unimolecular Charge Transfer by Exciting Bridge Vibrations, *J. Am. Chem. Soc.*, 2009, **131**(50), 18060–18062.
53 M. Delor, S. A. Archer, T. Keane, A. J. H. M. Meijer, I. V. Sazanovich, G. M. Greetham, *et al.*, Directing the path of light-induced electron transfer at a molecular fork using vibrational excitation, *Nat. Chem.*, 2017, **9**, 1099.
54 Z. Li, M. Smeu, A. Rives, V. Maraval, R. Chauvin, M. A. Ratner, *et al.*, Towards graphyne molecular electronics, *Nat. Commun.*, 2015, **6**, 6321.
55 H. Vazquez, R. Skouta, S. Schneebeli, M. Kamenetska, R. Breslow, L. Venkataraman, *et al.*, Probing the conductance superposition law in single-molecule circuits with parallel paths, *Nat. Nanotechnol.*, 2012, **7**, 663.
56 S. Richert, J. Cremers, I. Kuprov, M. D. Peeks, H. L. Anderson and C. R. Timmel, Constructive quantum interference in a bis-copper six-porphyrin nanoring, *Nat. Commun.*, 2017, **8**, 14842.
57 R. M. Young, S. M. Dyar, J. C. Barnes, M. Juríček, J. F. Stoddart, D. T. Co, *et al.*, Ultrafast Conformational Dynamics of Electron Transfer in ExBox4+⊂Perylene, *J. Phys. Chem. A*, 2013, **117**(47), 12438–12448.

58 Y. Shao, Z. Gan, E. Epifanovsky, A. T. B. Gilbert, M. Wormit, J. Kussmann, *et al.*, Advances in molecular quantum chemistry contained in the Q-Chem 4 program package, *Mol. Phys.*, 2015, **113**(2), 184–215.

59 D. Gosztola, M. P. Niemczyk, W. Svec, A. S. Lukas and M. R. Wasielewski, Excited Doublet States of Electrochemically Generated Aromatic Imide and Diimide Radical Anions, *J. Phys. Chem. A*, 2000, **104**(28), 6545–6551.

60 T. Okada, T. Fujita, M. Kubota, S. Masaki, N. Mataga, R. Ide, *et al.*, Intramolecular electron donor–acceptor interactions in the excited state of (anthracene)-(CH_2)*n*-(*N*,*N*-dimethylaniline) systems, *Chem. Phys. Lett.*, 1972, **14**(5), 563–568.

61 T. Okada, N. Mataga, W. Baumann and A. Siemiarczuk, Picosecond laser spectroscopy of 4-(9-anthryl)-*N*,*N*-dimethylaniline and related compounds, *J. Phys. Chem.*, 1987, **91**(17), 4490–4495.

62 M. Kasha, H. R. Rawls and M. Ashraf El-Bayoumi, The exciton model in molecular spectroscopy, *Pure Appl. Chem.*, 1965, 371.

63 Y. Wu, M. Frasconi, D. M. Gardner, P. R. McGonigal, S. T. Schneebeli, M. R. Wasielewski, *et al.*, Electron Delocalization in a Rigid Cofacial Naphthalene-1,8:4,5-bis(dicarboximide) Dimer, *Angew. Chem., Int. Ed.*, 2014, **53**(36), 9476–9481.

64 A. Weller, Photoinduced Electron Transfer in Solution: Exciplex and Radical Ion Pair Formation Free Enthalpies and their Solvent Dependence, *Z. Phys. Chem.*, 1982, 93.

65 A. Siemiarczuk and W. R. Ware, Complex excited-state relaxation in *p*-(9-anthryl)-*N*,*N*-dimethylaniline derivatives evidenced by fluorescence lifetime distributions, *J. Phys. Chem.*, 1987, **91**(13), 3677–3682.

66 B. M. Hoffman and M. A. Ratner, Reorganization energies and rate constants for electron reactions in glass-forming media and proteins, *Inorg. Chim. Acta*, 1996, **243**(1), 233–238.

67 R. A. Marcus, On the theory of electron-transfer reactions. VI. Unified treatment for homogeneous and electrode reactions, *J. Chem. Phys.*, 1965, **43**, 679–701.

68 *CRC Handbook of Chemistry and Physics*, CRC Press/Taylor & Francis, Boca Raton, FL, USA, 99th edn, 2018.

69 T. Furutsuka, T. Imura, T. Kojima and K. Kawabe, *Dielectric Constant of 2-Methyltetrahydrofuran*, Technology Reports of the Osaka University, 1974, vol. 24, p. 367.

70 A. Nitzan, *Chemical dynamics in condensed phases: relaxation, transfer, and reactions in condensed molecular systems*, Oxford University Press, New York, USA, 2006.

71 M. Zarea, M. A. Ratner and M. R. Wasielewski, Electron transfer in a two-level system within a Cole-Davidson vitreous bath, *J. Chem. Phys.*, 2014, **140**(2), 024110.

72 Z. E. X. Dance, S. M. Mickley, T. M. Wilson, A. B. Ricks, A. M. Scott, M. A. Ratner, *et al.*, Intersystem Crossing Mediated by Photoinduced Intramolecular Charge Transfer: Julolidine–Anthracene Molecules with Perpendicular π Systems, *J. Phys. Chem. A*, 2008, **112**(18), 4194–4201.

73 R. M. Young, S. C. Jensen, K. Edme, Y. Wu, M. D. Krzyaniak, N. A. Vermeulen, *et al.*, Ultrafast Two-Electron Transfer in a CdS Quantum Dot–Extended-Viologen Cyclophane Complex, *J. Am. Chem. Soc.*, 2016, **138**(19), 6163–6170.

Faraday Discussions

PAPER

Ultrafast transient absorption spectroscopy of doped P3HT films: distinguishing free and trapped polarons

Matthew G. Voss, D. Tyler Scholes, J. Reddy Challa and Benjamin J. Schwartz *

Received 27th November 2018, Accepted 14th December 2018
DOI: 10.1039/c8fd00210j

It is generally presumed that the vast majority of carriers created by chemical doping of semiconducting polymer films are coulombically trapped by the counteranion, with only a small fraction that are free and responsible for the increased conductivity essential for organic electronic applications. At higher doping levels, it is also possible for bipolarons to form, which are expected to be less conductive than single polarons. Unfortunately, there is no simple way to distinguish free polarons, trapped polarons and bipolarons using steady-state spectroscopy. Thus, in this work, we use ultrafast transient absorption spectroscopy to study the dynamics of polarons in 2,3,5,6-tetrafluoro-7,7,8,8-tetracyanoquinodimethane (F4TNCQ)-doped films of poly(3-hexylthiophene-2,5-diyl) (P3HT) as a function of dopant concentration and excitation wavelength. When exciting on the red side of the polaron P1 transition, our transient absorption spectra and kinetics match well with what is expected for free 2-D-delocalized polarons; the measurements are not consistent with a recent theory of doped conjugated polymer electronic structure that suggests that the half-filled state lies deeper in the conduction band rather than in the bandgap. As we tune the excitation wavelength to the blue, our measurements reveal an increasing amount of slower transient kinetics that are consistent with the presence of coulombically-trapped polarons rather than bipolarons. Taking advantage of their distinct ultrafast relaxation kinetics as a type of action spectroscopy, we are able to extract the steady-state absorption spectra of free and trapped polarons as a function of dopant concentration. By comparing the results to theoretical models, we determine that in F4TCNQ-doped P3HT films, trapped polarons sit ~0.4 nm away from the anion while free polarons reside between 0.7 and 0.9 nm from the counteranion. Perhaps counterintuitively, the ratio of trapped to free polarons increases at higher doping levels, an observation that is consistent with a plateau in the concentration-dependent conductivity of F4TCNQ-doped P3HT films.

Department of Chemistry and Biochemistry, University of California Los Angeles (UCLA), Los Angeles, CA 90095-1569, USA. E-mail: schwartz@chem.ucla.edu

1 Introduction

Like all organic semiconductors, conjugated polymers have several disadvantages relative to their inorganic counterparts when it comes to serving as functional electronic materials. The conductivity of conjugated polymers is generally poor, the result of intrinsically low charge mobilities and meager equilibrium carrier densities. As with inorganic materials, the carrier density and mobility can be improved by doping, which places additional charge carriers in the valence or conduction band. For inorganic semiconductors, atomic replacement of just a few parts per billion dopes these materials enough to become effective conductors. In contrast, the density of charge carriers needed to increase the conductivity of conjugated polymers by several orders of magnitude is quite high, and can be on the order of parts per thousand or even parts per hundred.[1–3]

Doping of plastic semiconductors is achieved by either oxidizing (for p-type doping) or reducing (for n-type doping) the conjugated polymer. The oxidation/reduction of conjugated polymers can be accomplished through a variety of means, including electrochemical doping,[4–6] which uses an applied bias and counterions from an electrolyte, or direct electrical methods, such as gate-doping in a field-effect transistor.[5,7] To permanently dope polymers, the best approach is chemical doping by adding a molecule to a conjugated polymer that is either a strong electron donor or acceptor, so that ground-state charge transfer reactions occur to produce equilibrium charge carriers.

Of course, conjugated polymers are full of defects (chain ends, oxidized or chemically altered monomers, kink sites, catalyst left over from the synthesis, *etc.*), that can trap or localize a charge carrier. Moreover, the low dielectric constant of organic semiconductors allows the carriers created by doping to have strong Coulomb interactions with the dopant counterions, interactions that are largely screened in doped inorganic materials. This means that at low doping concentrations in conjugated polymers, the carriers (also referred to as polarons) created are largely trapped. At higher concentrations, molecular doping is expected to create free carriers that contribute to improved charge transport and thus can be used as charge injection layers for organic–organic interfaces with good connections to organic active layers.[1] Applications of highly-doped semiconducting polymers include spintronics, thermoelectrics, and biological sensors as well as OLEDs and polymer-based solar cells.[8–12]

Much of the work studying conjugated polymer doping has focused on the prototypical semiconducting polymer, poly(3-hexylthiophene-2,5-diyl) (P3HT), whose chemical structure is pictured at the top left in Fig. 1.[1] One of the more common molecular dopants is 2,3,5,6-tetrafluoro-7,7,8,8-tetracyanoquinodimethane (F4TCNQ), whose chemical structure is shown in the upper right of Fig. 1. F4TCNQ has its LUMO level at 5.24 eV relative to vacuum, in a position, relative to the HOMO/valence band of P3HT at 5.2 eV,[13] to allow electrons to be transferred from P3HT to F4TCNQ, creating holes on P3HT.[2] F4TCNQ doping of P3HT dramatically increases the polymer's electrical conductivity by multiple orders of magnitude, although at the highest doping levels explored, the conductivity is observed to plateau with increased dopant loading.[3]

There are two main questions concerning the doping of polymers like P3HT with dopants like F4TCNQ that we will explore in this paper. First is the issue of

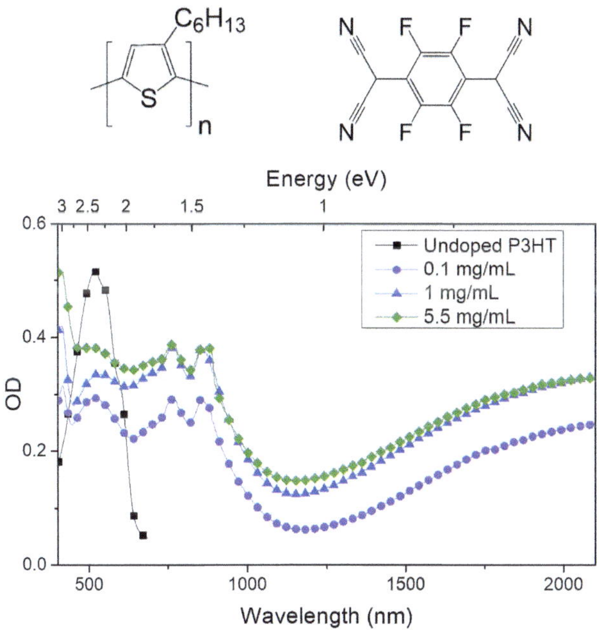

Fig. 1 Top Right: the chemical structure of the dopant F4TCNQ (2,3,5,6-tetrafluoro-7,7,8,8-tetracyanoquinodimethane). Top Left: the chemical structure of P3HT (poly(3-hexylthiophene-2,5-diyl). Lower Panel: steady-state UV-Vis-NIR absorption spectra of: an undoped P3HT film (black squares); an identical P3HT film after sequential doping using a 5.5 mg mL^{-1} solution of F4TCNQ in 70 : 30 THF : DCM (green diamonds); an identical P3HT film sequentially doped with a 1 mg mL^{-1} solution of F4TCNQ in DCM (blue triangles); and an identical P3HT film sequentially doped using a solution that was 0.1 mg mL^{-1} F4TCNQ in DCM (purple circles).

free *versus* trapped carriers in molecularly-doped conjugated polymer films. Previous work has estimated that ~95% of the polarons created by F4TCNQ-doping of P3HT at low concentrations remain trapped due to Coulomb binding with the F4TCNQ anion.[2] It is not clear, however, how this number changes with doping concentration or how Coulomb binding changes the spectroscopic and electrical properties of the polarons created by doping. Structural studies suggest that F4TCNQ resides in the lamellar region of P3HT crystallites, causing a reorganization of the P3HT crystal structure; F4TCNQ is also thought to increase order in the amorphous region of P3HT films.[14,15] Thus, the first main question we address in this paper is: can we use spectroscopic methods to separate the contributions of free and trapped carriers, and in doing so determine where the counterion resides relative to the polarons?

The second main issue we address in this work is the electronic nature of the charged carriers created by doping. The standard picture of doping for inorganic semiconductors, which has been extended to doped conjugated polymers, is shown in Fig. 2a and b.[16,17] The electronic structure associated with the polarons, however, can change if the holes pair at high concentrations to form bipolarons (Fig. 2c) or if the holes can delocalize to a neighboring P3HT chain as well as along the chain backbone (Fig. 2d).[18] Moreover, there are even competing theories as to the intrinsic electronic structure of a polaron, with a recent alternative model that

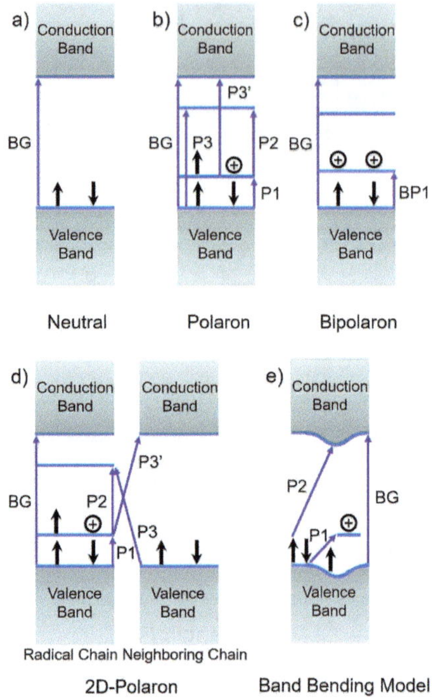

Fig. 2 (a) The band diagram for neutral semiconducting polymers, which shows only a single optical transition (labeled BG) and (b) the traditional, longstanding electronic structure model for polarons in semiconducting polymers and their corresponding optical transitions. The P3 and P3′ transitions are usually believed to be optically forbidden. (c) The traditional model for bipolarons and their allowed BP1 optical transition and (d) the model for 2-D-delocalized polarons, where optical transitions can occur both on the chain containing the majority of the radical cation character as well as diagonally between that chain and a neighboring, predominantly neutral chain. This makes the P3 and P3′ transitions become optically allowed and breaks their degeneracy. (e) A new band-bending model for the electronic structure of polarons on conjugated polymers that does not predict that a positively-charged polymer chain is easier to ionize than a neutral polymer chain.[19] There are no bipolarons or P3 or P3′ transitions in this model.

we refer to as the 'band-bending' model, which puts the polaronic states in the valence band rather than the bandgap, shown in Fig. 2e.[19–22] All of these different pictures predict different allowed transitions between the different polaronic electronic states. Thus, the second question we will address in this paper is: by studying the ultrafast relaxation dynamics of photoexcited polarons, can we distinguish between the various theoretical scenarios of polaron electronic structure outlined in Fig. 2?

To address both of these questions, we performed a series of ultrafast spectroscopy experiments examining the behavior of photoexcited polarons in F4TCNQ-doped P3HT films. We find clear evidence for two distinct polaron populations with different relaxation dynamics, which we assign to free and Coulombically-trapped polarons. By monitoring how the amplitudes of the two species change as the excitation wavelength is tuned, we are able to disentangle the spectra of the free and trapped polarons in our samples. We find that, perhaps counterintuitively, the number of trapped polarons increases with increasing doping levels, with the trapped polaron spectrum blue-shifting at higher doping concentrations. We also find that based on the features of the transient spectroscopy, there is little evidence for bipolaron formation at high doping levels. Instead, the transient spectroscopy appears to be consistent with that expected for 2-D-delocalized polarons, and not with band-bending models in which the electronic states of hole polarons lie in the valence band rather than the bandgap. Taken together, our results help to build a better understanding of how polarons in doped conjugated polymers contribute to electrical conductivity.

2 Methods

2.1 Sample preparation

All of the F4TCNQ-doped P3HT films in this work were prepared *via* solution sequential processing[3] to ensure high optical quality and control over the doping levels. Briefly, 120 nm thick films of undoped P3HT (purchased commercially from Rieke Metals) were fabricated by spin-coating a 20 mg mL^{-1} solution of P3HT in *o*-dichlorobenzene at 1000 rpm for 60 s onto glass substrates. The films were then doped following the solution sequential processing procedure outlined by Scholes *et al.*[3] by spinning solutions of F4TCNQ on top of the pristine P3HT films at 4000 rpm for 10 s. For lower doping levels ($\leq$1 mg mL^{-1}), the F4TCNQ solutions were prepared in dichloromethane, which is known to swell but not dissolve P3HT, allowing for infiltration of the dopant into the P3HT underlayer. For higher doping levels of 5.5 mg mL^{-1}, the F4TCNQ dopant was dissolved in a 70 : 30 v/v tetrahydrofuran : dichloromethane mixture, with the THF fraction needed to solubilize the higher concentration of dopant. All samples were stored under an inert argon atmosphere at 1 atm and all measurements were taken with the samples in an inert atmosphere within 72 hours of fabrication to limit any effects from F4TCNQ subliming or other de-doping mechanisms that could affect the sample over time.

2.2 Transient absorption spectroscopy

All of the ultrafast transient absorption experiments were performed using a commercial Ti:sapphire regenerative amplifier (Coherent, Inc.) and a Helios transient absorption spectrometer that together provide ~75 fs time resolution. The amplified laser system produced pulses of light at 800 nm with ~3 mJ of energy at a 1 kHz repetition rate. Part of the amplified 800 nm beam was used to pump an optical parametric amplifier, which was used to create ~10 μJ pulses at 1200, 1500, 1700, or 2000 nm to serve as the pump beam. The other part of the 800 nm pulse was focused into a sapphire plate to generate white light continuum in either the NIR (850 to 1300 nm) or visible regions (450 to 725 nm) to serve as the probe beam. We note that for both the NIR- and visible-probe experiments, scatter from the laser fundamental prevented transient absorption data from being collected in the region near 800 nm. The timing between the pump pulse and the probe pulse was modulated using a double-pass delay stage. Scans were averaged over at least 5 seconds per time point and at least three independent scans were taken.

2.3 Deconvoluting the transient absorption spectroscopy *via* differential kinetics

To determine the nature of the underlying and overlapping species in the ultrafast transient absorption measurements, we used the matrix decomposition technique Singular Value Decomposition. Singular Value Decomposition, which we applied according to methods elaborated by Doan *et al.*, factorizes a set of data into its significant components.[24] In our application, we took the data to be a function of wavelength on one dimension, and represented time steps, excitation wavelength and dopant concentration on the other dimension. Data was used from delays of 0.3 ps to 20 ps and for probe wavelengths from 500 to 720 nm and 900 to 1120 nm, including excitation of both 1 mg mL^{-1} and 5.5 mg mL^{-1} samples at 1200 nm, 1500 nm, 1700 nm and 2000 nm.

In our application of SVD, D holds the spectral components of our system, which is calculated from U, S, and V, which come from the Singular Value Decomposition of the data set in wavelength, excitation wavelength and concentration and time, along with F, which is a fitted model of how the measured absorption spectra change in time, excitation wavelength, and concentration:

$$\text{Data} = USV^{\text{T}} = DF^{\text{T}} \tag{1}$$

$$D = US(V^{\text{T}}F^{\text{T}+}) \tag{2}$$

We constructed the model F as a two-column matrix that described independent exponential decays for the two components. One row of F represents the fast component and the other the slow component. F is thus a function of the decay constants τ_1 and τ_2, which correspond to the fast and slow decay components, respectively, as well as the initial relative amplitudes of each component at each excitation wavelength and doping concentration. The columns of F run over time, describing the exponential decay of each component as a function of the initial amplitude of the spectral component and the time constant of its exponential decay.

We started by modeling the transient absorption data following 1200 nm excitation of the 1 mg mL^{-1} F4TCNQ-doped P3HT sample. After modeling this transient absorption decay, the matrix F would then be used to map out the decay following 1500 nm excitation, then that at 1700 nm excitation and so forth. This procedure was then repeated for the transient absorption data for different excitation wavelengths on the 5.5 mg mL^{-1} F4TCNQ-doped P3HT films. The two spectral components were constrained to have the same shape, regardless of excitation wavelength and doping concentration.

The difference between the transient absorption data and DF^{T} was minimized using a least squares technique to solve for the ratio of amplitudes, the spectral shape of the components, and the time constants of each component's exponential decay. Our attempts to include a third spectral component did not change the overall fit, as the first two components remained essentially unchanged and the third component had nearly zero amplitude.

Based upon initial fitting, the spectral shape and decay rate of each component were assumed to be constant for each excitation wavelength and the initial amplitude of each component was varied based upon the dopant concentration and excitation wavelength. We also explored several other models, such as one component decaying into the other or an inverse time dependence for the decay of one component, but none of the alternative models fit the data better, and most of the other models had more fitting parameters than the simple model of two independently-decaying transient species.

Error bars were generated by fixing the ratio of the slow and fast components' starting amplitudes for the corresponding excitation wavelength, and calculating the root-mean-square error of the fit for a particular fitting parameter. The ratio of the root-mean-square error of the constrained fit to the unconstrained fit compared well to an estimate of the general variance of the data constructed by fitting four time-dependent single-probe-wavelength traces for each excitation wavelength and concentration to the simple model and averaging their root-mean-square errors.

3 Results and discussion

3.1 The electronic structure of polarons and bipolarons in conjugated polymer films

One of the key motivators for the experiments we present here is that the nature of the electronic quasiparticle species present in doped semiconducting polymers has been the subject of debate in the literature.[2,16,19,20,22,25] The basic band structure of a neutral conjugated polymer is shown in Fig. 2a. When the neutral polymer becomes positively doped, the polymer relaxes from a benzoid to a quinoid structure to stabilize the charge, creating a polaron.[16] In the standard picture, this rearrangement shifts the occupied half-filled state from the valence band into the bandgap and lowers an equivalent state from the bottom of the conduction band, leading to the situation in Fig. 2b.[16] Within this picture, doping should lead to three new optical transitions in addition to the original bandgap transition: the P1 transition, which consists of exciting an electron from the valence band to the half-filled state, the P2 transition, which involves exciting the electron in the half-filled state to the other state inside the bandgap, and a degenerate pair of transitions known as P3 and P3′, which involve excitation of an electron between the valence band and the empty state or between the half-filled state and the conduction band, respectively.[16]

For an ideal, doped single conjugated polymer chain, the P3 and P3′ transitions are optically forbidden, but in a disordered system or a material in which there is partial delocalization of the polarons between neighboring polymer chains (so-called 2-D delocalization), one or both of these transitions may be observable.[18,25] In this 2-D delocalization picture, shown in Fig. 2d, one polymer chain carries most of the radical cation character, but there can be optical transitions involving electrons to or from neighboring chains.[18] When this happens, the P3 and P3′ transitions become symmetry-allowed and the P2 transition is expected to become partially optically forbidden.[18] In addition, with 2-D-delocalization the P3 and P3′ transitions are no longer required to be degenerate in energy.

At high doping levels, it is possible that polarons on conjugated polymers can pair, forming bipolarons. This is because it can be energetically less costly to put two like charges near each other than to distort the backbone of the conjugated polymer in two separate locations.[16] Bipolaron formation is particularly likely when the concentration of dopants is so high that the delocalization of polarons becomes limited by the total concentration of carriers.[16] Fig. 2c shows the standard electronic structure picture for bipolaron formation, in which there are two empty states within the bandgap, with the only optically-allowed transition going from the valence band to the lower-energy empty intragap state, termed the BP1 transition.[16,26,27]

Recently, however, several groups have proposed a new electronic structure scheme to explain polaron formation in doped conjugated polymers.[19,20] The new scheme, which is summarized in Fig. 2e, was motivated by the physical idea that it should be harder to ionize a positively-doped conjugated polymer chain than a neutral chain; the standard picture predicts that the half-filled polaronic state is closer to the vacuum level than the valence band.[19] In this new 'band-bending' model, which is based on density functional theory calculations, when a polaron forms, the state from which the electron was removed splits in half. After the state splits, a half-occupancy state containing the remaining electron bends below the bandgap into the valence band and the half-filled state containing the hole rises into the gap. The conduction band then bends down locally with the hole, forming another state that projects into the gap. The P1 transition would then result from promotion of an electron from the valence band to the intergap state and the P2 transition would arise when as valence band electron is promoted to the lowered part of the bent conduction band. Rather than being optically forbidden, there are no P3 or P3′ transitions in the band-bending model.[19,20] The band-bending model also provides no obvious mechanism for bipolaron formation.[19]

When a polaron is in close spatial proximity to its counteranion, the polaron can localize and become trapped in the Coulomb well of the counteranion.[2,28] Both experimental and theoretical work indicates that trapped polarons have a blue-shifted P1 peak due to their decreased delocalization and the presence of the Coulomb well.[23,29–31] Spano and his co-workers have pioneered calculations of polaron localization and trapping as a function of the distance of a counteranion from the doped polymer chain. These calculations show a clear shift to smaller wavelengths as well as a change in shape of the absorption band as the counteranion moves closer to the polaron, as shown in Fig. 3.[23] Of course, there may be an inhomogeneous distribution of free and weakly or more deeply trapped polarons in any given conjugated polymer film, but it is clear that polarons can absorb in different spectral regions depending on their degree of delocalization, which is directly related to their proximity to a dopant counterion.

Overall, there is still a great deal of complexity and controversy when it comes to describing polaronic states in conjugated polymers. There are multiple different pictures of the electronic structure of polarons, including the possibilities of traditional polarons, 2-D delocalized polarons and the new band-bending picture of polaron electronic structure. When this is combined with the fact that most polarons in chemically-doped conjugated polymer films are thought to be coulombically trapped rather than freely mobile, it is simply not at all clear how to best interpret the spectroscopy of doped conjugated polymer films. Our goal in this paper is to elucidate the spectroscopy of doped conjugated polymers by performing the ultrafast transient absorption experiments: as we will see below, the shape of

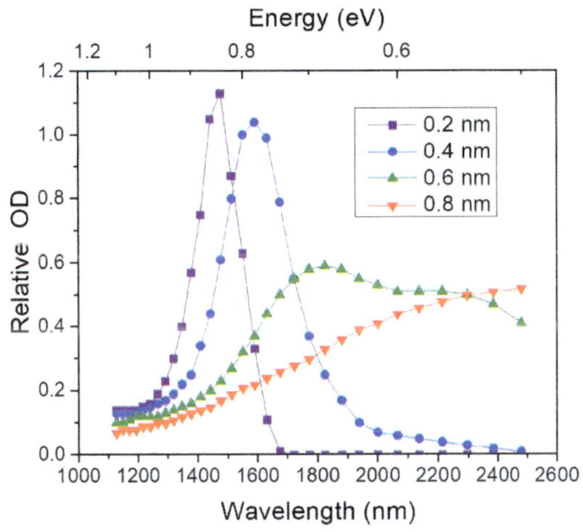

Fig. 3 A reproduction of theoretical calculations of the P3HT P1 polaron absorption band as a function of an anion distance from the P3HT chain, taken from ref. 23. As the anion distance decreases from 0.8 nm to 0.2 nm, the P1 peak shifts towards smaller wavelengths and narrows significantly.[23] The calculations used a 10-mer of P3HT with a point-charge counteranion situated at the given distance away from the center of the polymer chain, along with a dielectric constant and a Huang-Rhys parameter of 1.0.[23] The vibrational frequency was chosen to be 0.17 eV, the intrachain coupling to be −0.4 eV, the Gaussian disorder width to be 0.3 eV, and the homogeneous linewidth to be 0.03 eV.[23] See ref. 23 for details.

the excited-state absorption spectrum will tell us about polaron electronic structure, and the dynamics will tell us about the degree of coulombic polaron trapping.

3.2 The steady-state spectroscopy of F4TCNQ-doped P3HT films

We begin our study of the electronic structure of doped conjugated polymers in Fig. 1, which shows the steady-state optical absorption spectra of a neutral P3HT film (black squares) and P3HT films doped *via* sequential processing[3,32] with increasing concentrations of F4TCNQ (purple circles, blue triangles and green diamonds). The figure shows clearly that as neutral P3HT becomes doped, the absorption associated with the bandgap transition at 525 nm decreases in intensity, while new absorbing features grow in at 400, 700, 780, 880 and 2400 nm. All of these features have been observed in previous work and their assignments are well understood.[33–36] The peak near 2400 nm is assigned to the low-energy P1 transition of polarons on the doped P3HT. The new peaks that appear at 700, 780, and 880 nm are vibronic structure associated with the F4TCNQ anion, and the new peak at 400 nm is associated with neutral F4TCNQ.[33] We also expect that much of the absorption in the 650–900 nm region is the result of a P2 transition (and possibly P3 and/or P3′ transitions) that lie underneath the F4TCNQ anion absorption in this region.

Fig. 1 also shows that as the dopant concentration increases from 0.1 mg mL^{-1} to 1.0 mg mL^{-1}, the intensity of the doped P3HT P1 and P2/F4TCNQ anion absorption peaks increases. As the dopant concentration is further increased to 5.5 mg mL^{-1}, however, there is no further increase in the absorption intensity of these peaks; instead, a shoulder appears on the high-energy side of the broad P1 band in the region between 1200 and 1500 nm. This new absorption feature could be consistent with that expected for the BP1 transition of bipolarons that form at high dopant concentrations,[16] but it also is potentially consistent with the P1 transition of single polarons that are coulombically localized by their counterions.[14]

3.3 Connection between the electronic structure and expected ultrafast dynamics of doped conjugated polymer films

To better understand the spectral changes that take place upon doping P3HT with different concentrations of F4TCNQ, we performed a series of pump/probe transient absorption experiments. Most groups have used three-pulse pump-probe experiments to study polarons, in which a first laser pulse is used to create a polaron, a second, time-delayed pulse excites the polaronic species, and then a third pulse probes the resulting spectral dynamics.[37–41] These experiments suffer from being unable to cleanly excite polarons with the second laser pulse because of spectral congestion from other excited-state species (excitons, triplets, *etc.*).[37,42–44] Here, we excite already-formed polarons in our F4TCNQ-doped P3HT films, avoiding spectral congestion. Our idea was to perform one set of experiments exciting the low-energy P1 peak near 2400 nm (0.5 eV), and a second set of experiments exciting at wavelengths further to the blue to learn more about the nature of the absorbing species that gives rise to the blue shoulder at higher doping concentrations.

Fig. 4 shows roughly what can be expected in our pump/probe experiments based on the traditional model for the electronic structure of polarons and bipolarons in conjugated polymers,[16] with time flowing from left to right. For the case of polarons, shown in the upper part of the figure, exciting the low-energy P1 transition should bleach both the P1 peak and the neutral bandgap absorption since an electron is removed from the valence band. The addition of a second electron in the intra-gap state should also lead to an increase in absorption of the P2 peak. In this picture, one can think of exciting the P1 band as taking an electron from the valence band and using to fill the pre-existing polaronic hole, leaving a new hole somewhere else in the material; in other words, the P1 transition is essentially a charge-transfer band for moving the hole polaron from one place in the polymer to another. Once the polymer backbone relaxes around the new hole, which should take place relatively quickly, the electronic states will rearrange to recover the initial electronic configuration and thus the equilibrium absorption spectrum.

If we extend these ideas to the case of 2-D-delocalized polarons, in addition to all of the aforementioned changes in absorption, we would expect that exciting the P1 band of the polaron would also bleach the now-allowed P3 transition and cause a corresponding increase in absorption intensity for the P3′ transition. If the P3 and P3′ transitions are degenerate, then there would be no net change in absorbance at the P3 transition energy, but if the degeneracy is broken, this could produce an additional transient absorption signature.

This picture could change even further in the case of trapped polarons, where the P1 transition will occur at a higher energy, and the positions of the P2, P3, and P3′ peaks all will shift depending on how Coulomb trapping affects the

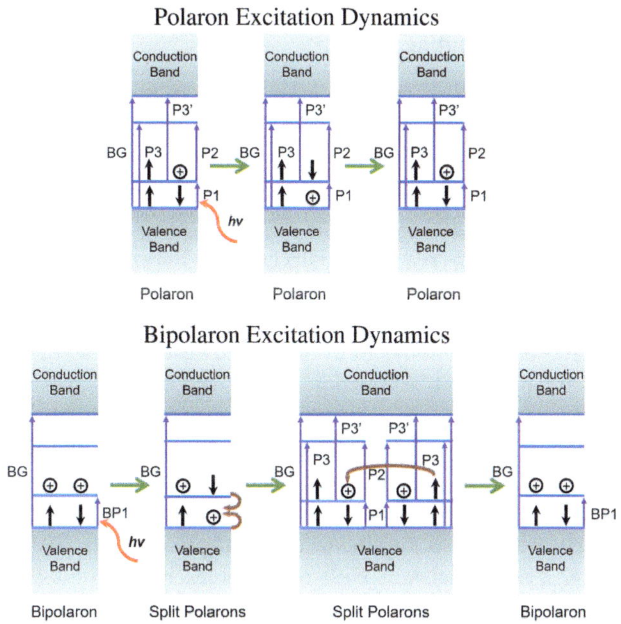

Fig. 4 Band diagrams explaining the dynamics expected following photoexcitation of polarons (top) and bipolarons (bottom) in the traditional conjugated polymer electronic structure picture; time flows from left to right. Following excitation of the low-energy polaron P1 transition, an electron is removed from the valence band and used to fill the pre-existing hole, creating a half-filled lower state and a filled P1 state. After rapid relaxation around the new hole, the system returns to the original ground-state polaron configuration. In contrast, when the BP1 transition of bipolarons is excited, an electron is taken from the valence band and used to fill one of the two holes, creating two single polarons in half-filled orbitals in different locations in the film. After any fast relaxation processes are complete, the two separated polarons must diffuse back together to reform the equilibrium bipolaron state, a process that is expected to occur on longer timescales.

corresponding energy levels. In the band-bending model of polaron electronic structure, the picture would be modified yet again, as exciting the P1 polaron band should temporarily bleach the P1, P2 and neutral bandgap transitions and there are no P3 or P3′ transitions to potentially bleach or increase in absorption intensity.

Finally, the lower part of Fig. 4 shows our expectations following excitation of the BP1 band of bipolarons. Here, one can think of the BP1 band as a charge-transfer transition that takes an electron from the valence band and moves it to fill one of the two original paired holes. This effectively dissociates the bipolaron into two single polarons, one of which remains in the original bipolaron location and the other of which appears somewhere else. Following excitation of the BP1 band, we expect both the BP1 and neutral bandgap transitions to be bleached by removal of the electron from the valence band. The two polaron-like states created will have new P1- and P2-like absorptions, and depending on the degree of delocalization between neighboring chains, there also could be new P3 and P3′ absorption transitions. As time progresses, the split single polarons will diffuse until they ultimately recombine back into an equilibrium bipolaron. Currently, there are no predictions regarding how bipolaron states might be affected upon excitation for the band-bending model.[19,20]

3.4 Pump-probe transient absorption spectroscopy of F4TCNQ-doped P3HT

To understand the nature of the steady-state optical absorption spectrum of F4TCNQ-doped P3HT shown in Fig. 1, we started by performing pump/probe experiments exciting near the peak of the P1 transition. Fig. 5 shows that when exciting the 1 mg mL^{-1} F4TCNQ-doped P3HT film at 2000 nm, there is a bleach of the P3HT neutral bandgap transition at 525 nm (2.4 eV), a second bleach that appears near 650 nm (1.9 eV), and a new, broad transient absorption that appears near 1000 nm (1.24 eV). The transient bleaches and absorption all decay at similar rates, suggesting that they all arise from the same electronic species.

These transient spectral signatures are not consistent with what is expected from the traditional polaron model outlined at the top of Fig. 4. In this picture, we would expect bleaching of the bandgap transition at 525 nm and an increase in absorption of the P2 transition near 1000 nm, but there is no way to explain the bleach that appears near 650 nm, below the bandgap of neutral P3HT. Instead, the data in Fig. 5 are consistent with the photoexcitation of 2-D delocalized polarons, with the 1000 nm transient absorption corresponding to the expected increase in the P2 transition and the 650 nm bleach corresponding to removal of the partially-allowed P3 transition. We might also expect 2-D delocalized polarons to show an increase in absorption due to the P3′ transition, but this is likely obscured by scatter from the laser fundamental at 800 nm (or possibly has significant overlap with the P3 and/or P2 bands). It is also worth noting that the bleach of the P3 transition at 650 nm is not consistent either with the

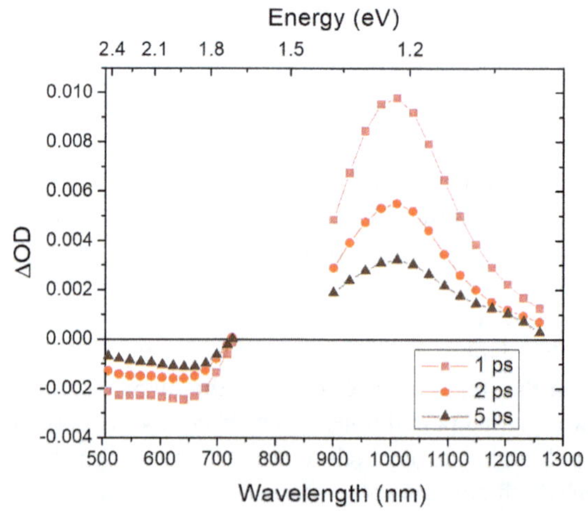

Fig. 5 Ultrafast transient absorption spectra of P3HT films sequentially doped with a 1 mg mL^{-1} F4TCNQ solution in DCM following excitation at 2000 nm; spectra are shown at delays of 1, 2, and 5 ps between excitation and collection. The data show two bleaches in the visible region, at 520 nm and 650 nm, and one broad transient absorption in the visible region near 1000 nm, which we assign to the bleach of neutral P3HT, the bleach of the P3 transition, and an increase of the P2 transition in the 2-D-delocalized polaron picture, respectively (*cf.* Fig. 2d).

photoexcitation of bipolarons, which would show increased absorption at this wavelength, or with the band-bending model, which does not predict any P3 or P3′ transitions.[19,20]

With the basic features of the ultrafast transient spectroscopy of polarons on P3HT established, we turn next to identifying the nature of the species that contributes to blue shoulder of the P1 band seen for P3HT at higher doping concentrations in Fig. 1. To this end, we repeated the experiment using a series of different excitation wavelengths spanning the range from 1200 to 2000 nm; the results for three such wavelengths are shown in Fig. 6. As we tune the excitation wavelength to the blue, we see clear shifts in the P2 transient absorption of P3HT that peaks near 1000 nm. In particular, the higher the energy of the excitation wavelength, the more blue-shifted the peak of the resulting P2 transient absorption. Moreover, higher-energy excitation also leads to a broadening on the blue side of the P2 transient absorption signature. But most importantly, the data in Fig. 6 show that transient absorption no longer decays uniformly with time: the blue side of the transient absorption that is enhanced when exciting further to the blue decays more slowly than the peak of the transient absorption band. All of these observations indicate that bluer excitation wavelengths are indeed accessing a new electronic species that was not significantly excited at 2000 nm.

To better understand the origin of the new electronic species, we also repeated our ultrafast transient absorption experiments on P3HT films as a function of dopant concentration. Fig. 7 shows the results when the bluest wavelength, 1200 nm, is used to excite P3HT films sequentially doped with F4TCNQ at concentrations of 1 mg mL^{-1} (red-colored curves) and 5.5 mg mL^{-1} (blue-colored curves). The data show that as the doping concentration is increased, the transient absorption spectrum shifts to the blue and shows a broadening on the blue

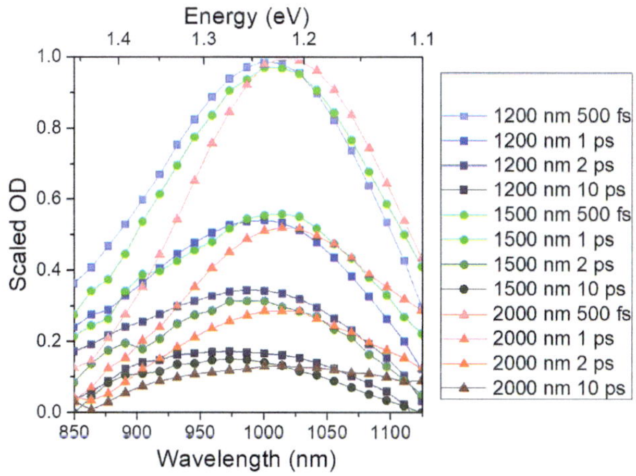

Fig. 6 Ultrafast transient absorption spectra of a P3HT film sequentially doped with a 1 mg mL^{-1} F4TCNQ solution in DCM following excitation at either 1200 nm (blue squares), 1500 nm (green circles) or 2000 nm (red triangles). Data are shown at 0.5, 1, 2, and 10 ps delays between excitation and collection, with darker colors used for later delay times. The observed transient absorption feature shifts to the blue, with the blue feature decaying more slowly, as the excitation wavelength is tuned to the blue.

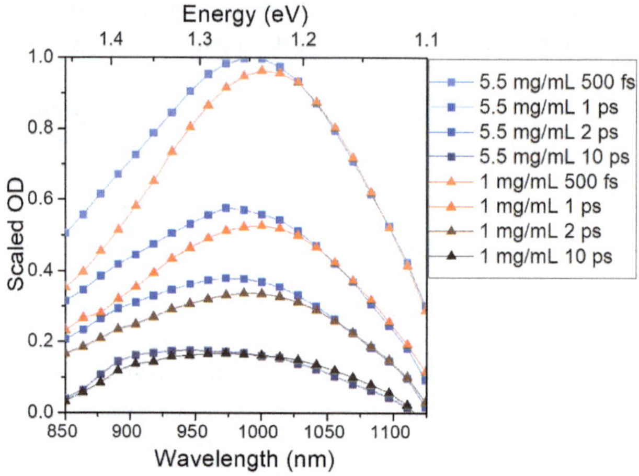

Fig. 7 Ultrafast transient absorption spectra of P3HT films that have been sequentially doped either with a 1 mg mL^{-1} solution of F4TCNQ in DCM (red triangles) or a 5.5 mg mL^{-1} solution of F4TCNQ in a 70 : 30 v/v mixture of THF : DCM (blue squares) following excitation at 1200 nm. Data are shown for 0.5, 1, 2, and 10 ps delays between excitation and collection, with darker colors corresponding to later delay times. The more highly-doped samples have the transient absorption shifted to the blue and the blue feature decaying more slowly than the lower-doped sample.

side, similar to what is seen when using higher-energy excitation. Moreover, also like what we saw above, the blue portion of the transient absorption spectrum decays more slowly than the peak. Thus, whatever the new electronic species is, its presence is manifest both when exciting at higher energies and when the doping concentration is increased.

3.5 Analysis of the ultrafast transient absorption spectra of F4TCNQ-doped P3HT films

To elucidate the nature of the second electronic species that is excited more at 1200 nm than 2000 nm and is more prevalent in P3HT films doped with F4TCNQ at 5.5 mg mL^{-1} than at 1 mg mL^{-1} doping level, we globally analyzed our transient absorption data across excitation wavelength and doping concentration. For our global analysis, we assumed a model in which there are precisely two species excited in the doped films. Each species in the model has a different absorbance at each excitation wavelength, is present in different amounts at different doping concentrations, and each has its own transient absorption spectrum and decay rate. We then used SVD, as described above, to extract the transient absorption spectra and decay times of each of the two species.

With the SVD procedure applied globally to all of our transient absorption data for the 1 and 5.5 mg mL^{-1}-doped samples (not just the subset of data shown in Fig. 5–7), we found that indeed SVD yielded a quickly-decaying (lifetime of ~0.37 ps), red-shifted transient absorption component with a larger initial amplitude and a more slowly-decaying (lifetime of ~15 ps), blue-shifted transient absorption component with a lower initial amplitude. The spectra of the two components are shown in Fig. 8. Based on their shapes, it makes sense to assign both of the

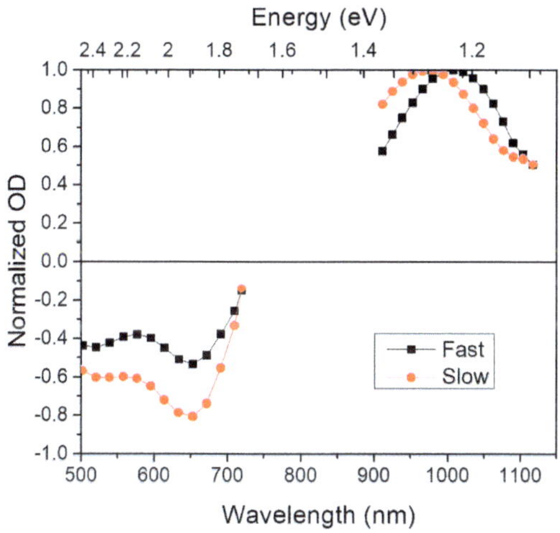

Fig. 8 The two spectral components extracted using SVD from our global fits to the transient absorption spectra of P3HT films sequentially doped with several different concentration solutions of F4TCNQ and excited at multiple wavelengths. The more slowly-decaying component (red circles) has a ~15 ps lifetime while the faster decaying component (black squares) has a 0.37 ps lifetime. Both components have two clear negative peaks in the visible region, at 530 nm and 650 nm, corresponding to bleaches of the BG and P3′ transitions. The slow component, which we assign to trapped polarons, shows a blue-shifted P2 peak in the NIR and more intense bleaching in the visible region relative to that of the fast component, which we assign to free polarons.

spectral components as arising from what would be expected following photo-excitation of a 2-D-delocalized polaron, with a bleach of the bandgap and P3 transitions and the transient appearance of P2 and P3′ peaks. Our hypothesis is thus that the two species can be assigned to free and coulombically-trapped 2-D-delocalized polarons: the species with the redder transient absorption spectrum and faster decay corresponds to that from free polarons, while the species with the blue-shifted transient absorption spectrum and slower recovery corresponds to trapped polarons.

To better characterize the SVD components, we fit each of the two SVD components that describe the transient spectroscopy to a sum of four Gaussian peaks, with the idea that each Gaussian peak reflects one of the expected transient 2-D-delocalized polaron transitions. Both the slow- and the fast-decaying spectral components, which we assign to trapped and free polarons, have two negative Gaussian peaks in the visible region, corresponding to the bleaches of bandgap and P3 transitions. Both transient spectra also have two positive Gaussian peaks, corresponding to newly-created P3′ and P2 absorptions. The Gaussian fitting parameters for both transient species are summarized in Table 1.

The Gaussian fits to the two SVD spectral components summarized in Table 1 make sense in terms our assignment of these as transitions arising from excited 2-D-delocalized polarons. As seen in Fig. 2d, the energy of the P2 transition plus twice the energy of the P1 transition should equal the optical bandgap energy, as should the sum of P3 and P3′ transition energies minus the P2 transition energy. In both cases, the energies of the fitted Gaussian peaks sum to within ~0.1 eV of the known ~2.2 eV bandgap of P3HT.[45]

Table 1 Fitted Gaussian peak locations for the free and trapped polaron spectral components

Assignment	Wavelength (nm)	Energy (eV)	Amplitude	Width (eV)
Free polarons				
Neutral	517	2.40	−0.45	0.22
P3′	658	1.89	−0.52	0.16
P3	765	1.62	0.12	0.17
P2	1000	1.24	1	0.11
P1	2400	0.52	N/A	N/A
Trapped polarons				
Neutral	519	2.39	−0.60	0.22
P3′	670	1.85	−0.86	0.19
P3	742	1.67	0.48	0.12
P2	969	1.28	1	0.11
P1	1500	0.83	N/A	N/A

With this assignment of the two transient species as arising from free and trapped polarons, the ~0.37 and ~15 ps time constants we observe imply that photoexcited free polarons in P3HT recover roughly forty times faster than the trapped polarons. This makes sense given what we expect from the standard 2-D-delocalized polaron picture. For free polarons, photoexcitation relocalizes the polaron and the system recovers to equilibrium as soon as the polymer backbone can relax to accommodate the hole in its new location, a process that can easily happen on sub-ps time scales. For coulombically-trapped polarons, it is reasonable to expect that the excited-state absorption is slightly blue-shifted from that of free polarons, much like the ground-state absorption spectra of free and trapped polarons shown above in Fig. 3. Moreover, we expect that photoexcitation of coulombically-bound polarons likely moves the hole away from the trap site, to where it must diffuse back to recover to equilibrium. Since diffusion is a much slower process (and since we expect polarons near traps to have lower mobilities than their free counterparts), it is not surprising that the excited-state recovery of trapped polarons is more than an order of magnitude slower than that of free polarons.

We turn next to exploring the relative amplitudes of the free and trapped polaron transient absorptions at each excitation wavelength and both doping levels. Fig. 9 shows the ratio of the slow (trapped) to fast (free) polaron components as a function of excitation wavelength for the 5.5 mg mL^{-1}-doped sample (upper panel) and the 1 mg mL^{-1}-doped sample (lower panel). The figure makes clear that more of the slow transient absorption component is excited near 1400 nm than at other wavelengths, particularly in the 5.5 mg mL^{-1}-doped sample. This fits perfectly with our expectation that trapped polarons absorb to the blue of free polarons, and that the blue shoulder observed on the P1 band of the 5.5 mg mL^{-1}-doped sample (Fig. 1) is indeed due to the presence of an increased number of trapped polarons. The fact that many of the polarons added at high doping concentrations are trapped fits well with the observation that doping at this level does not improve the overall film conductivity.[3,14]

Given our assignment of the two transient absorption species to free and trapped 2-D-delocalized polarons, one important question that needs to be

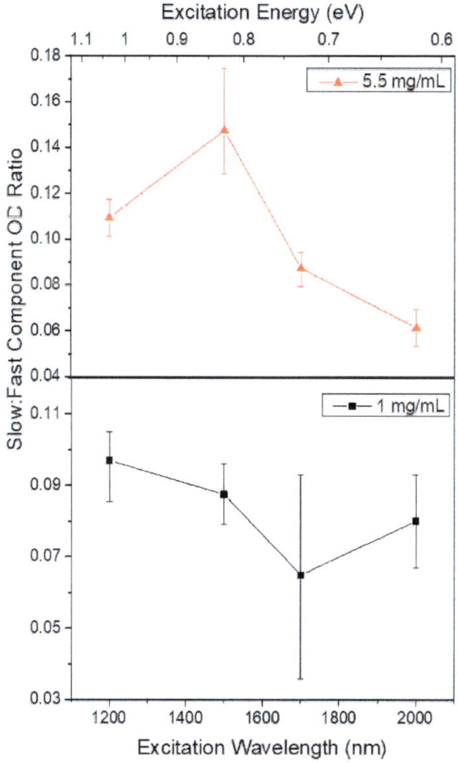

Fig. 9 The ratio of the initial optical densities of the extracted slow to fast components from the global fit of the transient absorption measurements as a function of excitation wavelength for: (a) a P3HT film sequentially doped with a 5.5 mg mL^{-1} F4TCNQ solution, and (b) a P3HT film sequentially doped with a 1 mg mL^{-1} F4TCNQ solution. It is clear that there is more of the slow component produced using 1200 and 1500 nm excitation wavelengths, particularly for the more highly-doped sample.

addressed is why does the number of trapped polarons appear to increase at higher doping concentrations? Previous work has shown that when P3HT is doped by F4TCNQ at low concentrations, the dopant intercalates into the lamellar regions of the P3HT crystallites, leading to a reorientation of the P3HT crystalline unit cell.[14,46–48] Because the F4TCNQ anion sits in the lamellar region, it lies relatively far from the P3HT backbone where the polaron resides, thus minimizing the Coulomb interaction between them. Once the crystalline reorganization is complete, there is no room for additional dopant in the P3HT crystallites, so at high doping concentrations F4TCNQ is forced to reside in the amorphous regions of the P3HT films. We speculate that when the anion is in the amorphous regions of the film, it can sit closer to the polymer backbone, thus increasing the Coulomb attraction and the degree of polaron trapping.

We also can analyze the relative amplitudes of free and trapped polarons in Fig. 9 in another way. If we follow the logic in the preceding paragraph, we can assume that the 0.1 mg mL^{-1}-doped P3HT sample has a negligible amount of trapped polarons. This means that the P1 band of this film is largely that of free polarons. If we then scale our measured ratio of trapped-to-free polarons by the

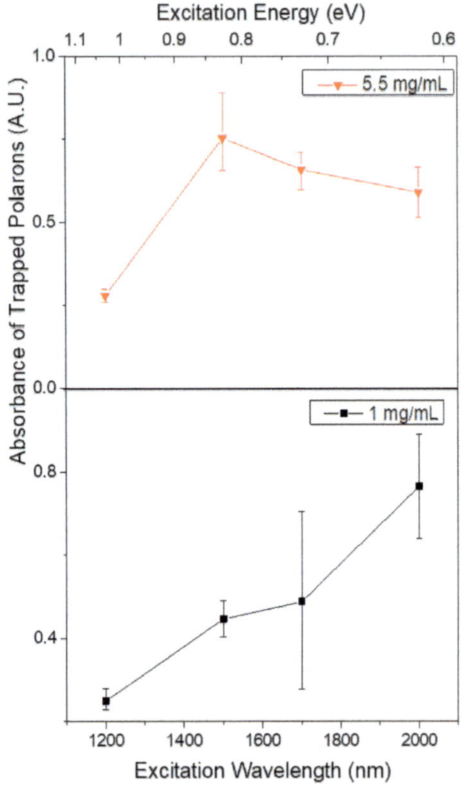

Fig. 10 The calculated action spectrum of trapped polarons for: (a) a P3HT film sequentially doped with a 5.5 mg mL^{-1} F4TCNQ solution, and (b) a P3HT film sequentially doped with a 1 mg mL^{-1} F4TCNQ solution. This spectrum was calculated by scaling the measured ratios of the slow to fast components in Fig. 9 by the absorption of a P3HT films sequentially doped with a 0.1 mg mL^{-1} F4TCNQ solution, which we expect has few trapped polarons. The data make clear that the more highly-doped sample has trapped polarons with a more blue-shifted absorption spectrum, indicating that these polarons are more highly trapped than those in the lower-doped sample.

0.1 mg mL^{-1}-doped film absorption spectrum, we can obtain the absorption spectrum of the trapped polarons in the higher-doped P3HT films. These spectra, which are one of the principal results of this work, are shown in Fig. 10. Essentially, these are reconstructed action spectra of the slow transient absorption component that we assign to the presence of trapped polarons.

The data in Fig. 10 make clear that the spectrum of trapped polarons is different at different doping concentrations. If we compare our extracted trapped polaron action spectra to what is expected from theory (Fig. 3, above),[14,23,29] we can estimate the average distance the polaron lies from its counterion at each doping concentration. We accomplished this by using a least squares regression to fit our measured action spectra of the trapped polaron in P3HT films doped with different F4TCNQ concentrations to a weighted sum of the simulated polaron spectra for varying anion distances,[23] as shown in Fig. 11. We conclude that the 'free' polarons in 0.1 mg mL^{-1}-F4TCNQ-doped P3HT films are still somewhat coulombically bound by their counterions, with an average polaron-to-anion

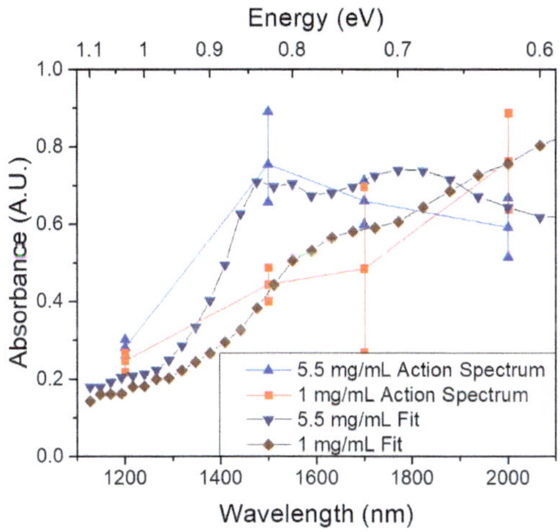

Fig. 11 The action spectra of trapped polarons from Fig. 10 fit to linear combinations of the anion-distance-dependent P1 spectra calculated by Spano and co-workers shown in Fig. 3.[23] The 5.5 mg mL^{-1} F4TCNQ-doped P3HT film (blue triangles) fits best to a sum of the theoretical spectra with anion distances calculated for 0.2, 0.4, and 0.6 nm spacing between the anion and the polymer chain with 20, 7, and 73% relative weightings, respectively. The 1 mg mL^{-1} F4TCNQ-doped P3HT film (red squares) fits best to a sum of the spectra for anion distances calculated for 0.4, 0.6, and 0.8 nm spacing between the anion and the polymer chain with 6, 4 and 90% relative weightings, respectively.

distance of 0.7 to 0.9 nm, consistent with the idea that the dopant counterions sit in the lamellar regions of the P3HT crystallites. We also see that the distance between the trapped polarons in the 5.5 mg mL^{-1}-doped P3HT films and their F4TCNQ counteranions is 0.2 nm to 0.6 nm, and the corresponding distance in the 1 mg mL^{-1}-doped films is 0.4 to 0.8 nm. This also fits well with our idea that at increased dopant concentrations, the dopant sits in the amorphous region of the film where it can reside closer, on average, to the polarons on the P3HT backbone.

As a check on the validity of our assignments of the peaks observed in the transient spectroscopy, we worked to reconstruct the 0.1 mg mL^{-1} UV-Vis spectrum (Fig. 1), which should result almost entirely from free polarons, by using the fitted peak positions and widths of the fast transient spectral component (Table 1), along with the known spectra of neutral P3HT and anionic F4TCNQ. The results, which also make use of a single Gaussian to represent the P1 absorption band, are shown in Fig. 12. Clearly, our ability to fit the steady-state UV-Visible absorption spectrum with only amplitudes as adjustable parameters is excellent. This lends credence to the idea that there are indeed 2-D-delocalized P3, P3′ and P2 transitions lying under the sharp F4TCNQ anion peaks in the UV-Vis of doped P3HT.

Finally, we are also able to check our understanding of the various polaronic transitions by working to reproduce the steady-state absorption spectrum of the 5.5 mg mL^{-1}-F4TCNQ-doped P3HT sample using the trapped polaron action spectrum extracted from our ultrafast spectroscopic analysis. Here, we started with the 0.1 mg mL^{-1}-doped-P3HT steady-state spectrum, and added additional absorption from the action spectrum for trapped polarons that we extracted from the slow transient absorption component. Fig. 13 shows that again, the results are

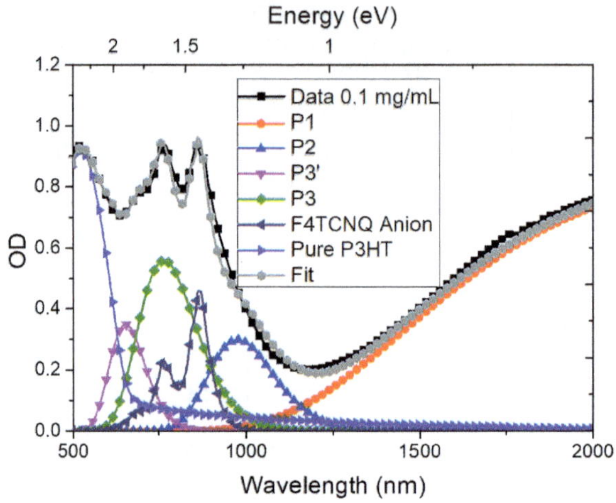

Fig. 12 Reconstruction of the absorption spectrum of a P3HT film sequentially doped with a 0.1 mg mL^{-1} solution of F4TCNQ in DCM using a linear combination of the known F4TCNQ anion spectra (taken from Wang *et al.*),[33] the known spectrum of neutral P3HT (Fig. 1), and the Gaussian fits of the peak positions of the fast component (which we assign to free polarons) extracted from the global model of our transient absorption measurements (Table 1); see text for details. The excellent agreement between the measured spectrum and the reconstructed fit shows that our transient absorption measurements, global fit, and steady-state absorption measurements are all internally consistent.

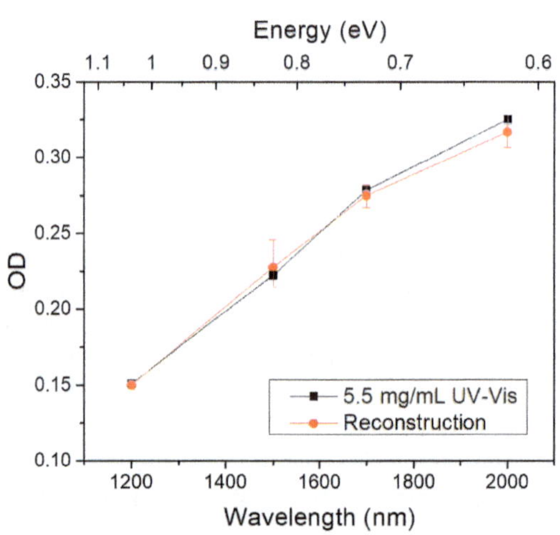

Fig. 13 Reconstruction of the absorption spectrum of a highly-doped P3HT film (created using a 5.5 mg mL^{-1} solution of F4TCNQ). The reconstructed spectrum uses the spectrum of a low-doped P3HT film (created using an 0.1 mg mL^{-1} F4TCNQ solution) and the action spectrum of trapped polarons that we extracted from the ultrafast transient absorption data in Fig. 10; see text for details. The excellent agreement provides confirmation that the action spectrum taken from our ultrafast measurements can be used to account for the spectral shape changes observed between the high- and low-doped P3HT films seen in Fig. 1.

excellent. Thus, the spectra of trapped polarons that we extracted from our ultrafast spectroscopy experiments are indeed internally-consistent with the measured steady-state spectra at each doping level.

4 Conclusions

In this paper we have found strong evidence for two populations of quasiparticles present in F4TCNQ-doped P3HT. The transient spectroscopy is consistent with F4TCNQ-doped P3HT consisting primarily of free and trapped polarons, both of which follow the traditional model for 2-D-delocalized polarons in terms of their observed absorption transitions. Neither the band-bending picture of polarons nor the presence of bipolarons accounts for the observed bleaches in our transient absorption measurements. We were able to distinguish free and trapped polarons by monitoring both the position and dynamics of their transient spectra: trapped polarons show blue-shifted features that decay more than an order of magnitude more slowly than free polarons, a direct result of their need to diffuse to return to equilibrium and their generally poor mobility.

By monitoring the relative amounts of free and trapped polarons produced at different excitation wavelengths, we were able to generate an action spectrum to disentangle the steady-state absorption of the two polaronic species. We found that the more highly-doped samples had greater amounts of polarons that were more tightly bound by their counterions. By comparing the action spectra of trapped polarons to theory, we found that the F4TCNQ dopant resides roughly 0.4 nm away from trapped polarons, compared to 0.7 to 0.9 nm for free polarons. For an F4TCNQ anion to be that close to a trapped polaron, it likely must pi-stack with the P3HT backbone, presumably in the amorphous regions of the film.

With an internally-consistent view of the electronic structure of free and trapped polarons in place, it may now be possible to design new materials to improve carrier mobility by decreasing the fraction of trapped polarons. For example, the use of dopants whose anions are electrostatically shielded from their corresponding holes provides a possible route to preventing the build-up of trapped polarons at high doping levels.[49,50] Another possibility would be to engineer the crystal structure of a polymer to better accommodate counterions into the lattice in positions far from the backbone, so that more dopants can be accommodated without having to produce significant numbers of trapped polarons. The hope is that with our new understanding, it should be possible to create new materials that do not suffer from increased polaron trapping at high doping concentrations.

Conflicts of interest

There are no conflicts of interest to declare.

Acknowledgements

This work was supported by the National Science Foundation under grant numbers CBET-1510353 and CHE-1608957. This material is based upon work supported by the National Science Foundation Graduate Research Fellowship under Grant No. DGE-1650604. Any opinion, findings, and conclusions or recommendations expressed in this material are those of the authors and do not necessarily reflect the views of the National Science Foundation.

References

1 F. Deschler, D. Riedel, A. Deák, B. Ecker, E. V. Hauff and E. Da, *Synth. Met.*, 2015, **199**, 381–387.

2 P. Pingel and D. Neher, *Phys. Rev. B: Condens. Matter Mater. Phys.*, 2013, **87**, 115209.

3 D. T. Scholes, S. A. Hawks, P. Y. Yee, H. Wu, J. R. Lindemuth, S. H. Tolbert and B. J. Schwartz, *J. Phys. Chem. Lett.*, 2015, **6**, 4786–4793.

4 C. Francis, D. Fazzi, S. B. Grimm, F. Paulus, S. Beck, S. Hillebrandt, A. Pucci and J. Zaumseil, *J. Mater. Chem. C*, 2017, **5**, 6176–6184.

5 E. Lim, K. A. Peterson, G. M. Su and M. L. Chabinyc, *Chem. Mater.*, 2018, **30**, 998–1010.

6 C. Enengl, S. Enengl, S. Pluczyk, M. Havlicek, M. Lapkowski, H. Neugebauer and E. Ehrenfreund, *ChemPhysChem*, 2016, **56**, 607–622.

7 J. Yamamoto and Y. Furukawa, *Org. Electron.*, 2016, **28**, 82–87.

8 M. Cox, E. H. M. van der Heijden, P. Janssen and B. Koopmans, *Phys. Rev. B: Condens. Matter Mater. Phys.*, 2014, **89**, 085201.

9 S. Y. Yang, B. N. Kim, A. A. Zakhidov, P. G. Taylor, J. K. Lee, C. K. Ober, M. Lindau and G. G. Malliaras, *Adv. Mater.*, 2011, **23**, 184–188.

10 G. Lu, J. Blakesley, S. Himmelberger, P. Pingel, J. Frisch, I. Lieberwirth, I. Salzmann, M. Oehzelt, R. Di Pietro, A. Salleo, N. Koch and D. Neher, *Nat. Commun.*, 2013, **4**, 1588.

11 P. Bujak, I. Kulszewicz-Bajer, M. Zagorska, V. Maurel, I. Wielgus and A. Pron, *Chem. Soc. Rev.*, 2013, **42**, 8895.

12 L. Sudha Devi, M. K. Al-Suti, C. Dosche, M. S. Khan, R. H. Friend and A. Köhler, *Phys. Rev. B: Condens. Matter Mater. Phys.*, 2008, **78**, 1–8.

13 P. P. Khlyabich, B. Burkhart and B. C. Thompson, *J. Am. Chem. Soc.*, 2011, **133**, 14534–14537.

14 D. T. Scholes, P. Y. Yee, J. R. Lindemuth, H. Kang, J. Onorato, R. Ghosh, C. K. Luscombe, F. C. Spano, S. H. Tolbert and B. J. Schwartz, *Adv. Funct. Mater.*, 2017, **27**, 1702654.

15 A. Hamidi-Sakr, L. Biniek, S. Fall and M. Brinkmann, *Adv. Funct. Mater.*, 2016, **26**, 408–420.

16 J. L. Bredas and G. B. Street, *Acc. Chem. Res.*, 1985, **18**, 309–315.

17 M. Nowak, S. D. D. V. Rughooputh, S. Hotta and A. J. Heeger, *Macromolecules*, 1987, **20**, 965–968.

18 D. Beljonne, J. Cornil, H. Sirringhaus, P. J. Brown, M. Shkunov, R. H. Friend and J.-L. Brédas, *Adv. Funct. Mater.*, 2001, **11**, 229–234.

19 G. Heimel, *ACS Cent. Sci.*, 2016, **2**, 309–315.

20 R.-Q. Png, M. C. Ang, M.-H. Teo, K.-K. Choo, C. G. Tang, D. Belaineh, L.-L. Chua and P. K. Ho, *Nat. Commun.*, 2016, **7**, 11948.

21 K. Kohlstedt, *ACS Cent. Sci.*, 2016, **2**, 278–280.

22 S. Kahmann, D. Fazzi, G. J. Matt, W. Thiel, M. A. Loi and C. J. Brabec, *J. Phys. Chem. Lett.*, 2016, **7**, 4438–4444.

23 R. Ghosh, A. R. Chew, J. Onorato, V. Pakhnyuk, C. K. Luscombe, A. Salleo and F. C. Spano, *J. Phys. Chem. C*, 2018, **122**, 18048–18060.

24 S. C. Doan and B. J. Schwartz, *J. Phys. Chem. C*, 2013, **117**, 4216–4221.

25 J. Cornil and J.-L. Brédas, *Adv. Mater.*, 1995, **7**, 295–297.

26 Y. H. Kim, S. Hotta and A. J. Heeger, *Phys. Rev. B: Condens. Matter Mater. Phys.*, 1987, **36**, 7486–7490.

27 K. E. Ziemelis, A. T. Hussain, D. D. C. Bradley and R. H. Friend, *Phys. Rev. Lett.*, 1991, **66**, 2231–2234.

28 V. I. Arkhipov, E. V. Emelianova, P. Heremans and H. Bässler, *Phys. Rev. B: Condens. Matter Mater. Phys.*, 2005, **72**, 235202.

29 A. R. Chew, R. Ghosh, Z. Shang, F. C. Spano and A. Salleo, *J. Phys. Chem. Lett.*, 2017, **8**, 4974–4980.

30 M. J. Bird, J. Bakalis, S. Asaoka, H. Sirringhaus and J. R. Miller, *J. Phys. Chem. C*, 2017, **121**, 15597–15609.

31 S. Kahmann, M. A. Loi and C. J. Brabec, *J. Mater. Chem. C*, 2018, **18**, 10270–10280.

32 I. E. Jacobs, E. W. Aasen, J. L. Oliveira, T. N. Fonseca, J. D. Roehling, J. Li, G. Zhang, M. P. Augustine, M. Mascal and A. J. Moulé, *J. Mater. Chem. C*, 2016, **4**, 3454–3466.

33 C. Wang, D. T. Duong, K. Vandewal, J. Rivnay and A. Salleo, *Phys. Rev. B: Condens. Matter Mater. Phys.*, 2015, **91**, 085205.

34 J. Gao, B. W. Stein, A. K. Thomas, J. A. Garcia, J. Yang, M. L. Kirk and J. K. Grey, *J. Phys. Chem. C*, 2015, **119**, 16396–16402.

35 F. M. McFarland, C. M. Ellis and S. Guo, *J. Phys. Chem. C*, 2017, **121**, 4740–4746.

36 J. Fuzell, I. E. Jacobs, S. Ackling, T. F. Harrelson, D. M. Huang, D. Larsen and A. J. Moulé, *J. Phys. Chem. Lett.*, 2016, **7**, 4297–4303.

37 O. G. Reid, R. D. Pensack, Y. Song, G. D. Scholes and G. Rumbles, *Chem. Mater.*, 2014, **26**, 561–575.

38 E. Lioudakis, I. Alexandrou and A. Othonos, *Nanoscale Res. Lett.*, 2009, **4**, 1475–1480.

39 A. Saeki, S. Seki, Y. Koizumi and S. Tagawa, *J. Photochem. Photobiol., A*, 2007, **186**, 158–165.

40 H. Li, A. Gauthier-Houle, P. Grégoire, E. Vella, C. Silva-Acuña and E. R. Bittner, *Chem. Phys.*, 2016, **481**, 281–286.

41 F. Provencher, N. Bérubé, A. W. Parker, G. M. Greetham, M. Towrie, C. Hellmann, M. Côté, N. Stingelin, C. Silva and S. C. Hayes, *Nat. Commun.*, 2014, **5**, 4288.

42 M. Menšík, J. Pfleger and P. Toman, *Chem. Phys. Lett.*, 2017, **677**, 87–91.

43 S. Das, P. P. Khlyabich, B. Burkhart, S. T. Roberts, E. Couderc, B. C. Thompson and S. E. Bradforth, *J. Phys. Chem. C*, 2014, **118**, 6650–6660.

44 H. Ohkita, S. Cook, Y. Astuti, W. Duffy, S. Tierney, W. Zhang, M. Heeney, I. McCulloch, J. Nelson, D. D. C. Bradley and J. R. Durrant, *J. Am. Chem. Soc.*, 2008, **130**, 3030–3042.

45 A. Guerrero, S. Loser, G. Garcia-Belmonte, C. J. Bruns, J. Smith, H. Miyauchi, S. I. Stupp, J. Bisquert and T. J. Marks, *Phys. Chem. Chem. Phys.*, 2013, **15**, 16456–16462.

46 A. Hamidi-Sakr, L. Biniek, J.-l. Bantignies, D. Maurin, L. Herrmann, N. Leclerc, P. Lévêque, V. Vijayakumar and N. Zimmermann, *Adv. Funct. Mater.*, 2017, **1700173**, 1–13.

47 D. T. Duong, C. Wang, E. Antono, M. F. Toney and A. Salleo, *Org. Electron.*, 2013, **14**, 1330–1336.

48 J. Hynynen, D. Kiefer, L. Yu, R. Kroon, R. Munir, A. Amassian, M. Kemerink and C. Müller, *Macromolecules*, 2017, **50**, 8140–8148.
49 Z. Liang, Y. Zhang, M. Souri, X. Luo, A. M. Boehm, R. Li, Y. Zhang, T. Wang, D.-Y. Kim, J. Mei, S. R. Marder and K. R. Graham, *J. Mater. Chem. A*, 2018, **6**, 16495–16505.
50 T. J. Aubry, J. C. Axtell, V. M. Basile, K. J. Winchell, J. R. Lindemuth, T. M. Porter, J.-Y. Liu, A. N. Alexandrova, C. P. Kubiak, S. H. Tolbert, A. M. Spokoyny and B. J. Schwartz, *Adv. Mater.*, 2019, **31**, 1805647.

Faraday Discussions

PAPER

Theoretical analysis of the inverted region in photoinduced proton-coupled electron transfer

Zachary K. Goldsmith, Alexander V. Soudackov and Sharon Hammes-Schiffer *

Received 14th December 2018, Accepted 7th January 2019

DOI: 10.1039/c8fd00240a

Photoinduced proton-coupled electron transfer (PCET) plays a key role in a wide range of energy conversion processes, and understanding how to design systems to control the PCET rate constant is a significant challenge. Herein a theoretical formulation of PCET is utilized to identify the conditions under which photoinduced PCET may exhibit inverted region behavior. In the inverted region, the rate constant decreases as the driving force increases even though the reaction becomes more thermodynamically favorable. Photoinduced PCET will exhibit inverted region behavior when the following criteria are satisfied: (1) the overlap integrals corresponding to the ground reactant and the excited product proton vibrational wavefunctions become negligible for a low enough product vibronic state and (2) the reaction free energies associated with the lower excited product proton vibrational wavefunctions contributing significantly to the rate constant are negative with magnitudes greater than the reorganization energy. These criteria are typically not satisfied by harmonic or Morse potentials but are satisfied by more realistic asymmetric double well potentials because the proton vibrational states above the barrier correspond to more delocalized proton vibrational wavefunctions with nodal structures leading to destructive interference effects. Thus, this theoretical analysis predicts that inverted region behavior could be observed for systems with asymmetric double well potentials characteristic of hydrogen-bonded systems and that the hydrogen/deuterium kinetic isotope effect will approach unity and could even become inverse in this region due to the oscillatory nature of the highly excited vibrational wavefunctions. These insights may help guide the design of more effective energy conversion devices.

1. Introduction

Photoinduced proton-coupled electron transfer (PCET) is an essential component of a wide range of energy conversion processes throughout chemistry and biology.

Department of Chemistry, Yale University, 225 Prospect Street, New Haven, Connecticut 06520, USA. E-mail: sharon.hammes-schiffer@yale.edu

Examples of such processes include natural and artificial photosynthesis, as well as photoelectrochemical reduction of carbon dioxide to carbon monoxide or hydrocarbons.[1–8] Given the complexity of these types of processes, which may include many different steps, simple model systems have been designed to enable the investigation of the fundamental underlying physical principles of photoinduced PCET.[9–21] In these model systems, PCET may be initiated by photoexcitation to a metal-to-ligand or ligand-to-metal charge transfer state[9,11,14,15,17,18] or to a locally excited state of an aromatic molecular component.[12,13,16,19–21] After this initial photoexcitation, the PCET reaction can occur through a variety of different mechanisms involving charge separation and charge recombination.

An interesting question that arises for photoinduced PCET is whether the Marcus inverted region behavior can be observed and, if so, under what conditions it is expected to be observed. In Marcus theory for electron transfer (ET), the rate constant depends exponentially on the free energy barrier,[22,23] which is expressed in terms of the reaction free energy $\Delta G^{\rm o}$ and the reorganization energy λ as follows:

$$\Delta G^{\dagger} = \frac{(\Delta G^{\rm o} + \lambda)^2}{4\lambda}. \tag{1}$$

Marcus theory predicts an inverted parabola for the dependence of the logarithm of the ET rate constant on the driving force $-\Delta G^{\rm o}$ (Fig. 1). The maximum rate constant is predicted to occur for $-\Delta G^{\rm o} = \lambda$ because the free energy barrier vanishes and the reaction becomes activationless. In the inverted region, where $-\Delta G^{\rm o} > \lambda$, the ET rate constant decreases as the driving force

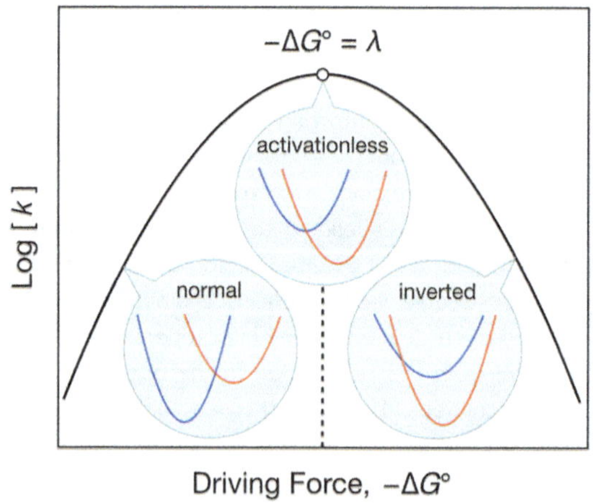

Fig. 1 Schematic plot of the logarithm of the rate constant k *versus* the driving force $-\Delta G^{\rm o}$ with the free energy barrier given by eqn (1). The dashed vertical line indicates the driving force equal to the reorganization energy λ. Encircled are depictions of the normal, activationless, and inverted regions, illustrated by the reactant (blue) and product (red) free energy parabolas as functions of a collective solvent coordinate. In the normal region, the free energy barrier decreases, and the rate constant correspondingly increases, as the driving force increases. The activationless region occurs when $-\Delta G^{\rm o} = \lambda$ and the barrier is zero. In the inverted region, the free energy barrier increases, and the rate constant correspondingly decreases, as the driving force increases.

increases (*i.e.*, as the reaction becomes more exoergic). The inverted region has been observed experimentally for electron transfer[24] and is technologically relevant because it has been proposed to be useful for slowing down charge recombination reactions to avoid wasting energy in energy conversion processes.[25,26] However, it can be significantly suppressed for systems with accessible excited intramolecular vibrational states[27–30] or completely disappear for electrochemical ET due to the available continuum of electronic states associated with the electrode.[31–33]

We have developed a general theory for PCET[34–41] that may be viewed as an extension of Marcus theory for ET. In this theory, the transferring proton, as well as the active electrons, is treated quantum mechanically, and the PCET reaction is described in terms of nonadiabatic transitions between mixed electron–proton vibronic states, corresponding to proton vibrational states for each electronic state (Fig. 2). The rate constant is calculated as the Boltzmann-weighted sum of the rate constants associated with transitions between all pairs of these electron–proton vibronic states, where each term depends exponentially on the corresponding free energy barrier and is proportional to the square of the overlap

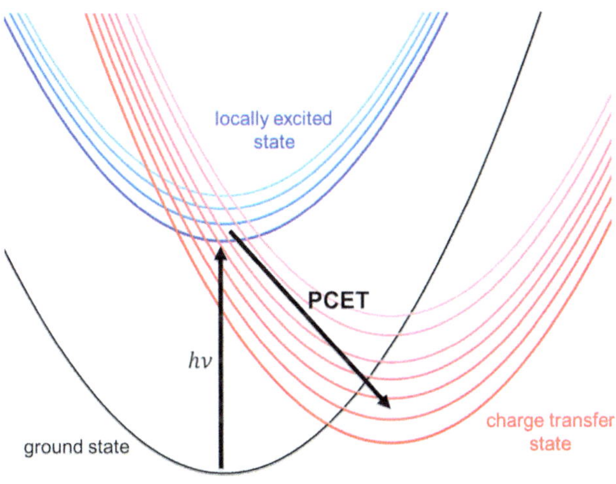

Fig. 2 Illustration of a reaction scheme for photoinduced PCET. Photoexcitation from the ground state (black) to a locally excited state (blue) is followed by PCET to a charge transfer state (red), which would subsequently decay to the ground state. The manifolds of stacked parabolas are representative of electron–proton vibronic states, which are essentially proton vibrational states for each electronic state, as functions of a collective solvent coordinate. The calculations presented herein are based on the assumption that vibrational cooling in the locally excited state is faster than PCET, and therefore the PCET reaction occurs predominantly from the ground vibronic state (lowest, darkest blue parabola). Moreover, PCET from this locally excited state to the charge transfer state involves transitions that span the normal (higher, lighter red parabolas) and inverted (lower, darker red parabolas) regions that are depicted in Fig. 1. This type of PCET process could exhibit inverted region behavior. An alternative situation could occur in which the PCET reaction from the charge transfer state back down to the ground state is more exoergic. In this case, PCET from mainly the ground vibronic state of the charge transfer state (lowest, darkest red parabola) to a manifold of proton vibrational states associated with the ground state (black stacked parabolas not shown here for simplicity) could exhibit inverted region behavior.

integral associated with the reactant and product proton vibrational wavefunctions. A previous study from our group[42] suggested that inverted region behavior is unlikely to be observed for PCET reactions because of the availability of excited proton vibrational states associated with the product (Fig. 2). As the driving force increases, the free energy barrier for the currently dominant product proton vibrational state increases, but the next excited proton vibrational state becomes nearly activationless and therefore becomes dominant. The accessibility of these excited proton vibrational states causes the rate constant to plateau at higher driving forces, and the inverted region behavior is not observed at experimentally relevant driving forces. However, this previous study was performed using harmonic proton potentials, which are not realistic representations of the bonds between the hydrogen and its donor or acceptor, especially for hydrogen-bonded systems.

Herein we investigate the dependence of the PCET rate constant on the driving force using more realistic proton potentials that are similar to those calculated with density functional theory (DFT) for experimentally studied PCET systems.[43–47] In particular, we perform a comprehensive comparison between the results for a model using Morse potentials and the results for a model using chemically motivated asymmetric double well potentials similar to those generated with DFT for real chemical systems. Our analysis illustrates that inverted region behavior is observed under certain conditions that depend on the shape of the proton potential energy curves. We also investigate the hydrogen/deuterium kinetic isotope effects (KIEs) for these model systems and explain why the KIEs become unity, or even slightly less than unity, at large driving forces. These results are also placed in the context of available experimental studies on photoinduced PCET.

2. Nonadiabatic PCET theory

A general theoretical framework has been developed to describe PCET reactions in solution and proteins.[34–41] In this PCET theory, the transferring proton, as well as the active electrons, are treated quantum mechanically. In the vibronically nonadiabatic regime, the PCET reaction is described in terms of nonadiabatic transitions between reactant and product electron–proton vibronic states. In practice, the proton vibrational states are computed for the reactant and the product diabatic electronic states, and each vibronic state is expressed as the product of one of the diabatic electronic states and an associated proton vibrational state. In this framework, the vibronically nonadiabatic PCET rate constant is a Boltzmann-weighted sum over the rate constants associated with nonadiabatic quantum transitions between all pairs of reactant and product electron–proton vibronic states.

This vibronically nonadiabatic PCET rate constant is expressed as a function of the distance R between the proton donor and acceptor as follows:[35]

$$k(R) = \frac{1}{\hbar}\sqrt{\frac{\pi}{\lambda k_B T}}\sum_{\mu,\nu} P_\mu \left|V_{\mu\nu}(R)\right|^2 \exp\left[-\frac{\Delta G^\dagger_{\mu\nu}}{k_B T}\right]. \tag{2}$$

Here, μ and ν correspond to the reactant and product vibronic states, respectively, P_μ is the Boltzmann population of reactant state μ, $V_{\mu\nu}(R)$ is the vibronic coupling between reactant and product states μ and ν, λ is the total reorganization energy for the reaction, and $\Delta G^{\dagger}_{\mu\nu}$ is the free energy barrier associated with the transition between reactant state μ and product state ν. The free energy barrier for reactant/product vibronic state pair (μ,ν) is

$$\Delta G^{\dagger}_{\mu\nu} = \frac{\left(\Delta G^{\mathrm{o}}_{\mu\nu} + \lambda\right)^2}{4\lambda}, \tag{3}$$

where $\Delta G^{\mathrm{o}}_{\mu\nu}$ is the reaction free energy associated with vibronic state pair (μ,ν). In the electronically nonadiabatic limit, the vibronic coupling $V_{\mu\nu}$ is expressed as the product of the electronic coupling V^{el} and the overlap integral $S_{\mu\nu}$ between the proton vibrational wavefunctions associated with reactant/product vibronic state pair (μ,ν):

$$V_{\mu\nu}(R) = V^{\mathrm{el}} S_{\mu\nu}(R). \tag{4}$$

Note that the electronic coupling is assumed to be independent of the proton donor–acceptor distance; however, the overlap integral depends strongly on this distance.[37]

The PCET rate constant is thermally averaged over the proton donor–acceptor distance to obtain the total rate constant, k_{PCET}. This thermal averaging is accomplished by weighting each rate constant $k(R)$ by the probability distribution $P(R)$ for sampling the proton donor–acceptor distance R and integrating over this distance:[38,43,44]

$$k_{\mathrm{PCET}} = \int_0^{\infty} \mathrm{d}R\, k(R) P(R). \tag{5}$$

In the present study, the proton donor–acceptor motion is assumed to be harmonic with an equilibrium proton donor–acceptor distance $\bar{R}$ and an effective force constant k_{eff}. Thus, $P(R)$ is a normalized Gaussian distribution function with the following form:

$$P(R) = \frac{\exp\left[-k_{\mathrm{eff}}(R-\overline{R})^2 / 2k_{\mathrm{B}}T\right]}{\int_{-\infty}^{\infty} \exp\left[-k_{\mathrm{eff}}(R-\overline{R})^2 / 2k_{\mathrm{B}}T\right]\mathrm{d}R}. \tag{6}$$

3. Model systems for PCET: Morse and asymmetric double well potentials

Two models for the diabatic reactant and product proton potential energy curves were used to probe the conditions for observing the inverted region for PCET. The first model describes the reactant and product proton potential energy curves as Morse potentials, which are nearly harmonic at the equilibrium bond length but also describe bond dissociation. The Morse potentials can be expressed as functions of the proton coordinate r along the proton donor–acceptor axis as follows:

$$U_1(r) = D_{\mathrm{e}}\left[1 - \mathrm{e}^{-\beta_1\left(r + \frac{\delta R}{2}\right)}\right]^2, \tag{7}$$

$$U_2(r) = D_e\left[1 - e^{\beta_2\left(r - \frac{\delta R}{2}\right)}\right]^2 \tag{8}$$

where D_e is the bond dissociation energy and β_1 and β_2 are parameters related to the frequency of the X–H bond for the proton donor and acceptor, respectively. Here δR is the distance between the minima of the two Morse potentials and is related to the proton donor–acceptor distance R by a constant defined as the sum of the two X–H equilibrium bond lengths. For this study, Morse parameters consistent with typical X–H bond frequencies of 3300 cm^{-1} and 3000 cm^{-1} for the proton donor and acceptor, respectively, were chosen. In particular, for the reactant proton potential, $D_e = 100$ kcal mol^{-1} and $\beta_1 = 1.14$ Å^{-1}, and for the product proton potential, $D_e = 100$ kcal mol^{-1} and $\beta_2 = 1.04$ Å^{-1}. These Morse potentials are depicted in Fig. 3A.

The second model for PCET describes the reactant and product proton potential energy curves as asymmetric double well potentials. Such double well potentials have been commonly observed for PCET systems previously studied by this group using DFT.[43,45–47] In particular, an asymmetric double well proton potential, or a proton potential with a significant shoulder, is expected to be

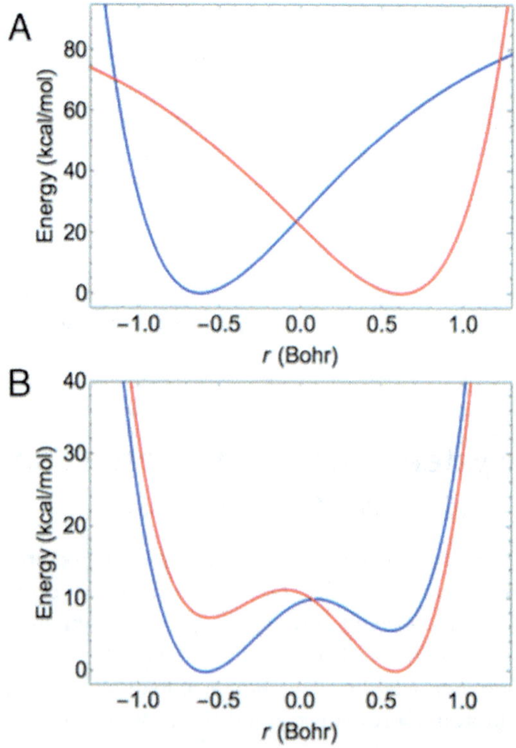

Fig. 3 (A) Morse potentials for the model reactant (blue) and product (red) proton potential energy curves associated with the diabatic electronic states. (B) Asymmetric double well potentials for the model reactant (blue) and product (red) proton potential energy curves associated with the diabatic electronic states.

found when the proton transfer occurs between a donor and acceptor interacting *via* a relatively strong hydrogen bond.

The asymmetric double well potentials used for this second model system were obtained from the reactant and product Morse potentials used in the first model system. The procedure for obtaining these asymmetric double well potentials at each value of R is as follows:

(1) The product Morse potential, $U_2(r)$, is shifted in energy by an asymmetry factor $\pm\Delta$ relative to the minimum of $U_1(r)$. For the reactant, $U_2(r)$ is raised by Δ, and for the product it is lowered by Δ.

(2) A 2 × 2 matrix, given in eqn (9), is constructed with the shifted Morse potentials as the diagonal elements and a coupling, V^{PT}, as the off-diagonal element:

$$\begin{bmatrix} U_1(r) & V^{\mathrm{PT}} \\ V^{\mathrm{PT}} & U_2(r) \pm \Delta \end{bmatrix}. \tag{9}$$

(3) The matrices described in eqn (9) are diagonalized, and the lowest eigenvalue for each matrix defines the reactant and product asymmetric double well potentials. For the regions of $|r| > 1.0$ Bohr, switching functions were used to transition smoothly from the lowest eigenvalue to the corresponding higher-energy diabatic potentials to circumvent unphysical avoided crossing points in these regions.

For this study, the energetic shift Δ was chosen to be 10 kcal mol^{-1}, and the coupling V^{PT} was chosen to be 30 kcal mol^{-1}. The reactant and product asymmetric double well potentials are depicted in Fig. 3B.

For each proton potential, the proton or deuteron vibrational wavefunctions and their energy levels were obtained by solving the one-dimensional Schrödinger equation analytically[48] for the Morse potentials or numerically[49] for the asymmetric double well potentials. The proton vibrational wavefunctions for the model systems described above are depicted in Fig. 4. This figure illustrates that

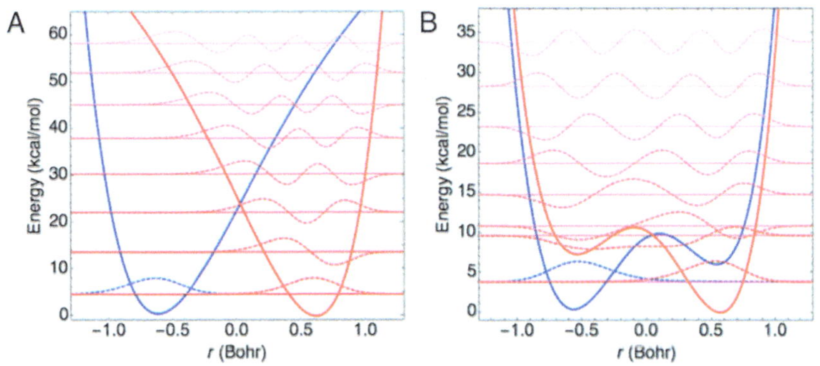

Fig. 4 Model reactant (blue, solid) and product (red, solid) proton potentials and the associated proton vibrational energy levels (horizontal lines) and wavefunctions (dashed curves) for the (A) Morse and (B) asymmetric double well potentials. Only the ground proton vibrational state for the reactant is shown along with a series of excited product proton vibrational states ($\nu = 0-7$). Note that the smaller energy scale in B indicates a much denser manifold of proton vibrational states (*i.e.*, smaller splittings between the energy levels) for the asymmetric double well potentials than for the Morse potentials.

the proton vibrational wavefunctions for the reactant and product Morse potentials are much more localized than those for the asymmetric double well potentials. This characteristic difference will have consequences in terms of the overlap integrals, $S_{\mu\nu}$, in the rate constant calculations. Furthermore, the proton vibrational states above the barrier for the asymmetric double well potentials exhibit significantly smaller energy level splittings than those for the Morse potentials, influencing the reaction free energies in the rate constant calculations. These issues will be discussed further below.

4. Relationship between driving force and PCET rate constant for model systems

The plausibility of observing the Marcus inverted region for PCET was investigated by calculating k_{PCET} over a range of driving forces ($-\Delta G^o$) from zero to values significantly larger in magnitude than λ for both model systems. The qualitative behavior is not altered by the specific values of the parameters entering the rate constant expression, and we chose values that are similar to those calculated for PCET systems previously studied by our group. The force constant associated with the harmonic proton donor–acceptor motion was chosen to be $k_{eff} = 0.051$ au, which is comparable to previously computed values for intramolecular, hydrogen-bonded PCET systems.[45] The reorganization energy was chosen to be $\lambda = 25$ kcal mol^{-1}, which is consistent with many calculated and experimental values for outer-sphere ET and PCET.[50] Lastly, the electronic coupling was chosen to be $V^{el} = 1$ kcal mol^{-1}. The electronic coupling is simply a constant scaling factor in the expression for k_{PCET} and does not impact the relationship between the rate constant and the driving force or the KIE. In all calculations, the number of electron–proton vibronic states included was fully converged, which required a significant number of product states (*i.e.*, up to 16 for the Morse potentials and 28 for the asymmetric double well potentials) for the highly exoergic reactions. The relationships between the driving force and k_{PCET}, as well as the KIE, for both types of model potentials are shown in Fig. 5.

As shown previously for harmonic potentials,[42] the inverted region for PCET is not observed for the model system using Morse potentials (Fig. 5A). Neglecting the excited proton vibrational state contributions to the rate constant (*i.e.*, including only the (0,0) vibronic state pair) yields a clear attenuation in the total rate constant at driving forces greater than λ (Fig. 5A, dashed gray line). In contrast, the results including the contributions from the excited proton vibrational states exhibit no apparent decrease in the rate constant up to $-\Delta G^o =$ 80 kcal mol^{-1}. In other words, the rate constant k_{PCET} approaches a plateau for the larger driving forces. We found that the computed k_{PCET} appears to start decreasing at *ca.* $-\Delta G^o = 90$–100 kcal mol^{-1}, but these rate constants may not be fully converged with respect to bound product proton vibrational states given the dissociation energy of 100 kcal mol^{-1} for the product Morse potential. Fig. 5 only shows the data for which the results are fully converged.

The KIEs calculated for the model with Morse potentials demonstrate a strong driving force dependence (Fig. 5B). At lower driving forces, corresponding to a virtually isoergic reaction, the KIE was computed to be nearly 30. As the driving force for the PCET reaction increases, this calculated KIE decreases nearly

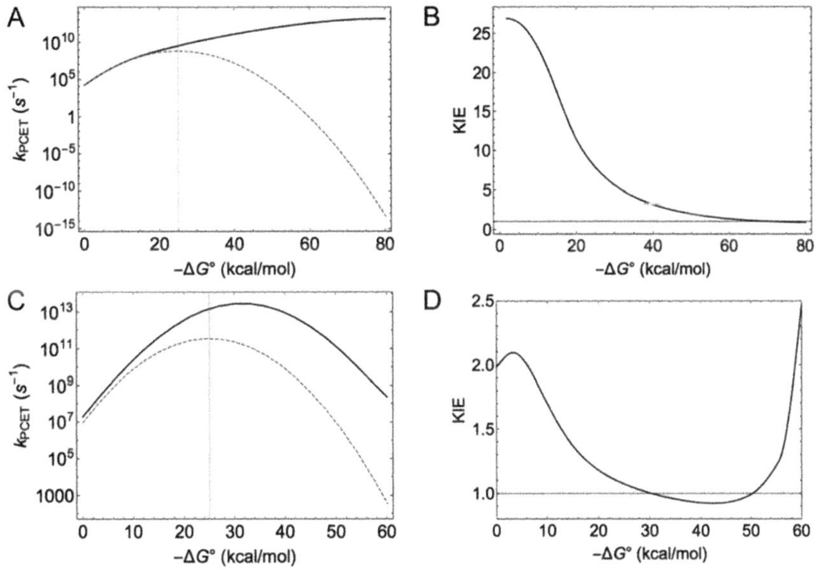

Fig. 5 Driving force dependence of the rate constant k_{PCET} and the KIE for Morse (A, B) and asymmetric double well (C, D) proton potentials. In (A and C), the dashed gray curves depict the rate constant calculated with only the transition from the ground reactant to the ground product vibronic state (*i.e.*, the (0,0) vibronic state pair), whereas the solid black curves depict the rate constant including contributions from the excited proton vibrational states. The solid gray vertical line in (A and C) corresponds to $\lambda = 25$ kcal mol^{-1}. In (B and D), the horizontal gray lines at KIE = 1 represent the situation in which the rate constants of PCET with hydrogen and deuterium are equal.

monotonically to unity. As will be demonstrated below, the highly delocalized nature of the proton vibrational wavefunctions predominantly contributing to the rate constant at higher driving forces attenuates the difference in overlap integrals between hydrogen and deuterium. Conversely, at lower driving forces, the ground-to-ground state transition is dominant, as illustrated by the overlap between the solid black and dashed gray lines for lower driving forces in Fig. 5A. The ground state proton vibrational wavefunctions are much more localized (Fig. 4A), and therefore the associated overlap integrals are extremely sensitive to the change in the mass of the transferring nucleus.

In contrast to the results with Morse potentials, as well as the previous results with harmonic potentials,[42] the inverted region can be clearly observed for the model using asymmetric double well proton potentials (Fig. 5C). Shortly after $-\Delta G^{o}$ exceeds λ, the value of the rate constant begins to decrease. The contributions from the excited proton vibrational states lead to asymmetry and shifting of the maximum of the inverted parabola describing the driving force dependence of the rate constant, as observed previously for ET reactions coupled to a vibrational quantum mode.[27,28]

The KIEs for the model with asymmetric double well potentials (Fig. 5D) are around 2 at low driving forces. These KIEs are significantly smaller than those obtained from the model with Morse potentials at low driving forces (Fig. 5B). As the driving force increases, the KIE decreases predominantly monotonically to unity and even slightly below unity until it starts to increase very rapidly for the most exoergic driving forces studied. The unexpected observation of inverse KIEs (*i.e.*, KIEs smaller than unity), which have also been observed experimentally,[15] and the increase in the KIE at higher driving forces[28] are due to a complex balance among the various contributions to the rate constants for hydrogen and deuterium. The inverse KIE most likely arises from greater vibrational wavefunction overlap integrals for deuterium than for hydrogen for the vibronic state pairs with dominant contributions to the corresponding rate constants. These issues will be analyzed in detail in the next section.

5. Analysis of vibronic state contributions to the rate constant

To further elucidate why the inverted region is or is not observed for these model systems, the contributions of individual reactant/product vibronic state pairs (μ,ν) to the total rate constant at significantly exoergic driving forces were analyzed. Thermal averaging over the proton donor–acceptor distance often avails larger proton vibrational wavefunction overlaps at smaller values of R, leading to dominant proton donor–acceptor distances (*i.e.*, the distance R corresponding to the maximum of the integrand in eqn (5)) that are shorter than the equilibrium proton donor–acceptor distance.[45] For both types of model potentials studied herein, however, the dominant proton donor–acceptor distance was determined to be very close to the equilibrium distance. This dominance of the equilibrium proton donor–acceptor distance is a consequence of significant contributions from highly delocalized wavefunctions to the total rate constant, thereby avoiding the necessity of sampling the smaller, less energetically favorable proton donor–acceptor distances. Thus, we analyzed the contributions from the individual reactant/product vibronic state pairs at the equilibrium proton donor–acceptor distance with $\delta R = 1.20$ Bohr.

We analyzed these contributions at a driving force of 60 kcal mol^{-1}. At this driving force, Fig. 5A and C indicate that the asymmetric double well model system is exhibiting inverted region behavior (*i.e.*, the rate constant is decreasing as the driving force increases), whereas the Morse system is not exhibiting this behavior (*i.e.*, the rate constant is slightly increasing as the driving force increases). The relative contributions of the individual reactant/product vibronic state pairs to the rate constant are determined mainly by a balance between the square of the overlap integral, $S_{\mu\nu}^2$, and the exponential of the free energy barrier, $\Delta G_{\mu\nu}^{\dagger}$. These and other associated quantities, as well as the percentage contribution to the total rate constant, are given in Tables 1 and 2 for the relevant vibronic state pairs (μ,ν). Note that the rate constant is computed by converging with respect to reactant and product vibronic states, but only the ground reactant vibronic state contributes significantly.

Table 1 demonstrates that the dominant contribution to the rate constant for the model with Morse potentials is from vibronic state pair (0,5), which constitutes ~57% of the total rate constant. The pairs (0,4) and (0,6) also each constitute ~20% of the total rate. Note that the value of $\Delta G_{\mu\nu}^{\dagger}$ is quite small for all three of these vibronic state pairs, and pair (0,4) is virtually activationless, with a free energy barrier of only 0.03 kcal mol^{-1}. The monotonic increase in the overlap

Table 1 Main contributions to the rate constant at $\Delta G^{\circ} = -60$ kcal mol^{-1} for Morse potentials with $\delta R = 1.20$ Bohr[a]

(μ,ν)	$\Delta G^{\circ}_{\mu\nu}$	$\Delta G^{\dagger}_{\mu\nu}$	$\exp[-\Delta G^{\dagger}_{\mu\nu}/k_BT]$	$S^2_{\mu\nu}$	% contrib.
(0,0)	−60.00	12.25	1.05×10^{-9}	7.27×10^{-7}	0.00
(0,1)	−51.00	6.76	1.11×10^{-5}	2.37×10^{-5}	0.00
(0,2)	−42.44	3.04	5.90×10^{-3}	3.69×10^{-4}	0.00
(0,3)	−34.33	0.87	2.30×10^{-1}	3.51×10^{-3}	0.86
(0,4)	−26.67	0.03	9.54×10^{-1}	2.18×10^{-2}	22.26
(0,5)	−19.45	0.31	5.94×10^{-1}	8.91×10^{-2}	56.71
(0,6)	−12.68	1.52	7.71×10^{-2}	2.31×10^{-1}	19.09
(0,7)	−6.35	3.48	2.83×10^{-3}	3.49×10^{-1}	1.06

[a] Free energies in kcal mol^{-1}.

Table 2 Main contributions to the rate constant at $\Delta G^{\circ} = -60$ kcal mol^{-1} for asymmetric double well potentials with $\delta R = 1.20$ Bohr[a]

(μ,ν)	$\Delta G^{\circ}_{\mu\nu}$	$\Delta G^{\dagger}_{\mu\nu}$	$\exp[-\Delta G^{\dagger}_{\mu\nu}/k_BT]$	$S^2_{\mu\nu}$	% contrib.
(0,0)	−60.00	12.25	1.05×10^{-9}	1.93×10^{-3}	0.00
(0,1)	−54.18	8.52	5.73×10^{-7}	5.90×10^{-1}	4.14
(0,2)	−52.96	7.82	1.85×10^{-6}	3.72×10^{-1}	8.44
(0,3)	−49.05	5.78	5.78×10^{-5}	3.40×10^{-2}	24.00
(0,4)	−45.18	4.07	1.03×10^{-3}	1.53×10^{-3}	19.33
(0,5)	−40.56	2.42	1.68×10^{-2}	1.08×10^{-5}	2.22
(0,6)	−35.47	1.10	1.57×10^{-1}	1.82×10^{-5}	34.99
(0,7)	−29.95	0.25	6.61×10^{-1}	7.96×10^{-7}	6.44
(0,8)	−24.05	0.01	9.85×10^{-1}	2.19×10^{-8}	0.26

[a] Free energies in kcal mol^{-1}.

integrals as the product proton vibrational state index ν increases results in the (0,5) pair contributing more than the (0,4) pair to the total rate constant. The dominant (0,5) pair is associated with a reaction free energy of $\Delta G^{\circ}_{\mu\nu} = -19.45$ kcal mol^{-1}, which is smaller in magnitude than λ. Thus, although the driving force, $-\Delta G^{\circ} = -\Delta G^{\circ}_{00}$ (*i.e.*, the reaction free energy for the (0,0) vibronic state pair), is much larger than λ, the reaction free energy for the dominant reactant/product vibronic state pair is in the normal region. Therefore, we observe that k_{PCET} is still increasing for the Morse potential model system at $\Delta G^{\circ} = -60$ kcal mol^{-1}. This trend continues as the driving force increases, with greater contributions from even higher excited product proton vibrational states because the overlap integrals continue to increase.

Table 2 indicates that more reactant/product vibronic state pairs contribute significantly to the total rate constant at this driving force for the model with asymmetric double well potentials. The largest contribution to the total rate constant for this model is from the (0,6) pair, which constitutes ~35% of the total rate constant. This pair is followed by pairs (0,3) and (0,4), which constitute ~24% and ~19%, respectively, of the total rate constant. Smaller contributions are associated with the (0,1), (0,2), (0,5), and (0,7) pairs as well. For each of these pairs, however, the value of $\Delta G^{\circ}_{\mu\nu}$ is greater in magnitude than λ. In other words, despite the participation of excited product proton vibrational states, those vibrational states that contribute significantly possess the key characteristic of the inverted region. Thus, at $\Delta G^{o} = -60$ kcal mol^{-1}, the asymmetric double well model system exhibits inverted region behavior.

The vibronic state pair in the asymmetric double well model system corresponding to an activationless transition is the (0,8) pair, which is associated with a free energy barrier of only 0.01 kcal mol^{-1}. This pair contributes negligibly to the rate constant, however, because the square of the proton vibrational wavefunction overlap integral, $S_{\mu\nu}^2$, is prohibitively small with a value of $\sim 10^{-8}$. All of the vibronic state pairs that contribute considerably to the total rate constant have values of $S_{\mu\nu}^2$ at least three orders of magnitude greater than that of the activationless (0,8) pair. Specifically, the dominant pairs, (0,6), (0,3), and (0,4), have values of $S_{\mu\nu}^2$ in the range of 10^{-2} to 10^{-5}, rather than 10^{-8}. For the model system with asymmetric double well potentials, the highly excited product proton vibrational states (*i.e.*, those corresponding to $\nu > 7$) become inaccessible because of the small overlap integrals (Fig. 6), which arise from phase cancellation between the ground reactant state proton vibrational wavefunction and the highly oscillatory excited state product proton vibrational wavefunctions. Thus, this model system exhibits inverted region behavior with the maximum of the inverted parabola shifted to a value slightly greater than λ due to contributions from excited proton vibrational states.

The overlap integrals of the ground reactant proton vibrational state with various product proton vibrational states are depicted in Fig. 6 for both model systems and for both hydrogen and deuterium. For the Morse potential model

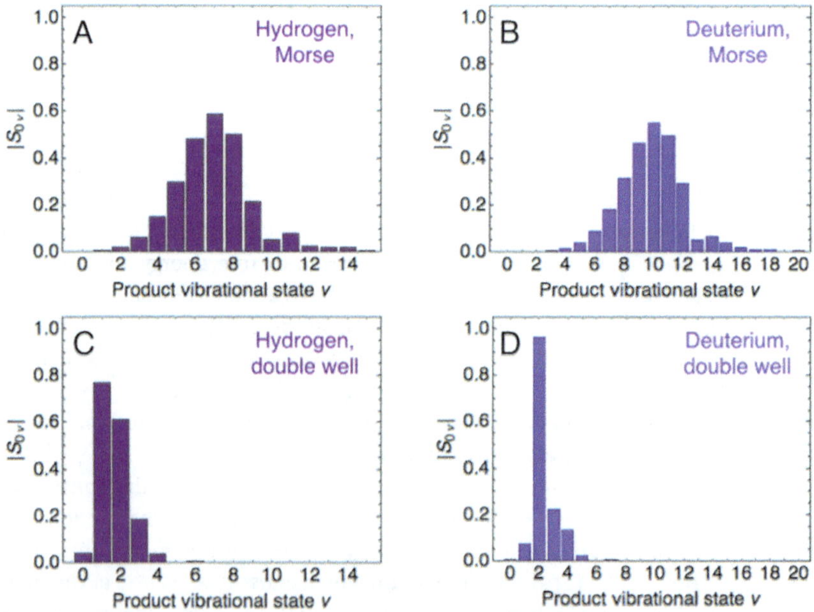

Fig. 6 Absolute overlap integrals between the ground reactant proton vibrational wavefunction and the product proton vibrational wavefunctions ν at $\delta R = 1.20$ Bohr for the Morse potentials (A and B) and asymmetric double well potentials (C and D) for hydrogen (A and C) and deuterium (B and D) transfer.

system (Fig. 6A), the overlap integrals $S_{0\nu}$ steadily grow as ν increases from zero to seven and subsequently decrease non-monotonically to small but not negligible values. This same trend is observed for deuterium in the Morse potential model (Fig. 6B), although it is spread over a larger range of product vibrational quantum numbers ν with the maximum shifted to a larger value of ν. Qualitatively different behavior of the overlap integral $S_{0\nu}$ is observed for the asymmetric double well model (Fig. 6C and D). For this model, the proton vibrational wavefunction overlap integrals increase very rapidly with ν, peaking at $\nu = 1$ for hydrogen and $\nu = 2$ for deuterium. As ν further increases, $S_{0\nu}$ decreases rapidly, becoming very small much more quickly than for the Morse potential model. Thus, the overlap integral becomes negligible for the asymmetric double well potential model at a quantum number ν for which the overlap is significant for the Morse potential model. As discussed further below, this difference is the main basis for the observation of the inverted region for the asymmetric double well potential model but not for the Morse potential model. Furthermore, the smaller overlap integrals for the lower product proton vibrational states in the Morse potential model provide an explanation for the much greater KIEs for the Morse potential model systems at lower driving forces, where these lower proton vibrational states dominate. In general, smaller overlap integrals for the dominant vibronic state pairs lead to larger KIEs.[39,41]

In addition, the proton vibrational states in the asymmetric double well potentials are much more energetically dense (*i.e.*, the energy level splittings are smaller) than are those in the Morse potentials (Fig. 4). The reason for the smaller energy level splittings in the asymmetric double well potentials is that the excited proton vibrational states are above the barrier and thus correspond to a softer potential than is associated with the Morse potentials. As a result, the reaction free energy associated with a given $(0,\nu)$ pair is more negative for the asymmetric double well potential model than for the Morse potential model, leading to inverted region behavior for the former and not for the latter.

The difference in the accessibility of highly excited product proton vibrational states for the asymmetric double well potentials compared to the Morse potentials explains why the inverted region behavior is observed for the former but not for the latter. Specifically, the overlap integrals associated with the relevant excited product proton vibrational states increase for the Morse potentials but become virtually zero for the asymmetric double well potentials. As a result, a progressively more excited product proton vibrational state becomes dominant as the driving force increases for the Morse potentials, preventing inverted region behavior. Because these excited product proton vibrational states are inaccessible for the asymmetric double well potentials due to the vanishing overlap integrals, the lower excited proton vibrational states continue to dominate as the driving force increases for the asymmetric double well potentials, leading to inverted region behavior.

6. Conclusions

The analysis presented herein indicates that inverted region behavior can be observed for photoinduced PCET reactions under certain conditions. The first condition is that the overlap integrals corresponding to the ground reactant and the product proton vibrational states decrease rapidly for higher excited product proton vibrational states. As a result, the highly excited proton vibrational states do not contribute to the rate constant, and the mid-level excited proton vibrational states remain dominant. The second condition is that the energy splittings between these mid-level excited proton vibrational states are small enough to ensure that the associated reaction free energies of the dominant states are negative with magnitudes greater than the reorganization energy. These conditions are typically not satisfied by harmonic and Morse potentials, and therefore inverted region behavior is not expected to be observed for these types of potentials. In contrast, these conditions were found to be satisfied by the chemically motivated asymmetric double well potentials studied herein because the mid-level excited proton vibrational states are more delocalized with smaller energy level splittings. Moreover, these excited proton vibrational wavefunctions are highly oscillatory, leading to destructive interference effects in the overlap with the ground reactant proton vibrational wavefunction. Thus, inverted region behavior is predicted to be observed for systems with these types of proton potentials, which are expected to be found in relatively strongly hydrogen-bonded systems. In addition, the KIE is expected to approach unity and could even become inverse as the driving force increases because the oscillatory nature of the highly excited vibrational wavefunctions produces overlap integrals that do not distinguish significantly between hydrogen and deuterium.

These calculations provide guidance in designing systems that could exhibit inverted region behavior. An underlying assumption of the PCET rate constant used herein is that the reactant is equilibrated in terms of the proton vibrational state populations and the solvent configuration. Thus, vibrational cooling and solvent relaxation are assumed to be faster than the lifetime of the reactant state. Examples of the types of experimentally attainable processes that could potentially exhibit inverted region behavior are photoexcitation to a locally excited state that undergoes exoergic PCET to a charge transfer state, or alternatively photoexcitation to a locally excited state, which relaxes to a charge transfer state that then undergoes exoergic PCET back down to the ground state. A wide range of other experimental scenarios are also possible. Observation of inverted region behavior is more likely for proton transfer across a hydrogen-bonded interface corresponding to asymmetric double well proton potential energy curves and also requires tuning of the reorganization energy and the optoelectronic properties of the photochemical system. These design principles may be helpful in the development of more effective and efficient energy conversion devices.

Conflicts of interest

There are no conflicts to declare.

Acknowledgements

This material is based upon work supported by the Air Force Office of Scientific Research under AFOSR Award No. FA9550-18-1-0134 and by the National Institutes of Health Grant GM056207.

References

1 C. Tommos and G. T. Babcock, *Biochim. Biophys. Acta, Bioenerg.*, 2000, **1458**, 199–219.

2 T. J. Meyer, M. H. V. Huynh and H. H. Thorp, *Angew. Chem., Int. Ed.*, 2007, **46**, 5284–5304.

3 L. Hammarström and S. Styring, *Philos. Trans. R. Soc., B*, 2008, **363**, 1283–1291.

4 A. J. Morris, G. J. Meyer and E. Fujita, *Acc. Chem. Res.*, 2009, **42**, 1983–1994.

5 S. C. Roy, O. K. Varghese, M. Paulose and C. A. Grimes, *ACS Nano*, 2010, **4**, 1259–1278.

6 L. Hammarström and S. Styring, *Energy Environ. Sci.*, 2011, **4**, 2379–2388.

7 D. R. Weinberg, C. J. Gagliardi, J. F. Hull, C. F. Murphy, C. A. Kent, B. C. Westlake, A. Paul, D. H. Ess, D. G. McCafferty and T. J. Meyer, *Chem. Rev.*, 2012, **112**, 4016–4093.

8 L. Hammarström, *Acc. Chem. Res.*, 2015, **48**, 840–850.

9 C. Turró, J. M. Zaleski, Y. M. Karabatsos and D. G. Nocera, *J. Am. Chem. Soc.*, 1996, **118**, 6060–6067.

10 J. P. Kirby, J. A. Roberts and D. G. Nocera, *J. Am. Chem. Soc.*, 1997, **119**, 9230–9236.

11 C. J. Gagliardi, B. C. Westlake, C. A. Kent, J. J. Paul, J. M. Papanikolas and T. J. Meyer, *Coord. Chem. Rev.*, 2010, **254**, 2459–2471.

12 J. N. Schrauben, M. Cattaneo, T. C. Day, A. L. Tenderholt and J. M. Mayer, *J. Am. Chem. Soc.*, 2012, **134**, 16635–16645.

13 T. T. Eisenhart and J. L. Dempsey, *J. Am. Chem. Soc.*, 2014, **136**, 12221–12224.

14 O. S. Wenger, *Coord. Chem. Rev.*, 2015, **282–283**, 150–158.

15 G. F. Manbeck, E. Fujita and J. J. Concepcion, *J. Am. Chem. Soc.*, 2016, **138**, 11536–11549.

16 E. W. Driscoll, J. R. Hunt and J. M. Dawlaty, *J. Phys. Chem. Lett.*, 2016, **7**, 2093–2099.

17 J. C. Lennox and J. L. Dempsey, *J. Phys. Chem. B*, 2017, **121**, 10530–10542.

18 J. C. Lennox, D. A. Kurtz, T. Huang and J. L. Dempsey, *ACS Energy Lett.*, 2017, **2**, 1246–1256.

19 S. D. Glover, G. A. Parada, T. F. Markle, S. Ott and L. Hammarström, *J. Am. Chem. Soc.*, 2017, **139**, 2090–2101.

20 M. A. Bowring, L. R. Bradshaw, G. A. Parada, T. P. Pollock, R. J. Fernández-Terán, S. S. Kolmar, B. Q. Mercado, C. W. Schlenker, D. R. Gamelin and J. M. Mayer, *J. Am. Chem. Soc.*, 2018, **140**, 7449–7452.

21 J. R. Hunt and J. M. Dawlaty, *J. Phys. Chem. A*, 2018, **122**, 7931–7940.

22 R. A. Marcus, *J. Chem. Phys.*, 1965, **43**, 2654–2657.

23 R. A. Marcus and N. Sutin, *Biochim. Biophys. Acta, Bioenerg.*, 1985, **811**, 265–322.

24 G. L. Closs and J. R. Miller, *Science*, 1988, **240**, 440.

25 H. Imahori, K. Tamaki, D. M. Guldi, C. Luo, M. Fujitsuka, O. Ito, Y. Sakata and S. Fukuzumi, *J. Am. Chem. Soc.*, 2001, **123**, 2607–2617.

26 D. I. Schuster, P. Cheng, P. D. Jarowski, D. M. Guldi, C. Luo, L. Echegoyen, S. Pyo, A. R. Holzwarth, S. E. Braslavsky, R. M. Williams and G. Klihm, *J. Am. Chem. Soc.*, 2004, **126**, 7257–7270.

27 J. Ulstrup and J. Jortner, *J. Chem. Phys.*, 1975, **63**, 4358–4368.

28 M. Bixon and J. Jortner, *Faraday Discuss.*, 1982, **74**, 17–29.
29 M. Bixon and J. Jortner, *J. Phys. Chem.*, 1991, **95**, 1941–1944.
30 M. D. Newton, *Coord. Chem. Rev.*, 2003, **238–239**, 167–185.
31 R. A. Marcus, *Annu. Rev. Phys. Chem.*, 1964, **15**, 155–196.
32 W. Schmickler and E. Santos, *Interfacial Electrochemistry*, Springer, 2nd edn, 1996.
33 A. Nitzan, *J. Phys. Chem. A*, 2001, **105**, 2677–2679.
34 A. Soudackov and S. Hammes-Schiffer, *J. Chem. Phys.*, 1999, **111**, 4672–4687.
35 A. Soudackov and S. Hammes-Schiffer, *J. Chem. Phys.*, 2000, **113**, 2385–2396.
36 S. Hammes-Schiffer, *Acc. Chem. Res.*, 2001, **34**, 273–281.
37 A. Soudackov, E. Hatcher and S. Hammes-Schiffer, *J. Chem. Phys.*, 2005, **122**, 014505.
38 S. Hammes-Schiffer, E. Hatcher, H. Ishikita, J. H. Skone and A. V. Soudackov, *Coord. Chem. Rev.*, 2008, **252**, 384–394.
39 S. Hammes-Schiffer and A. V. Soudackov, *J. Phys. Chem. B*, 2008, **112**, 14108–14123.
40 S. Hammes-Schiffer and A. A. Stuchebrukhov, *Chem. Rev.*, 2010, **110**, 6939–6960.
41 S. Hammes-Schiffer, *J. Am. Chem. Soc.*, 2015, **137**, 8860–8871.
42 S. J. Edwards, A. V. Soudackov and S. Hammes-Schiffer, *J. Phys. Chem. B*, 2009, **113**, 14545–14548.
43 B. Auer, L. E. Fernandez and S. Hammes-Schiffer, *J. Am. Chem. Soc.*, 2011, **133**, 8282–8292.
44 S. Horvath, L. E. Fernandez, A. V. Soudackov and S. Hammes-Schiffer, *Proc. Natl. Acad. Sci. U. S. A.*, 2012, **109**, 15663.
45 M. T. Huynh, S. J. Mora, M. Villalba, M. E. Tejeda-Ferrari, P. A. Liddell, B. R. Cherry, A.-L. Teillout, C. W. Machan, C. P. Kubiak, D. Gust, T. A. Moore, S. Hammes-Schiffer and A. L. Moore, *ACS Cent. Sci.*, 2017, **3**, 372–380.
46 P. Li, A. V. Soudackov and S. Hammes-Schiffer, *J. Am. Chem. Soc.*, 2018, **140**, 3068–3076.
47 E. R. Sayfutyarova, Z. K. Goldsmith and S. Hammes-Schiffer, *J. Am. Chem. Soc.*, 2018, **140**, 15641–15645.
48 H. Taseli, *J. Phys. A: Math. Gen.*, 1998, **31**, 779.
49 S. P. Webb and S. Hammes-Schiffer, *J. Chem. Phys.*, 2000, **113**, 5214–5227.
50 S. Ghosh, S. Horvath, A. V. Soudackov and S. Hammes-Schiffer, *J. Chem. Theory Comput.*, 2014, **10**, 2091–2102.

Faraday Discussions

PAPER

Symmetry breaking charge transfer as a means to study electron transfer with no driving force†

Michael Kellogg, Ali Akil, Daniel Sylvinson Muthiah Ravinson, Laura Estergreen, Stephen E. Bradforth * and Mark E. Thompson *

Received 22nd November 2018, Accepted 8th January 2019
DOI: 10.1039/c8fd00201k

Symmetry breaking charge transfer (SBCT) is a process where a symmetrically disposed pair of identical chromophores forms a charge transfer excited state with the hole and electron on different chromophores, *i.e.* chr–chr + $h\nu \rightarrow$ chr$^+$–chr$^-$. Herein we explore this process in two dipyrrin-based bichromophoric systems. One of these bisdipyrrins involved a pair of BODIPY chromophores linked by a single bond at their *meso*-positions (compound **1**) and the other involved two dipyrrin ligands coordinated in a tetrahedral geometry at the Zn^{2+} ion (compound **2**). Both compounds show rapid SBCT in polar solvents and only dipyrrin based emission in nonpolar solvents, the latter arising from a dipyrrin localized excited sate (LE). By "tuning" the solvent polarity the equilibrium between the LE and SBCT states can be shifted to favor either state. Ultrafast transient absorption spectroscopy (TA) was used to probe the kinetics of the charge transfer for **2** in solvents where the electron transfer is endergonic, exergonic and has a ΔG close to zero. Our TA derived rates were used to predict fluorescence efficiencies in each of the different solvent systems and showed a good correspondence to measured values. Detailed density functional theory (DFT) and time dependent DFT were used to model the ground states as well as the LE and SBCT states of **1** and **2**, in both polar and nonpolar media. The ground and LE excited states show small dipole moments, while the SBCT states show dipole moments of 16.4 and 20.3 D for **1** and **2**, respectively.

Introduction

The simplest way to form a charge separated excited state in a molecular chromophore is to link an electron donor to an acceptor so that an intramolecular

Department of Chemistry, University of Southern California, Los Angeles, CA 90089, USA. E-mail: met@usc.edu; stephen.bradforth@usc.edu

† Electronic supplementary information (ESI) available: The ESI contains (1) *xyz* coordinates for the ground state, LE and SBCT excited states for **1** and **2**, (2) the solvent compositions, $E_T(30)$ and dielectric values for all of the mixtures used here, (3) derivations of eqn (1) and (2), (4) TA spectra for **2** in all solvents and solvent mixtures (15 total). See DOI: 10.1039/c8fd00201k

charge transfer (ICT) state is formed upon irradiation. Strong coupling of D–A groups gives a good oscillator strength for the CT transition, but lowers the energy of the ICT state relative to the singlet energy of either the donor or acceptor. This limits the oxidization and reduction potential available from the ICT state to be well below that of the oxidized donor or reduced acceptor alone, respectively. Linking two dye molecules into a non-polar symmetric pair that are only weakly coupled is an alternative approach to promote charge separation (Fig. 1). An example of this occurs in the bacterial photosynthetic reaction centre, where two bacteriochlorophylls are spatially close, but weakly interacting, to form the "special pair". After bacteriochlorophylls are optically excited, ultrafast formation of an intradimer CT state is observed.[1] This approach leads to rapid (*i.e.* ps time scale) charge separation with an energy loss between the exciton and charge separated state of <100 mV.[2] This rapid formation of the SBCT state is especially useful for red and NIR excited states, such as those of bacteriochlorophylls, since this will efficiently outcompete fluorescence or nonradiative decay from the singlet excited state.

The small energy loss in charge separation in SBCT is in contrast to the typical offset of 500 mV or more between the singlet energy of the donor or acceptor and D^+/A^-, used to promote rapid intramolecular charge transfer. As previously noted by Rettig,[3] symmetry-breaking CT states facilitate charge separation with minimal

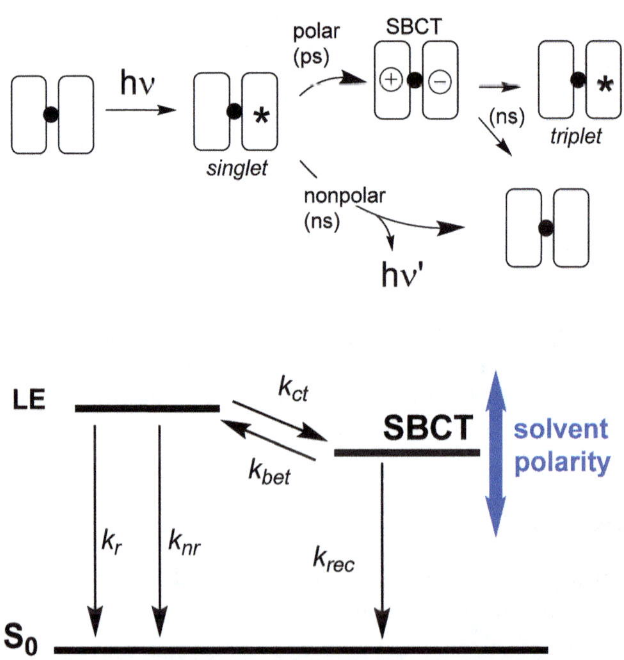

Fig. 1 (top) A schematic representation of the decay process for SBCT materials in polar and non-polar media. Two identical chromophores are represented by the open rectangles and the asterisk indicates an excited chromophore (singlet or triplet). (bottom) A kinetic scheme is shown for the SBCT process. The energy of the CT state is strongly solvent independent, while that of the chromophore localized singlet (labelled LE) is largely solvent independent.

energy loss, and simultaneously slow recombination rates for systems with orthogonal chromophores, both of which are beneficial in a wide range of applications, from photoelectrocatalysis and photovoltaics to photodynamic therapy. We have recently shown that SBCT can be used to significantly enhance the open circuit voltage of organic solar cells,[4] by increasing the rate of charge transfer between donor and acceptors layers in the devices. Symmetry-breaking charge transfer is proposed to occur in any symmetric molecular dyad provided: (1) the energy of the singlet state for the monomeric component of the dyad is nearly degenerate with or greater than that of redox gap ($\Delta E_{redox} = E^{ox}_{1/2} - E^{red}_{1/2}$), (2) the chromophores are spatially oriented to minimize both orbital and exciton coupling, and (3) the chromophores are close enough to undergo rapid electron transfer. Until recently, model compounds capable of SBCT have been largely confined to bi-acenes that absorb light at predominantly ultraviolet wavelengths.[3,5–7] The 9,9′-bianthryl molecule is the best studied system of this sort, and the emissive nature of its excited state provides a useful probe to study the nature of SBCT.[8–10]

The approach used to generate chromophore dimers suitable for SBCT has involved coupling two planar chromophores such that steric constraints force them to be orthogonal to each other. This geometry guarantees minimal overlap of their wavefunctions in the ground state or of the hole and electron formed in SBCT. The orthogonal geometry leads to a very small exchange energy for the singlet and triplet states and thus acts as an efficient intermediate state for the formation of the chromophore localized triplet (Fig. 1).[11] This efficient triplet formation has been used to efficiently sensitize singlet oxygen, *i.e.* chr–chr + $h\nu \rightarrow {}^{1}$chr–chr $\rightarrow {}^{1,3}$(chr^{+}–chr^{-}) $\rightarrow {}^{3}$chr–chr + $^{1}O_2 \rightarrow$ chr–chr + $^{3}O_2$ (chr = chromophore) and been proposed as a heavy metal free photodynamic therapy agent.[12–14] It is important to note that the SBCT state lives for nanoseconds before decaying to the chromophore localized triplet in the absence of oxygen and can be used to drive redox reactions at oxidation and reduction potentials close to the potentials of the isolated chr^{+} and chr^{-} ions, respectively.

Cyanine-type dyes such as difluoroboron dipyrrins (BODIPY) have the appropriate energetics to satisfy requirement (1) above for SBCT and absorb strongly at visible wavelengths.[15] The HOMO and LUMO orbitals are shown in Fig. 2 for a metal coordinated dipyrrin. The principal absorption band for these complexes involves a HOMO–LUMO transition, leading to strong absorption in the 500 nm range for a simple dialkyl-dipyrrin of the type pictured here ($\varepsilon = 10^{5}$ M^{-1} cm^{-1}). We have recently reported a meso-bridged BODIPY dimer dyad (**1** in Fig. 2) that undergoes SBCT upon excitation.[16] Excitation of **1** at 500 nm leads to rapid formation of the SBCT state (<1 ps) in polar solvents such as dichloromethane or acetonitrile. While we have not been able to get X-ray quality crystals of this compound, modeling predicts that the two dipyrrins are close to orthogonal (98°).

A second class of molecules that have the characteristics to exhibit SBCT are bis-dipyrrinato (dipy) zinc complexes such as **2**.[17,18] In these complexes the tetrahedral coordination geometry ensures an orthogonal relationship of the two dipyrrins, minimizing the orbital overlap and exciton coupling between the two ligands. The Zn complexes are readily synthesized, absorb strongly in the visible spectrum and do in fact undergo rapid charge separation (<5 ps), even in solvents of low polarity such as toluene.[17,19] Interestingly, SBCT occurs despite the fact that ΔE_{redox} is greater than the singlet energy in these complexes, seemingly counter to

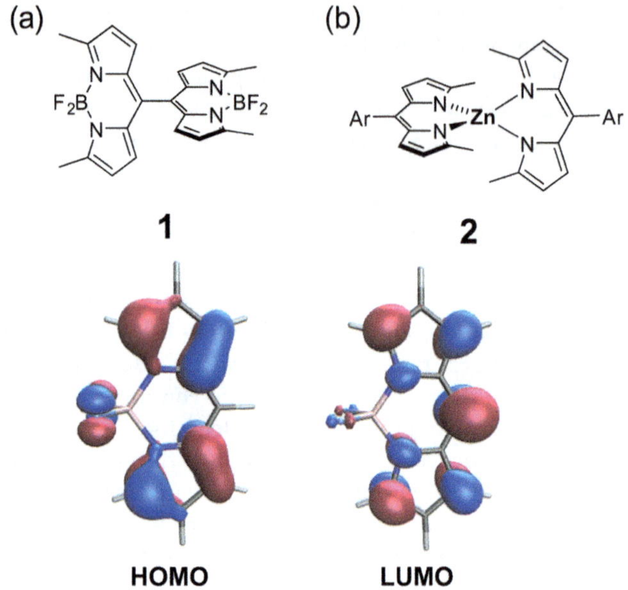

Fig. 2 (a) Dipyrrin structures used to study symmetry breaking charge transfer in this paper. (b) The HOMO and LUMO orbitals of BODIPY are shown. Ar = 1-mesityl. The surfaces are the same for Zn coordinated to the dipyrrin.

the thermodynamic requirement (1) given above. However, charge separation is favored in this case due to the close proximity of the chromophores and the additional stabilization from the screened interaction of the charge separated pair over the cation or anion by the solvent dielectric.

The 9,9′-bianthryl molecule is probably the best studied system exhibiting SBCT, and the emissive nature of its excited state provides a useful probe to study the nature of SBCT.[8–10] 9,9′-Bianthryl has been studied through time resolved microwave conductivity,[20] dc photocurrent measurements[21] and transient absorption measurements,[22–24] The combined studies infer that the charge transfer is occurring by a through-bond mechanism where the anthracene chromophores are initially coupled by activation of a torsional mode that has the chromophores oscillating between 60° and 120° with respect to one another.[24] This torsion results in an initial partial charge transfer in all solvents, largely at the meso-carbons joining the two anthryl units, followed by complete interchromophore charge transfer in polar solvents, where the CT state is further stabilized. Herein we will compare the rates of SBCT between the meso-bridged BODIPY dimer and an organometallic dimer system where the through-bond mechanism is inhibited.

The energy difference between the locally excited state on the chromophore and charge transfer state is strongly solvent dependent. Both **1** and **2** give high fluorescence efficiencies in nonpolar solvents, with emission spectra nearly identical to the isolated chromophore, consistent with the SBCT state lying well above the LE state. As the polarity of the solvent is increased, the SBCT state is stabilized to the point where it is the lowest energy excited state and its red-shifted and inefficient emission takes over (Fig. 1). Using the polarity of the solvent, we can create a situation where the LE and SBCT states become degenerate, so the free energy for the charge transfer reaction is zero. In this paper we discuss these experiments, where solvent mixtures were used to "fine tune" the solvent polarity. We have measured the kinetic parameters for all of the steps outlined in the scheme in Fig. 1 as a function of the solvent dielectric and will discuss them in the context of the charge transfer process in the absence of a driving force, *i.e.* $\Delta G^\circ = 0$. We will also discuss our molecular modeling of the LE and SBCT states for **1** and **2**, to better understand the structural and electronic changes that take place in this electron transfer process.

Results and discussion

Electron transfer at zero driving force

By adjusting the polarity of the solvent, we can shift the energy of the SBCT states to fall above or below the LE state of **2**. The same situation exists for **1**, but we will focus solely on **2** in this section. First, we will consider pure solvents: cyclohexane (CHX), toluene (TOL) tetrahydrofuran (THF) and acetonitrile (ACN). With cyclohexane as the solvent, the SBCT state is sufficiently high in energy that it is not populated in the lifetime of the LE state, so the fluorescence efficiency Φ_{fl} is high ($\Phi_{fl} \sim 0.9$). In Fig. 3 we show the measured Φ_{fl} values for **2** as a function of the solvent dielectric, for both pure and mixed solvents (the mixed solvent dielectric is taken as the weighted average of the dielectric constants of the two pure solvents). Shifting to toluene decreases the Φ_{fl} to ~0.2. This reduction in Φ_{fl} is due to the non-emissive SBCT state dropping below the LE state and rate of populating the SBCT state outcompeting fluorescence.[17] Shifting from toluene to more polar solvents further degrades Φ_{fl}, such that fluorescence is barely measurable in THF and is not observed in acetonitrile. In order to look at more subtle effects of the solvent polarity on the photophysical properties we have investigated Φ_{fl} of **2** in solvent mixtures of TOL–CHX, TOL–THF and CHX–THF. The Φ_{fl} values in solvent mixtures fall monotonically with the increasing dielectric constant of the mixture.

The Φ_{fl} for the TOL–THF and CHX–THF mixtures seem contradictory. While the solvent ratios chosen here (volume ratios were TOL–THF: 70/30–85/15 and

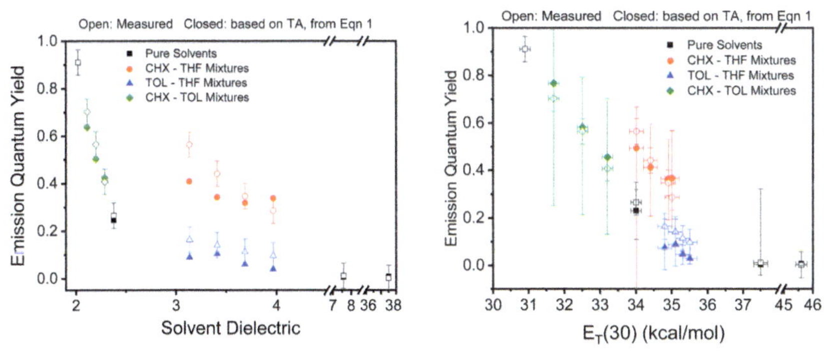

Fig. 3 Photoluminescence quantum yields in a range of solvents and solvent mixtures. Both measured (open symbols) and values calculated from the TA measurements using eqn (1) (closed symbols) are shown. The Φ_{fl} values are plotted *versus* both solvent dielectric (left) and solvent polarity, using the $E_T(30)$ scale (right).

CHX–THF: 65/35–80/20) have similar dielectric values, the measured Φ_{fl} values differ markedly. It appears that the solvent dielectric does not accurately represent the solvent effects in these two solvent mixtures. An alternative way to treat the solvent effects is to consider the polarity of the solvent, using the $E_T(30)$ scale.[25] This solvent polarity scale is based on the absorption spectrum of a charge transfer dye (Reichardt's dye: 2,6-diphenyl-4-(2,4,6-triphenylpyridin-1-ium-1-yl) phenolate) dissolved in the solvent or a solvent mixture. Polar solvents hypsochromically shift the absorbance and less polar solvents lead to bathochromic shifts. Reichardt used this method to develop a polarity scale for common organic solvents,[26] which was later extended to solvent mixtures.[27] The absorption band used in deriving the $E_T(30)$ values for each solvent or mixture involved a phenoxide to pyridinium charge transfer, so many of the same solvent effects captured in the $E_T(30)$ scale should be similar for the electronic transitions being studied here. When the Φ_{fl} values are replotted against the $E_T(30)$, the values of the solvent mixtures overlap less (Fig. 3). While there are still some contradictions at the points where the TOL–THF and CHX–THF mixtures have similar $E_T(30)$ values, the $E_T(30)$ scale captures the solvent effects better than the solvent dielectric, so we will use $E_T(30)$ in our discussion hereafter.

The Φ_{fl} values in TOL–CHX mixtures are perhaps the most interesting. In this case, the non-emissive SBCT appears to smoothly transition from not being significantly populated in the excited state to being the dominant excited state. Assuming that cyclohexane gives an LE state and toluene gives a predominantly SBCT state, this mixed solvent system allows us to find the conditions where $\Delta G^\circ = 0$ for the LE $\rightarrow$ SBCT transition and study the rates of charge transfer and recombination as the system flips from endergonic to exergonic. In order to probe this solvent region, the rates of forward and backward charge transfer were experimentally determined using femtosecond transient absorption spectroscopy (TA). The TA data was fit to the three-state model proposed in Fig. 1 to determine the rate constants. In order to determine the best fit for each of the samples, the fit for the bands at 380 nm and 538 nm, the absorption maxima assigned to the CT state, were monitored. Assigning error bars for the fast rate constants, k_{bet} and k_{ct}, was achieved by incrementing one value by hand, until the quality of the fit in the time and spectral domains was obvious as a systematic deviation in the residuals. As the time-range probed in the TA experiment is not sensitive to $(k_r + k_{nr})$, this value was fixed throughout to the inverse lifetime measured for the complex in cyclohexane. With the values of k_r and k_{nr} from the cyclohexane solution in hand we can use the lifetime and quantum efficiencies in other solvent systems to solve for k_{rec} (see ESI, eqn S3 and S4†). For the pure solvents, THF and acetonitrile, we find that the TA data does not allow robust assignment of k_{ct} and k_{bet}. Even after assigning rate constants ~5 times the best fit value, no deviation in any of the time traces was detected. The ratio of these values are better pinned by the quantum yield measurement. All derived rate parameters from the TA analysis were used to construct an LE state fluorescence quantum yield (QY) utilizing eqn (1), and were found to show a good correlation with the experimentally measured values (Fig. 3).[28,29]

$$\Phi_{fl} = \frac{k_r(k_{bet} + k_{rec})}{(k_r + k_{nr})(k_{bet} + k_{rec}) + k_{ct}k_{rec}} \tag{1}$$

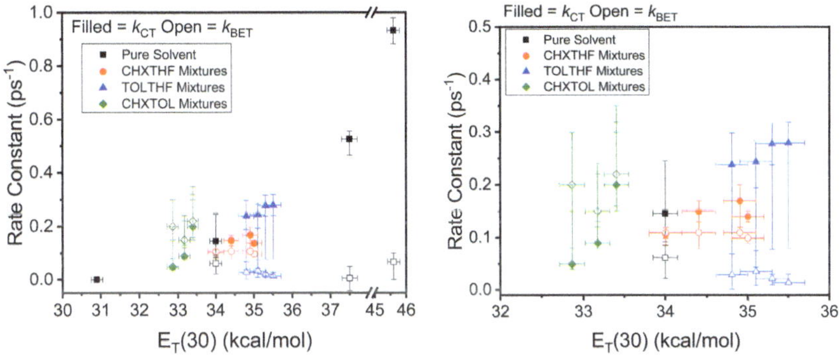

Fig. 4 (left) Rate constants from TA experiments are shown for a range of solvents, from nonpolar to polar. (right) Rate constants from TA experiments are shown for both pure solvent sand solvent mixtures, focusing on the less polar solvents and mixtures. Filled symbols are k_{ct} and open symbols are k_{bet} (see Fig. 1).

Fig. 4 shows the rate constants as a function of solvent polar for pure solvents. In a highly polar medium, the charge transfer state is greatly stabilized and as such, the SBCT state is populated faster than for a nonpolar medium ($k_{ct} > k_{bet}$). For the solvents acetonitrile and THF, k_{ct} is much faster than k_{bet}, so much so that the TA data can be fit even by simplifying to $k_{bet} = 0$. This indicates that for these solvents, the equilibrium established is so far to the side of charge separation that the TA experiment no longer detects the small LE population. On the other hand, the SBCT state in cyclohexane is too destabilized to support any form of SBCT so the rates for the forward reaction is assumed to be zero and the rate constants involved on the SBCT side of Fig. 4 are now undefined. The mixed solvent systems allow us to probe media with polarities between toluene, where $k_{ct} > k_{bet}$ and cyclohexane, where $k_{ct} < k_{bet}$, showing a clear crossing over of the rates of forward and backward charge transfer. For a nonpolar medium, the backward charge transfer rate outcompetes the forward rate.

Using the values for k_{ct} and k_{bet}, the equilibrium constant and then ΔG of formation of the SBCT state from the LE state was calculated at room temperature (Fig. 5). From the calculation of ΔG, we see that all mixtures of CHX and TOL lead to a positive free energy change. Thus, for a solvent less polar or polarizable than toluene, the CT state is destabilized but not enough to eliminate the population of this pathway during the lifetime of fluorescence. In addition to the TOL/CHX mixtures, the 20/80 THF–CHX mixture gives a ΔG value close to zero. The $E_T(30)$ solvent parameters for the CHX–THF mixed solvent that are >60% CHX may be impacted by the low solubility of Reichardt's dye in CHX. We are exploring other solvent models for the mixed solvent systems used here to find a better match to the experimental trends.

Modeling LE and SBCT states

The first step in the modeling studies was to optimize the geometries of the ground states for **1** and **2**. Both structures were optimized using Density Functional Theory (DFT) with the B3LYP functional and 6-31+G* basis set. A crystal structure for **1** is not available for comparison, but the geometry optimized structure of **2** matches the experimental one closely with no more than 0.03 Å

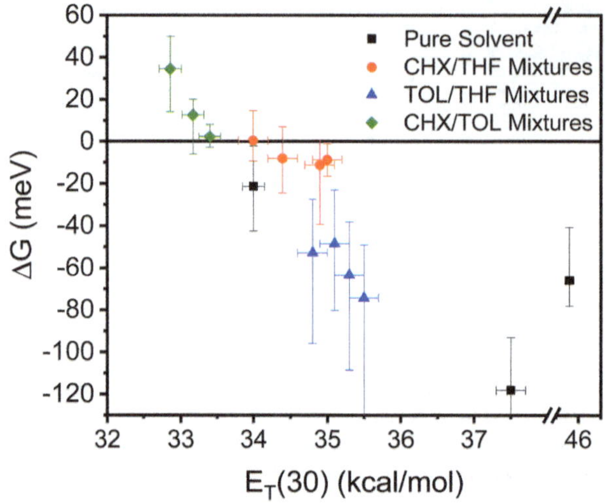

Fig. 5 ΔG for the equilibrium $S_1 \rightleftarrows$SBCT for **2** as a function of solvent $E_T(30)$. A line has been added at $\Delta G = 0$ as a guide.

difference in corresponding C–C or C–N bond lengths between the two structures. The bond lengths and angles of the dipyrrin ligands of both **1** and **2** are very similar to other BODIPY and zinc-dipyrrin compounds, respectively. The geometry optimized structure for the meso-bridged **1** has a dihedral angle of 81° while the geometry optimized structure of the zinc-based compound **2** is 90°. The dipole moments for the two compounds are very similar and quite low, <0.2 D. The DFT derived MOs of **1** and **2** (Fig. 6) show the top two filled orbitals in both compounds are dipyrrin π-orbitals, and the bottom two unfilled orbitals are comprised of the dipyrrin π*-orbitals, (see Fig. 2(b) for π and π* orbitals). The LUMO and LUMO+1 orbitals of **1** also show substantial overlap of the p_z orbitals of the two meso-carbons, enabled by the dihedral angle of 81°.

The key to the photophysics of these molecules is their solvent dependence. For this reason, we carried out time dependent DFT (TDDFT) calculations both in

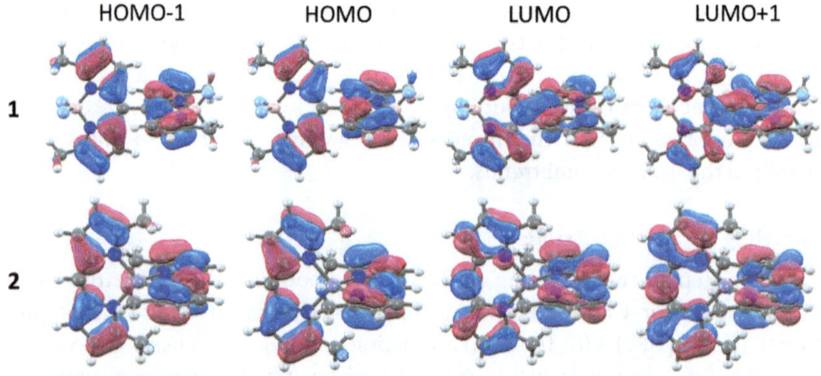

Fig. 6 Molecular orbitals for **1** and **2** (DFT: B3LYP functional, 6-31+G* basis).

a vacuum and with an implicit solvent continuum to probe the excited states. The B3LYP functional, 6-31G basis set was used for geometry optimization of excited states and 6-31+G* was used for electronic structure calculations with the optimized structures. These TDDFT calculations were carried out with a nonequilibrium implicit continuum solvation model. More specifically, the conductor-like screening model, COSMO,[30] is used to qualitatively model the character of the excited states in polar and nonpolar environments. COSMO is a self-consistent reaction field method. In COSMO, the solute cavity takes the shape of the molecule and is constructed using atom-centered spheres. The exact electron density of the solute is used to polarize the cavity surface into discrete point charges. COSMO also includes a correction for outlying charges, which is the part of the solute electron density that extends beyond the cavity. It should be noted, however, that COSMO is better suited for modeling high dielectric solvents. Thus, there is a sacrifice of quantitative accuracy in favor of a qualitative description.

For these COSMO modeling studies, we chose the dielectric of cyclohexane, a solvent where we see only the LE state, and acetonitrile, where the SBCT state dominates. At both a low and high dielectric we see the lowest four excited states being composed of two LE states and two SBCT states. The dihedral angles between the dipyrrin planes of **1** and **2** in both the LE and SBCT states are within 1° of their ground state values, *i.e.* 81° and 90°, respectively. The natural transition orbitals (NTOs) for the LE and SBCT states for **1** and **2** are illustrated in Fig. 7. The

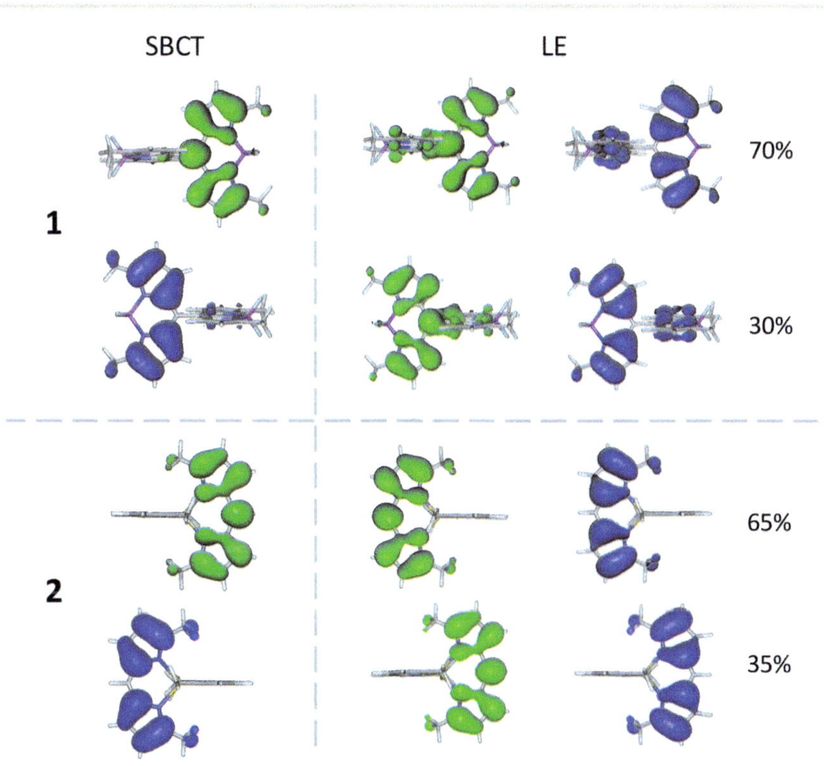

Fig. 7 SBCT and LE natural transition orbitals (NTOs) for **1** and **2**. The hole orbitals are in blue and the electron orbitals are in green.

two SBCT states have a hole on one dipyrrin and an electron on the other, *i.e.* (dpy$^+$)Zn(dpy$^-$) and (dpy$^-$)Zn(dpy$^+$). The LE state consists of linear combinations of the dipyrrin localized sates on the two dipyrrins. The LE states for **1** shown in Fig. 7 has NTO contributions (hole + electron) that are 70% from one dipyrrin and 30% from the other. The other LE state (not shown) has the opposite composition. The LE state does not lead to any charge build up in the molecule, and as such the LE excited states have dipole moments close to those of the ground state, 0.6 D for both **1** and **2**, with a calculated hole–electron separation of <0.1 Å. The low ground state and LE dipole moments are consistent with the low degree of solvatochromism for the absorption and emission spectra for $S_0 \rightarrow S_1$/LE. It is noteworthy that the electron NTO for the LE state of **1** has character at both meso-carbons, consistent with the non-orthogonal orientation of the two dipyrrins. No such "bridging" character is seen for **2**. In contrast to the LE states, the SBCT states for **1** and **2** give very large dipole moments of 16.4 and 20.3 D and hole–electron separations of 3.4 and 4.4 Å, respectively, consistent with the high degree of charge separation in the SBCT state.

The SBCT state of **2** shows a significant distortion of the coordination sphere around the Zn ion (Fig. 8). The ground state structure of **2** has the zinc ion equally spaced between the two dipyrrin ligands, as judged by the zinc to meso-carbon distance (a and b in Fig. 8). The Zn^{2+} ion in the SBCT structure shifts to be closer to the reduced dipyrrin and further from the oxidized one. This shift in the position of the Zn^{2+} ion is dependent on the dielectric of the continuum, with a nonpolar solvent (low dielectric) giving a 0.24 Å difference in the Zn to meso-carbon distances, while the shift in a high dielectric medium is 0.16 Å. This is consistent with more efficient screening of charge in the high dielectric medium. A related, albeit much smaller shift is observed for **1**. In the SBCT state for **1** the B atom is closer to the meso-carbon of the dipyrrin with the electron than it is to the meso-carbon of the dipyrrin with the hole, but the difference in distances is only 0.06 Å.

A problem with the dielectric continuum model presented above is that while the orbital make-up and transitions are consistent with the experimental data, the energies of the states are not well matched to experiments. In both high and low

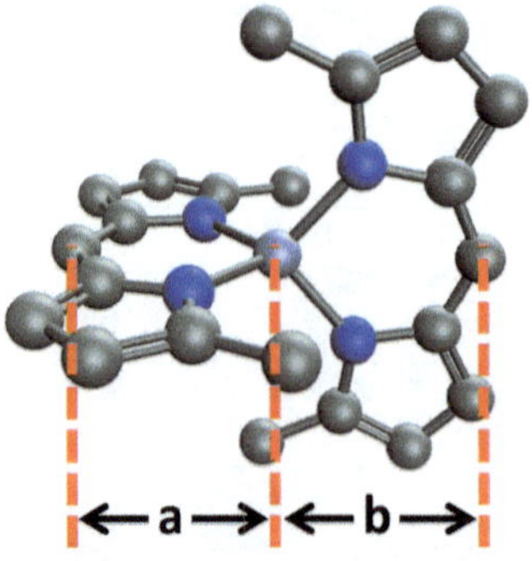

	Medium	a (Å)	b (Å)	Δ(Å)
Ground State	Gas	3.30	3.30	0
	CHX	3.30	3.30	0
	ACN	3.31	3.31	0
LE	Gas	3.30	3.30	0
	CHX	3.33	3.37	0.04
	ACN	3.33	3.36	0.03
SBCT	Gas	3.48	3.19	0.29
	CHX	3.46	3.22	0.24
	ACN	3.41	3.25	0.16

Fig. 8 Geometry optimized excited state structures of **2** differ in media with different dielectric constants. The structural change here is largely a displacement of the Zn ion along the axis containing the two meso-carbons and the Zn ion, given by a and b in the image. The effect is marked in the SBCT state, where the Zn ion moves closer to the dipyrrin carrying the electron. The effect is most pronounced in nonpolar media.

dielectric media the SBCT state falls below the LE state in energy. This is also true in the vacuum calculations of both **1** and **2**. In order to better model the system, a system with explicit solvent molecules may be needed. Moreover, it has been found that using common hybrid functionals, such as B3LYP, severely underestimate the energies of charge transfer states,[31–33] whereas long-range corrected functionals, such as CAM-B3LYP, ωPBE and ωB97xD, better match the observed energies,[34–37] and will be used in future studies of **1** and **2**. To incorporate an explicit solvent model a hybrid multiscale approach that employs classical Molecular Dynamics (MD) simulations in conjunction with the aforementioned higher level TDDFT method is being explored. In this hybrid scheme, the formal atomic charge forcefield parameters of the chromophore are replaced by the CHELPG (Charges from Electrostatic Potentials using a Grid-based method) charges calculated for the state of interest using DFT. MD simulations are then performed with these modified parameters. Single-point TDDFT calculations are performed on snapshots of the MD equilibrated cell with the solvent atoms replaced by their corresponding CHELPG atomic charges to serve as a polarizing influence on the chromophore. The results of these studies will be reported in due course.

Comparing the SBCT process in meso-BODIPY (1) and Zn(dpy)$_2$ (2)

It is instructive to compare the CT and charge recombination rates for compounds **1** and **2**, shown in Table 1. The two compounds have very different connections between the two dipyrrin units, which lead to very different CT rates. The meso-bridged dimer shows a CT rate that is nearly an order of magnitude faster than the zinc dipyrrin complex. The two dipyrrins are orthogonal to each other in both structures, but the connections between them in **1** and **2** are very different. The closest contact between the two dipyrrins in **1** is 1.49 Å (C–C bond), while the closest contact between the two dipyrrins in **2** is 3.39 Å (interligand N–N contact). It is also important to stress that the differences go beyond just distance. The closest contact for **1** is at the meso-carbon of each dipyrrin, which is a node in the HOMO and has substantial orbital density in the LUMO (Fig. 2(b)). In contrast, the closest dipyrrin–dipyrrin contacts in **2** are the nitrogen atoms of the ligand. Both the HOMO and LUMO orbitals have substantial nitrogen character, but the symmetries of the two are opposite. Twisting of the ligands away from an orthogonal relationship leads to a very weak but positive overlap for the LUMOs of the two ligands and no constructive overlap for the HOMOs.

Looking closely at the NTOs for the LE states of **1** and **2**, one can see that there is no interaction between the hole levels of the dipyrrin ligands for either compound. The same is seen for the electron levels for **2**, where the NTOs remain

Table 1 Kinetic parameters for compounds **1** and **2**

	Solvent	τ_{ct} (ps)	τ_{rec} (ns)	Φ_{fl}
1	Acetonitrile	0.18	0.65	<0.01
	Cyclohexane	—	—	0.8
2	Acetonitrile	1.1	0.9	<0.01
	Cyclohexane	—	—	0.9

fully localized on single dipyrrins. However, the electron levels for **1** show a weak but positive interaction between the two dipyrrin ligands at the bridging meso-carbons. This is enabled by the 80° dihedral angle observed for **1**. This meso–meso orbital overlap is also observed in the LUMO orbitals of the ground state of **1**. A similar orbital overlap is seen for bianthryl and has been suggested as a bridge state for the LE → SBCT transition in that system.[22] In the case of **1** and **2** we postulate that it is this LE bridge state in **1** that increases the rate of charge transfer to the SBCT state by an order of magnitude, relative to charge transfer rate for **2**. In the SBCT states of both **1** and **2** no such bridge state exists; the hole and electron NTOs are fully localized in single dipyrrin ligands. Thus, it is not surprising that the recombination rates (k_{rec}) are nearly the same for both **1** and **2**, with lifetimes of 0.65 and 0.9 ns.

Another interesting comparison is that the emissive state in polar solvents of **1** is from the CT state, like 9,9′-bianthryl, meaning that radiative recombination to the ground state is weakly allowed from the SBCT state,[16] whereas **2** displays only emission from the S_1 of the BODIPY chromophore.[17] In order for recombination from the SBCT state to radiatively relax back to the ground, sufficient orbital overlap and dipolar coupling are necessary. The bridging group in **1** allows for sufficient torsion where the BODIPY chromophores can become less orthogonal, increasing orbital overlap and dipole coupling.

SBCT in **1** displays charge transfer rates like that of 9,9′-bianthryl (hundreds of femtoseconds) while in **2** the CT rate is an order of magnitude slower. Like 9,9′-bianthryl, **1** shows an electron density localized on the bridge. This implies that the BODIPY dimer complex transfers charge through-bond and may also require activation of the torsion mode as noted for bianthryl, allowing for charge transfer to occur rapidly, relative to **2**. An interesting question is what moves in the LE → SBCT transition, the hole or the electron? Either carrier could be transferred to form the SBCT state. While the coupling of the two meso-carbons in the LUMO and electron NTO suggest it is the electron that will be transferred, the data we have presented here does not allow us to state which carrier moves unequivocally. We are currently exploring the use of ultrafast transient absorption anisotropy measurements to answer this question.

Conclusion

Marcus theory relates the rate of electron transfer to the electronic coupling between the initial and final states (H_{AB}), the solvent and molecular reorganization energies (λ) and the free energy of the electron transfer process (ΔG).[38,39] For a case where $\Delta G = 0$, the forward and backward rates are determined by only H_{AB}, λ and the temperature. The system reported here allows us to potentially determine the electronic coupling directly, if the reorganization energy can be estimated. Experimental approaches to determining λ using the Stokes shift[40–42] are not useful since the SBCT state does not emit. We could alternatively estimate λ theoretically, however, the dielectric continuum calculations discussed above are not accurate enough for this. We are currently carrying out QMMM modeling studies with explicit solvent molecules, at a markedly high level of quantum mechanical theory for the excited chromophores than used above. Deriving a reliable λ value in mixed solvent systems with this QMMM approach may be problematic, but should give us good values in pure solvents. Thus, we will use

QMMM derived λ and our experimental values for ΔG and k_{ct} in toluene, THF and acetonitrile to measure the electron coupling in each of these solvents. Solvent dependent structural differences between the LE and SBCT structures will affect λ and may impact the electronic coupling as well.

Methods

Compounds 1 and 2 were prepared *via* literature procedures.[16,17]

Photoluminescence quantum yield measurements. All absorption measurements were taken on a Cary 50 UV-VIS. Steady state relative QY measurements were performed using a fluorescence QY standard. The standard used was fluorescein in 0.1 M NaOH. Each of the pure solvents and solvent mixtures were tested. First, zDIP2 was dissolved in its respective solvent or solvent mixture to obtain an absorbance of near or below 0.1 OD at 496 nm, the excitation wavelength used in fluorescence measurements in a cuvette with a pathlength of 4 mm. This wavelength was chosen because it was the wavelength used for fluorescein, the quantum yield reference, as well as being close to the maximum absorption of zDIP2. The emission spectra were as taken. Fluorescence spectra were obtained on a Horiba Jobin Yvon FluoroMax-3 fluorometer using DataMax v2.2 software. The parameters used for fluorescence measurements were 0.25 nm interval, 1 nm slit width, and 100 ms integration time. The reference fluorescein was run under the same absorption and emission conditions as **21**. The total area of the fluorescence spectrum was then calculated after subtracting off the dark background. The fluorescence QY in reference to fluorescein was calculated using eqn (2):

$$QY_S = \frac{E_S f_R n_S^2}{E_R f_S n_R^2} QY_R. \tag{2}$$

Here, S and R subscripts refer to the sample and reference respectively, E_x is the area under the emission curve, n_x is the refractive index of the solvent or mixed solvent, QY_R is the reference QY of fluorescein (95%)[5] and f_x is the absorption correction factor where $f_x = 1 - 10^{-A_x}$ where A_x is the absorbance at the excitation wavelength used in the emission experiment. The refractive indices of the pure solvents were taken from the literature and the solvent mixture refractive indices were just calculated as a weighted sum of the pure solvents, according to the Arago–Biot equation.[43,44] This is not the ideal case for a binary mixture but the other errors in the QY measurement due to absorbance and fluorescence are assumed to be larger.

Ultrafast transient absorption spectroscopy. Pump probe experiments were done after taking the output of a Ti:sapphire regenerative amplifier (Coherent Legend Elite, 1 kHz, 3.8 mJ, 35 fs) with both single color pumping and broadband probing the sample. For the pure solvents and the CHX–THF mixtures, the excitation pulses were generated from a type-II OPA (Spectra Physics OPA-800CF), centered around 500 nm with a spectral bandwidth of 10–13 nm.

For the TOL–THF and CHX–TOL mixtures, a homebuilt noncollinear optical parametric amplifier (NOPA) was used to pump the sample instead of the commercial OPA mentioned previously. Here the residual 800 nm pump from the amplifier was sent through a type I BBO to frequency double the pump which was then combined with a white light continuum generated by a sapphire disk to generate 500 nm. The signal output was then sent through a double prism compressor to temporally shorten the pulses.

The probe pulses were generated after taking the residual 800 nm amplifier output, passing it through a $\lambda/2$ waveplate, and focusing it onto a rotating 2 mm CaF_2 window. This white light supercontinuum (320 nm to 950 nm) was then collimated and focused at the sample using a pair off-axis parabolic mirrors. After passing through the sample, the supercontinuum was then collimated and focused at the slit of a Czerny–Turner monochromator. The probe was then dispersed onto a 256-pixel silicon diode array (Hamamatsu) for multiplexed detection of the probe. The probe polarization was set at magic angle (54.7°) with respect to the pump to avoid any contribution to the observed signal from orientational dynamics.

After passing through a chopper tuned to 500 Hz, the pump was focused before the sample using a 25 cm lens. The cross-correlation of the pump and probe pulses through a 1 mm quartz cuvette was determined to have a $<$150 fs instrument response for the OPA and 80 fs for the NOPA. The pump was blocked after exciting the sample.

Theoretical methods. All Density Functional Theory (DFT) calculations reported in this work were performed using the Q-Chem (5.1) software package.[30] Ground state geometries of the molecules were optimized at the B3LYP/6-31+G* level in a vacuum and using the conductor-like screening model[45] (COSMO) for cyclohexane and acetonitrile. Vertical excitation energies were computed using TDDFT on the corresponding ground state optimized structures in a vacuum and using the COSMO linear response solvation approach in the non-equilibrium limit. Excited state geometry optimization calculations were performed using TDDFT at the B3LYP/6-31G level.

Conflicts of interest

There are no conflicts of interest to report.

Acknowledgements

The authors would like to acknowledge the Office of Basic Energy Sciences at the Department of Energy (DE-SC0016450) for supporting this work. The authors have no competing financial interest in this work. The authors would also like to acknowledge Kyro Grace for performing the first set of photoluminescence experiments.

References

1 J. R. Norris, M. E. Druyan and J. J. Katz, *J. Am. Chem. Soc.*, 1973, **95**, 1680–1682.
2 W. W. Parson, A. Scherz and A. Warshel, in *Antennas and Reaction Centers of Photosynthetic Bacteria*, ed. M. E. Michel-Beyerle, Springer-Verlag, Berlin, 1985, pp. 122–130.
3 W. Rettig, *Angew. Chem., Int. Ed. Engl.*, 1986, **25**, 971–988.
4 A. N. Bartynski, M. Gruber, S. Das, S. Rangan, *et al.*, *J. Am. Chem. Soc.*, 2015, **137**, 5397–5405.

5 Z. R. Grabowski, K. Rotkiewicz and W. Rettig, *Chem. Rev.*, 2003, **103**, 3899–4031.
6 J. M. Giaimo, A. V. Gusev and M. R. Wasielewski, *J. Am. Chem. Soc.*, 2002, **124**, 8530–8531.
7 J. M. Giaimo, J. V. Lockard, L. E. Sinks, A. M. Scott, T. M. Wilson and M. R. Wasielewski, *J. Phys. Chem. A*, 2008, **112**, 2322–2330.
8 F. Schneider and E. Lippert, *Ber. Bunsen-Ges.*, 1968, **72**, 1155–1160.
9 F. Schneider and E. Lippert, *Ber. Bunsen-Ges.*, 1970, **74**, 624–630.
10 F. C. Grozema, M. Swart, R. W. J. Zijlstra, J. J. Piet, L. D. A. Siebbeles and P. T. van Duijnen, *J. Am. Chem. Soc.*, 2005, **127**, 11019–11028.
11 S. Das, W. G. Thornbury, A. N. Bartynski, M. E. Thompson and S. E. Bradforth, *J. Phys. Chem. Lett.*, 2018, **9**, 3264–3270.
12 S. Duman, Y. Cakmak, S. Kolemen, E. U. Akkaya and Y. Dede, *J. Org. Chem.*, 2012, **77**, 4516–4527.
13 L. Wang, J. Cao, J.-w. Wang, Q. Chen, A.-j. Cui and M.-y. He, *RSC Adv.*, 2014, **4**, 14786–14790.
14 Y. Cakmak, S. Kolemen, S. Duman, Y. Dede, *et al.*, *Angew. Chem., Int. Ed.*, 2011, **50**, 11937–11941.
15 A. B. Nepomnyashchii and A. J. Bard, *Acc. Chem. Res.*, 2012, **45**, 1844–1853.
16 M. T. Whited, N. M. Patel, S. T. Roberts, K. Allen, P. I. Djurovich, S. E. Bradforth and M. E. Thompson, *Chem. Commun.*, 2012, **48**, 284–286.
17 C. Trinh, K. O. Kirlikovali, S. Das, M. E. Ener, H. B. Gray, P. I. Djurovich, S. Bradforth and M. E. Thompson, *J. Phys. Chem. C*, 2014, **118**, 21834–21845.
18 I. V. Sazanovich, C. Kirmaier, E. Hindin, L. H. Yu, D. F. Bocian, J. S. Lindsey and D. Holten, *J. Am. Chem. Soc.*, 2004, **126**, 2664–2665.
19 C. Trinh, K. Kirlikovali, S. Das, M. E. Ener, H. B. Gray, P. Djurovich, S. E. Bradforth and M. E. Thompson, *J. Phys. Chem. C*, 2014, **118**, 21834–21845.
20 J. J. Piet, W. Schuddeboom, B. R. Wegewijs, F. C. Grozema and J. M. Warman, *J. Am. Chem. Soc.*, 2001, **123**, 5337–5347.
21 S. N. Smirnov and C. L. Braun, *Rev. Sci. Instrum.*, 1998, **69**, 2875–2887.
22 T. Takaya, H. O. Hamaguchi and K. Iwata, *J. Chem. Phys.*, 2009, **130**, 014501.
23 K. Nishiyama, T. Honda and T. Okada, *Acta Phys. Pol., A*, 1998, **94**, 847–856.
24 S. Hashimoto, A. Yabushita, T. Kobayashi, K. Okamura and I. Iwakura, *Chem. Phys.*, 2018, **512**, 128–134.
25 C. Reichardt, *Chem. Rev.*, 1994, **94**, 2319–2358.
26 C. Reichardt, *Angew. Chem., Int. Ed. Engl.*, 1979, **18**, 98–110.
27 P. M. E. Mancini, A. Terenzani, M. G. Gasparri and L. R. Vottero, *J. Phys. Org. Chem.*, 1995, **8**, 617–625.
28 D. Veldman, S. M. Chopin, S. C. Meskers and R. A. Janssen, *J. Phys. Chem. A*, 2008, **112**, 8617–8632.
29 W. R. Ware, D. Watt and J. D. Holmes, *J. Am. Chem. Soc.*, 1974, **96**, 7853–7860.
30 Y. H. Shao, Z. T. Gan, E. Epifanovsky, A. T. B. Gilbert, *et al.*, *Mol. Phys.*, 2015, **113**, 184–215.
31 A. Dreuw, J. L. Weisman and M. Head-Gordon, *J. Chem. Phys.*, 2003, **119**, 2943–2946.
32 A. Dreuw and M. Head-Gordon, *J. Am. Chem. Soc.*, 2004, **126**, 4007–4016.
33 D. J. Tozer, R. D. Amos, N. C. Handy, B. O. Roos and L. Serrano-Andres, *Mol. Phys.*, 1999, **97**, 859–868.
34 J. D. Chai and M. Head-Gordon, *J. Chem. Phys.*, 2009, **131**, 174105.

35 D. Jacquemin, E. A. Perpete, G. E. Scuseria, I. Ciofini and C. Adamo, *J. Chem. Theory Comput.*, 2008, **4**, 123–135.
36 D. Jacquemin, V. Wathelet, E. A. Perpete and C. Adamo, *J. Chem. Theory Comput.*, 2009, **5**, 2420–2435.
37 T. Tsuneda and K. Hirao, *Wiley Interdiscip. Rev.: Comput. Mol. Sci.*, 2014, **4**, 375–390.
38 R. A. Marcus, *Angew. Chem., Int. Ed. Engl.*, 1993, **32**, 1111–1121.
39 Theory of Electron Transfer Reactions: Insights and Hindsights, in *Progress in Inorganic Chemistry*, ed. S. J. Lippard, 1983, pp. 441–498, DOI: 10.1002/9780470166314.ch9.
40 E. L. Mertz, V. A. Tikhomirov and L. I. Krishtalik, *J. Phys. Chem. A*, 1997, **101**, 3433–3442.
41 M. Halder, S. Datta, P. Bolel, N. Mahapatra, S. Panja, H. Vardhan, S. Kayal, D. K. Khatua and I. Das, *Anal. Methods*, 2016, **8**, 2805–2811.
42 W. W. Parson, *Modern Optical Spectroscopy: With Exercises and Examples from Biophysics and Biochemistry*, Springer Berlin Heidelberg, 2015.
43 S. C. Bhatia, N. Tripathi and G. P. Dubey, *Indian J. Chem., Sect. A: Inorg., Bio-inorg., Phys., Theor. Anal. Chem.*, 2002, **41**, 266–269.
44 S. O. Isehunwa, E. B. Olanisebe, O. O. Ajiboye and S. A. Akintola, *Afr. J. Pure Appl. Chem.*, 2015, **10**, 58–64.
45 A. Klamt and G. Schuurmann, *J. Chem. Soc., Perkin Trans. 2*, 1993, 799–805, DOI: 10.1039/p29930000799.

Faraday Discussions

PAPER

Understanding the potential for efficient triplet harvesting with hot excitons

T. Northey, T. Keane, J. Eng and T. J. Penfold

Received 12th November 2018, Accepted 17th December 2018
DOI: 10.1039/c8fd00174j

Excited state energy transfer in disordered systems has attracted significant attention owing to the importance of this phenomenon in both artificial and natural systems that operate in electronically excited states. Of particular interest, especially in the context of organic electronics, is the dynamics of triplet excited states. Due to their weak coupling to the singlet manifold they can often act as low energy trapping sites and are therefore detrimental to device performance. Alternatively, by virtue of their long lifetime they can lead to enhanced diffusion lengths important for organic photovoltaics (OPV). Herein, we explore the triplet energy transfer mechanism from dichlorobenzene to thioxanthone in methanol solution. We rationalise previous experimental observations as arising from preferential population transfer into the lowest triplet state rather than the higher lying triplet state that is closer in energy. The reason for this is a delicate balance between the electronic coupling, reorganisation energy and the energy gap involved. The present results provide the understanding to potentially develop a hot exciton mechanism in materials for organic light emitting diodes (OLED) to achieve higher device efficiencies.

1 Introduction

Triplet–triplet energy transfer (T-TET) has attracted considerable interest during the past decades[1–5] owing to the importance of triplet excitons in both artificial and natural systems that operate in electronically excited states. This includes emerging organic electronics technologies, such as organic photovoltaics (OPV) and organic light emitting diodes (OLEDs). Importantly, in these cases, the weak coupling between the singlet and triplet manifolds means that triplet states are often considered as undesirable low energy trap states and therefore are detrimental to device performance. However, the lifetime of triplet states are also considerably longer than their singlet counterparts, and therefore have far more desirable diffusion length scales.[6] Despite their importance, the description of triplet energy transfer pathways are often over simplified or ignored during the analysis of device performance or the design of new components.

Chemistry-School of Natural and Environmental Sciences, Newcastle University, Newcastle upon Tyne, NE1 7RU, United Kingdom. E-mail: tom.penfold@ncl.ac.uk

For OLEDs, which are rapidly emerging in display[7] and illumination[8] markets, the excited states responsible for emission are generated by charge recombination (electrons and holes) within the active amorphous guest–host layer. Owing to the spin statistics of charge recombination, this leads to a 25% generation of emissive singlet states and 75% generation of non-emissive triplets.[9] Consequently, the efficiency of OLEDs crucially depends on guests that are able to make these triplet states emissive. This has most commonly been achieved using guests that exhibit either efficient phosphorescence[10] or thermally activated delayed fluorescence (TADF)[11–13] (see Fig. 1).

For the latter, TADF, even in the presence of the spin-vibronic coupling[14,15] and/or the multiple singlet–triplet state pairs approach[16] to promote efficient triplet harvesting, a major drawback of the small exchange interaction required to minimise the energy gap between the low lying singlet and triplet states is that it also gives rise to small radiative rate constants of the fluorescence.[17] This renders non-radiative processes competitive and can reduce the luminescence efficiency. Crucially, this limitation is based upon the assumption that it is the S_1 and T_1 states which are exclusively populated during the electronic excitation and involved during the triplet harvesting. While examples exist where this is proposed not to be the case,[18,19] these do not provide any specific evidence other than high device external quantum efficiencies. Consequently, detailed understanding of the process of exciton generation on the guest remains incomplete making it challenging to design molecules which achieve this.

To overcome this limitation, Adachi and co-workers proposed the hyper-fluorescence approach.[20] This utilises a fluorophore alongside the TADF material. It is proposed that recombination occurs at the TADF co-host molecules. However, the co-host is not used for emission, but instead for purely converting the triplet excitons into singlets. It can therefore be optimised for the maximum rate of reverse intersystem crossing at the sacrifice of the oscillator strength. Emission occurs *via* singlet energy transfer from the TADF materials (co-host) to the fluorophore *via* Förster resonant energy transfers (FRET). While this is a promising approach, the extra step involved in the emission increases the complication of these devices. In addition, the energetics make developing the hyper-fluorescence concept for efficient blue emitters extremely challenging.

Alternatively, as shown in Fig. 1c, it is potentially possible to avoid these problems by relying on upper excited triplet states, *e.g.* T_n. Provided that T_n is strongly coupled to the singlet manifold it could be possible to achieve efficient

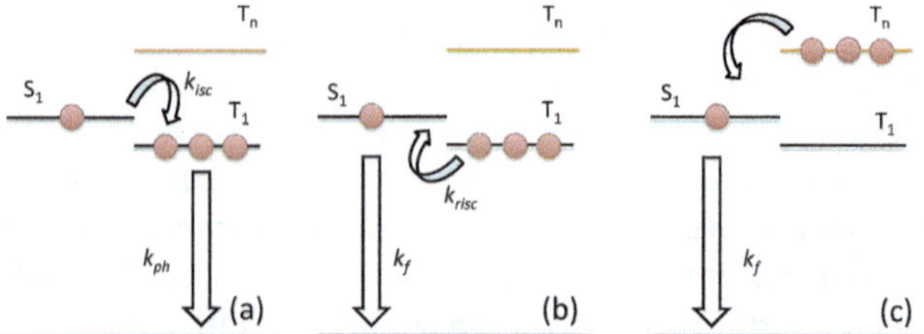

Fig. 1 Triplet harvesting through (a) phosphorescence, (b) Thermally Activated Delayed Fluorescence (TADF) and (c) the hot exciton mechanism. The latter is the focus of the present work.

triplet harvesting with large radiative rates. Recently, Gilch and co-workers[21] have proposed that simple aromatic carbonyls, such as thioxanthone (TX), could be used to achieve direct and controlled access to higher lying excited states. To illustrate this they used a model system composed of a methanol solution containing 1,4-dichlorobenzene (DCB, triplet sensitizer) and TX (harvester). They showed that upon addition of the triplet sensitizer, an enhanced emission was observed because of T-TET from the lowest triplet state of DCB into the isoenergetic T_2 state of TX. Previous work[22–27] has shown that this state exhibits rapid reverse intersystem crossing (rISC) into the bright $S_1(\pi\pi^*)$ state leading to the enhanced emission. However despite the fact that the T_1 state of DCB and the T_2 state of TX are isoenergetic, only a small enhancement was observed, and this remains unexplained.

In this contribution we study this triplet harvesting mechanism. Using a combination of quantum chemistry and molecular and quantum dynamics we rationalise the T-TET between TX and DCB. We reveal that the origin of the weak fluorescence enhancement is due to a strong preference for transfer into the T_1 state of TX, which is significantly lower in energy than the S_1 state and therefore acts as a nonradiative trap site for triplet excitons. The reason for this is a delicate balance between the electronic coupling, reorganisation energy and the energy gap involved which provides the understanding to potentially control this process in OLEDs.

2 Methods and computations

2.1 Quantum chemistry calculations

The geometries of TX and DCB in their ground and relevant excited states were calculated using density functional theory (DFT) and linear response time-dependent density functional theory (LR-TDDFT) within the Tamm–Dancoff approximation and the PBE0 exchange and correlation functional[28,29] as implemented within the Q-chem quantum chemistry package.[30] A Def2-TZVP[31] basis set was used throughout and the solvent was described using a conductor-like polarisable continuum model[32] using the dielectric constant of methanol.

The electronic coupling responsible for T-TET population transfer from the DCB to the TX fragments was calculated using the fragment-excitation difference (FED) method[33] as implemented within the Q-chem quantum chemistry package.[30] Here the coupling is expressed:

$$V_{\mathrm{DA}}^{\mathrm{EET}} = \frac{(E_2 - E_1 \Delta\overline{x}_{12})}{\sqrt{(\Delta x_{11} - \Delta x_{22})^2 + 4\Delta\overline{x}_{12}{}^2}} \tag{1}$$

where $E_2 - E_1$ is the energy difference between the TX and DCB excited triplet states. Δx_{12} is defined:

$$\Delta x_{12} = \int_r \rho_{\mathrm{ex}}{}^{(12)}(r)\mathrm{d}r - \int_r \rho_{\mathrm{ex}}{}^{(21)}(r)\mathrm{d}r \tag{2}$$

where

$$\rho_{\mathrm{ex}}{}^{(12)}(r) = \rho_{\mathrm{hole}}{}^{(12)}(r) + \rho_{\mathrm{elec}}{}^{(12)}(r) \tag{3}$$

This expresses the excitation density as the sum of the attachment (ρ_{elec}) and detachment (ρ_{hole}) densities created in an excitation. In contrast to dipole-induced coupling, the coupling of triplet energy transfer is an exchange mechanism and therefore requires wavefunction overlap between the donor and acceptor fragments.[34] Consequently, the coupling strength decays exponentially with distance between the fragments.

2.2 Classical molecular dynamics

The classical molecular dynamics (CMD) were performed using the Amber MD package[35] with TX and dichlorobenzene solvated in methanol. The force fields for the molecular dynamics were generated using the general Amber force-field (GAFF).[36] DFT(PBE0)/6-31G* determined Mulliken charges were used to define the charges of TX and DCB. The force fields for DCB and methanol were taken from ref. 37. All force fields were defined based upon ground state structures. In the context of the T-TET coupling, the initial conditions of the mixture should correspond to DCB in its triplet state. However the coupling elements are most strongly determined by the distance between the two fragments rather than the specific structure of the DCB. Consequently, we retained the ground state structure of DCB obtained from previous parameterisations.[37]

The simulations consisted of a box containing a TX molecule and 20 dichlorobenzene molecules. These were solvated with methanol in a box with initial dimensions of 49 Å^3. The system was then equilibrated using a constant pressure of 1 atmosphere and a temperature of 300 K until all properties (energy, temperature and pressure) were converged. The dynamics were propagated for a further 200 ps. All of the properties were simulated from this 200 ps of dynamics.

2.3 *Ab initio* molecular dynamics

The *ab initio* molecular dynamics of TX in its electronic ground state and DCB in its lowest triplet state was also performed using Q-chem quantum chemistry package.[30] The potential energy and forces was calculated using DFT(PBE0) and a Def2-TZVP basis set. The effect of the environment was included using a conductor-like polarisable continuum model[32] using the dielectric constant of methanol. Snapshots from these 10 ps of MD were used to compute the energy and corresponding distribution for the TX and DCB molecules.

2.4 Model Hamiltonian for TX and TX-DCB

The ISC dynamics of TX and the energy transfer dynamics of TX-DCB were studied using a model Hamiltonian operator based upon the Linear Spin-Vibronic Coupling Hamiltonian[38] ($\boldsymbol{H}^{\text{SO-vib}}$). This Hamiltonian is the sum of a non-relativistic vibronic coupling Hamiltonian matrix ($\boldsymbol{H}^{\text{vib}}$) and spin–orbit ($\boldsymbol{H}^{\text{SO}}$) Hamiltonian matrices:

$$\boldsymbol{H}^{\text{SO-vib}} = \boldsymbol{H}^{\text{vib}} + \boldsymbol{H}^{\text{SO}} \tag{4}$$

$\boldsymbol{H}^{\text{SO}}$ is consisted of off-diagonal spin–orbit coupling terms, which can either be Q-dependent[39] or independent,[40] where Q represents the dimensionless (mass-frequency scaled) normal mode coordinates. In this present work, due to the rigidity of TX, the spin–orbit coupling terms are assumed to be Q-independent and the coupling at the Franck–Condon geometry used.

The vibronic coupling Hamiltonian matrix is expressed:

$$\boldsymbol{H}^{\text{vib}} = (T_{\text{N}} + V_0)1 + \boldsymbol{W}^{(0)} + \boldsymbol{W}^{(1)} \tag{5}$$

T_{N} is the kinetic energy operator and V_0 is the ground state potential. V_0 is defined as a harmonic oscillator with vibrational frequencies ω_i for dimensionless normal coordinate Q_i. As shown in Table 1, for the TX only model ω_i corresponds to 15 normal modes included in the Hamiltonian and calculated at the ground state optimised geometry of TX. For the TX-DCB, 5 additional modes (1d–5d, Table 1), corresponding to the normal mode frequencies of the T_1 state of DCB at its optimised geometry, were included. $\boldsymbol{W}^{(0)}$ is the zeroth order diabatic matrix that contains the excited state energies at Q_0, the Franck–Condon geometry. $\boldsymbol{W}^{(1)}$ is the first-order linear coupling matrix written:

$$W_{ij}^{(1)} = \sum_{\alpha} \left\langle \Phi_i(Q_0) \left| \frac{\partial \hat{\mathcal{H}}_{\text{el}}}{\partial Q_\alpha} \right| \Phi_j(Q_0) \right\rangle Q_\alpha \tag{6}$$

where the on-diagonal and off-diagonal terms are simplified to:

Table 1 The expansion coefficients used in the model Hamiltonians for TX and energy transfer from DCB to TX[a]

Mode	ω (eV)	$\kappa_{T_{1(\pi\pi^*)}}$	$\kappa_{T_{2(n\pi^*)}}$	$\kappa_{S_{1(\pi\pi^*)}}$	$\lambda_{T_1\text{–}T_2}$
ν_{11}	0.056	0.000	0.000	0.000	0.0006
ν_{14}	0.061	0.000	0.000	0.000	0.0000
ν_{19}	0.085	0.010	0.030	0.030	0.0000
ν_{24}	0.098	0.020	0.050	0.050	0.0000
ν_{33}	0.133	0.000	−0.039	−0.044	0.0000
ν_{38}	0.144	0.000	−0.041	−0.040	0.0000
ν_{39}	0.149	0.000	0.040	0.043	0.0000
ν_{41}	0.150	0.000	0.035	0.045	0.0000
ν_{43}	0.161	0.000	−0.030	−0.030	0.0000
ν_{46}	0.173	0.000	0.037	0.042	0.0000
ν_{47}	0.185	0.000	0.040	0.039	0.0000
ν_{50}	0.189	0.000	0.000	0.000	0.0000
ν_{52}	0.205	0.000	0.000	0.000	0.0000
ν_{54}	0.208	0.000	−0.073	−0.067	0.0000
ν_{55}	0.219	0.140	0.200	0.200	0.0001
ν_{1d}	0.008	−0.040	−0.040	−0.040	0.000
ν_{2d}	0.022	0.050	0.050	0.050	0.000
ν_{3d}	0.046	−0.050	−0.050	−0.050	0.000
ν_{4d}	0.062	0.070	0.070	0.070	0.000
ν_{5d}	0.079	0.100	0.100	0.100	0.000

[a] The modes without the label d are normal modes of TX, while the last 5 modes including the d label are for the triplet state of DCB. The energies of the TX states at the Franck–Condon geometry: $T_1(\pi\pi^*)$ = 2.80 eV, $T_2(\pi\pi^*)$ = 3.50 eV and $S_1(\pi\pi^*)$ = 3.20 eV. The energy for the T_1 state of DCB is 3.45 eV. Additional parameters include spin–orbit coupling between the $S_1(\pi\pi^*)$–$T_2(n\pi^*)$ = 17.62 cm^{-1}. The couple between DCB and TX is included in the same way at the spin–orbit coupling, with T_1(DCB)–T_1(TX) = 0.0025 eV and T_1(DCB)–T_2(TX) = 0.0002 eV.

$$W_{ii}^{(1)} = \sum_{\alpha} \kappa_{\alpha}^{(i)} Q_{\alpha} \tag{7a}$$

$$W_{ij}^{(1)} = \sum_{\alpha} \lambda_{\alpha}^{(i,j)} Q_{\alpha} \tag{7b}$$

where α refers to a specific vibrational degree of freedom. $W^{(1)}$ contains on- (κ) and off-diagonal (λ) coupling elements which are Q dependent.[38] The on-diagonal elements are the forces acting within an electronic surface and are responsible for structural changes of excited-state potentials compared to the ground state. The off-diagonal elements are the nonadiabatic couplings responsible for transferring wavepacket population between different excited states. Here, because we truncate the expansion of the Hamiltonian to first order, we are using the linear vibronic coupling (LVC) model.[41]

The potential and couplings used to describe TX were obtained by performing a fit of the linear vibronic coupling equations to TDDFT(PBE0) calculations at geometries along the important normal modes, so called diabatisation by *ansatz*. The normal modes were identified from those exhibiting the largest linear coupling constants and were in good agreement with modes previously identified in ref. 24. The total Hamiltonian included 15 normal modes and 3 excited states. All of the parameters are provided in Table 1.

The Hamiltonian describing the TX-DCB triplet energy transfer included an additional state, corresponding to the T_1 state of DCB. An additional five normal modes associated with the relaxation of the DCB in its triplet geometry to ground state geometry, which occurs upon energy transfer, were also included. The coupling between DCB and TX, which was assumed to be Q independent, were included as off-diagonal elements in the same way as H^{SO}.

Table 2 Computational details for the MCTDH simulations of the TX and TX-DCB model[a]

	Modes	N_i,N_j,N_k	n_i,n_j,n_k,n_l
TX only	ν_{11}, ν_{14}	6,6	4,10,2
	ν_{19}, ν_{24}, ν_{33}	6,6,6	4,10,2
	ν_{38}, ν_{39}, ν_{41}	6,6,6	4,10,2
	ν_{43}, ν_{46}, ν_{47}	6,6,6	4,10,2
	ν_{50}, ν_{52}, ν_{54}	6,6,6	4,10,2
	ν_{55},	61	4,10,2
TX-DCB	ν_{11}, ν_{14}	6,6	6,6,2,2
	ν_{19}, ν_{24}, ν_{33}	6,6,6	6,6,2,2
	ν_{38}, ν_{39}, ν_{41}	6,6,6	6,6,2,2
	ν_{43}, ν_{46}, ν_{47}	6,6,6	6,6,2,2
	ν_{50}, ν_{52}, ν_{54}	6,6,6	6,6,2,2
	ν_{55}	61	6,6,2,2
	ν_{1d}	121	6,6,2,2
	ν_{2d}	81	6,6,2,2
	ν_{3d}, ν_{4d}, ν_{5d}	6	6,6,2,2

[a] N_i,N_j,N_k refers to the number of primitive harmonic oscillator discrete variable representation (DVR) basis functions used to describe each mode. n_i,n_j,n_k and n_l are the number of single-particle functions used to describe the wavepacket on each state. This is 3 in the case of the TX only model and 4 for the TX-DCB model.

All dynamics were performed using the multi-configuration time-dependent Hartree (MCTDH) method as implemented in the Quantics quantum dynamics package.[42,43] For the ISC simulations, the initial wavefunction in the ground state, built using one-dimensional harmonic oscillator functions with zero initial momentum, was vertically excited into the S_1 state of TX at the FC geometry ($Q = 0$). For the energy transfer simulations, the initial wavepacket was obtained from a relaxation of the wavepacket on the T_1 state of DCB. The calculation details are provided in Table 2.

3 Results

3.1 Intersystem crossing: thioxanthone

The ISC of TX has previously been investigated experimentally[22,23] and theoretically.[24–27] This work has shown that in protic solvents, the bright $S_1(\pi\pi^*)$ state is isoenergetic and in equilibrium with the $T_2(n\pi^*)$ state. The structure of TX and the orbitals involved in these excited states are shown in Fig. 2. This equilibration is established within ~10 ps and consequently, ISC and rISC must also occur on this timescale. Importantly, this equilibrium persists into the nanoseconds, due to the slow internal conversion between the $T_1(\pi\pi^*)$ and $T_2(n\pi^*)$ states.

Using perturbation theory, Marian and co-workers[27] reported that for the $S_1(\pi\pi^*)$–$T_2(n\pi^*)$ equilibrium, $k_{ISC} = 3 \times 10^9\ s^{-1}$ and $k_{rISC} = 4 \times 10^{10}\ s^{-1}$, although k_{ISC} is somewhat sensitive to the hydrogen bonding between the oxygen on the TX and methanol solvent. This was in agreement with the rate determined experimentally[23] of $k_{ISC} = 7 \times 10^{10}\ s^{-1}$ and $k_{rISC} = 1 \times 10^{11}\ s^{-1}$. This previous work is used throughout the remainder of this section to benchmark present results and the model for the excited states of TX before studying the energy transfer between TX and DCB.

Fig. 3 shows the TX population kinetics during the first 14 ps after vertical excitation into the $S_1(\pi\pi^*)$ state using quantum dynamics with the previously described model Hamiltonian. The dynamics included the $T_1(\pi\pi^*)$, however due

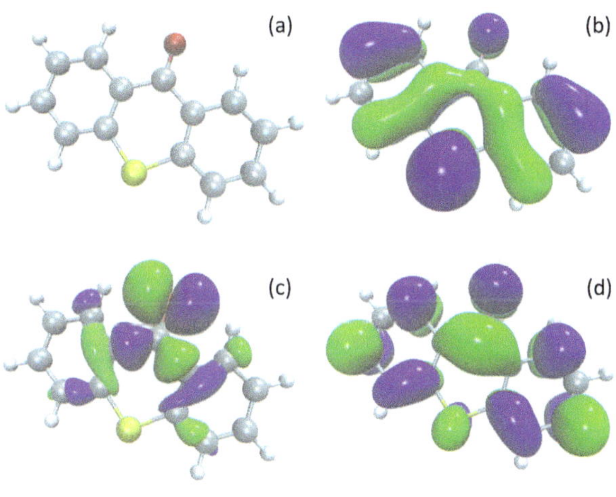

Fig. 2 The (a) ground state equilibrium structure and (b) HOMO−2, (c) HOMO and (d) LUMO orbitals of thioxanthone.

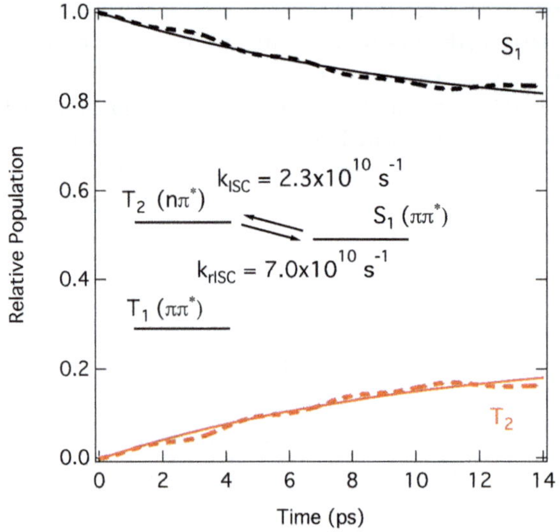

Fig. 3 Relative diabatic state populations of the S_1 (black) and T_2 (red) states of thioxanthone for 14 ps following excitation into the S_1 state. The solid lines of fits of the population kinetics derives from eqn (8).

to the very weak coupling (see Table 1), the population of this state is negligible on the timescale considered and therefore is not shown. These populations kinetics show a fast ISC from the $S_1(\pi\pi^*)$ into the $T_2(n\pi^*)$ state, which at the ground state geometry is 0.3 eV higher than the $S_1(\pi\pi^*)$. Although, they become degenerate with excited state structural distortions. At later times (>10 ps) an equilibrium between the two states is reached. To establish k_{ISC} and k_{rISC}, the $S_1(\pi\pi^*)$ populations kinetics are fit using the rate expression:

$$[S_1(t)] = 1 - \left[\frac{k_{ISC}}{k_{ISC} + k_{rISC}} \times \left(1 - \exp^{-\{k_{ISC}+k_{rISC}\}t}\right)\right] \tag{8}$$

As shown in Fig. 3, this yields a $k_{ISC} = 2.3 \times 10^{10}\ s^{-1}$ and a $k_{rISC} = 7.0 \times 10^{10}\ s^{-1}$ in good agreement with the previous experiments and theory, validating the model for TX used herein. This also illustrates that this process is a direct spin–orbit ISC process, driven by the El-Sayed's allowed spin–orbit coupling ($\sim$18 cm^{-1}) and small energy gap between the $S_1(\pi\pi^*)$ and $T_2(n\pi^*)$ states.

3.2 Triplet-triplet energy transfer: thioxanthone and dichlorobenzene

We now turn our attention to the T-TET from DCB to TX. The energy transfer between these two molecules was recently studied by Torres Ziegenbein *et al.*[21] in methanol solution. Owing to the concentration of TX and DCB and the exponential distance dependence of the exchange coupling associated with T-TET, the energy transfer step in this system is diffusion controlled quenching. However, herein we focus upon establishing the origin for the preferential population transfer into the $T_1(\pi\pi^*)$ over the $T_2(n\pi^*)$ state. In the context of triplet harvesting, this has crucial implications as population entering the T_1 state becomes trapped and cannot undergo rISC to the S_1 state due to the weak coupling and large energy gap.

In the perturbative limit, the rate of triplet energy transfer can be approximated using Fermi's golden rule, which within the semi-classical limit can be expressed:

$$k_{\text{T-TET}} = \frac{2\pi}{\hbar}\left|\langle\psi_f|\hat{\mathcal{H}}_{\text{T-T}}|\psi_i\rangle\right|^2 \frac{1}{\sqrt{4\pi\lambda k_B T}} \exp\left[-\frac{(\Delta G+\lambda)^2}{4\lambda k_B T}\right] \tag{9}$$

The first term is responsible for the electronic coupling between the initial (ψ_i) and final (ψ_f) triplets states. The other terms correspond to the Franck–Condon weighted density of states. Eqn (7) shows that there are three components that can control the rate: the electronic coupling ($\langle\psi_f|\hat{\mathcal{H}}_{\text{T-T}}|\psi_i\rangle$), the free-energy difference between the states (ΔG) and the reorganisation energy (λ).

To understand the role of ΔG, Fig. 4 shows the energy distribution of the $T_1(\pi\pi^*)$ and $T_2(n\pi^*)$ states of TX and the $T_1(\pi\pi^*)$ state of DCB. For both molecules, the geometries were sampled from 10 ps of *ab initio* molecular dynamics of DCB in the lowest triplet state and TX in its electronic ground state, mimicking the starting electronic structure of the system before T-TET. The distribution of energies for DCB is very broad owing to its flexibility[44] in the $T_1(\pi\pi^*)$ state. Importantly, the energies and their spread are in reasonable agreement with experimentally recorded phosphorescence spectra.[45–47]

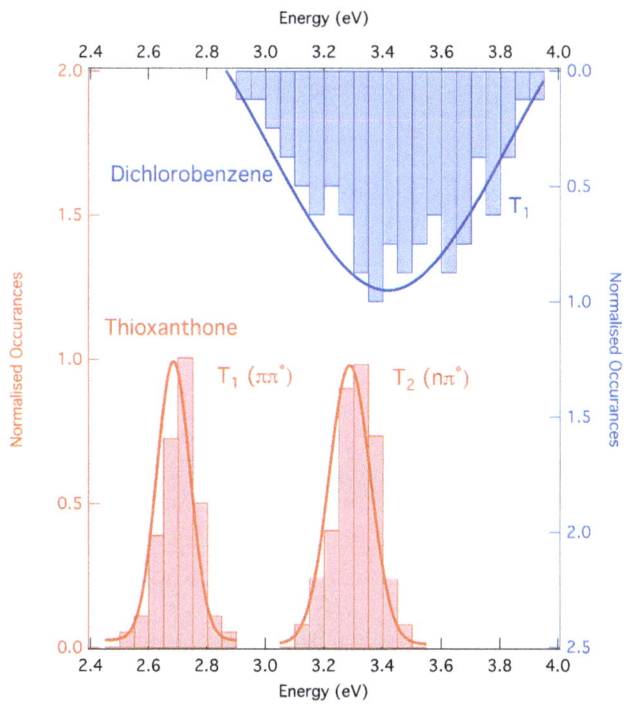

Fig. 4 A distribution of the energies of the $T_1(\pi\pi^*)$ and $T_2(n\pi^*)$ states of TX and the $T_1(\pi\pi^*)$ state of DCB extracted from 100 configurations sampled from 10 ps of *ab initio* molecular dynamics.

For TX, the average energies of $T_1(\pi\pi^*)$ and $T_2(n\pi^*)$ are in good agreement with previous high level calculations performed by Rai-Constapel *et al.*[26] It is emphasised that in the present study, the effect of the solvent, methanol, is accounted for implicitly using a continuum model. In ref. 27, the authors assessed the effect of explicit hydrogen bonding between the carbonyl on the TX and methanol. This increased the energy of the $T_2(n\pi^*)$ up by ~0.1 eV. We have observed a similar effect when using specific snapshots from the MD simulations including an explicit description of a methanol molecule. Given the relative position of the T_1 state in DCB and the T_2 for TX, this will increase the number of configurations for which it is energetically unfavourable for energy transfer from the DCB to the T_2 of TX to occur. However, this is only a small fraction and therefore cannot explain the lack of energy transfer into the T_2 state of TX reported experimentally.

As demonstrated in eqn (9), the reorganisation energy (λ) also plays a crucial role in the rate of triplet energy transfer. To establish the relative rates of the two channels we have calculated λ following ref. 48. For the T-TET into the T_2 of TX λ = 2.1 eV, while for transfer into the T_1, λ = 1.9 eV. In both cases the vast majority of this reorganisation energy (1.8 eV) derives from the structural change of DCB when moving from its triplet to ground state geometry. Importantly, this large reorganisation energy favours population transfer to states with a larger ΔG, as shown in eqn (9).

Fig. 5 shows the electronic coupling between the T_1 of DCB and the T_1 of TX (red trace) and T_2 of TX (blue trace). Important in the context of the present work is that the coupling to the $T_1(\pi\pi^*)$ of TX is consistently larger at all distances between TX and DCB. The origin of this difference is due to the orbitals involved

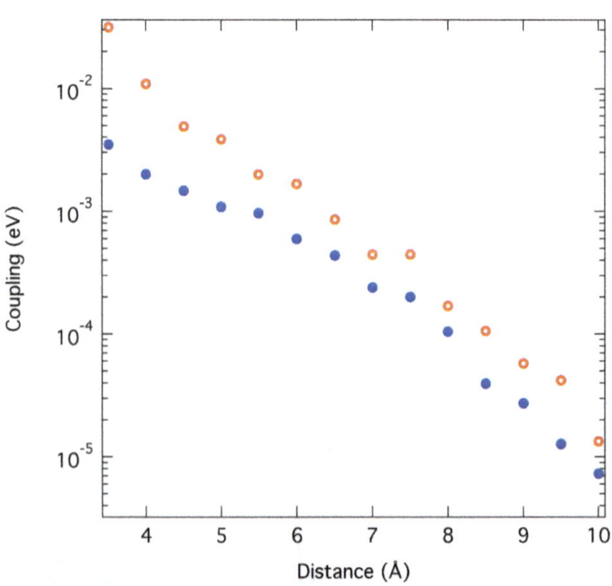

Fig. 5 The T-TET electronic coupling as a function of distance between the centroids of DCB and TX, calculated using the fragment-excitation difference (FED) method[33] as implemented within the Q-Chem quantum chemistry package,[30] between the $T_1(\pi\pi^*)$ of DCB and the $T_1(\pi\pi^*$, red) and $T_2(n\pi^*$, blue) of TX. Each point is the average of many MD sampled couplings assigned to 0.5 Å distance bins in the range 3.5–10 Å.

in the excitation of the two interacting fragments. In the case of T_1(DCB)–T_1(TX) coupling, the transitions are both involving π and π* orbitals which are spread over the entire molecule. This ensures the largest overlap and therefore coupling for a particular separation between the fragments. However, for the T_1(DCB)–T_2(TX) coupling, the T_2 state of TX is an (nπ*) transition, and therefore is more localised on the C=O bond, which reduces the overlap and therefore the coupling.

Using the aforementioned parameters in conjunction with eqn (9), Table 3 shows the rate of population transfer from the DCB to the TX. This shows a strong preference for transfer to the T_1 state with a rate of 2.34×10^8 s^{-1} compared to 4.35×10^5 s^{-1} for transfer into the T_2 state of TX. This difference derives from the $\Delta G + \lambda$ term. However, using these rates within a kinetic model, the large difference between the two pathways leads to negligible population in the T_2 state (0.2%) and therefore cannot explain the enhanced fluorescence observed. However, this analysis uses a single geometry and therefore cannot account for distributions in geometry and how this affects the rates. Indeed, as shown in Fig. 4, DCB is quite flexible and therefore accounting for a range of geometries is important to achieve a full description of the rates of population transfer.

Fig. 6a shows quantum dynamics simulations of the energy transfer from DCB to TX. The model Hamiltonian is based upon the one used previously for TX, but includes 5 additional degrees of freedom associated with the DCB. These modes are primarily responsible for the structural changes between triplet to the ground state of DCB which occurs upon energy transfer. The populations kinetics, like the Fermi's Golden Rule calculations, shows a strong preferences for transfer to the T_1(ππ*) state rather than the T_2(nπ*) state, with rates of 9.9×10^8 s^{-1} and 5.2×10^7 s^{-1} for the T_1 and T_2 states, respectively.

Using these rates, Fig. 7 shows a kinetic model similar to that presented in ref. 13. The solid lines correspond to the T_2 state of TX, while the dashed lines correspond to the T_1 state. The red lines use the rates calculated using the Fermi's Golden Rule approach, and yield a final population distribution of <0.2% in the T_2(nπ*) state. However, when using the rates extracted from the quantum dynamics (black trace), we observed ~4.5% population of the T_2(nπ*) state, in excellent agreement with the ~3% reported by Gilch and co-workers.[21] In this case, the quantum dynamics provides a better description of the ratio between the two rates as the inherent width of the wavepacket more effectively samples the range of configurations possible, especially in the T_1 state of DCB. However, it is emphasised that the ratio between these two pathways can be modified quite easily. Fig. 6b shows the ratio of the rates as a function of the coupling to the T_2 state. The coupling to the T_1 state is kept fixed at 0.0025 eV. This shows an

Table 3 The electronic coupling ($\langle\psi_f|\hat{\mathcal{H}}_{T\text{-}T}|\psi_i\rangle$), free-energy difference between the states (ΔG) and the reorganisation energy (λ) used to calculate $k^{FGR}_{T\text{-}TET}$ between the two channels using eqn (9). The same rates are also calculated using quantum dynamics for comparison

	$\langle\psi_f|\hat{\mathcal{H}}_{T\text{-}T}|\psi_i\rangle$ (eV)	λ (eV)	ΔG (eV)	$k^{FGR}_{T\text{-}TET}$ (s^{-1})	k^{QD}_{TETT} (s^{-1})
$T_1^{DCB} \rightarrow T_1^{TX}$	0.015	2.10	−0.70	2.34×10^8	9.90×10^8
$T_2^{DCB} \rightarrow T_2^{TX}$	0.005	1.90	−0.30	4.35×10^5	5.21×10^7

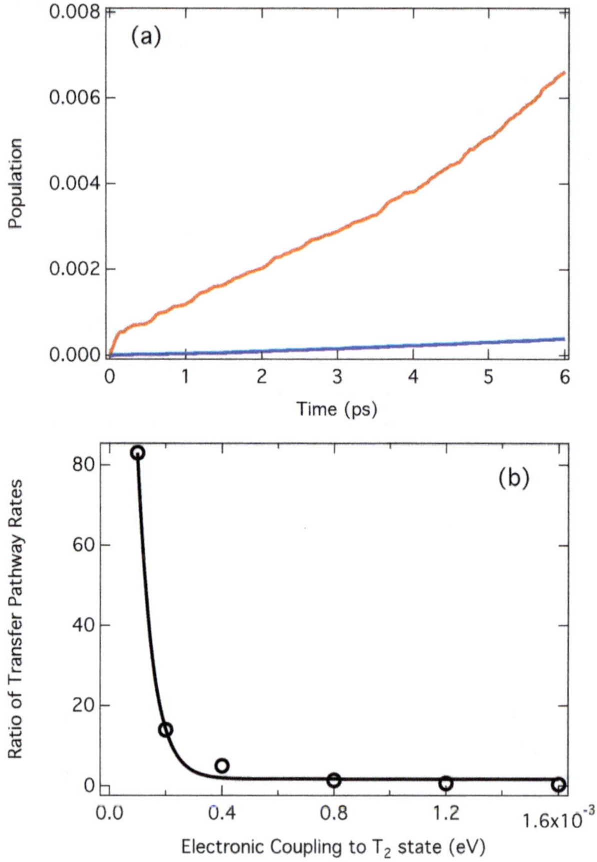

Fig. 6 (a) Relative diabatic state populations of the T_1 (red) and T_2 (black) states of thioxanthone for 5 ps, following initial wavepacket population into the T_1 state of DCB. (b) The ratio of the T-TET rates (k_{T_1}/k_{T_2}) into the T_1 and T_2 states of thioxanthone as a function of the electronic coupling between the T_1 state of DCB and the T_2 state of TX.

exponential dependence on the relative rates of the two channels, with population transfer to the T_2 state preferred if the T-TET coupling exceed 0.0010 eV.

4 Discussion and conclusions

In the previous sections we have simulated the ISC and T-TET dynamics to shed more insight onto the energy transfer between DCB and TX. In agreement with extensive previous experimental and theoretical work, the ISC of TX only occurs mainly from the S_1 into the T_2 state with a rate of $\sim 10^{10}$ s^{-1}. This is driven by the El-Sayed's allowed spin orbit coupling between the two states. In addition, the near degeneracy of these states means that an equilibrium is formed in $\sim$10 ps. Internal conversion between the T_2 and T_1 states is slow due to the large energy gap between them and the weak coupling. In the context of triplet energy transfer this creates two separate channels, $T_1^{DCB} \rightarrow T_1^{TX}$ and $T_1^{DCB} \rightarrow T_2^{TX}$. As recently

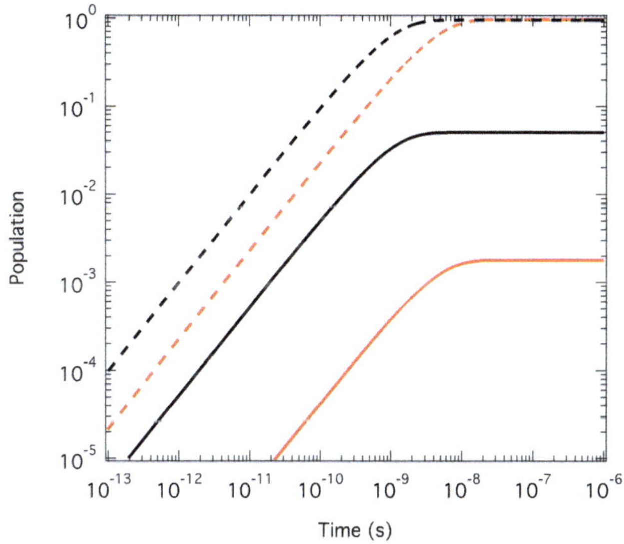

Fig. 7 Time-dependent population of the $T_1(\pi\pi^*$, dashed) and $T_2(n\pi^*$, solid) state using a kinetic model and the rates calculated using Fermi's golden rule (red) and quantum dynamics (black).

reported experimentally,[21] only the latter results in an enhancement of the fluorescence yield of TX, due to the rapid rISC arising from the small energy gap.

Our simulations of the energy transfer from DCB to TX show, in agreement with ref. 21, that population transfer into the T_1 state of TX is strongly preferred. This is a consequence of the larger coupling of this state to the T_1 state of DCB and the combination of the large energy gap with a large reorganisation energy, derived principally from DCB, which creates a favourable term $(\Delta G + \lambda)$. Interestingly, we have shown that this is a rather delicate balance and relatively small variations in the coupling strength between the two fragments can significantly alter this, providing an opportunity for controlling this process in the context of OLEDs. Indeed, host molecules are designed to have minimal reorganisation energy, which according to the $(\Delta G + \lambda)$ term will favour energy transfer between states that are almost isoenergetic. For the present case this would correspond to the T_2 of TX, which results in fluorescence enhancement. This is nicely illustrated by the work of Tabachnyk *et al.*,[4] who reported that triplet energy transfer from pentacene to PbSe nanocrystals could only be achieve in band gap states of the nanocrystals which were close to resonant with the triplet state of pentacene. This is because pentacene has a low reorganisation energy upon energy transfer.

Importantly, the intermolecular energy transfer from the DCB sensitiser to TX is rate determining and can be assumed to be diffusion controlled. Indeed, owing to the low concentration of DCB[21] and the exponential decay of the coupling constant with distance, population transfer can only occur when DCB comes close to TX. As shown in ref. 21, the rate of population transfer can be estimated from the rate of diffusion of DCB in methanol. This is somewhat different to the situation expected from OPV or OLEDs, which exist in the solid state. In the case of OLEDs, guest and host molecules will be in much closer proximity, facilitating triplet energy transfer. However, crucially this approach to generating higher lying excited states requires exciton generation, *i.e.* recombination of charges, on the host rather than the emitter. But triplet states generated on the host are generally not desirable as these are forced to diffuse through the host region to the emitter and tend to annihilate. Consequently, to achieve higher device efficiencies it would appear preferable to ensure exciton generation on the emitter.[3]

While this would appear to make the proposed hot exciton mechanism unfavourable, it is not quite so straightforward. von Borczyskowski and co-workers[49,50] studied the triplet energy transfer in dichloro- and dibromobenzene crystals. In contrast to the primarily diffusion driven mechanisms of triplet energy transfer from DCB in solution, in the solid state, the transfer rates in the solid state exhibit a stretched exponential kinetics arising from the disorder of sites. Indeed, they reported two mechanism for energy transfer namely thermally activated pathways or energy funnels. For the latter the dichlorobenzene (monomers and dimers) lower the energy levels of the surrounding dibromobenzene molecules generating a funnel promoting energy transfer to the dichlorobenzene units. This is similar to the charge transport recently simulated by van Voorhis and co-workers.[51] Exciton formation in the region of the guest molecules, promoted by this funnelling effect could lead each to hot exciton formation through the mechanism proposed herein, although it is likely that this is hard to control.

Alternatively, it is important to consider that the formation of higher lying excited states on the emitter is still possible if the exciton is generated directly on the emitter, which is usually considered preferable for OLEDs. Indeed the energetics of guest and host frontier orbitals mean that electron and hole transfer on to the emitter does not necessarily have to occur into the HOMO or LUMO, but can populate higher (lower) lying orbitals. In these cases, there will be a competition between the orbital relaxation and the timescale taken to form the exciton, *i.e.* arrival of the second charge carrier. The latter will clearly strongly depend on the current density. Due to the longer timescales of relaxation (>10 ns) we have not been able to study this process in the present work. However, recently, the competition between deep hole transfer and relaxation has been studied in DNA hairpins.[52] In this work, the authors proposed deep-hole transfer along a chain of nucleobases, which involved long-range migration of a hole through lower-lying electronic states of the nucleobases. They reported hole relaxation from deep lying occupied orbitals into the HOMO occurring on the 10–100 ns timescale. Importantly, as trapping of charges usually occurs on the μs timescale, this would appear to allow sufficient time for relaxation[53] before formation of the exciton. As the second charge transfer event forming the exciton can also populate higher (lower) lying orbitals, this creates a subspace of states which can be accessed: HOMO → LUMO+n in the case that the hole is trapped first, or HOMO−n → LUMO in the case that the electron is trapped first. This is shown schematically in Fig. 8. Exploring this further is the objective of future work.

In conclusion, we study a triplet harvesting mechanism, arising from T-TET into higher lying states on an emitting molecule. Using a combination of quantum chemistry, molecular and quantum dynamics we have rationalised previous experimental results for T-TET between TX and DCB in solution. Our simulations show that the origin of the weak fluorescence enhancement is due to a strong preference for transfer into the T_1 state of TX. In the broader context of

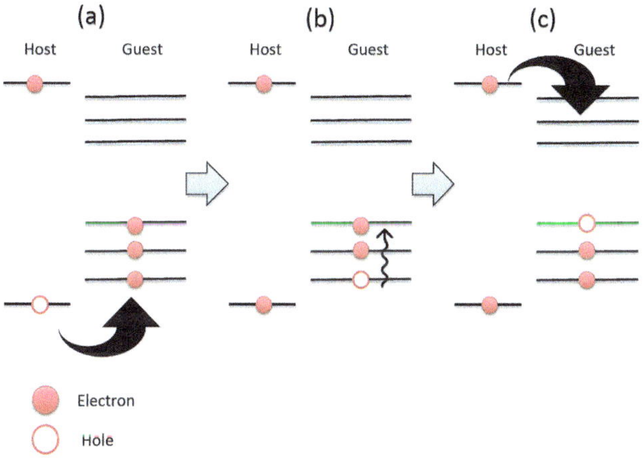

Fig. 8 A schematic of a potential mechanism for hot exciton transfer. In this case it is applied to hole trapping occurring first, but is equally valid for electron trapping. (a) The hole is trapped on the guest in the HOMO−2 orbital. (b) Relaxation occurs so that the hole relaxes into the HOMO orbitals. (c) The electron is transferred and trapped in the LUMO+*n* orbitals forming a subset of the complete range of valence excited states.

organic electronics, this preference derives from a large reorganisation energy of DCB, and therefore as materials in OLEDs tend to minimise this, it would be possible to control energy transfer into higher lying triplet states. However, crucially this requires exciton formation on the host and subsequent diffusion to the emitter, which is often detrimental to device performance. Consequently, while promising, further insight into the mechanism of exciton formation on the guest and the formation of higher lying excited states through this mechanism needs to be achieved before establishing the best approach for controlling hot exciton processes in organic electronics.

Conflicts of interest

No conflict of interest.

Acknowledgements

We gratefully acknowledge support from the EPSRC through grants EP/N028511/1, EP/R021503/1 and EP/P012388/1.

References

1 C. Mongin, S. Garakyaraghi, N. Razgoniaeva, M. Zamkov and F. N. Castellano, Direct observation of triplet energy transfer from semiconductor nanocrystals, *Science*, 2016, **351**, 369–372.

2 G. L. Closs, M. D. Johnson, J. R. Miller and P. Piotrowiak, A connection between intramolecular long-range electron, hole, and triplet energy transfers, *J. Am. Chem. Soc.*, 1989, **111**, 3751–3753.

3 M. A. Baldo and S. R. Forrest, Transient analysis of organic electrophosphorescence: I. transient analysis of triplet energy transfer, *Phys. Rev. B: Condens. Matter Mater. Phys.*, 2000, **62**, 10958.

4 M. Tabachnyk, B. Ehrler, S. Gélinas, M. L. Böhm, B. J. Walker, K. P. Musselman, N. C. Greenham, R. H. Friend and A. Rao, Resonant energy transfer of triplet excitons from pentacene to PbSe nanocrystals, *Nat. Mater.*, 2014, **13**, 1033–1038.

5 T. S. Lee, Y. L. Lin, H. Kim, R. D. Pensack, B. P. Rand and G. D. Scholes, Triplet energy transfer governs the dissociation of the correlated triplet pair in exothermic singlet fission, *J. Phys. Chem. Lett.*, 2018, **9**, 4087–4095.

6 D. N. Congreve, J. Lee, N. J. Thompson, E. Hontz, S. R. Yost, P. D. Reusswig, M. E. Bahlke, S. Reineke, T. Van Voorhis and M. A. Baldo, External quantum efficiency above 100% in a singlet-exciton-fission–based organic photovoltaic cell, *Science*, 2013, **340**, 334–337.

7 J.-F. Tremblay, The rise of OLED displays, *Chem. Eng. News*, 2016, **94**, 29–34.

8 H. Sasabe and J. Kido, Development of high performance OLEDs for general lighting, *J. Mater. Chem. C*, 2013, **1**, 1699–1707.

9 M. A. Baldo, D. F. O'Brien, M. E. Thompson and S. R. Forrest, Excitonic singlet-triplet ratio in a semiconducting organic thin film, *Phys. Rev. B: Condens. Matter Mater. Phys.*, 1999, **60**, 14422.

10 M. A. Baldo, S. Lamansky, P. E. Burrows, M. E. Thompson and S. R. Forrest, Very high-efficiency green organic light-emitting devices based on electrophosphorescence, *Appl. Phys. Lett.*, 1999, **75**, 4–6.

11 Q. Zhang, B. Li, S. Huang, H. Nomura, H. Tanaka and C. Adachi, Efficient blue organic light-emitting diodes employing thermally activated delayed fluorescence, *Nat. Photonics*, 2014, **8**, 326–332.

12 F. B. Dias, T. J. Penfold and A. P. Monkman, Photophysics of thermally activated delayed fluorescence molecules, *Methods Appl. Fluoresc.*, 2017, **5**, 012001.

13 T. J. Penfold, F. B. Dias and A. P. Monkman, The theory of thermally activated delayed fluorescence for organic light emitting diodes, *Chem. Commun.*, 2018, **54**, 3926–3935.

14 J. Gibson, A. P. Monkman and T. J. Penfold, The importance of vibronic coupling for efficient reverse intersystem crossing in TADF molecules, *ChemPhysChem*, 2016, **17**, 2956–2961.

15 J. Gibson and T. J. Penfold, Nonadiabatic coupling reduces the activation energy in thermally activated delayed fluorescence, *Phys. Chem. Chem. Phys.*, 2017, **19**, 8428–8434.

16 P. L. dos Santos, J. S. Ward, D. G. Congrave, A. S. Batsanov, J. Eng, J. E. Stacey, T. J. Penfold, A. P. Monkman and M. R. Bryce, Triazatruxene: A rigid central donor unit for a d–a3 thermally activated delayed fluorescence material exhibiting sub-microsecond reverse intersystem crossing and unity quantum yield *via* multiple singlet–triplet state pairs, *Adv. Sci.*, 2018, **5**, 1700989.

17 T. J. Penfold and J. Gibson, The role of vibronic coupling for intersystem crossing and reverse intersystem crossing rates in TADF molecules, *Highly Efficient OLEDs. Materials Based on TADF*, 2018.

18 W. Li, Y. Pan, L. Yao, H. Liu, S. Zhang, C. Wang, F. Shen, P. Lu, B. Yang and Y. Ma, A hybridized local and charge-transfer excited state for highly

efficient fluorescent OLEDs: Molecular design, spectral character, and full exciton utilization, *Adv. Opt. Mater.*, 2014, **2**, 892–901.

19 R. Chen, Y. Tang, T. Wan, T. Chen, C. Zheng, Y. Qi, Y. Cheng and W. Huang, Promoting singlet/triplet exciton transformation in organic optoelectronic molecules: Role of excited state transition configuration, *Sci. Rep.*, 2017, **7**, 6225.

20 H. Nakanotani, T. Higuchi, T. Furukawa, K. Masui, K. Morimoto, M. Numata, H. Tanaka, Y. Sagara, T. Yasuda and C. Adachi, High-efficiency organic light-emitting diodes with fluorescent emitters, *Nat. Commun.*, 2014, **5**, 4016.

21 C. Torres Ziegenbein, S. Fröbel, M. Glöß, R. S. Nobuyasu, P. Data, A. P. Monkman and P. Gilch, Triplet harvesting with a simple aromatic carbonyl, *ChemPhysChem*, 2017, **18**, 2314–2317.

22 D. Wöll, S. Laimgruber, M. Galetskaya, J. Smirnova, W. Pfleiderer, B. Heinz, P. Gilch and U. E. Steiner, On the mechanism of intramolecular sensitization of photocleavage of the 2-(2-nitrophenyl) propoxycarbonyl (nppoc) protecting group, *J. Am. Chem. Soc.*, 2007, **129**, 12148–12158.

23 T. Villnow, G. Ryseck, V. Rai-Constapel, C. M. Marian and P. Gilch, Chimeric behavior of excited thioxanthone in protic solvents: I. experiments, *J. Phys. Chem. A*, 2014, **118**, 11696–11707.

24 V. Rai-Constapel, M. Kleinschmidt, S. Salzmann, L. Serrano-Andrés and C. M. Marian, Thioxanthone: on the shape of the first absorption band, *Phys. Chem. Chem. Phys.*, 2010, **12**, 9320–9327.

25 R. Mundt, T. Villnow, C. Torres Ziegenbein, P. Gilch, C. Marian and V. Rai-Constapel, Thioxanthone in apolar solvents: ultrafast internal conversion precedes fast intersystem crossing, *Phys. Chem. Chem. Phys.*, 2016, **18**, 6637–6647.

26 V. Rai-Constapel, S. Salzmann and C. M. Marian, Isolated and solvated thioxanthone: a photophysical study, *J. Phys. Chem. A*, 2011, **115**, 8589–8596.

27 V. Rai-Constapel, T. Villnow, G. Ryseck, P. Gilch and C. M. Marian, Chimeric behavior of excited thioxanthone in protic solvents: Ii. theory, *J. Phys. Chem. A*, 2014, **118**, 11708–11717.

28 J. P. Perdew, K. Burke and M. Ernzerhof, Generalized gradient approximation made simple, *Phys. Rev. Lett.*, 1996, **77**, 3865–3868.

29 C. Adamo and V. Barone, Toward reliable density functional methods without adjustable parameters: The PBE0 model, *J. Chem. Phys.*, 1999, **110**, 6158–6170.

30 Y. Shao, Z. Gan, E. Epifanovsky, A. T. B. Gilbert, M. Wormit, J. Kussmann, A. W. Lange, A. Behn, J. Deng, X. Feng, *et al.* Advances in molecular quantum chemistry contained in the Q-Chem 4 program package, *Mol. Phys.*, 2015, **113**, 184–215.

31 F. Weigend, Accurate coulomb-fitting basis sets for H to Rn, *Phys. Chem. Chem. Phys.*, 2006, **8**, 1057–1065.

32 J. Tomasi, B. Mennucci and R. Cammi, Quantum mechanical continuum solvation models, *Chem. Rev.*, 2005, **105**, 2999–3094.

33 C.-P. Hsu, Z.-Q. You and H.-C. Chen, Characterization of the short-range couplings in excitation energy transfer, *J. Phys. Chem. C*, 2008, **112**, 1204–1212.

34 Z.-Q. You and C.-P. Hsu, Theory and calculation for the electronic coupling in excitation energy transfer, *Int. J. Quantum Chem.*, 2014, **114**, 102–115.

35 D. A. Case, D. A. Pearlman, J. W. Caldwell, T. E. Cheatham, J. Wang, W. S. Ross, C. L. Simmerling, T. A. Darden, K. M. Merz, R. V. Stanton, A. L. Cheng,

J. J. Vincent, M. Crowley, V. Tsui, H. Gohlke, R. J. Radmer, Y. Duan, J. Pitera, I. Massova, G. L. Seibel, U. C. Singh, U. K. Weiner, and P. A. Kollman, *Amber 14*, University of California, San Francisco.

36 J. Wang, R. M. Wolf, J. W. Caldwell, P. A. Kollman and A. A. Case, Development and testing of a general amber force field, *J. Comput. Chem.*, 2004, **25**, 1157–1174.

37 F.-Y. Dupradeau, A. Pigache, T. Zaffran, C. Savineau, R. Lelong, N. Grivel, D. Lelong, W. Rosanski and P. Cieplak, The red. tools: Advances in resp and esp charge derivation and force field library building, *Phys. Chem. Chem. Phys.*, 2010, **12**, 7821–7839.

38 T. J. Penfold, E. Gindensperger, C. Daniel and C. M. Marian, The spin vibronic mechanism for intersystem crossing, *Chem. Rev.*, 2017, **118**, 6975–7025.

39 G. Capano, M. Chergui, U. Rothlisberger, I. Tavernelli and T. J. Penfold, A quantum dynamics study of the ultrafast relaxation in a prototypical Cu(i)–phenanthroline, *J. Phys. Chem. A*, 2014, **118**, 9861–9869.

40 J. Eng, C. Gourlaouen, E. Gindensperger and C. Daniel, Spin-vibronic quantum dynamics for ultrafast excited-state processes, *Acc. Chem. Res.*, 2015, **48**, 809–817.

41 H. Köppel, W. Domcke and L. S. Cederbaum, Multimode molecular dynamics beyond the Born-Oppenheimer approximation, *Adv. Chem. Phys.*, 1984, **57**, 59–246.

42 H.-D. Meyer, U. Manthe and L. S. Cederbaum, The multi-configurational time-dependent Hartree approach, *Chem. Phys. Lett.*, 1990, **165**, 73–78.

43 M. H. Beck, A. Jäckle, G. A. Worth and H.-D. Meyer, The multiconfiguration time-dependent Hartree method: A highly efficient algorithm for propagating wavepackets, *Phys. Rep.*, 2000, **324**, 1–105.

44 O. Rubio-Pons, B. Minaev, O. Loboda and H. Ågren, Ab initio calculations of vibronic activity in phosphorescence microwave double resonance spectra of p-dichlorobenzene, *Theor. Chem. Acc.*, 2005, **113**, 15–27.

45 T. Takemura, Y. Yamada and H. Baba, Dual phosphorescence of halogenated benzenes in rigid-glass solution, *Chem. Phys.*, 1982, **68**, 171–177.

46 M. Kinoshita, N. Goto and N. Iwasaki, Phosphorescence spectra of p-dichlorobenzene in various matrices at 1.4 k, *Bull. Chem. Soc. Jpn.*, 1983, **56**, 2591–2597.

47 A. P. Marchetti and D. R. Kearns, Investigation of singlet-triplet transitions by the phosphorescence excitation method. iv. the singlet-triplet absorption spectra of aromatic hydrocarbons, *J. Am. Chem. Soc.*, 1967, **89**, 768–777.

48 K. Anna and H. Bässler, What controls triplet exciton transfer in organic semiconductors?, *J. Mater. Chem.*, 2011, **21**, 4003–4011.

49 S. A. Gilbert, T. Kirski, H. Brenner and C. von Borczyskowski, Low temperature tunneling of triplet excitation energy, *Chem. Phys. Lett.*, 1991, **178**(4), 379–387.

50 T. Kirski, J. Grimm and C. von Borczyskowski, Concentration dependent nondiffusive triplet energy transport on a disordered lattice in the chemically mixed crystal dichlorobenzene–dibromobenzene, *J. Chem. Phys.*, 1987, **87**, 2062–2066.

51 T. Zhu and T. Van Voorhis, Charge recombination in phosphorescent organic light-emitting diode host–guest systems through qm/mm simulations, *J. Phys. Chem. C*, 2016, **120**, 19987–19994.

52 N. Renaud, M. A. Harris, A. P. N. Singh, Y. A. Berlin, M. A. Ratner, M. R. Wasielewski, F. D. Lewis and F. C. Grozema, Deep-hole transfer leads to ultrafast charge migration in dna hairpins, *Nat. Chem.*, 2016, **8**, 1015.
53 M.-J. Tsai and H.-F. Meng, Electron traps in organic light-emitting diodes, *J. Appl. Phys.*, 2005, **97**, 114502.

Faraday Discussions

PAPER

Decomposing electronic and lattice contributions in optical pump – X-ray probe transient inner-shell absorption spectroscopy of CuO†

Johannes Mahl,[ab] Stefan Neppl,‡[a] Friedrich Roth,[cd] Mario Borgwardt,[a] Catherine Saladrigas,[ae] Benjamin Toulson,[a] Jason Cooper,[f] Tahiyat Rahman,[a] Hendrik Bluhm,[ag] Jinghua Guo,[g] Wanli Yang,[g] Nils Huse,[bd] Wolfgang Eberhardt[dg] and Oliver Gessner*[a]

Received 10th December 2018, Accepted 8th January 2019

DOI: 10.1039/c8fd00236c

Electronic and lattice contributions to picosecond time-resolved X-ray absorption spectra (trXAS) of CuO at the oxygen K-edge are analyzed by comparing trXAS spectra, recorded using excitation wavelengths of 355 nm and 532 nm, to steady-state, temperature-dependent XAS measurements. The trXAS spectra at pump-probe time-delays ≥150 ps are dominated by lattice heating effects. Insight into the temporal evolution of lattice temperature profiles on timescales up to 100s of nanoseconds after laser excitation are reported, on an absolute temperature scale, with a temporal sensitivity and a spatial selectivity on the order of 10s of picoseconds and 10s of nanometers, respectively, effectively establishing an "ultrafast thermometer". In particular, for the 532 nm experiment at ~5 mJ cm^{-2} fluence, both the initial sample temperature and its dynamic evolution are well captured by a one-dimensional thermal energy deposition and diffusion model. The thermal conductivity $k = (1.3 \pm 0.4)$ W m^{-1} K^{-1} derived from this model is in good agreement with the literature value for CuO powder, $k_{powder} =$ 1.013 W m^{-1} K^{-1}. For 355 nm excitation, a quantitative analysis of the experiments is hampered by the large temperature gradients within the probed sample volume owing to the small UV penetration depth. The impact of the findings on mitigating or utilizing

[a]Chemical Sciences Division, Lawrence Berkeley National Laboratory, Berkeley, California, USA. E-mail: ogessner@lbl.gov

[b]Physics Department Universität Hamburg, Hamburg, Germany

[c]Institute for Experimental Physics, TU Bergakademie Freiberg, Germany

[d]Center for Free-Electron Laser Science DESY, Hamburg, Germany

[e]Department of Chemistry, University of California, Berkeley, California, USA

[f]Joint Center for Artificial Photosynthesis, Lawrence Berkeley National Laboratory, Berkeley, California, USA

[g]Advanced Light Source, Lawrence Berkeley National Laboratory, Berkeley, California, USA

† Electronic supplementary information (ESI) available. See DOI: 10.1039/c8fd00236c

‡ Present Address: Helmholtz-Zentrum Berlin für Materialien und Energie, Berlin, Germany.

photoinduced lattice temperature changes in future X-ray free electron laser (XFEL) experiments is discussed.

1. Introduction

As research in the field of ultrafast X-ray science progresses toward the use of high repetition rate X-ray free-electron lasers[1,2] and high-order harmonic generation (HHG) light sources,[3,4] it is crucial to understand the impact of corresponding high-repetition rate excitation processes on non-replenishing samples that are typical for many applications in materials and chemical sciences. Thermal excitations including phase transitions and their X-ray spectroscopic fingerprints play a particularly important role as they are a ubiquitous, often unintended consequence of virtually all photon absorption processes in condensed phase materials through coupling of electronic and nuclear motion. Understanding and, in many cases, mitigating the impact of thermal excitations on ultrafast X-ray spectroscopy and scattering signals is crucial for the success of a large range of experiments. This applies, in particular, to dynamics that are strongly coupled to thermal excitations[5] and/or that extend beyond the first ~100 fs after photoexcitation where signal contributions from lattice motion due to electron–phonon coupling may become relevant.[6] Despite their importance, systematic studies of the relative contributions and dynamic trends of electronic and lattice excitations in time-resolved X-ray spectroscopy data are scarce to date. Hayes *et al.* found that photoinduced changes in the iron K-edge X-ray absorption spectrum of hematite (α-Fe_2O_3), recorded 100 ps after super-bandgap excitation with 3.5 eV photons, are entirely dominated by lattice heating effects.[7] Lin *et al.* used femtosecond time-resolved XUV absorption spectroscopy to investigate photoinduced dynamics in PbI_2.[8] The transient XUV spectra in this study are well reproduced by lattice temperature dependent changes for pump-probe delays greater than ~4 ps. In an attosecond time-resolved XUV transient absorption experiment on silicon, Schultze *et al.* observed lattice-motion induced band-gap modifications as early as ~60 fs after photoexcitation.[6] Evidently, signal contributions due to lattice motion are an integral part of time-resolved X-ray spectroscopy that can have decisive impact on the feasibility of experiments. It is essential to characterize these signal contributions and to understand their dynamics based on the underlying physics.

Here, we study photoinduced dynamics in the transition metal oxide semiconductor CuO (cupric oxide) using picosecond time-resolved laser-pump – X-ray-probe transient absorption spectroscopy (trXAS). CuO is a p-type semiconductor with possible applications in photocatalytic degradation of organic pollutants,[9,10] hydrogen production from water,[11–13] and as an anode material in lithium batteries.[14] In many of these applications, CuO is used in conjunction with wide band gap semiconductors, such as ZnO and TiO_2, to cover a large portion of the solar spectrum. In the experiments presented here, we study a CuO film that is generated by thermal surface oxidation of a sheet of copper. The trXAS experiments are performed at the oxygen K-edge near ~540 eV using a previously described optical-pump – X-ray-probe picosecond trXAS setup at beamline 8 of the Advanced Light Source (ALS).[15] The pump pulse repetition rate of 127 kHz is comparable to the current 100 kHz design target for the first stage of the Linac Coherent Light Source upgrade (LCLS II). Pump-probe experiments are performed at two different photon energies of 2.33 eV and 3.49 eV. Band gap values for CuO have been reported across a range of ~1.0–2.1 eV,[16–23] thus, both laser wavelengths induce super-bandgap excitations. Relaxation to the conduction band minimum *via* electron-phonon coupling occurs on timescales much smaller than the 70 ps time resolution of the experiment. Therefore, the main difference between measurements with the two different wavelengths is the transfer of an additional 1.2 eV into phonons before relaxation across the band gap, *i.e.*, a potentially higher lattice temperature after excitation with 355 nm, assuming the same number of excitations per unit sample volume (in the following referred to as excitation density). Thus, comparing measurements for both excitation wavelengths (a) using the same excitation density and (b) using the same total excitation energy within the sample volume V $\int_V (\text{photon energy} \times \text{excitation density})$ provides insight into the relative impact of electronic and lattice contributions to the trXAS spectra. Relaxation dynamics are monitored across a pump-probe time-delay range of ~350 ns.

The time-dependent optical-pump – X-ray-probe experiments are complemented by steady-state, temperature-dependent XAS (XAS_{temp}) measurements at sample temperatures between 22 °C and 145 °C, providing a reference for purely temperature-induced effects. The photoinduced changes in the CuO oxygen K-edge trXAS spectra recorded ≥150 ps after photoexcitation with either 355 nm or 532 nm are well reproduced by effects due to increased lattice temperatures. The XAS_{temp} measurements are used to create a "temperature ruler" for the trXAS data, yielding picosecond time-resolved lattice temperature trends. Temperatures for both excitation wavelengths at comparable pump laser fluences of ~5 mJ cm^{-2} exhibit an initial fast decay within ~10s of nanoseconds, followed by a slower cooling trend on ~100s of nanosecond timescales. These dynamics are well reproduced by a one-dimensional thermal diffusion model that explicitly takes into account the depth-dependent intensity distributions of both the UV/vis pump pulses as well as the X-ray probe pulse. Comparison of the 355 nm and 532 nm results illustrates that a detailed modeling of these depth-dependent intensity profiles is crucial for a quantitative understanding of the physics underlying the laser-induced heating and subsequent cooling processes and their impact on the recorded trXAS spectra. The measurements yield a thermal conductivity of the CuO film of $k = (1.3 \pm 0.4)$ W m^{-1} K^{-1}, which is in good agreement with the literature value for CuO powder of 1.013 W m^{-1} K^{-1} in ref. 24 and approximately an order of magnitude smaller than the smallest literature value for bulk CuO of 18 W m^{-1} K^{-1} in ref. 25. The findings are rationalized by surface morphology modifications during the sample preparation by thermal annealing, which have previously been documented using temperature-dependent atomic force microscopy (AFM).[26]

2. Experimental details

2.1 Time-resolved XAS

For the trXAS experiment, a mobile laser-pump – X-ray-probe setup is employed at beamline 8.0.1.2 of the ALS. The setup and its operation have previously been described in detail.[15] Briefly, CuO samples are excited with 10 ps full-width-at-half-maximum (FWHM) laser pulses of either 532 nm (2.33 eV) or 355 nm (3.49 eV) wavelength (photon energy) and interrogated with 70 ps long X-ray pulses from the

ALS. Time-dependent oxygen K-edge absorption spectra are recorded in total fluorescence yield (TFY) mode with a time-sensitive micro-channel plate detector.[27] The detector is capable of distinguishing TFY signals from individual ALS pulses, enabling the simultaneous recording of trXAS spectra at multiple laser-pump – X-ray-probe time delays. The laser system is synchronized to the ALS and operated at ~127 kHz repetition rate. Excited state spectra ("pumped") are compared to ground state reference spectra ("unpumped") by creating difference spectra for each pump-probe delay, $\Delta trXAS(\Delta t) = trXAS_{pumped}(\Delta t) - trXAS_{unpumped}$. The sample is raster-scanned with 250 μm s^{-1} scan speed and a 500 μm spacing between neighboring lines to mediate sample damage and thermally induced phase transitions.

2.2 Selection of pump laser fluences

Experiments are performed using pump laser photon energies (wavelengths) of 3.49 eV ($\lambda = 355$ nm) and 2.33 eV ($\lambda = 532$ nm), both of which are beyond the band gap of CuO ($\approx$1.0–2.1 eV (ref. 16–23)). For each laser wavelength, measurements are performed at two different fluences, 1.8 mJ cm^{-2} and 4.6 mJ cm^{-2} for 355 nm as well as 1.7 mJ cm^{-2} and 4.8 mJ cm^{-2} for 532 nm. Fluences are estimated based on the assumption that the FWHM laser focus area contains 50% of the pulse energy. The two higher fluence values are chosen such that both the fluence and the total energy deposited in the sample are the same at each laser wavelength (with the laser spot sizes approximately the same for both colors). The 1.8 mJ cm^{-2} fluence for 355 nm excitation is chosen such that the average laser excitation density within the probed volume is expected to be approximately comparable to that for 532 nm and 4.8 mJ cm^{-2}. The measurement with 1.7 mJ cm^{-2} fluence at 532 nm pump wavelength has been performed as an additional reference point to gauge the impact of the laser fluence on the sample response.

2.3 Temperature-resolved XAS

The trXAS measurements are complemented by temperature-dependent X-ray absorption spectra (XAS_{temp}) recorded at ALS beamline 11.0.2. The XAS_{temp} spectra are acquired in TFY mode using a photodiode covered with an Al filter to suppress background signals from ambient light. The sample is mounted on a button heater stage to control the sample temperature, which is monitored with a thermocouple attached to the front side of the sample. All measurements are performed under 3 mTorr of oxygen, in order to increase the temperature at which the phase transition of CuO to Cu_2O starts to occur.[28,29] Temperature-induced difference spectra ΔXAS_{temp} are calculated by subtracting room temperature (RT) spectra from spectra acquired at elevated temperatures $\Delta XAS_{temp}(\Delta T = T - RT) = XAS_{temp}(T) - XAS_{temp}(RT)$. For direct comparison of the $\Delta XAS_{temp}(\Delta T)$ and the $\Delta trXAS(\Delta t)$ spectra, RT = 28 °C was used as prolonged laser exposure leads to a slightly increased equilibrium temperature of the sample (see Section 4 for details).

Further details on the sample preparation, trXAS technique, and choice of pump laser fluences are given in the ESI.†

3. Results

Fig. 1a shows static X-ray absorption spectra acquired in total fluorescence yield (TFY, black) and total electron yield (TEY, blue) mode. The spectra agree well with

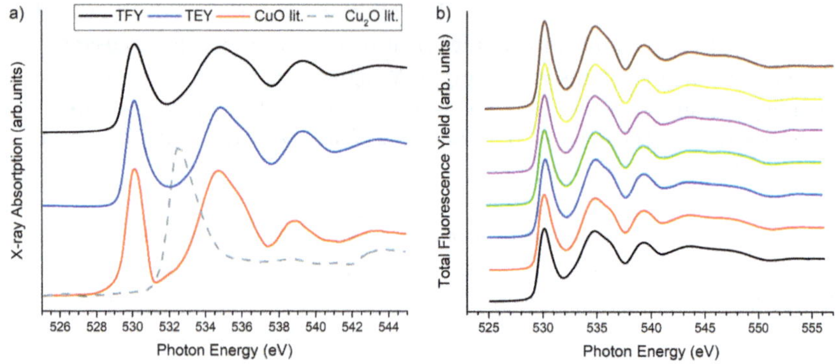

Fig. 1 (a) Comparison of static oxygen K-edge XAS spectra of CuO acquired in TFY (black) and TEY (blue) mode with CuO (red) and Cu_2O (gray, dashed) reference spectra from Jiang *et al.*[30] (b) Static TFY spectra from all samples used for time-resolved measurements. Spectra in (a) and (b) have been offset vertically for improved clarity.

TEY measurements by Jiang *et al.*, shown in red.[30] For comparison, the Cu_2O spectrum from the same study is also included (gray, dashed). The absence of the main Cu_2O resonance feature at 532.5 eV in the black and blue spectra indicates a high purity of the prepared CuO film. The slightly better agreement, concerning the amplitude, between the TEY spectra compared to the TFY spectrum is likely due to saturation and self-absorption effects in the TFY signal.[31] These are minimized by choosing a geometry with a normal X-ray incidence angle and grazing angle X-ray fluorescence detection ($\sim 0°$ and $\sim 82°$, respectively, relative to the sample surface normal). Fig. 1b shows static TFY spectra from all samples used for the time-resolved measurements presented here, demonstrating very good reproducibility of the sample preparation and experimental conditions. All spectra are normalized to the same increase between pre- and post-edge signals.

Fig. 2 illustrates the characteristic ΔtrXAS effects of the CuO sample for various pump-probe time delays using the 355 nm, 4.6 mJ cm^{-2} data set as an example. Fig. 2a covers the full spectral range of the measurement, while Fig. 2b provides a magnified view of the first resonance. The static XAS spectrum is overlaid in gray

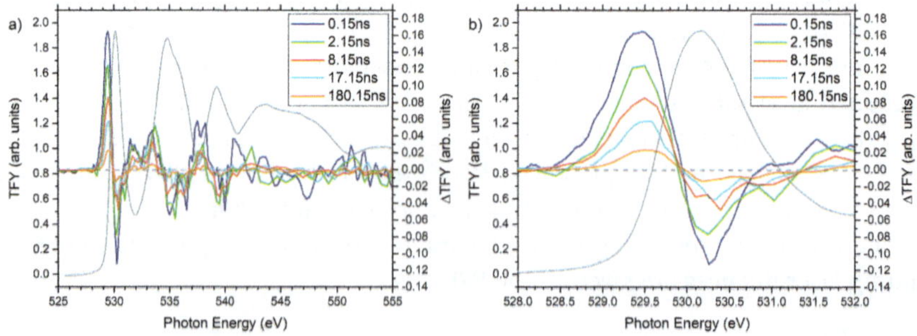

Fig. 2 (a) Overview of the 355 nm 4.6 mJ cm^{-2} trXAS data set. The reference spectrum (gray, left ordinate) is shown along with ΔtrXAS spectra at several pump-probe delays (right ordinate). (b) The lowest energy resonance at 530.1 eV displays the largest relative changes and is analysed in more detail. See text for details.

for reference. The left and right ordinates correspond to XAS and ΔtrXAS spectra, respectively. While the XAS units are arbitrary, they are the same for both the XAS and ΔtrXAS spectra. The blue (0.15 ns) and green (2.15 ns) curves correspond to data from the first and second X-ray pulse after excitation, respectively. In order to investigate smaller differences at longer pump-probe delays, where the ΔtrXAS signals are smaller, signals from multiple ALS bunches are averaged for this figure. The red curve (8.15 ns) is an average of three pulses ($\Delta t - 6.15$–10.15 ns), the cyan curve (17.15 ns) of six pulses ($\Delta t =$ 12.15–22.15 ns) and the orange (180.15 ns) of 81 pulses ($\Delta t =$ 100.15–260.15 ns). For each of the averaged spectra, the indicated delay is the mean delay of the averaged spectra. The by far strongest effects are observed in the region of the first resonance at ~530.1 eV (Cu e_g ref. 30) which is shown in more detail in Fig. 2b. The difference spectra exhibit a new feature emerging at the low-energy side of the static absorption spectrum (gray) with a peak at ~529.4–529.5 eV. Concurrently, signal is depleted near the static peak maximum with a maximum depletion at ~530.3 eV. The ΔtrXAS spectra exhibit an isosbestic point near ~529.9 eV, which separates the signal enhancement and depletion ranges. At first glance, the difference curves resemble differentials of the static XAS spectrum, which would be an indication that they may be described by a simple energy shift (ΔE) of the XAS spectrum upon photoexcitation, *i.e.* $\Delta\text{trXAS} = \alpha \times [\text{trXAS}(E - \Delta E) - \text{trXAS}(E)]$. However, no combination of parameters α (the fraction of reference spectrum trXAS that is shifted) and ΔE (the shift in X-ray photon energy) can be found that describes the difference spectra reasonably well. This is mostly due to the asymmetry of trXAS signal depletion and emergence, readily apparent in Fig. 2b.

As a first metric to investigate the relaxation dynamics of the photoexcited system, the area underneath the positive peak of the ΔtrXAS spectra in Fig. 2b is plotted as a function of pump-probe delay in Fig. 3. All data sets have been recorded during the same beamtime except for the 355 nm, 1.8 mJ cm^{-2} data. In that case, an additional 355 nm, 4.6 mJ cm^{-2} scan was performed to ensure it matches the data from the previous experiment, and inevitable changes from beamtime to beamtime, such as slight variations of the laser spot size, did not affect the measured signals within the scatter of the data. Each data point in Fig. 3 corresponds to a single pump-probe delay. The amplitudes and dynamic trends of

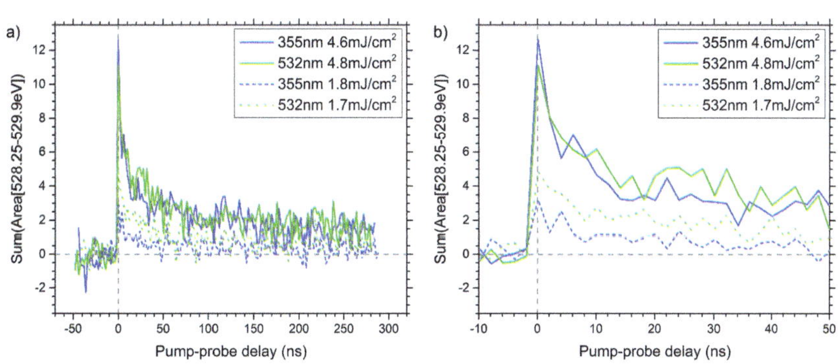

Fig. 3 (a) Time-dependent intensity of the positive ΔtrXAS feature, integrated between 528.25 eV and 529.9 eV (Fig. 2b) for all trXAS data sets. (b) Magnified view for delays up to 50 ns.

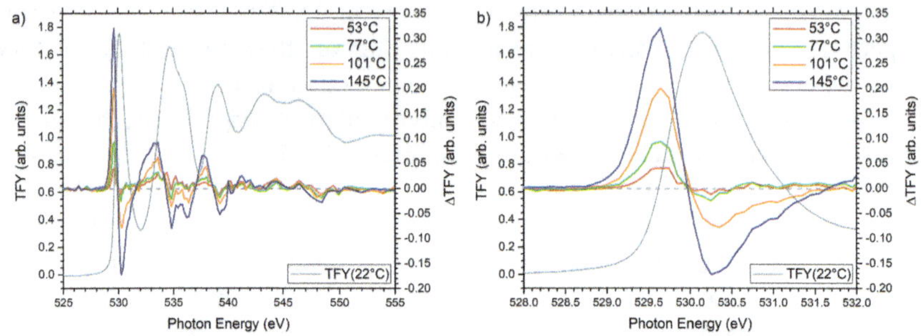

Fig. 4 (a) Temperature-dependent difference spectra ΔXAS_{temp} (colored, right ordinate) together with the reference spectrum XAS_{temp}(22 °C) measured at room temperature (gray, left ordinate). (b) Magnified view near the Cu e_g resonance feature at ~530.1 eV.

the new low-energy feature vary for different excitation wavelengths and fluences. For the highest fluences (4.6–4.8 mJ cm^{-2}), however, they are very similar for both 355 nm and 532 nm excitation. Each of these two data sets display a transient behavior marked by two decay timescales of roughly $t_1 = 10$ ns and $t_2 = 300$ ns. Evidently, the dynamic processes underlying the trXAS signals cannot be characterized by a simple, first-order rate equation. Instead, a more complex description is required that will be derived in Section 4 within the framework of a thermal diffusion model.

Fig. 4 shows the results of the temperature-dependent XAS_{temp} measurements. Fig. 4a shows the room temperature reference spectrum acquired at 22 °C in gray (left ordinate). This reference spectrum is subtracted from spectra acquired at elevated temperatures of 53 °C, 77 °C, 101 °C and 145 °C to generate the color coded ΔXAS_{temp} difference spectra as indicated in the figure legend (right ordinate). Fig. 4b shows a magnified view of the region near the Cu e_g resonance.[30] A comparison of Fig. 2 and 4 reveals striking similarities between the temperature-induced difference spectra $\Delta XAS_{temp}(\Delta T)$ and the photoinduced $\Delta trXAS(\Delta t)$ data. For every resonance peak of the XAS spectrum (~530 eV, ~535 eV and 539 eV), a signal enhancement feature appears at its low-energy side (529.65 eV, 533.50 eV and 537.75 eV) and a depletion feature at its high-energy side (530.3 eV, 535.5 eV and 539.3 eV). Both the positions and relative heights of these features appear to be very similar in both types of measurements. The second resonance around 535 eV has mainly Cu4s and some Cu3d character[30] and consists of two strongly overlapping peaks. Correspondingly, the ΔXAS_{temp} spectra at higher temperatures exhibit a double peaked depletion feature, likely due to overlapping absorption features. As in the case of the ΔtrXAS data, the most distinct features appear in the spectral range of the Cu e_g resonance. Interestingly, the ΔXAS_{temp} spectra also exhibit an isosbestic point separating the positive and negative difference signal for all measured temperatures. It is located at ~530.0 eV, very close to the ~529.9 eV value for the ΔtrXAS curves.

4. Data analysis

Different data sets were recorded with slightly different energy resolutions. In particular, the XAS_{temp} data (Fig. 4) exhibit better energy resolution than the trXAS data. In order to facilitate a quantitative comparison of all observed difference spectra, the energy resolution mismatch is compensated by identifying the ground state XAS spectrum with the lowest energy resolution (532 nm, 4.8 mJ cm^{-2}, further on referred to as XAS$_{ref}$), and convoluting all other data sets with Gaussian functions that effectively impose a common energy resolution on all ground state and RT spectra. Data analysis is restricted to the energy region between 528.25 and 531.75 eV containing the Cu e$_g$ peak at 530.1 eV. This region exhibits the strongest difference features and small differences in the energy scaling across more extended NEXAFS ranges make a quantitative comparison across larger energy ranges challenging. Due to the asymmetric peak shape, the convolution changes the peak position as well. Therefore, for each ground state spectrum, two free fit parameters are used to match it to XAS$_{ref}$, the FWHM of the Gaussian convolution function and a rigid shift of the entire spectrum. In order to prevent artifacts due to small peak height variations, all spectra are scaled to the same Cu e$_g$ peak height of 1. The FWHM, energy shift and scaling values derived using the ground state spectra are then applied to both the ground and corresponding excited state spectra to calculate the ΔtrXAS and ΔXAS$_{temp}$ difference spectra. Fig. 5a shows the ground state (unpumped) reference spectrum XAS$_{ref}$ (gray, left ordinate) along with ΔtrXAS difference spectra for both excitation wavelengths, 355 nm in blue and 532 nm in green, and two pump laser fluences each at a pump-probe delay of 150 ps (right ordinate). Fig. 5b shows a magnified view of the energy range used for the fit (528.25–531.75 eV). While all data sets show qualitatively similar trends, the best quantitative match is found between the 355 nm, 4.6 mJ cm^{-2} and the 532 nm, 4.8 mJ cm^{-2} data sets. These comparable difference spectra together with the matching peak areas and similar decay behavior illustrated in Fig. 3 strongly suggest that the observed XAS dynamics are primarily defined by the laser fluence, not the photon energy. In contrast, the excitation density within the probed volume appears to play a less prominent role, as evidenced by the significant differences between the traces associated with the 355 nm, 1.8 mJ cm^{-2} and 532 nm, 4.8 mJ cm^{-2} data sets shown in Fig. 3 and 5. In other words, the deposited energy per unit area rather than the number of excitations per unit volume correlates most strongly with the observed transient XAS response, which is a strong indication that thermal effects, rather than the

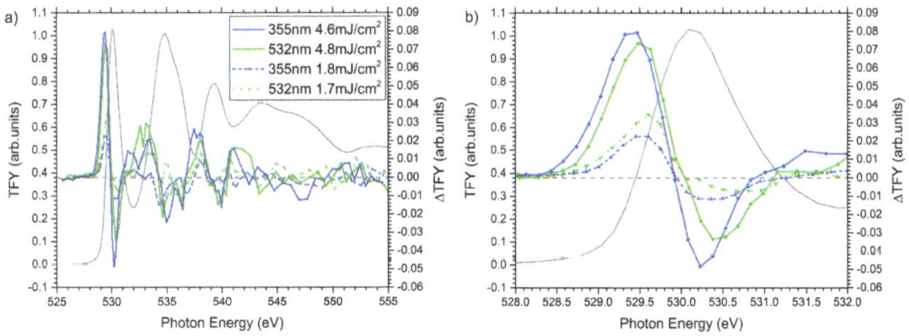

Fig. 5 (a) ΔtrXAS difference spectra at a pump-probe delay of 150 ps for both excitation wavelengths and various pump laser fluences. (b) Spectral region of the lowest energy feature, Cu e$_g$, which is used to adjust all data sets toward a common energy resolution. Data is shown after the convolution procedure is applied. The ground state reference spectrum XAS$_{ref}$ is shown in gray in both panels.

number of excited electrons, generate the dominant contributions to the ΔtrXAS data.

In order to test this hypothesis, a detailed comparison between the two highest fluence ΔtrXAS data (355 nm, 4.6 mJ cm^{-2} and 532 nm, 4.8 mJ cm^{-2}) and the ΔXAS_{temp} data is performed for each pump-probe time delay. The comparison is used to generate time-dependent sample temperature profiles for the two trXAS measurements, which are interpreted with the aid of a thermal diffusion model. As demonstrated below, the model reproduces the non-trivial dynamic trends indicated in Fig. 3 and provides access to time-dependent surface- and bulk-temperatures as well as the thermal conductivity of the sample.

In a first step, a linear interpolation of the XAS_{temp} spectra measured at 22 °C, 53 °C, 77 °C, 101 °C and 145 °C is performed in 1 °C steps in order estimate the shape of the spectra on a finer temperature grid. Spectra for temperature values >145 °C are estimated by a corresponding extrapolation using the 101 °C and 145 °C data. ΔXAS_{temp} spectra similar to Fig. 4 are generated for all interpolated and extrapolated data beyond 28 °C by subtracting the (interpolated) reference spectrum at 28 °C from the higher temperature spectra. The slightly elevated reference temperature (compared to the nominal 22 °C room temperature) is chosen to account for a slight increase in average temperature during the trXAS measurements that is determined with a thermocouple. The 28 °C spectrum is also used to determine the convolution parameters described above to match the energy resolution of the temperature-dependent and time-dependent spectra.

Two different fit procedures are applied to derive the time-dependent temperature profiles of the sample volumes probed by the trXAS measurements. The first method is easier to implement, but partly neglects inhomogeneous temperature effects, which are fully included in the second method. In the following, we discuss the results achieved by both methods in order to illustrate which effects need to be taken into account and which approximations are applicable when modeling thermal effects in trXAS spectra of condensed phase samples.

In the first method ("homogeneous fit"), the fit routine determines the temperature T for each time delay Δt, for which the absolute difference between the $\Delta XAS_{temp}(T)$ and the $\Delta trXAS(\Delta t)$ spectra (integrated over the energy range of the Cu e_g resonance) is minimal. Fig. 6a shows the result of this fit procedure for the 532 nm, 4.8 mJ cm^{-2} data set at a pump-probe delay of 150 ps. Depicted are the ΔtrXAS(150 ps) spectrum (green) and the best fit (red) corresponding to the curve $\Delta XAS_{temp}(73\ °C) = XAS_{temp}(101\ °C) - XAS_{temp}(28\ °C)$. The residual of the fit is shown as a dashed, black curve. Fig. 6b illustrates the convergence of the fit. The black trace is derived by integrating the absolute value of the residual across the fit energy range for each interpolated $\Delta XAS_{temp}(T)$ spectrum. The minimum position corresponds to the best fit shown in Fig. 6a. The best fit quality varies for different time delays and the corresponding residual curve (as shown in Fig. 6b) is flatter for longer time delays. Nevertheless, the fit always converges to a single temperature to best reproduce the measured $\Delta trXAS(\Delta t)$ spectrum. Note that the temperature T is the only free parameter of the fit in this method. For any fixed values of T and Δt, the amplitudes of the $\Delta XAS_{temp}(T)$ and $\Delta trXAS(\Delta t)$ spectra are determined by normalizing the Cu e_g peak heights of the $XAS_{temp}(28\ °C)$ and the ground state trXAS spectra to 1.

Performing this fit for all pump-probe time-delays leads to the time-dependent temperature estimates displayed in Fig. 7. Fig. 7a shows the results for the

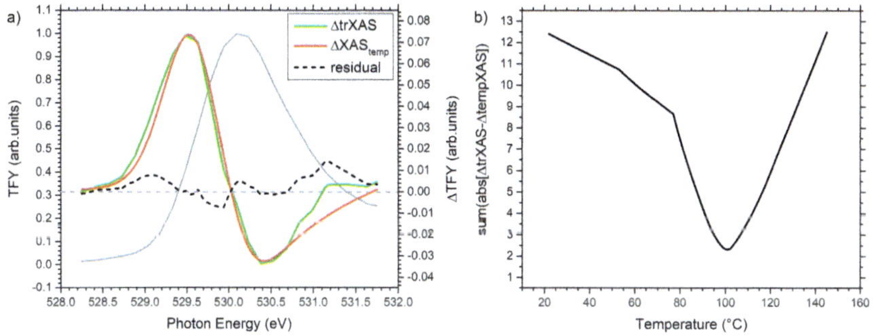

Fig. 6 (a) Photoinduced oxygen K-edge trXAS difference spectrum (ΔtrXAS, green) upon excitation of CuO with 532 nm, ~5 mJ cm^{-2} pulses, probed after 150 ps. The best fit by a temperature-dependent difference spectrum (ΔXAS_{temp}) is shown in red, the fit residual in black (dashed). For reference, the ground state XAS spectrum is shown in gray. All difference spectra refer to the right ordinate, the ground state spectrum to the left ordinate. (b) Integral over the absolute value of the fit residual curve in (a) as a function of temperature. The fit routine chooses the minimum of this curve to determine the best ΔXAS_{temp} fit to the ΔtrXAS data.

355 nm, 4.6 mJ cm^{-2} data (blue diamonds), Fig. 7b the corresponding results for the 532 nm, 4.8 mJ cm^{-2} data (green circles). Similar to the traces shown in Fig. 3, the dynamic trends are marked by a fast initial decay within 10s of nanoseconds, followed by a much slower relaxation on 100s of nanosecond timescales. The data are interpreted with the aid of a thermal diffusion model represented by the solid curves in Fig. 7.

The model is based on the assumption that a certain amount of thermal energy is deposited by the pump pulse within a near-surface volume and then propagates further into the bulk by thermal diffusion. Heat transport by convection is neglected, since the experiment is conducted at a pressure of $<5 \times 10^{-9}$ mbar. A simple estimate shows that radiative heat loss is negligible, too. According to the Stefan–Boltzmann law, the total radiated power of a black body at 120 °C is

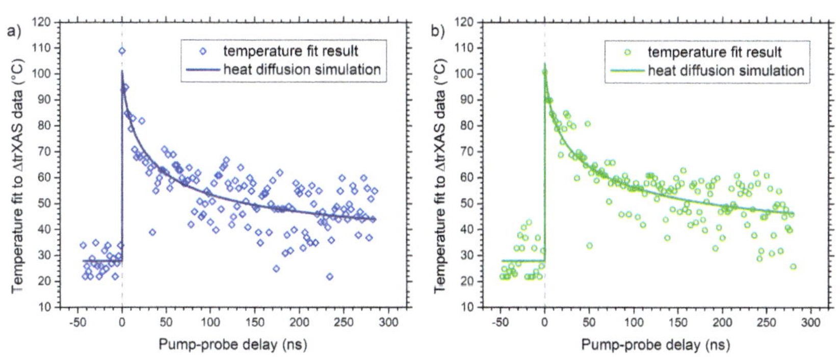

Fig. 7 Time-dependent sample temperatures for excitation with (a) 355 nm (blue diamonds) and (b) 532 nm (green circles) pulses at fluences of ~5 mJ cm^{-2}. Temperatures are derived by a homogeneous temperature fit model that determines a single effective temperature for the probed sample volume. The solid curves are fit results, describing the data within a one-dimensional thermal diffusion model. See text for details.

$\sim$1.4 kW m^{-2}. Using a FWHM laser beam size of roughly 250 × 250 μm^2 and a pulse energy of 3 μJ, the fraction of pulse energy emitted within the first 300 ns is 7×10^{-6}, which is negligible within the precision of the experiment. Even when allowing for substantially higher temperatures within the first 150 ps before the first probe pulse arrives, the order of magnitude of this estimate would be unaffected. Heat conduction is modeled as a one-dimensional diffusion process along the surface normal, z. For CuO, the optical penetration depths of 21 nm (λ = 355 nm) and 125 nm (λ = 532 nm)[32,33] as well as the effective X-ray probing depth of 43 nm (ref. 34) (see ESI†) are three to four orders of magnitude smaller than the FWHM of the laser spot (150–250 μm). Furthermore, the diameter of the area probed by the X-ray beam is only about half of the laser spot size. Therefore, the excited and probed sample volumes can be thought of as thin, concentric disks. The characteristic temperature gradients in and around the probed sample volume are thus orders of magnitude larger along the surface normal compared to radial directions parallel to the surface. Consequently, heat transport within the sample may be modeled by diffusion along an insulated cylinder with no lateral heat loss as described by[35–37]

$$\frac{\delta Q}{\delta t} = -k \times A \frac{\delta T(t,z)}{\delta z} . \tag{1}$$

Here, δQ is the amount of thermal energy transferred through a cross-sectional area A per time δt, k is the thermal conductivity and T is the temperature profile as a function of time t and depth z relative to the sample surface. The thermal conductivity is assumed to be constant for simplicity and because slight temperature dependencies would be negligible compared to the overall precision of the analysis presented here. The temperature distribution at time zero $T(t = 0,z)$ is proportional to the derivative of the depth-dependent laser intensity $I_{\text{laser}}(z) = \exp[-z/\delta_{\text{laser}}(\lambda)]$, where δ_{laser} is the laser penetration depth. The amplitude of the resulting exponential function is the initial temperature increase after laser excitation at the surface $\Delta T(t = 0,z = 0)$. This boundary condition leads to

$$T(t=0,z) = \frac{(1-R(\lambda))E_{\text{pulse}}}{2\rho A c(T)\delta(\lambda)_{\text{laser}}} \mathrm{e}^{-\frac{z}{\delta(\lambda)_{\text{laser}}}} + T_{\text{ref}} , \tag{2}$$

where $R(\lambda)$ is the sample reflectivity, E_{pulse} is the laser pulse energy, $\delta_{\text{laser}}(\lambda)$ the 1/e laser penetration depth, ρ the sample density, A is the FWHM laser focus area and $c(T)$ the specific heat capacity.[38] The sample equilibrium temperature $T_{\text{ref}} = 28$ °C is added to the laser-induced temperature jump to convert to an absolute temperature scale $T(t = 0,z)$. Using eqn (2) and the experimental parameters given above, the initial surface temperature estimates are $T(0,0) = 747$ °C for 355 nm (4.6 mJ cm^{-2}) and $T(0,0) = 157$ °C for 532 nm (4.8 mJ cm^{-2}) excitation, respectively.

The subsequent thermal diffusion process is modeled numerically by an iterative computer simulation. Based on an initial temperature profile $T(t = 0,z)$ according to eqn (2), it calculates the amount of heat dQ transferred to and from each volume element $A \times$ dz within a short time step dt using eqn (1). The slightly modified thermal energy distribution is used to calculate a new temperature profile $T(t + \mathrm{d}t,z)$, which in turn is employed to calculate the depth-dependent heat transfer rates during the next time step and so on. Typical time and depth step sizes are dt = 10 ps and dz = 3 nm, respectively. Convergence of the simulation is tested by varying these step sizes and ensuring that the results are not affected within the overall precision of the method.

For the first, basic temperature analysis presented in Fig. 7, the effective temperature $T_{\text{eff}}(t)$ probed by the X-ray pulse is estimated by multiplying the simulated temperature profile $T(t,z)$ with the probe intensity distribution $I_{\text{X-rays}}(z) \sim \exp[-z/\delta_{\text{X-rays}}]$ and integrating the result over z. Here, $\delta_{\text{X-rays}}$ is the effective 1/e X-ray escape depth described in the ESI.† The free parameters varied during the fit are the amplitude of the initial temperature distribution $T(t=0,z=0)$, *i.e.*, the initial surface temperature, and the thermal conductivity k. The best fit results, shown as solid curves in Fig. 7, correspond to thermal conductivities of $k_{355\ \text{nm}} = (0.16 \pm 0.06)\ \text{W m}^{-1}\ \text{K}^{-1}$ and $k_{532\ \text{nm}} = (1.3 \pm 0.5)\ \text{W m}^{-1}\ \text{K}^{-1}$ and initial surface temperatures of $T(0,0)_{355\ \text{nm}} = (190 \pm 20)\ °\text{C}$ and $T(0,0)_{532\ \text{nm}} = (120 \pm 11)\ °\text{C}$. While a substantial difference in the initial surface temperatures is to be expected for the two laser colors due to the factor of ~6 difference in absorption depths, the thermal conductivities should ideally be the same within the precision of the method. In order to improve the temperature dynamics estimates and to get a better understanding of factors affecting the precision of the fit results, a second more advanced temperature modeling and dynamics fit procedure is applied ("inhomogeneous fit").

A key challenge is that each of the XAS_{temp} spectra is acquired at a single temperature that is constant throughout the entire sample, while temperatures in the trXAS experiments vary significantly within the probed sample volume. The first, homogeneous fit method described above hinges on the assumption that an effective temperature T_{eff} can be defined for the probed volume by a weighted average of an extended temperature profile and that the measured $\Delta\text{trXAS}(t)$ spectrum emerging from this volume is well represented by the corresponding $\Delta\text{XAS}_{\text{temp}}(T_{\text{eff}})$ spectrum measured with a homogeneous temperature T_{eff} across the entire sample. However, the temperature dependence of the $\Delta\text{XAS}_{\text{temp}}$ spectra is nontrivial (see Fig. 4) and, generally, the (weighted) average spectrum $\Delta\text{XAS}_{\text{temp}}^{\text{avg}} = \sum_i w_i \Delta\text{XAS}_{\text{temp}}(T_i) \Big/ \sum_j w_j$ of the probed sample volume does not have to be well approximated by the corresponding difference spectrum $\Delta\text{XAS}_{\text{temp}}(T_{\text{eff}})$ measured at the average temperature $T_{\text{eff}} = \sum_i w_i T_i \Big/ \sum_j w_j$. The quality of the approximation is expected to depend strongly on the range of temperatures within the probed volume. For the 532 nm data, the optical penetration depth of 125 nm is significantly larger than the X-ray probing depth of 43 nm, leading to a relatively homogeneous temperature distribution within the probed sample volume and the approximation is expected to hold fairly well. In contrast, for 355 nm excitation, the penetration depth is only 21 nm, much smaller than the X-ray probing depth, resulting in a strongly inhomogeneous temperature distribution within the probed volume, at least for short pump-probe delays. In this case, the approximation is expected to break down.

In order to avoid potential shortcomings of the effective homogeneous temperature approach, a modified inhomogeneous fit procedure is applied that explicitly takes into account both the time- and depth-dependent temperature profiles as well as the complex correlation between local temperatures and corresponding ΔXAS_{temp} profiles. Starting with a thermal conductivity value k and an initial surface temperature $T(0,0)$, the 2D matrix $T(t,z)$ of a time- and depth-dependent temperature profile is calculated using the numerical model based on eqn (1). This provides, in particular, a temperature distribution $T_N(z)$ at the time of arrival of each probe pulse N, *i.e.*, $T_N(z) = T(t_N = t(\text{bunch } N), z)$. The simulated spectrum $\Delta trXAS_{sim}(E, t_N)$ for each bunch N is then calculated as the weighted average of the $\Delta XAS_{temp}(E, T_N(z))$ spectra using the X-ray intensity distribution $I_{\text{X-rays}}(z) \sim \exp[-z/\delta_{\text{X-rays}}]$ as the weight function:

$$\Delta trXAS_{sim}(E, t_N) = \frac{\sum_z I_{\text{X-rays}}(z)\Delta XAS_{temp}(E, T_N(z))}{\sum_{z'} I_{\text{X-rays}}\left(z'\right)}. \tag{3}$$

During the fit procedure, the values of k and $T(0,0)$ are varied such that the sum of all residuals $R_{total} = \sum_N R_N = \sum_N \sum_E |\Delta trXAS_{sim}(E, t_N) - \Delta trXAS(E, t_N)|$ for all energies E within the range 528.25–531.75 eV and for all times t_N is minimized. This inhomogeneous fit procedure leads to best fit values of $k_{355\ nm} = (0.29^{+0.11}_{-0.07})$ W m^{-1} K^{-1} and $k_{532\ nm} = (1.3 \pm 0.4)$ W m^{-1} K^{-1} for the heat conductivities and $T(0,0)_{355\ nm} = (230 \pm 30)$°C, $T(0,0)_{532\ nm} = (123 \pm 10)$ °C for the initial surface temperatures at the two excitation wavelengths. Uncertainties correspond to one sigma standard deviations computed from the fit residuals according to eqn (12) in ref. 39. As expected, the values for 532 nm are very similar to the results derived with the homogeneous, effective temperature fit model, while the values for 355 nm differ considerably.

Fig. 8 shows a comparison of the results from the two modeling procedures for both excitation wavelengths. Gray circles are the result of the homogeneous, single effective temperature fits (same as the data points in Fig. 7) while the black solid lines result from the inhomogeneous fit routine. Similar to the homogeneous fit curves in Fig. 7, the inhomogeneous fit curves in Fig. 8 are generated by multiplying the simulated temperature profiles $T(t,z)$ with the probe intensity distribution $I_{\text{X-rays}}(z) \sim \exp[-z/\delta_{\text{X-rays}}]$ and integrating the results over z. In order to illustrate the effect of varying the thermal conductivity parameter k in the inhomogeneous fit routine, two dashed lines are added, corresponding to the temperature transients computed with half (blue) and double (green) the best fit value of k, while keeping the surface temperatures $T(0,0)$ constant. The curves indicate a pronounced sensitivity of the temperature trends to the absolute value of the thermal conductivity.

Note that the homogeneous and inhomogeneous fit procedures lead to almost identical volume-averaged temperature trends (solid curves in Fig. 7, black curves in Fig. 8). The best fit values for k and $T(0,0)$, however, agree for both fit procedures only for 532 nm excitation, while they differ significantly for the 355 nm data. For the latter, the reconstructed surface temperatures also differ from the volume-averaged ones while they agree reasonably well for the 532 nm measurement. This is illustrated by the red and violet (dotted) curves in Fig. 8, which show the surface temperature trends, $T(t,0)$, for the inhomogeneous and homogeneous fits, respectively. For 532 nm, the two surface temperature curves and the black curve are very similar (Fig. 8b). In contrast, all three curves differ

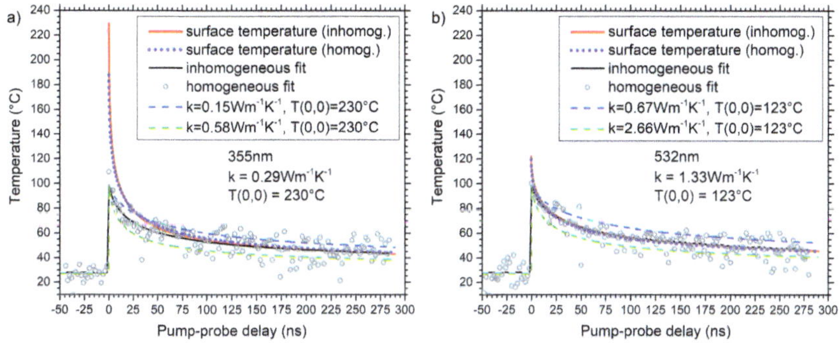

Fig. 8 Temperature dynamics resulting from the two fit models as a function of pump-probe delay for excitation with (a) 355 nm and (b) 532 nm. Singular data points (gray) are the result of the effective, homogeneous temperature model, the black line the result of the inhomogeneous fit model. Red and violet (dotted) lines show the surface temperature dynamics derived from the inhomogeneous and homogeneous fit models, respectively. Blue and green dashed lines indicate the variation of the inhomogeneous fit model results when varying the heat conductance by factors of 1/2 and 2, respectively, compared to the best fit values, while keeping the surface temperature constant.

notably from each other for the 355 nm data set (Fig. 8a). The large differences in the fit results for the 355 nm and 532 nm data are rooted in the different absorption depths of the two excitation wavelengths in relation to the X-ray escape depth. This is illustrated in more detail in Fig. 9, which shows the temperature distributions at time zero (gray lines) and for the first nine probe pulses (colored lines) after excitation, reconstructed with the inhomogeneous fit model. Fig. 9a and b show the results for 355 nm and 532 nm excitation, respectively. For comparison, the X-ray escape profile, offset by the sample equilibrium temperature and scaled to the initial surface temperature, is overlaid as a solid black line in both panels. Note the different ranges of *z*-values in Fig. 9a and b. As expected from the different optical penetration depths, the temperature profiles have much

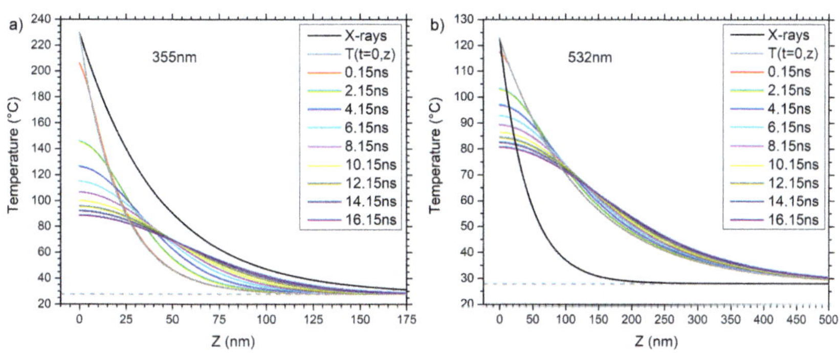

Fig. 9 Time-dependent temperature distributions for (a) 355 nm and (b) 532 nm excitation as a function of distance *z* from the surface along the surface normal. The gray line corresponds to $t \approx 0$ and colored lines to the times of arrival of the first nine X-ray pulses following the laser excitation. For reference, the depth-dependent X-ray probe intensity, offset by the sample equilibrium temperature and scaled to the initial surface temperature, is shown in black.

higher gradients and spread across a larger range of values for the 355 nm experiment compared to the 532 nm experiment. These details, however, are largely obscured in the volume-averaged temperatures "seen" in the experiment. Thus, the striking similarities between the observed volume-averaged temperature dynamics for 355 nm and 532 nm excitation displayed in Fig. 7 and 8 conceal significant differences in the underlying physical phenomena that need to be recovered through careful analysis and modeling. We note that the appearance of significantly higher surface temperatures in the 355 nm experiment compared to the 532 nm experiment is independently corroborated by a more efficient conversion of CuO to Cu_2O upon UV excitation of the sample compared to the visible excitation. This leads to the emergence of an additional signal in the CuO XAS spectrum at the location of the ~532 eV white line of the Cu_2O spectrum (see Fig. 1a), which needs to be suppressed by scanning every sample spot only once in the 355 nm experiment.

5. Discussion

The results presented above strongly suggest that the photoinduced changes in the XAS spectra of CuO beyond ~100 ps pump-probe delay are dominated by thermal effects. This observation is in agreement with the results of Hayes *et al.*[7] and Lin *et al.*[8] for super-bandgap excited α-Fe_2O_3 and PbI_2, respectively. Lin *et al.* excited ~70 nm thick polycrystalline films of PbI_2 (deposited on 50 nm thick silicon nitride) with 400 nm radiation and probed the induced dynamics with femtosecond time-resolved XUV absorption at the iodine 4d level near ~50 eV.[8] Good agreement was found between the photoinduced XUV absorption change after 40–100 ps and a temperature-dependent absorption change induced by heating the sample from 20 °C to 120 °C. A complementary ultrafast electron diffraction experiment by the same group corroborated the finding that temperature effects start to dominate the pump-probe difference spectra after ~4 ps.[8] Hayes *et al.* performed oxygen and iron K-edge trXAS measurements on 20 nm thick films of hematite (α-Fe_2O_3, deposited on 100 nm thick silicon nitride) excited with 400 nm and 351 nm pulses, respectively.[7] The Fe K-edge ΔtrXAS spectrum at 100 ps pump-probe delay is well reproduced by a ΔXAS_{temp} spectrum generated by subtracting a steady-state XAS spectrum recorded at ~30 °C from one acquired at ~230 °C. Corresponding theoretical modeling of the spectra using multiple scattering calculations confirmed that the dominant effects in the ΔtrXAS spectrum are indeed due to thermal effects with contributions from both thermal disorder and lattice distortions.[7]

The study presented here reproduces a predominance of lattice temperature induced effects in picosecond trXAS for another transition metal oxide semiconductor. In addition, it provides further insight into the temporal evolution of lattice temperature profiles after laser excitation on an absolute temperature scale with a temporal sensitivity and a spatial selectivity on the order of 10s of picoseconds and 10s of nanometers, respectively. The results presented in Fig. 7 and 8 indicate that the heat transport dynamics may be approximated by a one-dimensional diffusion model as previously employed in picosecond time-resolved electron diffraction studies of laser-heated metal surfaces.[35] However, comparison of the trXAS results for 355 nm and 532 nm excitation also reveals that a quantitative interpretation of the underlying physics requires detailed understanding and modeling of the relation between the temperature distribution within the sample and the resulting X-ray spectroscopic signatures. In order to understand and predict thermal effects in time-dependent X-ray absorption spectra, it should be ensured that the X-ray probing depth is small compared to the penetration depth of the optical excitation. Otherwise, signal contributions from regions with significantly different temperatures need to be disentangled. This effect is documented by the significant fit-model dependence of both the reconstructed temperature profiles as well as the thermal conductivity k in the 355 nm experiment. In contrast, the corresponding results for the 532 nm measurements are largely model-independent.

Additional evidence that the 532 nm experiment is better understood on a quantitative level is provided by a comparison of the fit results with surface temperature predictions based on eqn (2). The calculated surface temperature after absorption of a 532 nm pump pulse with a fluence of 4.8 mJ cm^{-2} is $T(0,0)_{532\ \mathrm{nm}}(\mathrm{calc}) = 157$ °C, which agrees reasonably well with the outcome of the inhomogeneous fit $T(0,0)_{532\ \mathrm{nm}}(\mathrm{fit}) = 123$ °C. In contrast, the corresponding values for 355 nm excitation are $T(0,0)_{355\ \mathrm{nm}}(\mathrm{calc}) = 747$ °C and $T(0,0)_{355\ \mathrm{nm}}(\mathrm{fit}) = 230$ °C, *i.e.*, the calculated and reconstructed temperature jumps differ by a factor of ~3.6. The reason for this large discrepancy is currently unknown. One contributing factor may be that, despite our best efforts, the convolution methods described in Section 4 could provide an inaccurate modeling of the ΔtrXAS response when the sampled region exhibits a steep-gradient temperature distribution. Another potential issue is that the $\Delta\mathrm{XAS}_{\mathrm{temp}}$ spectra for temperatures above 145 °C have been estimated based on a linear extrapolation of the measurements at 101 °C and 145 °C, which is most likely increasingly inaccurate with increasing temperatures. Interestingly, Hayes *et al.* also observed a significant discrepancy between the estimated temperature rise in their hematite sample based on the absorbed photon energy (~470 °C) and the steady-state temperature rise needed to reproduce the photon-induced change in X-ray absorption (200 °C).[7] In this case, temperature gradients should not play a major role due to the small sample thickness of only 20 nm. Note that in both cases, the hematite experiment using 351 nm excitation and the CuO experiment at 355 nm excitation, the temperature rise estimated by a complete conversion of the absorbed laser pulse energy into heat is much larger than the measured temperature increase based on a comparison to steady-state temperature-dependent $\Delta\mathrm{XAS}_{\mathrm{temp}}$ spectra. Additionally, the estimated temperatures in both cases were significantly higher than the highest temperatures during the steady-state measurements used to determine the $\Delta\mathrm{XAS}_{\mathrm{temp}}$ response. Thus, it could be that this response saturates toward higher temperatures, which would limit their impact on the XAS fingerprint. Further studies with higher reference temperatures are required to gain a better understanding of this regime.

Given the challenges associated with the quantitative analysis of the 355 nm experiment, we restrict a more detailed discussion of the fit results to the 532 nm measurement. The good agreement between the observed and reconstructed temperature dynamics illustrated in Fig. 7b, between the absolute temperature values derived from the absorbed laser pulse energy and the dynamics fit, as well as the model-independence of the fit results for 532 nm excitation suggest that heat transport within the CuO sample is well represented by a thermal diffusion constant $k = (1.3 \pm 0.4)$ W m^{-1} K^{-1}. Literature values for the thermal conductivity of CuO vary across almost two orders of magnitude from 1.013 W m^{-1} K^{-1} in ref. 24 to 18 W m^{-1} K^{-1} in ref. 25 and 33 W m^{-1} K^{-1} in ref. 40 and 78 W m^{-1} K^{-1} in ref. 41. We note, however, that an extensive literature search did not yield any original reference that would describe the experiments used to derive any of these values. Thus, it is not entirely clear whether the large variance is primarily the result of different sample preparation techniques or of uncertainties in the measurements. Notably, however, the smallest value, reported by Samsonov[24] is the only one for which the sample form is indicated as powder. Given that this value is remarkably close to the one derived here and that all other literature values are more than an order of magnitude larger, it appears that the thermal conductivity of the CuO film prepared in this study most closely resembles that of a CuO powder.

At first glance, this result may be surprising as the sample is prepared by annealing a solid sheet of copper. Note, however, that the CuO sample consists of a relatively thin film at the Cu surface and the oxidation process is likely accompanied by significant changes in the surface morphology. Gong *et al.* used annealed sputtering to deposit thin films of copper at temperatures between 200 °C and 400 °C in air and studied the resulting surfaces using X-ray diffraction (XRD), Raman spectroscopy, and atomic force microscopy (AFM).[26] It was concluded that, below 300 °C, Cu is oxidized to Cu_2O and, upon further temperature increase ≥300 °C, CuO is formed. In particular, the AFM images reveal substantial surface morphology variations during the phase transitions. During the Cu_2O formation, the surface roughness increases from ~10s to ~100s of nanometer while grains of Cu_2O form and coalesce to typical grain sizes of ~150 nm. At 400 °C, CuO forms in lamellar secondary grains with a thickness of ~5 nm, which are distributed within the larger Cu_2O grains. Based on these observations, it is conceivable that the CuO film prepared here by heating of a Cu surface in air to 500 °C is also marked by a nano-structured morphology rather than a crystalline bulk structure. In this case, the thermal conductivity of the film may be expected to lie closer to the value for a CuO powder than a crystalline bulk sample. Further characterization of the CuO film using alternative techniques may provide additional insight in this regard.

6. Conclusion and outlook

Photoinduced changes in the oxygen K-edge X-ray absorption spectrum of a thin film of CuO have been studied by picosecond time-resolved XAS and complementary steady-state temperature-dependent XAS. The results show that for pump-probe delays ≥150 ps, the trXAS spectra are dominated by lattice temperature induced effects for both 355 nm and 532 nm excitation. The laser-induced heating and subsequent sample cooling dynamics at pump laser fluences of ~5 mJ cm^{-2} are well described by a one-dimensional thermal diffusion model that explicitly takes into account the depth-dependence of both the optical excitation and the X-ray probing. A comparison of the measurements with different pump laser wavelengths shows that a detailed modeling of these depth-dependent profiles is crucial for a quantitative understanding of the thermal dynamics in terms of the material properties. *Vice versa*, predicting the impact of thermal excitations on future trXAS studies will require a correspondingly well-developed understanding of the sample properties and nanoscale intensity profiles of pump- and probe-pulse intensities. The thermal conductivity of the

CuO sample used in this study, $k = (1.3 \pm 0.4)$ W m^{-1} K^{-1}, is in good agreement with the literature value for CuO powder, k_{powder} = 1.013 W m^{-1} K^{-1},[24] and approximately an order of magnitude smaller than the smallest literature value for bulk CuO, k_{bulk} = 18 W m^{-1} K^{-1}.[25] This underlines the importance of proper sample characterization prior to a trXAS experiment. A prediction of, for example, how fast thermal excitations in the CuO film may be removed between pump-pulses of a high repetition-rate trXAS experiment would lead to a far too optimistic estimate if it were based on the bulk properties of cupric oxide. Moreover, a predictive understanding of thermal dynamics and their spectral fingerprints is not only important to mitigate sample damage and unwanted signal contributions but also to perform experiments where photoinduced thermal excitations are the actual trigger for the dynamics under investigation.[5] The results presented here demonstrate that a quantitative modeling of thermal effects in trXAS spectra is possible but also that it requires great dedication to detail regarding the material properties and experimental conditions of the pump-probe experiment to make quantitatively accurate predictions.

Conflicts of interest

There are no conflicts to declare.

Acknowledgements

This work was supported by the U.S. Department of Energy, Office of Science, Office of Basic Energy Sciences, Chemical Sciences, Geosciences and Biosciences Division, through Contract No. DE-AC02-05CH11231. This research used resources of the Advanced Light Source, which is a DOE Office of Science User Facility under contract no. DE-AC02-05CH11231. J. M. acknowledges partial support from the ALS Doctoral Fellowship in Residence Program (2017). SN and MB acknowledge support by the Alexander von Humboldt foundation. F. R. acknowledges financial support by VI 419 of the Helmholtz Association. FR and WE were supported by the Helmholtz Research Association (Germany) within the framework program of 'structure of matter'. WE also would like to acknowledge the hospitality of the ALS during some extended visits.

References

1 M. L. Grünbein, J. Bielecki, A. Gorel, M. Stricker, R. Bean, M. Cammarata, K. Dörner, L. Fröhlich, E. Hartmann, S. Hauf, M. Hilpert, Y. Kim, M. Kloos, R. Letrun, M. Messerschmidt, G. Mills, G. Nass Kovacs, M. Ramilli, C. M. Roome, T. Sato, M. Scholz, M. Sliwa, J. Sztuk-Dambietz, M. Weik, B. Weinhausen, N. Al-Qudami, D. Boukhelef, S. Brockhauser, W. Ehsan, M. Emons, S. Esenov, H. Fangohr, A. Kaukher, T. Kluyver, M. Lederer, L. Maia, M. Manetti, T. Michelat, A. Münnich, F. Pallas, G. Palmer, G. Previtali, N. Raab, A. Silenzi, J. Szuba, S. Venkatesan, K. Wrona, J. Zhu, R. B. Doak, R. L. Shoeman, L. Foucar, J.-P. Colletier, A. P. Mancuso, T. R. M. Barends, C. A. Stan and I. Schlichting, *Nat. Commun.*, 2018, **9**, 3487.

2 R. W. Schoenlein, *et al.*, *LCLS-II Science Opportunities, Report SLAC-R-1053*, SLAC National Accelerator Laboratory, Menlo Park CA, https://

portal.slac.stanford.edu/sites/lcls_public/Lists/Publications/DispForm.aspx?ID=1004&ContentTypeId=0x0100D2D2C16F06384A44A845B10E6FC226F600DC7CB9184F2FB34484887A99919A7390, accessed 1 December 2018.

3 S. M. Teichmann, F. Silva, S. L. Cousin, M. Hemmer and J. Biegert, *Nat. Commun.*, 2016, **7**, 11493.

4 D. Popmintchev, B. R. Galloway, M.-C. Chen, F. Dollar, C. A. Mancuso, A. Hankla, L. Miaja-Avila, G. O'Neil, J. M. Shaw, G. Fan, S. Ališauskas, G. Andriukaitis, T. Balčiunas, O. D. Mücke, A. Pugzlys, A. Baltuška, H. C. Kapteyn, T. Popmintchev and M. M. Murnane, *Phys. Rev. Lett.*, 2018, **120**, 093002.

5 H. Öström, H. Öberg, H. Xin, J. LaRue, M. Beye, M. Dell, J. Gladh, M. L. Ng, J. A. Sellberg, S. Kaya, G. Mercurio, D. Nordlund, M. Hantschmann, F. Hieke, D. Kühn, W. F. Schlotter, G. L. Dakovski, J. J. Turner, M. P. Minitti, A. Mitra, S. P. Moeller, A. Föhlisch, M. Wolf, W. Wurth, M. Persson, J. K. Nørskov, F. Abild-Pedersen, H. Ogasawara, L. G. M. Pettersson and A. Nilsson, *Science*, 2015, **347**, 978–982.

6 M. Schultze, K. Ramasesha, C. D. Pemmaraju, S. A. Sato, D. Whitmore, A. Gandman, J. S. Prell, L. J. Borja, D. Prendergast, K. Yabana, D. M. Neumark and S. R. Leone, *Science*, 2014, **346**, 1348–1352.

7 D. Hayes, R. G. Hadt, J. D. Emery, A. A. Cordones, A. B. F. Martinson, M. L. Shelby, K. A. Fransted, P. D. Dahlberg, J. Hong, X. Zhang, Q. Kong, R. W. Schoenlein and L. X. Chen, *Energy Environ. Sci.*, 2016, **9**, 3754–3769.

8 M.-F. Lin, M. A. Verkamp, J. Leveillee, E. S. Ryland, K. Benke, K. Zhang, C. Weninger, X. Shen, R. Li, D. Fritz, U. Bergmann, X. Wang, A. Schleife and J. Vura-Weis, *J. Phys. Chem. C*, 2017, **121**, 27886–27893.

9 S. Harish, J. Archana, M. Sabarinathan, M. Navaneethan, K. D. Nisha, S. Ponnusamy, C. Muthamizhchelvan, H. Ikeda, D. K. Aswal and Y. Hayakawa, *Appl. Surf. Sci.*, 2017, **418**, 103–112.

10 R. Saravanan, S. Karthikeyan, V. K. Gupta, G. Sekaran, V. Narayanan and A. Stephen, *Mater. Sci. Eng., C*, 2013, **33**, 91–98.

11 S. Xu, A. J. Du, J. Liu, J. Ng and D. D. Sun, *Int. J. Hydrogen Energy*, 2011, **36**, 6560–6568.

12 J. Yu, Y. Hai and M. Jaroniec, *J. Colloid Interface Sci.*, 2011, **357**, 223–228.

13 J. Bandara, C. P. K. Udawatta and C. S. K. Rajapakse, *Photochem. Photobiol. Sci.*, 2005, **4**, 857.

14 W. Zhang, G. Ma, H. Gu, Z. Yang and H. Cheng, *J. Power Sources*, 2015, **273**, 561–565.

15 S. Neppl, J. Mahl, A. S. Tremsin, B. Rude, R. Qiao, W. Yang, J. Guo and O. Gessner, *Faraday Discuss.*, 2016, **194**, 659–682.

16 N. Serin, T. Serin, Ş. Horzum and Y. Çelik, *Semicond. Sci. Technol.*, 2005, **20**, 398–401.

17 B. Balamurugan and B. R. Mehta, *Thin Solid Films*, 2001, **396**, 90–96.

18 F. Marabelli, G. B. Parravicini and F. Salghetti-Drioli, *Phys. Rev. B: Condens. Matter Mater. Phys.*, 1995, **52**, 1433–1436.

19 M. T. Nair, L. Guerrero, O. L. Arenas and P. Nair, *Appl. Surf. Sci.*, 1999, **150**, 143–151.

20 S. C. Ray, *Sol. Energy Mater.*, 2001, **68**, 307–312.

21 C. E. Ekuma, V. I. Anisimov, J. Moreno and M. Jarrell, *Eur. Phys. J. B*, 2014, **87**, 23.

22 M. Dahrul, H. Alatas and Irzaman, *Procedia Environ. Sci.*, 2016, **33**, 661–667.
23 J. F. Pierson, A. Thobor-Keck and A. Billard, *Appl. Surf. Sci.*, 2003, **210**, 359–367.
24 G. V. Samsonov, *The Oxide Handbook*, IFI/Plenum, New York, 1973.
25 A. L. Edwards, *A Compilation Of Thermal Property Data For Computer Heat-Conduction Calculations*, Report No. UCRL-50589, 1969.
26 Y. S. Gong, C. Lee and C. K. Yang, *J. Appl. Phys.*, 1995, **77**, 5422–5425.
27 A. S. Tremsin, O. H. Siegmund, J. S. Hull, J. V. Vallerga, J. B. McPhate, J. Soderstrom, J.-W. Chiou, J. Guo and Z. Hussain, *IEEE Nucl. Sci. Symp. Conf. Rec.*, 2006, **2**, 735–739.
28 S. Poulston, P. M. Parlett, P. Stone and M. Bowker, *Surface and Interface Analysis: An International Journal devoted to the development and application of techniques for the analysis of surfaces, interfaces and thin films*, 1996, vol. 24, pp. 811–820.
29 M. Heinemann, B. Eifert and C. Heiliger, *Phys. Rev. B: Condens. Matter Mater. Phys.*, 2013, **87**, 115111.
30 P. Jiang, D. Prendergast, F. Borondics, S. Porsgaard, L. Giovanetti, E. Pach, J. Newberg, H. Bluhm, F. Besenbacher and M. Salmeron, *J. Chem. Phys.*, 2013, **138**, 024704.
31 S. Eisebitt, T. Böske, J.-E. Rubensson and W. Eberhardt, *Phys. Rev. B: Condens. Matter Mater. Phys.*, 1993, **47**, 14103–14109.
32 D. Tahir and S. Tougaard, *J. Phys.: Condens. Matter*, 2012, **24**, 175002.
33 W. K. C. Yung, B. Sun, Z. Meng, J. Huang, Y. Jin, H. S. Choy, Z. Cai, G. Li, C. L. Ho, J. Yang and W. Y. Wong, *Sci. Rep.*, 2016, **6**, 39584.
34 CuO – X-ray attenuation length, http://henke.lbl.gov/optical_constants/atten2.html, accessed 28 November 2018.
35 H. E. Elsayed-Ali and J. W. Herman, *Appl. Phys. Lett.*, 1990, **57**, 1508–1510.
36 J. H. Bechtel, *J. Appl. Phys.*, 1975, **46**, 1585–1593.
37 J. Crank, *The Mathematics of Diffusion*, Oxford University Press, Oxford, England, end edn, 1975.
38 Copper oxide – National Insitute for Standards and Technology, https://webbook.nist.gov/cgi/cbook.cgi?ID=C1317380&Type=JANAFS&Table=on, accessed 28 November 2018.
39 P. J. Cumpson and M. P. Seah, *Surf. Interface Anal.*, 1992, **18**, 345–360.
40 M. Liu, M. Lin and C. Wang, *Nanoscale Res. Lett.*, 2011, **6**, 297.
41 D. Anandan and K. S. Rajan, *Asian J. Sci. Res.*, 2012, **5**, 218–227.

Faraday Discussions

DISCUSSIONS

Photo-induced electron transfer: general discussion

Michael Ashfold, Jon Bender, David N. Beratan, Stephen Bradforth, Jeffrey Cina, Animesh Datta, Jahan Dawlaty, Ryan Dill, Amro Dodin, Marta Duchi, Laura Estergreen, Graham Fleming, Terry Frankcombe, Gregory Gate, Oliver Gessner, Naomi Ginsberg, Christopher Grieco, Michael Haggmark, Sharon Hammes-Schiffer, Vanessa Huxter, Michael Kellogg, Nadia Korovina, Yumin Lee, Johannes Mahl, Karen Morenz, Jennifer Ogilvie, Thomas A. A. Oliver, Tom Penfold, Petter Persson, Benjamin Schwartz, Minjung Son, Vasilios Stavros, Collin Steen, Mark Thompson, Michael Wasielewski, Emily Weiss and Jack Woolley

DOI: 10.1039/C9FD90029B

Nadia Korovina opened discussion of the paper by David N. Beratan: Does the excited state energy of the bridge affect the coupling between the donor and the acceptor? How does the rate of energy transfer through the bridge scale with respect to the bridge energy?

David N. Beratan replied: With Dexter energy transfer reactions, as with electron-transfer reactions, the donor–acceptor coupling and the transfer rate are expected to grow as the states of the bridge become nearly resonant with the donor and acceptor states. In the case of Dexter energy transfer, it is the excitation energy of the bridge states compared to those of the donor/acceptor species that should be considered. The exponential distance decay constant, typically called beta, will become smaller as the bridge electronic transitions approach resonance with the donor and acceptor transitions. Interestingly, we have found that the nature of the mediating virtual states changes as the bridge exciton state energies become closer to being resonant with the donor and acceptor exciton energies. For large excitation energy gaps with the bridge states, charge-transfer virtual states dominate the Dexter coupling. However, as the bridge virtual states approach resonance with the donor and acceptor states, virtual exciton states enter the Dexter mediation. We have described this transition in the Dexter coupling mechanism in ref. 1. Of course, as coupling to the bridge becomes very large, the two-state quantum dynamics formulation and the Fermi's golden rule rate treatment will fail.

1 S. S. Skourtis, C. Liu, P. Antoniou, A. M. Virshup, and D. N. Beratan, *Proc. Natl. Acad. Sci. U. S. A.*, 2016, **113**, 8115–8120.

Tom Oliver commented: I believe that your model is purely electronic and does not include any vibrational degrees of freedom. Is this a strictly valid approximation, and how would your model change if these were included in your model?

David N. Beratan answered: There are two families of vibronic effect to consider: those arising from the donor and acceptor chromophores and those arising from the bridge. Our studies assume bridge electronic excited states are far off of resonance with the electronic transitions of the donor and acceptor species. In this regime, neglect of vibronic effects that could arise from the bridge is justified since bridge mixing is so weak. As bridge electronic transitions become closer to degenerate with the donor and acceptor transitions, the bridge, donor, and acceptor states will be more strongly mixed and bridge vibronic effect could become quite rich and significant. This is analogous to typical considerations in electron transfer reactions where bridge motion could produce inelastic transport as well as non-Condon effects on the bridge-mediated couplings. In our current treatment of Dexter interactions, we have assumed a Condon approximation and have not explored the (donor and acceptor) nuclear coordinate dependence of the Dexter couplings.

Tom Penfold remarked: You mentioned that for Ru and Os complexes, the picture of distinct singlet and triplet states breaks down. Do you have an idea at what point it becomes important to consider the breakdown of spin-classification? For example, at what point can triplet energy transfer be driven by a Förster mechanism arising from a mixing of singlet character into the triplet state?

David N. Beratan responded: The balance between the Förster and Dexter mechanisms depends on the extent of spin–orbit mixing among the spin states, the magnitude of the transition dipole moments for the donor and acceptor chromophores, and the strong distance dependence of the Dexter coupling. In the Ru-metal organic framework structures that are doped with Os exciton traps,[1] weak transition dipole moments tend to disfavor Förster transfer and produce a greater role for Dexter transfer than might be expected from the admixture of singlet and triplet character in the excited states (the Dexter rate is independent of the transition dipoles of the donor and acceptor chromophores). As such, one expects that the relative balance of Förster and Dexter contributions to be distance dependent (the Förster component will have a small 'prefactor' due to the transition dipole effect, but will have a much softer distance decay than the Dexter component). It would be interesting to develop systems in which the donor–acceptor distance could be varied over a wide range; in such a case with mixed spin states one might expect to find exponential distance dependence (Dexter mechanism transfer) at shorter distances and $1/R^6$ dependence at longer distances. Studies that access a wide range of transfer distances could probably be carried out in disordered glasses, where there is a long history of studying both exciton and electron transfer reactions (see classic studies of Inokuti, Hirayama, Hopfield, and Miller[2–4]).

1 J. Lin, X. Hu, P. Zhang, A. Van Rynbach, D. N. Beratan, C. A. Kent, B. P. Mehl, J. M. Papanikolas, T. J. Meyer, Thomas, W. Lin, S. S. Skourtis and M. Constantinou, *J. Phys. Chem. C*, 2013, **117**, 22250–22259.
2 M. Inokuti and F. Hirayama, *J. Chem. Phys.*, 1965, **43**, 1978–1989.
3 D. G. Thomas, J. J. Hopfield and W. M. Augustyniak, *Phys. Rev.*, 1965, **140**, A202–A220.
4 J. V. Beitz and J. R. Miller, *J. Chem. Phys.*, 1979, **71**, 4579.

Emily Weiss asked: How does one think about triplet energy transfer from a donor for which the eigenstates are not pure singlets and triplets (due, say, to spin–orbit coupling) to an acceptor in which the eigenstates are pure singlets and triplets?

David N. Beratan responded: Please see the answer to the previous question for a discussion of this issue in the context of Ru and Os excitation energy transfer. For chromophores that include transition metals, the spin orbit interaction mixes the spin states and one can expect a mixture of both Förster and Dexter mechanisms to become available, with the balance among them set by the strength of the relevant transition dipole matrix elements (Förster component) and the Dexter couplings. The dominance of one mechanism over another is expected to depend on distance; it would be very interesting to look at distance dependent quenching in such systems to identify the crossover point between the Förster and Dexter mechanisms.

Petter Persson commented: In the scheme for alternative Dexter energy transfer pathways in Donor–Bridge–Acceptor (D–B–A) systems that you show in Fig. 2 of your paper (DOI: 10.1039/c9fd00007k) and explore throughout the paper you distinguish clearly between one-particle and two-particle interactions. Can you elaborate any further on any significant trends that you have observed for the energy transfer processes in terms of state energies and coupling elements relating to the fundamentally different nature of the (virtual) states involved in the two different types of pathways as either 'charge-transfer' or 'excitonic' in nature, respectively? Do you think that the difference in the nature of these states could manifest itself in terms of different sensitivity to factors that could be varied experimentally such as a surrounding solvent polarity (or similar), and could this potentially be used to perform controlled energy transfer experiments on relevant molecular systems in which the relative contributions for the different types of pathways to the net energy transfer could be significantly influenced?

David N. Beratan replied: The clearest observation is that bridges with electronic transitions closer to resonance to the donor and acceptor electronic transitions will tend to favor mediation *via* excitonic virtual states, while bridges with higher energy electronic transitions will favor charge-transfer state mediation. This means that the latter mediation mechanism will more closely resemble the mechanism suggested by Closs, Fleming and co-workers[1,2] cited in our paper (where the Dexter coupling is approximated by a product of electron and hole superexchange interactions), while the former case will be more complex. This is described in greater detail in ref. 3. An exciting challenge is to design chemically bridged donor acceptor systems in which electron, hole, and Dexter energy transfer can all be studies so that these mechanistic interrelationships can be probed in detail.

1 Closs, Fleming and co-workers.
2 S. S. Skourtis, C. Liu, P. Antoniou, A. M. Virshup, and D. N. Beratan, *Proc. Natl. Acad. Sci. U. S. A.*, 2016, **113**, 8115–8120.

Jeffrey Cina commented: As illustrated in Fig. 2 of your paper (DOI: 10.1039/c9fd00007k), the donor–bridge–acceptor systems you are considering exhibit multiple electronic pathways between donor-excited and acceptor-excited states mediated by Dexter transfer. While the initial and final adiabatic electronic states are the same in all these pathways, the different pathways consist of distinct sequences of electron and hole propagations. Using the analogy of what occurs in a multiple slit experiment, you predict that these pathways can interfere quantum mechanically at the level of amplitudes, and that the sum of these transfer amplitudes must be summed before squaring to obtain the transfer probability. My question is whether the differing response of the 'environment' of the electronic system, comprising internal and external nuclear degrees of freedom, and possibly the local dielectric response of the surroundings, might provide a form of 'which-path' information, which tends to 'observe' through which of these multiple 'slits' the excitation is transferred? Could these effects perhaps be tuned by varying solvents, isotopes, *etc.* to control the net extent of interference between the multiple pathways?

David N. Beratan answered: The diagram shown in Fig. 2 of DOI: 10.1039/c9fd00007k elaborates the electronic coupling pathways that exist during the fleeting existence of the transition state for the energy transfer reaction. In the cases studied in our paper, the virtual states that mediate the Dexter coupling are far out of energy resonance with the initial and final states. As such, on the timescale of the transition state (fs) and the electronic dynamics, the nuclei are essentially frozen and the coupling pathways interfere coherently. In the simpler case of single-electron transfer, orbital symmetry effects are know to exert large effects on the transport kinetics, and this influence is a manifestation of the coherent interference among tunneling pathways.[1,2]

As a result of the coherent interference of coupling pathways in bridge-mediated Dexter energy transfer, one could create 'which way' experiments. Indeed, we have developed this concept and studied it in the context of coherent vibrationally-perturbed electron-transfer reactions. These concepts and designs map very directly onto Dexter energy transfer experiments, as suggested. We have explicitly considered using vibrational markers (based on isotopic and weakly perturbing chemical substitution) to reveal the path taken by an electron, and thus to change the nature of the quantum interferences and the electron transfer kinetics. We believe that multi-slit and 'which way' experiments at the molecular scale provide a rich area for further study. Please see: ref. 3-6 for more detail. Prof. Julia A. Weinstein at the University of Sheffield is also pursuing important studies in a similar spirit.

1 S. S. Skourtis and D. N. Beratam, *Adv. Chem. Phys.*, 1999, **106**, 377–452.
2 S. S. Skourtis, C. Liu, P. Antoniou, A. M. Virshup and D. N. Beratan, *Proc. Natl. Acad. Sci. U. S. A.*, 2016, **113**, 8115–8120.
3 S. S. Skourtis, D. H. Waldeck and D. N. Beratan, *J. Phys. Chem. B*, 2004, **108**, 15511–15518.
4 D. Xiao, S. S. Skourtis, I. V. Rubtsov and D. N. Beratan, *Nano Lett.*, 2009, **9**, 1818–1823.
5 Z. Lin, C. M. Lawrence, D. Xiao, V. V. Kireev, S. S. Skourtis, J. L. Sessler, D. N. Beratan and I. V. Rubtsov, *J. Am. Chem. Soc.*, 2009, **131**, 18060–18062.
6 Z. Ma, Z. Lin, C. M. Lawrence, I. V. Rubtsov, P. Antoniou, S. S. Skourtis, P. Zhang and D. N. Beratan, *Chem. Sci.*, 2018, **9**, 6395–6405.

Jon Bender asked: Within this theoretical framework, is it possible to explore systems in which the acceptor and bridges are of the same identity in order to explore differences in triplet exciton transfer/transport in different packing motifs, *i.e.* brick-and-mortar (two bridges) compared to face-to-face slip stacking (one bridge)? Studies such as these could be beneficial for establishing design guidelines for crystals with enhanced triplet exciton transport.

David N. Beratan responded: In fact, we have explored multi-step Dexter hopping among degenerate sites in the following study, including structures that differ in 3D architecture.[1] In addition, Fig. 9 of our Faraday Discussions paper in the present volume (DOI: 10.1039/c9fd00007k) shows how shifting the napthalene–napthalene stacking influences the Dexter couplings. When donor and bridge species interact very strongly, there are additional issues of how delocalized (trapped) the initially prepared donor exciton state will be, as this characteristic will influence the subsequent dynamics.

1 J. Lin, X. Hu, P. Zhang, A. Van Rynbach, D. N. Beratan, C. A. Kent, B. P. Mehl, J. M. Papanikolas, T. J. Meyer, T. W. Lin, S. S. Skourtis and M. Constantinou, *J. Phys. Chem. C*, 2013, **117**, 22250–22259.

Karen Morenz commented: I was really interested in the role of the molecular orientations or positions in the constructive and destructive interference you see here, but unfortunately in real life I can not just arbitrarily slide my molecules along exactly as far as I want to, so what experiment would you suggest to experimentally verify this?

David N. Beratan answered: Similar issues of molecular orientation, distance, bridge structure, and quantum interference arise in the field of electron transfer photochemistry (see, *e.g.* ref. 1). As the theoretical studies of Dexter energy transfer mature, and the nature of the quantum interferences are better understood, we hope to target specific covalently bonded or self-assembled structures for study. Such structures can "look in" geometries of specific interest.

1 V. Balzani, *Electron Transfer in Chemistry*, vols. *I–V*, Wiley-VCH, 2001.

Karen Morenz asked: You mentioned that perhaps one could build some rigid covalently bonded structures in order to hold the molecules in a particular orientation—are you planning to do some calculations on such a system so that we could compare experimental observations to that, or how would we connect the current work to an experiment on such a system?

David N. Beratan answered: Considerable attention has been placed on building rigid covalently bonded structures in which to study the distance and bridge-structure dependence of Dexter energy transfer rates.[1,2] As we come to understand how the Dexter pathways interfere with each other and how the couplings scale with the number of parallel bridge pathways, we hope to suggest some compelling new molecular designs.

1 G. L. Closs, P. Piotrowiak, J. M. Macinnis and G. R. Fleming, *J. Am. Chem. Soc.*, 1988, **110**, 2652–2653.
2 S. Speiser, *Chem. Rev.*, 1996, **96**, 1953–1976.

Stephen Bradforth opened a discussion of the paper by Michael R. Wasielewski: Your results present interesting evidence for quantum coherence in electron transfer pathways. Can the temperature dependence of the rate ratios presented be connected to a simple theory of how quickly the bath should damp out the coherence?

Michael Wasielewski answered: Yes, the connection to theory can be found in ref. 1. In the context of this theory, we are operating in the large free energy drop, low temperature limit, where damping is modest. This also correlates with our own work.[2]

1 A. J. Leggett *et al.*, *Rev. Mod. Phys.*, 1987, **59**, 1–85.
2 D. D. Powell *et al.*, *J. Phys. Chem. B*, 2017, **121**, 7190–7203.

Benjamin Schwartz asked: To better understand the fluctuations that could be destroying coherence in the multi-pathway electron transfer, would it be possible to build a similar molecule that does not do photoinduced charge transfer so that detailed spectroscopy could be performed on the NDI acceptors? In this way, it might be possible to get a handle on the nature of all the fluctuations and coupling in the system, which could lead to a better understanding of the interference that contributes to the overall photoinduced charge transfer.

Michael Wasielewski replied: This is a matter of timescales. We have examined an NDI-NDI cyclophane in which one of the NDI molecules is reduced with a single electron. On an EPR timescale this electron is rapidly hopping between the two NDIs, but on an electronic timescale, it is not. However, this experiment does not properly answer the question because there is a counterion that actually determines how fast the electron can hop between the two NDIs. In other words, it is difficult to generate a radical anion that can probe the environment on a fast enough timescale in the presence of other ancillary effects interfering.

Minjung Son commented: The DMA-An and NDI moieties are the only donor–acceptor pair for charge transfer considered in this work, for both 1NDI and 2NDI. In the case of 2NDI, however, could there be symmetry-breaking charge transfer between the two NDIs (such as reported in ref. 1 and 2) that might interfere with the signal observed here? Have you considered this possibility at all?

1 E. Vauthey, *ChemPhysChem*, 2012, **13**, 2001–2011.
2 N. Banerji *et al.*, *J. Phys. Chem. A*, 2009, **113**, 8202–8212.

Michael Wasielewski replied: We do not excite NDI, so that symmetry-breaking charge separation is not an option.

Tom Oliver queried: If you directly photoexcite the localised excited states of 1NDI or 2NDI instead of the charge-transfer state, does this change the observed charge separation dynamics?

Michael Wasielewski responded: No, energy transfer from excited NDI to the DMA-An CT state is faster than we can measure.

Jahan Dawlaty asked: Is it possible to photoexcite the split state? The gap is probably somewhere in the mid IR, and it seems reasonable that it should be optically allowed. Therefore it should be possible to study it with a UV-vis pump and mid-IR probe experiment.

Michael Wasielewski replied: The gap between the states is about 1450 cm^{-1}, so it should be possible to see a mid-IR transition provided that other vibrational features do not overlap severely.

Jon Bender said: Do you observe any solvent dependence of the 2-NDI excited state dynamics at room temperature?

Michael Wasielewski replied: We have not performed an extensive survey of solvents.

Jennifer Ogilvie addressed Michael Wasielewski and David N. Beratan: In your paper (DOI: 10.1039/c8fd00218e) you mention the use of selective vibrational labels to probe quantum interference effects. This is perhaps a question to both you and David N. Beratan. Can you please elaborate on how such labels could be used? Might they provide 'which way' information or would their use destroy interference effects by making the pathways distinguishable?

Michael Wasielewski responded: This would depend, of course, on how great a perturbation the change would be. For example, it may be possible to substitute a single deuterium atom at a C–H bond of one NDI of the two, which would have a small effect on the electronic symmetry. Observation of the C–D vibrations would be required. Whether this perturbation is sufficiently small would need to be determined experimentally.

David N. Beratan replied: In our most optimistic scenario, vibrational labels could indeed provide 'which way' information. This pathway manipulation, although a manifestation of purely coherent quantum dynamics, could have profound implications for electron transfer kinetics and yields. For example, imagine that the electron transfer process is symmetry forbidden in the absence of the vibronic effect described above. The process that excites (or de-excites) a localized vibrational mode would turn on the electron transfer process by eliminating destructive interference between the pathways. We have described these effects in some detail in previous work.[1,2] A complexity is that the key experiments may need to be carried out at very low temperatures to avoid thermal fluctuations that could also lower symmetry and turn on otherwise 'forbidden' processes. Experimental efforts aimed at realizing vibrationally perturbed electron transfer have most likely manipulated the reaction coordinate motion, rather than strongly perturbing the electronic dynamics.[3] Synthetic, spectroscopic, and theoretical collaborations are underway to refine the molecular designs in order to realize the novel double-slit quantum dynamics at the nanoscale that has been predicted by theory. We have not fully extended the double-slit concepts to Dexter exciton transfer, but the coherent bridge-mediated nature of the Dexter coupling suggests that entirely analogous physics should be accessible.

1 S. S. Skourtis, D. H. Waldeck and D. N. Beratan, *J. Phys. Chem. B*, 2004, **108**, 15511–15518.
2 D. Xiao, S. S. Skourtis, I. V. Rubtsov, and D. N. Beratan, *Nano Lett.*, 2009, **9**, 1818–1823.
3 Z. Ma, Z. Lin, C. M. Lawrence I. V. Rubtsov, P. Antoniou, S. S. Skourtis, P. Zhang and D. N. Beratan, *Chem. Sci.*, 2018, **9**, 6395–6405.

Tom Oliver asked: Instead of using an electric field to bias electron transfer to a specific NDI acceptor, could you place the molecule inside an asymmetric protein, and in analogy with photosynthetic reaction centres,[1] make the electron transfer directional without changing the chemical structure of electron donors/acceptors?

1 M. A. Steffen, *Science,* 1994, **264**, 810.

Michael Wasielewski responded: In principle, yes, provided an appropriate protein could be found with a large enough binding site to accommodate 2NDI.

Oliver Gessner questioned: Could you provide an intuitive, physical picture for the significant increase in charge transfer (CT) rates due to quantum interference between the two CT pathways beyond mathematical expressions? Parallels to the famous double-slit analogue seem obvious at first glance but, in the case of CT, questions arise about how to translate the fundamental ideas of local signal enhancements into rate enhancements. Is there 'a price to pay' for the increased CT rate, for example, in terms of the amount of charge that may be transferred within a given time interval? Could the basic concepts, principally, be extended 'indefinitely', *i.e.*, scaled to an arbitrarily large number of interference pathways with correspondingly increasing CT rates (neglecting practical issues associated with chemical synthesis)?

Michael Wasielewski replied: The result of using more than two slits (interfering pathways) can be described as a sum of two-slit experiments (Born's rule) as has been recently validated by experiment.[1] We would expect a rate increase limited by the time resolution of the experiment (and in the limit by relativistic effects). Also it is likely that dephasing will be faster with more sites since there are more system–bath interactions, so the experiment becomes far more challenging. However, there should be no effect on the overall yield of electron transfer.

1 J. P. Cotter, *et al., Sci. Adv.*, 2017, **3**, e1602478.

Nadia Korovina remarked: The quantum interference effect that results in a >2× increase in charge transfer when two cofacial acceptor compounds are attached to the donor is due to substantial orbital overlap of the two acceptor choromophores. If the two acceptors were attached to the donor in such a way that the orbital overlap was minimized between them, would the increase in the rate of charge transfer be exactly two upon addition of the second acceptor? Do you have any experimental evidence of this?

Michael Wasielewski answered: We expect that the quantum interference effect would diminish as the electronic coupling between the two acceptors also decreases. While there is no experimental evidence, we have examined this case in ref. 1.

1 D. D. Powell, M. R. Wasielewski and M. A. Ratner, *J. Phys. Chem. B,* 2017, **121**, 7190–7203.

Jon Bender asked: Can you imagine a scenario in which rigid or alkyl functional groups containing cyano Stark effect probes at the imide positions of NDIs within your NDI dimer could be sensitive enough to changes in the individual NDI moiety charge states/density while being able to resolve changes in the electric field about each NDI monomer individually? Such a scenario may prove to be a fruitful test of your experimental interpretation.

Michael Wasielewski answered: Addition of CN or other functional groups having characteristic vibrations would have to be done on each NDI to maintain symmetry. We are currently exploring whether the vibrations of the carbonyl groups of NDI can offer similar information.

Vasilios Stavros remarked: From your TA spectra of 1- and 2-NDI, do you see any evidence of vibrational cooling from the spectral narrowing of, say, the ground state bleach signature?

Michael Wasielewski answered: No, there is no significant evidence of vibrational cooling in these systems on the timescale of our measurements.

Vanessa Huxter opened discussion of the paper by Benjamin J. Schwartz: Although it may be difficult to achieve experimentally, could you comment on the possibility of controlling disorder as a function of dopant choice to modify the ratio of trapped to free polarons.

Benjamin Schwartz answered: This is a very interesting question and one that we are still actively pursuing. Our main conclusion in this paper is that disorder in the distribution of dopant counterions leads to increased polaron trapping and thus poorer carrier mobility. We have been working on controlling this disorder in two different ways. In ref. 14 of our paper (DOI: 10.1039/c8fd00210j), we describe work in which we take advantage of our sequential doping method to control the crystallinity of P3HT doped with F4TCNQ. We find that, as expected, more crystalline P3HT has less disorder and thus higher carrier mobilities, but this is primarily because there are more places to put the dopant counterions far from the polarons on the polymer backbone rather than because crystallinity allows for greater polaron delocalization. With this in mind, in ref. 50 of the paper we used dodecaborane cluster-based dopants to isolate the counterion from the polaron. Even though these large dopants completely destroyed crystallinity in our doped P3HT films, the carrier mobilities we observed were higher than in our much more crystalline films doped with F4TCNQ, verifying that controlling the interactions between the polaron and counterion is more important than controlling crystallinity when it comes to increasing the ratio of free to trapped polarons. We are continuing to work with new dopant/polymer combinations in order to better understand all the interactions between polymer crystallinity, counterion proximity to the polaron and the general nature of disorder in determining carrier mobility.

Tom Oliver said: In your paper you illustrate that only specific polaron transitions are optically bright. What are the selection rules for polaron absorptions?

Benjamin Schwartz responded: Based on Fig. 2b in our paper (DOI: 10.1039/c8fd00210j), the standard picture of polaron optical transitions is that P1 and P2 are optically allowed while the P3 and P3′ transitions are forbidden by symmetry. However, if the polaron has a significant degree of two-dimensional delocalization, as illustrated in Fig. 2d of our paper, then the P3 and P3′

transitions become more optically allowed (and possibly non-degenerate) while the P2 transition loses oscillator strength. Since polarons in P3HT have a partial 2-D character, this means that P1, P2, P3 and P3′ generally can all be observed. There is an excellent discussion of polaron optical selection rules in ref. 18 of our paper.

Tom Oliver asked: Would I be right in saying that the symmetry of the polymer itself then dictates the selection rules and allowed polaron absorptions?

Benjamin Schwartz answered: Yes; most conjugated polymers can be roughly classified as having electronic states of Ag and Bu symmetry, and this is what gives rise to the basic selection rules for polaronic optical transitions.

Tom Oliver asked: Due to polaron formation in your doped P3HT films, do you observe any infrared active vibrational (IRAV) modes?[1] And are the IRAV modes similarly modulated by the presence of bipolarons?

1 A. J. Heeger, *Rev. Mod. Phys.*, 1988, **60**, 781.

Benjamin Schwartz responded: We do observe IRAV bands associated with our polarons, which result from the fact that the backbone of the polymer is reorganized in the presence of the positively-charged polaron (in other words, a P3HT radical cation has different vibrations than a neutral P3HT chain). These bands occur at energies lower than 0.5 eV (wavelengths redder than 2400 nm), so we were unable to probe their dynamics in the experiments presented in our paper. We have studied them with steady-state spectroscopy in previous work, however, because the ratio of the intensity of the IRAV bands to the P1 optical transition provides a nice handle on the degree of polaron trapping and localization; see ref. 14 and 50 of our paper (DOI: 10.1039/c8fd00210j) for more details.

Christopher Grieco queried: In this study was the P3HT regio-regular or regio-random in chemical structure? If it is regio-regular, the polymer should form both amorphous and crystalline domains in thin films. The crystalline structure is typically lamellar, in which multiple chains pack together through pi-stacking. It has been reported in the past that 'delocalized polarons' can form in these crystalline domains, in which the (hole) polaron delocalizes through space across multiple polymer chains. These species have absorption bands that are distinct from those of localized polarons. In literature that I am most familiar with (*e.g.* that of Z. Valy Vardeny and coworkers), the absorption bands associated with localized polarons are designated as 'P', and those associated with delocalized polarons are designated as 'DP' (for example, see Fig. 4 and 6 in ref. 1). Why don't you observe 'DP' bands in your spectra?

1 Jiang *et al.*, *Adv. Funct. Mater.*, 2002, **12**, 587–597.

Benjamin Schwartz answered: The P3HT studied in this work is >98% regioregular, purchased commercially from Rieke Metals. Thus, as you point out, our films are semicrystalline, with a crystalline morphology that has the lamellar stacking edge-on to the substrate and the pi-stacking direction in the plane of the film (see, *e.g.*, ref. 14 in our paper (DOI: 10.1039/c8fd00210j) for a more detailed structural analysis of our P3HT films doped with F4TCNQ by the sequential processing method that we also use in this paper). Indeed, it is whether or not the dopant counterion is located in a crystallite, where it sits in the lamellar regions relatively far from the polaron on the polymer backbone, or in the amorphous regions where it can be much closer to the positive charge, that determines whether or not the polaron is free (delocalized in your language) or trapped. The whole point of our paper is that ultrafast transient absorption can be used as a type of action spectroscopy to detangle the optical transitions of free and trapped polarons, whose absorption spectra strongly overlap in the P1 region. Thus, we do observe free polarons and trapped polarons with different spectra in our experiments, and what we observe is well aligned with the theory we developed in ref. 14 of the paper. The 'DP' band you refer to from the work of Vardeny and co-workers is at a much lower energy (0.1–0.2 eV) than the P1 transition we probe in our paper (0.4–0.7 eV); unfortunately we do not have access to ultrafast pulses in this energy range, so we were not able to probe this band in our samples.

Laura Estergreen remarked: Referring to the UV-vis spectrum that was presented (ref. 3 in DOI: 10.1039/c8fd00210j), as the dopant concentration is increased from 0.001 to 1 mg mL^{-1}, there is a nice isosbestic point right around 2.0 eV. However, at 1 mg mL^{-1} it appears to deviate. Why is that? Will this impact the transient absorption features and interpretation, as, in the paper, the primary dopant concentrations compared are 1 and 5.5 mg mL^{-1}?

Benjamin Schwartz responded: We believe that this is a direct reflection of the fact that the F4TCNQ sits primarily in the P3HT crystallites at lower doping concentrations but resides in the amorphous P3HT regions at higher doping concentrations. The spectrum of P3HT is well known to depend on its degree of crystallinity, with enhanced vibronic structure on the red edge of the absorption as the degree of order increases. At low and mid-doping concentrations, this vibronic structure is clearly present in the residual neutral P3HT absorption, indicating that there are still undoped crystalline regions remaining in the film. Since only the crystalline P3HT spectrum is bleached at these concentrations, there is a nice isosbestic point, as you point out. At the highest doping concentration, however, the F4TCNQ dopant has filled the crystalline regions, and the residual P3HT absorption that remains no longer has any pronounced vibronic structure and thus is more reflective of amorphous material. Since the amorphous P3HT has a different spectrum than the crystalline P3HT, the fact that the bleached P3HT spectrum changes with doping concentration is what causes the loss of the isosbestic point at high doping concentrations.

Tom Oliver commented: P3HT is commonly used in bulk heterojunction thin films in conjunction with C60 electron acceptors, which obviously changes the thin film composition and morphology compared to F4NCQ doped P3HT. Do you anticipate this would alter the polaron and bipolaron formation?

Benjamin Schwartz answered: The P3HT samples in our paper are doped in their electronic ground state, as F4TCNQ is a strong enough oxidizing agent to pull an electron directly out of the P3HT valence band. Thus, our situation is different from that in bulk heterojunction solar cells, where polarons on the polymer are created by photoinduced charge transfer to fullerenes from the excited state. Since the bandgap transition of P3HT is largely suppressed when the material is doped (because electrons have been removed from the valence band), there wouldn't be an easy way to create excitons to do photoinduced charge transfer to fullerenes if fullerenes and F4TCNQ were somehow present in the same P3HT film.

Tom Oliver returned to the discussion of the paper by Michael Wasielewski: How can we retrieve more information (*e.g.* phase) about the superposition of charge-transfer states formed upon coherent charge-separation in Mike Wasielewski's 2NDI molecule? Would a mid-infrared laser probe pulse resonant with the energy splitting between the two acceptor states (A_+ and A_-) be sufficient? Or could transient X-ray experiments provide additional key information?

Michael Wasielewski replied: I do not think that the application of a mid-infrared pulse or the X-ray experiment would provide the phase information. Use of 2D electronic/vibronic spectroscopy would be more informative. We are currently pursuing these experiments.

Oliver Gessner responded: While a mid-IR experiment would probe the existence of a certain energy level spacing, which would be the acceptor state energy splitting here, X-rays would principally be able to provide a local probe of the valence electron dynamics from the perspective of specific reporter atoms. To probe differences between the charge transfer pathways with X-rays, however, one would need to be able to distinguish them, *i.e.*, one would need different X-ray absorption or emission energies, which would require some kind of elemental labeling of the different channels. This would imply, however, that the two pathways would not be strictly equivalent anymore. In this case, one may speculate whether reporter atoms with very similar valence electronic structures and chemical characteristics but different X-ray transitions may be used as 'minimally invasive' probes.

Naomi Ginsberg answered: Perhaps a technique such as time resolved scanning tunneling microscopy would be most useful because it is able to directly map the wavefunction of the molecule.

David N. Beratan responded: If a coherent superposition of N identical acceptor species is formed as the product, one expects 1/N of an electron to be located on each of the acceptor units. As such, spectroscopic probes that are diagnostic of the net electron charge on an individual acceptor would be useful. Probes of the acceptor species superposition states are also potentially accessible, although such superposition states would need to be probed on ultrafast timescales before medium polarization traps a full electron on one on the many accessible A sites. Delocalizing charge coherently over many acceptors in the product state will reduce the reorganization energy for electron transfer. The reorganization energies for transfer to a coherent superposition of acceptor states is expected to scale inversely with the number of participating acceptors participating in the coherent state. This scaling of reorganization energy with N should have kinetic consequences and, given a family of homologous species with different N values with intervalence (charge transfer) transitions, one could look for the signatures of this expected reorganization energy scaling with N in the energies of charge transfer bands.

Johannes Mahl asked: In the example of donor and acceptor connected with two bridge molecules: while any probe of which of the two bridges the charge took on its way from donor to acceptor will collapse the quantum mechanical wave function, could you think of other characteristics of charge transfer that can be accessed, using a less invasive/more indirect probe technique, while preserving the system's quantum mechanical properties?

Michael Wasielewski replied: Multi-dimensional electronic/vibronic spectroscopy should be able to reveal the presence of coherences that result from the two electron transfer pathways.

Naomi Ginsberg addressed Michael Wasielewski and David N. Beratan: I wonder if there might be a way to do a form of quantum non-demolition measurement to be able to measure the wavefunction amplitude in the different branches of your molecular interferometer. The phase of the wavefunction would presumably be oscillating at optical frequencies. Could one couple to a different degree of freedom, *e.g.* using a spin label and a nitrogen vacancy?

Or take for example the simpler situation in which we might ask a 'which way' question in a molecular system like benzene. One could in principle ask whether energy (information) is directed along one side of the ring or the other in order to reach the *para* position. But we are describing a single molecule that has wavefunctions that we can compute, *e.g.* using Huckel theory. The wavefunction is an eigenstate of the whole molecule, as would be the case for the more complex molecular interferometer in question. So to localize the wavefunction amplitude on one side of the benzene molecule one could create a superposition of wavefunctions, but because this is quantum mechanics can we truly talk about different routes around the ring? The closest quantum mechanical analogy I can think of would be a wavepacket of some sort (*i.e.* a superposition of states that enables localisation of the wavefunction). One might consider the creation of a Rydberg atom, which we can think of as having a very high energy state that localises the electronic wavefunction far away from the nucleus. Of course this sort of state is approaching a classical limit in which the electron becomes more particle-like. Is this a good analogy or does it obviate the need for an interferometer?

Michael Wasielewski responded: I think that once the interaction between the two acceptors is negligible relative to system–bath interactions, one is in the single particle limit as you suggest for the Rydberg state, so that the idea of an interferometer becomes moot.

David N. Beratan responded: One is free to pose the 'which way' or pathway interference analysis in the language of eigenstates or of localized orbitals (natural bond orbitals, CC sigma bond, CC pi bonds, *etc.*). The coupling pathway language has roots in Dyson expansions of quantum mechanical amplitudes and proves very convenient and intuitive in understanding multi-pathway quantum interferences in the tunneling regime (compared to an eigenstate language). Some discussion of these complementary frameworks and the utility of the tunneling pathway framework appear in ref. 1–3.

1 S. S. Skourtis and D. N. Beratan, *Adv. Chem. Phys.*, 1999, **106**, 377–452.
2 S. S. Skourtis, D. H. Waldeck and D. N. Beratan, *J. Phys. Chem. B*, 2004, **108**, 15511–15518.
3 J. N. Onuchic and D. N. Beratan, *J. Am. Chem. Soc.*, 1987, **109**, 6771–6778.

Animesh Datta responded: Quantum mechanical interference is observed when an event can occur by any of several alternate pathways. If an experiment is performed in which it is possible, even in principle, to determine which of the alternate pathways was actually taken, then the interference is lost. Almost never explicitly stated, asking a 'which way' question assumes that these alternate pathways are macroscopically distinct and experimentally identifiable. It seems to me that this may be unlikely for 'different routes around the ring' of benzene using far-field optical methods. So the 'different routes around the ring' may not be macroscopically distinct. But maybe some near-field method can identify these 'routes' after which we can start talking of quantum interference in a 'benzene interferometer'. After that, of course, one needs to identify quantum excitation that will move along these 'routes' and make sure they do not encounter anything that distinguishes which of the 'routes' was taken. There are 'quantum erasure' ideas in quantum optics that can suppress distinguishing aspects, but its unclear to me to what extent they can be transferred to chemical/molecular systems.

Naomi Ginsberg returned to the discussion of the paper by Benjamin J. Schwartz: I wonder if one could use transient reflectivity instead of transient absorption in order to help decouple the presence of the many excited species in the spectrum. There could be areas where examining the real, rather than the imaginary, part of the electronic response aids in discerning one feature from another.

Benjamin Schwartz answered: There would definitely be additional information in adding transient reflectivity experiments, but it is not immediately obvious to me how they could be interpreted. This is because I am not aware that the steady-state index of refraction of doped polarons has ever been measured out in the infrared where the P1 band absorbs, so that even if the transient changes in the real and imaginary parts of the index could be deconvoluted, it would be challenging to understand the information carried in the real part of the index.

David N. Beratan opened the discussion of the paper by Sharon Hammes-Schiffer: Please explain why the high frequency nuclear mode associated with proton transfer does not lead to washing out of the Marcus inverted effect in the proton-coupled electron transfer reaction that was studied.

Sharon Hammes-Schiffer responded: For electron transfer reactions coupled to a vibrational mode, the Marcus inverted region may not be exhibited due to contributions from the excited vibrational states, leading to a plateau as the driving force increases. A similar effect is observed for PCET reactions with harmonic or Morse proton potentials. In these cases, the Marcus inverted region is not exhibited because of contributions from excited proton vibrational states that are nearly activationless. The excited proton vibrational states contribute significantly because the overlap integral corresponding to the lowest reactant proton vibrational state and an excited product proton vibrational state increases as the quantum number for the product state increases (see Fig. 4A from our paper, DOI: 10.1039/c8fd00240a, where the tails of the product wavefunctions extend further toward the reactant for higher excited vibrational states). Fig. 6A from our paper also illustrates this increase in the overlap integrals up to around the seventh excited state, after which the overlap starts to decrease.

In contrast, PCET reactions can exhibit the Marcus inverted region behavior for asymmetric double well proton potentials. In this case, the excited proton vibrational states become delocalized over both wells, as shown in Fig. 4B from our paper, where the third excited state is already over the barrier and therefore delocalized. The delocalized, oscillatory nature of the highly excited proton vibrational wavefunctions leads to phase cancellation and therefore negligible overlap integrals associated with the lowest reactant and highly excited product proton vibrational states. Fig. 6C from our paper depicts the fast decay of the overlap integral after the first few excited product proton vibrational states. Because the square of the overlap integral is a prefactor for each term in the non-adiabatic PCET rate constant expression (eqn 2 in the paper), the vanishing overlap integral eliminates the contributions from these excited proton vibrational states. In this case, for highly exoergic PCET reactions, the contributions from the pairs of reactant/product vibronic states that are activationless are multiplied by zero. Thus, the mid-level excited product proton vibrational states are the dominant contributors, and because the splittings are also relatively small, these pairs of vibronic states are still in the inverted region in that the driving force is greater than the reorganization energy. In this case, the nuclear mode associated with proton transfer does not lead to washing out of the Marcus inverted effect.

Karen Morenz queried: And so to clarify, would you always see the inverse kinetic isotope effect in a proton coupled energy transfer system like this, or is that something that just happens to be inherent to this particular system?

Sharon Hammes-Schiffer answered: The inverse kinetic isotope effect (KIE) will not always be observed in these types of PCET systems. The rate constants for hydrogen and deuterium are computed by summing over all of the reactant and product vibronic states. The percentage contribution from each reactant/product vibronic state pair is determined by a complex balance among the Boltzmann population for the reactant state, which favors the lowest proton vibrational state, the free energy barrier, and the overlap integral associated with the reactant and product proton vibrational wavefunctions. The relative contributions from the vibronic state pairs are different for hydrogen and deuterium due to differences in the vibrational wavefunctions, as well as the smaller energy level splittings for deuterium. The observation of an inverse KIE will depend on the contributions from the vibronic state pairs for hydrogen and deuterium, which in turn depend on a complex balance among several factors inherent to the specific system. Thus, it is difficult to predict an inverse kinetic isotope effect without computing the full rate constants for hydrogen and deuterium, although it is more likely to occur for large driving force.

Karen Morenz said: This is another sort of clarification question. I was interested in how you achieve the inverse kinetic isotope effect—I mean, I understand that it comes from the vibrational states of deuterium *vs.* hydrogen but is there something particular about the deuterium vibrational states overlap in the proton coupled energy transfer system that leads to this inverse kinetic isotope effect?

Sharon Hammes-Schiffer replied: The inverse KIE most likely arises from greater vibrational wavefunction overlap integrals for deuterium than for hydrogen for the vibronic state pairs with dominant contributions to the corresponding rate constants.

Michael Ashfold asked: Can you please expand a bit on how I should picture photoinduced proton coupled electron transfer in, for example, a hydrogen bonded phenol-pyridine complex using asymmetric double well potentials like those shown in Fig. 4B of your paper (DOI: 10.1039/c8fd00240a).

Sharon Hammes-Schiffer answered: For a triad composed of anthracene, phenol, and pyridine, photoexcitation to a local excited state of the anthracene may be followed by concerted PCET involving electron transfer from the phenol to the anthracene and proton transfer from the phenol to the pyridine.[1] The resulting charge-separated state can then decay to the ground state by another concerted PCET involving electron transfer from the anthracene to the phenol and proton transfer from the pyridinium to the phenoxyl radical. For this system, the one-dimensional asymmetric double well potentials shown in Fig. 4B of our paper (DOI: 10.1039/c8fd00240a) can be generated by moving the proton from the oxygen of the phenol to the nitrogen of the pyridine with all other nuclei fixed. In practice, the proton coordinate axis is determined by connecting the equilibrium positions of the proton bonded to the oxygen or the nitrogen. In the charge-separated state, the lower-energy minimum is near the nitrogen, whereas in the ground state, the lower-energy minimum is near the phenol. This flipping of the asymmetry of the double well potential can be explained in terms of electrostatic interactions between the electron(s) and the proton; when the electron transfers away from the phenol, the proton is more stable on the nitrogen of the pyridine, but when the electron transfers back to the phenol, the proton is more stable on the oxygen of the phenol. This behavior of electron transfer flipping the asymmetric double well proton potentials, or in some cases shifting of the minima of single well potentials that may exhibit shoulders, is characteristic of PCET reactions in general.

1 G. A. Parada, Z. K. Goldsmith, S. Kolmar, B. P. Rimgard, B. Q. Mercado, L. Hammarström, S. Hammes-Schiffer and J. M. Mayer, *Science*, 2019, **364**, 471–475.

Jahan Dawlaty said: As you very nicely showed, the ground state vibrational wavefunction will have vanishing overlap with higher excited states of the acceptor, resulting into the inverted region.

Is it possible to think of a system in which the higher vibrational states of the donor overlaps with higher vibrational states of the acceptor? In that case, the overlap factor will not go to zero (at least not as quickly). This will involve transfer prior to vibrational cooling in the donor. Basically it can be phrased as, is it possible to imagine cases where transfer occurs prior to vibrational cooling?

Sharon Hammes-Schiffer replied: Our calculations in this paper (DOI: 10.1039/c8fd00240a) assume that vibrational cooling is faster than the PCET reaction, and therefore the reactant state is represented by a Boltzmann distribution among the proton vibrational states. In the non-adiabatic PCET rate constant expression, each term in the summation over reactant and product vibronic states is weighted by the Boltzmann population of the reactant vibronic state. If the system were not at equilibrium prior to PCET, then this rate constant expression would no longer be valid. In this case, we could perform non-adiabatic dynamics simulations to explore the non-equilibrium dynamics of the PCET process. We have developed methods for these types of non-adiabatic dynamics simulations and have applied them to PCET in solution.[1,2] From the experimental perspective, this situation will occur when photoexcitation leads to non-equilibrium populations of the excited vibrational states, presumably weighted by the Franck–Condon factors, and then the PCET reaction occurs prior to vibrational cooling. If the solvent equilibration were much faster than both vibrational cooling and PCET, then the non-adiabatic PCET rate constant expression presented in the paper would still be valid but would include averaging over the non-equilibrium populations of the vibrational states in certain regimes. Further calculations would be required to determine if these various scenarios could exhibit inverted region behavior.

1 P. Goyal, C. A. Schwerdtfeger, A. V. Soudackov and S. Hammes-Schiffer, *J. Phys. Chem. B*, 2015, **119**, 2758–2768.
2 P. Goyal and S. Hammes-Schiffer, *ACS Energy Lett.*, 2017, **2**, 512–519.

Tom Oliver remarked: In your bidirectional photoinduced PCET study, you observe both an inverted region and inverse kinetic isotope effect as a function of increasing the driving force relative to the reorganisation energy. Do you anticipate this will also occur for photoinduced H-atom transfer reactions where the proton and electron are transferred to a common acceptor?

Sharon Hammes-Schiffer answered: Hydrogen atom transfer (HAT), in which the electron and proton transfer between the same donor and acceptor, is a subset of PCET. In general, PCET encompasses reactions in which the electron and proton transfer between the same or different donors and acceptors and in the same or different directions. My group has published several papers[1–3] providing diagnostics that distinguish HAT, which is electronically adiabatic but often vibronically non-adiabatic, from the alternative electron-proton transfer (EPT) mechanism, which is typically vibronically and electronically non-adiabatic. Another distinction is that typically EPT reactions involve substantial charge redistribution, while HAT reactions do not involve such charge redistribution. In principle, the inverted region and inverse kinetic isotope effect could be observed in either regime. In practice, we have observed this behavior in theoretical models that are vibronically and electronically non-adiabatic but not yet in models that are vibronically non-adiabatic and electronically adiabatic. Moreover, we do not expect to observe inverted region behavior for fully vibronically adiabatic reactions, which are defined to occur on the ground vibronic state. However, further exploration is required to fully understand the limits of this behavior.

1 J. H. Skone, A. V. Soudackov and S. Hammes-Schiffer, *J. Am. Chem. Soc.*, 2006, **128**, 16655–16663.
2 A. Sirjoosingh and S. Hammes-Schiffer, *J. Phys. Chem. A*, 2011, **115**, 2367–2377.
3 S. Hammes-Schiffer, *Energy Environ. Sci.*, 2012, **5**, 7696–7703.

Amro Dodin commented: Does this study provide any insight into what properties of the potential energy surfaces lead to the presence or disappearance of the Marcus inverted regime? In particular, is there a specific type of anharmonicity or curvature of the potential that determines this?

Sharon Hammes-Schiffer responded: The experimental accessibility of the Marcus inverted regime depends on a balance of many factors in the rate constant expression. In particular, the rate constant depends on the Boltzmann population of the reactant state, the vibronic coupling (which depends on the electronic coupling and the overlap between the proton vibrational wavefunctions in the electronically non-adiabatic regime), and the free energy barrier (which depends on the reorganization energy, the driving force, and the proton vibrational energy level splittings). For each system, we would need to compute these quantities and use them to calculate the PCET rate constant as a function of driving force. However, one trend that emerged from our calculations on model systems is that the Marcus inverted regime is more likely to be observed for proton potentials that are asymmetric double wells for which the lowest few proton vibrational states are below the barrier, but the remaining vibrational states are above the barrier and therefore are delocalized over both wells. The oscillatory wavefunctions delocalized over both wells typically exhibit negligible overlap with the reactant proton vibrational wavefunction, thereby allowing observation of the inverted region behavior. As shown in our Faraday Discussion paper (DOI: 10.1039/c8fd00240a), the Marcus inverted region behavior is more likely to be observed if this overlap decreases quickly with product state quantum number, as observed for the asymmetric double well proton potentials. For the Morse proton potentials, where this overlap decreases more gradually (*i.e.*, does not become negligible until higher quantum numbers) the inverted region is inaccessible. Thus, the shape of the proton potentials plays a significant role in determining whether or not the Marcus inverted region will be observed. In terms of the physical properties of the system, asymmetric double well proton potentials are common in systems for which the proton transfers across a hydrogen bond but are also found in other types of systems.

Jack Woolley said: The model system that you presented displays an intramolecular hydrogen bond. Could this type of double well potential model situations of intermolecular proton transfer, in particular those between solute and solvent, and if so what sort of additional considerations would you expect to play a role?

Sharon Hammes-Schiffer replied: This PCET theory is also applicable to systems exhibiting intermolecular proton transfer, including proton transfer between the solute and a solvent molecule. My group has computed similar asymmetric double well potentials or shifted single well potentials for these types of systems. The main additional consideration is the formation of a hydrogen-bonded complex prior to PCET. In this case, the observed rate constant is the product of the equilibrium constant for the formation of this hydrogen-bonded complex and the PCET rate constant. In terms of observing inverted region behavior for such systems, the experimental challenge would be ensuring that the hydrogen-bonded complex remains stable during the photoexcitation, charge separation, and charge recombination processes.

Collin Steen opened discussion of the paper by Mark E. Thompson: Thanks for this very nice contribution. It is exciting to see such a clear dependence of the propensity for symmetry-breaking charge transfer on the environment, specifically solvent polarity. How general is this phenomenon, especially with regard to fluorescence quenching in photosynthetic systems? Do you think there are other factors beyond environmental polarity that need to be considered in more complicated systems such as light-harvesting antenna complexes?

Mark Thompson responded: The polarizability of the medium is also very important. We see SBCT being fast in toluene and benzene and in doped polystyrene thin films. In these cases the media have low or zero polarity, but are polarizable.

Yumin Lee said: Thank you for the great work and presentation. I just wondered that what happens if the dihedral angle changes, for example from pi-extension of the rings or using another metal instead of Zn, how can that alter the kinetics regarding charge-transfer? In your insight, do you expect to see the formation of exciton coupling by the angle change could occur and compete with the charge transfer?

Mark Thompson replied: The dihedral angle is very important. If you twist too far away from 90 degrees exciton coupling this will lower the excited energy too low to engage in charge transfer. There are examples of zinc based materials that deviate significantly from 90 degrees and they do not give SBCT states, but non-radiatively decay to the ground state directly from the S_1.

Graham Fleming asked: Concentration quenching is an almost universal phenomenon with no clear microscopic explanation. Could symmetry breaking charge transfer be the common cause?

Mark Thompson answered: I think that is a great suggestion. This could well be the case, but is hard for us to examine. In the concentration range we need to look for quenching the TA signal will be swamped out. Also, with the rapid tumbling of the molecules in solution, the lifetime of the SBCT state will be very short (and lead to extensive quenching as observed). We will continue to think about this and see if we can design an experiment to test this hypothesis.

Nadia Korovina queried: The spectral signature of your assigned charge transfer state looks very similar to the triplet induced absorption spectrum in analogous BODIPY compounds. Is there spectroelectrochemical data to confirm the charge transfer assignment?

Mark Thompson answered: The spectrum does not in fact look like the triplet. Our *J. Phys. Chem. C* paper[1] shows both spectroelectrochemical data and the TA spectra of the $Zn(dpy)_2$ triplet and you can see that the signatures we see in the TA spectra between 0.5 and 900 ps do not match the triplet.

1 C. Trinh, K. Kirlikovali, S. Das, M. E. Ener, H. B. Gray, P. Djurovich, S. E. Bradforth and M. E. Thompson, *J. Phys. Chem. C*, 2014, **118**(38), 21834–21845.

Gregory Gate asked: Why did you also study a Zn dimer, as both dimers appeared to have similar dynamics?

Mark Thompson replied: The coupling between the two chromophores are quite different. One pair is coupled through a C–C bond at the *meso* positions of the two dyes and the other is coupled through a zinc ion. The C–C bond coupling gives rise to a 10× faster rate of charge transfer than the pair coupled through a zinc ion.

Marta Duchi addressed Mark Thompson and Michael Kellogg: Are the observed oscillations in the time-domain transient absorption data a signature of excited state wavepacket dynamics?

Michael Kellogg responded: I do not believe that the oscillations observed in time traces for 2 for the pure solvents are a result of wavepacket motion. There are a few reasons for this. If I plot the time traces for the other CT band at 379 nm, there are no definitive oscillations like there are in the 538 nm band. I would expect to see oscillations for both CT bands, not just one of them. The time traces for the CT only extend out to 20 ps after excitation. If I extend the same CT band out to the entire time window (950 ps), I still see oscillations of the same amplitude which I don not believe are real. Actual oscillations would not live that long. This is for pure solvents. It is important to note that each of the samples for the pure solvents were done on the same day. So, if there was a laser problem like power instability or compression issues, it would likely be the same for all samples. This is why each of the pure solvents seem to have these oscillations. If I plot the CT band for different solvents, say for the CHXTHF mixtures, I do not see any oscillations in the time trace for any wavelength. There is noise in these wavelengths that has the same amplitude as the oscillations in the pure solvents, but they are not a set of periodic oscillations, like the pure solvents. I am not sure what the specific identity of the oscillations are. They are not due to noise because our signal is above the noise floor. I have not repeated these experiments save for ACN and TOL, which I did a few weeks later. And for 538 nm for these new solvents, I do not see these oscillations. Likely, the oscillations were experimental artefacts, but I do not know the origin of them.

Mark Thompson replied: No, the oscillations are not that. They are due to 150 fs time constant for the apparatus. Those oscillation are purely instrumental.

Tom Oliver commented: At the end of your paper (DOI: 10.1039/c8fd00201k) you discuss the relative rate of charge recombination for the two dipyrrin structures, and note that rate constants are similar. Given the structural differences between the compounds and presumably different re-organisation energies associated with the respective charge transfer states, would one not expect the recombination rates to be quite different?

Mark Thompson responded: Note that he forward CT is facilitated by the bridging MO at the *meso* carbons of the BODIPY dimer and no such state exists for the $Zn(dpy)_2$ molecule. Since neither SBCT state has a bridge state and the driving force for the two dimers are the same, we believe that the direct recombination is largely controlled by the hole–electron distances, which are similar in the two complexes. Note that, while we say they are similar, the BODIPY dimer is 50% faster than the $Zn(dpy)_2$ complex.

Ryan Dill opened the discussion of the paper by Thomas. J. Penfold: In your paper you mention that thioxanthone undergoes a spin-forbidden equilibration with a time constant of 10 picoseconds. Why is it so fast? It contains a sulfur, but I have also seen that xanthone, with no sulfur, has a similar intersystem crossing rate, so the heavy-atom can not be the important factor here. I understand that it is El-Sayed allowed, but it still seems faster than I would have expected.

Tom Penfold responded: The S_1–T_2 spin–orbit coupling in thioxanthone is El-Sayed allowed, and a value of 18 cm^{-1} is found. However the key aspect here is the fact that these two states are degenerate. The mixing between two states scales as the coupling divided by the energy gap and so the small energy gap helps drive rapid intersystem crossing.

Mark Thompson said: Thioxanthone has a really fast ISC, but is a rarity. For internal conversion to be slow relative to intersystem crossing is even rarer. What are your thoughts on how to slow down internal conversion?

Tom Penfold replied: The ISC of thioxanthone is fast, but similar rates can be achieved in organic molecules exhibiting El-Sayed's allowed spin–orbit coupling. Another well-known example is benzophenone. But as you correctly identify, the challenge in the proposed mechanism is restricting the internal conversion rate. This is slow in thioxanthone because of the large energy gap between the T_1 and T_2 states and consequently for this approach to work, this gap would need to be maintained. A counter challenge, in the context of OLEDs would be that the lower lying T_1 state would act as a non-radiative trap. This would likely cause problems with efficiency and lifetime. To truly establish the ability for this approach to be effective we have to understand how the excited states are formed upon the thioxanthone and if we can control this to avoid T_1 becoming populated. This work is currently underway.

Tom Oliver asked: In Fig. 3 of the paper (DOI: 10.1039/c8fd00201k), the time evolving normalised diabatic state populations of S_1 and T_2 are displayed. There appears to be a small oscillatory component to these dynamics—can you comment on its origin?

Tom Penfold responded: Very small! The timescale of the oscillation is ~3 ps which corresponds to ~11 cm^{-1} which is similar to the spin–orbit coupling between the S_1 and T_2 states. I could not expect this to be visible experimentally, and will be largely because the reduced space model will be slightly over-coherent, *i.e.* no bath modes or damping effect of the environment.

Tom Oliver asked: The coupling between the T_2 and T_1 electronic states is very small. Is simply because of the large energy difference between states, or are there additional important factors?

Tom Penfold replied: It is twofold. For thioxanthone the molecule is fairly rigid and therefore any coupling that is activated by vibrational degrees of freedom will be small. In addition, as you say, the gap between the states is very large, which will reduce the effect of any coupling.

Tom Oliver queried: On a more general point, what are the key requirements for efficient triplet–triplet energy transfer? It is apparent the efficiency of such processes are important precursor to enhancing triplet–triplet annihilation up-conversion for solar energy applications.[1,2]

1 K. A. Phillips *et al.*, *Chem. Eur. J.*, 2018, **24**, 8577–8588.
2 S. Baluschev *et al.*, *Phys. Rev. Lett.*, 2005, **97**, 143903.

Tom Penfold responded: Triplet–triplet energy transfer (TTET), as with all transfer mechanisms governed by the Fermi Golden rule are dependent (in the semi-classical limit) on the coupling, the reorganisation energy and the free energy released upon transfer. For the coupling, TTET is a Dexter mechanism and is therefore driven by the exchange interaction. This therefore requires some overlap between the two wavefunctions. This is why the coupling to the T_2 of thioxanthone is smaller as it is a more localised π* excitation and therefore there is less wavefunction overlap with the T_1 of dichlorobenzene. It is expected that the TTET rates decay exponentially as distances increases, as observed in our work.

Mark Thompson commented: Has the T_2 of thioxanthone been directly sensitized? It would be useful to show that you get highly efficient photoluminescence from the T_2.

Tom Penfold answered: In the experiments performed in ref. 1 and simulated in our present contribution (DOI: 10.1039/c8fd00201k), the dichlorbenzene is photoexcited and used to transfer energy to the triplet states of thioxanthone. For enhanced fluorescence to be observed, the triplet-energy transfer has to pass from the T_1 of dichlorobenzene to the T_2 of thioxanthone. However, as observed experimentally and in our contribution, only a small fraction passes through the T_2 state, instead the T_1 is preferred. This can be understood from calculating the coupling between the states, free energy and reorganisation energy upon population transfer, as shown in the paper.

1 C. Torres Ziegenbein, S. Fröbel, M. Glöß, R. S. Nobuyasu, P. Data, A. Monkman and P. Gilch, *ChemPhysChem*, 2017, **18**(17), 2314–2317.

Jon Bender opened discussion of the paper by Oliver Gessner†: Have you explored changing the steady-state temperature of the copper oxide and underlying substrate to tune their relative thermal conductivities? Or have you explored changing the underlying substrate identity to one with a drastically different thermal conductivity in order to explore the role of the substrate in heat dissipation?

† Oliver Gessner's paper was presented by Johannes Mahl, Lawrence Berkeley National Laboratory, U. S. A.

Oliver Gessner replied: We have not. However, we do not expect the substrate to play a significant role in this particular experiment for the following reasons. The time-averaged substrate temperature during the pump–probe experiment was monitored with a thermocouple and increased by only a few degrees C compared to the ambient room temperature as a result of the laser exposure. Thus, most of the laser-induced heat from each pump pulse dissipates before the arrival of the next laser pulse, as otherwise a much higher average heating would be observed. Our spatiotemporal temperature modeling is consistent with this finding and strongly indicates that the dissipation of heat is largely accomplished within several hundred nanometers from the surface. Therefore, the impact of the substrate is likely negligible within the precision of this experiment.

Jon Bender said: I am really surprised by the large initial change in temperature that you observe in this system. I have primarily experienced these effects in organic molecular solids personally, which tend to achieve much more modest lattice temperatures. Can you please comment on how this is achieved? Additionally, is it electronic or lattice temperature interested in?

Oliver Gessner replied: Principally, we are interested in both, the electronic and lattice temperatures of the material after photoexcitation. However, as the coupling between electrons and phonons typically occurs within tens of picoseconds or faster, we expect our experiment to be predominantly sensitive to the lattice temperature. The observed temperature jumps are either in reasonable agreement (for 532 nm excitation) or significantly below (for 355 nm excitation) the expected temperature changes, which were calculated using eqn (2) and the experimental laser pulse parameters and material characteristics, as discussed in sections 4 and 5 of the manuscript. We don't know why the experimentally determined temperature jump for 355 nm excitation is much lower than the calculated value but point out that a similar phenomenon was reported by Chen and co-workers in a transient X-ray absorption study of photoexcited hematite (ref. 7 of the paper).

Michael Haggmark said: Your study suggests that the vast majority of the energy is thermalized. Generally speaking, how well characterized are the photodynamics of copper oxide? Another way of asking this, is the quantum yield of fluorescence quite low for powdered copper oxide?

Oliver Gessner replied: The timescales of charge carrier recombination found in the literature are mostly below the temporal resolution of the study presented here. Born *et al.*[1] observed three timescales in the decay dynamics of CuO nanocrystals following 1.6 eV excitation in an optical pump–probe study. A 330 to 630 fs (pump fluence dependent) decay was attributed to carrier-carrier scattering and exciton-exciton annihilation, a 2 ps decay to carrier-phonon scattering and a 50 ps time constant to trapping and recombination in trap states. Othonos *et al.*[1] observed similar timescales of 0.4 and 2.1 ps in the decay of photo-generated carriers in CuO nanowires.

1 B. Born *et al.*, *ACS Photonics*, 2016, **3**, 2475.
2 A. Othonos *et al.*, *Nanoscale Res. Lett.*, 2011, **6**, 622.

Michael Haggmark remarked: This study nicely followed the thermal effects. Are you planning on studying something with a long-lived excited electronic state as a follow up?

Oliver Gessner responded: A study where both thermal and electronic signatures are present would be an ideal next step for this project. In fact, this was the primary motivation for this Faraday Discussion contribution. However, while the methodology developed within this Faraday Discussion article (DOI: 10.1039/c8fd00236c), in principle, allows to disentangle thermal and electronic effects, it should be noted that, so far, any effort to study photo-induced effects with X-ray spectroscopies on picosecond and longer timescales has found that thermal effects dominated the photo-induced difference spectra (see ref. 7 and 8 of the paper). A promising advantage of acquiring data at multiple pump–probe delays as with the setup used in our study is that thermal and non-thermal effects with similar spectral fingerprints are distinguishable, as long as the timescales of their decay are sufficiently different.

Karen Morenz asked: This is very related to the previous question from Michael Haggmark, and perhaps also a little more off-topic, but I know that a lot of people are really interested in heating and even melting in quantum dots, so could you connect your work up with that situation and tell us how these results can inform us about that sort of system?

Oliver Gessner answered: Time-resolved X-ray absorption may very well be suitable for studying heating and phase transitions in quantum dots, in particular, since the element specificity allows to target the quantum dot and the substrate independently (assuming they consist of different materials). Thus, in addition to the analysis presented here, heterogeneous heat transfer from the quantum dot to the substrate could also be studied by probing two core levels that are exclusively connected to the quantum dot and the substrate, respectively. One may also expect significant changes in the X-ray absorption spectrum upon melting, see *e.g.* data on silicon by Johnson *et al.*,[1] and germanium by Stebel *et al.*[2]

1 S. L. Johnson *et al.*, *Phys. Rev. Lett.*, 2003, **91**, 157403.
2 L. Stebel *et al.*, *Rev. Sci. Instrum.*, 2011, **82**, 123109.

Benjamin Schwartz remarked: Although your picture of the transient dynamics as resulting from thermal effects makes good sense, I was wondering if it would be possible to independently measure the thermal conductivity of the same sample using a different ultrafast technique. David Cahill's group at UIUC has developed a method known as time-resolved thermoreflectance, in which a short laser pulse is used to heat a metal placed on one side of the sample, and then the change in reflectivity of the metal is monitored as a function of time, allowing one to extract the thermal conductivity of the material under the metal. In your case, there is already copper metal on the bottom side of the sample, which means it would be straightforward to perform this type of measurement on the exact same sample with a similar excitation pulse but a visible-pulse probe, providing a nice way to examine the thermal effects with two different techniques.

Oliver Gessner responded: This is a very interesting cross-reference to a technique that accesses similar characteristics of the sample as the experimental technique described in our paper. The targets used by Cahill's group consist of well-defined double-layer structures with the sample material deposited on a thin metal foil with thicknesses on the order of 100 nm or less. The metal substrate used in our study is ~2 mm thick, which would render a similar experiment unfeasible since the observed dynamics would be entirely dominated by the substrate. However, if a CuO sample could be grown in a better defined way and on a thin film of Cu or other metallic surface, the method presented here could most likely be applied as well as time-domain thermoreflectance measurements.

Terry Frankcombe commented: When I look at the determinations of the temperatures from the experimental data in Fig. 7 of your paper (DOI: 10.1039/c8fd00236c), there seems to be a very large spread. For example, the determined temperatures for long delays in Fig. 7a spread from the low 20s to more than 60 degrees. This is much larger than the spread in temperatures shown in the pre-pump background to the left of time zero. Can you comment on the source of such a large spread of values, and what the uncertainty of each temperature determination should be?

Oliver Gessner answered: The data shown in Fig. 7 of the paper result from the homogeneous fit method, which approximates the ΔtrXAS data at every pump–probe delay individually by identifying one steady-state temperature difference spectrum that best fits the ΔtrXAS data. Generally, three pump–probe time delay regions may be distinguished, in which different effects dominate the uncertainty of the data: the pre-pump region with $\Delta t \leq 0$, early dynamics with $\Delta t \approx 0$–50 ns delay, and dynamics at longer delays $\Delta t \geq 50$ ns. In the pre-pump region at nominally negative delays, any laser-induced effect has decayed and the scatter in the inferred temperatures is essentially defined by the reproducibility of the (tr) XAS spectra at room temperature. In the second region, $\Delta t \approx 0$–50 ns, the temperature uncertainty is relatively small due to the high absolute sample temperatures. In this regime, the XAS spectra are particularly temperature sensitive, *i.e.*, the temperature-induced differences ΔXAS_{temp} per change in temperature ΔT are largest at higher temperatures, as can be seen in Fig. 4. Conversely, this leads to relatively small temperature uncertainties when fitting ΔtrXAS data corresponding to higher ΔT. In contrast, the temperature sensitivity of the XAS spectra decreases for low to intermediate temperatures, leading to larger temperature uncertainties for longer delays. Furthermore, for positive delays we cannot exclude potential contributions of laser-induced effects that are not captured by describing the ΔtrXAS data with ΔXAS_{temp} spectra. We have tried to identify systematic trends beyond the temperature induced effects but were not able to isolate any within the signal-to-noise ratio of the experiment.

It should also be mentioned, that the results of the second (inhomogeneous) fit procedure we applied are subject to much smaller variations since this method simultaneously fits entire temperature profiles and their temporal evolution based on the one-dimensional heat transport model. This global fit routine inherently produces smooth temperature evolution curves as can be seen in Fig. 8 of the paper (DOI: 10.1039/c8fd00236c).

Conflicts of interest

There are no conflicts to declare.

Faraday Discussions

PAPER

Carotenoid–chlorophyll energy transfer in the fucoxanthin–chlorophyll complex binding a fucoxanthin acyloxy derivative

Hristina Staleva-Musto,[a] Robert West,[a] Marco Trathnigg,[a] David Bína,[ab] Radek Litvín[ab] and Tomáš Polívka*[ab]

Received 19th November 2018, Accepted 29th November 2018

DOI: 10.1039/c8fd00193f

The fucoxanthin–chlorophyll *a* protein from *Emiliania huxleyi* (E-FCP) is a member of the LHC family of light-harvesting proteins. It has a rather unusual pigment composition as its binds more Chl-*c* than Chl-*a*, and 19′-hexanoyloxyfucoxanthin (hFx) as the main carotenoid instead of fucoxanthin (Fx) typically found in various FCP complexes. The presence of a hexanoyloxy tail in hFx suppresses the charge transfer character of the S_1/ICT state resulting in almost no effect of polarity on the excited state dynamics of hFx, strongly contrasting with the excited-state properties of Fx. Here we report on the dynamics of the energy transfer between hFx and Chl in E-FCP, and we compare it with Fx–Chl energy transfer in the FCP complex from *Phaeodactylum tricornutum*. In both complexes, the excited hFx (Fx) transfers energy from the S_2 state with a sub-100 fs time constant and no effect of the hexanoyloxy tail on the efficiency of the S_2 route was found. The energy transfer *via* the S_1/ICT state has in E-FCP two channels characterized by 1.5 and 11 ps time constants, while for FCP these two channels operate with time constants of 0.8 and 4.5 ps. Thus, minimizing the charge transfer character of S_1/ICT in hFx results in about twice slower energy transfer *via* the S_1/ICT state, underlining the importance of the ICT state in facilitating carotenoid–Chl energy transfer in systems utilizing keto carotenoids as energy donors.

Introduction

The efficient conversion of solar radiation energy into chemical energy carried out by photosynthetic organisms heavily relies on the performance of their light-harvesting systems. The role of light-harvesting (or antenna) complexes is to capture sunlight and to transfer the absorbed energy to reaction centers, in which the excitation energy is converted into chemical energy.[1,2] The structure and

[a]*Faculty of Science, University of South Bohemia, Branišovská 1760, 370 05 České Budějovice, Czech Republic. E-mail: tpolivka@jcu.cz; Tel: +420 387 776 259*

[b]*Institute of Plant Molecular Biology, Biological Centre, Czech Academy of Sciences, České Budějovice, Czech Republic*

function of reaction centers remained mostly conserved throughout their evolution and only two types of reaction centers can be found in photosynthetic organisms. In contrast, the light harvesting complexes evolved into a large variety of systems containing various pigments carrying out the light-harvesting function.[1–3] The diversity of light-harvesting systems is especially pronounced in photosynthetic microorganisms which, depending on the ecological niches they are living in, employ different pigments (typically carotenoids and chlorophylls) resulting in a broad range of light-harvesting efficiencies.[1–3]

Besides the light-harvesting function, antenna complexes also have an important regulatory role as they are able to quench the Chl excited state under high-light conditions when reaction centers are not able to accept and use all absorbed energy.[4] Without regulatory mechanisms, the unexploited energy of an excited Chl would generate a Chl triplet that could in turn react with oxygen resulting in production of deleterious singlet oxygen. Thus, photoprotective mechanisms are clearly needed. It is known that the molecules crucial for photoprotection are carotenoids.[5] Using carotenoids, photosynthetic organisms have developed a three-layer photoprotection consisting of the direct quenching of singlet states of Chl,[6,7] quenching of Chl triplet states formed by intersystem crossing,[8,9] or scavenging singlet oxygen generated through sensitization from Chl triplets.[10]

Thus, carotenoids in light-harvesting complexes significantly affect the overall efficiency of the light-harvesting and photoprotective processes in a particular photosynthetic organism. Numerous studies addressed the specific excited-state properties of carotenoids which, in the first approximation, may be described with a three-state model consisting of the ground state (S_0), the lowest excited state S_1, which is forbidden for one-photon transitions from the ground state, and the strongly absorbing S_2 state.[11] After excitation of a carotenoid into the S_2 state, it promptly relaxes to the S_1 state and, when embedded in light-harvesting complexes, both the S_1 and S_2 excited states can serve as energy donors in carotenoid–Chl energy transfer.[2,3] The actual efficiency of either channel depends mainly on the carotenoid structure.[3] The role of carotenoid excited states in photoprotection is less clear, but at least in some light-harvesting complexes it has been proven that the carotenoid S_1 state can accept energy from excited Chl-*a*.[6,12–14]

A special family of carotenoids contain a conjugated C=O group in their structure: the keto-carotenoids. They exhibit specific excited-state properties as the conjugated keto group introduces an intramolecular charge transfer (ICT) state, which, *via* coupling to the S_1 state, makes their spectroscopic properties dependent on polarity.[15–18] Keto-carotenoids often occur in the light-harvesting complexes of photosynthetic microorganisms, *e.g.* dinoflagellates and diatoms,[19] and it has been shown that their excited-state properties are potentially tunable by the binding site. This is an important feature, allowing the highly efficient carotenoid–Chl energy transfer that is typically observed in these systems. The prototypical light-harvesting complexes containing keto-carotenoids are the peridinin–chlorophyll-complex (PCP) from dinoflagellates and fucoxanthin–chlorophyll-protein (FCP) from diatoms, which contain keto-carotenoids peridinin and fucoxanthin (Fx), respectively. In both PCP and FCP, the keto-carotenoids transfer the absorbed energy to Chl-*a* with ~90% efficiency,[20,21] and the ICT state, specifically its coupling to the S_1 state,[22,23] plays a key role in achieving such high energy transfer efficiency.

Fucoxanthin in FCP transfers energy to Chl-*a* from both S_2 and S_1/ICT states;[21,23–26] the S_1/ICT pathway consists of two channels that are likely associated with two different Fx–Chl pairs in FCP. These two S_1/ICT channels yielded time constants of 0.6 and 2.5 ps for FCP from *Cyclotella meneghiniana*[21,24,25] and 0.8 and 4.5 ps for FCP from *Phaeodactylum tricornutum*.[23] In both FCP complexes, there is also a long-lived Fx (25 and 38 ps for FCP from *C. meneghiniana* and *P. tricornutum*, respectively) that does not transfer energy to Chl. Interestingly, Chl-*a* in isolated FCP has a lifetime of ~4.5 ns, indicating that the singlet excited states of Chl *a* in the isolated FCP are not quenched. Instead, the photoprotection mechanism in FCP efficiently quenches the Chl triplets by triplet–triplet energy transfer to Fx.[27] Except for the keto-carotenoid Fx, both FCPs also bind small amounts of diadinoxanthin and/or diatoxanthin: the two carotenoids of the diatoms xanthophyll cycle. However, their amount in complexes vary significantly depending on the light conditions during cultivation as well as on the isolation procedure.[28–30]

Another FCP complex is found in the coccolithophore *Emiliania huxleyi*. Marine algae of the group Coccolithophyceae (phyllum Haptophyta) are among the most significant contributors to both geologic and present global carbon cycles.[31] Haptophyte algae most likely represent an independent branch of eukaryotic life and are not closely related to either diatoms or red algae.[32] On the other hand, the major light harvesting antenna proteins of these algae are closely related to diatom FCP.[33] In contrast to FCPs studied so far, the FCP complex from *E. huxleyi* (hereafter denoted as E-FCP) binds fucoxanthin acyloxy derivative 19′-hexanoyloxyfucoxanthin (hFx) as the major carotenoid.[34] Under high-light or nutrient starvation conditions, the amount of hFx increases at the expense of Fx,[35] suggesting that the acyloxy moiety may provide some advantage for protection. Interestingly, even though the hexanoyloxy group is not conjugated, we have recently shown that its presence significantly affects the excited-state properties. Namely, the presence of a hexanoyloxy tail nearly completely disables the characteristic polarity-dependent behavior of fucoxanthin, and all the typical spectroscopic markers of the ICT state disappear in hFx.[36] Therefore, the ICT-assisted fucoxanthin–Chl energy transfer reported for FCPs from *C. meneghiniana* or *P. tricornutum* should not work for E-FCP. Here we show that binding hFx to E-FCP indeed slows down the carotenoid–Chl energy transfer compared to the FCP complex from *P. tricornutum*, underscoring the importance of the ICT state in facilitating energy transfer between a keto-carotenoid and Chl.

Materials and methods

Diatom FCP light-harvesting complexes from *P. tricornutum* were obtained as described previously.[23] Cells of *E. huxleyi* strain CCMP1516 were used as a source material for obtaining E-FCP samples. *E. huxleyi* was batch-grown in an artificial seawater medium with f/2 nutrient addition[37] in 5 L Erlenmeyer flasks. The cultures were illuminated by a metal halide lamp at an intensity of 300 μmol per photons per m^2 per s^1 (15 h light, 9 h dark). Cells were pelleted by centrifugation at $8000 \times g$ and stored at −80 °C until further use. The E-FCP antenna was isolated as described previously for FCP.[23,38] Briefly, cells were resuspended in a breaking buffer (50 mM HEPES pH 7.5, 2 mM KCl, 10 mM EDTA) and broken by a high-pressure cell disruptor (EmulsiFlex-C5, Avestin Inc., Canada) by several passages at 800 bar. Thylakoid membranes were pelleted by centrifugation at 150000×*g*, solubilized using 2% *n*-dodecyl β-D-maltoside and pre-purified using linear sucrose gradient ultracentrifugation. The darkest zone, which contained the antenna proteins, was collected from the gradient and further purified using anion exchange chromatography on a 2 mL column packed with DEAE Sepharose CL-6B (Sigma-Aldrich, St. Louis, USA) using a linear NaCl gradient of 5 400 mM (elution buffer: 50 mM HEPES pH 7.5, 2 mM KCl, and 0.03% *n*-dodecyl β-D-maltoside). The fractions eluted at 180–220 mM NaCl were selected for spectroscopic measurements based on the simultaneous criteria of lowest protein/Chl-*a* (A270/A674) and highest carotenoid/Chl-*a* (A480/A674) ratios. The collected fractions were desalted and stored at −80 °C. Prior to experiments, the isolated E-FCP was dissolved in a buffer (50 mM HEPES pH 7.5, 2 mM KCl, and 0.03% *n*-dodecyl β-D-maltoside) to yield an optical density of 0.4 in a 2 mm path length cuvette at the maximum of the Q_y band of Chl-*a*.

The pigment composition of the samples was determined using an HPLC analysis as described previously.[39] Pigments were identified by their absorption spectra, elution times and by comparison with known standards (*e.g.* FCP pigment extract was used as a fucoxanthin and diadinoxanthin standard). Pigment molar ratios were estimated from areas of their HPLC peaks corrected by their respective extinction coefficients.[40] The extinction coefficient of Fx was used for the quantification of hFx.

Steady-state absorption spectra were measured using an Agilent 8453 UV-VIS spectrophotometer. Ultrafast transient absorption measurements were performed with a regenerative chirped pulse amplification system (Spectra-Physics) producing a 100 fs pulse width and 800 nm central wavelength pulses at a repetition rate of 1 kHz at 4 W average power. Beam splitters directed pulses to an optical parametric amplifier, TOPAS Prime (Light Conversion), which generated pump pulses at 510 nm. A third beam path led to a 3 mm sapphire plate, which produced a supercontinuum for probing at the repetition rate of the source. Additionally, this supercontinuum beam was split into a reference beam, which passed outside the sample and probe beam, which overlapped the pump beam in the sample. The timing of the probe pulses relative to the pump pulses was adjusted using a computer-controlled delay line allowing measurement up to 6 ns. Near the sample space, the pump beam was focused with a 350 mm focal-length fused silica lens. The white-light continuum was collimated by an off-axis parabolic mirror, and, after splitting into reference and probe beams, the probe beam was focused with a 300 mm spherical mirror to minimize chirp. The resulting spot-sizes, where the beams overlapped in the sample space, were approximately 300 and 120 μm for pump and probe, respectively. The resulting pump photon flux at the sample space was adjusted using a neutral density filter and kept below 10^{14} photons per pulse per cm^2. A Berek compensator placed into the pump beam fixed the pump beam polarization to 54.7 degrees to the probe beam. Time-resolved spectra obtained from reference and probe signals were detected using a double CCD array system (Pascher Instruments).

Microsecond triplet-minus-singlet (TmS) spectra of fucoxanthin were measured using the home-built microsecond flash-photolysis setup.[41] Carotenoid triplet states were generated using Chl-*a* as photosensitizer. The contribution of the Chl-*a* triplet-minus-singlet spectrum to the signal was removed by subtracting the pure Chl-*a* solution spectrum acquired in a separate experiment, scaled to the bleaching of the Chl-*a* Q_y band.

The spectro-temporal datasets from the femtosecond transient absorption experiments were fitted globally using the CarpetView fitting software (Light Conversion). To visualize the excited state dynamics, for simplicity we assumed a sequential, irreversible scheme. In this model, the spectral profiles of the individual excited-state species are called evolution-associated difference spectra (EADS), and although they do not correspond to pure spectra of individual excited states, they provide valuable information about the time evolution of the whole system.[42]

Results

The HPLC chromatogram of E-FCP presented in Fig. 1 shows the complex pigment composition. The E-FCP complex binds three types of Chl-*c*: Chl-c_3, Chl-c_2 and Chl-c_2 esterified with monogalactosyl diacylglycerol (Chl-c_2-MGDG), in an approximately 3 : 2 : 1 ratio. Altogether, Chl-*c* comprises almost 60% of the chlorophyll content of E-FCP. This contrasts with diatom FCP, where forms of Chl-*c* typically account for less than 25% of all chlorophyll content, and Chl-c_2-

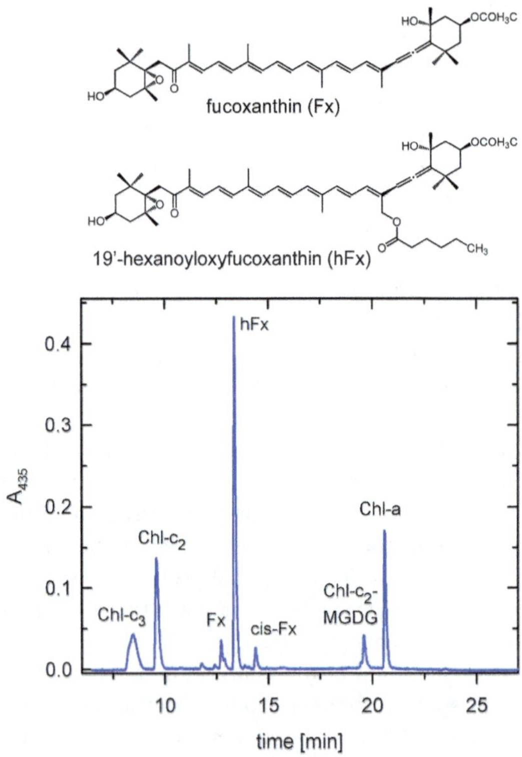

Fig. 1 HPLC analysis of the pigment content of the purified antenna of *E. huxleyi* (E-FCP). The chromatogram was detected at 435 nm. The major pigments are labelled at their respective peaks; Fx – fucoxanthin; hFx – 19′-hexanoyloxyfucoxanthin; Chl – chlorophyll. The molecular structures of Fx and hFx are also shown.

MGDG is completely absent.[28–30] The pigment stoichiometry of E-FCP is summarized in Fig. 2. The carotenoid pool of E-FCP is dominated by hFx (over 85%), accompanied by ~8% of Fx. Other carotenoids identified include a *cis*-isomer of Fx, diadinoxanthin, and a trace amount (<1%) of β-carotene. The small, substoichimetric amount of the xanthophyll cycle carotenoid diadinoxanthin is not surprising and is in line with diatom FCP, where the amount of this carotenoid is known to vary between preparations depending on the purification procedure.[28,29,43] In spite of the significant difference in Chl-*a*/*c* content between E-FCP and FCP, the two complexes maintain a total carotenoid : total Chl ratio close to 1.

The absorption spectra of the E-FCP and FCP complexes are shown in Fig. 3. This comparison underlines the significantly larger amount of pigments in E-FCP. The Q_y band of Chl-*a* is the same in both complexes, peaking at 672 nm, but the significantly larger amount of Chl-*c* in E-FCP is manifested by the characteristic Chl-*c* peak at 590 nm. Further, the domination of Chl-*c* over Chl-*a* in E-FCP is also reflected in the Soret band peaking at 467 nm, the maximum corresponding to Chl-*c*, with a Chl-*a* shoulder at 440 nm. A reverse ratio of these peaks is observed for FCP. Due to the higher hFx(Fx)/Chl-*a* ratio in E-FCP (Fig. 2), the carotenoid band is more intense in E-FCP. Even though E-FCP binds hFx and FCP Fx, there is no significant difference in the carotenoid absorption bands; this agrees with the absorption data on Fx/hFx showing that the absorption spectra are very similar.[36] Yet, the red-edge of the carotenoid absorption is slightly red-shifted in FCP (560 nm), which is the first hint that the hexanoyloxy group of hFx minimizes the characteristic features of keto-carotenoids. The extension of the absorption to long wavelengths in a polar environment is typical for fucoxanthin.[16]

To gain information about a possible effect of the hexanoyloxy group of hFx on carotenoid–Chl energy transfer, we have excited Fx (in FCP) as well as hFx (in E-

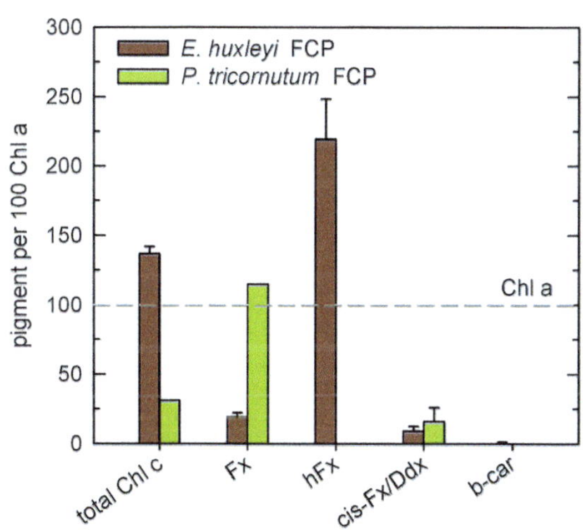

Fig. 2 Comparison of the pigment composition E-FCP from *E. huxleyi* and FCP of the pennate diatom *P. tricornutum*. The pigment stoichiometry is expressed in mol/100 mol Chl-*a* (mean ± standard deviation of three replicates). For simplicity, all Chl-*c* forms are grouped, as are the minor carotenoids *cis*-Fx and diadinoxanthin.

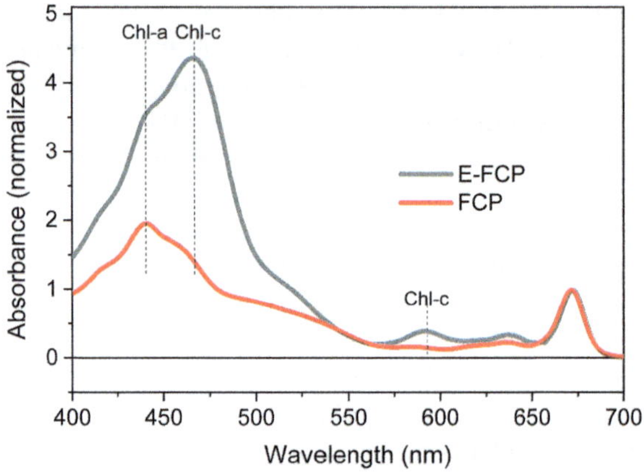

Fig. 3 Absorption spectra of the E-FCP and FCP complexes.

FCP) at 510 nm. Transient absorption spectra taken at 1 ps after excitation are depicted in Fig. 4. In the carotenoid bleaching region, the Fx bleaching signal in FCP extends further to the red compared to hFx in E-FCP, confirming the observation from the absorption spectra. The positive signal is due to a S_1–S_n excited state absorption which, in agreement with comparison of hFx and Fx in solution,[36] peaks at the same wavelength of 560 nm for both complexes. The major differences are in the Chl bleaching region. First, the magnitude of the Chl bleaching signal is markedly larger for FCP, indicating faster energy transfer from Fx to Chl-*a* in FCP compared to hFx to Chl-*a* transfer in E-FCP. Second, the positive signal in the 630–700 nm spectral region, which is due to the ICT-S_n transition overlapping with the Chl-*a* Q_y bleaching, is also larger in FCP. This shows that the ICT character of the coupled S_1/ICT state is diminished for hFx bound to E-FCP, mirroring the situation reported for the excited-state dynamics of hFx in solution.[36]

The differences between Fx and hFx bound to antenna complexes are further exhibited in the energy transfer dynamics. The kinetics were measured at the

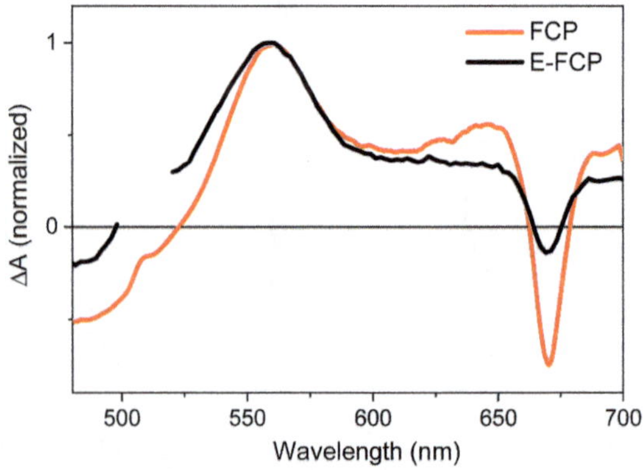

Fig. 4 Transient absorption spectra measured at 1 ps after excitation of E-FCP (black) and FCP (red) complexes. Data are normalized to the S_1–S_n maximum.

maximum of Chl-*a* bleaching at 672 nm to monitor the arrival of energy to Chl-*a* after excitation of Fx/hFx at 510 nm (Fig. 5a). The rise of the Chl-*a* bleaching signal is markedly slower in E-FCP, showing that hFx is less efficient in energy transfer to Chl-*a* compared to Fx in FCP. The excited Chl-*a* populated by energy transfer from Fx/hFx has the same lifetime in FCP and E-FCP (Fig. 5a, inset), implying no role of the particular fucoxanthin structure (Fx, hFx) in the possible quenching of excited Chl-*a*. The Chl-*a* lifetime in both complexes corresponds to essentially unquenched (lifetime close to 4 ns, see below) Chl-*a*. The less efficient energy transfer in E-FCP is also reflected in the decay of the hFx S_1/ICT state which is slower than that of Fx in FCP (Fig. 5b), further confirming the effect of the hexanoyloxy group of hFx on the energy transfer and excited-state dynamics.

Detailed information about the dynamics of carotenoid–Chl energy transfer in both complexes is provided in Fig. 6, which shows the results of global fitting in terms of the EADS corresponding to individual compartments in the sequential fitting model. Note that these compartments do not reflect differential spectra of true excited states of the system as they mix carotenoid and Chl states, yet they provide useful information about the energy flow through the system after excitation of Fx/hFx in both complexes. The initial EADS (black) correspond to the

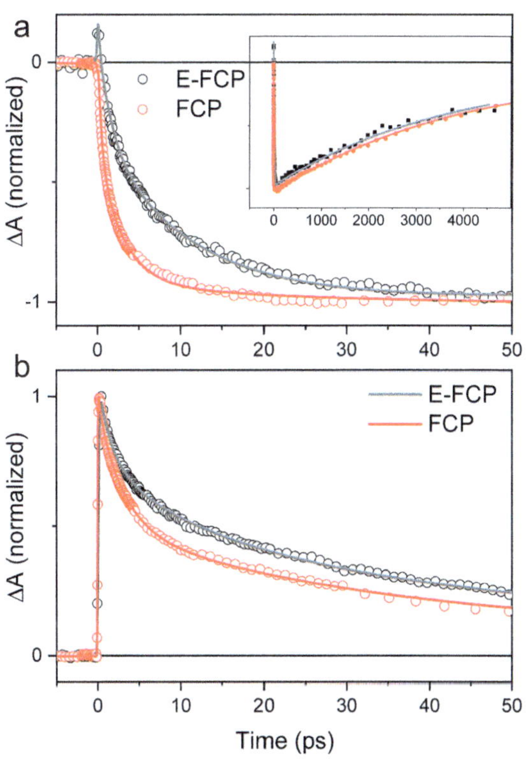

Fig. 5 (a) Kinetics measured at the maximum of Chl-*a* bleaching at 672 nm. Inset shows the long time scale monitoring the decay of excited Chl-*a*. (b) Decay of the S_1/ICT state monitored at the maximum of the S_1–S_n band at 560 nm. The excitation wavelength was 510 nm, and the data are normalized either to Chl-*a* bleaching minimum (a) or S_1–S_n maximum (b). E-FCP: black, FCP: red.

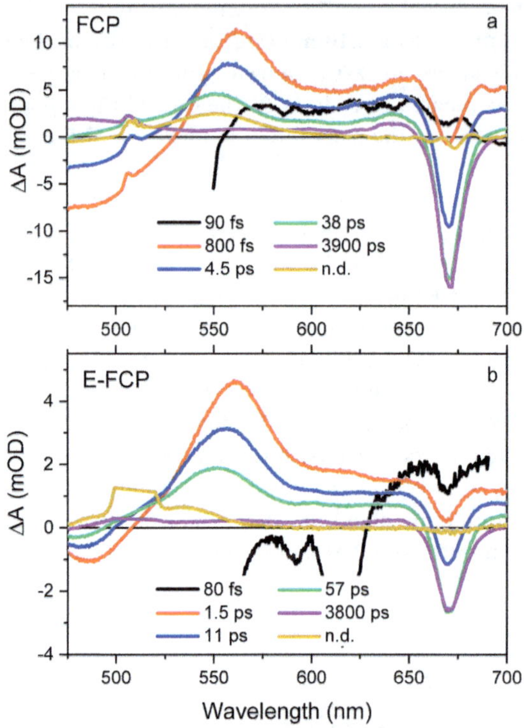

Fig. 6 EADS obtained from global fitting the data measured for (a) FCP and (b) E-FCP complexes. The excitation wavelength was 510 nm, and the short-wavelength (<570 nm) part of the first EADS is omitted due to strong scattering from excitation.

spectrum of the initially excited S_2 state of Fx/hFx. In both FCP and E-FCP, the Fx/hFx S_2 state decays on a sub-100 fs time scale to populate S_1/ICT of Fx/hFx and Chl-*a*. This is evidenced by the characteristic spectroscopic markers in the red EADS: the S_1–S_n band of Fx/hFx peaking at 560 nm and Chl-*a* bleaching at 672 nm. The process characterized by transformation of the red to blue EADS is accompanied by a significant increase of Chl-*a* bleaching and decay of the S_1–S_n band. The same occurs during the decay of the blue EADS; therefore, these two processes represent two channels of energy transfer *via* the S_1/ICT state. For FCP from *P. tricornutum*, these two channels have lifetimes of 0.8 and 4.5 ps, but they are slower for E-FCP, 1.5 and 10.8 ps. The prolongation of the S_1/ICT-mediated energy transfer in E-FCP is clearly visible in the kinetics of the Chl-*a* rise in Fig. 5a.

Thus, the first three EADS in Fig. 6 are related to three energy transfer channels from Fx/hFx to Chl-*a*. The fastest channel (<100 fs) characterizes the S_2-mediated transfer; the other two are due to the S_1/ICT route. No other energy transfer channels were identified, and the subsequent dynamics is due to decay of the S_1/ICT state of Fx/hFx which does not transfer energy to Chl-*a* (green EADS). This EADS has a lifetime of 38 ps (FCP) and 57 ps (E-FCP). During the decay, the S_1–S_n signal disappears completely, but no change in the amplitude of Chl-*a* bleaching is observed, confirming that this process is related solely to the decay of Fx/hFx. The difference in lifetimes of the non-transferring Fx (FCP) and hFx (E-FCP) agrees well with the reported S_1/ICT lifetimes of Fx and hFx in solution.[36] The 3.9 ns (FCP) and 3.8 ns (E-FCP) EADS reflect the decay of excited Chl-*a* and the formation of the carotenoid triplet state (the final EADS), which has a lifetime in the microsecond timescale, far beyond the time window accessible in our experiment.

Discussion

The specific pigment composition in E-FCP, the major carotenoid being the acyloxy fucoxanthin derivative hFx instead of the fucoxanthin occurring in other FCP antenna, clearly affects the dynamics of the energy transfer between carotenoids and Chl. While the position of the S_1–S_n transition remains essentially unaltered when switching from Fx to hFx, the energy transfer rates associated with the S_1/ICT state slow down for E-FCP binding hFx (Fig. 5a and 6). To trace the origin of this effect we recall the recent report of the spectroscopic properties of hFx in solution which demonstrated that the non-conjugated hexanoyloxy group of hFx eliminates the polarity-dependent behavior characteristic of Fx, and the spectroscopic markers of the ICT state are nearly absent in hFx.[36]

It is possible to demonstrate that comparable behavior occurs when hFx is bound to E-FCP. In Fig. 7, we show the EADS associated with the fast energy transfer channel *via* the S_1/ICT route with the contribution of Chl-*a* subtracted. The EADS in Fig. 7 thus represent an estimation of the 'pure' transient spectra of Fx and hFx bound to FCP and E-FCP, respectively. It is obvious that, when normalized to the S_1–S_n maximum, the characteristic ICT signal in the 630–700 nm spectral region is enhanced for Fx in FCP, confirming the assumption of the lack of ICT character of the S_1/ICT state of hFx in E-FCP. We cannot exclude some involvement of the ICT state in hFx since some ICT character has been observed in alcohols,[36] suggesting that hydrogen bonding to the conjugated keto-oxygen of hFx may be a trigger for the appearance of ICT-related features. Since we do not know the structure of either FCP or E-FCP, hydrogen bonding to a keto-oxygen in some hFx cannot be excluded. Yet, our analysis unequivocally shows

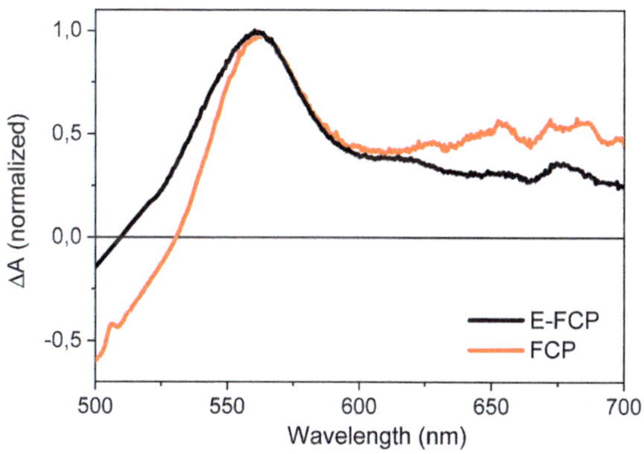

Fig. 7 Normalized EADS corresponding to the fast S_1/ICT energy transfer channel in E-FCP (black, 1.5 ps) and FCP (red, 0.8 ps). The Chl-*a* bleaching contribution was removed by subtracting the appropriately scaled Chl-*a* EADS (3.9 ns).

that the involvement of the ICT state, thus also its coupling to the S_1 state, is significantly diminished for hFx in E-FCP.

The near absence of the ICT features of hFx in E-FCP translates to the significantly slower energy transfer *via* the S_1/ICT state, underscoring the importance of the ICT state for energy transfer between keto-carotenoids and Chl-*a*. The concept that the ICT state is a key factor in maintaining high energy transfer efficiency was suggested to explain the peridinin–Chl energy transfer in the peridinin–chlorophyll-complex (PCP)[20,22] and later in the FCP complex.[21,23] However, comparison of FCP and E-FCP provides the final proof, as we can compare two fucoxanthins having an identical structure of the conjugated backbone; yet one of them, hFx, lacks the ICT-related features. As a result, both energy transfer channels involving the S_1/ICT state of hFx are about twice as slow as those utilizing Fx. Since ICT-related features are nearly absent in hFx, this result also supports the hypothesis that the S_1 part of the S_1/ICT potential surface is the energy donor in Fx/hFx–Chl energy transfer.[23] The S_1-ICT coupling is the mechanism for tuning the energy transfer, mostly by enhancing the transition dipole moment of the donor, resulting in a faster energy transfer rate.

It is important to note, however, that even though the energy transfer rate is slower when hFx is the energy donor, the overall efficiency of the S_1/ICT route is not much affected. Since we know that the S_1/ICT lifetime of hFx in solution is 60 ps, even for the slower energy transfer channel characterized by the 10.8 ps S_1/ICT lifetime of hFx in E-FCP (Fig. 6b), the efficiency is still ~80% (the energy transfer time is 13.5 ps). The same calculation for the slower S_1/ICT channel in FCP, assuming the intrinsic S_1/ICT lifetime of Fx is about 35 ps (based on the lifetime of the non-transferrin Fx), yields an efficiency of 87%, only slightly larger than that for hFx in E-FCP. Therefore, although the absence of the ICT state in hFx slows down the energy transfer, the energy transfer efficiency decreases only marginally.

However, the importance of the ICT state for tuning the energy transfer rate is obvious. This is also supported by nearly identical rates for the S_2-channel that is not affected by the charge-transfer character of the S_1/ICT state. In both FCP and E-FCP are Fx/hFx molecules which do not transfer energy to Chl-*a* (Fig. 5, green EADS). The lifetimes of the non-transferring carotenoids are 38 ps (Fx) and 57 ps (hFx), somehow mirroring the S_1/ICT lifetimes obtained for these molecules in solution. For Fx, the S_1/ICT lifetime of 38 ps indicates a mildly polar environment as the S_1/ICT lifetimes of Fx span the 20–60 ps range, depending on the polarity and proticity of the environment.[16,44] The 58 ps S_1/ICT lifetime of non-transferring hFx in E-FCP agrees with the 60 ps S_1/ICT lifetime in nearly all solvents, regardless of polarity.[36] The only parameter that could decrease the S_1/ICT lifetime of hFx in solution is the proticity of the solvent, as hydrogen bonding in alcohols decreased the lifetime to 40 ps.[36] This suggests that no hydrogen bonding to the keto-oxygen of hFx occurs in E-FCP. It is also important to note that, based on the amplitude of the S_1–S_n band in the EADS corresponding to the non-transferring Fx/hFx, in both complexes the non-transferring carotenoids represent about one third of the Fx/hFx population excited at 510 nm. This implies that even though there are substantially more carotenoids per Chl-*a* in E-FCP than in FCP (Fig. 2), the fraction of hFx involved in energy transfer remains approximately the same as for Fx in FCP. Thus, the additional hFx molecules must be also involved in energy transfer.

The decay of the Chl-*a* excited state also remains unaffected by the exchange of Fx for hFx since the lifetime in both complexes is very similar: 3.8 and 3.9 ns, respectively. This shows that presence of hFx does not induce any quenching; the Chl-*a* remains unquenched in both FCP and E-FCP. This is not surprising, because in both quenching mechanisms, proposed for direct quenching of excited Chl-*a* by carotenoids, energy transfer to the S_1 state[6,12–14] or electron transfer forming carotenoid radical cations[7] are not favorable for Fx or hFx. The S_1 energy of Fx is around 16 500 cm^{-1},[16] which is far too high to quench the Q_y band of Chl-*a*, having an energy of ~14 900 cm^{-1}. The spectroscopic properties of hFx, namely the identical maximum of the S_1–S_n band, suggests the S_1 energy of hFx is comparable to Fx. For electron transfer quenching, the redox properties of keto carotenoids are not favorable as they are much harder to oxidize than their non-keto counterparts.[45] Thus, similar to other complexes containing keto-carotenoids such as PCP, FCP and E-FCP complexes, mechanisms are not employed for direct quenching of the excited singlet state of Chl-*a*. Instead, they use very efficient quenching of the Chl-*a* triplet state *via* triplet–triplet energy transfer to the carotenoid.[27]

The normalized triplet spectra of Fx in FCP and hFx in E-FCP are depicted in Fig. 8. Interestingly, although the S_1–S_n transition is identical for Fx and hFx, the

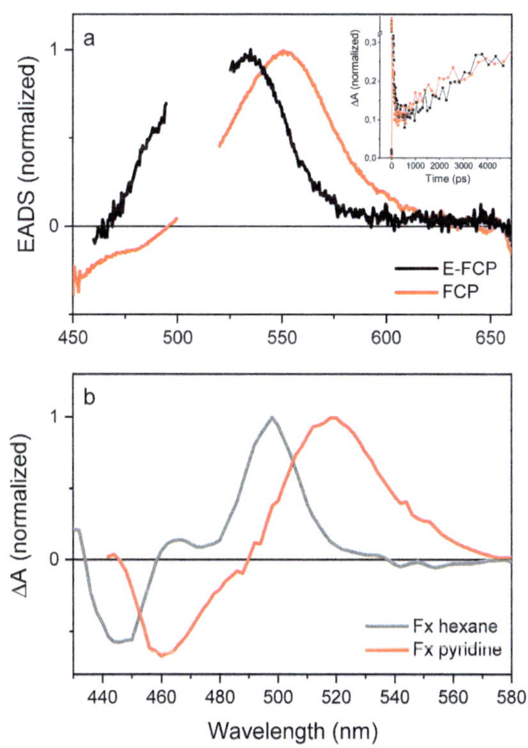

Fig. 8 (a) Normalized EADS of the triplet state of hFx in E-FCP (black) and Fx in FCP (red). The part of the spectrum around the excitation wavelength (510 nm) is omitted due to scattering. The inset shows the triplet rise at 535 nm (E-FCP) and 550 nm (FCP). (b) Triplet-minus-singlet spectra of fucoxanthin in solution in *n*-hexane and pyridine.

spectra of the triplet state generated *via* quenching of the Chl-*a* triplet are different for the two fucoxanthin derivatives. While the T–T maximum of Fx in FCP is at 552 nm, hFx has its T–T maximum at 533 nm. This shift of the T–T spectrum is likely another manifestation of the different response of the two fucoxanthin derivatives to the polarity of the environment. A detailed study of the triplet spectra of the keto-carotenoid peridinin in solvents with different polarity showed that the maximum of the T–T band depends on the solvent polarity: it shifts from 513 nm in the non-polar *n*-hexane to 525 nm in the polar acetonitrile.[46] We have measured the triplet spectra of Fx in two organic solvents, the non-polar *n*-hexane and polar pyridine (Fig. 8b), confirming the trend reported for peridinin, though part of the red-shift of the T–T spectrum of Fx in pyridine is related to the high polarizability of this solvent. We showed that the spectroscopic properties of hFx in essentially all solvents are reminiscent of Fx in a non-polar solvent, and this feature remains also for hFx bound to E-FCP. Then, the reason for the blue shift of the T–T spectrum of hFx in E-FCP is just a different response to the environment polarity. The rise of the triplet state signal is nearly identical in FCP and E-FCP, mirroring the Chl-*a* lifetime. This shows that, as in essentially all light-harvesting complexes studied so far, the rate limiting step in formation of the carotenoid triplet is the singlet-triplet conversion in Chl-*a*. The Chl-*a* to carotenoid triplet–triplet transfer is much faster, taking place on the time scale of 10–100 ps.[9,47]

Conclusions

The light-harvesting complex from *E. huxleyi*, E-FCP, is yet another example of how various photosynthetic organisms adapt to the light conditions in which they live. E-FCP binds the fucoxanthin derivative hFx that belongs to the family of keto-carotenoids. In contrast to its parent molecule, however, hFx suppresses the formation of the ICT state.[36] This gives significantly slower hFx–Chl energy transfer *via* the S_1/ICT route in E-FCP, providing clear proof that the magnitude of the S_1-ICT coupling is the key feature determining the energy transfer rate. However, minimizing the S_1-ICT coupling in hFx also prolongs the lifetime of the energy donor, the coupled S_1/ICT state. This situation results in only a small decrease of energy transfer efficiency of the S_1/ICT route as the slower energy transfer rate is compensated by the longer lifetime of the energy donor. Thus, even though E-FCP exhibits the slowest carotenoid–Chl energy transfer rate *via* the S_1 or S_1/ICT state reported so far, it still maintains an energy transfer efficiency of around 80%. The regulatory effect of using hFx instead of Fx is therefore only marginal and it cannot serve as a tool to regulate energy flow through the photosynthetic antenna systems.

Conflicts of interest

There are no conflicts to declare.

Acknowledgements

The research was funded from the Czech Science Foundation grant 16-10417S, and by institutional support RVO:60077344.

References

1 R. Croce and H. van Amerongen, *Nat. Chem. Biol.*, 2014, **10**, 492–501.
2 T. Mirkovic, E. E. Ostroumov, J. M. Anna, R. van Grondelle, Govindjee and G. D. Scholes, *Chem. Rev.*, 2017, **117**, 249–293.
3 T. Polivka and H. A. Frank, *Acc. Chem. Res.*, 2010, **43**, 1125–1134.
4 A. V. Ruban, *J. Exp. Bot.*, 2015, **66**, 7–23.
5 C. D. P. Duffy and A. V. Ruban, *J. Photochem. Photobiol., B*, 2015, **152**, 215–226.
6 A. V. Ruban, R. Berera, C. Ilioaia, I. H. M. van Stokkum, J. T. M. Kennis, A. A. Pascal, H. van Amerongen, B. Robert, P. Horton and R. van Grondelle, *Nature*, 2007, **450**, 575–578.
7 N. E. Holt, D. Zigmantas, L. Valkunas, X. P. Li, K. K. Niyogi and G. R. Fleming, *Science*, 2005, **307**, 433–436.
8 M. Mozzo, L. Dall'Osto, R. Hienerwadel, R. Bassi and R. Croce, *J. Biol. Chem.*, 2008, **283**, 6184–6192.
9 Z. Kvicalova, J. Alster, E. Hofmann, P. Khoroshyy, R. Litvin, D. Bina, T. Polivka and J. Psencik, *Biochim. Biophys. Acta, Bioenerg.*, 2016, **1857**, 341–349.
10 R. Schmidt, *J. Phys. Chem. A*, 2004, **108**, 5509–5513.
11 T. Polivka and V. Sundstrom, *Chem. Rev.*, 2004, **104**, 2021–2071.
12 H. Staleva, J. Komenda, M. K. Shukla, V. Slouf, R. Kana, T. Polivka and R. Sobotka, *Nat. Chem. Biol.*, 2015, **11**, 287–U296.
13 D. M. Niedzwiedzki, T. Tronina, H. J. Liu, H. Staleva, J. Komenda, R. Sobotka, R. E. Blankenship and T. Polivka, *Biochim. Biophys. Acta, Bioenerg.*, 2016, **1857**, 1430–1439.
14 Y. Hontani, M. Kloz, T. Polivka, M. K. Shukla, R. Sobotka and J. T. M. Kennis, *J. Phys. Chem. Lett.*, 2018, **9**, 1788–1792.
15 H. A. Frank, J. A. Bautista, J. Josue, Z. Pendon, R. G. Hiller, F. P. Sharples, D. Gosztola and M. R. Wasielewski, *J. Phys. Chem. B*, 2000, **104**, 4569–4577.
16 D. Zigmantas, R. G. Hiller, F. P. Sharples, H. A. Frank, V. Sundstrom and T. Polivka, *Phys. Chem. Chem. Phys.*, 2004, **6**, 3009–3016.
17 K. Redeckas, V. Voiciuk and M. Vengris, *Photosynth. Res.*, 2017, **131**, 119.
18 R. G. West, M. Fuciman, H. Staleva-Musto, V. Sebelik, D. Bina, M. Durchan, V. Kuznetsova and T. Polivka, *J. Phys. Chem. B*, 2018, **122**, 7264–7276.
19 C. Buchel, *J. Plant Physiol.*, 2015, **172**, 62–75.
20 D. Zigmantas, R. G. Hiller, V. Sundstrom and T. Polivka, *Proc. Natl. Acad. Sci. U. S. A.*, 2002, **99**, 16760–16765.
21 E. Papagiannakis, I. H. M. van Stokkum, H. Fey, C. Buchel and R. van Grondelle, *Photosynth. Res.*, 2005, **86**, 241–250.
22 K. Redeckas, V. Voiciuk, D. Zigmantas, R. G. Hiller and M. Vengris, *Biochim. Biophys. Acta, Bioenerg.*, 2017, **1858**, 297–307.
23 R. G. West, D. Bina, M. Fuciman, V. Kuznetsova, R. Litvin and T. Polivka, *Biochim. Biophys. Acta, Bioenerg.*, 2018, **1859**, 357–365.
24 N. Gildenhoff, J. Herz, K. Gundermann, C. Buchel and J. Wachtveitl, *Chem. Phys.*, 2010, **373**, 104–109.
25 N. Gildenhoff, S. Amarie, K. Gundermann, A. Beer, C. Buchel and J. Wachtveitl, *Biochim. Biophys. Acta, Bioenerg.*, 2010, **1797**, 543–549.

26 A. Gelzinis, V. Butkus, E. Songaila, R. Augulis, A. Gall, C. Buchel, B. Robert, D. Abramavicius, D. Zigmantas and L. Valkunas, *Biochim. Biophys. Acta, Bioenerg.*, 2015, **1847**, 241–247.

27 M. Di Valentin, C. Buchel, G. M. Giacometti and D. Carbonera, *Biochem. Biophys. Res. Commun.*, 2012, **427**, 637–641.

28 M. Herbstova, D. Bina, R. Kana, F. Vacha and R. Litvin, *Sci. Rep.*, 2017, **7**, 11976.

29 B. Lepetit, D. Volke, M. Szabo, R. Hoffmann, G. Z. Garab, C. Wilhelm and R. Goss, *Biochemistry*, 2007, **46**, 9813–9822.

30 L. Premvardhan, B. Robert, A. Beer and C. Buchel, *Biochim. Biophys. Acta, Bioenerg.*, 2010, **1797**, 1647–1656.

31 B. A. Read, J. Kegel, M. J. Klute, A. Kuo, S. C. Lefebvre, F. Maumus, C. Mayer, J. Miller, A. Monier, A. Salamov, J. Young, M. Aguilar, J. M. Claverie, S. Frickenhaus, K. Gonzalez, E. K. Herman, Y. C. Lin, J. Napier, H. Ogata, A. F. Sarno, J. Shmutz, D. Schroeder, C. de Vargas, F. Verret, P. von Dassow, K. Valentin, Y. Van de Peer, G. Wheeler, J. B. Dacks, C. F. Delwiche, S. T. Dyhrman, G. Glockner, U. John, T. Richards, A. Z. Worden, X. Y. Zhang, I. V. Grigoriev, A. E. Allen, K. Bidle, M. Borodovsky, C. Bowler, C. Brownlee, J. M. Cock, M. Elias, V. N. Gladyshev, M. Groth, C. Guda, A. Hadaegh, M. D. Iglesias-Rodriguez, J. Jenkins, B. M. Jones, T. Lawson, F. Leese, E. Lindquist, A. Lobanov, A. Lomsadze, S. B. Malik, M. E. Marsh, L. Mackinder, T. Mock, B. Mueller-Roeber, A. Pagarete, M. Parker, I. Probert, H. Quesneville, C. Raines, S. A. Rensing, D. M. Riano-Pachon, S. Richier, S. Rokitta, Y. Shiraiwa, D. M. Soanes, M. van der Giezen, T. M. Wahlund, B. Williams, W. Wilson, G. Wolfe, L. L. Wurch and A. Emiliania Huxleyi, *Nature*, 2013, **499**, 209–213.

32 F. Burki, M. Kaplan, D. V. Tikhonenkov, V. Zlatogursky, B. Q. Minh, L. V. Radaykina, A. Smirnov, A. P. Mylnikov and P. J. Keeling, *Proc. R. Soc. B*, 2016, **283**, 10.

33 G. E. Hoffman, M. V. S. Puerta and C. F. Delwiche, *BMC Evol. Biol.*, 2011, 11.

34 F. T. Haxo, *J. Phycol.*, 1985, **21**, 282–287.

35 G. N. Harris, D. J. Scanlan and R. J. Geider, *J. Phycol.*, 2005, **41**, 851–862.

36 H. Staleva-Musto, V. Kuznetsova, R. G. West, G. Kesan, B. Minofar, M. Fuciman, D. Bina, R. Litvin and T. Polivka, *J. Phys. Chem. B*, 2018, **122**, 2922–2930.

37 R. R. Guillard and J. H. Ryther, *Can. J. Microbiol.*, 1962, **8**, 229–239.

38 Z. Gardian, R. Litvin, D. Bina and F. Vacha, *Photosynth. Res.*, 2014, **121**, 79–86.

39 M. Herbstova, D. Bina, P. Konik, Z. Gardian, F. Vacha and R. Litvin, *Biochim. Biophys. Acta, Bioenerg.*, 2015, **1847**, 534–543.

40 S. W. Jeffrey, R. F. C. Mantoura and S. W. Wright, *Phytoplankton pigments in oceanography: Guidelines to modern methods, Monographs on Oceanographic Methodology*, UNESCO Publishing, Paris, 2nd edn, 2005, vol. 10.

41 D. Bina, R. Litvin, F. Vacha and P. Siffel, *Photosynth. Res.*, 2006, **88**, 351–356.

42 I. H. M. van Stokkum, D. S. Larsen and R. van Grondelle, *Biochim. Biophys. Acta, Bioenerg.*, 2004, **1657**, 82–104.

43 A. Beer, K. Gundermann, J. Beckmann and C. Buchel, *Biochemistry*, 2006, **45**, 13046–13053.

44 G. Kesan, M. Durchan, J. Tichy, B. Minofar, V. Kuznetsoya, M. Fuciman, V. Slouf, C. Parlak and T. Polivka, *J. Phys. Chem. B*, 2015, **119**, 12653–12663.

45 D. Z. Liu, Y. L. Gao and L. D. Kispert, *J. Electroanal. Chem.*, 2000, **488**, 140–150.
46 M. Fuciman, M. M. Enriquez, S. Kaligotla, D. M. Niedzwiedzki, T. Kajikawa, K. Aoki, S. Katsumura and H. A. Frank, *J. Phys. Chem. B*, 2011, **115**, 4436–4445.
47 M. E. Tejeda-Ferrari, C. L. Brown, G. C. Coutinho, G. A. Gomes de Sá Julio, L. Palma, M. J. Llansola-Portoles, G. Kodis, V. Mujica, J. Ho, D. Gust, T. A. Moore and A. L. Moore, *Photochem. Photobiol.*, 2019, **95**, 211–219.

Faraday Discussions

PAPER

Can we use on-the-fly quantum simulations to connect molecular structure and sunscreen action?†

Gareth W. Richings, Christopher Robertson and Scott Habershon *

Received 4th December 2018, Accepted 18th January 2019
DOI: 10.1039/c8fd00228b

The ground-up design of new molecular sunscreens, with improved photostability, absorbance and spectral coverage, stands as a challenge to fundamental chemical science. Correlating sunscreen molecular structure and function requires detailed insight into the relaxation pathways available following photoexcitation; however, the complex coupled electron/nuclear dynamics in these systems stands as a tough challenge to computational chemistry. To address this challenge, we have recently developed efficient and accurate simulation methods to model non-adiabatic dynamics of general molecular systems as a route to correlating photoinduced dynamics and potential sunscreen activity. Our approach, combining the multi-configuration time-dependent Hartree (MCTDH) method with PESs generated using machine-learning, represents a new "on-the-fly" strategy for accurate wavefunction propagation, with the potential to provide a new "black box" strategy for interrogating ultrafast dynamics in general photoinduced energy transport processes. Here, we illustrate our attempts to apply this methodology to study the ultrafast photochemistry of four compounds derived from mycosporine-like amino acids (MAAs), compounds which are believed to act as microbial sunscreens in micro-organisms such as algae. Specifically, we investigate how the choice of active vibrational space and diabatic electronic states in MCTDH strongly influences the predictions of ultrafast dynamic relaxation in representative MAAs. Our results serve to demonstrate that "on-the-fly" quantum dynamics using MCTDH is increasingly viable, but important barriers relating to coordinate choices and diabatisation remain.

1 Introduction

The direct solution of the time-dependent Schrödinger equation (TDSE) can provide a wavefunction describing the electronic and nuclear motion in

Department of Chemistry and Centre for Scientific Computing, University of Warwick, Coventry, CV4 7AL, UK. E-mail: S.Habershon@Warwick.ac.uk

† Electronic supplementary information (ESI) available. See DOI: 10.1039/c8fd00228b

a molecular system undergoing either adiabatic or non-adiabatic dynamics. As such, wavefunction propagation simulations are an invaluable tool for interrogating ultrafast (*i.e.* a few hundred fs) chemical dynamics in molecular systems, with applications including calculation of absorption and emission spectra, determination of reactive fluxes, prediction of reaction mechanisms, and prediction of non-linear or multidimensional spectroscopies.

Among the wide range of methods proposed for propagating wavefunctions using the TDSE, the multi-configuration time-dependent Hartree (MCTDH) method occupies a central position.[1–3] In MCTDH, the time-dependent wavefunction is written as a completely general tensor product of basis functions (termed single-particle functions [SPFs] in MCTDH) for each degree-of-freedom, with each Hartree product of SPFs being associated with a complex expansion coefficient. In MCTDH, both the SPFs and the expansion coefficients are time-dependent, and the Dirac–Frenkel variational principle is used to derive equations-of-motion which are propagated to produce a time-dependent wavefunction. To date, MCTDH (and extensions such as multi-layer MCTDH,[4–6] aimed at treating systems with greater numbers of degrees-of-freedom) has been used to study an enormous range of molecular chemistry, with representative examples including non-adiabatic dynamics in organic molecules such as pyrazine,[7–9] pyrroles[10] and DNA bases,[11] to complex materials such as dye chromophores adsorbed on TiO_2 nanoparticles[12,13] or spin–orbit coupling in organometallic complexes.[14]

However, a key computational drawback of MCTDH has played an important role in preventing wider uptake of this powerful approach; in particular, for efficient numerical implementation, MCTDH requires that the potential energy surface (PES) of the molecular system of interest be available in the form of a sum-of-products. While one can always fit a PES with a functional form of the appropriate sum-of-products structure required for efficient MCTDH simulations,[15–18] this fitting process can itself be cumbersome and inaccurate; for example, while approximate models such as the vibronic coupling Hamiltonian are useful in describing small-amplitude molecular dynamics, more complex processes, such as hydrogen dissociation, can be difficult to describe correctly. In addition, fitting sum-of-products PES forms to *ab initio* data can itself present challenges, notably in the choice of reference data. Technically, MCTDH requires a *global* PES describing all accessible molecular geometries which might feature in a wavefunction propagation simulation; however, this requirement is contrary to the *local* nature of *ab initio* electronic structure calculations.

In our recent work,[19–22] we have made a number of contributions which are aimed at implementing an "on-the-fly" or "direct" MCTDH simulation strategy. We have shown how one can construct both adiabatic and non-adiabatic PESs describing complex chemical systems during the course of an MCTDH wavefunction propagation, in contrast to the standard approach of fitting a global PES *prior* to propagation. Our approach, described below, uses kernel ridge regression (KRR) to automatically generate a simple functional representation of the PES upon which the MCTDH wavefunction is evolving,[23] in a similar manner to the GROW methodology proposed by Collins and co-workers.[24,25] Combined with an additional two-body or three-body decomposition of the resulting KRR PES, our scheme enables automatic generation of a PES in tandem with wavefunction propagation, providing an overall wavefunction propagation method that does not require pre-fitting of a global PES. Our direct-dynamics MCTDH (DD-MCTDH) scheme offers a new computational tool for modelling quantum chemical dynamics while circumventing some of the computational roadblocks that have stood in the way of direct implementations of MCTDH in the past.

One particularly important emerging application domain for direct quantum dynamics simulations is the design of new molecular sunscreens.[26–29] With the increasing prevalence of health issues arising from over-exposure to the sun, notably melanoma and other cancers, as well as increased awareness of the dangers of sun exposure, there is a globally growing demand to develop molecular sunscreens that outperform current market products in regards to higher UVA (315–400 nm) or UVB (280–315 nm) absorption strength, minimal photoproduct formation, maximal photostability and low risk of photosensitization or allergy. However, the *de novo* design of new organic sunscreen molecules is hampered because the molecular mechanism of action of many sunscreen filters is simply not well enough understood at the molecular level; while general principles regarding the desired properties of sunscreens, such as strong UVA/UVB absorption, photostability and the absence of toxic photoproducts, are clearly defined, the connection between molecular structure, electronic state manifolds and photodynamic relaxation in governing the ultimate efficiency of molecular sunscreens remains difficult to predict.

In this paper, we present the first application of our DD-MCTDH scheme to directly investigate the correlation between molecular electronic structure, molecular quantum photodynamics, and emergent sunscreen potential. In particular, we investigate the photodynamics in four different organic molecules which are derived from a category of natural products known as mycosporine-like amino acids (MAAs; Fig. 1). MAAs comprise a core moiety of either a cyclohexenone or cyclohexenimine with a variety of peripheral functional groups (typically alcohols or carboxylic acids).[30,31] These molecules are commonly found

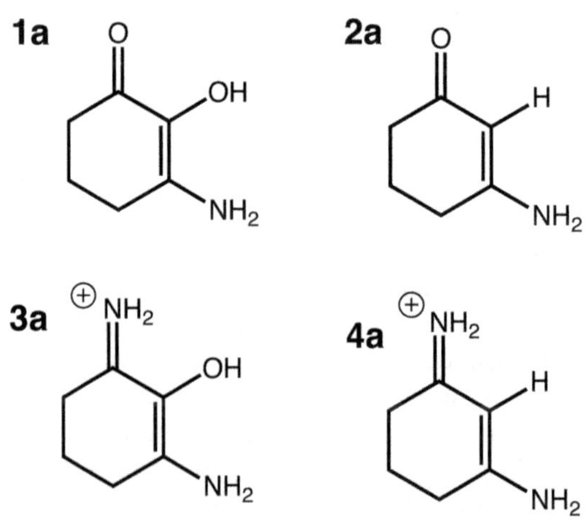

Fig. 1 MAA analogue structures studied here. Structures **1a–4a** are closely related to MAAs that have been studied previously by experiment and surface-hopping simulations; the only difference, as noted below, is that methyl groups on the amines/imines have been replaced with hydrogen in order to reduce computational complexity.

in high-UV natural environments, and have been discovered in microbes, algae and fungi; MAAs are strongly associated with a function as microbial sunscreens, with the expectation that the cyclic conjugated core structure acts as a strong UV absorber. In addition to their absorption properties, MAAs exhibit a high degree of photostability, as well as other desirable properties for sunscreen formulations, such as water solubility.

In a recent experimental and computational study of sunscreen performance for MAAs, Losantos and co-workers used a series of quantum chemical calculations for several MAA-like compounds, as well as direct synthesis and absorption spectra characterization, in an attempt to rationally design molecular compounds with improved sunscreen action.[28] Excited-state minimum energy path (MEP) calculations for **1a** and **3a** suggested that **3a** is a better sunscreen candidate than **1a**, with a first excited-state (S_1) PES which leads to fast relaxation to the S_0 ground-state through a conical intersection. This conclusion is supported by non-adiabatic trajectory surface-hopping simulations of **3a**, which illustrated ultrafast relaxation to the ground-state in around 200 fs. Building on these studies of **3a**, this prior work then proposed a number of modified molecular structures in the search for improved sunscreen function building on the structure of **3a**, with promising sun-protection factors measured for formulations containing five-membered ionic compounds which are analogous to the cyclohexenimine core.

The purpose of this contribution is to investigate whether quantum dynamics simulations, using our newly-developed DD-MCTDH strategy, can provide direct predictive insight of sunscreen function in MAA-like compounds; in other words, we will investigate whether the predictions of DD-MCTDH simulations align with previous insights into MAA properties obtained from MEP calculations, and to what extent DD-MCTDH can be used to directly correlate molecular structure and sunscreen action. In addition, we note that direct application of DD-MCTDH to the complex molecular species of Fig. 1 remains extremely difficult, even using our DD-MCTDH approach; a key challenge in applying DD-MCTDH to such many-atom molecules is in identifying the set of vibrational coordinates which are essential in describing relaxation and energy dissipation following photoexcitation. To investigate this problem in the context of DD-MCTDH, this article also proposes and tests a simple computational scheme, based on finding vibrational modes with strong non-adiabatic coupling terms (NACTs) and/or large excited-state forces, to identify the set of active vibrational modes within which DD-MCTDH propagation should be performed.

The remainder of this paper is organised as follows. In Section 2, we first outline the standard MCTDH and DD-MCTDH methods, before describing approaches for identifying an active vibrational coordinate-space within which to perform quantum dynamics, based on projection of excited-state forces and non-adiabatic couplings onto the set of ground-state normal modes. In Section 3, we then highlight the application of DD-MCTDH, along with the newly-proposed active coordinate selection, to study non-adiabatic dynamics in four MAA derivatives. Our quantum dynamics results are analysed in the context of prior computational and experimental work by Losantos and co-workers, with emphasis on direct non-adiabatic quantum dynamics as a route to connecting molecular structure to the rapid non-radiative relaxation, a desirable feature in molecular sunscreens. Disappointingly, we find that our DD-MCTDH simulations cannot reproduce the ultrafast dynamics of the selected model MAAs; however, the underlying reasons for this are explored, and are found to be revealing in pointing the way towards making progress in applying direct quantum dynamics to the correlation of structure and dynamics.

2 Method

In this section, we outline the key aspects of our computational approach to non-adiabatic quantum dynamics; further details have been published elsewhere.[19–21] In addition, we also discuss simple strategies that can be used to identify a set of vibrational coordinates that might be expected to be important in the non-adiabatic relaxation dynamics of given molecular systems; such automated coordinate identification methods will be key to the further application of DD-MCTDH to larger molecules.

2.1 DD-MCTDH using a KRR PES

The MCTDH method[1–3] provides a numerically-exact method for solving the TDSE, yielding a time-dependent wavefunction for the evaluation of dynamic expectation values and observables such as absorption spectra. The MCTDH method has been, and continues to be, applied to study wavefunction dynamics in a wide range of molecular systems (typically containing a few tens of active degrees-of-freedom), while the development of the multi-layer formulation has pushed the upper limit of system size which is treatable within this framework.[4–6]

Assuming a molecular system with f degrees-of-freedom, the MCTDH wavefunction *ansatz* uses a sum-of-products of basis functions,

$$\Psi^{(s)}(Q_1,\cdots,Q_f,t)=\sum_{j_1=1}^{n_1}\cdots\sum_{j_m=1}^{n_m}A^{(s)}_{j_1,\ldots,j_m}(t)\prod_{\kappa=1}^{m}\varphi^{(s,\kappa)}_{j_\kappa}(Q_\kappa,t)=\sum_J A^{(s)}_J(t)\Phi^{(s)}_J(Q,t). \quad (1)$$

Here, $\varphi^{(s,\kappa)}_{j_\kappa}(Q_\kappa,t)$ is the j_κ-th basis function on electronic state s and Q_κ indicates the corresponding degree(s)-of-freedom. Each Hartree product $\Phi_J^{(s)}(Q,t)$ is associated with a complex expansion coefficient $A_J^{(s)}(t)$, where we have introduced a compound index, $J=j_1,\ldots,j_f$. At this point, it is worth highlighting the fact that our work on developing direct-dynamics MCTDH schemes has focussed exclusively, to date, on using mass-frequency scaled normal-mode coordinates as the degrees-of-freedom defining molecular configuration space; such coordinates have, so far, been adequate for our simulations, with the added advantage of removing the challenges associated with non-diagonal kinetic energy operators. However, we note that our computational approach can, in principle, be generalised to arbitrary coordinate systems.

In MCTDH, the expansion coefficients and basis functions are time-dependent; in this way, the number of terms necessary for accurate wavefunction representation can be minimised, enabling treatment of larger system-sizes than more standard grid-based quantum dynamics schemes. Using the Dirac–Frenkel variational principle,[32] one can derive equations-of-motion for the coefficients and time-dependent basis functions (termed single-particle functions [SPFs] in MCTDH terminology). The resulting equations-of-motion, for a system with N_s electronic states, are

$$i\hbar\dot{A}_J^{(s)} = \sum_{u=1}^{N_s}\sum_L \left\langle \Phi_J^{(s)} \middle| \hat{H}^{(su)} \middle| \Phi_L^{(u)} \right\rangle A_L^{(u)}, \tag{2a}$$

$$i\hbar\dot{\varphi}^{(s,\kappa)} = \left(1 - \hat{P}^{(s,\kappa)}\right)\left(\rho^{(s,\kappa)}\right)^{-1} \sum_{u=1}^{N_s} \left\langle \hat{H}^{(su)} \right\rangle^{(\kappa)} \varphi^{(u,\kappa)}. \tag{2b}$$

In eqn (2b), $\hat{P}^{(s,\kappa)}$ is a projection operator onto the SPF space along mode κ, and $(\rho^{(s,\kappa)})^{-1}$ is the inverse of the density matrix associated with κ. The mean-field matrix is given by $\langle \hat{H}^{(su)} \rangle_{jl}^{(\kappa)} = \langle \Psi_j^{(s,\kappa)} | \hat{H}^{(su)} | \Psi_l^{(u,\kappa)} \rangle$, where $\Psi_l^{(s,\kappa)} = \sum_{J^\kappa} A_{J_l^\kappa}^{(s)} \Phi_{J^\kappa}^{(s)}$ defines a single-hole function written in terms of functions $\Phi_{J^\kappa}^{(s)}$, corresponding to SPF Hartree products in all modes apart from κ.

The SPFs, $\varphi^{(s,\kappa)}$, are functions of a small set (1–4) of the degrees-of-freedom in the system, $Q_\kappa = (q_{\kappa_1}, \ldots, q_{\kappa_p})$. The SPFs are expanded in a time-independent basis set, here taken to be a discrete variable representation (DVR) basis,

$$\varphi_{j_\kappa}^{(s,\kappa)}(Q_\kappa, t) = \sum_{i_\kappa=1}^{N_\kappa} c_{i_\kappa}^{(s,\kappa,j_\kappa)}(t) X_{i_\kappa}^{(\kappa)}(Q_\kappa), \tag{3}$$

with $X_{i_\kappa}^{(\kappa)}(Q_\kappa)$ identifying the i_κ-th basis function for coordinate Q_κ.

A key computational challenge in applying MCTDH lies in evaluating the necessary integrals over the Hamiltonian operator (eqn (2)). When working in rectilinear coordinate systems, as we do here using vibrational normal-mode coordinates, integrals over kinetic energy operators can be evaluated straightforwardly as one-dimensional terms along each degree-of-freedom. The integrals over the PES, on the other hand, are far from trivial, formally requiring integration over the full f-dimensional space; as a result, a prerequisite for the *standard* MCTDH algorithm is the availability of a global PES function. In addition, for efficient computation, it is typically required that the global PES function is in the form of a "sum-of-products". This is a stringent requirement which acts as a roadblock to "on-the-fly" implementations.

In recent work, we have shown how kernel ridge regression (KRR) can be used to generate a PES in tandem with MCTDH wavefunction propagation; in other words, we have shown how one can achieve direct-dynamics simulations using MCTDH and KRR. In our scheme,[19–21] the PES is given as a weighted-sum of Gaussian kernel functions, with the following form:

$$V(q) = \sum_{i=1}^{M} w_i k_i(q_i, q).$$

As described below, we typically use a set of normal-mode vibrational coordinates q to describe molecular configuration-space; q_i is the normal-mode coordinates of KRR reference point i. In our initial work, the kernel functions were chosen to be f-dimensional Gaussian products centred at a set of reference-points q_i, namely

$$k_i(q_i, q) = \prod_{j=1}^{f} \mathrm{e}^{-\alpha_j\left(q_i^{(j)} - q^{(j)}\right)^2},$$

where $q_i^{(j)}$ is the value of normal-mode coordinate j for reference point i. Subsequently, we have demonstrated that so-called additive kernels of the form

$$k_i(q_i, q) = \sum_{j=1}^{f} \mathrm{e}^{-\alpha_j\left(q_i^{(j)} - q^{(j)}\right)^2} + \sum_{j=1}^{f}\sum_{k=1}^{f} \mathrm{e}^{-\alpha_j\left(q_i^{(j)} - q^{(j)}\right)^2 - \alpha_k\left(q_i^{(k)} - q^{(k)}\right)^2} + \ldots$$

offer a much more computationally-convenient choice of basis function as a result of the increased configuration-space volume associated with each additive kernel function. Regardless of which kernel function is selected, the KRR weights are found by solving a linear system of equations, given by

$$\boldsymbol{K}\boldsymbol{w} = \boldsymbol{b},$$

where $\boldsymbol{w}$ is the vector of weights, the vector $\boldsymbol{b}$ contains the set of known PES values evaluated at the reference points (for example, using *ab initio* electronic structure methods),

$$\boldsymbol{b}_i = V(q_i),$$

and $\boldsymbol{K}$ is an $M \times M$ covariance matrix (for M reference points) whose elements contain the kernel evaluated for each reference-point pair, and

$$\boldsymbol{K}_{ij} = \boldsymbol{k}(q_i, q_j) + \gamma^2\delta_{ij}$$

where γ^2 is a small, positive coefficient used to regularise the matrix. The result of this KRR scheme is the generation of a global PES, which is naturally tailored to be more accurate in regions of wavefunction evolution. In addition, the KRR PES generated described above is naturally in the sum-of-products form required for efficient MCTDH simulation. However, we note that further efficiency improvements are possible, for example, we have demonstrated that an additional many-body decomposition of the KRR PES onto the underlying DVR grid of the MCTDH wavefunction results in a much more efficient propagation scheme overall.[21] This approach is used extensively below.

Two important remaining implementation issues relate to the choice of reference points and the adaptation of our scheme to account for non-adiabatic effects. In our DD-MCTDH approach,[19–21] we select KRR reference points based on the natural dynamics of the time-evolving wavefunction, as described previously. Here, we generate trial configurations using a uniform sampling function centred on the position expectation value of the time-dependent wavefunction, and with a sufficiently large width to span the relevant region of configuration space. The use of Sobol sequences to select trial geometries within this region ensures an even sampling. At these trial configurations, the variance of the current KRR PES is evaluated. If this variance is larger than a user-defined tolerance, the KRR PES at the selected configuration is judged to be insufficiently accurate and an *ab initio* energy evaluation is performed; the configuration and the corresponding energy are then added to the database of KRR reference points, the KRR weights are updated, and propagation continues. This accept/reject test is performed continually throughout wavefunction propagation (typically every fs), such that the KRR database constantly adapts as new regions of configuration space are explored. At the end of the DD-MCTDH simulation, once a large database of reference points is generated, we usually repeat the dynamics calculation on the KRR PES generated using the full database, without further *ab initio* calculations.

In order to account for non-adiabatic effects, our DD-MCTDH scheme uses one of two different routes to generate diabatic electronic states and energies; we then perform the same KRR fitting scheme described above for each diabatic state (and coupling surface). To date, we have shown that this diabatisation can be achieved using either (i) propagation of an approximate diabatisation matrix using line-integration between nearby points in configuration space,[19,21,33,34] or (ii) by projecting a complete active space self-consistent field (CASSCF) wavefunction expansion onto a reference expansion.[35] We have found so far that these two approaches are equally applicable in DD-MCTDH, but with distinct advantages; for example, the projection diabatisation scheme does not require calculation of non-adiabatic coupling matrix elements, but is less rigorous than the propagation diabatisation. However, in both cases, we have found that calculated dynamic properties are qualitatively similar, and we focus on the projection diabatisation scheme here.

2.2 Selecting active vibrational modes

The discussion of DD-MCTDH above has been predicated on the availability of a set of active vibrational coordinates (or normal modes) which adequately describe the molecular configuration space; but how does one go about choosing the *active* vibrational coordinates *a priori*? For typical organic molecular systems undergoing non-adiabatic dynamics, one might reasonably expect that a few important vibrational coordinates will be most strongly responsible for photo-relaxation, while many other vibrational modes might be simply "spectators", undergoing only simple vibrational motion. Identifying the "active" vibrational coordinates is clearly an important task in successful application of MCTDH; using too many vibrational degrees-of-freedom is computationally-expensive, whereas not using enough might risk "trapping" the time-dependent wavefunction in regions of space from which access to the ground-state becomes hindered.

As a result, several schemes have been proposed to identify active coordinates.[36–38] For example, symmetry considerations are often useful in this regard, and are commonly exploited in MCTDH simulations where relevant, although symmetry is not helpful for complex organic molecules lacking higher-symmetry elements. Identification of active coordinates using classical trajectory simulations has also been proposed and tested successfully, but not in the context of MCTDH. Finally, we note that the approach taken by Lasorne and coworkers,[36] is most similar in spirit to that employed below, focussing on building a simple analytical model which provides orthonormal coordinates pointing along the 'most important' coordinate vectors for ultrafast dynamics; our approach is similar in spirit, but with the ambition to be simpler to implement and avoiding additional PES fitting.

As an alternative step in this direction, we consider here how one might choose a set of active normal-mode coordinates for propagation in DD-MCTDH *without* relying on prior dynamical knowledge or symmetry properties. Our approach can be summarised as follows:

(1) We perform a geometry optimization calculation on the ground electronic state, giving a reference geometry $\mathbf{r}_0$,

(2) At the reference geometry, we perform a normal-mode analysis, giving the set of ground-state normal-modes, $\{\mathbf{q}_0\}$. From this set of normal-modes, we would like to choose the "active" set which are most likely to be relevant to the subsequent excited-state dynamics.

(3) We then calculate the force arising from the *excited-state PES* along each ground-state normal-mode; this is achieved simply using the ground-state normal-mode transformation matrix and the Cartesian forces calculated on the excited-state PES at the Franck–Condon (FC) point;

(4) Simultaneously, we calculate the non-adiabatic coupling matrix element between the ground-state and excited-state projected along each of the ground-state normal modes.

(5) Finally, from $\{\mathbf{q}_0\}$, we then select the N normal-modes which have either the largest excited-state force components, the largest non-adiabatic coupling components, or both.

The above scheme is clearly motivated by a simply physical picture; those vibrational coordinates which have the largest gradients in the excited-state PES would be expected to be most relevant to dynamics in a 'steepest descent' picture while, at the same time, those coordinates which are most likely to lead towards conical intersections, along strong directions of non-adiabatic coupling, are also most likely to be essential in capturing the non-adiabatic dynamics.

The advantage of this scheme is clearly its simplicity; however, it does rely heavily on the underlying idea that the set of ground-state normal-mode coordinates and the available diabatic-states picture are both appropriate to capturing the overall dynamical effects relevant to photodynamics. As we see below, this is often not the case.

3 Application, results and discussion

In this section, we use *ab initio* electronic structure calculations (all carried out with Molpro[39–41]), the DD-MCTDH method described above, and our newly-proposed strategy for selecting dynamically-active vibrational coordinates, to investigate the photodynamics of four MAA-like molecules in an attempt to assess the extent to which wavefunction propagation can be used to predict sunscreen function. We focus here on modelling the molecules shown in Fig. 1; we note that these four molecules are essentially the same as the initial MAA-like molecules investigated by Losantos and co-workers,[28] with the minor change in replacing the methyl groups on the amine/imine moieties with hydrogen atoms. This change is implemented in order to expedite the required *ab initio* electronic structure calculations, and our preliminary investigations indicate that it does not significantly alter the important molecular electronic characteristics which are so vital to photodynamics.

3.1 Single-point energy calculations and coordinate selection

For each of the four molecules considered (Fig. 1), we first performed exploratory CIS calculations using Hartree–Fock natural orbitals. In each case, bright states were located that agreed with the previous work for closely-related MAAs; initially-excited states were determined as S_2 from molecules **1a** and **2a**, and S_1 for molecules **3a** and **4a**. These initial CIS calculations were subsequently used to identify orbital contributions to excited-states in order to define the active space for subsequent CASSCF calculations.

The electronic basis used for molecules **1a** and **2a** was 6-31G*, and for molecules **3a** and **4a** it was 6-31G* on carbon and hydrogen, with 6-31+G* on nitrogen and oxygen. A smaller basis was used for molecules **1a**/**2a** due to difficulties with stability in the active space, while for **3a**/**4a** the augmented set was used for O and N as this is usually deemed desirable.

State-averaged-CASSCF (SA-CASSCF) calculations were performed to optimise molecular geometries on S_0. Normal-mode analyses were then performed at the S_0 optimised geometries to generate mass-frequency-scaled normal-mode coordinates for the dynamics calculations. Calculations were also performed at the FC geometries for each molecule to determine forces and Hessians for the bright states, and also to get the non-adiabatic coupling vectors between the bright state and the state immediately below it in energy. Using the normal-modes and frequencies calculated above, the gradients, Hessians and NACTs could be transformed from molecular Cartesian coordinates to mass-frequency scaled normal modes.

We then proceeded to select active vibrational modes for dynamics calculations based on the magnitude of the forces and NACTs at the FC point in the bright state for each molecules. The results of these simulations are shown in Fig. 2 for molecule **1a**. For the modes illustrated, it is clear that there is a range of magnitudes for both the excited-state gradients and the NACTs; in fact, we have only illustrated the largest magnitude normal-modes, with the remainder having

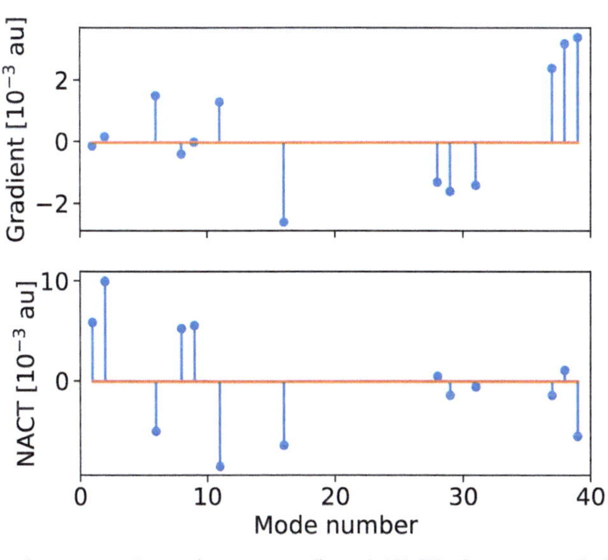

Fig. 2 Excited-state gradients (upper panel) and NACTs (lower panel) for selected ground-state normal-mode numbers for molecule **1a**. The lengths of the lines indicate the magnitude of the forces along each normal-mode in the excited-state, or the NACTs along that normal-mode; in each case, the normal-mode vectors come from geometry optimization and normal-mode analysis on the ground-state.

negligible contributions. From the gradients, it is clear that some modes have large force contributions in the excited-state, and it is relatively easy to select potentially-active modes from this set. In addition, there are clearly modes which have large NACTs. Perhaps most interesting and useful, it is found that these sets of modes do not necessarily overlap; modes 29–31 would not be selected on the basis of NACTs alone, but are selected once the gradients are also accounted for.

Based on the gradient and NACT information for each molecule, we selected appropriate normal-modes for subsequent dynamics (see ESI†). In addition, we also examined off-diagonal terms in the transformed Hessian in the bright state for each molecule; large absolute values indicate strong coupling between the modes, and strongly-coupled modes were subsequently grouped together in the MCTDH calculations as combined modes to reduce the effort required in the dynamics. By increasing the number of normal-modes which are included for each molecule (*i.e.* by decreasing the threshold on forces and NACTs below which modes are included), we can generate models for MCTDH simulations with increasing complexity in order to test the influence of coupled vibrational modes on the photodynamics.

3.2 MCTDH dynamics simulations

Following the identification of sets of active vibrational coordinates for each molecule, we proceeded to set up DD-MCTDH simulations. As a first step, we performed one-dimensional wavepacket propagation simulations for each active vibrational coordinate for each molecule; these simulations were performed to select an appropriate KRR width parameter, as well as identify an appropriate number of DVR basis functions to support the DD-MCTDH wavefunction. With the width parameters and DVR basis set sizes selected, we proceeded to perform f-dimensional dynamics simulations, with increasing numbers of coordinates f for each molecule.

To begin, 6-D DD-MCTDH calculations were run for all four molecules (see ESI† for details of SPFs, *etc.*); in all cases, the multi-set formulation of MCTDH was used.[2] Initial wavefunctions were taken to be a product of Gaussian functions along each active mode, with width $1/\sqrt{2}$, centred at the FC point on S_2 for **1a**/**2a** and on S_1 for **3a**/**4a**. The default variable mean field integration scheme was used to solve the MCTDH EOMs, employing the Adams–Bashforth–Moulton integrator of order 6, accuracy 10^{-5} and initial step of 10^{-4} fs.

For the generation of the KRR PES, 10^3 molecular geometries *per* state were sampled every 1 fs, within 3 SD of the centres of the wavepackets on each state. These geometries were discarded if the KRR PES variance estimate was below 10^{-3} au. If, instead, the sampled geometries had a variance greater than 10^{-3} au, SA-CASSCF calculations were performed, the resulting energies were diabatised and the results added to the growing KRR PES database. As noted above, to accelerate propagation, a many-body expansion of the KRR PES (truncated at third-order), was employed throughout.

The diabatic state populations for **1a** are shown in Fig. 3, as is the absorption spectrum, calculated as the Fourier transform of the wavepacket time-correlation function. Most disappointingly, for molecules **1a**, **3a** and **4a**, we see negligible population transfer from the initially-excited state; the diabatic state populations for **3a** and **4a** behave in the same way as **1a**, and are not shown. In the case of

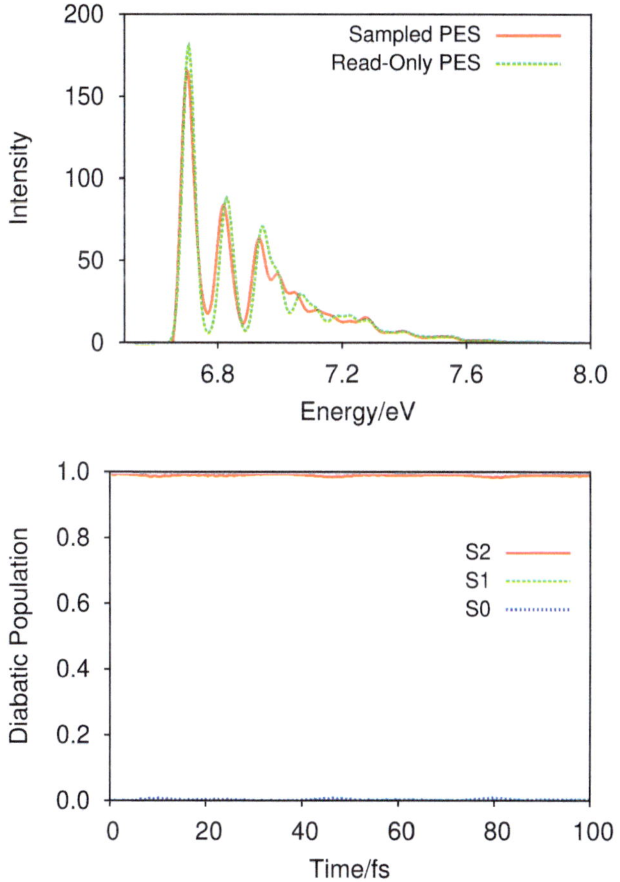

Fig. 3 Upper panel: absorption spectra of **1a** after vertical excitation to the S_2 state, calculated using DD-MCTDH with six normal-modes chosen according to their gradients and non-adiabatic couplings at the FC point. The red, solid line is the result from a calculation with the PES being built on-the-fly whilst the green, dashed line is the result from a second calculation using the pre-computed database to construct the PES. Lower panel: diabatic state populations from DD-MCTDH simulations calculated on the fully-sampled KRR PES.

molecules **3a** and **4a**, rapid transfer to S_0 is expected, as was observed for similar molecules by Losantos and coworkers in both experiments and surface-hopping simulations.[28] In particular, these previous investigations found an excited-state lifetime of 186 $\pm$ 49 fs for the methylated version of **3a.**

However, for molecule **2a**, we do indeed observe population transfer from the initially-excited S_2 state to the S_1 and S_0 states (Fig. 4); this seems to be achieved by a very fast (<20 fs) time-scale, followed by a much longer decay, such that relaxation is far from complete after the 100 fs simulation time. Furthermore, the absorption spectrum for **2a** is also indicative of ultrafast decay dynamics, with a broad peak with some weaker vibrational peaks superimposed.

Overall, these results are somewhat disappointing; we have performed challenging DD-MCTDH simulations, but we do not observe the expected ultrafast photodynamics for molecules **1a**, **3a** and **4a**. The approach being taken in our

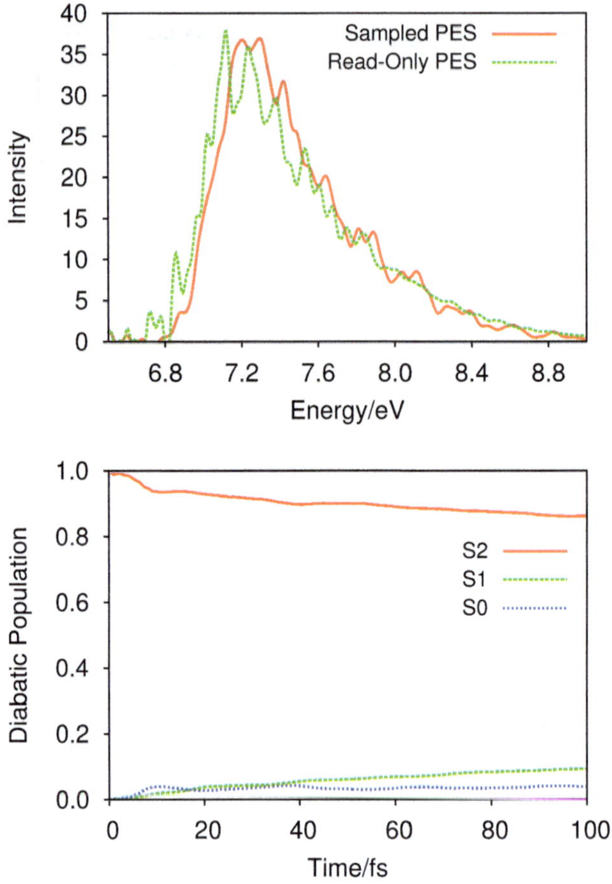

Fig. 4 Upper panel: absorption spectra of molecule **2a** after vertical excitation to the S_2 state, calculated using DD-MCTDH with six normal-modes chosen according to their gradients and non-adiabatic couplings at the FC point. The red, solid line is the result from a calculation with the PES being built on-the-fly whilst the green, dashed line is the result from a second calculation using the pre-computed database to construct the PES. Lower panel: diabatic state populations from DD-MCTDH simulations calculated on the fully-sampled KRR PES.

simulations and the previous investigations of these molecules are very similar in regards to electronic structure; as a result, we conclude that there are important differences between the dynamics generated by our DD-MCTDH simulations and the previous surface hopping calculations. To assess whether this difference arises due to the reduced dimensionality of our DD-MCTDH calculations (compared to the full-dimensional surface hopping simulations of previous work), we performed a further set of simulations with more active vibrational coordinates. In particular, we have performed additional simulations using 14 active vibrational modes; this represents the largest DD-MCTDH simulations we have performed to date. However, very disappointingly, we found that this made little difference to the dynamics for any of the molecules; in other words, it is not the limited dimensionality of our model systems which is preventing relaxation to the ground-states. There is more to this story, and we believe that the culprits may be the choice of coordinate system and/or the identity of missing diabatic states.

To begin to unravel exactly why we do not see the expected excited-state dynamics in our DD-MCTDH simulations, we focussed on molecule **4a**, the smaller of those molecules expected to demonstrate fast decay to the ground-state. To assess whether our choice of normal-mode coordinates was hindering approach to the conical intersection (CI), and hence decay to the ground-state, we modified our choice of normal-mode coordinates in an attempt to drive dynamics towards the CI. First, the optimized geometry of **4a** on S_1 and the CI were determined; these calculations found that the S_1 optimized geometry was very close to the CI, and could be reached from a "steepest descent"-like path on S_1. We then tried to find a new set of "pseudo" normal-mode coordinates which are constructed by generating linear combinations of the original set of normal-mode coordinates *plus* additional vectors that connect two or more geometries of interest; the procedure for achieving this is described in the ESI.† The result of this procedure is a new set of rectilinear normal-mode-like coordinates, which contain the original set of normal-mode coordinates, plus additional coordinates that are designed to span the coordinate space connecting the specified geometries of interest; in our simulations, we chose these geometries to be the FC geometry on S_1 and the corresponding CI.

Using this approach, we then performed DD-MCTDH propagation for 1 ps using a 6-D Hamiltonian constructed from the five modes with the largest forces at the FC geometry, plus the additional "pseudo" normal mode connecting the FC geometry on S_1 to the CI. Despite providing a direct coordinate connecting the FC configuration and the CI, we still do not observe ultrafast decay dynamics; in fact, only around 0.2% of the initial S_1 population is transferred to S_0 over the course of 1 ps. Expanding this approach, we then added an additional "pseudo" normal-mode which was designed to connect the FC point to a geometry which was visited during geometry optimization towards the CI on S_1. Again, this additional coordinate gave no further improvement in the predicted excited-state dynamics. Our only conclusion is that the choice of normal-mode coordinates is not significantly hindering excited-state dynamics. When we include additional modes specifically designed to connect key "way-points" along the trajectory from the FC point to the CI, we do not see improvement in the expected dynamics.

Instead, the answer to this problem is suggested in Fig. 5, which shows slices through the diabatic and adiabatic PESs along the "pseudo" normal-mode coordinate connecting the FC point to the CI for molecule **4a**. The upper panel in Fig. 5 shows the S_0 and S_1 adiabatic potential energy curves along the "pseudo" normal mode between the FC point (at coordinate $q = 0$) and the CI (seen at around coordinate $q = 21$). Two features are important here: (i) the CI is significantly lower in energy than the S_1 state at the FC point, consistent with the earlier work by Losantos *et al.* where the CI was reached very quickly after initial photoexcitation, and (ii) a large energy barrier exists between the FC and CI geometries along this path (around 2.5 eV), suggesting that additional modes must be included for the wavepacket to reach the CI by a barrierless process.

However, the plot in the lower panel of Fig. 5, illustrating the diabatic PESs along the same coordinate as that used to generate the adiabatic PES plots, indicates another problem. The curves in the lower panel of Fig. 5 indicate that the diabatic curves match the adiabatic curves quite closely between the FC point

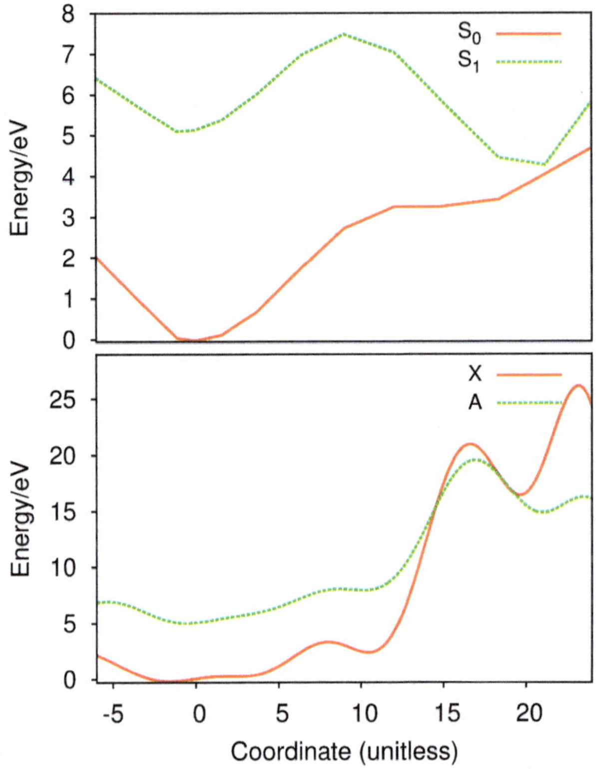

Fig. 5 Upper panel: adiabatic potential energy curves along the vector between the FC and CI geometries of molecule **4a** calculated during a DD-SM wavepacket propagation. The red, solid line is the energy of the S_0 electronic state whilst the green, dashed line is the energy of the S_1 state. Lower panel: diabatic potential energy curves generated from the adiabatic curves in the upper panel. The red, solid line is the *X* state which corresponds to the adiabatic S_0 at the FC point whilst the green, dashed line is the *A* state which corresponds to the S_1 adiabatic state at the same point.

and about coordinate $q \simeq 10$ (*i.e.* approaching the top of the adiabatic barrier). However, beyond this point and towards the CI, the diabatic and adiabatic curves diverge significantly. As the projection diabatisation scheme used here relies on following the configuration interaction expansion vector across configuration space so that a particular diabatic state has a consistent expansion at all geometries, the curves in Fig. 5 indicate that the configuration defining the S_0 and S_1 states at the CI is not included in the reference electronic states selected at the FC point. The implication is that, even by including ever more normal modes in the dynamics calculations, the CI will not be reached because not enough diabatic states have been included to allow the PESs to take the form necessary for the wavepacket to access the CI; the wavepacket is "diabatically trapped" close to the FC point. The projection diabatisation scheme, in principle, allows the inclusion of configurations defined at more than one geometry so that the electronic states near CI could be described correctly; this requires additional work in our dynamics code, which has not yet been implemented. However, one encouraging point to note is that the active space natural orbitals correspond closely at both the FC and CI geometries, so the problem is not the "breaking" of the active space. More work will have to be done to resolve this issue.

4 Conclusions

On one hand, the conclusions of this paper are somewhat disappointing but, on the other hand, they are very encouraging! We have been able to run direct non-adiabatic wavefunction propagations completely "on-the-fly" using our DD-MCTDH scheme; the ability to run such simulations for systems containing more than 10 degrees-of-freedom is very exciting, and opens up all sorts of possible future applications. However, in the particular examples chosen here, we have found that application of DD-MCTDH was hindered by unanticipated obstacles. Most importantly, we have found that the choice of diabatisation scheme can have a profound effect on the emergent dynamics, to the extent that expected ultrafast decay dynamics is completely absent. We have rationalised this as arising from a poor choice of diabatic states which were not sufficient to describe the approach towards the CI for our selected molecules, leading to "diabatic trapping".

Fortunately, there is a relatively easy fix – we simply need to adapt our diabatisation scheme so that many more initial adiabatic states are incorporated into our projection scheme for diabatisation. Here, states that may be very high in energy at the FC geometry, but might become much lower energy at geometries near the CI, could readily be included at little additional cost to the MCTDH propagation. By including these states, the description of the electronic PESs across all sampled geometries would be expected to be much more accurate, and one might anticipate agreement with previous experiments and surface-hopping simulations would emerge. This is ongoing work, which should be complete soon, enabling us to then meet our initial goal of connecting sunscreen molecular structure and ultrafast dynamics.

As a final point, we note that further studies performed using the original propagation diabatisation scheme,[19,21,33,34] (as opposed to the projection diabatisation scheme considered here) combined with additional normal-mode-like coordinates as described above, have begun to provide promising results for the ultrafast dynamics of MAAs. A full analysis of these emerging results will be published in due course.

Conflicts of interest

There are no conflicts to declare.

Acknowledgements

We gratefully acknowledge funding from the Leverhulme Trust (RPG-2016-055), the Engineering and Physical Sciences Research Council (EP/R020477/1), and the Energy Global Research Priority at the University of Warwick. We also acknowledge Simon Holland and Vasilios Stavros (University of Warwick) for useful discussions relating to MAA photodynamics.

Notes and references

1 H.-D. Meyer, U. Manthe and L. Cederbaum, *Chem. Phys. Lett.*, 1990, **165**, 73–78.

2 M. H. Beck, A. Jäckle, G. A. Worth and H. D. Meyer, *Phys. Rep.*, 2000, **324**, 1–105.

3 *Multidimensional quantum dynamics: MCTDH theory and applications*, ed. H.-D. Meyer, F. Gatti and G. A. Worth, Wiley, Weinheim, Germany, 2009.

4 H. Wang and M. Thoss, *J. Chem. Phys.*, 2003, **119**, 1289–1299.

5 H. Wang, *J. Phys. Chem. A*, 2015, **119**, 7951–7965.

6 O. Vendrell and H.-D. Meyer, *J. Chem. Phys.*, 2011, **134**, 044135.

7 G. Worth, H.-D. Meyer and L. Cederbaum, *J. Chem. Phys.*, 1996, **105**, 4412–4426.

8 G. Worth, H.-D. Meyer and L. Cederbaum, *J. Chem. Phys.*, 1998, **109**, 3518–3529.

9 A. Raab, G. Worth, H.-D. Meyer and L. Cederbaum, *J. Chem. Phys.*, 1999, **110**, 936–946.

10 S. P. Neville and G. A. Worth, *J. Chem. Phys.*, 2014, **140**, 034317.

11 G. M. Krishnan and O. Kühn, *Chem. Phys. Lett.*, 2007, **435**, 132–135.

12 J. Li, I. Kondov, H. Wang and M. Thoss, *J. Phys.: Condens. Matter*, 2015, **27**, 134202.

13 J. Li, H. Wang, P. Persson and M. Thoss, *J. Chem. Phys.*, 2012, **137**, 22A529.

14 M. Fumanal, E. Gindensperger and C. Daniel, *J. Chem. Theory Comput.*, 2017, **13**, 1293–1306.

15 L. S. Cederbaum, H. Köppel and W. Domcke, *Int. J. Quantum Chem.*, 1981, **15**, 251–267.

16 Q. Meng and H.-D. Meyer, *J. Chem. Phys.*, 2013, **138**, 014313.

17 M. Fumanal, F. Plasser, S. Mai, C. Daniel and E. Gindensperger, *J. Chem. Phys.*, 2018, **148**, 124119.

18 R. G. McKinlay, J. M. Żurek and M. J. Paterson, *Theoretical and Computational Inorganic Chemistry*, Academic Press, 2010, vol. 62, pp. 351–390.

19 G. W. Richings and S. Habershon, *Chem. Phys. Lett.*, 2017, **683**, 228–233.

20 G. W. Richings and S. Habershon, *J. Chem. Theory Comput.*, 2017, **13**, 4012–4024.

21 G. W. Richings and S. Habershon, *J. Chem. Phys.*, 2018, **148**, 134116.

22 G. W. Richings, C. Robertson and S. Habershon, *J. Chem. Theory Comput.*, 2019, **15**, 857–870.

23 C. Williams, *Handbook of Brain Theory and Neural Networks*, The MIT Press, Cambridge, Massachusetts, 2002, pp. 466–470.

24 M. A. Collins and D. H. Zhang, *J. Chem. Phys.*, 1999, **111**, 9924–9931.

25 J. Ischtwan and M. A. Collins, *J. Chem. Phys.*, 1994, **100**, 8080–8088.

26 L. A. Baker, S. E. Greenough and V. G. Stavros, *J. Phys. Chem. Lett.*, 2016, **7**, 4655–4665.

27 L. A. Baker, M. D. Horbury, S. E. Greenough, F. Allais, P. S. Walsh, S. Habershon and V. G. Stavros, *J. Phys. Chem. Lett.*, 2016, **7**, 56–61.

28 R. Losantos, I. Funes-Ardoiz, J. Aguilera, E. Herrera-Ceballos, C. García-Iriepa, P. J. Campos and D. Sampedro, *Angew. Chem., Int. Ed.*, 2017, **56**, 2632–2635.

29 J. M. Woolley, M. Staniforth, M. D. Horbury, G. W. Richings, M. Wills and V. G. Stavros, *J. Phys. Chem. Lett.*, 2018, **9**, 3043–3048.

30 W. M. Bandaranayake, *Nat. Prod. Rep.*, 1998, **15**, 159–172.

31 Q. Gao and F. Garcia-Pichel, *Nat. Rev. Microbiol.*, 2011, **9**, 791–802.
32 J. Broeckhove, L. Lathouwers, E. Kesteloot and P. Van Leuven, *Chem. Phys. Lett.*, 1988, **149**, 547–550.
33 G. Richings and G. Worth, *J. Phys. Chem. A*, 2015, **119**, 12457–12470.
34 G. W. Richings and G. A. Worth, *Chem. Phys. Lett.*, 2017, **683**, 606–612.
35 C. Robertson, J. Gonzalez-Vazquez, S. Diaz-Tendero, I. Corral and C. Diaz, *J. Comput. Chem.*, 2019, **40**, 794–810.
36 B. Lasorne, F. Sicilia, M. J. Bearpark, M. A. Robb, G. A. Worth and L. Blancafort, *J. Chem. Phys.*, 2008, **128**, 124307.
37 S. Manzhos and T. Carrington, *J. Chem. Phys.*, 2008, **129**, 224104.
38 P. Tavadze, G. A. Franco, P. Ren, X. Wen, Y. Li and J. P. Lewis, *J. Am. Chem. Soc.*, 2018, **140**, 285–29039.
39 H.-J. Werner, P. J. Knowles, G. Knizia, F. R. Manby, M. Schutz, P. Celani, W. Gyorffy, D. Kats, T. Korona, R. Lindh, A. Mitrushenkov, G. Rauhut, K. R. Shamasundar, T. B. Adler, R. D. Amos, S. J. Bennie, A. Bernhardsson, A. Berning, D. L. Cooper, M. J. O. Deegan, A. J. Dobbyn, F. Eckert, E. Goll, C. Hampel, A. Hesselmann, G. Hetzer, T. Hrenar, G. Jansen, C. Koppl, S. J. R. Lee, Y. Liu, A. W. Lloyd, Q. Ma, R. A. Mata, A. J. May, S. J. McNicholas, W. Meyer, T. F. Miller III, M. E. Mura, A. Nicklass, D. P. O'Neill, P. Palmieri, D. Peng, K. Pfluger, R. Pitzer, M. Reiher, T. Shiozaki, H. Stoll, A. J. Stone, R. Tarroni, T. Thorsteinsson, M. Wang and M. Welborn, *MOLPRO, version 2015.1, a package of ab initio programs*, 2015, http://www.molpro.net.
40 H.-J. Werner and P. J. Knowles, *J. Chem. Phys.*, 1985, **82**, 5053–5063.
41 P. J. Knowles and H.-J. Werner, *Chem. Phys. Lett.*, 1985, **115**, 259–267.

Faraday Discussions

PAPER

Two-photon excitation spectroscopy of photosynthetic light-harvesting complexes and pigments

Alexander Betke †[a] and Heiko Lokstein ‡*[ab]

Received 21st November 2018, Accepted 17th January 2019

DOI: 10.1039/c8fd00198g

In addition to (bacterio)chlorophylls, (B)Chls, light-harvesting complexes (LHCs) bind carotenoids, and/or their oxygen derivatives, xanthophylls. Xanthophylls/carotenoids have pivotal functions in LHCs: in stabilization of the structure, as accessory light-harvesting pigments and, probably most importantly, in photoprotection. Xanthophylls are assumed to be involved in the not yet fully understood mechanism of energy-dependent (qE) non-photochemical quenching of Chl fluorescence (NPQ) in higher plants and algae. The so called "xanthophyll cycle" appears to be crucial in this regard. The molecular mechanism(s) of xanthophyll involvement in qE/NPQ have not been established, yet. Moreover, excitation energy transfer (EET) processes involving carotenoids are also difficult to study, due to the fact that transitions between the ground state (S_0, $1^1A_g^-$) and the lowest excited singlet state (S_1, $2^1A_g^-$) of carotenoids are optically one-photon forbidden ("dark"). Two-photon excitation spectroscopic techniques have been used for more than two decades to study one-photon forbidden states of carotenoids. In the current study, two-photon excitation profiles of LHCII samples containing different xanthophyll complements were measured in the presumed $1^1A_g^- \rightarrow 2^1A_g^-$ ($S_0 \rightarrow S_1$) transition spectral region of the xanthophylls, as well as for isolated chlorophylls *a* and *b* in solution. The results indicate that direct two-photon excitation of Chls in this spectral region is dominant over that by xanthophylls. Implications of the results for proposed mechanism(s) of qE/NPQ will be discussed.

1. Introduction

Photosynthetic pigment–protein complexes, *i.e.*, the photoactive reaction centres (RCs) and light-harvesting complexes (LHCs), bind besides (bacterio)chlorophylls ((B)Chls) also carotenoids (carotenes and/or their oxygen derivatives,

[a] *Institut für Physik und Astronomie, Universität Potsdam, D-14476 Potsdam-Golm, Germany*

[b] *Institut für Biochemie und Biologie, Universität Potsdam, D-14476 Potsdam-Golm, Germany*

† Present address: Sophie-Charlotten-Str. 86, D-14059 Berlin, Germany.

‡ Present address: Charles University, Department of Chemical Physics and Optics, Ke Karlovu 3, 121 16 Prague, Czech Republic. *E-mail:* lokstein@karlov.mff.cuni.cz, Tel: +42 95155 1396.

xanthophylls). The major light-harvesting complex (LHCII) of plants (Fig. 1) binds – in addition to Chls *a* and *b* – a variety of xanthophylls per monomeric subunit of the trimeric complex: two luteins, one neoxanthin, and one violaxanthin (and/or its de-epoxidation products).[1,2]

Carotenoids, including xanthophylls, have pivotal functions in LHCs: in stabilization of the structure, in light-harvesting and in photoprotection.[2] Xanthophylls are assumed to be involved in energy-dependent (qE) non-photochemical quenching (NPQ) of Chl fluorescence.[2] The xanthophyll cycle, the enzymatic interconversion of violaxanthin *via* antheraxanthin into zeaxanthin and back, appears to be crucial in this regard.[3] However, the molecular mechanism(s) of xanthophyll involvement in qE/NPQ have so far not been established.

Carotenoids share similarities with all-trans polyenes which belong to the idealized C_{2h} symmetry point group. Such molecules possess a center of symmetry and electronic transitions have to obey the Laporte selection rule: they require a change in parity for one-photon transitions to be allowed, since the electric dipole transition moment is vanishing if both participating states share the same parity. Transitions, such as g → g or u → u, if appearing at all, are rather weak and, thus, denoted as "optically forbidden" or "dark". Complementary, if the transition simultaneously involves two photons, it is only allowed if the parity does not change. For carotenoids both the ground (S_0) and the first excited (S_1) singlet state are of the same symmetry, A_g^-, and, hence, S_0–S_1 transitions have to comply with the Laporte rule. Excitation energy transfer (EET) processes involving the lowest excited electronic states (S_1) of carotenoids and (B)Chls are, therefore, difficult to study with conventional (one-photon) spectroscopic techniques, but it

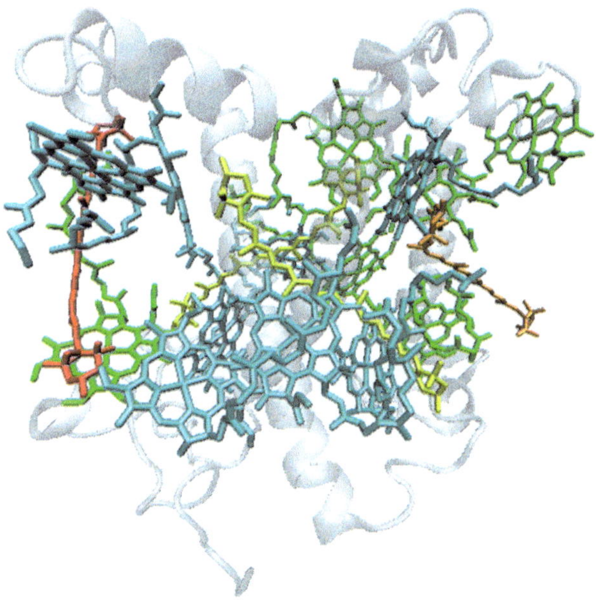

Fig. 1 Structure of LHCII (monomeric subunit). Chls *a* are in cyan and Chls *b* in green. The xanthophylls are displayed as follows: luteins – yellow; neoxanthin – orange; violaxanthin (or its de-epoxidation products – red).

is commonly assumed that they can be investigated by means of (simultaneous) two-photon excitation spectroscopy (TPE).

Two-photon excitation (TPE) spectroscopic techniques have extensively been used previously to study the energetic position and processes involving the one-photon forbidden S_1 ($2^1A_g^-$) states of carotenoids and xanthophylls.[4–11] These (simultaneous) TPE techniques rely on excitation with short laser pulses of in the order of 100 fs duration, tunable in the near infrared region. The effect – simultaneous two-photon absorption – has to be distinguished from step-wise (resonant) two-photon absorption, *via* the S_1 states of (B)Chls to higher excited states, which can occur under similar excitation conditions in the ~650 to 700 nm spectral region.[12]

It is generally assumed that TPE in the ~1000 to 1500 nm spectral region is exclusively (or at least to a large extent) due to transitions to the "dark" S_1 states of the respective carotenoids.[4–11] Subsequent EET to the S_1 state of adjacent (B)Chls is assumed, and the intensity of the resultant (B)Chl fluorescence is being monitored. These processes are schematically depicted in Fig. 2. A sizeable direct TPE of (B)Chls is usually not assumed to occur in the respective spectral region.[4–11] Employing TPE spectroscopy, Krueger *et al.* reported the observation of the S_1 state of spheroidene in the purple bacterial peripheral light-harvesting complex, LH2.[5] Even EET from "hot" S_1 states of carotenoids onto the S_1 state(s) of Chls *a* was suggested to occur in LHCII based on TPE experiments.[8] Moreover, using a sophisticated combination of TPE and pulse-amplitude modulated fluorescence measurements, a mechanism for NPQ based on changes in excitonic interactions between certain Chls and xanthophylls was proposed.[11]

However, a study comparing carotenoid-less and carotenoid-containing purple bacterial LH2 complexes strongly suggested that at least in some purple bacterial antenna complexes a significant proportion of TPE is due to direct excitation of BChls, and not to carotenoids.[13] Subsequent further TPE studies with LH2 complexes corroborated and further extended the findings of Krikunova and co-workers,[13] and suggested that TPE in purple bacterial LH2 complexes may be essentially due to excitonic states of strongly coupled BChl molecules.[14,15] Moreover, a very recent TPE study of core-antenna complexes (LH1-RC) from the purple bacterium *Rhodospirillum rubrum* and their subunit B820 suggested that the

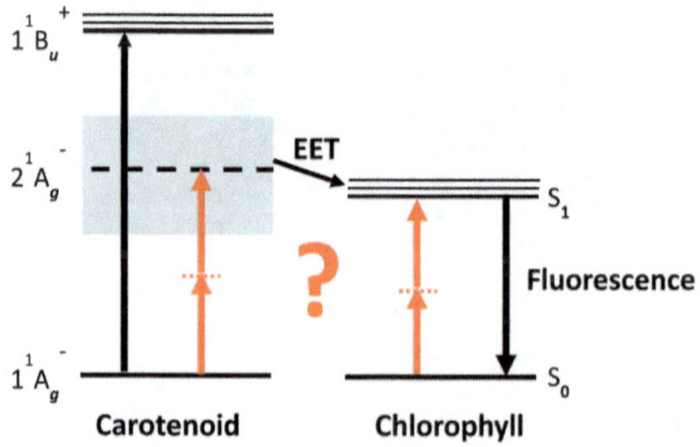

Fig. 2 Energy level schemes to explain two-photon excited (TPE) fluorescence of LHCII. Red arrows indicate possible TPE events, and fluorescence emanates from Chls *a* of LHCII, either following EET from xanthophylls, or upon direct TPE of Chls.

"dark" states addressed by TPE in the ~675 nm range might be due to coupling of charge-transfer states to excitonic states.[16] Unluckily, TPE experiments comparing carotenoid-containing and carotenoid-less samples cannot be performed with Chl-containing LHCs, since these complexes are not stable in the absence of carotenoids. However, the question remains to what extent direct Chl TPE occurs in Chl-containing photosynthetic pigment–protein complexes, including LHCs, RCs and entire photosystems.

To address this long-standing question, TPE profiles of LHCII samples containing different xanthophyll complements were measured in the presumed $1^1A_g^- \rightarrow 2^1A_g^-$ ($S_0 \rightarrow S_1$) transition region of the xanthophylls. Additionally, TPE profiles of Chls *a* and *b* in solution were measured. The results indicate that two-photon absorption by Chls in the spectral region ~1000 to 1400 nm (corresponding to a 500–700 nm one-photon excitation range) is vastly dominant over that of carotenoids/xanthophylls.

2. Experimental

2.1. Biological materials

LHCII samples containing different xanthophyll complements were obtained as follows: first, LHCII from dark-adapted spinach leaves was isolated according to the procedure described in ref. 17. Zeaxanthin-enriched spinach thylakoids were obtained according to a procedure described previously[18] with some modifications: spinach thylakoids were suspended at ~300 μg Chl *a*+*b* in 50 mM MES buffer, pH 5.0, containing 50 mM ascorbate, for 2 h. Subsequently, LHCII was isolated as above,[17] except that the Triton-X 100 concentration was increased to 0.8% instead of 0.7% for initial solubilisation of the thylakoids. Additionally, monomeric LHCII was obtained from the lutein-deficient *lut2*-mutant of *Arabidopsis thaliana*,[19] using sucrose density gradient ultracentrifugation. Chls *a* and *b* were prepared freshly from LHCII by HPLC, according to a previously published protocol.[20] TPE profiles of the thus obtained pure Chls *a* and *b* fractions were measured immediately thereafter. Water soluble chlorophyll protein (WSCP) from *Lepidium virginicum*[21] was received as a kind gift from Professor A. Uchida, Toho University, Tokyo, Japan.

2.2. Experimental setup

For TPE measurements, a femtosecond near-infrared tunable laser system was used. It comprised a Ti:Sapphire femtosecond regenerative amplifier (Coherent Legend Elite) that was seeded by a Coherent Mira 900 femtosecond oscillator with 800 nm at 1 kHz constant repetition rate. The regenerative laser amplifier provided a 1.2 W power optical pump to feed a wavelength tunable optical parametric amplifier (Coherent OperA Solo). For the generation of a near infrared pulsed excitation laser beam the parametrical amplified signal wave was used within the range of 1100–1450 nm, and the second harmonic of the idler wave within 900–1100 nm wavelength range. Possibly interfering residues of the complementary downconverted as well as fundamental wave have been filtered out using dichroitic mirrors. An achromatic half wave plate in combination with a Glan laser polarizer served as a variable laser attenuator enabling the pulse energy to be set at a certain level (approximately 0.5 μJ for all excitation wavelengths) during the TPE measurements. The fluorescence yield showed a quadratic dependence on laser intensity over the entire wavelength range (see Fig. 3A). The excitation laser beam was focused onto the samples in a rotating cuvette.

The chlorophyll fluorescence was collected by an aspheric lens and coupled into a Czerny–Turner monochromator that selectively detected it at 685 nm using a photomultiplier tube (PMT). A bandpass filter in front of the monochromator entrance slit (Semrock FF01-685/40-25) in combination with a Schott KG-shortpass filter additionally suppressed possible stray light. Fig. 3B shows a typical fluorescence spectrum that was generated by two-photon excitation of LHCII.

The PMT and photodiode signals were recorded using an 8 channel sample and hold PC card (Becker & Hickl SHM-180) at 1 kHz. That allowed for accounting on small pulse-to-pulse energy fluctuations before averaging the fluorescence signal. The whole setup was computer operated and semi automatized in order to speed up the TPE spectra measurements. Linear absorption spectra were measured before and remeasured following the two-photon excitation measurements in order to exclude sample degradation.

To obtain the TPE spectra, fluorescence yields have to be compared at the same level of photon flux density for all excitation wavelengths. As the photon flux density sensitively depends on wavelength, beam spot diameter, pulse duration and pulse energy fluctuations, all these laser parameters had to be carefully determined. A Hamamatsu TG-Series NIR-Mini spectrometer allowed for precise laser wavelength determination. The pulse energy was measured by means of an InGaAsP integrated photo diode module (Becker & Hickl PDI 400) that was calibrated prior to the TPE measurement using an Ophir Nova II pulse energy meter.

Laser pulse widths, $\tau(\lambda)$, as full widths at half-maximum (FWHM), were extracted from an interferometric autocorrelation trace measured with a self-built Mach–Zehnder interferometer. A nonlinear signal for the autocorrelation trace was obtained from silicon (above 1350 nm) and InAsP (below 1350 nm) photo diodes within their respective nonlinear spectral response ranges (requiring photon energies below the material bandgap).

The determination of the laser beam width at the sample position was carried out using a knife-edge technique with a home built setup. The laser beam was found to be slightly elliptical with principal axes along x and y. A blade has been moved along the x and y directions and the beam diameters $\Delta x(\lambda)$ and $\Delta y(\lambda)$ were extracted by fitting the transmissive laser signal data with the complementary error function: $\frac{1}{2}\,\mathrm{erfc}(\sqrt{2}\,(r - r_0)/w)$ in which $r = x, y$ and $2w = \Delta x$ and Δy.

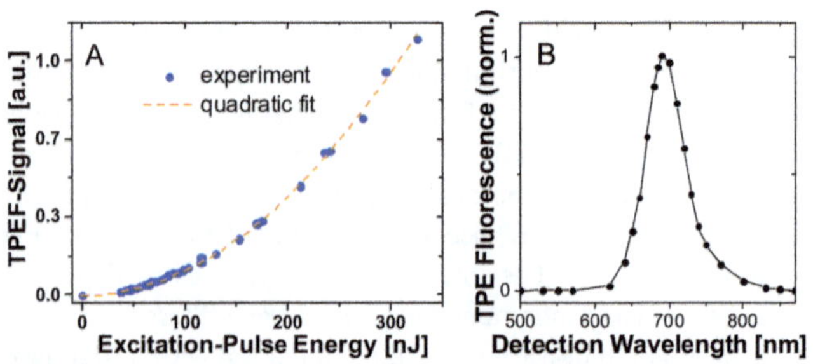

Fig. 3 (A) Intensity dependence of the two-photon excited (TPE) fluorescence signal. (B) Fluorescence emission spectrum of LHCII, obtained upon two-photon excitation.

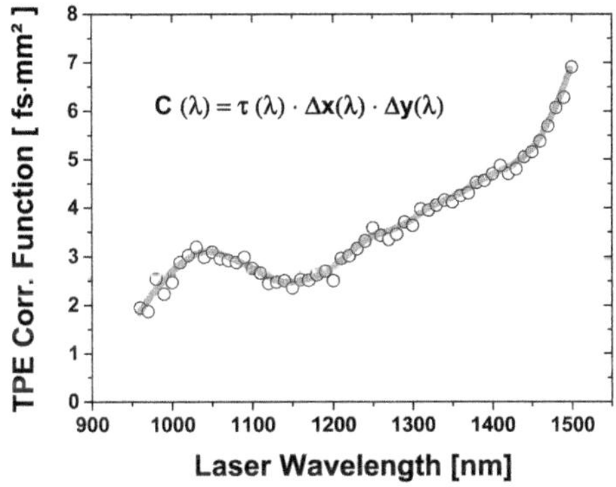

Fig. 4 Correction function for the two-photon excited (TPE) fluorescence signal at different excitation wavelengths.

The photon flux density ϕ_{photon} of the laser pulses that are penetrating the sample in the beam focus can be determined as: $\phi_{photon} = E_{pulse}(\lambda)\cdot\lambda/(h\cdot c\cdot\pi\cdot\Delta x(\lambda)\cdot\Delta y(\lambda)\cdot\tau(\lambda))$. According to previous reports[22] the fluorescence raw data F^{raw} have to be multiplied by a correction function $C(\lambda) = \Delta x(\lambda)\cdot\Delta y(\lambda)\cdot\tau(\lambda)$ and $1/(\lambda\cdot E_{pulse}(\lambda))^2$ in order to receive a corrected fluorescence signal S^{TPE} that is proportional to the two-photon absorption cross section of the sample in solution:

$$S^{TPE} = F^{raw}C(\lambda)\cdot 1/(\lambda\cdot E_{pulse}(\lambda))^2\alpha\sigma_2(\lambda).$$

The correction function $C(\lambda)$, which comprises all wavelength dependencies of the laser parameters is shown in Fig. 4. According to the course of $C(\lambda)$ one would incur an error of more than 300% within the respective spectral range without considering these wavelength dependent effects. For all samples, TPE profiles were measured in the presumed xanthophyll $S_0 \rightarrow S_1$ absorption region, from 1000 to 1400 nm (corresponding to a one-photon scale of 500 to 700 nm).

2.3. Two-photon polarization ratio measurements

The two-photon polarization ratio (Ω) was measured by introducing an achromatic quarter-wave plate into the excitation laser beam, directly in front of the focussing lens. By rotation of the angle between the wave plates fast axis and the electric field vector of the incident linear polarized beam it was possible to set the beam to the desired elliptical polarization. In particular, it was feasible to change the beam polarization from linear to circular and back by 45° rotation of the wave plate around the optical axis. The two-photon polarization ratio, $\Omega = F(\text{circ,circ})/F(\text{lin,lin})$, was then obtained from the two photon excited fluorescence yields: $F(\text{circ,circ})$ and $F(\text{lin,lin})$, recorded for circular (circ) and linear (lin) co-polarized excitation, respectively, under otherwise identical measuring conditions.

3. Results and discussion

3.1. Two-photon excitation measurements with LHCII

Initially, the TPE profile of dark-adapted LHCII (containing only violaxanthin as the sole xanthophyll cycle pigment) was measured. The TPE profile (Fig. 5A, lower panel) was consistent with previously measured ones, except probably for a somewhat better resolution.[7] The TPE curve features several peaks in a spectral region where the linear absorption of LHCII (Fig. 5A, upper panel) is low and rather unstructured. This has previously been taken as an indicator that the TPE profiles are more or less exclusively originating from carotenoid contributions. In the current study, additional TPE profiles were obtained for zeaxanthin-enriched LHCII, as well as LHCII from the *lut2*-mutant. Within the range of error (on the order of 1%) the TPE spectra are virtually indistinguishable (Fig. 5A, lower panel). Obviously, alteration of the xanthophyll contents does not have any impact on the TPE spectra of LHCII. This observation suggests that the contribution of different xanthophylls to the TPE profiles may be much less pronounced than previously assumed. Moreover, significant direct TPE of the Chls in LHCII can thus be inferred.

To verify the two-photon excitation origin of the observed LHCII fluorescence emission, the dependence of the fluorescence signal on excitation pulse intensity was measured (Fig. 3A). The fluorescence signal intensity follows a strict quadratic dependence over the entire intensity range as well as for all excitation wavelengths, indicating a two-photon excitation origin of the measured signals. Additionally, the fluorescence spectrum, generated by TPE of dark-adapted LHCII, was measured (Fig. 3B). The obtained spectral profile with an emission maximum at ~680 nm is typical for LHCII, being proof that the monitored fluorescence signal indeed emanates from LHCII-bound Chls *a*.

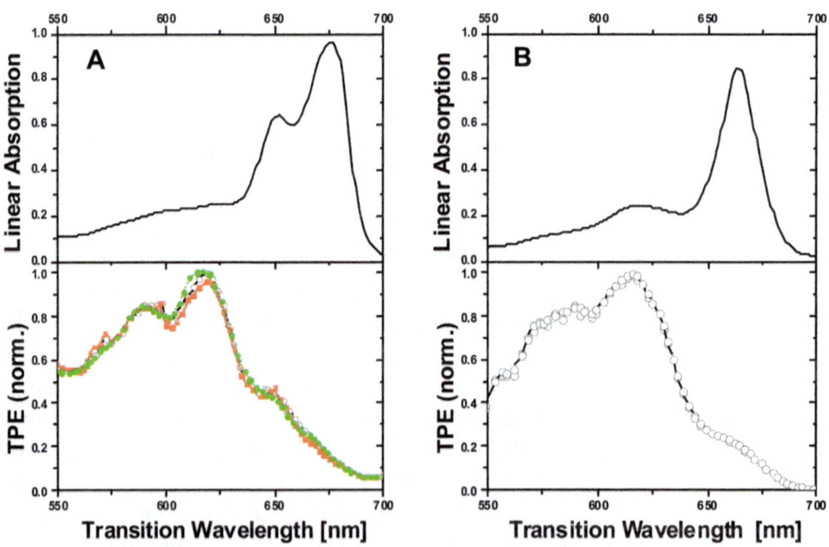

Fig. 5 Linear (one-photon) absorption spectra (upper panels) and TPE spectra (lower panels) of LHCII samples with different xanthophyll contents (A) and of WSCP from *Lepidium virginicum* (B). In (A), lower panel, the black curve refers to LHCII from dark-adapted spinach, the red curve to zeaxanthin-enriched LHCII, and the green curve to LHCII from the lutein-deficient *lut2*-mutant of *Arabidopsis thaliana*. In (A), upper panel, only the linear absorption spectrum for dark-adapted LHCII is shown (the different LHCII samples were almost indistinguishable).

3.2. Two-photon excitation measurements with isolated chlorophylls in solution

To assess the direct contributions of Chls *a* and *b* to TPE of LHC II, freshly prepared solutions of these Chls were measured. In Fig. 6 the one-photon (linear absorption, upper panels) and TPE spectra (lower panels) of Chls *a* (A) and *b* (B) are displayed. Both the one-photon and TPE spectra are shown on a transition wavelength scale (*i.e.*, half the excitation laser wavelength in the case of TPE) for direct comparison.

Chls *a* and *b* have a molecular structure comprising a chlorin macrocycle. Thus, they share some basic symmetry properties with the C_{2v} point group. They do not possess a pronounced center of symmetry and, hence, do not conform to the Laporte selection rule, but would rather show non-vanishing transition dipole strengths for both one- and two-photon transitions. The measured TPE spectra, however, appear to be remarkably different to the one-photon absorption spectra of the respective Chls upon first inspection (Fig. 6). Taking a closer look, most of the bands in the one-photon spectra appear also in the TPE spectra, but with rather different amplitudes. Those bands which have a large amplitude in the one-photon (linear absorption) spectra are low in the TPE spectra – and *vice versa*. In the TPE spectra the higher vibrationally excited Q_y states are more pronounced, while in one-photon (linear) absorption $S_0 \rightarrow S_{1,0}$ is the strongest transition. This phenomenon seems not to be unusual for tetrapyrrolic compounds, such as Chls[23] and is, to the best of our knowledge, not yet fully understood. As a possible explanation, a so called "vibrational intensity borrowing" has been suggested. This is caused by coupling between the involved electronic and vibrational degrees of freedom. The phenomenon can be theoretically explained within the framework of the Herzberg–Teller approximation.[24] A thorough discussion of this

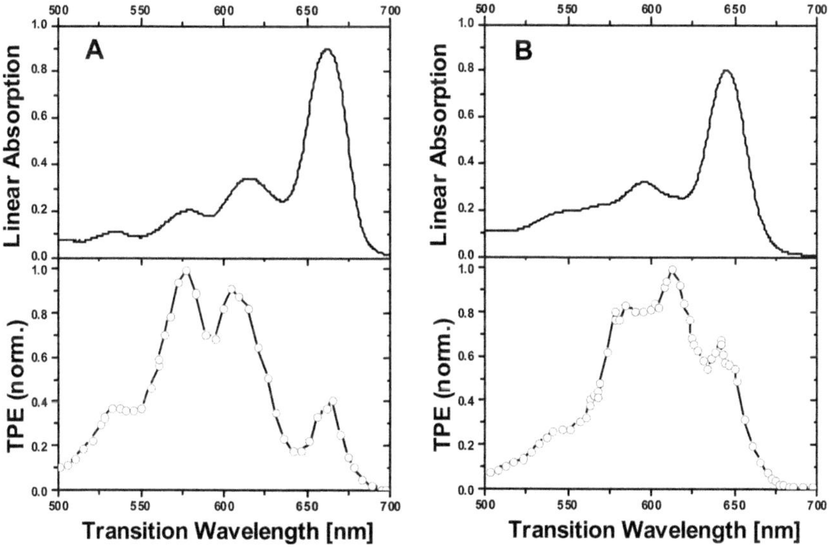

Fig. 6 Linear (one-photon) absorption spectra (upper panels) and TPE spectra (lower panels) of Chls *a* (A) and *b* (B) in acetonitrile/ethyl acetate/water.

phenomenon and related effects in tetrapyrrolic compounds is beyond the scope of the present paper and can be found elsewhere.[23,25,26]

To further quantify the contributions of Chls *a* and *b* to the TPE spectra of LHCII, a linear combination of the Chl *a* and *b* TPE spectra in solution was compared to the TPE spectra of LHCII. Interestingly, this linear combination of the Chl *a* and *b* TPE spectra (together with a slight red shift) appears to be sufficient to reconstruct the LHCII spectra (Fig. 7). No further contribution of other components (*i.e.* xanthophylls spectra) would be required.

WSCP from *Lepidium virginicum*[21] is a Chl *a*/*b*-binding protein of still not fully understood function.[21,27] Most remarkably, WSCP does not contain carotenoids.[21] However, TPE profiles of WSCP strongly resemble those of LHCII (Fig. 5, lower panel, B). Minor differences in the intensities of the TPE band structure can be explained by the lower Chl *b* contents of WSCP as compared to LHCII. The strong similarity between the TPE profiles of LHCII and WSCP is further evidence that a contribution of carotenoids/xanthophylls to the TPE of LHCII is negligible (Fig. 5, lower panels).

3.3. The two-photon polarization ratio Ω

As compared to linear absorption, in two-photon absorption the polarization of the exciting light matters, even for randomly oriented molecules.[28] TPE was recorded for linear (lin) and circular (circ) co-polarized excitation with otherwise identical measuring conditions, leading to different fluorescence yields, F(lin,lin) and F(circ,circ), respectively (Fig. 8A).

Thus, the two-photon polarization ratio $\Omega = F(\text{circ,circ})/F(\text{lin,lin})$ can be used as an indicator for the symmetry of the molecules under investigation.[28,29] For carotenoids (as well as for C_{2h} symmetric linear long-chain polyenes in general) $1^1A_g^- \rightarrow 2^1A_g^-$ ($S_0 \rightarrow S_1$) transitions, a value below Ω ~0.8 would be expected.[29] Our polarization resolved measurements with LHCII yielded Ω values between 1 and 1.2 for the spectral range between 610 and 690 nm, and slightly less than 1 (but very close to 1) below 600 nm (Fig. 8B). This would suggest that the TPE excited molecules have, unlike carotenoids, a molecular symmetry different from C_{2h}. Furthermore, Ω has also been measured in the range between 660 to 680 nm

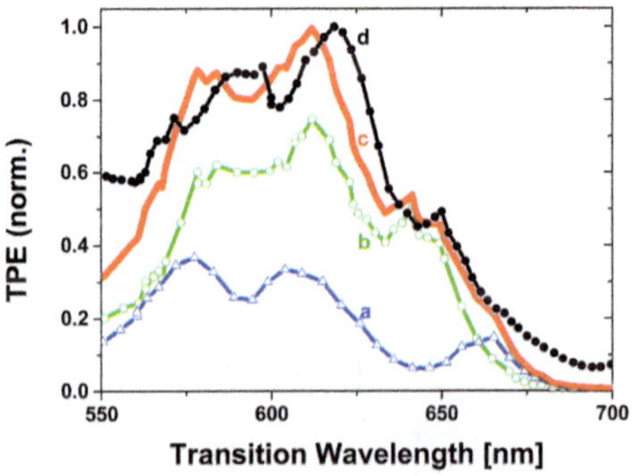

Fig. 7 TPE spectra of Chls *a* and *b* and their linear combination, in comparison to the TPE spectrum of LHCII. Blue (a) Chl *a*, green (b) Chl *b*, red (c) linear combination of TPE spectra of Chls *a* and *b*, and black (d) LHCII.

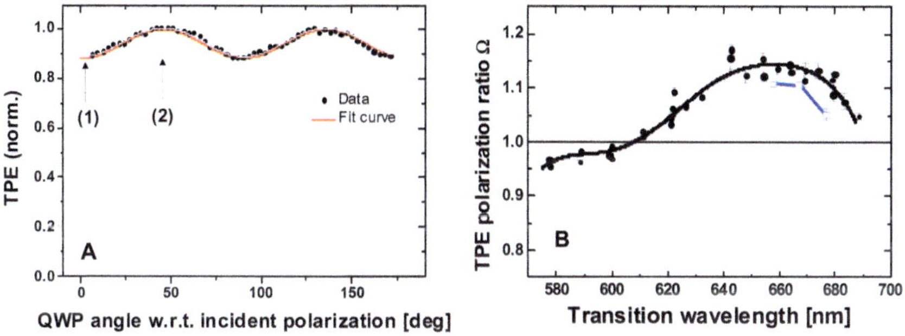

Fig. 8 (A) Polarization dependent variation of LHCII-TPE upon excitation at 1330 nm. Quarter wave plate angle positions leading to linear and circular polarized excitations are indicated by (1) and (2), respectively. (B) TPE polarization ratio Ω for LHCII (black circles) in comparison to the one for Chl *a* (open blue squares) in the spectral regions of 580 to 690 nm and 660 to 680 nm, respectively.

for Chl *a* and was found to be comparable to the Ω-values of LHCII under the same conditions (Fig. 8B). This, in turn, is further and independent evidence that the molecules, of which the fluorescence has been two-photon-excited, are rather not xanthophylls, but Chls.

The presented results are consistent with previous TPE experiments comparing carotenoid-containing and carotenoid-less light-harvesting complexes (LH2) from *Allochromatium minutissimum*. These earlier experiments suggested prevailing direct TPE of BChls in the spectral region where the carotenoid S_1 state was expected.[13] Further TPE studies comparing carotenoid-containing and carotenoid-less LH2 samples similarly came to the conclusion that excitonic states of BChls and/or mixing of excitonic states with charge transfer states of BChls in LH1-RC were the origin of TPE in the near infrared spectral region, rather than carotenoid dark states.[14–16] These experimental results are also corroborated by recent theoretical investigations which suggest significant two-photon absorption by (B)Chls.[30]

However, our results do not rule out the possibility of TPE by carotenoid "dark" states, including S_1, at all. In fact, TPE of isolated carotenoids (*e.g.*, in solution) can be readily achieved and detected using the well-known excited state absorption from S_1.[8] Nevertheless, our results clearly indicate that direct TPE of (B)Chls in the respective spectral region appears to be at least about two orders of magnitude stronger (determined by the error levels of the measurements) than that of carotenoids, when TPE is monitored by (B)Chl fluorescence. This, in turn, suggests that fluorescence-detected TPE spectroscopy is not a suitable technique to investigate carotenoid "dark" states in (photosynthetic) pigment–protein complexes that contain (B)Chls. However, TPE techniques have proven to be very useful for characterizing optically "dark" excitonic and charge transfer states of (B)Chls, that cannot be properly investigated otherwise in photosynthetic pigment–protein complexes.[14–16]

Moreover, the results also bear important implications on the interpretation of previously published TPE data and the proposed mechanism(s) of xanthophyll involvement in qE/NPQ[11] and EET between carotenoids and (B)Chls.[8] Apparently the interpretations of these previous studies will have to be re-assessed.

We have to emphasize that we do not claim that the mechanism for NPQ, as proposed by Walla and co-workers,[11] would be disproved by the results of the current study. In fact, previous work, using non-linear polarization spectroscopy in the frequency domain (NLPF),[31] resulted in rather similar conclusions as suggested by Bode and co-workers.[11] NLPF has been shown previously to be a uniquely suited technique to investigate carotenoid–Chl interactions, *e.g.*, in the peridinin-Chl *a*-protein (PCP).[32]

NLPF spectra of LHCII obtained by tuning the pump laser beam across the Chl *a*/*b* Q_y-region and probing at selected wavelengths in the xanthophyll $2^1B_u^+$ (S_2) absorption region clearly indicated dramatic changes in xanthophyll–Chl interactions upon altering the aggregation state (and, consequently, the fluorescence yield) of LHCII, supposed to be related to NPQ.[31] Obviously a structural change occurs: a xanthophyll, most likely the violaxanthin in the respective binding pocket, seems to undergo a transition from close interaction with a Chl *b* (with an absorption maximum at 650 nm) to interaction with Chl *a* (with long wavelength absorption at about 682 nm) which appears to be part of the so-called terminal emitter cluster of Chls in LHCII.[33]

4. Conclusions

TPE profiles of LHCII samples containing different xanthophyll complements were measured in the presumed $1^1A_g^- \rightarrow 2^1A_g^-$ ($S_0 \rightarrow S_1$) transition region of xanthophylls. Additionally, TPE profiles of Chls *a* and *b* in solution and WSCP, which does not contain carotenoids, were measured. The results indicate that two-photon absorption by Chls in this spectral region is dominant over that of carotenoids, with negligible contributions of the latter. These results suggest the re-evaluation and, possibly, re-interpretation of previously published TPE data obtained with photosynthetic pigment–protein complexes containing (B)Chls and carotenoids.

Conflicts of interest

There are no conflicts to declare.

Acknowledgements

Long-term financial support by the Deutsche Forschungsgemeinschaft (SFB 429) as well as by the Czech Science Foundation (project P501/12/G055) is gratefully acknowledged. We would like to thank Prof. A. Uchida, Toho University, Tokyo, Japan, for the kind gift of WSCP.

References

1 Z. Liu, H. Yan, K. Wang, T. Kuang, J. Zhang, L. Gui, X. An and W. Chang, *Nature*, 2004, **428**, 287–292.

2 J. Standfuss, A. C. Terwischa van Scheltinga, M. Lamborghini and W. Kühlbrandt, *EMBO J.*, 2004, **24**, 919–928.

3 P. Jahns and A. R. Holzwarth, *Biochim. Biophys. Acta*, 2011, **1817**, 182–193.

4 A. P. Shreve, J. K. Trautman, T. G. Owens and A. C. Albrecht, *Chem. Phys. Lett.*, 1990, **170**, 51–56.
5 B. P. Krueger, J. Yom, P. J. Walla and G. R. Fleming, *Chem. Phys. Lett.*, 1998, **310**, 57–64.
6 P. J. Walla, P. A. Linden, C. P Hsu, G. D. Scholes and G. R. Fleming, *Proc. Natl. Acad. Sci. U. S. A.*, 2000, **97**, 10808–10813.
7 P. J. Walla, J. Yom, B. P. Krueger and G. R. Fleming, *J. Phys. Chem. B*, 2000, **104**, 4799–4806.
8 P. J. Walla, P. A. Linden, K. Ohta and G. R. Fleming, *J. Phys. Chem. A*, 2002, **106**, 1909–1916.
9 A. Wehling and P. J. Walla, *J. Phys. Chem. B*, 2005, **109**, 24510–24516.
10 M. Hilbert, A. Wehling, E. Schlodder and P. J. Walla, *J. Phys. Chem. B*, 2004, **108**, 13022–13030.
11 S. Bode, C. C. Quentmeier, P.-N. Liao, N. Hafi, T. Barros, L. Wilk, F. Bittner and P. J. Walla, *Proc. Natl. Acad. Sci. U. S. A.*, 2009, **106**, 12311–12316.
12 D. Leupold, K. Teuchner, J. Ehlert, K. D. Irrgang, G. Renger and H. Lokstein, *Biophys. J.*, 2002, **82**, 1580–1585.
13 M. Krikunova, A. Kummrow, B. Voigt, M. Rini, H. Lokstein, A. Moskalenko, H. Scheer, A. Razjivin and D. Leupold, *FEBS Lett.*, 2002, **528**, 227–229.
14 I. Stepanenko, V. Kompanetz, Z. Makhneva, S. Chekalin, A. Moskalenko and A. Razjivin, *J. Phys. Chem. B*, 2009, **113**, 11720–11723.
15 I. Stepanenko, V. Kompanetz, Z. Makhneva, S. Chekalin, A. Moskalenko and A. Razjivin, *J. Phys. Chem. B*, 2012, **116**, 2886–2890.
16 A. Razjivin, A. Solovév, V. Kompanets, S. Chekalin, A. Moskalenko and H. Lokstein, *Photosynth. Res.*, 2019, DOI: 10.1007/s11120-018-0602-0.
17 Z. Krupa, N. P. A. Huner, J. P. Williams, E. Maissan and D. James, *Plant Physiol.*, 1987, **84**, 19–24.
18 A. Färber, A. J. Young, A. V. Ruban, P. Horton and P. Jahns, *Plant Physiol.*, 1997, **115**, 1609–1618.
19 H. Lokstein, L. Tian, J. Polle and D. DellaPenna, *Biochim. Biophys. Acta*, 2002, **1553**, 93–103.
20 A. V. Shpilyov, V. V. Zinchenko, S. V. Shestakov, B. Grimm and H. Lokstein, *Biochim. Biophys. Acta*, 2005, **1706**, 195–203.
21 D. Horigome, H. Satoh, N. Itoh, K. Mitsunaga, I. Oonishi, A. Nakagawa and A. Uchida, *J. Biol. Chem.*, 2007, **273**, 6525–6531.
22 N. S. Makarov, M. Drobizhev and A. Rebane , *Opt. Express* , 2008 , **16** , 4029–4047.
23 A. Karotki, M. Drobizhev, M. Kruk, C. Spangler, E. Nickel, N. Mamardashvili and A. Rebane, *J. Opt. Soc. Am. B*, 2003, **20**, 321–332.
24 G. Herzberg and E. Teller, *Z. Phys. Chem., Abt. B*, 1933, **21**, 410–446.
25 J. A. Greco, S. Shima, N. L. Wagner, J. R. McCarthy, K. Atticks, C. Brückner and R. R. Birge, *J. Phys. Chem. C*, 2015, **119**, 3711–3724.
26 J. L. Hughes, B. Conlon, T. Wydrzynski and E. Krausz, *Phys. Procedia*, 2010, **3**, 1591–1599.
27 S. Damaraju, S. Schlede, U. Eckhardt, H. Lokstein and B. Grimm, *J. Plant Physiol.*, 2011, **168**, 1444–1451.
28 W. M. McClain, *J. Chem. Phys.*, 1971, **55**, 2789–2796.
29 R. R. Birge and B. M. Pierce, *J. Chem. Phys.*, 1979, **70**, 168–178.

30 A. Graczyk, J. M. Żurek and M. J. Paterson, *Photochem. Photobiol. Sci.*, 2014, **13**, 103–111.

31 B. Voigt, M. Krikunova and H. Lokstein, *Photosynth. Res.*, 2008, **95**, 317–325.

32 M. Krikunova, H. Lokstein, D. Leupold, R. G. Hiller and B. Voigt, *Biophys. J.*, 2006, **90**, 261–271.

33 H. Rogl, R. Schödel, H. Lokstein, W. Kühlbrandt and A. Schubert, *Biochemistry*, 2002, **41**, 2281–2287.

Faraday Discussions

PAPER

Role of charge transfer states into the formation of cyclobutane pyrimidine dimers in DNA

Wook Lee [a] and Spiridoula Matsika *[b]

Received 17th November 2018, Accepted 5th December 2018
DOI: 10.1039/c8fd00184g

Photoinduced charge transfer between neighboring bases plays an important role in DNA. One of its important effects is shown in its ability to affect the photochemical yields of the formation of cyclobutane pyrimidine dimer (CPD) products between adjacent pyrimidine bases. In this work we examine how the energies of charge transfer states depend on the sequences of oligonucleotides using a hybrid quantum and molecular mechanics (QM/MM) methodology combined with the algebraic diagrammatic construction through a second order electronic structure method for excited states. Specifically, we examine 10 sequences with guanine being on the 5′ or 3′ position of two pyrimidine bases. The results show that the energies of charge transfer states are affected by the nature of the donor acceptor pair, by the distance between them, and by other electrostatic effects created by the surrounding environment.

1 Introduction

Even though DNA is quite photostable, photochemical reactions can occur with low yield.[1,2] The most common photochemical product after absorption of UV radiation by DNA is the formation of cyclobutane pyrimidine dimers (CPDs), formed by cycloaddition [2 + 2] between the C5–C6 double bonds of two pyrimidines.[1] Although all pyrimidine dimers can form these products, it has been determined that Thymine–Thymine (TT) is the most photoreactive sequence followed by Thymine–Cytosine (TC), whereas lower amounts of damage are produced at Cytosine–Thymine (CT) and Cytosine–Cytosine (CC) sites.[3]

The quantum yield of thymine dimer formation is significantly influenced by adjacent bases, and a correlation between the degree of quenching and the oxidation potential of the flanking base has been observed experimentally.[4–7] A theoretical explanation for these observations was provided by our previous work, which demonstrated that electron transfer from the flanking base to thymine,

[a]Department of Chemistry, Pohang University of Science and Technology (POSTECH), Pohang 37673, Korea
[b]Department of Chemistry, Temple University, Philadelphia, PA 19122, USA. E-mail: smatsika@temple.edu; Tel: +1 215 204 7703

facilitated by a charge transfer state between these two bases, can provide a decay pathway for the population to escape from dimer formation, which eventually leads to the formation of an exciplex.[8]

In addition to the quenching effect, it was also found that, in some cases, the charge transfer states can have an opposite effect and actually enhance the reactivity. This pathway also involves the formation of an exciplex which facilitates intersystem crossings. In this case CPDs are formed through triplet states.[9] This mechanism is not expected to be widely operative. Its relevance depends on the characteristics of the exciplex formed through the charge transfer states. Specifically, the exciplex should lead to diradicals with two radical centers separated by relatively large distances, and, as a result, the singlet and triplet states are degenerate, facilitating intersystem crossing. Formation of CPDs through this mechanism occurs through sequential steps and their existence is supported by the experimental detection of triplet intermediates.[10]

Overall, the importance of electron transfer in the reactivity of oligonucleotides has been highlighted in these previous studies. In the singlet mechanism, charge transfer states quench CPD formation while in the triplet mechanism they enhance it. Examining how sequence affects the charge transfer states is crucial in order to understand reactivity, and this is the scope of the current work. We will only examine guanine as a flanking base because it has the lowest oxidation potential among the four standard nucleic acid bases, allowing easy access to the charge transfer states, and was shown to have the greatest quenching effect. We will focus on three effects: initially, we will examine how guanine affects the formation of CPDs in the different pyrimidine dimers, TT, CT, TC, and CC. Subsequently, we will examine the effect of placing guanine on the 5′ position or the 3′ position in the sequence. Finally, the effect of methylation of cytosine on charge transfer states will be examined.

2 Computational details

A total of 10 model systems, where the core sequences were inserted in the middle of duplex DNA (18-mer for the TTG sequence and 17-mer for the rest), were created (see Fig. 1). The initial structures of duplex DNA for each model system were built using the NAB module[11] in AMBER Tools Ver. 1.5. The AMBER ff99 parameter set[12] with bsc0 corrections[13] was used for standard nucleic acids and the general AMBER force field (GAFF) parameter set[14] was used for 5-methylcytosine residue. The atomic partial charges for 5-methylcytosine were fitted using the restrained electrostatic potential (RESP) method.[15] Here, the required potentials were computed using HF/6-31G* on the structure optimized with B3LYP/6-31G*, and these calculations were done using Gaussian03.[16] The initial structures were then solvated in a truncated octahedral box filled with TIP3P water molecules[17] in which the minimum distance between any atom in solute and the edge of the box is 8 Å. The system was neutralized by adding Na^+ counter ions. All model systems were minimized for 500 steps to avoid any possible steric crashes, and then the temperature was gradually increased from 0 to 300 K with positional restraints on the solute. After reaching 300 K, the restraints were gradually removed, and equilibration was carried out for 1 ns followed by 2 ns of productive Molecular Dynamics (MD) run. All MD simulations were performed using the sander module of AMBER 11.[18] The temperature was controlled using

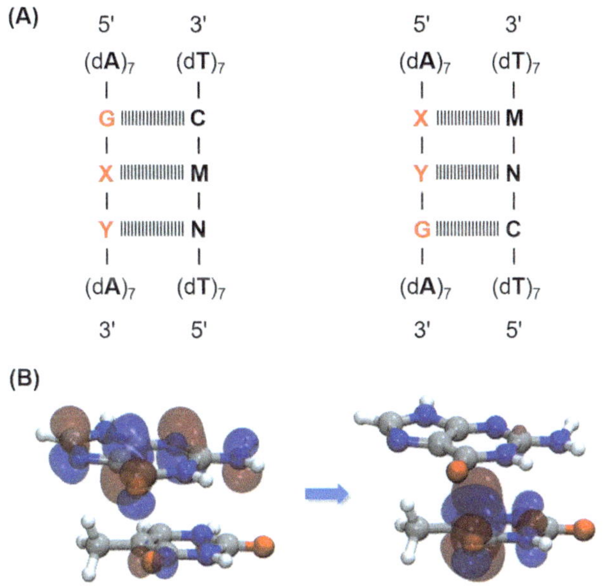

Fig. 1 Schematic representation of the duplex DNA model systems examined in this work (A) and an exemplary natural transition orbital of a charge transfer state (B). X and Y shown in (A) are the sequences of interest and M and N are their complementary bases, respectively.

the Langevin thermostat, and the pressure was kept at 1 bar with Berendsen's weak-coupling algorithm.[19] The SHAKE algorithm was applied on all bonds including hydrogen atoms,[20] and a time step of 2 fs was used. Periodic boundary conditions were employed, and the cut-off distance of 8 Å was adopted for non-bonded interactions. For long-range electrostatic interactions, the particle mesh Ewald summation method was used.[21]

For QM/MM calculations, the Chemshell suite[22] was used to provide an interface between QM and MM calculations. The Turbomole program package[23] and DL-POLY code[24] were used for QM and MM calculations, respectively. The link atom scheme was employed at the boundary of QM and MM,[22] and the charge shift scheme was applied to avoid overpolarization.[25] A total of 20 snapshots separated by an equal interval were extracted from 2 ns of MD simulation trajectory for each model system, and water molecules farther than 25 Å away from the central base of duplex DNA in all snapshots were removed for computational efficiency. All snapshots then underwent ground state energy minimization with BLYP/TZVPP[26–29] in conjunction with Grimme's dispersion correction[30] before vertical excitation energy calculations, and all residues farther than 10 Å away from the central base were frozen during these minimizations. Since we are focusing on only the charge transfer state and also dealing with many sequences, the QM region includes only two bases, namely, the flanking guanine and its neighboring pyrimidine, for computational efficiency. Comparisons of the current results with the previous study where we included three bases[8] shows that the difference of the average of charge transfer states is about 0.1 eV, and it should not affect any of our conclusions.

Vertical excitation energies were computed using the algebraic diagrammatic construction through the second order (ADC(2)) method[31,32] with the def2-SVP basis set.[29] VMD[33] was used to visualize natural transition orbitals[34] and to measure distances.

In order to generate the potential energy profile of TC, geometry optimizations were performed with the constraint on the distance *d* between the middle point of the C5–C6 double bonds of the two pyrimidine bases. After the constrained geometry optimizations with dispersion corrected BLYP/TZVPP at each point, vertical excitation energy calculations with ADC(2)/def2-SVP were followed. Further constrained optimization on the ground state leads to the CPD product. The potential energy profile of TT was adopted from our previous study[8] and redrawn.

3 Results and discussion

3.1 Mechanism for CPD formation

This work is motivated by previous findings that charge transfer states play an important role in the quantum yields of the photochemical formation of CPDs. In previous work we examined the formation of CPDs in TT sequences.[8] It was shown that the lower energy exciton state formed from the ππ* state in thymine is responsible for the formation of CPDs. The reactive state becomes obvious when we plot the excited states as a function of the distance between the double bonds

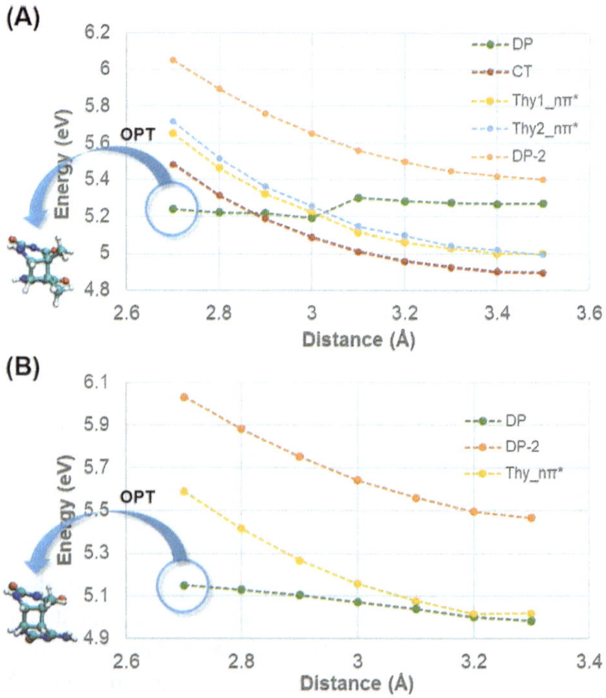

Fig. 2 Potential energy profiles of (A) TT and (B) TC along the distance *d* between the two pyrimidine bases.

in thymine (see Fig. 2A). While all other states are destabilized when the two thymine bases approach each other, one of the exciton states remains stable. In previous work we proved that optimization along that surface eventually leads to the CPD product (depicted in Fig. 2A). Therefore, we will refer to this state as the dimer precursor (DP) state and the other state from excitonic coupling as the DP-2 state. In addition, two $n\pi^*$ excitations of each constituting thymine are denoted as Thy1_$n\pi^*$ and Thy2_$n\pi^*$, and the charge transfer state is denoted as CT in Fig. 2A. In order to check the generality of these arguments we calculated similar potential energy profiles for the TC dimer. Fig. 2B shows the plot for the TC dimer where we observe the same behavior for the reactive state as we saw in TT, *i.e.* the lower exciton state is stable with respect to the distance d between the two pyrimidine bases. This validates our argument that the photochemistry between two pyrimidine bases should proceed through similar pathways, and that the importance we observed of the interplay between the charge transfer and DP states in TT holds true for the other dimers. Several theoretical studies examining the mechanism for CPD formation have been published, ranging from studies of gas phase bases to more complicated oligonucleotides.[35–46] The nature of the state responsible for CPD formation has been debated in the literature. While many studies agree with our conclusions that the lowest exciton is the reactive state[45,46] an alternative explanation has been given, namely that a doubly excited state is responsible.[44] Doubly excited states are too high in energy to be populated directly by absorption, however. Furthermore, when the distance between the two pyrimidine bases becomes short the pathway is the same regardless of the initial state. Doubly excited states cannot be described by single reference methods, such as ADC(2), but a study employing both single and multireference methods agrees that the DP state is responsible for the photoproduct.[45]

Based on our previous work, it is expected that the lower the energy of the charge transfer states the stronger the quenching effect will be. For instance, the quenching will take place if the excited population transfers from DP to CT through non-adiabatic transitions around the area of maximum coupling, which is shown in our 1-dimensional picture in Fig. 2A to be around 2.9 Å where the two states cross. Therefore, after we establish that the DP state should be similar in the different pyrimidine sequences (please note that this work is not examining any inherent differences in the reactivity of the pyrimidine bases), we mainly focus on the energy of charge transfer states, and systematically examine the factors affecting the energetic location of the charge transfer states in the context of various sequences. To this end, we create model systems where the sequence of interest is inserted in the middle of duplex DNA (Fig. 1A) and compute the excitation energies of the charge transfer state in each model system. The sequence of interest is comprised of two pyrimidine bases and one flanking guanine at either the 5′ or 3′ side. Fig. 1B shows the natural transition orbitals describing a charge transfer state and shows that electron density is shifted from the flanking guanine to an adjacent pyrimidine base.

3.2 Factors affecting the energy of charge transfer states

3.2.1 Effect of sequence: GTT, GTC, GCT, GCC. We first examine the four pyrimidine dimers, TT, CT, TC, and CC with guanine placed at the 5′ position. The difference in reactivity between TT and CC has been studied experimentally before for isolated dimers.[39] The intrinsic differences in the reactivity of the pyrimidines are not examined in this work, where we only focus on the charge transfer states between guanine and the pyrimidine next to it.

Fig. 3 shows the energy of charge transfer states for all 20 snapshots for two of the systems we compare in this section (GTT and GCC) as a representative example of how the charge transfer states fluctuate based on the fluctuations of the environment. The charge transfer states depend on the environment and there are strong variations in all cases. The standard deviation is 0.2–0.4 eV for all systems studied in this work. From now on we will use the average energies produced from these snapshots since they represent a more compact way to analyze the results.

The average energies of charge transfer states for all systems are shown in Fig. 4A. The energy of the charge transfer states changes significantly for the four sequences, GTT, GTC, GCT, and GCC. The charge transfer states in GTT have the lowest energy, with an average of 5.1 eV, followed by GTC with an average of 5.4 eV. On the other hand, GCT and GCC both have charge transfer states with average energies at 5.6 eV (5.6335 and 5.6379 eV, respectively). This pattern shows that when guanine is π-stacked with thymine the charge transfer states are stabilized compared to when guanine is π-stacked with cytosine. The main effect is caused by the first neighbors, *i.e.* whether guanine is next to thymine or cytosine. There is however a small secondary effect caused by the second neighboring base. This is evident when comparing GTT to GTC or GCC to GCT. For GTT and GTC for example, guanine is on the 5′ position of thymine, but the difference is the pyrimidine base next to thymine, which is either thymine or cytosine. This secondary effect is <0.3 eV compared to the main effect, which is as large as 0.6 eV.

These results indicate that the quenching of the reactivity to form CPDs should be more important for GTT and GTC but less important for GCC and GCT.

3.2.2 Effect of guanine being on the 5′ *vs.* 3′ position. We next examine the effect of placing guanine on the 5′ *vs.* 3′ position. Fig. 4A shows the average energies of the charge transfer states for five pairs of sequences. In each pair guanine has been placed either on the 5′ or the 3′ position of reactive pyrimidine dimer. In the first two pairs (GTT, TTG, GTC, and CTG), the neighboring pyrimidine next to guanine is always thymine, while it is cytosine in the next two pairs (GCT, TCG, GCC, and CCG). In the last pair, methylated cytosine is adjacent to

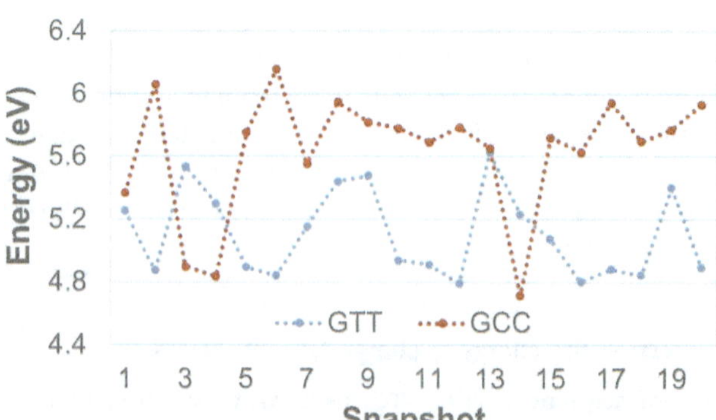

Fig. 3 Energy of charge transfer states for GTT and GCC for 20 snapshots.

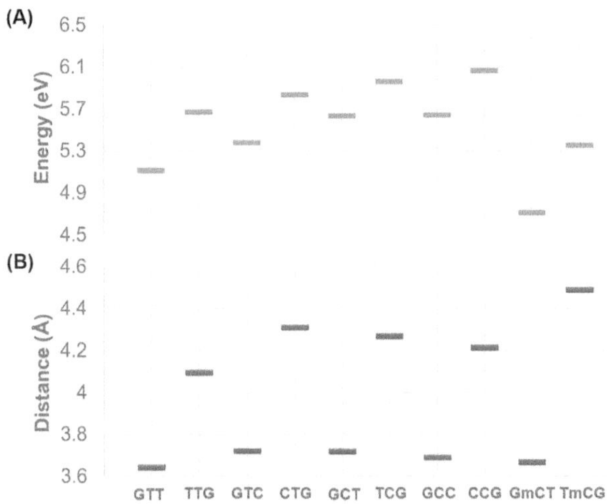

Fig. 4 (A) Average charge transfer energies obtained from 20 snapshots and (B) average distances between two bases involved in charge transfer obtained from 200 snapshots for all sequences.

guanine. In the two pairs in the middle (GTC, CTG, GCT, and TCG), the two pyrimidines are asymmetric. Since we have observed that the effect of guanine is very different in T compared to C, in order to exclude the effect of different bases and focus only on the 5′ *vs.* 3′ effect, we compare GTC(GCT) with CTG(TCG), where in both cases G is adjacent to T(C). For similar reasons the fifth pair involves the sequences GmCT and TmCG.

In all five pairs, it is clearly shown that the energy of the charge transfer states is stabilized when guanine is on the 5′ position compared to the 3′ position. The effect is not equivalently strong in all cases. In GmCT the effect is the strongest with the charge transfer state blueshifted by 0.64 eV moving from the 5′ to the 3′ position, and the effect is getting weaker in the order of GTT, GTC, GCC, and GCT.

In this case, we can again relate the energy of charge transfer states to the quenching of CPD formation. Our results indicate that guanine on the 5′ position is a stronger quencher of reactivity compared to guanine on the 3′ position. The change in the quenching behavior of guanine for TT based on the position (GTT *vs.* TTG) has been indeed observed experimentally in single strands and hairpins[6] and in double strands.[7]

3.2.3 Effect of methylation. Finally, we examine the effect of methylation. It has been observed that methylation has a pronounced effect on CPD formation. Comparisons of TCG and TmCG have shown that the yield for dimerization increases considerably when cytosine is methylated.[5] In our previous study we found that the differences in energies of charge transfer states may be responsible for this.[9] Here we explore further the charge transfer states of sequences with cytosine and methylated cytosine. Specifically we compare two pairs of systems, GmCT *vs.* GCT and T5mCG *vs.* TCG. Fig. 5A shows the average energies of charge transfer states for GmCT, GCT, and TmCG *vs.* TCG. It is obvious that methylation affects the charge transfer states considerably. When cytosine is methylated in GCT the energy of the charge transfer state is stabilized by 0.9 eV, while methylation of TCG leads to a stabilization of 0.6 eV.

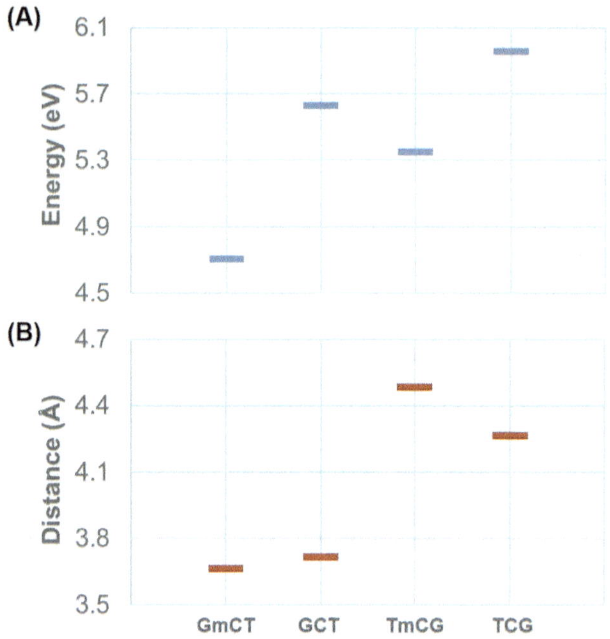

Fig. 5 (A) Average charge transfer energies obtained from 20 snapshots and (B) average distances between two bases involved in charge transfer obtained from 200 snapshots for methylated (GmCT, TmCG) and non-methylated sequences (GCT, TCG).

3.3 Reasons for changes in charge transfer energies

The main reasons that can affect the energy of charge transfer states are either intrinsic properties such as difference in the reduction/oxidation potential between the two neighboring bases, or environmental effects such as conformational restriction exerted by the helical structure of DNA or electrostatic interactions with their surroundings. We examine how these effects have determined the energies of charge transfer states observed in this work.

3.3.1 Redox potentials of bases involved. In our previous work we examined the effect of different flanking bases, guanine, adenine, and cytosine, where the neighboring pyrimidine was always thymine. The redox properties of these bases are quite different, and it was shown that the oxidation potential of flanking bases correlates with the energy level of the charge transfer state, thereby determining whether the charge transfer state intersects with the state that can lead to dimer formation. In the current work the flanking base is always guanine but the pyrimidine base can be either thymine or cytosine, and we observed that there is a big effect on the charge transfer states. It turns out that this effect is also correlated with the energies of charge transfer states. The reduction potentials of bases, shown in Table 1, indicate that thymine is the easiest to reduce.[47] The reduction potential of thymine compared to cytosine indicates that reduction of thymine is more favorable than cytosine. Thus there is a stronger driving force for electron transfer from guanine to thymine. This explains the energies of charge transfer states that we see in our calculations. However, this cannot explain the case of methylated cytosine because its reduction potential is more negative than

Table 1 One-electron reduction potentials for DNA Bases (V *vs.* NHE) calculated at M06-2X/6-31++G(d,p) level, taken from ref. 48. Experimental values taken from ref. 49

	E_{red}/V	E_{red}/V exp.
Guanine	−2.97	<−2.76
Cytosine	−2.36	−2.35
Thymine	−2.31	−2.18
5-Methyl-cytosine	−2.40	

cytosine and thymine, and it should not be reduced as readily. We will return to this issue later.

3.3.2 Conformational restriction by the helix structure of DNA. Besides the redox potentials mentioned above, another factor that affects the energy of the charge transfer states in general is environment effects, which can be divided into electrostatic interactions and conformational restriction. In the case of the positional effect of a flanking guanine at the 5′ or 3′ side on charge transfer state energies, the difference must originate from the environment since the constituting bases are identical. In order to probe the reason for this positional effect, we investigated two pairs of sequences, namely, GTT, TTG, GCC and CCG. We chose snapshots from each sequence whose charge transfer energy is close to the average charge transfer energy and removed the MM environments to perform gas phase QM calculations with the constituting bases involved in CT. The results (Table 2) show that the charge transfer energy difference depending on the position of a flanking guanine is maintained even after removing the MM environments, indicating that the difference is not attributed to electrostatic interactions with the surroundings. Therefore, the only reason left is a conformational restriction exerted by the helical structure of DNA. The simplest parameter that can significantly affect the charge transfer state energy is the distance between the

Table 2 Energies of charge transfer states (in eV) for a representative snapshot calculated using the whole model system (QM and MM) and only the QM region. The energy difference Δ between the two previously listed sequences is also shown

	QM/MM	QM
GTT	5.158	5.414
TTG	5.675	6.142
$\Delta E_{TTG-GTT}$	0.517	0.728
GCC	5.629	5.271
CCG	6.082	5.854
$\Delta E_{CCG-GCC}$	0.453	0.583
GmCT	4.7	5.731
GCT	5.649	5.231
$\Delta E_{GCT-GmCT}$	0.949	−0.5
TmCG	5.262	6.103
TCG	5.942	5.803
$\Delta E_{TCG-TmCG}$	0.68	−0.3

donor and acceptor, so we collected the distances between the geometric centers of two bases involved in charge transfer from MD trajectories for all sequences. Averages of the distances from 200 snapshots are shown in Fig. 4B, and the average distance between the centers of guanine and pyrimidine is 3.6–3.7 Å when guanine is on the 5′ position, while the average distance is 4.1–4.5 Å when guanine is on the 3′ position. The difference in the charge transfer energies due to the position of guanine at the 5′ or 3′ side (Fig. 4A) exhibits clear correlation with the difference in the distances (Fig. 4B), indicating that the charge transfer energy difference resulting from the position of guanine at the 5′ or 3′ side is attributed to the distance between the bases involved in CT.

Fig. 6 shows the actual data used to produce the distance averages for GTT and TTG. It is clear that there is a great variation in the distances, but the distances of TTG are almost always higher than those of GTT. Thus, the average is a very good indicator of the behavior.

3.3.3 Electrostatic interactions with surroundings. The charge transfer energy difference between sequences including methylated cytosine and its non-methylated counterpart should also originate from environmental effects since the redox potential of methylated cytosine is comparable to that of the non-methylated one (Table 2). However, the cause of this charge transfer energy difference is not due to the difference in the distances between the two bases involved in charge transfer, since the average distance observed in TCG is shorter than that of TmCG while the charge transfer energy of TmCG is lower than that of TCG (Fig. 5B). In the case of GmCT, whose charge transfer energy is much lower than that of GCT, the distance is shorter than that of GCT, but the difference is very small (0.05 Å). We also compared the distance between methylated cytosine and a neighboring thymine to that of the non-methylated sequence, but the distance differences were negligible. Therefore, no correlation is observed between the charge transfer energies and distances in this case, so we conclude that the conformational restriction exerted by helical structure is not the reason of the difference in charge transfer energies here.

As a next step, we performed gas phase calculations without the MM environment to evaluate the effect of electrostatic interactions with the surroundings (Table 2). Interestingly, the energetic order of charge transfer state between the methylated sequence and its counterpart becomes reversed after turning off the charges from the environment, and the charge transfer energy of the methylated

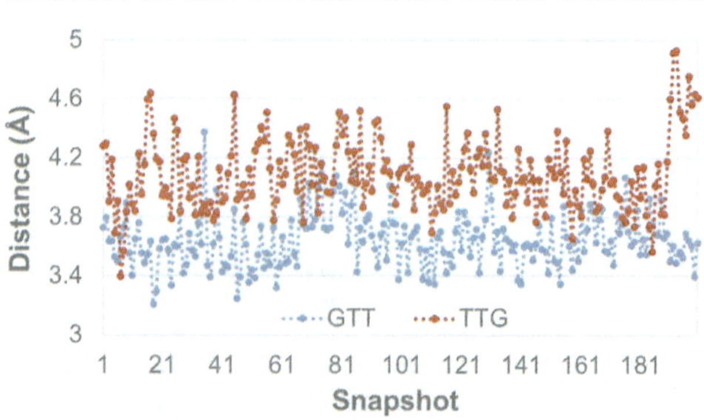

Fig. 6 The distance between the geometric centers of two bases.

sequence is now higher than that of the non-methylated counterpart. This is clearly opposite from the case observed in the charge transfer energy difference caused by the position of a flanking guanine. The bulky methyl group of cytosine seems to distort the surrounding structures, resulting in differences in structures from those of non-methylated one, and it is quite interesting that this distortion leads to the rearrangement of surrounding charges in such a way as to stabilize the charge transfer energies.

4 Conclusions

We used QM/MM calculations to examine how the energies of charge transfer states change for 10 different sequences that include a pair of pyrimidine dimers which can be photo-dimerized to form CPDs. All sequences include guanine as the flanking base. In the first tetrad (GTT, GTC, GCT, GCC) we examined how the different pyrimidines can affect charge transfer states, and we found that when guanine is next to thymine the charge transfer states are low in energy, while they are destabilized when guanine is neighboring cytosine. The secondary base has a much smaller effect. The second effect we examined was how placement of guanine at the 5′ *vs.* 3′ side changes the charge transfer states, and we found that guanine at the 3′ position destabilizes the charge transfer states. Finally, methylation of cytosine was found to lead to extra stabilization of charge transfer states. Overall, this work shows that the energies of charge transfer states are affected by (i) the nature of the donor acceptor pair, (ii) the distance between them and (iii) other environmental effects created by the surrounding environment. Updated information on the methylation effects and their sensitivity to force field parameters can be found in the comments section of this Faraday Discussion. The reader should read these comments for a better understanding.

Conflicts of interest

There are no conflicts of interest to declare.

Acknowledgements

SM acknowledges funding from the National Science Foundation (grant number: CHE-1800171). This research includes calculations carried out on Temple University's HPC resources and thus was supported in part by the National Science Foundation through major research instrumentation grant number 1625061 and by the US Army Research Laboratory under contract number W911NF-16-2-0189.

References

1 J. Cadet, S. Mouret, J.-L. Ravanat and T. Douki, *Photochem. Photobiol.*, 2012, **88**, 1048–1065.
2 J. Cadet, A. Grand and T. Douki, *Top. Curr. Chem.*, 2015, **356**, 249.
3 T. Douki and J. Cadet, *Biochemistry*, 2001, **40**, 2495–2501.
4 M. R. Holman, T. Ito and S. E. Rokita, *J. Am. Chem. Soc.*, 2007, **129**, 6–7.
5 V. J. Cannistraro and J.-S. Taylor, *J. Mol. Biol.*, 2009, **392**, 1145–1157.

6 Z. Pan, M. Hariharan, J. D. Arkin, A. S. Jalilov, M. McCullagh, G. C. Schatz and F. D. Lewis, *J. Am. Chem. Soc.*, 2011, **133**, 20793–20798.

7 Y. K. Law, R. A. Forties, X. Liu, M. G. Poirier and B. Kohler, *Photochem. Photobiol. Sci.*, 2013, **12**, 1431.

8 W. Lee and S. Matsika, *Phys. Chem. Chem. Phys.*, 2015, **17**, 9927–9935.

9 W. Lee and S. Matsika, *ChemPhysChem*, 2018, **19**, 1568–1571.

10 B. M. Pilles, D. B. Bucher, L. Liu, P. Clivio, P. Gilch, W. Zinth and W. J. Schreier, *J. Phys. Chem. Lett.*, 2014, **5**, 1616–1622.

11 T. Macke and D. A. Case, Modeling Unusual Nucleic Acid Structures, in *Molecular Modeling of Nucleic Acids*, ed. N. B. Leontes and J. J. SantaLucia, American Chemical Society, Washington, DC, 1998, ch. 24, pp. 379–393.

12 J. Wang, P. Cieplak and P. Kollman, *J. Comput. Chem.*, 2000, **21**, 1049–1074.

13 A. Peréz, I. Marchán, D. Svozil, T. E. I. I. I. Cheatham, C. A. Laughton and M. Orozco, *Biophys. J.*, 2007, **92**, 3817–3829.

14 J. Wang, R. M. Wolf, J. W. Caldwell, P. Kollman and D. A. Case, *J. Comput. Chem.*, 2004, **25**, 1157–1174.

15 C. I. Bayly, P. Cieplak, W. Cornell and P. Kollman, *J. Phys. Chem.*, 1993, **97**, 10269–10280.

16 M. J. Frisch, G. W. Trucks, H. B. Schlegel, G. E. Scuseria, M. A. Robb, J. R. Cheeseman, J. A. Montgomery Jr, T. Vreven, K. N. Kudin, J. C. Burant, J. M. Millam, S. S. Iyengar, J. Tomasi, V. Barone, B. Mennucci, M. Cossi, G. Scalmani, N. Rega, G. A. Petersson, H. Nakatsuji, M. Hada, M. Ehara, K. Toyota, R. Fukuda, J. Hasegawa, M. Ishida, T. Nakajima, Y. Honda, O. Kitao, H. Nakai, M. Klene, X. Li, J. E. Knox, H. P. Hratchian, J. B. Cross, V. Bakken, C. Adamo, J. Jaramillo, R. Gomperts, R. E. Stratmann, O. Yazyev, A. J. Austin, R. Cammi, C. Pomelli, J. W. Ochterski, P. Y. Ayala, K. Morokuma, G. A. Voth, P. Salvador, J. J. Dannenberg, V. G. Zakrzewski, S. Dapprich, A. D. Daniels, M. C. Strain, O. Farkas, D. K. Malick, A. D. Rabuck, K. Raghavachari, J. B. Foresman, J. V. Ortiz, Q. Cui, A. G. Baboul, S. Clifford, J. Cioslowski, B. B. Stefanov, G. Liu, A. Liashenko, P. Piskorz, I. Komaromi, R. L. Martin, D. J. Fox, T. Keith, M. A. Al-Laham, C. Y. Peng, A. Nanayakkara, M. Challacombe, P. M. W. Gill, B. Johnson, W. Chen, M. W. Wong, C. Gonzalez and J. A. Pople, *Gaussian 03, Revision C.02*, 2004.

17 W. Jorgensen, J. Chandrasekhar, J. Madura and M. Klein, *J. Chem. Phys.*, 1983, **79**, 926–935.

18 D. A. Case, T. A. Darden, T. E. Cheatham III, C. L. Simmerling, J. Wang, R. E. Duke, R. Luo, R. C. Walker, W. Zhang, K. M. Merz, B. Roberts, B. Wang, S. Hayik, A. Roitberg, G. Seabra, I. Kolossvry, K. F. Wong, F. Paesani, J. Vanicek, J. Liu, X. Wu, S. R. Brozell, T. Steinbrecher, H. Gohlke, Q. Cai, X. Ye, M.-J. Hsieh, G. Cui, D. R. Roe, D. H. Mathews, M. G. Seetin, C. Sagui, V. Babin, T. Luchko, S. Gusarov, A. Kovalenko and P. A. Kollman, *Amber 11*, University of California, San Francisco, 2010.

19 H. J. C. Berendsen, J. P. M. Postma, W. F. van Gunsteren, A. DiNola and J. R. Haak, *J. Chem. Phys.*, 1984, **81**, 3684–3690.

20 J. P. Ryckaert, G. Ciccotti and H. J. C. Berendsen, *J. Comput. Phys.*, 1977, **23**, 327–341.

21 U. Essmann, L. Perera, M. L. Berkowitz, T. Darden, H. Lee and L. G. Pedersen, *J. Chem. Phys.*, 1995, **103**, 8577.

22 P. Sherwood, A. H. de Vries, M. F. Guest, G. Schreckenback, C. R. A. Catlow, S. A. French, A. A. Sokol, S. T. Bromley, W. Thiel, A. J. Turner, S. Billeter, F. Terstegen, S. Thiel, J. Kendrick, S. C. Rogers, J. Casci, M. Watson, F. King, E. Karlsen, M. Sjovoll, A. Fahmi, A. Schäfer and C. Lennartz, *J. Mol. Struct.: THEOCHEM*, 2003, **632**, 1–28.
23 *TURBOMOLE V6.5*, A development of University of Karlsruhe and Forschungszentrum Karlsruhe GmbH, 1989–2007, TURBOMOLE GmbH, since 2007, available from, http://www.turbomole.com.
24 W. Smith and T. R. Forester, *J. Mol. Graphics*, 1996, **14**, 136–141.
25 P. Sherwood, A. H. de Vries, S. J. Collins, S. P. Greatbanks, N. A. Burton, M. A. Vincent and I. H. Hillier, *Faraday Discuss.*, 1997, **106**, 79–92.
26 A. D. Becke, *Phys. Rev. A: At., Mol., Opt. Phys.*, 1988, **38**, 3098–3100.
27 W. Y. C. Lee and R. G. Parr, *Phys. Rev. B: Condens. Matter Mater. Phys.*, 1988, **37**, 785–789.
28 B. Miehlich, A. Savin, H. Stoll and H. Preuss, *Chem. Phys. Lett.*, 1989, **157**, 200–206.
29 F. Weigend and R. Ahlrichs, *Phys. Chem. Chem. Phys.*, 2005, **7**, 3297–3305.
30 S. Grimme, *J. Comput. Chem.*, 2006, **27**, 1787–1799.
31 A. B. Trofimov and J. Schirmer, *J. Phys. B: At., Mol. Opt. Phys.*, 1995, **28**, 2299.
32 J. H. Starcke, M. Wormit, J. Schirmer and A. Dreuw, *Chem. Phys.*, 2006, **329**, 39–49.
33 W. Humphrey, A. Dalke and K. Schulten, *J. Mol. Graphics*, 1996, **14**, 33–38.
34 R. L. Martin, *J. Chem. Phys.*, 2003, **118**, 4775–4777.
35 B. Durbeej and L. A. Eriksson, *J. Photochem. Photobiol., A*, 2002, **152**, 95–101.
36 R. B. Zhang and L. A. Eriksson, *J. Phys. Chem. B*, 2006, **110**, 7556.
37 M. Boggio-Pasqua, G. Groenhof, L. V. Schäfer, H. Brubmüller and M. A. Robb, *J. Am. Chem. Soc.*, 2007, **129**, 10996–10997.
38 L. Blancafort and A. Migani, *J. Am. Chem. Soc.*, 2007, **129**, 14540–14541.
39 J. J. Serrano-Pérez, I. González-Ramírez, P. B. Coto, M. Merchán and L. Serrano-Andrés, *J. Phys. Chem. Lett.*, 2008, **112**, 14096–14098.
40 T. Climent, I. González-RamŠrez, R. González-Luque, M. Merchán and L. Serrano-Andrés, *J. Phys. Chem. Lett.*, 2010, **1**, 2072–2076.
41 S. Rössle, J. Friedrichs and I. Frank, *ChemPhysChem*, 2010, **11**, 2011–2015.
42 M. Barbatti, *ChemPhysChem*, 2014, **15**, 3342–3354.
43 J. I. Mendieta-Moreno, D. G. Trabada, J. Mendieta, J. P. Lewis, P. Gomez-Puertas and J. Ortega, *J. Phys. Chem. Lett.*, 2016, **7**, 4391–4397.
44 C. Rauer, J. J. Nogueira, P. Marquetand and L. González, *J. Am. Chem. Soc.*, 2016, **138**, 15911–15916.
45 I. Conti, L. Martinez-Fernandez, L. Esposito, S. Hofinger, A. Nenov, M. Garavelli and R. Improta, *Chem.–Eur. J.*, 2017, **23**, 15177–15188.
46 L. Martínez-Fernández and R. Improta, *Photochem. Photobiol. Sci.*, 2018, **17**, 586.
47 C. E. Crespo-Hernandez, D. M. Close, L. Gorb and J. Leszcynski, *J. Phys. Chem. B*, 2007, **111**, 5386–5395.
48 Y. Paukku and G. Hill, *J. Phys. Chem. A*, 2011, **115**, 4804–4810.
49 C. A. M. Seidel, A. Schulz and M. H. M. Sauer, *J. Phys. Chem.*, 1996, **100**, 5541–5553.

Faraday Discussions

PAPER

Probing eumelanin photoprotection using a catechol:quinone heterodimer model system†

Christopher Grieco, Jennifer M. Empey, Forrest R. Kohl and Bern Kohler *

Received 6th December 2018, Accepted 9th January 2019

DOI: 10.1039/c8fd00231b

Eumelanin is a natural pigment with photoprotective and radical scavenging characteristics, which are vital for a multitude of living organisms. However, the molecular mechanisms behind these functions remain obscure, in part because eumelanin is a heterogeneous polymer composed of a complex assortment of structural and chemical domains. Despite uncertainty about its precise structure, the functional units of eumelanin are thought to include quinones in various oxidation states. Here, we investigate the photochemistry of a catechol : *o*-quinone heterodimer as a model system for uncovering the photoprotective roots of eumelanin. Ultrafast transient absorption measurements in the UV to near-IR spectral regions are used to identify the photochemical processes that follow selective excitation of the *o*-quinone in the heterodimer using 395 nm light. We find that both singlet and triplet *o*-quinone excited states induce hydrogen atom transfer from the catechol, forming semiquinone radical pairs that persist beyond 2.5 ns, which is the upper time limit accessible by our instrument. Furthermore, the hydrogen atom transfer reaction was found to occur 1000 times faster *via* the singlet channel. Excited state pathways such as these may be important in eumelanin, where similar hydrogen-bonded interfaces are believed to exist between catechol and *o*-quinone functional groups.

1. Introduction

Eumelanin is a redox-mutable biopolymer with a photoprotective and radical scavenging character that is widely distributed throughout nature.[1,2] However, its excited state deactivation processes are not well understood due to the heterogeneity of its physical and chemical structure, which may give rise to a number of different deactivation pathways. Studies of model systems containing substructures proposed to occur in eumelanin have led to important insights, but much of

Department of Chemistry and Biochemistry, The Ohio State University, 100 West 18th Avenue, Columbus, Ohio 43210, USA. E-mail: kohler.40@osu.edu

† Electronic supplementary information (ESI) available. See DOI: 10.1039/c8fd00231b

this work has focused on intramolecular processes that occur within monomers and oligomers of 5,6-dihydroxyindole (DHI) and 5,6-dihydroxyindole carboxylic acid (DHICA), or intermolecular processes with the solvent (see Scheme 1).[3–7] Very little is known about how intermolecular interactions between DHI and/or DHICA units in different redox states affect excited state relaxation processes in eumelanin.[8]

Two major functional groups present in eumelanin, *ortho*-positioned dihydroxy (catechol) and quinone groups, are believed to interact through π-stacking and/or intermolecular hydrogen bonding.[9–11] Because the indole units present in eumelanin bearing these groups spontaneously polymerize in solution to form eumelanin, we consider the chemically stable 3,5-di-*tert*-butyl-substituted catechol (C) and *o*-quinone (Q) molecules as an alternative model system. To our knowledge, only one account[12] attempts to time-resolve the photochemistry of heterodimers of this pair. Using nanosecond transient absorption spectroscopy, Van Anh *et al.*[12] determined that a semiquinone radical pair is produced upon selective photoexcitation of Q. The authors concluded that this species must be a neutral radical pair formed by proton-coupled electron transfer, and not a radical ion pair formed by electron transfer, based on a 20 nm shift in the transient absorption band. However, the mechanism for radical pair formation could not be time-resolved with their instrumentation, and additional spectral support for this assignment is desirable.

In this work, we expand on the results of Van Anh *et al.*[12] by time-resolving the dynamics immediately following selective excitation of the quinone in hydrogen-bonded heterodimers of Q and C in cyclohexane. Our results support their assignment of neutral radical pair formation, and reveal the complex excited state

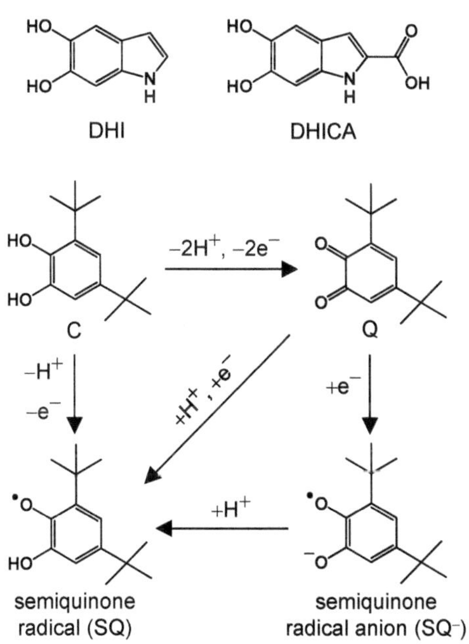

Scheme 1 Eumelanin building blocks (DHI and DHICA) along with the chemical structures discussed in the text.

pathways that precede formation of this species. In particular, the singlet excited state of Q may either induce hydrogen atom transfer from C to form a semiquinone neutral radical pair, or undergo ultrafast intersystem crossing. This bifurcated deactivation process is controlled by the hydrogen bond joining Q and C. The triplet state of Q also drives hydrogen atom transfer from C to form a semiquinone neutral radical pair, but on a much slower timescale. Our results reveal a hydrogen atom transfer interface that serves as a light-driven radical generation center that may be present in eumelanin.

2. Methods

2.1. Materials and solution preparation

All chemicals were purchased and used as received. Cyclohexane was from Acros Organics (HPLC grade, 99.8%), and 2-propanol was from Sigma Aldrich (ACS reagent grade, >99.5%). The 3,5-di-*t*-butylcatechol (C) and 3,5-di-*t*-butyl-*o*-quinone (Q) were from Sigma Aldrich (98% purity each). For UV/Vis and FTIR spectroscopy measurements, cyclohexane solutions of Q and C were prepared in concentrations ranging from 5–100 mM. To prepare heterodimer solutions, Q and C were dissolved in cyclohexane to yield equal initial concentrations (1 : 1 mole ratio). Hereafter, the concentration of every Q + C solution mentioned in the text refers to the initial concentration of Q, which is equal to the initial concentration of C. Steady-state spectroscopy measurements were performed on Q + C solutions with concentrations ranging from 5–100 mM. For femtosecond transient absorption measurements, solutions of Q, C, and Q + C were all 30 mM. Additional measurements were performed with 2-propanol.

2.2. UV/Vis spectroscopy

UV/Vis spectra of all samples were recorded using a Cary 5000 UV/Vis/NIR Spectrometer (Agilent; Santa Clara, CA). Solutions were measured in a custom-made demountable liquid cell with Teflon spacers and CaF_2 windows. The pathlengths used ranged from 25–500 μm depending on the sample concentration in order to keep the maximum absorbance above 300 nm below 1.0. The spectral resolution was 2 nm.

2.3. FTIR spectroscopy

Vibrational spectroscopy was performed using a FTIR spectrometer (FT/IR-4200, JASCO; Easton, MD). All samples were measured in the same liquid cell used for UV/Vis spectroscopy, but with a pathlength of 100 μm for solutions containing monomer concentrations exceeding 50 mM, and 200 μm for all other samples. The spectral resolution was 0.5 cm^{-1}.

2.4. Femtosecond UV/Vis and Vis/NIR pump-probe spectroscopy

Femtosecond UV/Visible measurements were performed using an ultrafast UV/Vis pump-probe spectrometer described previously.[8] For all experiments, the excitation wavelength was 395 nm, and the incident laser fluence calculated assuming Gaussian spatial and temporal profiles was approximately 500 μJ cm^{-2}. For each sample scan, about 4 mL of solution was recirculated through a demountable liquid flow cell (TFC-M25-3, Harrick Scientific Products; Pleasantville, NY) with

CaF_2 windows and a 500 μm Teflon spacer. The absorbance of each sample at 395 nm was approximately 0.6. A prism spectrograph was used to disperse the probe light after transmitting the sample on a Si CCD detector, and the detected spectral range was 320–630 nm.

Probe pulses for femtosecond visible/near-infrared (Vis/NIR) broadband measurements between 600 and 1100 nm were obtained from a white light supercontinuum that was generated by focusing 1400 nm pulses in a 5 mm thick sapphire window. The 1400 nm pulses were from the signal field produced by driving an optical parametric amplifier (TOPAS Prime, Coherent; Santa Clara, CA) using a portion of the output of a ~100 fs Ti:Sapphire laser amplifier (Astrella, Coherent; Santa Clara, CA). Measurements carried out on cyclohexane solutions were conducted using the same Si CCD detector used in the femtosecond UV/Visible broadband experiments. The measurements on the 2-propanol solution were carried out using an InGaAs array detector, an upgrade to our NIR detection system that was completed during the course of these experiments.

The total transient absorption spectra between 320 nm and 1100 nm were measured piecewise using the femtosecond UV/Vis and Vis/NIR spectrometers. To connect the spectra, a single scaling factor was applied to the entire Vis/NIR dataset until the mean square deviations in signal intensities for both the UV/Vis and Vis/NIR datasets were minimized. This correction factor, which never exceeded a value of 1.315, accounts for slight differences in the excitation fluence used between each set of measurements. After scaling, the signals measured in the region of overlap between 600 nm and 630 nm were averaged. The connected transient absorption spectra were modeled with target analysis using the Glotaran software package (version 1.5.1).[13]

3. Results

The FTIR spectra of a series of equimolar Q + C solutions in cyclohexane (Fig. 1a) change dramatically with concentration. This is easily seen when the spectra are presented as the apparent molar absorption coefficient, $\varepsilon_{\text{apparent}}$, which is defined as the measured absorbance divided by the product of the initial concentration of Q (or C) and the sample pathlength. For the 5 mM Q + C solution (the black trace in Fig. 1a), two intense and narrow peaks appear at 3561 cm^{-1} and 3626 cm^{-1} due to O–H stretching by C (see the FTIR spectra of the separate Q and C solutions in Fig. S1†). A broader, lower-frequency O–H stretching peak is also observed at 3460 cm^{-1}, which is assigned to the heterodimer formed between Q and C (hereafter designated by Q : C). As the concentration increases, the two sharp peaks decrease while the 3460 cm^{-1} peak increases and a new narrow peak grows in at 3540 cm^{-1}. These spectra were decomposed to obtain the Q : C heterodimer FTIR spectrum, which is compared to the measured Q and C monomer spectra in Fig. 1b. The Q : C heterodimer spectrum lacks the 3626 cm^{-1} band of the C monomer. In addition to a broad band at 3460 cm^{-1}, the heterodimer has a narrow band at 3540 cm^{-1}, which is red shifted compared to the 3561 cm^{-1} band of C. These spectra, which will be discussed below in detail, were used to deduce the hydrogen bonded structures shown in Fig. 1c.

Fig. 2a shows the UV/Vis spectra recorded for cyclohexane solutions of Q, C, and Q + C. Each solution has the same concentration as that used in the femtosecond transient absorption experiments. The $\pi \rightarrow \pi^*$ transition of Q appears at

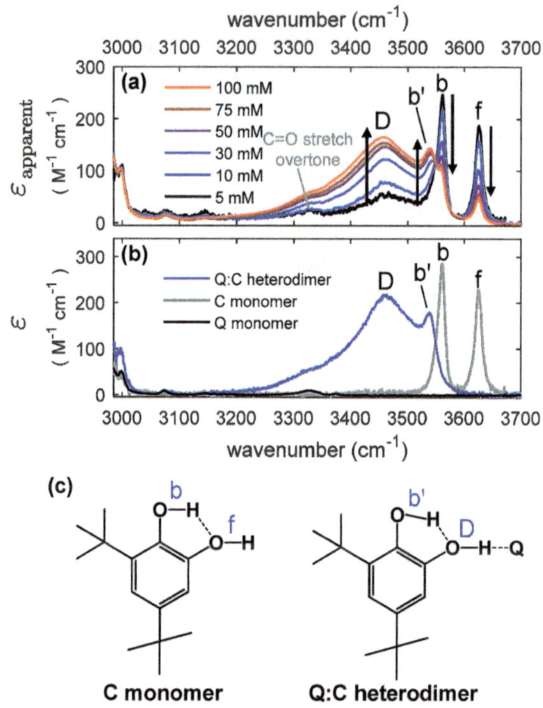

Fig. 1 (a) FTIR spectra of equimolar mixtures of Q + C in cyclohexane. The legend lists the initial concentration of C (or Q) added to the solution. The peaks are identified as free OH (f), intramolecularly bonded OH (b), perturbed intramolecularly bonded OH (b′), and heterodimer (D). (b) The IR absorption spectra of the Q : C heterodimer and C and Q monomers in cyclohexane. (c) Structures of C with and without complexation to Q. O–H bonds are labeled by their type of hydrogen bonding interaction. $\varepsilon_{apparent}$ is the apparent molar absorption coefficient, and ε is the molar absorption coefficient.

386 nm and 393 nm for the Q and Q + C solutions, respectively. UV/Vis spectra recorded in a sufficiently wide spectral window to see the weak n → π* transition of Q at approximately 595 nm are shown in Fig. S3.† Assignment of the n → π* transition is supported by the observed blue shifting of this band as the solvent polarity is increased (Fig. S5 and S6†). UV/Vis spectra of Q + C solutions varying between 5 mM and 100 mM are shown in Fig. S7† along with difference spectra. Based on the FTIR spectra of the 30 mM solutions, the UV/Vis spectrum of the equimolar Q + C solution is decomposed to reveal the contributions of the Q monomers, C monomers, and Q : C heterodimers in Fig. 2b. The contributions to the total absorption at 395 nm by the populations of the Q monomer and the Q : C heterodimer are used later to isolate their individual transient absorption spectra.

Transient absorption spectra were recorded for Q and Q + C solutions in cyclohexane (Fig. 3) using a pump wavelength of 395 nm to excite the π → π* band of Q. The transient spectra for the Q + C solution in Fig. 3 have been corrected by subtracting signal contributions from the Q monomer, as discussed in the ESI.† The raw, uncorrected transient spectrum for the Q + C solution is shown in Fig. S11.† The 2D maps in Fig. 3a and b are plotted on a linear photon energy scale, while the spectral slices in Fig. 3c and d are graphed on a linear wavelength

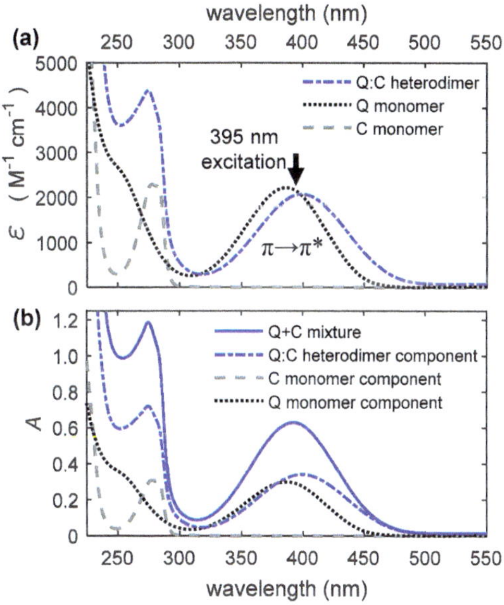

Fig. 2 (a) UV/Vis spectra for cyclohexane solutions of the Q monomer, C monomer, and Q : C heterodimer. ε is the molar absorption coefficient. (b) Spectral decomposition of an equimolar mixture of Q + C based on the results from FTIR spectroscopy, showing the absorbance contributions of the C monomer, Q monomer, and Q : C heterodimer.

axis. For both Q monomers and Q : C heterodimers, intense photoinduced absorptions (PIAs) are observed at the UV wavelengths for all time delays, while weaker PIAs are observed at the visible wavelengths. A negative signal assigned to a ground state bleach (GSB) feature is observed in the vicinity of the $\pi \rightarrow \pi^*$ peak of the electronic ground state (see inverted ground state absorption spectra, as indicated by the gray curves in Fig. 3c and d).

The 0.5 ps spectrum of the Q monomer (Fig. 3c) has a broad peak at 580 nm, a shoulder at 462 nm, and an additional broad peak around 1100 nm (see Fig. S15† for an extended spectrum). As discussed later, these peaks can be identified with electronic transitions predicted from UV/Vis spectroscopy, which are included in Fig. 3c. Over time, the transient spectrum evolves into a narrower band with clear vibronic structure centered at approximately 533 nm (see dark blue curve). The 0.5 ps spectrum of the Q : C heterodimer (Fig. 3d) somewhat resembles that of the Q monomer. However, it possesses a peak at approximately 610 nm that lacks the shoulder seen in the Q monomer at 462 nm, and has a significantly broadened tail extending into the NIR region. After several picoseconds, this spectrum evolves into one with peaks at approximately 550 nm and 710 nm. By 2 ns, the peak at 550 nm has completely decayed, while the peak at 710 nm continues to grow in intensity and a distinct and narrow peak is seen at 370 nm (see red arrow in Fig. 3d).

The early-time transient absorption spectrum of the Q : C heterodimer averaged from 300 to 700 fs is presented in Fig. 4a and b. This spectrum differs from the transient spectra recorded for the Q monomer in cyclohexane and the hydrogen-bonding solvent, 2-propanol (Fig. 4a). Target analysis of the Q : C

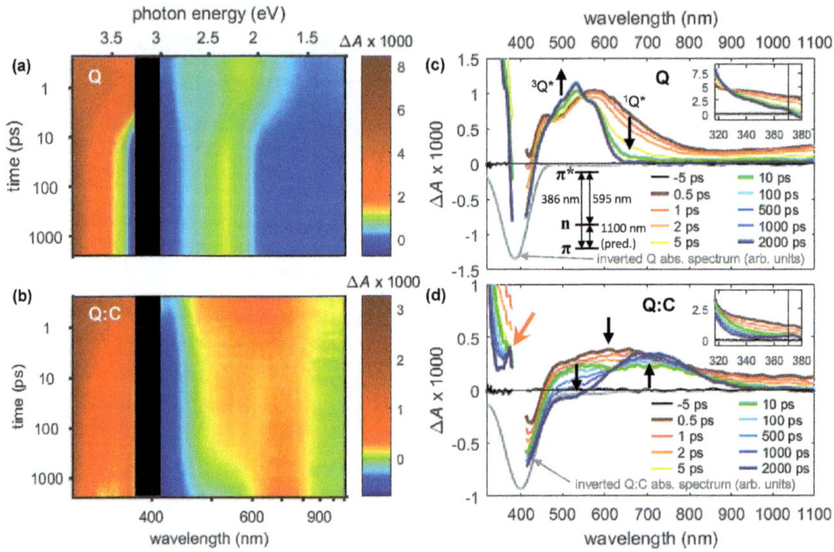

Fig. 3 (a, b) 2D maps of the transient absorption spectrum of (a) Q monomers and (b) Q : C heterodimers in cyclohexane, measured using 395 nm excitation. The black rectangles mask the scatter appearing in the vicinity of the excitation wavelength. (c, d) Transient absorption spectra at select time delays for (c) Q monomers and (d) Q : C heterodimers. The gray traces in (c) and (d) are the inverted absorption spectra of Q monomers and Q : C heterodimers, respectively. The insets in (c) and (d) show the UV portion of the spectra. The vertical lines at 370 nm are included for reference. The energy level diagram shown in (c) includes electronic transitions for Q in cyclohexane as determined from UV/Vis spectroscopy. The red arrow in (d) emphasizes the sharp feature at 370 nm.

heterodimer spectrum in Fig. 3b reveals that additional absorbing species contribute to the early spectrum of the Q : C heterodimer (Fig. 4b). One of these species produces a peak at 710 nm (pink curve in Fig. 4b), which matches the peak position seen in the 2000 ps spectrum in Fig. 3d. Fig. 4d compares the species associated difference spectrum (SADS) of this species, SADS 2, to that of the longest-lived species, SADS 4. The two spectra share a peak location of 710 nm, but SADS 4 appears narrower. This reflects the dynamic peak narrowing observed in the transient absorption spectrum of the Q : C heterodimer (Fig. 4c).

Transient absorption kinetics for the Q monomer and the Q : C heterodimer are compared in Fig. 5. The kinetics were obtained by averaging the transient spectra in Fig. 3 over the indicated wavelength ranges, which were selected to correspond to peaks in the transient spectra highlighted above. For Q monomers (Fig. 5a), the kinetic signals averaged over 650–1100 nm decay within ~25 ps, but include a component that extends beyond 2 ns. The kinetics averaged over 475–580 nm exhibit a rapid decay over the first few picoseconds, followed by a slower growth in intensity over the first ~25 ps. This signal does not decay by 2 ns.

For the Q : C heterodimer (Fig. 5b), the kinetic signal averaged over 950–1100 nm exhibits a faster initial decay than the 650–1100 nm trace for Q monomers, followed by a slower decay during the first 25 ps. The kinetics averaged over 500–530 nm exhibit a fast decay during the first few picoseconds followed by

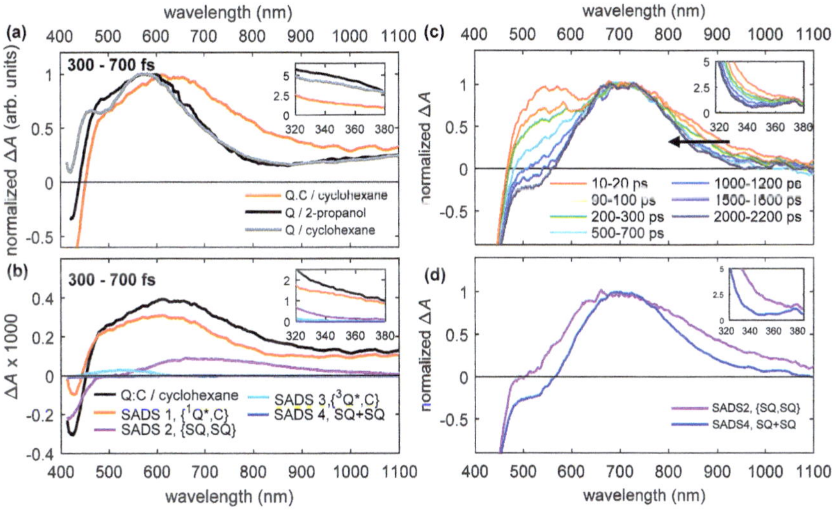

Fig. 4 (a) Comparison of the normalized transient absorption spectra of Q monomers in either cyclohexane or 2-propanol with the Q : C heterodimer spectrum averaged from 300 to 700 fs. (b) The Q : C heterodimer spectrum from (a), decomposed using the target analysis results. (c) A normalized transient absorption spectrum of Q : C heterodimers averaged over the indicated time ranges from 10 to 2200 ps. (d) Normalized species associated difference spectra (SADS) for the two semiquinone (SQ) states. The insets of all graphs show the UV portion of the spectra.

a slower growth over the next several picoseconds. Unlike the 475–580 nm trace for Q monomers, the 500–530 nm trace for Q : C heterodimers decays completely by 2 ns. The kinetics averaged over 650–800 nm contain a rapid decay component like the 950–1100 nm trace. However, the signal does not decay further beyond 5 ps, but instead persists until 25 ps, after which it slowly grows during the remainder of the measured time window. This growth mirrors the decay of the 500–530 nm trace.

Fig. 5c shows the kinetics averaged over 415–425 nm for Q monomers and Q : C heterodimers in cyclohexane. These signals contain overlapping PIA signals that partially obscure the pure bleaching kinetics. In both systems, the intensity of the GSB traces increases in magnitude over the first few picoseconds, mirroring the decay of the 650–1100 (black squares in Fig. 5a) and 950–1100 nm (black squares in Fig. 5b) traces for the Q monomers and Q : C heterodimers, respectively, but then remains constant over the next 100 ps. Thereafter, the GSB trace for the Q monomer decays very slightly toward zero, while that for the Q : C heterodimer becomes more negative, mirroring the late growth of the 650–800 nm trace in Fig. 5b.

4. Discussion

4.1. Heterodimer formation

Intermolecular hydrogen bonding drives the association of the Q and C molecules in concentrated cyclohexane solutions. We chose the weakly-perturbing solvent, cyclohexane, to isolate the interaction between Q and C from interactions of each with the solvent. Structural sensitivity of the O–H stretching modes in hydrogen-

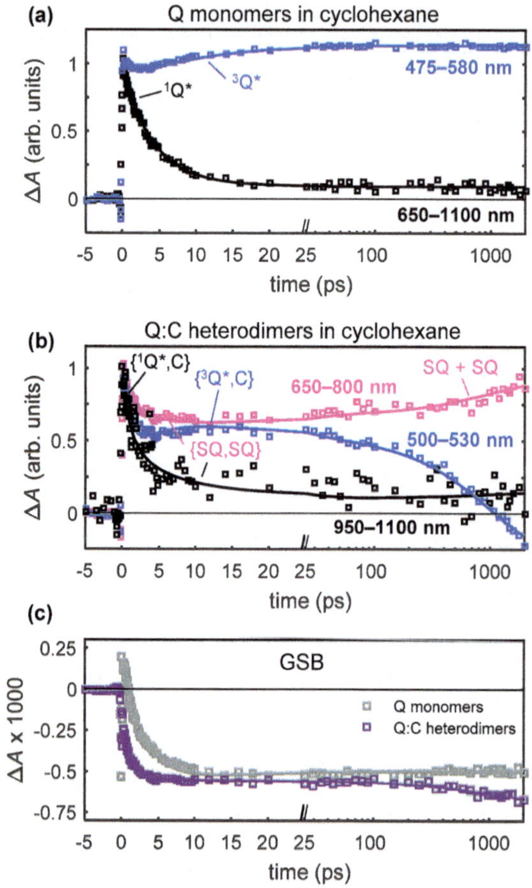

Fig. 5 Transient absorption kinetics averaged over the indicated wavelengths for (a) Q monomers and (b) Q : C heterodimers. (c) Ground state bleach (GSB) kinetics averaged between 415 and 425 nm. Solid lines are fits to the data from target analysis as described in the text.

bonded systems makes vibrational spectroscopy a useful method for characterizing their structure.[8,14,15] As such, FTIR spectroscopy of the O–H stretching modes in C can be used to characterize the hydrogen bonding structure of Q : C heterodimers that form in Q + C mixture solutions.

The sharp O–H stretching peaks at 3561 cm^{-1} and 3626 cm^{-1} in Fig. 1a were previously assigned to the intramolecularly hydrogen bonded (b) and free (f) O–H groups of the C monomer (see left structure in Fig. 1c).[8,14,15] Indeed, these peaks are present for C solutions ranging in concentration from 5 mM to 100 mM (Fig. S1a†). The redshifted and broadened peak at 3460 cm^{-1} is not present in the C-only solution and must originate from intermolecular hydrogen bonding between Q and C molecules. The growth of this peak with increasing Q and C concentration is associated with the appearance of the new narrow peak at 3540 cm^{-1} (peak b′ in Fig. 1a). This spectral behavior has been previously observed in solutions of catechol in aprotic solvents containing hydrogen bond-accepting bases.[15] Similar observations have been made for 3,5-di-*t*-butylcatechol.[14] In these studies, the catechol molecule forms complexes with base molecules through intermolecular hydrogen bonding with the free O–H group. This complexation causes the intensity of the f band to decrease and is accompanied by the growth of a broader, downshifted band below 3500 cm^{-1} due to the intermolecular hydrogen bond. The intermolecular hydrogen bond causes the intramolecular hydrogen bond in catechol to strengthen *via* cooperativity,[14,15] inducing the b band to change into the lower frequency b′ band. Based on this work, we conclude that a similar complexation mechanism occurs between the C and Q molecules in the heterodimer, as illustrated in Fig. 1c. In particular, one O–H group in C donates a hydrogen bond to Q, while the other remains intramolecularly hydrogen bonded.

The magnitude of the frequency shift of the b band of catechol and its derivatives is well correlated with the strength of the intermolecular hydrogen bond, or proton acceptor ability of the base.[14,15] In particular, Varfolomeev *et al.* have demonstrated that this shift[15] is linearly correlated to the hydrogen bond formation enthalpy of methanol with proton acceptors.[16] Using this correlation, the 21 cm^{-1} downshift of the b band in Fig. 1 indicates a hydrogen bond formation enthalpy between C and Q of approximately −13.3 kJ mol^{-1}, most similar to that for the catechol : dimethylsulfoxide complex.[15] Bifurcation of the hydrogen bond, which would produce a 3-centered interaction between the O–H group in C and the C=O groups in Q, is unlikely because the resulting 5-membered ring structure would possess an angular geometry (120°) unfavorable to hydrogen bond formation.[17] We therefore conclude that the free O–H group in C hydrogen bonds to only one of the C=O groups in Q.

The concentration of Q : C heterodimers in the Q + C solutions was quantified as a first step toward interpreting the transient absorption spectra, assuming a simple equilibrium model:

$$Q + C \leftrightarrow Q:C,\ K_{eq} = 91\ M^{-1}. \tag{1}$$

From the FTIR spectra, the concentration of C monomers, [C], can be determined from the decrease in extinction of the free O–H peak at 3626 cm^{-1} (see the ESI† for further details).[8] The concentration of the Q : C heterodimers, [Q : C], can be determined by subtracting [C] from the initial C concentration, $[C]_{init}$:

$$[Q:C] = [C]_{init} - [C]. \tag{2}$$

Using eqn (2), the concentration of Q : C heterodimers in the 30 mM Q + C solution used in the ultrafast spectroscopy measurements was determined to be 17 mM. Using eqn (1) and (2), the equilibrium constant, K_{eq}, for the dimerization reaction was estimated to be 91 M^{-1}. Additionally, the molar absorption spectrum of the Q : C heterodimer could be determined (Fig. 1b).

4.2. UV/Vis spectroscopy and discriminating the transient absorption spectra produced by monomers and heterodimers

The UV/Vis spectra in Fig. 2a are dominated at wavelengths longer than 300 nm by the $\pi \rightarrow \pi^*$ transition of Q, which is assigned based on the magnitude of the molar extinction coefficient (2100 M^{-1} cm^{-1}) and previous reports.[18–20] The $n \rightarrow \pi^*$ transition is located at 595 nm and has a molar extinction coefficient of approximately 40 M^{-1} cm^{-1} (Fig. S3b†). The redshift observed for Q + C mixtures (see Fig. 2a and S7†) reflects the change in environmental polarity experienced by Q upon hydrogen bonding with C. Indeed, such a redshift of the $\pi \rightarrow \pi^*$ peak (and simultaneous blueshift of the $n \rightarrow \pi^*$ peak) of Q originates from solvatochromism (see Fig. S5†) as reported previously by Van Anh *et al.*[12] In fact, we find that the $\pi \rightarrow \pi^*$ and $n \rightarrow \pi^*$ peak positions correlate linearly with the Gutmann acceptor number of the solvent (Fig. S6†). Ravi Kumar *et al.* reported a similar phenomenon for such transitions in benzophenone solutions.[21]

Based on these observations, the gradual redshift of the $\pi \rightarrow \pi^*$ peak observed for the Q + C mixtures as their concentrations are increased (Fig. S7†) could be explained by a gradual change in the polarity experienced by the Q molecules. However, we find that a two-state model better describes the gradual redshift of the Q + C mixtures, in which the total measured absorbance of Q + C (A_{Q+C}) can be described as the sum of the Q monomer spectrum (A_Q) and the Q : C heterodimer spectrum ($A_{Q:C}$) for all solution concentrations:

$$A_{Q+C} = A_Q + A_{Q:C}. \tag{3}$$

The hydrogen bond in the Q : C heterodimer is expected to redshift its absorbance to a distinct position due to solvatochromism (Fig. S5†), which would produce the gradual redshift observed as the Q population converts to Q : C (Fig. S7a†). The results published in the ESI section of Van Anh *et al.* show that the difference absorption spectrum of Q does not shift as the mole ratio of Q : C is increased from 1 : 10 to 1 : 80 in methylcyclohexane.[12] Likewise, we observe no significant shifting in difference spectra as the total concentration in an equimolar mixture of Q + C is increased to 100 mM (Fig. S7b†). These observations support the two-state model. Notably, the redshift of the $\pi \rightarrow \pi^*$ peak upon increasing solution concentration (Fig. S7a†) approaches that of dilute solutions of covalently tethered Q and C molecules, which is located at 400 nm.[12]

As Fig. 2b indicates, the 395 nm excitation wavelength used in the transient absorption measurements excites both Q monomers and Q : C heterodimers present in the Q + C solution. Because the total absorption spectrum can be described using the 2-state model (eqn (3)), we can decompose the transient absorption signal for the Q + C mixture, $\Delta A_{Q+C}(t)$, into components due to the Q monomers and Q : C heterodimers (see ESI Section S.3† for full details),

$$\Delta A_{Q+C}(t) = \Delta A_{Q:C}(t) + \Delta A_Q(t), \tag{4}$$

where $\Delta A_{Q:C}(t)$ and $\Delta A_Q(t)$ are the transient absorption signals produced by the population of Q : C heterodimers and Q monomers, respectively. Knowledge of the absorbance fraction of each species excited allows the interfering transient absorption signal produced by Q monomers to be subtracted from the total measured spectrum (ΔA_{Q+C}). This is because, in a mixture, the transient absorption signal of each component contributes in proportion to the ratio of the absorbance of that species to the total absorbance at the pump wavelength (see ESI†). In eqn (4), cross terms describing signals arising from interactions among the various species excited are not needed because bimolecular encounters occur beyond the time window measured by our instrument (2.5 ns). For example, diffusion-limited reactions occur in cyclohexane at a rate of approximately 7.1 ×

10^9 M^{-1} s^{-1}, implying that a bimolecular encounter requires approximately 240 ns for a solute concentration of 60 mM.

The absorption spectrum of the Q + C mixture solution can be decomposed using the fraction of Q : C heterodimers determined from FTIR spectroscopy of the same solution (Fig. 2b). As a check on the accuracy of the decomposition procedure, the Q : C heterodimer peak appears at ~400 nm (Fig. 2b) in excellent agreement with the peak location of covalently tethered Q and C molecules.[12] The absorption intensities of the decomposed Q monomer and Q : C heterodimer bands at 395 nm are then used to determine how much of the Q monomer transient absorption spectrum to subtract from the total measured transient absorption spectrum of the Q + C mixture. Further details on this subtraction procedure are provided in the ESI.† The results of this process are shown in Fig. 3b and d and will be discussed below.

4.3. Excited state dynamics of the Q monomer

Understanding of the photophysics of the Q monomer in cyclohexane is needed to interpret the results for the Q : C heterodimer. The 0.5 ps spectrum in Fig. 3c is assigned to an excited singlet state of Q (^{1}Q*) based on previous observations for other quinones.[22] Additionally, the peaks at 580 nm and 1100 nm can be assigned to the n → π* and π → n transitions of the ^{1}nπ* state of Q, respectively, as predicted from UV/Vis spectroscopy (see energy level diagram in Fig. 3c and S15†). As in other quinones,[22] xanthones,[23] and benzophenone,[24] the excited singlet state undergoes ultrafast intersystem crossing in several picoseconds to yield a distinctive, blue-shifted triplet absorption band (see the blue curves in Fig. 3c). The triplet state of Q (^{3}Q*) does not decay during the observation window.

The ^{1}Q* and ^{3}Q* population kinetics are extracted by averaging over their respective bands at 650–1100 nm and 475–580 nm (Fig. 5a). Due to substantial overlap between their spectra, only portions of their bands are averaged to obtain their kinetics. This still captures their true population dynamics because no significant changes in their lineshapes are observed. The kinetics cannot be described as a single unimolecular singlet → triplet intersystem crossing reaction as previously reported for chloranil,[22] but instead require a more complicated reaction scheme like those reported for other ketones.[23,24] The triplet growth kinetics are often described using a sequential kinetic model, in which the initial ^{1}nπ* state rapidly intersystem crosses in about 1 ps to form the ^{3}ππ* state, followed by a slower internal conversion step to produce the ^{3}nπ* state in about 10 ps. When exciting to the ^{1}ππ* state, a branched kinetic scheme is sometimes invoked,[23] in which the ^{1}ππ* state is able to undergo either intersystem crossing to the ^{3}nπ* state or internal conversion to the ^{1}nπ* state.

Because the ^{1}nπ* state is observed as early as 500 fs for Q (Fig. 3c and S15†), the sequential model more accurately describes intersystem crossing in Q. Using the sequential kinetic model, target analysis was performed on the transient absorption spectrum of Q monomers in cyclohexane (Fig. S8†). The formation of ^{3}Q* is described by the time constants of 3.8 ps and 7.7 ps. The latter value agrees well with the 8 ps triplet formation time constant reported for chloranil in dichloromethane.[22]

4.4. Excited state dynamics of Q : C heterodimers

Over a few picoseconds, the 0.5 ps spectrum of the Q : C heterodimer in Fig. 3d, which resembles that of the Q monomer, decays to reveal ^{3}Q* absorption around 533 nm and an additional peak around 710 nm (green trace, Fig. 3d). Over time, the ^{3}Q* peak decays completely, while the 710 nm band increases in intensity and the sharp 370 nm feature is revealed (see 2000 ps trace in Fig. 3d). The spectrum at 2000 ps matches that expected for the 3,5-di-*t*-butyl-substituted neutral semiquinone (SQ) radical.[12,25,26] Although it is difficult to use the visible band to distinguish the neutral and anionic forms of the SQ radical, the 370 nm band is a clear signature of the neutral form; the anionic form contains a distinct peak at 350 nm,[25] which is absent from our spectrum at all times. Therefore, we conclude that neutral SQ radicals are produced in Q : C heterodimers upon excitation at 395 nm *via* a net hydrogen atom transfer (HAT) reaction from C to Q. Considering that the ^{3}Q* signal of the heterodimer sample in Fig. 5b completely decays by 2 ns due to the formation of SQ radicals, we conclude that the ^{3}Q* state mediates a HAT reaction from C. Note that the complete decay of the ^{3}Q* state in the heterodimer rules out the possibility that the heterodimer dissociates following photoexcitation. If dissociation were to occur, then the long-lived ^{3}Q* state should persist as seen for the Q monomer (blue trace in Fig. 5a).

The 0.5 ps spectrum of Q : C heterodimers in Fig. 3d does not completely match the spectrum of the ^{1}Q* state in Fig. 3c, so we investigated whether additional absorbing species are present. We considered whether the intermolecular hydrogen bond between C and Q could perturb the ^{1}Q* state, giving rise to a redshifted and broadened spectrum. This effect would be similar to the solvatochromic shifts observed for the ground state absorption spectrum of Q in Fig. S5.† To test this, the transient absorption spectrum of Q was measured in 2-propanol (Fig. 4a), which is a solvent expected to hydrogen bond to Q and that redshifts the n → π* band of Q to the same location in the Q : C heterodimer (Fig. S4†). The ^{1}Q* spectrum averaged between 300 and 700 ps qualitatively matches that of Q in cyclohexane, indicating that excited state hydrogen bonding alone does not produce the 300–700 ps spectrum seen in the Q : C heterodimer and that additional absorbing species are present.

The prominent absorption band at 710 nm in the green trace in Fig. 3d reveals that a neutral SQ radical species is present as early as 10 ps. At this time, the triplet population is not yet quenched, indicating that SQ radicals are also formed from ^{1}Q* states. As such, both singlet- and triplet-mediated HAT reactions are incorporated into a target model used to describe the Q : C heterodimer spectrum in Fig. 3d (Scheme 2a and Fig. S14a†). In the target model, it was necessary to include two separate SQ radical states to capture the dynamic narrowing of the 710 nm band seen in Fig. 4c, which we will return to later. Fig. 4b shows the decomposition of the Q : C heterodimer spectrum averaged between 300 and 700 fs in terms of the SADS from the target model, confirming that there are significant contributions from both the ^{1}Q* and SQ states at this time.

We now consider the bifurcation of the excited state deactivation of ^{1}Q* between the intersystem crossing and HAT channels. From previous work on benzophenone,[21,24] we expect the hydrogen bonded and non-hydrogen bonded Q molecules to absorb closer to the lower-energy and higher-energy edge, respectively, of the π → π* band. Consequently, the 395 nm excitation likely produces

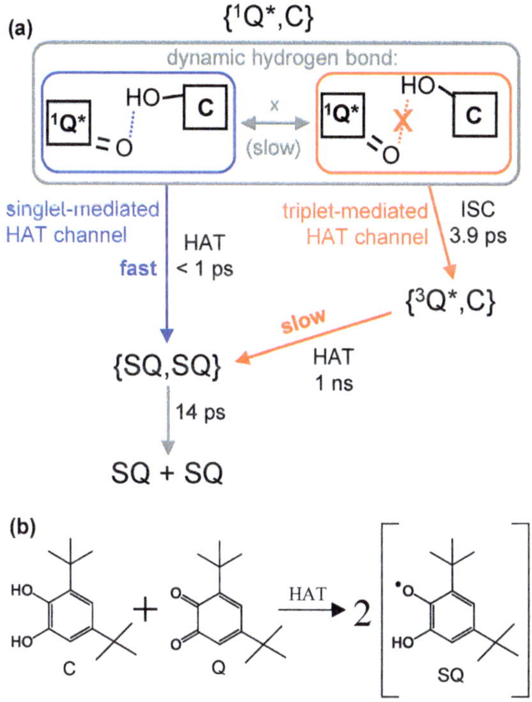

Scheme 2 (a) Excited state decay pathways determined in this study for the Q : C heterodimer excited at 395 nm. Braces indicate contact pairs. The time constants were obtained from target analysis of the transient absorption spectra. (b) Comproportionation reaction by excited state hydrogen atom transfer (HAT) leading to a pair of neutral semiquinone radicals.

three excited state populations in the Q + C solution: the first (trivial case) is a population of Q* monomers that is accounted for as discussed above. The second is a population of Q : C heterodimers in which the ^{1}Q* molecule is hydrogen bonded with a C molecule at the instant of excitation. The third is a population of Q : C heterodimers in which the hydrogen bond between Q* and C is broken at the instant of excitation. Hydrogen bonding is known to influence electronic excited state dynamics.[27] For example, Venkatraman *et al.* have reported recently how hydrogen bonding slows down the rate of ultrafast intersystem crossing in benzophenone due to excited state solvation effects.[24] Consequently, the binary distribution of hydrogen bonded and non-hydrogen bonded populations of ^{1}Q* in Q : C heterodimers could be responsible for the bifurcation of its excited state deactivation.

The hydrogen bond lifetime of the Q : C heterodimer is expected to be at least several picoseconds based on MD simulation estimates for benzophenone in methanol and ethanol by Ravi Kumar *et al.*[21] The expected hydrogen bond lifetime is longer than the longest ^{1}Q* decay time constant, which is 3.8 ps and 3.9 ps in Q monomers and Q : C heterodimers, respectively (Fig. S8 and S14†). We therefore propose that the dynamic hydrogen bond between Q and C in the ground state gates the excited state dynamics. In other words, ^{1}Q* molecules that are not hydrogen bonded with C at the instant of photoexcitation undergo intersystem crossing, while those engaged in a hydrogen bond induce HAT to form the SQ singlet RP state. This is illustrated in Scheme 2a.

The SQ radical formation reaction by $^1Q^*$ takes place approximately 1000 times faster than by $^3Q^*$ (Scheme 2a). Such a difference between singlet and triplet reactivity of Q highlights the general importance of its complexation to C for enhancing the SQ radical formation efficiency. Nau *et al.* similarly reported a significantly enhanced hydrogen abstraction reactivity, yet lower efficiency, for bimolecular reactions of excited singlet states of acetone with several different hydrogen bond donors.[28]

We now discuss the necessity of using two states in our target model to describe the SQ radicals formed by photoexcitation of the Q : C heterodimers (Scheme 2a). The dynamic peak narrowing of the SQ band in Fig. 4c, which is reflected by the difference in peak width of the two SADS describing the SQ radicals in Fig. 4d, indicates that their electronic states are perturbed at early time. It is likely that this perturbation results from the electronic coupling[29] between the two proximal SQ radicals that are formed from the Q : C heterodimer. As such, the SADS 2 is assigned to a contact SQ radical pair state, {SQ,SQ}, and the SADS 4 is assigned to a solvent-separated SQ radical pair state, SQ + SQ. In accordance with the target model in Scheme 2a, the contact SQ pairs undergo diffusional separation with a time constant of 14 ps. Such spectral changes from diffusional separation of the SQ radical pairs are reminiscent of the observations made for correlated triplet pair states occurring in molecular films of singlet fission materials, in which the vibronic peak intensity ratios of $T_1 \rightarrow T_n$ PIA bands were found to be sensitive to whether or not the correlated triplet pairs are spatially separated, due to changes in excitonic coupling.[30]

Radical recombination of the contact SQ pairs in the Q : C heterodimers by back HAT is not observed. In particular, the kinetic signal averaged over the 710 nm band (pink data in Fig. 5b) does not decay during the tens of picoseconds timescale. In addition, the GSB signal does not decay toward zero at any time (purple data in Fig. 5c). Note that the rise of the GSB signal does overlap with positive transient absorption features; the early growth of the GSB signal is likely associated with the decay of the overlapping $^1Q^*$ peak, and the continued growth of the GSB signal following 200 ps is likely associated with the decay of the overlapping $^3Q^*$ peak. Instead of restoring the ground state, the contact SQ pairs separate diffusionally, causing the dynamic narrowing of the SQ band in Fig. 4c. The recombination of the contact SQ pairs would require a HAT event from one SQ to the other. While such a disproportionation reaction is quite efficient for semiquinones in general,[31–34] we propose that it is less efficient for the SQs in this study. Because the intramolecular hydrogen bond in an *ortho*-semiquinone is twice as strong as in catechol,[31,35,36] it stabilizes the radical photoproduct. Additionally, if the intramolecular hydrogen bond in SQ is favored over an intermolecular hydrogen bond between two SQs, the diffusional separation of the contact pair may be facilitated.

5. Conclusions

An *o*-quinone : catechol heterodimer was used as a model system for investigating how the intermolecular hydrogen bond between quinone- and catechol-containing indolic units in eumelanin may influence its photochemistry. Excitation of the quinone using 395 nm generates singlet excited states that are capable of inducing ultrafast hydrogen atom transfer from the catechol. Some of this population instead undergoes ultrafast intersystem crossing to form triplet states that induce hydrogen atom transfer on a late picosecond timescale. The resulting semiquinone neutral radical pair photoproducts do not recombine during the full time range (2.5 ns) of these measurements. Furthermore, we propose that the dynamic hydrogen bond between Q and C controls the bifurcated deactivation of the singlet excitations.

The results of our model system study show how the specificity of the interfaces between intermolecular hydrogen-bonded indolic units in eumelanin may control their photochemistry. In particular, the *o*-quinone species are capable of inducing excited state hydrogen atom transfer with catechol groups when they are hydrogen bonded, leading to long-lived semiquinone radical pairs. The chemical heterogeneity proposed for eumelanin[1,37,38] intrinsically sets up a wide distribution of intermolecular interactions between indolic units with different redox states. Our study reveals that the hydrogen bonded quinone–catechol interaction sets up a photoinduced hydrogen atom transfer interface that can serve as an ultrafast radical generation center in eumelanin, acting as a source for SQ radicals. Therefore, radical formation may be facilitated in eumelanin structures that are rich in units containing the UVA- and visible-absorbing *o*-quinone functional group. Note that our measurements do not reveal the timescale during which the SQ radicals disproportionate. However, nanosecond transient absorption measurements by Van Anh *et al.* of both a Q + C mixture and a covalently-tethered Q : C compound reveal long-lived SQ radicals with half-lives of approximately 10 μs and 300 ns, respectively.[12] Therefore, neutral SQ radicals generated at hydrogen-bonded Q : C interfaces are expected to have sufficient time to undergo further radical chemistry. Our study presents a possible photoinduced radical generation pathway in eumelanin that aligns with proposals[39–41] that eumelanin is not only photoprotective, but also a harmful radical source.

Conflicts of interest

There are no conflicts of interest to declare.

Acknowledgements

This work was supported in part by start-up funding from the Ohio State University. The broadband femtosecond transient absorption spectrometer was constructed with partial support from NSF grant CHE-1800471.

References

1 P. Meredith and T. Sarna, *Pigm. Cell Res.*, 2006, **19**, 572–594.

2 M. d'Ischia, K. Wakamatsu, A. Napolitano, S. Briganti, J. C. Garcia-Borron, D. Kovacs, P. Meredith, A. Pezzella, M. Picardo, T. Sarna, J. D. Simon and S. Ito, *Pigm. Cell Melanoma Res.*, 2013, **26**, 616–633.

3 A. Corani, A. Pezzella, T. Pascher, T. Gustavsson, D. Markovitsi, A. Huijser, M. d'Ischia and V. Sundstrom, *J. Phys. Chem. Lett.*, 2013, **4**, 1383–1388.

4 A. Corani, A. Huijser, T. Gustavsson, D. Markovitsi, P. A. Malmqvist, A. Pezzella, M. d'Ischia and V. Sundstrom, *J. Am. Chem. Soc.*, 2014, **136**, 11626–11635.

5 M. Gauden, A. Pezzella, L. Panzella, A. Napolitano, M. d'Ischia and V. Sundstrom, *J. Phys. Chem. B*, 2009, **113**, 12575–12580.

6 A. Huijser, A. Pezzella, J. K. Hannestad, L. Panzella, A. Napolitano, M. d'Ischia and V. Sundstrom, *ChemPhysChem*, 2010, **11**, 2424–2431.

7 J. J. Nogueira, A. Corani, A. El Nahhas, A. Pezzella, M. d'Ischia, L. González and V. Sundström, *J. Phys. Chem. Lett.*, 2017, **8**, 1004–1008.

8 C. Grieco, F. R. Kohl, Y. Zhang, S. Natarajan, L. Blancafort and B. Kohler, *Photochem. Photobiol.*, 2019, **95**, 163–175.

9 D. R. Dreyer, D. J. Miller, B. D. Freeman, D. R. Paul and C. W. Bielawski, *Langmuir*, 2012, **28**, 6428–6435.

10 E. D. Głowacki, M. Irimia-Vladu, S. Bauer and N. S. Sariciftci, *J. Mater. Chem. B*, 2013, **1**, 3742.

11 L. Panzella, G. Gentile, G. D'Errico, N. F. Della Vecchia, M. E. Errico, A. Napolitano, C. Carfagna and M. d'Ischia, *Angew. Chem., Int. Ed.*, 2013, **52**, 12684–12687.

12 N. Van Anh and R. M. Williams, *Photochem. Photobiol. Sci.*, 2012, **11**, 957–961.

13 J. J. Snellenburg, S. P. Laptenok, R. Seger, K. M. Mullen and I. H. M. van Stokkum, *J. Stat. Software*, 2012, **49**, 1–22.

14 M. C. Foti, L. R. C. Barclay and K. U. Ingold, *J. Am. Chem. Soc.*, 2002, **124**, 12881–12888.

15 M. A. Varfolomeev, D. I. Abaidullina, A. Z. Gainutdinova and B. N. Solomonov, *Spectrochim. Acta, Part A*, 2010, **77**, 965–972.

16 B. N. Solomonov, V. B. Novikov, M. A. Varfolomeev and N. M. Mileshko, *J. Phys. Org. Chem.*, 2005, **18**, 49–61.

17 I. Rozas, I. Alkorta and J. Elguero, *J. Phys. Chem. A*, 1998, **102**, 9925–9932.

18 J. W. Sidman, *J. Am. Chem. Soc.*, 1956, **78**, 4567–4572.

19 H. Bettermann and I. Dasting, *Chem. Phys.*, 1995, **196**, 531–541.

20 A. Berger, P. Hertl and A. Rieker, in *The chemistry of quinonoid compounds*, ed. S. Patai and Z. Rappoport, John Wiley & Sons, Chinchester, 1st edn, 1988, vol. 2, ch. 2, pp. 29–86.

21 V. Ravi Kumar, C. Verma and S. Umapathy, *J. Chem. Phys.*, 2016, **144**, 064302.

22 S. M. Hubig, T. M. Bockman and J. K. Kochi, *J. Am. Chem. Soc.*, 1997, **119**, 2926–2935.

23 H. Satzger, B. Schmidt, C. Root, W. Zinth, B. Fierz, F. FKrieger, T. Kiefhaber and P. Gilch, *J. Phys. Chem. A*, 2004, **108**, 10072–10079.

24 R. K. Venkatraman, S. Kayal, A. Barak, A. J. Orr-Ewing and S. Umapathy, *J. Phys. Chem. Lett.*, 2018, **9**, 1642–1648.

25 S. V. Jovanovic, K. Konya and J. C. Scaiano, *Can. J. Chem.*, 1995, **73**, 1803–1810.

26 P. P. Levin, A. B. Belyaev and V. A. Kuzmin, *Bull. Acad. Sci. USSR, Div. Chem. Sci.*, 1987, **36**, 406–408.

27 G. Zhao and K. Han, *Acc. Chem. Res.*, 2012, **45**, 404–413.

28 W. M. Nau, F. L. Cozens and J. C. Scaiano, *J. Am. Chem. Soc.*, 1996, **118**, 2275–2282.

29 I. R. Gould, R. H. Young, R. E. Moody and S. Farid, *J. Phys. Chem.*, 1991, **95**, 2068–2080.

30 R. D. Pensack, E. E. Ostroumov, A. J. Tilley, S. Mazza, C. Grieco, K. J. Thorley, J. B. Asbury, D. S. Seferos, J. E. Anthony and G. D. Scholes, *J. Phys. Chem. Lett.*, 2016, **7**, 2370–2375.
31 J. J. Warren, T. A. Tronic and J. M. Mayer, *Chem. Rev.*, 2010, **110**, 6961–7001.
32 S. K. Wong, W. Sytnyk and J. K. S. Wan, *Can. J. Chem.*, 1972, **50**, 3052–3057.
33 F. G. Bordwell and J. Cheng, *J. Am. Chem. Soc.*, 1991, **113**, 1736–1743.
34 A. E. Alegria, M. Lopez and N. Guevara, *J. Chem. Soc., Faraday Trans.*, 1996, **92**, 4965.
35 H. Korth, M. I. de Heer and P. Mulder, *J. Phys. Chem. A*, 2002, **106**, 8779–8789.
36 J. S. Wright, *J. Am. Chem. Soc.*, 2001, **123**, 1173–1183.
37 M. d'Ischia, A. Napolitano, A. Pezzella, P. Meredith and T. Sarna, *Angew. Chem., Int. Ed.*, 2009, **48**, 3914–3921.
38 R. Micillo, L. Panzella, M. Iacomino, G. Prampolini, I. Cacelli, A. Ferretti, O. Crescenzi, K. Koike, A. Napolitano and M. d'Ischia, *Sci. Rep.*, 2017, **7**, 41532.
39 F. Solano, *Polym. Int.*, 2016, **65**, 1276–1287.
40 H. Z. Hill, *BioEssays*, 1992, **14**, 49–56.
41 A. B. Mostert, S. B. Rienecker, C. Noble, G. R. Hanson and P. Meredith, *Sci. Adv.*, 2018, **4**, eaaq1293.

Faraday Discussions

DISCUSSIONS

Photo-protection/photo-damage in natural systems: general discussion

Michael Ashfold, Shuming Bai, Stephen Bradforth, Pavel Chabera, Jeffrey Cina, Carlos E. Crespo-Hernández, Natércia das Neves Rodrigues, Marta Duchi, Graham Fleming, Christopher Grieco, Scott Habershon, Michael Haggmark, Sharon Hammes-Schiffer, Shou-Ting Hsieh, Bern Kohler, Heiko Lokstein, Andrew Marcus, Todd Martinez, Spiridoula Matsika, Thomas A. A. Oliver, Luis Ortiz-Rodríguez, Tomas Polivka, Minjung Son, Vasilios Stavros, Collin Steen, Matthew Turner, Peter Jomo Walla and Jack Woolley

DOI: 10.1039/C9FD90031D

Collin Steen opened discussion of Tomas Polivka's paper: It would be very interesting to study the excited state dynamics of this E-FCP protein in the context of the whole organism. Dynamics and behavior are frequently altered in an *in vivo* sample compared to an isolated one. As an example, I'd like to present some recent results from Graham Fleming's group at UC Berkeley in which we apply *in vivo* spectroscopy to study the photoprotective quenching mechanisms in intact cells of the heterokont algae *Nannochloropsis oceanica*.[1] *N. oceanica* is a small marine algae, 2–3 μM in size, which results in minimized light scattering compared to larger organisms such as *E. huxleyi*. It also possesses a relatively simple membrane structure and pigment composition including the higher plant xanthophyll cycle carotenoids (violaxanthin-antheraxanthin-zeaxanthin), one type of chlorophyll, and Chl *a*-binding LHCX proteins, which are closely related to the FCP proteins discussed in the present contribution (DOI: 10.1039/c8fd00193f). We have also made use of two knock-out mutants of *N. oceanica*: *vde*, which is unable to synthesize Zea in high light, and *lhcx1*, which lacks the LHCX1 protein.

Fluorescence lifetime snapshots revealed that wild type cells undergo a significant decrease in the average chlorophyll fluorescence lifetime (t_{avg}) during 10 min of high-light exposure, which is indicative of the onset of quenching. However, both mutants lack any substantial decrease in t_{avg} (Fig. 1B in Park *et al.*[1]), indicating that VDE and LHCX1 are both required for photoprotection in this species of algae.

In order to gain insight into the mechanism behind the dissipation of excess excitation energy, we employed *in vivo* Transient Absorption (TA) spectroscopy on *N. oceanica* cells that had been dark acclimated (unquenched) and after exposure to high light for 30 min (quenched). The decay of the signal at 980 nm following excitation of Chl *a* was clearly different in the unquenched and quenched states (Fig. 2C in Park *et al.*[1]). Given that this spectral region corresponds to $Car^{•+}$ ESA, this observed difference provides evidence for charge transfer (CT) quenching in which energy is transferred from the Q_y state of Chl to a heterodimer CT state consisting of $Car^{•+}$ and $Chl^{•-}$. Probing at 540 nm also revealed differences between the unquenched and quenched conditions (Fig. 2A in Park *et al.*[1]), thereby confirming the presence of excitation energy transfer (EET) quenching from the Chl Q_y to the Car S_1 state. Interestingly, the *vde* and *lhcx1* mutants exhibited no difference between unquenched and quenched samples probed at either 980 nm or 540 nm (Fig. 3 in Park *et al.*[1]). This strongly suggests that VDE (and thus Zea) as well as LHCX1 are required for the CT and EET photoprotective quenching mechanisms in intact *N. oceanica*.

So in summary: (1) We observe Chl-Car energy transfer processes in living cells of the algae *Nannochloropsis oceanica*; (2) these energy transfer processes involve the $Car^{•+}$ (CT) and Car S_1 (EET) states; (3) this energy transfer specifically requires VDE (Zea) as well as the LHCX1 protein, which is closely related to FCP. I understand that it is not possible to directly compare energy transfer dynamics in intact *N. oceanica* and isolated E-FCP proteins, but what do you make of the differences between your results and ours? Additionally, do you have plans to study energy transfer dynamics of the FCP protein *in vivo*? What differences would you anticipate between FCP and LHCX1 in the context of a fully intact system?

1 S. Park, C. J. Steen, D. Lyska, A. L. Fischer, B. Endelman, M. Iwai, K. K. Niyogi and G. R. Fleming, *Proc. Natl. Acad. Sci. U. S. A.*, 2019, **116**, 3385–3390.

Tomas Polivka responded: We do not fully understand the energy transfer pathways in isolated E-FCP, especially their response to high- or low-light conditions. We know that the carotenoid composition (Fx to hFx ratio) varies with light conditions, but we do not know how this may affect the energy transfer pathways. So we first plan to explore the isolated protein in detail, later we may focus on an *in vivo* system containing E-FCP. Concerning the differences between E-FCP and an intact system from *N. oceanica*, the key difference is the carotenoid composition. The direct Chl-Car quenching observed in *N. oceanica* requires carotenoids with low S_1 energy, such as zeaxanthin. Since such carotenoids are not present in E-FCP, the absence of quenching is rather expected. Fucoxanthin is obviously unable to quench the singlet states of Chl-a, because its S_1 energy is too high to do it.

Tomas Polivka addressed Collin Steen: The kinetics you showed in your presentation certainly suffers from annihilation. Then, when normalizing to the long-time plateau of your kinetics, if the amount of annihilation is different in the quenched and unquenched sample, such a difference must affect your results. How do you deal with this?

Collin Steen replied: Thanks for bringing up the topic of Chl*–Chl* annihilation as it does influence the observed kinetics. The signals measured at 540 nm are a combination of Car S_1 ESA and Chl ESA. High-light exposure activates a variety of de-excitation pathways, resulting in a reduced overall amplitude of Chl ESA in the quenched sample. However, the kinetics of the ESA signals were observed to be kinetically equivalent at time delays ≥40 ps where Car S_1 states are nearly fully de-populated. In order to extract the contribution of Car S_1 ESA from the overall Chl ESA, we scale the kinetic profile measured in the dark (mostly Chl ESA) to match that in high light based on signals at long time delays and subtract the two.

This normalization procedure of the EET kinetic traces at long time delays is based on the assumption that the amount of annihilation is consistent during the transition from dark to high light. If the amount of annihilation were different under the quenched and unquenched conditions, we would expect to see a significant change in the Chl ESA and GSB dynamics. However, in a previous paper, we found that the decay of the Chl ESA and GSB signals probed at 620 and 680 nm respectively were identical for dark- and high-light-exposed spinach thylakoid membranes.[1] Therefore, it is reasonable to assume that Chl*–Chl* annihilation is consistent during the transition from dark to high light and thus the decay of Chl ESA is kinetically equivalent. This is further supported by the *N. oceanica vde* mutant that lacks Zea, which upon scaling exhibits identical decays at 540 nm under unquenched and quenched conditions.[2]

1 S. Park, A. L. Fischer, C. J. Steen, M. Iwai, J. M. Morris, P. J. Walla, K. K. Niyogi and G. R. Fleming, *J. Am. Chem. Soc.*, 2018, **140**, 11965–11973.
2 S. Park, C. J. Steen, D. Lyska, A. L. Fischer, B. Endelman, M. Iwai, K. K. Niyogi and G. R. Fleming, *Proc. Natl. Acad. Sci. U. S. A.*, 2019, **116**, 3385–3390.

Natércia das Neves Rodrigues addressed Collin Steen: This is a question regarding your TA experiments on living cells, which are of course rather complex samples. I am curious to know how much this was a problem experimentally, *i.e.* how difficult it was to generate signal, if there were problems with excessive scatter or any other experimental challenges that arose from using a complex sample like this. Moreover, I am interested in knowing how you addressed these, both experimentally – *e.g.* did you have to avoid translating the sample, or were there other things you had to adjust experimentally – and in terms of analysis, *e.g.* did you have problems with over-complicated spectra, how you addressed these and how you isolated the spectral features of interest?

Collin Steen answered: Experimentally, it is important that the sample is continuously translocated during the experiment and bubbled regularly between measurements to prevent sample damage. To minimize scattering of the pump we use a polarization filter set to the probe polarization. To ensure a clean probe signal we also add a notch filter specific to the pump wavelength prior to our detector. In terms of analysis, we are able to identify the molecular species involved in the quenching of excitation energy based on the features of the difference kinetic decay profile, which is taken as the difference between the sample in its quenched (high-light) state and its unquenched (dark) state. By fitting this difference profile, we obtain rise and decay time constants for the species of interest, which can be compared to literature values for the lifetimes for various pigments. Verification of the molecular species of interest is also based on comparing the observed excited-state absorption spectrum with the literature. Additionally, the use of mutants or chemical treatments that target specific molecules in the organism can provide further insight into assignment of the spectra and decay profiles.

Pavel Chabera asked Tomas Polivka: It would seem that the synthetic pathway for producing carotenoid with an extra hexanol group is rather complicated. Why does the organism (plankton) undertake this challenge? You indicate in the paper (DOI: 10.1039/c8fd00193f) that this is not happening for protection against excess light – why? Is there any better plausible explanation?

Tomas Polivka replied: Our results show that the overall efficiency of carotenoid-chlorophyll energy transfer is not affected by fucoxanthin (Fx) - hexanoylfucoxanthin (hFx) exchange, though the energy transfer is slower if hFx is the donor. Similarly, there is no additional quenching of Chl singlet excited states if hFx is present. Yet, the triplet quenching, which has not been studied in detail in our paper, could still be affected by Fx-hFx exchange. The enhanced synthesis of hFx under high-light conditions is clearly proven, so there should be some advantage of having hFx in the antenna under high-light conditions. But the data we have collected so far do not give any clear answers to what this advantage actually is.

Pavel Chabera followed: In a "standard carotenoid model", it is believed that only groups attached to the conjugated backbone of a carotenoid can influence its excited state properties. How do you explain that the hexanol attached to the molecule but *not* conjugated to carotenoid's conjugated system can influence its properties? In particular, that it can switch off the ICT state?

Tomas Polivka answered: Yes, it is correct that we have always considered the groups that are not in conjugation to be spectroscopically silent. On the other hand, we know that spectroscopic properties of keto-carotenoids are strongly affected by solvent polarity. Thus, there is clearly a way to modify spectroscopic properties by something that is not a part of the conjugated system. Here, the attached hexanoyloxy tail obviously affects the electron distribution along the conjugated backbone, preventing stabilization of the ICT state.

Graham Fleming asked: Can you add a bit more detail on why you say charge transfer is necessary for rapid energy transfer?

Tomas Polivka responded: The coupling of the ICT state to the S_1 state modifies the transition dipole moment of the donor transition. It presumably enhances the transition dipole moment (as evidenced by *e.g.* increased S_1 fluorescence quantum yield of keto-carotenoids), making the donor–acceptor coupling stronger, resulting in more efficient energy transfer.

Heiko Lokstein remarked: There has been a lot of discussion recently about the so-called S* states of carotenoids. I wonder whether you have observed an S* state in either one of your investigated carotenoids, or, more specifically, whether there was any difference in S* transient absorption spectra between them.

Tomas Polivka answered: No, we have not observed any spectral features attributable to S* signal in these carotenoids. It is not so surprising as the S* signal is typically reported for carotenoids with long conjugation length, let's say about N=11 or longer. Fucoxanthin and its derivatives are much shorter.

Vasilios Stavros said: Tomas, you present some really interesting data. I was wondering whether you had considered looking at the influence of temperature on the excited state dynamics of hFx, as conformational flexibility may influence the role of S_1/ICT on excited state energy transfer.

Tomas Polivka responded: No, we have not studied temperature dependence so far though I agree it could provide valuable information as it indeed can reveal some extra information about the role of the hexanoyloxy tail in hFx. In the future, we plan to run experiments at both low temperature (down to about 200 K to keep the solution liquid) and high (up to about 325 K to prevent degradation).

Shuming Bai opened discussion of the paper by Scott Habershon: As I understand, you apply the machine learning diabatization to obtain an "on-the-fly" PES for the MCTDH method. My first question: In principle, other effective diabatization methods should be applicable in your approach, right? My second question: The universal diabatization method is still quite challenging for the systems with more than several atoms. For "on-the-fly" applications it can be more difficult because different configurations of state can join in along the dynamics, as you have found. Is it possible to monitor situations like this, and provide warning information during the calculation?

Scott Habershon replied: For the first question, the answer is yes; in principle, our approach is compatible with any diabatization scheme.

For the second part of the question, we agree that some sort of diabatization monitoring is essential; we are already thinking about this somewhat. In the projection diabatization scheme which we employ, for example, one can keep track of the energies of the different adiabatic states throughout the sampled configurations in our on-the-fly scheme; if adiabatic states change in energy such that they become relevant to the diabatization scheme (as in the example considered in our paper), then we can of course go back and re-run the dynamics with these additional states included. This procedure could be repeated until no further states are required. As an alternative, we are also investigating how one might include more than one set of reference electronic states in our projection diabatization scheme; this is ongoing work, however.

Bern Kohler asked: Have you considered applying your on-the-fly simulations to uracil, a nucleobase with well-understood internal conversion pathways?

Scott Habershon answered: Yes - that is exactly the sort of molecular system we'd like to be in a position to directly simulate using these on-the-fly methods. With 12 atoms, so a maximum of 30 active degrees-of-freedom (3N-6), this is at the upper-end of the system sizes we can currently model; however, the development of more efficient potential energy surface sampling schemes and improved software parallelisation should help us push towards this limit.

Natércia das Neves Rodrigues said: In the model you describe, do you have to restrict and select which coordinates to include or are you able to use the full 3N dimensional space?

Scott Habershon responded: Although the methods we describe can in principle be used to describe the full 3N-6-dimensional space of an N-atom molecule, the expense of *ab initio* calculations and propagation of the MCTDH equations-of-motion mean that one usually has to make a selection of the "active" vibrational coordinates. Various approaches towards this have been suggested previously, including the one in our contribution (DOI: 10.1039/c8fd00228b), but this aspect remains something of an open problem.

Natércia das Neves Rodrigues remarked: In my gas-phase experiments, after I pump my system with one photon, I then ionize the system in order to probe it. This ionization is effectively a second transition from the excited state to a given cationic state. I have found that the Franck–Condon factor for this transition—which changes in time as the wavepacket samples the excited state surface and is projected onto different parts of the cationic state surface—affects my experimental observable. In particular, the auto-correlation function which describes the time-dependent wavepacket behavior alone does not successfully model my experimental observations, as it does not account for the changing probe step Franck–Condon factor. Is it realistic that your models may be expanded to model this time-dependent projection of the excited state population onto the cationic state? Alternatively, do you know of any other methods which may model this behavior?

Scott Habershon answered: Yes, we believe that wavefunction methods like MCTDH could be used to model processes such as that described, although we expect it would be very challenging. For example, one could imagine introducing a time-dependent term to the Hamiltonian to simulate excitation to the ionic states at different stages during the MCTDH evolution. However, accounting for the configuration-dependent Franck–Condon factors seems like it would require further work; performing these "jumps" for trajectory-based methods might be more straightforward using Franck–Condon factors calculated at each configuration for each trajectory, but performing a similar simulation for a grid-based method would be much more demanding.

It is already possible within the MCTDH framework, using pre-fitted potential energy surfaces, to model time-resolved photoelectron spectroscopy experiments, so such an extension to the on-the-fly method should be possible. The cation potential energy surfaces can be fitted as for neutral states, and time-dependent fields can be applied to model laser pulses. So, assuming that the transition dipole (and polarisability, *etc.*, if necessary) matrix elements are also fitted, then excitations to the cation states my also be modelled. It is also necessary to account for the energy of the ejected electron; this can be achieved by introducing an extra discrete variable representation-type grid into which the electron wavefunction is excited, a so-called discretised continuum.

Jeffrey Cina asked: Could you please comment on the relative computational overhead of the MCTDH and electronic structure portions of these calculations?

Scott Habershon replied: In the initial calculation, where one is running a MCTDH simulation while simultaneously sampling configurations at which to perform *ab initio* electronic structure calculations, the computational burden lies towards the electronic structure calculations. In follow-up calculations, where one can perform an MCTDH simulation on the constructed potential energy surface, the computational expense lies firmly with the propagation of the MCTDH equations-of-motion.

It is also worth noting that the distribution of computational expense will be highly problem-dependent. For example, problems with many strongly-coupled degrees-of-freedom might require more *ab initio* electronic structure calculations in order to generate an accurate representation of the potential energy surface, such that this stage may be extremely costly. In addition, the exponential scaling of the MCTDH algorithm itself will eventually become the bottleneck in the simulations as the system-size increases. In the initial dynamics calculations reported here, using the normal modes chosen using forces and couplings, the overwhelming computational cost was in the electronic structure calculations. For each of the molecules studied in this paper (DOI: 10.1039/c8fd00228b), the percentage CPU times for the dynamics were: 0.35% (1a), 4.6% (2a), 0.11% (3a) and 0.5% (4a).

For the simulations of 4a using the 5 "steepest" normal modes, as well as the "false" normal mode directed towards the conical intersection, MCTDH accounted for about 40% of the total computational cost. This much larger proportion was due to the much larger DVR grids being used.

In addition to the point about highly-correlated potential energy surfaces being more difficult to fit, we also note that such systems also require more single-particle functions in the MCTDH calculation to account for the motion of the coupled modes.

Sharon Hammes-Schiffer questioned: How do you generate the initial conditions, and how sensitive are the results to them? Does MCTDH properly describe zero-point energy and the transfer of energy between vibrational modes?

Scott Habershon answered: The initial wavepacket is a Hartree product of ground-state harmonic oscillator eigenfunctions, centered at the minimum-energy geometry of the electronic ground-state. As such, harmonic zero-point energy is included in the initial wavefunction.

The MCTDH equations-of-motion are derived using the time-dependent variational principle. As such, they correctly describe zero-point energy conservation and other quantum-mechanical effects such as tunnelling. In addition, energy transfer between different vibrational modes is also captured (as long as this coupling is also described by the underlying potential energy surface, of course).

We have not investigated the effects of alternative initial conditions in this paper, although this is possible in the MCTDH method. For example, an initial imaginary-time relaxation of the wavepacket on the ground-state would account for any effects of anharmonicity on the shape of the initial wavepacket. This wavepacket can then be vertically promoted to the excited electronic state or, pushing towards higher accuracy, a laser pulse can be modelled as a time-dependent field which, allied with a representation of the transition-dipole moment surface, allows a more realistic representation of the excitation. To adopt this more accurate representation of the excitation process in our on-the-fly method requires more work to model the transition dipoles, but this should be feasible.

Spiridoula Matsika asked: In this study you discovered that there was a problem with the chosen diabatization scheme because the DD-MCTDH predictions did not agree with the known ultrafast dynamics. Is there a way to test/know if your calculations predict the correct behavior if this behavior is not known *a priori*?

Scott Habershon responded: Yes - one can, in principle, compare the features of the diabatic and adiabatic states as an indicator of potential problems in diabatization. In addition, in one of the diabatization schemes used in this paper (projection diabatization) (DOI: 10.1039/c8fd00228b), one can monitor the adiabatic states as they change in energy across the range of sampled configuration space, allowing one to self-consistently update the manifold of states used in the diabatization scheme. Finally, although we have not investigated fully, we believe that the projection and propagation diabatization schemes certainly have room for improvement, for example by combining the two different approaches to seek a "best-of-both-worlds" approach.

Jack Woolley commented: Given that certain vibrational modes are required to traverse conical intersections, could this method, knowing the correct active space, be used as a black box software to explore how different chemical functionalities associated with the two nitrogen atoms influence the excited state dynamics? Recent experimental data on a model MAA and the MAA Shinorine shows subtle differences (see Fig. 1 below) suggesting that careful thought behind choice of functionality will be crucial in developing next generation, nature inspired sunscreens, and hence theory will be very important, to aid with functionalisation of core structures, like those discussed in the contribution.

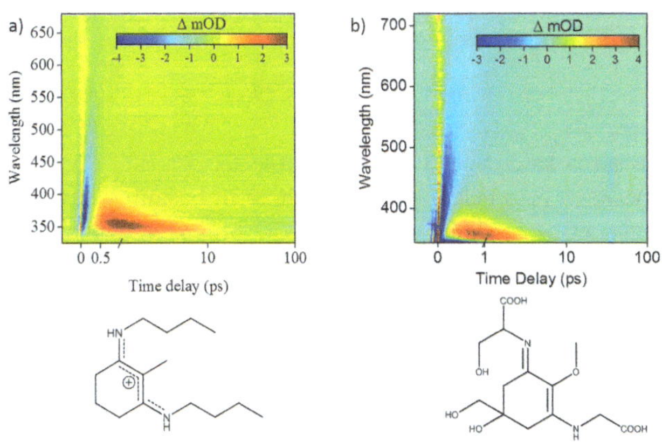

Fig. 1 (a) False colour heat map of transient absorption data, reported by Woolley *et al.*[1] for a similar system to those studied by Habershon and co-workers. (b) False colour heat map of transient absorption date of the MAA Shinorine.

1 J. M. Woolley, M. Staniforth, M. D. Horbury, G. W. Richings, M. Wills and V. G. Stavros, *J. Phys. Chem. Lett.*, 2018, **9**, 3043–3048.

Scott Habershon replied: Yes, the ultimate (if ambitious) aim of this line of research is to try to move towards a more "black box" method for modelling quantum chemical dynamics as a route to exploring the impact of different chemical functionality on molecular photochemistry. Such a method would enable design of photo-functional molecules such as sensors or switches. However, as noted in the question, barriers remain to this, such as the accuracy and expense of *ab initio* electronic structure calculations and methodological aspects such as the appropriate choice of active vibrational modes. Knowing which vibrational modes are essential to capture the experimentally-observed dynamics is very challenging, and may require a multi-method approach, such as using trajectory-surface hopping to inform the set-up of MCTDH simulations.

Todd Martinez asked: Is the diabatization process that you are using dependent on the path taken? If so, how do you deal with the resulting ambiguity?

Scott Habershon answered: In the propagation diabatization scheme, we propagate the diabatization matrix outwards from some initial geometry (usually the minimum-energy geometry on the lowest electronic state). In this scheme, the diabatization matrix is propagated using line integrals along the shortest-path between configurations (defined using simple Euclidean distance). As a result, there is some ambiguity here; if one changes the set of sampled configurations, then the propagation of the diabatization matrix will also change, and so the diabatic states will also change.

To date, we have not noticed this becoming a problem in our simulations. Our impression is that the density of configurations which must be sampled to generate an accurate potential energy surface implicitly means that there is little variation between the propagation of the diabatization matrix between different simulation runs. In addition, we have recently begun using Sobol grid sampling to generate configurations which are subsequently used to construct the diabatic potential energy surfaces and inter-state couplings; this means that the MCTDH dynamics is less susceptible to "jitter" in the constructed potential energy surfaces which might arise from differences in sampled configurations. This point is discussed more in a previous publication.[1]

1 G. W. Richings, C. Robertson and S. Habershon, *J. Chem. Theory Comput.*, 2019, 15, 857–870.

Vasilios Stavros said: You discuss possible factors that could influence the agreement between your calculations and those of Losantos and coworkers. One is the inclusion of more diabatic states in the projection scheme, which would lower the barrier along a relaxation coordinate. Do your calculations include solvation effects and if not, is it possible that the disagreement may also be due to the absence of these effects?

Scott Habershon replied: Our calculations do not include solvation effects, and are performed in vacuum. It is possible that solvent effects can change the character of the electronic states, and that could, of course, influence the accuracy of comparison with experiment.

Including solvation effects directly into our on-the-fly scheme is the next frontier here. We can see some ways forward, such as using an Ehrenfest-type scheme for mixed quantum/classical dynamics whereby the solvent molecules are modelled using classical molecular dynamics, but the on-the-fly nature of our approach might enable alternatives; we are actively investigating these aspects.

Peter Jomo Walla opened discussion of the paper by Heiko Lokstein: I would like to make a comment about the relative contributions of chlorophyll two-photon excitation and carotenoid two-photon excitation to the two-photon excitation spectra of LHC II and other pigment-protein complexes. In 2017 we addressed this question in a paper in which we explored relative cross-sections in two-photon excitation spectra of chlorophylls (Chls) as well as carotenoids.[1] After recognizing Dr Lokstein's contribution (DOI: 10.1039/c8fd00198g) I contacted him because, based on his reported results, he concluded that there is almost no contribution from carotenoid two-photon excitation to the two-photon spectra of LHC II. However, these results did not consider the two-photon excitations spectra of carotenoids as observed by us.[1] We therefore agreed to join forces and directly compare all two-photon results that we both have obtained in our previous studies. I have prepared a slide that summarizes the major conclusions from these comparisons (shown in Fig. 2).

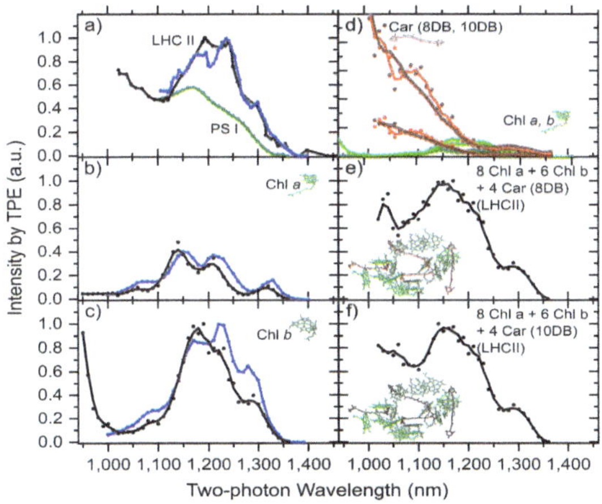

Fig. 2 Comparison of experimental two-photon excitation spectra of LHC II, PS I and isolated pigments. (a) Two-photon excitation spectra of LHC II by Walla *et al.*[2] (black curves) and Betke and Lokstein (blue) (DOI: 10.1039/c8fd00198g), as well as, PS I by Wehling *et al.*[3] (green). (b) and (c) Two-photon excitation spectra of isolated Chl a and b from Gacek *et al.*[1] and Betke and Lokstein (blue) (DOI: 10.1039/c8fd00198g). (d) Two-photon excitation spectra of carotenoids with 8 and 10 conjugated double bonds from Gacek *et al.*[1] (red) in comparison to two-photon excitation spectra of Chls *a* and *b*. (green). The relative magnitudes of the spectra in (d) reflect the relative contributions of the individual pigments to two-photon spectra in ref. 1. (e) and (f) LHC II two-photon excitation spectra reconstructed from summing experimental two-photon excitation spectra of 8 Chls *a*, 6 Chls *b* and 4 carotenoids (xanthophylls) from the individual pigment two-photon excitation spectra shown in (d). The protein and pigment structures were created with VMD: Visual molecular dynamics,[4] from a data set of Su *et al.*[5]

In general, the two-photon spectra observed in our work are in good agreement with those reported by Dr Lokstein. Fig. 2a link accordingly demonstrates that the LHC II spectrum reported in Betke and Lokstein (blue) agrees quite well with the one we reported in 2000 in Walla *et al.*,[2] (black in Fig. 2a) only that the spectral range reported in our publication is larger. For comparison, Fig. 2a also shows the two-photon excitation spectrum of PS I that we reported in 2005, for example.[3] Also the agreement of the two-photon spectra of Chl *a* reported in Gacek *et al.*[1] as well as Betke Lokstein (DOI: 10.1039/c8fd00198g) in solution (Fig. 2b, black and blue, respectively) and of Chl *b* in solution (Fig. 2c) is quite reasonable. Minor differences are likely due to different solvents that were used in both studies. A surprising result observed in Gacek *et al.* was that Chl *b* has, in general, a significantly larger two-photon absorption cross-section than Chl *a*.[1]

However, the larger spectral range reported for LHC II in Walla *et al.*[1] (Fig. 2a, black) demonstrates that there are more contributions from pigments other than Chls since the Chls two-photon absorption cross section drastically drops in the spectral range below 1100 nm (Fig. 2b and c). These additional signals from other pigments to the two-photon excitation spectra of LHC II can be readily explained by significant contributions from carotenoid dark states since the two-photon excitation spectra of carotenoids reported in Gacek *et al.*[1] are quite large in this spectral range (Fig. 2d).

Relative magnitudes of the two-photon excitation spectra of individual carotenoids and Chls were determined in Gacek *et al.*[1] Thus, it appears to be reasonable to reconstruct a LHC II two-photon excitation spectrum by using the sum of 8 Chls *a*, 6 Chls *b* and 4 carotenoids corresponding to the stoichiometric ratio of these pigments in LHC II. In fact, the agreement of these reconstructed spectra (Fig. 2e and f) with that of the experimental two-photon spectrum of LHC II (Fig. 2a) is quite reasonable.

The spectrum of PS I (green curve in Fig. 2a) demonstrates that the relative contribution of carotenoids may be even larger here. PS I contains exclusively Chl *a* (besides β-carotene), hence, this is in good agreement with the observation that Chl *a* has a smaller two-photon cross-section than Chl *b*.

The observations made by Dr Lokstein (DOI: 10.1039/c8fd00198g) that LHC II, containing different complements of carotenoids having also different numbers of conjugated double bonds, have very similar two-photon excitation spectra agrees well with our observation (Gacek *et al.*[1]), that carotenoids with different numbers of conjugated double bonds have surprisingly similar two-photon excitation spectra (Fig. 2d).

In summary, carotenoid two-photon excitation contributes significantly to the two-photon excitation spectra of PS I and LHC II below a spectral range of around 1100 -1200 nm. This observation also provides evidence that in both PS I and LHC II, excitation energy transfer occurs from carotenoid dark states to Chls; otherwise no Chl fluorescence would be observed after carotenoid two-photon excitation. Dr Lokstein and we agree to join forces and further explore our observations, in order to provide more quantitative information about what the contributions of individual pigments, Chls and carotenoids (xanthophylls) at various excitation wavelengths in two-photon excitation and absorption spectra of photosynthetic pigment-protein complexes are.

Dr Lokstein might also elucidate further, to what extend these insights as well as further observations obtained by his group support current suggestions about the molecular mechanisms relevant in the regulation of photosynthetic light-harvesting.

1 D. A. Gacek, A. L. Moore, T. A. Moore and P. J. Walla, *J. Phys. Chem. B*, 2017, **121**, 10055.
2 P. J. Walla, J. Yom, B. P. Krueger and G. R. Fleming, *J. Phys. Chem. B*, 2000, **104**, 4799.
3 A. Wehling and P. J. Walla, *J. Phys. Chem. B*, 2005, **109**, 24510.
4 W. Humphrey, A. Dalke and K. Schulten, *J. Mol. Graph.*, 1996, **14**, 33.
5 X. Su, J. Ma, X. Wei, P. Cao, D. Zhu, W. Chang, Z. Liu, X. Zhang and M. Li, *Science*, 2017, 357, 815.

Heiko Lokstein responded: I became aware of Dr Walla's 2017 paper[1] when I was contacted by him following the online publication of our current Faraday Discussions paper (DOI: 10.1039/c8fd00198g). We both do agree that our experimental data are rather similar, but our interpretations of these were considerably differing, namely regarding the relative contributions of carotenoids and chlorophylls (Chls) to the two-photon excitation (TPE) spectra of various photosynthetic pigment-protein complexes, and, in particular those of plant major light-harvesting complex II (LHC II). In fact, we concluded that there is (almost) no contribution from carotenoids to two-photon excitation spectra of LHC II. Moreover, we pointed out that direct TPE of (Bacterio)Chls in the presumed spectral region of the forbidden carotenoid S_1 ($2^1A_g^-$) state(s) was largely neglected in previous publications.

However, after discussions with Dr Walla and careful re-examination of our own data we both came to the conclusion that our previous bold statements have to be reconsidered. We therefore agreed to join forces and directly compare all the two-photon excitation results that we had obtained.

Indeed, Walla and co-workers had previously reported TPE data for a more extended spectral range, extending below 1100 (550 nm) for LHC II. These data may indicate that there may be contributions from pigments other than Chls (namely from dark states of carotenoids), since the Chls two-photon absorption cross section apparently drops in the TPE range below 1100 (550 nm) nm, possibly carotenoid contributions may even be apparent already at 1200 (600) nm.

Summarizing, we do agree that, apparently, carotenoid two-photon excitation can significantly contribute to the two-photon excitation spectra of photosynthetic pigment protein complexes, including PS I and LHC II, in particular in the spectral range below 1100 (550 nm), possibly even already at 1200 (600) nm, however this would need to be further quantified.

Dr Walla and we agreed to further examine our observations, in order to provide a more quantitative view of the contributions of individual pigments, Chls and carotenoids (xanthophylls), to two-photon excitation at various excitation wavelengths in photosynthetic pigment-protein complexes.

Moreover, we would like to stress that the mechanism of photoprotective non-photochemical Chl-fluorescence quenching (NPQ) proposed by Dr Wallas group, based on a combination of TPE and pulse-amplitude modulated fluorescence (PAM) experiments[2] would be consistent with our own observations: Non-linear polarization spectroscopy in the frequency domain (NLPF) experiments with LHC II in different states of aggregation (commonly serving as a model system for NPQ) indicated altered xanthophyll-Chl interactions: Upon pumping in the Chl *a*+*b* Qy spectral region (640–680 nm) and probing at selected wavelengths in the energetically much higher allowed xanthophyll absorption region (490–510 nm) it was found that a xanthophyll molecule absorbing at ~510 nm strongly interacts with a Chl *b* absorbing at about 650 nm.[3] Upon aggregation of LHC II the picture changes completely: now the same xanthophyll molecule, probed at 510 nm, interacts with Chl(s) *a* absorbing at ~682 nm, comprising the terminal emitter in LHC II.[4]

1 D. A. Gacek, A. L. Moore, T. A. Moore and P. J. Walla, *J. Phys. Chem. B*, 2017, **121**, 10055.
2 S. Bode, C. C. Quentmeier, P. N. Liao, N. Hafi, T. Barros, L. Wilk, F. Bittner and P. J. Walla, *Proc. Natl. Acad. Sci. U. S. A.*, 2009, **106**, 12311.
3 B. Voigt, M. Krikunova and H. Lokstein, *Photosynth. Res.*, 2008, **95**, 317.
4 H. Rogl, R. Schödel, H. Lokstein, W. Kühlbrand and A. Schubert, *Biochemistry*, 2002, **41**, 2281.

Tomas Polivka remarked: Your detection of two-photon absorption relies on carotenoid-chlorophyll energy transfer. Can the absence of any carotenoid contribution to the two-photon signal in LHCII simply mean there is no carotenoid-chlorophyll energy transfer *via* the S_1 state in LHCII?

Heiko Lokstein responded: A very good question. Indeed, it seems that excitation energy transfer onto chlorophyll S_1 following supposed two-photon excitation of carotenoid S_1 states is somewhat inefficient (or even negligible) as also suggested by previous studies with purple bacterial LH2 complexes (*e.g.*, Krikunova *et al.*[1]) – but further experimental work is clearly required.

1 M. Krikunova, A. Kummrow, B. Voigt, M. Rini, H. Lokstein, A. Moskalenko, H. Scheer, A. Razjivin and D. Leupold, *FEBS Lett.*, 2002, **528**, 227–229.

Tomas Polivka addressed Peter Jomo Walla and Heiko Lokstein: From the data presented by Peter Walla it seems that there is onset of carotenoid contribution to the two-photon absorption in LHCII at significantly higher energies than expected. Do you think this imply that the energy threshold for carotenoid-chlorophyll S_1-mediated energy transfer channel is much higher than expected? Only hot S_1 contributes to energy transfer? Or even some other state?

Peter Jomo Walla responded: We were also surprised by this observation and it might indeed indicate energy transfer from hot carotenoid S_1 states. We have already speculated about this possibility in a paper in 2002 (Walla *et al.*[1], see also A. Wehling and P. J. Walla[2]). When the 0-0-transition of the carotenoid S_1 state is significantly higher than that of the tetrapyrrole acceptor, the contributions of the carotenoid S_1 0-0-transtion relative to hot states might be larger in two-photon excitation spectra. On the other hand, the reported two-photon excitation spectra (Gacek *et al.*[3]) might also indicate the presence of further dark states between the carotenoid S_1 and carotenoid S_2 state such as the putative carotenoid S_x or carotenoid S* states proposed by others. We currently cannot disentangle the exact nature of these states but the results provide evidence that there are dark carotenoid states, either hot S_1, S_x or S* states, that can efficiently transfer energy to tetrapyrroles and contribute to two-photon excitation in dyads and pigment-protein complexes.

1 P. J. Walla, P. A. Linden, K. Otha and G. R. Fleming, *J. Phys. Chem. B*, 2002, **106**, 1909.
2 A. Wehling and P. J. Walla, *J. Phys. Chem. B*, 2005, **109**, 24510.
3 D. A. Gacek, A. L. Moore, T. A. Moore and P. J. Walla, *J. Phys. Chem. B*, 2017, **121**, 10055.

Heiko Lokstein replied: It seems, taking also the data of Dr Walla and co-workers into account, that obviously two-photon absorption in LHC II starts at significantly higher energies than previously expected, (*i.e.*, for S_1 of the xanthophylls). In fact, we cannot exclude that the energy threshold for carotenoid-to-chlorophyll excitation energy transfer is much higher, so that possibly hot S_1 states or, even, some other, yet unassigned, state(s) contribute… Clearly, further experimental work is required.

Minjung Son addressed Heiko Lokstein and Peter Jomo Walla: Could you make some general comments on the sensitivity of this experiment as to whether it can distinguish between contributions from different carotenoid molecules, *i.e.*, whether the carotenoid is a β-carotene or a lutein, or whether the carotenoid has eight conjugated bonds or nine, for example?

Peter Jomo Walla replied: The two-photon excitation spectra of carotenoid-tetrapyrrole dyads containing carotenoids with different numbers of conjugated double bonds looked surprisingly similar (Gacek *et al.*[1]). Again, dominant contributions from hot carotenoid S_1 energy transfer might be an explanation in this case. Then, the cut off-wavelength in two-photon excitation spectra and energy transfer for each vibrational donor level would rather depend on the limiting energy level of the acceptor and not so much on varying 0-0-transtion energies of the carotenoid donors themselves. However, our studies also indicate that the overall efficiency of energy transfer depends on the number of conjugated double bonds or the carotenoid used, *i.e.* the relative contribution of the carotenoid donors to two-photon spectra varies compared to that of the tetrapyrrole acceptor. Two-photon excitation spectra of pigment-protein complexes reconstituted with different carotenoids that include more of the spectral range dominated by carotenoid dark states might provide further insights into that.

1 D. A. Gacek, A. L. Moore, T. A. Moore and P. J. Walla, *J. Phys. Chem. B*, 2017, **121**, 10055.

Heiko Lokstein answered: It appears that – regarding basically our data (see also Fig. 5 of the current paper (DOI: 10.1039/c8fd00198g)) – we cannot distinguish contributions from different carotenoids, or, whether the carotenoid has eight conjugated bonds or nine, or even more… This may, at least in part, also be related to inefficient excitation energy transfer to chlorophylls.

Stephen Bradforth asked: As your experiment uses two photons from the same beam, this limits somewhat the polarization information available. However, in your presentation you showed polarization ratio data comparing two circular polarized photons to two linearly polarized photons. Can you interpret what that ratio information is revealing?

Heiko Lokstein answered: The polarization-dependent experiment does not aim at obtaining information about (relative) orientations of the molecules. However, unlike in linear absorption, for two-photon absorption the polarization of the exciting light matters, even for randomly oriented molecules.[1] Thus, two-photon excited fluorescence was recorded for linear and circular excitation with otherwise identical measuring conditions.

The two-photon polarization ratio can be an indicator for the symmetry of the molecules investigated.[1] For the $S_0 \rightarrow S_1$ transition of a carotenoid (or linear polyene), a value of ~ 0.8 would be expected.[2] In our polarization measurements with LHCII, however, values between 1 and 1.2 were obtained for the spectral range of 610–690 nm and close to 1 below 600 nm. This would suggest that the excited molecules have, unlike carotenoids, a molecular symmetry different from the C_{2h} point group. Furthermore, the polarization ratio has been measured for pure chlorophyll *a* in the range between 660 to 680 nm. It was found to be comparable to the values for LHCII. These observations, in turn, can be taken as further indication that the two-photon-excited molecules in this spectral region are not carotenoids, but chlorophylls in LHCII.

1 W. M. McClain, *J. Chem. Phys.*, 1971, **55**, 2789–2796.
2 R. R. Birge and B. M. Pierce, *J. Chem. Phys.*, 1979, **70**, 168–178.

Vasilios Stavros addressed Heiko Lokstein and Peter Jomo Walla: In terms of the (simplified) energy level scheme presented in Fig. 2 (DOI: 10.1039/c8fd00198g), I would anticipate large spectral changes in the excited state absorption (ESA) following excited state energy transfer from the carotenoid to the chlorophyll. From the previous studies that you cite in your paper,[1–4] is there any evidence that such large spectral shifts are observed in, say, transient absorption studies? If so, I would expect this to be indicative of initial TPE of the carotenoid.

1 B. P. Krueger, J. Yom, P. J. Walla and G. R. Fleming, *Chem. Phys. Lett.*, 1998, **310**, 57–64.
2 P. J. Walla, P. A. Linden, C. P Hsu, G. D. Scholes and G. R. Fleming, *Proc. Natl. Acad. Sci. U. S. A.*, 2000, **97**, 10808–10813.
3 P. J. Walla, J. Yom, B. P. Krueger and G. R. Fleming, *J. Phys. Chem. B*, 2000, **104**, 4799–4806.
4 P. J. Walla, P. A. Linden, K. Ohta and G. R. Fleming, *J. Phys. Chem. A*, 2002, **106**, 1909–1916.

Peter Jomo Walla responded: Actually, there is also some evidence for carotenoid two-photon excitation from two-photon ESA studies. When exciting pure carotenoids at two-photon wavelengths corresponding to energies significantly below the allowed carotenoid S_2 absorption ($\lambda_{TPE}\sim$1200–1310 nm), carotenoid dark state ESA decay kinetics ($\lambda_{det}\sim$550 nm) are observed corresponding to the lifetime of the dark carotenoid S_1 state in the absence of any energy transfer (~9–15 ps). When repeating the same experiment with the carotenoids in bacterial light-harvesting complexes the carotenoid dark state ESA decay shifts to 2 ps in *Rb. sphaeroides*,[1,2] confirming the time constant for carotenoid dark state to bacteriochlorophyll energy transfer previously reported by Zhang *et al.*[3] In complexes with less overall carotenoid-chlorophyll energy transfer (*Rps. acidophila*) two-photon ESA constants are observed that are closer to the carotenoid S_1 lifetime in the absence of energy transfer (~ 7 ps). Also, different ESA kinetics are observed when exciting the same state energy by two- as well as one-photon excitation in green plants or algae. For example, the fast decay kinetics (~ 0.8 ps) observed in the spectral range of the carotenoid dark state ESA ($\lambda_{det}\sim$550 nm) with a two-photon excitation wavelength of $\lambda_{TPE}\sim$1150 nm is not observed in the same complex with one-photon excitation at $\lambda_{OPE}\sim$575 nm.[4] In both cases identical state energies are excited with both types of excitation but different kinetics are observed. This excludes that the same pigment, *i.e.* only Chlorophylls, is excited in both cases as exciting the same pigment at the same state energy should result in the same ESA kinetics. Again, the faster kinetics observed after two-photon excitation in the spectral range of the carotenoid dark state ESA that is not observed after chlorophyll one-photon excitation of the same state energies corresponds to the time scales for energy transfer from forbidden carotenoid states to chlorophylls.

1 P. J. Walla, P. A. Linden, C. P. Hsu, G. D. Scholes and G. R. Fleming, *Proc. Natl. Acad. Sci. U. S. A.*, 2000, **97**, 10808–10813.
2 Walla, Fleming *et al.*, Ultrafast Phenomeny XII, 671–673, (2000).
3 J.-P. Zhang, R. Fujii, P. Qian, T. Inaba, T. Mizoguchi, Y. Koyama, K. Onaka and Y. Watanabe, *J. Phys. Chem. B*, 2000, **104**, 3683.
4 A. Wehling and P. J. Walla, *J. Phys. Chem. B*, 2005, **109**, 24510.

Heiko Lokstein answered: A very good question, indeed. I have to admit that we have not performed ourselves excited state absorption (ESA) studies following two-photon excitation of carotenoids, except for β-carotene in solution. However, as apparent from the discussion, it seems that excitation energy transfer to (bacterio)chlorophylls following two-photon excitation of carotenoids, at least in the spectral region above 1200 (600) nm, is inefficient (or even negligible). Thus, the origin of the previously observed ESA signals following (supposed) two-photon excitation of carotenoids in pigment-protein complexes may well have to be reconsidered, as also suggested by our current paper.

Graham Fleming asked: The potential surfaces of carotenoids are very complex and there may be multiple minima on a single electronic state surface such as S_1. In fact, in experiments we did some time ago we found that the transient infrared spectra of β-apo carotenal obtained by either exciting S_2 or two photon excitation of S_1 were not the same and the lifetime was different also.[1] Thus representing the energy levels by a single line can lead to significant ambiguity in my view.

1 Y. Pang, M. A. Prantil, A. J. Van Tassle, G. A. Jones and G. R. Fleming, *J. Phys. Chem. B*, 2009, **113**, 13086–13095.

Heiko Lokstein replied: Indeed, we have to admit that the (generally) used energy level schemes for carotenoids are somewhat over-simplified. There is also theoretical evidence that the states reached by different excitation schemes are not the same. However, the allotted presentation time precluded introduction of any more sophisticated representations!

Andrew Marcus opened a discussion of the paper by Spiridoula Matsika: You found that in general the distance between adjacent pyrimidine bases is one of the key factors that can affect the energies of charge transfer states and the rate of cyclobutane pyrimidine dimer (CPD) formation. Is CPD formation similarly sensitive to the local conformation (or relative orientation) of adjacent pyrimidine bases?

Spiridoula Matsika answered: This question has been addressed before by many studies. Studies by Schatz and co-workers examined that questions several years ago[1,2] and found that while the orientation plays some role, the most important factor is the distance between the two thymine double bonds. More recent studies have looked at many geometrical parameters in more detail,[3] and they find other geometrical parameters affecting CPD formation but the distance is still the most significant.

1 M. McCullagh, M. Hariharan, F. D. Lewis, D. Markovitsi, T. Douki and G. C. Schatz, *J. Phys. Chem. B*, 2010, **114**, 5215–5221.
2 Z. Z. Pan, M. McCullagh, G. C. Schatz and F. D. Lewis, *J. Phys. Chem. Lett.*, 2011, **2**, 1432–1438.
3 I. Conti, L. C. Martinez Fernandez, L. Esposito, S. Hofinger, A. Nenov, M. Garavelli and R. Improta, *Chem.–Eur. J.*, 2017, **23**, 15177–15188.

Luis Ortiz-Rodríguez commented: It is known that DNA can adopt different conformations such as A-, B-, or Z-forms. In your work, it is mentioned that conformational restrictions exerted by the helical structure of DNA are involved in the stabilization of the charge transfer (CT) states. One parameter highlighted in the paper (DOI: 10.1039/c8fd00184g) is the distance between the stacking bases, which is very sensitive to the conformation of DNA. Could you comment on the DNA conformation selected in the investigation? Do other parameters such as the sugar pucker, intrastrand phosphate-phosphate distance and twist angle play an important role in the stabilization of the CT-states?

Spiridoula Matsika answered: The DNA conformation that we adopted in this study is a regular B DNA form. Other parameters may play some role in the stabilization of the CT states, but we did not examine them in detail. The distance however seems to explain the trends we observe, and there is a direct correlation between the distance and energy of the CT states, so our results indicate that distance is the most important parameter.

Marta Duchi remarked: McCamant's group have reported that the AMBER ff14SB and ff14 force fields designed for double stranded DNA do not accurately predict the conformations of single-stranded thymine due to the over-stabilisation of 5′-thymine in a *syn* conformation.[1] Do you observe similar effects in your molecular dynamics simulations?

1 C. Nganou, S. D. Kennedy and D. W. McCamant, *J. Phys. Chem. B*, 2016, **120**, 1250–1258.

Spiridoula Matsika responded: The study by McCamant focused on single stranded thymine strands and dinucleotides, and it was concluded that the current AMBER force field does not accurately predict the conformation of single-stranded thymine. Our work considers double stranded oligonucleotides which is expected to be much better represented by AMBER force fields. That said, we do not have a direct way to compare our results to experiments to verify the validity of the simulations. Our results, however, indirectly agree with experiment, and we believe this is a form of validation.

Bern Kohler said: You argue persuasively that a low-lying CT state can quench CPD formation by depopulating a dimer precursor (DP) state that would otherwise form the photoproduct. My question has to do with whether the probability that an initial excited state can reach the DP state in the first place could also be an important factor in determining sequence-dependent CPD quantum yields. In your earlier study,[1] you argue that the energetic proximity of the bright exciton state formed from the two $\pi\pi^*$ states of the two T bases to the DP state will promote internal conversion to the latter state. However, it seems that the initial excited state could also decay to a non-DP state that is lower in energy such as an $n\pi^*$ state localized on either T. In addition, there is a bright $n\pi^*$ state localized on G that could decay to the DP state, possibly even resulting in some CPD formation. In a sequence like GTTA, one could also ask whether excitons associated with GT or TA mediate dimerization by the central thymines. Can you say anything about whether the many internal conversion pathways that seem to exist for the various kinds of initial excitations make it more or less likely to reach the DP state that is subject to quenching by the CT mechanism?

1 W. Lee and S. Matsika, *Phys. Chem. Chem. Phys.*, 2015, **17**, 9927–9935.

Spiridoula Matsika responded: It is absolutely true that there are many pathways present, and initial excitation can decay in many different ways instead of reaching the DP state. That is a reason that CPD formation in general has a very low quantum yield. In our studies we assume that these other pathways will be similar in the different sequences we studied, and in that case, what differentiates the sequences is the probability to decay from the DP state to the CT state. This seems like a reasonable assumption since the most important difference between the sequences is the energy of the CT states. A more complete study would have to examine all the different pathways and their competition, but that is a formidable task.

Vasilios Stavros stated: I have two questions/comments:

1. Am I correct in assuming that the role of the cation is simply as a spectator, or does it influence the energy of the charge transfer state?
2. Can you say a little more about solvation, specifically the size of the octahedral box used in your simulations, and what guided your choice of box size?

Spiridoula Matsika responded: 1. The cations can influence the energy of the CT state depending on their positions. However, they are expected to be found around negatively charged backbone in general (because they are positively charged), so their positions are very unlikely to be sequence dependent. Therefore, it is not likely that the cations make any significant contribution to the CT energy differences observed in our comparisons. In this sense, I think that, in our comparisons, it is safe to consider the role of the cation as a spectator.

2. For solvation, water molecules were added to make the minimum distance between any atom of DNA and the edge of the octahedral box be 8 ångstroms.

The number of water molecues added is different depending on the sequences, but they are usually around 8000–9000 molecules. For example, a total 8732 of water molecules were added for the TmCG sequence.

Stephen Bradforth commented: Your calculations show very subtle differences in the CT state energies. Have you performed a sensitivity analysis with respect to the solvation model on your results?

Spiridoula Matsika answered: That is a very good point. We have not performed a sensitivity analysis with respect to the solvation model. It is correct however that the solvation model can affect the exact energies we see. We do not expect this to alter the trends we see and the conclusions we have made, but if we are interested in very accurate energetics we should explore the solvation model further.

Tom Oliver commented: Echoing the earlier comment made by Bern Kohler, the nature of the initially photoexcited state can greatly alter the yield of charge-transfer in DNA strands.[1] In our recent study of 2'-deoxyguanosine 3'-monophosphate 5'-thymidine, d(GpT) we observed the relative product branching into d(G^+pT^-) products was wavelength independent between 273–257 nm, and determined that the initial Franck–Condon excited states of the dinucleotide are significantly delocalised across both nucleobases.[2] Could you comment on the nature of the initially excited states predicted by your ADC(2) calculations, and any whether $^1\pi\pi^*$ or $^1n\pi^*$ states preferentially couple to charge-transfer states.

1 Y. Zhang, J. Dood, A. A. Beckstead, X.-B. Li, K. V. Nguyen, C. J. Burrows, R. Improta and B. Kohler, *Proc. Natl. Acad. Sci. U. S. A.*, 2014, **111**, 11612–11617.
2 M. Duchi, M. P. O'Hagan, R. Kumar, S. J. Bennie, M. Carmen Galan, B. F. E. Curchod and T. A. A. Oliver, *Phys. Chem. Chem. Phys.*, 2019, DOI: 10.1039/C8CP07864E.

Spiridoula Matsika replied: The initially excited states in the Franck–Condon region in our calculations had various degrees of delocalization along two bases. Some of them were localized on one base, and some were delocalized over two bases, either the two pyrimidine bases, or in some cases between thymine and the adjacent guanine base. In agreement with the above ref. 2, we also see that the nature of the vertical excited states is very dependent on the specific geometry and relative orientation of the bases. The charge transfer states we see involve π orbitals, so they are mainly coupled to the $\pi\pi$ states.

Carlos E. Crespo-Hernández remarked: According to the results presented in Fig. 3 in your paper (DOI: 10.1039/c8fd00184g), the energy of the charge transfer state of a given sequence varies by more than 1 eV depending on the specific molecular dynamics snapshot used in the calculation. Can reliable inferences be made using an average charge transfer energy that is based on 20 snapshots given that the standard deviation is 0.2 to 0.4 eV for each of the sequences studied in this work? In other words, is the averaging of 20 snapshots accurate enough to represent the average charge transfer energy of a given sequence relative to the average energy of the other sequences? How sensitive are the average charge transfer energies, and the standard deviations, to the number of snapshots used in the calculation?

Spiridoula Matsika replied: The strong variation of the energy of the CT states with the snapshot is physical and insightful since it shows how sensitive the CT states are to the environment. We believe the average to be qualitatively correct, especially for this type of study, which compares different systems. So we expect the trends observed in this study to be preserved even if we do larger simulations with more than 20 snapshots. The average is probably not statistically converged, but given the expense of the calculations this is the best that can be done at the present time.

Carlos E. Crespo-Hernández followed: The reported one-electron reduction potentials for cytosine and thymine shown in Table 1 are practically identical taking into consideration the theoretical and experimental accuracy of the methods used in ref. 48 and 49 in your paper (DOI: 10.1039/c8fd00184g), respectively. Then, the significantly different average charge transfer energy predicted for the GTC/GCT and GTT/GCC pair of sequences seems surprising, given that the average distance between the two bases involved in charge transfer does not vary by more than 0.1 ångstroms according to Fig. 4 in your paper. Are other base stacking geometric parameters (such as tilt, twist, shift, slide, *etc.*) important in the determination of the average charge transfer energies?

Spiridoula Matsika responded: The difference in the experimental reduction potentials between thymine and cytosine is actually substantial. And it is well established that thymine is reduced more readily than cytosine. This we believe is one component leading to the difference in the energies of CT states. The distance seems to be another parameter since shorter distances are observed between GT compared to GC. So, these two effects contribute in the same direction of affecting the energy of CT states. We think the additive effect makes the shift on the CT states stronger. Other geometrical parameters may play some small role, but we have not explored that in this work. It is worth examining in more detail in the future.

Shou-Ting Hsieh communicated: The research is interesting, and sorry for such a naive question. I just think that our DNA sequence was determined when we were born (neglecting the probability of mutation). Therefore, even though we know some people have a sequence that may cause CPD easily, it is hard to do any "modification". I would like to know what kinds of method may be developed based on this research. I guess that one day in the future, maybe we can predict the probability of CPD formation based on the DNA sequence. As precision medicine becomes more important, this may allow people to know with how much care they need to prevent UV light.

Spiridoula Matsika communicated in reply: We are still far away from being able to predict the probability of CPD formation based on someone's DNA sequence and offer solutions. But progress is made in small steps, and we hope that the current fundamental research such as ours and that of others can help in making progress towards this direction. By the way, there are experimental groups that try to determine experimentally the effect of sequence on CPD formation, such as the group of Prof. John-Stephen Taylor at Washington University.

Tom Oliver opened discussion of Christopher Grieco's paper: What is the mechanism for the very fast intersystem crossing you observe in the quinone moiety?

Christopher Grieco replied: The very fast intersystem crossing process appears to occur from the $n\pi^*$ singlet state of the quinone to the triplet manifold. Although we are directly exciting to the $\pi\pi^*$ singlet state in our experiments, the observation of the two excited state absorption bands described in the paper at the earliest resolvable time delay is evidence that internal conversion from the $\pi\pi^*$ singlet state to the $n\pi^*$ singlet state occurs very rapidly. In accordance to the El-Sayed rules, the intersystem crossing process likely occurs from this state to a $\pi\pi^*$ triplet state. Because of the rapid time constant for this process, we consider that the energy levels of these two states may also be close. However, we currently do not have data to support this. We believe quantum mechanical calculations would provide much further insight into the intersystem crossing mechanism.

Matthew Turner said: In a previous paper[1] you suggested that, when in cyclohexane, 4-*t*-butylcatechol could aggregate through two possible pathways; a "linear" and a "bifurcated" hydrogen bonding structure. In contrast, in this current paper (DOI: 10.1039/c8fd00231b) you discuss the dimer of 3,5-di-*t*-butylcatechol (C) + 3,5-di-*t*-butyl-*o*-quinone (Q) but only consider the "linear" structure. Is it possible the "bifurcated" dimer could form and if so how would this effect the interpretation of your results?

1 C. Grieco, F. R. Kohl, Y. Zhang, S. Natarajan, L. Blancafort and B. Kohler, *Photochem. Photobiol.*, 2019, **95**, 163–175.

Christopher Grieco responded: In the paper, we studied intermolecular hydrogen bonding interactions that occur within 4-*tert*-butylcatechol aggregates in cyclohexane solutions. The "bifurcation" refers to an intermolecularly hydrogen-bonded O-H group in a catechol molecule that can simultaneously donate an intermolecular hydrogen bond to another neighboring catechol molecule (see Fig. 3a). In this *Faraday Discussions* paper (DOI: 10.1039/c8fd00231b), we considered a linearly hydrogen-bonded structure between Q and C. In this case, we argue (see the text of the paper) that a single O-H group is unlikely to hydrogen bond to both of the carbonyl groups in Q simultaneously (see Fig. 3b). We consider that a bifurcated structure in which the intramolecularly hydrogen-bonded O-H group in C can also donate a hydrogen bond to a carbonyl group in Q is also unlikely (see Fig. 3c). This is because forming a heterodimer of C and C causes the free O-H band in the FTIR spectrum to decrease in intensity, which signifies that the free O-H group must be donating a hydrogen bond.

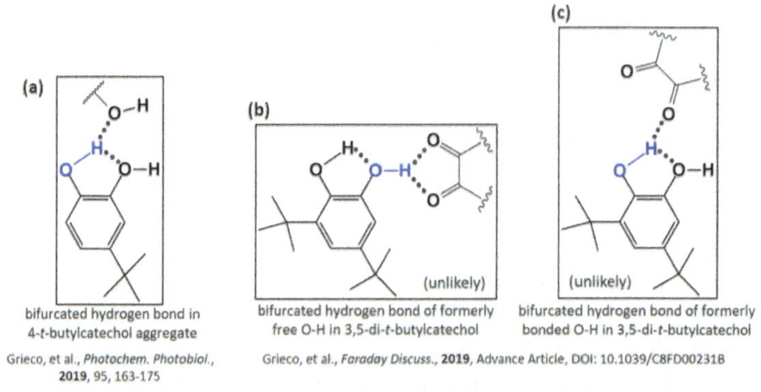

Fig. 3 (a) Possible bifurcated hydrogen bond structure in a 4-*t*-butylcatechol aggregate. The O–H group that intramolecularly hydrogen bonds to the other O–H group in the catechol also intermolecularly hydrogen bonds to an O–H group of a neighboring molecule. (b) Bifurcation of the intermolecular hydrogen bond between an O–H group of 3,5-di-*t*-butylcatechol and two carbonyl groups in a neighboring 3,5-di-*t*-butyl-*o*-quinone molecule. (c) Bifurcated hydrogen bond structure for the O–H group in a 3,5-di-*t*-butylcatechol molecule that retains its intramolecular hydrogen bond while simultaneously donating an intermolecular hydrogen bond to one of the carbonyl groups of a 3,5-di-*t*-butyl-*o*-quinone molecule.

1 C. Grieco, F. R. Kohl, Y. Zhang, S. Natarajan, L. Blancafort and B. Kohler, *Photochem. Photobiol.*, 2019, **95**, 163–175.

Carlos E. Crespo-Hernández said: Fig. 2 in your paper (DOI: 10.1039/c8fd00231b) shows that the Q : C heterodimer can be excited selectively from the Q and C monomer components in the Q + C mixture at wavelengths longer than 450 nm. Even though the two-state model that you describe for discriminating the transient absorption spectra produced by the monomers and heterodimers is reasonable, I wonder if you had considered exciting the mixture at 450 nm or longer wavelengths in order to measure directly the spectral and excited-state dynamics of the Q : C heterodimer independently of those of the Q monomer?

Christopher Grieco answered: The process of discriminating the transient absorption spectra produced by the monomers (Q+C) and heterodimers Q : C is an effective, but undesirable method for studying the excited state processes in the Q : C heterodimer. This is because it requires multiple transient absorption measurements to be made, which can intrinsically carry error into the final spectrum determined for Q : C due to the spectral subtraction process involved. A more desirable measurement is as you describe, and it appears possible if exciting with 450 nm. This experiment is worth trying. However, there is a small amount of Q monomer absorption present at 450 nm, and it may still contribute to the transient absorption signal. In particular, the absorption % by Q monomers in the Q+C mixture at 450 nm is ~17%. Using a longer wavelength, such as 500 nm, would be challenging because of possible resonance with the n → pi* transition. Measuring a higher concentration of Q+C would shift the solution equilibrium to Q : C and favorably change the relative absorption of Q and Q : C at 450 nm (*i.e.* > 30 mM Q + 30 mM C). Provided that other experimental conditions can be accommodated, such as a smaller sample path length, the experiment using 450 nm could then be ideal.

Carlos E. Crespo-Hernández asked: In the introductory remarks of your presentation, you described that eumelanin exhibits a complex chemical heterogeneity in which stacking and hydrogen-bonding interactions are important. Could you comment on how stacking and other intermolecular interactions in eumelanin may modulate the excited-state dynamics and the formation of neutral semiquinone radicals you elucidated in the Q : C heterodimer model system?

Christopher Grieco answered: Intermolecular interactions, such as π-stacking, that are present in eumelanin are expected to modulate the excited state dynamics and the formation of neutral semiquinone radicals. For example, alternative, competitive excited state pathways may turn on due to the presence of these different intermolecular interactions. We are actively studying these processes, and believe that the rapid (~10 ps) decay of the excited states in eumelanin materials may involve deactivation pathways that are distinct from what we observe in the Q+C model system studied here (photophysical relaxation *versus* photochemical change). Regardless of the differences between eumelanin and our model system, we believe that it is possible that some of the excited states in eumelanin follow a similar hydrogen atom (net) transfer reaction as in the model

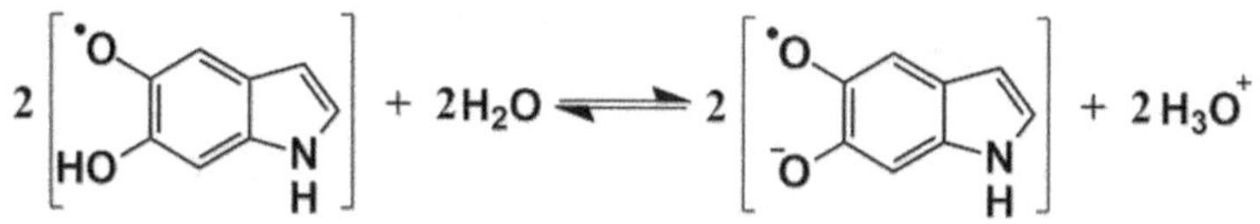

Fig. 4 Redox equilibrium of the neutral and anionic semiquinone forms of 5,6-dihydroxyindole in aqueous solution.

system. To add to this discussion, the neutral semiquinone of the eumelanin precursor molecules are known to undergo a subsequent proton transfer event that could lead to a metastable semiquinone radical ion, which may play an additional role (see Fig. 4). This can occur in eumelanin because its structure is likely to be more polar than the Q and C compounds, and is suspended or dissolved in a polar solvent (water).

Michael Ashfold commented: The photochemistry reported for the benzoquinone:catechol (Q : C) heterodimer is reminiscent of that displayed by other hydrogen-bonded organic chromophores, where Q is the hydrogen-bond acceptor and C the hydrogen-bond donor. [1-3] UV excitation typically populates a $\pi\pi^*$ state, in which the π and π^* orbitals are both localized on either Q or C –forming a configuration that is commonly termed a locally excited (LE) state. The UV absorption data shows that the orbitals involved in the $\pi^*\leftarrow\pi$ transition in the Q : C heterodimer are localized on Q (DOI: 10.1039/c8fd00231b). Analogy with the benzoquinone-water (Q:H_2O) complex, for example, suggests that this $\pi\pi^*$ LE state will evolve to a $\pi\pi^*$ charge-transfer (CT) state with the π-hole localized on C and a π^* electron on Q.[4]

This photoinduced charge separation affects the subsequent hydrogen migration from C to Q.[1] But, as noted elsewhere in this Discussion (DOI: 10.1039/c8fd00240a), the nature of the migrating species (*i.e.* proton *vs.* hydrogen atom) is sensitive to the non-adiabatic coupling between the LE and CT states and thus to the topography of the adiabatic excited state potential energy surface (PES). If the diabatic PESs of the LE and CT states associated with the minimum energy reaction path (MERP) are within the Marcus normal region, then the state reached by $\pi^*\leftarrow\pi$ excitation at the Franck–Condon (FC) geometry has dominant LE character and non-adiabatic coupling between the LE and CT states is weak. The migrating species is likely to be hydrogen atom-like: the H atom 'carries' the electron from C towards Q, and the excited adiabatic PES only gains appreciable CT character after the O-H bond has started extending. In contrast, if the diabatic PESs of the LE and CT states associated with the MERP are barrierless or within the Marcus-inverted region, excitation at the FC geometry samples a region of strong non-adiabatic coupling between the LE and CT states and the migrating species will tend to be proton-like. In this case, an intermolecular Q^-–C^+ charge separation is established at very early time, which is then neutralized by proton-transfer from the O-H group of C to the O atom of Q.

In the case of Q:H_2O, the calculated LE and CT $^1\pi\pi^*$ states show a Marcus inverted evolution along the Q....H–OH.... coordinate.[2] The analogous singlet reaction path in Q : C can be expected to be Marcus inverted also, since catechol is more acidic than water and thus a stronger proton-donor. The calculated $^3\pi\pi^*$ reaction path in Q:H_2O is more reminiscent of normal Marcus behavior.[2] The barriered nature of this process is likely to extend to Q : C also and offers a possible explanation for the deduced ns timescale of the hydrogen migration on the triplet PES. The present data for Q : C heterodimer can be expected to inspire future computational studies of this and related systems.

1 T. Schultz, E. Samovlova, W. Radloff, I.V. Hertel, A.L. Sobolewski and W. Domcke, *Science*, 2004, **306**, 1765–1768.
2 M.Barbatti, *J. Am. Chem. Soc.*, 2014, **136**, 10246–10249.
3 T. Kumpulainen, B. Lang, A. Rosspeintner and E. Vauthey, *Chem. Rev.*, 2017, **117**, 10826–10939, and references therein.
4 T.N.V. Karsili, D. Tuna, J. Ehrmaier and W. Domcke, *Phys. Chem. Chem. Phys.*, 2015, **17**, 32183–32193.

Christopher Grieco responded: We thank you for your in-depth, insightful analysis. We look forward to future computational studies of the processes you describe. Your discussion also inspires future experimental studies in different solvent environments that may perturb the CT state energy in a predictable way, forming an additional way to test the models you discuss.

Some additional responses: does the charge separation process compete with internal conversion from $\pi\pi^* \rightarrow n\pi^*$ in Q? The $n\pi^*$ state is possibly lower in energy than the $\pi\pi^*$ state like in xanthone (for example, see Satzger *et al.*[1]). Our transient absorption spectra of Q monomers in different solvents show peaks that match optical transitions expected for the $n\pi^*$ state of Q. Future computational studies may be useful for determining whether these spectral assignments are correct and to help determine how the $n\pi^*$ state plays a role in the excited state pathways in Q.

1 H. Satzger, B. Schmidt, C. Root, W. Zinth, B. Fierz, F. Krieger, T. Kiefhaber and P. Gilch, *J. Phys. Chem. A*, 2004, **108**, 10072–10079.

Todd Martinez asked: Is there any experimental information about these processes in protic solvents?

Christopher Grieco replied: Such experimental information would be very useful, but no experiments on Q : C heterodimers have been made in polar solvents yet. This appears difficult to study because the dimerization constant is expected to depend on solubility, which changes with solvent polarity. For example, our unpublished results show that the Q : C heterodimers form in much smaller yields in a polar solvent, chloroform, compared to in cyclohexane. For other polar, protic solvents, it is difficult to establish whether a significant population of heterodimers form because such solvents typically contain O–H or N–H groups whose IR spectrum interferes with that of the solute. These groups also compete with hydrogen bonding to the Q molecule and can also lead to lower Q : C heterodimer yields. Increasing the Q : C formation yield may require exceptionally high concentrations of solute (*e.g.* $\gg$ 100 mM) that may not be feasible in experiments. We note that changes in the UV-vis spectrum may be used to quantify Q : C heterodimer formation yields in protic solvents because the Q spectrum is sensitive to changes in solvent polarity, but this is a less direct method.

Alternatively, the problem of reduced Q : C heterodimer formation yield in polar solvent environments may be solved by future studies of covalently linked Q and C dimer compounds, such as the ones studied by Van Ahn and Williams.[1] This is because spatially constraining the Q and C pair may favor intermolecular hydrogen bonding between them.

1 N. Van Anh and R. M. Williams, *Photochem. Photobiol.*, 2012, **11**, 957–961.

Andrew Marcus remarked: Your model for the photo protection and radical scavenging mechanisms of eumelanin focus on the chemical interactions between DHI and DHICA. In a living organism, could these molecular interactions be mediated in a significant way by the presence of proteins or other biochemical species?

Christopher Grieco answered: While the presence of proteins may affect the molecular interactions in eumelanin in living organisms, we believe they may not generally be significant. The eumelanin structure is believed to consist of large agglomerates of nanoparticles that are hundreds of nanometers in size. These nanoparticles are believed to consist of aggregated structures containing the DHI and DHICA moieties. For example, see Fig. 4 in Watt *et al.*[1] These structures are expected to predominantly include intermolecular interactions between eumelanin chromophores that likely give rise to many of its photoproperties. However, it is possible that the photochemistry at the surface of the particles, which may experience interactions with the aqueous solvent environment and surrounding proteins, could be modulated and would be interesting to study. Indeed, proteins have been shown to bind to eumelanin.[2]

1 A. A. R. Watt, J. P. Bothma and P. Meredith, *Soft Matter*, 2009, **5**, 3754–3760.
2 S. Sharma, S. Wagh and R. Govindarajan, *Pigment Cell Res.*, 2002, **15**, 127–133.

Michael Haggmark commented: In the last scheme of the paper, there is a lifetime associated with the complex separating. I was wondering if the melanin polymer is confined and as such unable to separate, what dynamics would follow. That is to say, would it be a more faithful model system due to the confinement? Have you considered extending this experiment to more confined conditions, perhaps in a glassy matrix? It would make a very interesting follow up project.

Christopher Grieco responded: The likelihood for solvent-separation of the semiquinone (SQ) radical pair is certainly a process that can change the rate of decay of the radicals to restore Q and C. We agree that a model in which the SQ radicals cannot spatially separate would be a more faithful system for studying the characteristics of the SQ radical pair. However, in our paper we are primarily interested in the dynamics that precede formation of the SQ radical pair, and believe that the Q+C system is a faithful model system for these processes. We agree that it would very interesting to follow up on the work in this paper by studying Q+C in a more confined condition. Two approaches come to mind: (1) study the complex in a glassy matrix (as you've suggested), such as a glass-forming solvent; (2) study a covalently tethered Q+C system, such as those studied by Van Ahn and Williams.[1] Each system has its limitations: in the first, one would need to ensure that heterodimers form, and may be limited to experiments using lower temperatures; in the second one, the molecules may still be able to spatially separate (become solvent separated). Regardless, we believe both systems would likely provide valuable information about the reactivity of the SQ radical pair.

1 N. Van Anh and R. M. Williams, *Photochem. Photobiol.*, 2012, **11**, 957–961.

Vasilios Stavros remarked: Chris, you show some really beautiful data. My question follows from the previous comment regarding restricting the position of the C/Q pair. Have you considered increasing the viscosity of your solvent? I would expect the time for separation of your contact pair to increase from 14 ps due to the imposed friction.

Christopher Grieco answered: We have considered increasing the viscosity of the solvent to see if we observe an increase of the separation time of the contact semiquinone (SQ) radical pair, but we have not yet designed or attempted any experiments. The chosen solvent would not only need higher viscosity, but must also support the formation of Q : C heterodimers in large enough quantities to measure. Such an experiment would be very insightful and help provide better evidence for spatial separation of the SQ radical pair. Your question reminds me of an alternative model we previously considered (not discussed in the paper) that may explain the spectral narrowing of the SQ absorption band: intersystem crossing. For example, after the net hydrogen transfer event has occurred, the total multiplicity of the contact SQ pair may be singlet. It is possible for a subsequent spin-flip to occur to form a triplet SQ radical pair state. As such, this process may be confused with spatial separation of the SQ radical pair. We currently favor the latter process, but are interested in future experiments that can help support our interpretation of the spectral narrowing.

Vasilios Stavros followed: Am I correct in saying that you do not see any sharpening of the radical feature in Fig. 4d in the paper (due to vibrational cooling) (DOI: 10.1039/c8fd00240a) because of the much slower formation of a neutral radical from the triplet-mediated HAT channel (>1 ns)?

Christopher Grieco answered: What you are saying is expected because the sharpening of the radical feature at 370 nm would occur with a much faster rate than the formation of the neutral radicals from the triplet-mediated hydrogen atom transfer channel. In other words, our resolution of spectral sharpening would be poor. However, it would be interesting if we could observe spectral sharpening of the radical feature after it forms from the singlet-mediated hydrogen atom transfer channel, which occurs much more quickly. Unfortunately, strong signal overlap from excited singlet and triplet states of Q also absorb in the same spectral region that make it difficult to discern the peak at 370 nm. We envision that a time-resolved vibrational spectroscopy experiment could be used to resolve vibrational cooling of the neutral radical by monitoring dynamic shifts in the vibrational frequency of a marker band (if one could be found).

Conflicts of interest

There are no conflicts to declare.

Faraday Discussions

PAPER

Ultrafast photoinduced energy and charge transfer: concluding remarks

Bern Kohler

Received 30th May 2019, Accepted 30th May 2019
DOI: 10.1039/c9fd90030f

The ability to characterize and control the energy and charge transfer events triggered by the photoexcitation of molecules and materials is of fundamental importance to many fields, including the sustainable capture and conversion of solar energy. This article summarizes the papers that were presented and discussed at the recent Faraday discussion meeting on ultrafast photoinduced energy and charge transfer. Ultrafast laser spectroscopy and theory were at the center of discussions on photoinduced phenomena in biological and nanoscale systems of interacting absorbers. Many of the questions that motivate this field of science have occupied scientists for many decades, as a look back to a Faraday discussion meeting that took place 60 years earlier reveals.

1. Introduction

The meeting on Ultrafast Photoinduced Energy and Charge transfer, which is recorded in *Faraday Discussions* volume 216, took place April 8–10, 2019 in Ventura. It was the first time that a Faraday discussion meeting had been held in California, and the abundant sunshine of the region was a fitting accompaniment to a meeting focused on photoinduced phenomena. Although "It never rains in Southern California", at least according to an old song, it had clearly rained a great deal leading up to the meeting, and participants were rewarded with views of verdant, flower-dotted hillsides on the drive to the conference hotel—a further treat for the many fans of photosynthesis in attendance.

As the title indicates, the meeting was focused on the dynamical phenomena of energy and charge (*i.e.*, electron and/or proton) transfer initiated by the photoexcitation of molecules and materials. The word "ultrafast" is an indication of the rapid timescales involved, and most of the authors of the experimental papers discussed at the meeting reported results obtained using a variety of ultrafast laser spectroscopies. The theoretical frameworks that are crucial for interpreting the experimental observations were also at the core of this meeting that attracted the participation of many theorists. The organizing committee encouraged investigators to attend, who use spectroscopic and computational approaches to

Department of Chemistry and Biochemistry, The Ohio State University, 100W. 18th Ave., Columbus, OH 43210, USA. E-mail: kohler.40@osu.edu

learn about fundamental events that play out in molecules and in their nanoscale assemblies.

Although biology is not mentioned in the meeting title, biomolecules and their assemblies were discussed in nearly half of the papers presented, and most of the remaining papers were inspired to a lesser or greater degree by the photodynamics of natural systems. The natural photosynthesis that powers life on Earth makes extensive use of excitation energy transfer and this constituted another theme of the meeting, as will be discussed below. While the ability to probe events on ultrafast timescales continues to improve, nature remains the master of results, meaning that human-engineered systems for efficiently and sustainably generating electricity or chemical fuels from sunlight still leave much to be desired. Possible strategies for achieving advances in man-made systems were discussed frequently.

With only a few exceptions, the systems presented in the many excellent papers and posters involved two or more interacting light absorbers. Only a handful of papers were concerned with solids with long-range translational symmetry and delocalized states. Most of the studies addressed systems in the fascinating middle ground between small molecules and extended solids. It is an area of special interest and opportunity for physical chemists, who bring their familiarity with localized states of molecular units to complex systems in which interactions among multiple chromophores over nanoscale distances give rise to rich new phenomena in energy and charge transfer.

2. Excited states through the lens of time

The meeting attracted many excellent contributions that spanned an impressively broad range of topics, as even cursory inspection of this volume indicates. Distilling these diverse contributions into a coherent summary is a nearly impossible task. The scientific heterogeneity on display at the meeting is both a hallmark and strength of a field that seeks to characterize and exploit the multiple paths that lead from excited electronic states back to the ground state or to photoproducts.

Many of the themes and questions addressed at the meeting have old roots that are worth reflecting on. As the introduction to Prof. Fleming's paper in this volume (DOI: 10.1039/c8fd00190a) indicates, efforts to understand energy transfer in photosynthetic organisms began in earnest more than 70 years ago. Given the focus on fundamental light–matter interactions, a theme with great longevity in physical chemistry circles, it is hardly a surprise that a substantial number of preceding Faraday discussion meetings have had at least some topical overlap with the 2019 meeting. There have been, for example, multiple Faraday discussions on electron and proton transfer and ones on photochemistry topics date back to at least 1925.

There is, however, one Faraday discussion meeting from the past that is a particularly interesting precursor to the April 2019 meeting. This discussion, which was titled "Energy Transfer with special reference to Biological Systems", took place almost exactly 60 years earlier at Nottingham University from April 14–16, 1959. Sixty is of course an appropriately round number, but as the number of seconds in a minute, and the number of minutes in an hour, it is an irresistible anniversary for celebrating a field that revolves around time-domain observations.

The 1959 meeting featured many notable scientists working in spectroscopy and photoinduced energy transfer. Participants included a past (Albert Szent-Györgyi) and a future (George Porter) Nobel laureate. Szent-Györgyi received the 1937 Nobel Prize in Physiology or Medicine, while George Porter would receive a Nobel Prize in 1967. Other distinguished scientists present in Nottingham whose names would be recognizable to many of the 2019 participants included: John B. Birks, Edmund J. Bowen, Henry Eyring, Theodor Förster, H. Christopher Longuet-Higgins, John P. Simons, Gregorio Weber, Eugene Rabinowitch, Joseph Weiss, and Albert Weller. Most of these individuals and many of the other scientists in attendance were younger than 50 in 1959, and the paper presenters were on average younger than their counterparts in 2019. To underscore this youthful dynamism, there was even an Olympic champion among the attendees (American Britton Chance, inventor of the stopped flow technique, and a sailing gold medalist at the 1952 games in Helsinki).

That energy and charge transfer usually play out on ultrafast time scales was already well known sixty years ago. With a simple and direct elegance, Nobel laureate Albert Szent-Györgyi succinctly described this challenge in the opening sentences of his written contribution,[1]

"If a photon, ejected by the sun, interacts with an electron of a molecule on our globe, then the electron is raised to a higher energy level to drop back, as a rule within 10^{-8}–10^{-9} s, to its ground state… Life has shoved itself between the two processes, catches the electron in its high-energy state and lets it drop back to the ground level within its machinery, using the energy thus released for its maintenance…"

The business of understanding this "machinery" and how it functions has occupied physical chemists and other scientists for many years now. The lack of techniques for directly observing energy transfer over the fastest time scales on which it takes place was an obvious barrier to progress six decades ago. Because lasers and the ultrashort light pulses they produce had not yet been invented, observing excited state dynamics, particularly of short-lived singlet excited states, required an extra measure of ingenuity and effort in the late 1950s.

Two of the 25 papers from the 1959 meeting include figures that depict transient measurements of excited electronic states. Fittingly, one of the papers is by George Porter, who would go on to receive the 1967 Nobel Prize in Chemistry together with Manfred Eigen and George Norrish for pioneering studies of fast reactions. At the 1959 meeting, Porter and Wright described flash photolysis measurements with microsecond time resolution, which they used to determine the lifetime of the lowest triplet state of naphthalene in various solvents.[2]

The second paper by Burton and Dreeskamp reported measurements of luminescence decay times from scintillators excited by 11 ns X-ray pulses.[3] Their apparatus permitted measurements of lifetimes as short as a few nanoseconds, a remarkable, state-of-the-art achievement in 1959. Although the time delays in their experiments were laboriously generated by manually swapping out coaxial cables cut to just the right length to delay an electrical gate pulse by the desired amount of time, the sampling approach of performing many repeated measurements at discrete values of the time after photoexcitation is in essence the same one used in the many pump-probe experiments discussed in the 2019 meeting.

The ability to directly observe excited state dynamics using ultrafast lasers has been revolutionary. Lasers are mentioned explicitly in all but a few of the experimental papers at the 2019 meeting. Ultrafast laser sources and related technology have matured greatly since the first lasers that directly generated femtosecond laser pulses became widely available in the early 1980s, and it is now possible to generate sub-picosecond laser pulses across a broad swath of the electromagnetic spectrum from THz radiation to X-rays. The remarkable range of observations that are now possible using ultrafast laser techniques was well documented in the introductory lecture by Prof. Majed Chergui (DOI: 10.1039/c9fd00036d).

The papers discussed at the 2019 meeting explored many themes in energy and charge transfer studies at a level of microscopic detail that would have been hardly imaginable in 1959. The organizing committee organized the meeting into four sessions each with a distinct theme. These are briefly presented in the following followed by some concluding remarks that look back as well as ahead.

3. Energy transfer and charge transfer in natural photosynthesis

A special feature of Faraday discussion meetings has always been the exchanges among participants that are preserved for posterity. Even though both questions and answers acquire some additional luster through editing by questioner and questionee after the meeting, they nevertheless serve up freshly baked (and occasionally half-baked) science at the frontiers of a field, as it unfolded in real time. One of the more interesting comments from the 1959 meeting came from Prof. Rufus Lumry from the University of Minnesota (italics added):

"The comments of several contributors to this discussion *indicate a yearning for the appearance in biological systems of exotic processes depending upon the quantum mechanics of highly ordered* systems and the like. While one should not at present dismiss these in any confident manner, it seems desirable not to neglect unusual mechanisms for which we already have evidence and which might be reasonably expected to exist in biological systems."

This comment captures the longstanding interest in (and occasional uneasiness about) quantum effects in biology that continues to the present day.[4] The papers in this first session of the meeting dealt with experimental and theoretical advances in understanding transient processes relevant to natural photosynthesis and the "exotic processes" mentioned by Lumry. While most of the presentations investigated events that occur in natural systems, Prof. Leggett described how light harvesting complexes isolated from spinach can be deposited on arrays of gold nanostructures and used to observe and control plasmon–exciton couplings (DOI: 10.1039/c8fd00241j).

The other presentations in this session sought insight into how the electronic and vibrational coherences observed in natural systems influence energy and charge transfer dynamics. Since the report that quantum coherences can be detected in experiments that monitor the transport of excitation energy in photosynthetic complexes,[5] there has been great interest in the physics underlying this phenomenon,[6–8] and how it might be used to design improved synthetic systems for solar energy harvesting.[9]

Prof. Olaya-Castro reported on fundamental quantum effects that appear when two coupled oscillators become synchronized (DOI: 10.1039/c9fd00006b). Theoretical treatments like this one are important because they can ask questions about the functioning of isolated or altered subsystems that may be difficult or impossible to probe experimentally. The paper presented by Dr Fingerhut explored the effects of intramolecular vibrations on the coherent dynamics in a model system based on a bacterial reaction center (DOI: 10.1039/c8fd00189h). The calculations suggest that charge transfer proceeds non-adiabatically at room temperature, following the equilibration of excitonic coherences.

In a further theoretical study, Prof. Cheng investigated the photosystem II supercomplex using large-scale molecular dynamics simulations and quantum chemical calculations that modeled excitonic couplings among all 74 pigments (DOI: 10.1039/c8fd00205c). A coarse-grained version of the model reproduces key excitation energy transfer pathways and shows promise for simulating dynamics in large networks. In his contribution, Prof. Fleming (DOI: 10.1039/c8fd00190a) described the advantages of two-dimensional electronic-vibrational spectroscopy for studying photosynthetic energy transfer. This recently developed technique is advantageous for studying correlations between electronic and vibrational degrees of freedom.

4. Photovoltaics and bioinspired light harvesting

The papers in this session are notable for their diversity of topics. Prof. Persson reported on his latest efforts to understand the photophysics of iron carbene complexes, which have remarkably long ^{3}MLCT state lifetimes (DOI: 10.1039/c8fd00232k). Prof. Weiss discussed in her presentation the dynamics of singlet fission in aggregates of pentacene derivatives that form spontaneously on the surface of PbS quantum dots (DOI: 10.1039/c8fd00157j). The rate of singlet fission was observed to increase by a factor of five, suggesting a possible strategy for increasing the efficiency of photovoltaic devices.

An artificial light harvesting system in which porphyrin chromophores are arranged on the 'struts' of a metal organic framework was discussed by Prof. Morris together with her team's observations of exciton migration using time-resolved spectroscopy (DOI: 10.1039/c8fd00194d). Prof. Marcus discussed how interchromophoric couplings among a dimer of dye molecules can be used to reveal dynamically fluctuating conformations at junctions between single- and double-stranded DNA (DOI: 10.1039/c8fd00245b). This work illustrates how spectroscopy and theory can yield powerful insights into fluctuating structures on time scales that are not accessible using other techniques.

The debate about what allows a photogenerated electron and hole to form free charge carriers in an organic photovoltaic device despite their strong mutual coulombic attraction of $\sim$0.5 eV was addressed by Prof. Bittner (DOI: 10.1039/c8fd00182k). Entropy has been shown to play an important role in reducing the free energy barrier to charge separation even in the face of a significant potential energy barrier.[10,11] The formalism presented by Prof. Bittner was used to calculate the degree of quantum entanglement and the Shannon entropy of excitonic and charge transfer states in a model heterojunction. Finally, the light-driven movement of protons and the factors that influence the photobasicity of organic molecules were the focus of Prof. Dawlaty's contribution (DOI: 10.1039/c8fd00215k).

5. Photoinduced electron transfer

When two or more pathways connect initial and final states, quantum interference can enhance and suppress rates. This fundamental aspect of quantum systems has recently motivated attempts to control rate processes. Quantum interference among pathways that couple two molecules that undergo Dexter energy transfer was the subject of the contribution from Prof. Beratan (DOI: 10.1039/c9fd00007k). Prof. Wasielewski presented results of spectroscopic experiments showing a non-statistical rate enhancement for intramolecular electron transfer in a compound containing one *vs.* two electron acceptors at cryogenic temperatures (DOI: 10.1039/c8fd00218e). However, bath fluctuations at room temperature spoil the coherent coupling between the two acceptors, resulting in a statistical rate.

Three papers discussed further issues of importance in the field of organic photovoltaic devices. Prof. Schwartz discussed a new spectroscopic approach for distinguishing free polarons from trapped polarons in doped organic polymers (DOI: 10.1039/c8fd00210j). The role played by the solvent in influencing charge separation was studied at the fundamental level of two identical chromophores in a paper presented by Prof. Thompson (DOI: 10.1039/c8fd00201k). The contribution from Prof. Penfold examined factors affecting triplet–triplet energy transfer (DOI: 10.1039/c8fd00174j). Triplet states can be low energy traps, but their long lifetimes can potentially allow them to diffuse over great distances.

Biology often couples proton and electron transfers to efficiently separate charge using light. Marcus[12] first predicted theoretically the so-called inverted region behavior in which electron transfer rates decrease even as the reaction exergonicity increases—an effect that was subsequently observed experimentally in the 1980s.[13] Prof. Hammes-Schiffer described in her presentation the conditions necessary for observing inverted region behavior in photoinduced proton-coupled electron transfer (PCET) (DOI: 10.1039/c8fd00240a), which has now been seen in very recent experiments by Parada *et al.*[14]

A final paper in this session presented an analysis of thermal effects in high repetition rate ultrafast X-ray studies. Prof. Gessner and his team studied the lattice heating that results from electron-phonon coupling in the transition metal oxide semiconductor CuO (DOI: 10.1039/c8fd00236c). While time-resolved X-ray experiments are still performed overwhelmingly on inorganic systems, the thermal effects described by Gessner will also be important in systems of organic chromophores, like the ones that dominated the presentations in the meeting.

6. Photoprotection/photodamage in natural systems

The final session addressed the downside to using energy-rich excited electronic states. Even minute quantum yields of photoproduct formation or other off-pathway events pose a significant challenge to the long-term viability of photoactive systems, both natural and man-made. The generation of triplet excited states poses a particularly grave threat due to their long lifetimes and their ability to form reactive oxygen species like singlet oxygen. The specific pathways that give rise to photodamage can be as varied as the chromophores involved, and papers in the final session covered a range of biomolecular absorbers that included carotenoids, melanin model compounds, mycosporine-like amino acids, and nucleobases. These chemically diverse chromophores absorb from the visible to the deep UV and are all thought to contribute in one way or another to enhancing the photostability of the structures they occur in.

Carotenoids play a key role in the non-photochemical quenching mechanism that protects the photosynthetic machinery from high light levels,[15,16] but they also act as accessory absorbers capable of transferring energy to chlorophyll. Prof. Polívka described how energy transfer from fucoxanthin keto-carotenoids to chlorophyll is slowed when chemical modifications eliminate the intramolecular charge transfer state of fucoxanthin (DOI: 10.1039/c8fd00193f). In spite of this change, overall energy transfer efficiency is not compromised because slower energy transfer is compensated by a longer S_1 lifetime.

The linear polyenes that are the chromophores of carotenoids have long fascinated spectroscopists because the lowest singlet excited state is a dark state.[17] Although one-photon excitation to the S_1 state is forbidden, this state can be readily reached *via* an allowed two-photon transition. The paper presented by Dr Lokstein (DOI: 10.1039/c8fd00198g) addressed an important issue in studies of light harvesting complexes that contain two or more chromophores that can be excited by two-photon absorption. Polarization-dependent emission spectra revealed that the two-photon excitation of chlorophyll using wavelengths between 1000 and 1500 nm is two orders of magnitude more probable than the two-photon excitation of a xanthophyll carotenoid when two-photon excitation is detected by bacteriochlorophyll fluorescence.

Prof. Matsika explored in her paper how base sequence and the methylation state of a nucleobase affect the energies of charge transfer states in DNA strands (DOI: 10.1039/c8fd00184g). Charge transfer states are of great importance in DNA strands because the short distances between stacked nucleobases promote their coupling. Depending on spin multiplicity, these states can quench or enhance the formation of the cyclobutane pyrimidine dimer photoproduct formed in DNA by UV radiation.

Mycosporine-like amino acids (MAAs) are UVB sunscreens found in algae. Prof. Habershon used MAA analogs to test "on-the-fly" wavefunction propagation methods (DOI: 10.1039/c8fd00228b). Quantum dynamical simulations like these can directly simulate ultrafast excited state dynamics and help to bridge the gap between experiments that are generally focused on kinetics and theoretical treatments that are most often focused on energy landscapes.

Melanin is a ubiquitous biopolymer found in organisms throughout the tree of life, but the lack of a molecular structure has severely hampered structure–function studies. Dr Grieco (DOI: 10.1039/c8fd00231b) discussed photoinduced hydrogen atom transfer in catechol:quinone dimers that are thought to model some of the molecular motifs present in this polymer.

7. Outlook

It is easy to imagine that the 1959 participants would have greatly enjoyed the 2019 meeting due, firstly, to themes that have obvious connections to the science that they discussed 60 years earlier, and, also to the delight that they would certainly feel at the stunning technological advances that have transformed how experiments (lasers) and theoretical studies (vastly more powerful computers) are performed today. It is also worth noting that several of the participants at the 1959 meeting coined terms like excimer (B. Stevens in 1960 (ref. 18)) and exciplex (R. Lumry in 1966 (ref. 19)) that are now part of the vocabulary of the field and that were invoked in Ventura on multiple occasions.

The fact that many of the questions discussed in 2019 have roots that can be easily recognized in the program from a meeting from sixty years earlier (a near eternity given the pace of modern science) might give the impression that there is "nothing new under the sun". However, it would be a mistake to confuse this thematic continuity with the steady and dramatic progress that has taken place in a field focused on complex phenomena. There are plenty of reasons to believe that fundamental studies of excited states will continue to be important for years to come, but what ideas imported from other fields will fuel new advances? What new technology will be as indispensable and transformative to the field as the laser has been beginning in the early 1960s?

Answering these questions seems just as foolhardy today as predicting the state of the field in 2019 would have been in 1959, but a new direction poised to occupy scientists in the field, at least in the short term, was discussed in Ventura. Several speakers commented on the new frontier in quantum information science and the contributions that workers in the field of ultrafast energy and charge transfer are expected to make in this emerging area. Prof. Fleming was frank in his assessment of the difficulties faced by experimentalists who will need to work with quantum light sources. However, he offered ultimately an optimistic view of the future when he reminded the audience of the difficulties that previous generations of scientists overcame to break into the regime of picosecond and femtosecond optical studies. This journey, which has largely been completed in the sixty years since the 1959 meeting, clearly illustrates the folly of predicting the trajectory or limits of scientific discovery.

The diversity of the science presented in Ventura in 2019 is also a platform for future success. Science that advances incrementally along a predictable trajectory established by small numbers of insiders is often science in a rut. Newcomers and researchers from other fields bring unorthodox approaches and occasionally 'heretical' perspectives that have driven advances in unexpected directions time and time again. These scientist immigrants have the special mixture of courage and naïveté needed to break out from the well-worn paths laid down in a field. It seems certain that new impulses from unexpected corners will move the field forward into the future.

The strength of the field today is measured not just by the diversity of the themes, but also by the diversity of the participants. The use of first name initials in the discussion proceedings makes it difficult to tell with certainty whether there were any female participants at the 1959 meeting, but the striking imbalance in gender pronouns in the published general discussions tells its own story:

there are plenty of instances of 'he', 'his', and 'Mr', zero mentions of 'she' or 'her', and only a solitary 'Miss' in a comment by a Dr Redfern about work conducted with a Miss A. M. Pumphrey, who may or may not have been at the meeting. In 2019, women scientists were present in large numbers that will hopefully increase further in the future.

The papers and posters presented in Ventura document the exquisitely fine probing of excited electronic states that is now possible. Yet, for all of the splendid advances in theory and ultrafast laser spectroscopy, the ability to synthesize chemicals at will using solar photons—the ability to do what plants do with efficiency and using earth-abundant materials—remains elusive. Many decades of effort testify to the magnitude of the difficulty, but it nevertheless seems reasonable to predict that there will be a breakthrough in artificial photosynthesis that will end the currently unsustainable and climate-damaging energy economy of fossil fuels.

The fact that life was able to shove itself into the niche in time between photon absorption and excited-state deactivation as Szent-Györgyi wrote in 1959 indicates that the problem of how to store solar energy in chemical bonds for later retrieval is not an intractable one. Although the desirability of accomplishing this has been discussed for many years,[20] the urgency of finding a solution has never felt greater. A rapidly degrading climate, mass extinction of species, deforestation and damage to Earth's ocean ecosystems are all reasons to move swiftly toward a solution with all of the optimism that science can muster. When push comes to shove, it is a grand challenge too important to cede to the future.

Conflicts of interest

There are no conflicts to declare.

Acknowledgements

The author thanks Professors Michael Ashfold and Stephen Bradforth and the members of the Organizing Committee for the invitation to present the Concluding Remarks lecture and this written contribution. The author is grateful to the National Science Foundation for research support (CHE-1800471).

References

1 A. Szent-Győrgyi, *Discuss. Faraday Soc.*, 1959, **27**, 111–114.
2 G. Porter and M. R. Wright, *Discuss. Faraday Soc.*, 1959, **27**, 18–27.
3 M. Burton and H. Dreeskamp, *Discuss. Faraday Soc.*, 1959, **27**, 64–73.
4 N. Lambert, Y. N. Chen, Y. C. Cheng, C. M. Li, G. Y. Chen and F. Nori, *Nat. Phys.*, 2013, **9**, 10–18.
5 G. S. Engel, T. R. Calhoun, E. L. Read, T. K. Ahn, T. Mancal, Y. C. Cheng, R. E. Blankenship and G. R. Fleming, *Nature*, 2007, **446**, 782–786.
6 N. Christensson, H. F. Kauffmann, T. Pullerits and T. Mancal, *J. Phys. Chem. B*, 2012, **116**, 7449–7454.
7 V. Tiwari, W. K. Peters and D. M. Jonas, *Proc. Natl. Acad. Sci. U. S. A.*, 2013, **110**, 1203–1208.
8 A. Chenu and G. D. Scholes, *Annu. Rev. Phys. Chem.*, 2015, **66**, 69–96.

9 G. D. Scholes, G. R. Fleming, A. Olaya-Castro and R. van Grondelle, *Nat. Chem.*, 2011, **3**, 763–774.
10 N. R. Monahan, K. W. Williams, B. Kumar, C. Nuckolls and X. Y. Zhu, *Phys. Rev. Lett.*, 2015, **114**, 5.
11 S. N. Hood and I. Kassal, *J. Phys. Chem. Lett.*, 2016, **7**, 4495–4500.
12 R. A. Marcus, *J. Chem. Phys.*, 1956, **24**, 966–978.
13 J. R. Miller, L. T. Calcaterra and G. L. Closs, *J. Am. Chem. Soc.*, 1984, **106**, 3047–3049.
14 G. A. Parada, Z. K. Goldsmith, S. Kolmar, B. Pettersson Rimgard, B. Q. Mercado, L. Hammarström, S. Hammes-Schiffer and J. M. Mayer, *Science*, 2019, **364**, 471.
15 A. V. Ruban, M. P. Johnson and C. D. P. Duffy, *Biochim. Biophys. Acta, Bioenerg.*, 2012, **1817**, 167–181.
16 P. Jahns and A. R. Holzwarth, *Biochim. Biophys. Acta, Bioenerg.*, 2012, **1817**, 182–193.
17 B. S. Hudson and B. E. Kohler, *J. Chem. Phys.*, 1973, **59**, 4984–5002.
18 B. Stevens and E. Hutton, *Nature*, 1960, **186**, 1045–1046.
19 M. S. Walker, T. W. Bednar and R. Lumry, *J. Chem. Phys.*, 1966, **45**, 3455–3456.
20 G. Ciamician, *Science*, 1912, **36**, 385–394.

ADDITIONS AND CORRECTIONS

Role of charge transfer states into the formation of cyclobutane pyrimidine dimers in DNA

Wook Lee and Spiridoula Matsika*

Faraday Discuss., 2019, DOI: 10.1039/C8FD00184G. ***Amendment published 1st July 2019.***

After this paper (Role of charge transfer states into the formation of cyclobutene pyrimidine dimers in DNA) was accepted we found out that the partial charges we had used for methylated cytosine (5mC) are extremely sensitive to the details of obtaining them. In order to be able to compare directly cytosine and methylated cytosine the charges should be as similar as possible. So, we recalculated the charges on 5mC in the exact same protocol as the ones for cytosine. The details of how this was done and how the results changed are outlined below. We want to emphasize, however, that we see this as a cautionary tale. The partial charges are extremely sensitive to details, such as the number of grid points, and this is unsettling. So, one has to be very careful in obtaining these charges, and in our opinion benchmarks have to be made to decide which charges are best. Ideally, quantum mechanical calculations should be carried out to decide which answer is closer to reality. Furthermore, when charge transfer states are calculated, the charges can change the energies significantly, so even further caution is needed.

The RESP procedure we carried out to obtain the 5mC charges has some small deviations from the one originally used to obtain AMBER force field charges for nucleobases. The first deviation is that we used far more grid points than the default value used in Gaussian for fitting point charges (thinking mistakenly that this would make it more accurate), and the second is that the fragment we used for generating RESP charges is different from the one used for generating standard AMBER force field.[1] In standard AMBER force field, the whole nucleoside was included in the calculations when atomic partial charges were derived, while we only included the base and part of a sugar ring. The main difference between the charges in C and 5 mC that affects our results is the charge at C8 atom, which is at the boundary between the QM and MM region (see Figure 1). This charge was rather large in our original calculations. After correcting for the two factors described above, we could obtain a small positive charge of C8 atom as shown in Table 1.

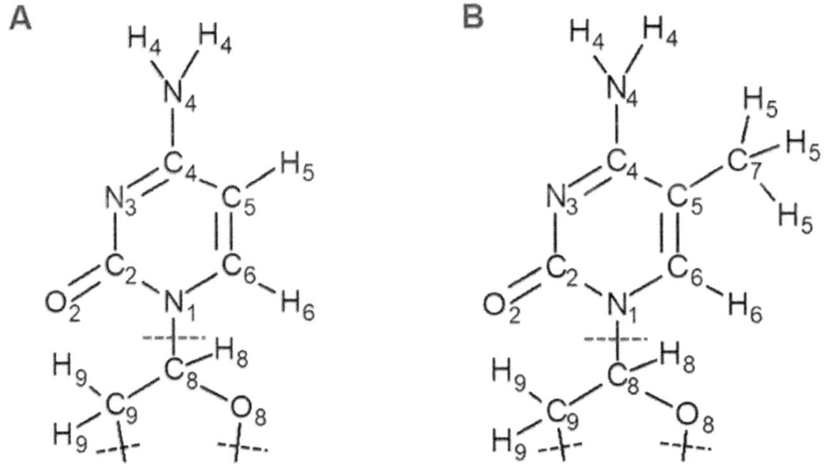

Figure 1. The chemical structure of cytosine (left) and 5-methylcytosine (right) with the labeling of the atoms. The red dotted lines indicate QM/MM boundaries, and the atoms beyond the black dotted lines are omitted because the charges of those atoms are the same between cytosine and 5-methylcytosine.

Table 1. Atomic partial charges of cytosine from standard AMBER force fields (ff99) and the RESP charges of 5mC from the original paper and new calculations

	Cytosine (AMBER ff99)	**Original charges of 5 mC**	**New charges of 5 mC**
N_1	-0.0339	-0.3408	-0.0398
C_2	0.7959	0.8749	0.7063
O_2	-0.6548	-0.6594	-0.6288
N_3	-0.7748	-0.7944	-0.6882
C_4	0.8439	0.7942	0.6341
N_4	-0.9773	-0.9514	-0.8770
H_4	0.4314	0.4159	0.3943
C_5	-0.5222	-0.1936	-0.1182
H_5	0.1863	0.1033	0.0767

C_6	-0.0183	-0.0126	-0.1343
H_6	0.2293	0.1972	0.2350
C_8	-0.0116	0.5571	0.0484
H_8	0.1963	0.0421	0.1963
O_8	-0.3691	-0.4738	-0.3691
C_9	-0.0854	-0.1901	-0.0854
H_9	0.0718	0.0718	0.0718
C_7		-0.3239	-0.2309

With these new charges of 5mC, we re-computed the vertical excitation energy of charge transfer states for the sequences involving 5mC, GmCT and TmCG. Figures 4 and 5 in the original paper need to be updated with the new energies. The updated figures are shown below.

The main effect of the new charges is the description of methylation effects in Section 3.2.3 of the manuscript. According to the new results methylation does not have a pronounced effect on the charge transfer states. In fact, methylation causes a small effect in the energies of charge transfer states, which follows the trends of the distances between donor and acceptor. The main idea described in Section 3.3.3, i.e. that the electrostatic interactions with the surroundings will affect the charge transfer states, should still be valid. But there is no direct evidence of that in our study. This is shown in the updated section of Table 2 below. This shows that the energy of charge transfer states is not dramatically different between QM and QM/MM calculations, indicating that the surroundings do not play such a big role as we originally thought.

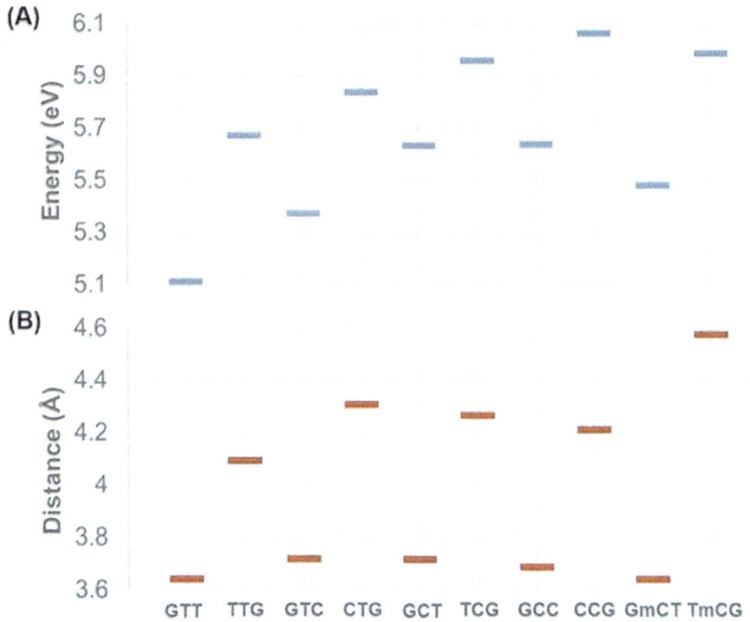

Updated Figure 4: (A) Average charge transfer energies obtained from 20 snapshots and (B) average distances between two bases involved in charge transfer obtained from 200 snapshots for all sequences. The only different results here compared to the original figure are in part (A) the last two energies (corresponding to GmCT and TmCG).

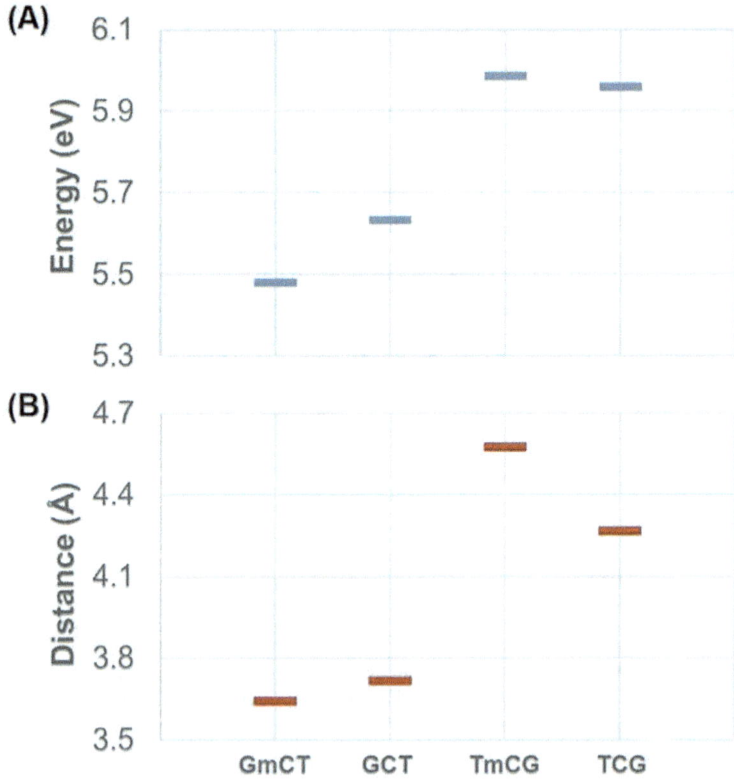

Updated Figure 5: (A) Average charge transfer energies obtained from 20 snapshots and (B) average distances between two bases involved in charge transfer obtained from 200 snapshots for methylated (GmCT, TmCG) and non-methylated sequences (GCT, TCG). The only different results here compared to the original figure are in part (A) the energies of the methylated systems.

Updated Table 2: Energies of charge transfer states (in eV) for a representative snapshot calculated using the whole model system (QM and MM) and only the QM region. The energy difference ΔE between the two previously listed sequences is also shown.

	QM/MM	**QM**
GmCT	5.452	5.384
GCT	5.649	5.231
$\Delta E_{GCT\text{-}GmCT}$	0.197	-0.153
TmCG	5.978	5.920
TCG	5.942	5.803
$\Delta E_{TCG\text{-}TmCG}$	-0.036	-0.117

Reference

1 P. Cieplak, W. D. Cornell, C. Bayly, P. A. Kollman, *J. Comput. Chem.* 1995, **16**, 1357–1377.

The Royal Society of Chemistry apologises for these errors and any consequent inconvenience to authors and readers.

Poster titles

Determination of secondary species through use of explicit-solvent TDDFT, **M. Turner, M.Horbury, V. Stavros, N. Hine,** *University of Warwick, United Kingdom*

Chlorophyll-carotenoid energy and charge transfer mechanisms in nonphotochemical quenching in green plants and microalgae, **C. Steen, S. Park, A. Fischer, D. Lyska, M. Iwai, K. Niyogi, G. Fleming,** *University of California, Berkeley, USA*

Quantum interferences among Dexter energy transfer pathways, **S. Bai, P. Zhang, P. Antoniou, S. Skourtis, D. Beratan,** *Duke University, U.S.A*

Cavity-enhanced transient absorption spectroscopy of singlet fission processes in molecules, **N. D. Cooper, R. Basnet, K. Warren, M. A. R. Reber,** *The University of Georgia, U.S.A.*

Determining zero driving force of electron transfer for a zinc dipyrrin in various solvent polarities using ultrafast spectroscopy, **M. Kellogg, A. Akil, D. Sylvinson, L. Estergreen, S. Bradforth, M. Thompson,** *University of Southern California, U.S.A.*

Dynamics of photoinduced charge transfer processes in FeII and FeIII carbene photosensitizers, **P. Chabera, K. S. Kjær, L. A. Fredin, N. Rosemann, J. Uhlig, L. Lindh, O. Prakash, K. Wärnmark, A. Yartsev, V. Sundström, P. Persson,** *Lund University, Sweden*

Liquid-microjet photoelectron spectroscopy of biological chromophores, **J. Riley, B. Wang, A. Henley, M. A. Parkes, H. Fielding,** *University College London, United Kingdom*

Micro-focused MHz pink beam for X-ray emission spectroscopy, **M.-F. Tu, G. Doumy, A. Al Haddad, A. M. March, S. Southworth, Y. Kumagai, D. Walko, L. Young, A. DiChiara, C. Bostedt,** *Argonne National Laboratory, U.S.A.*

Direct visualization of long-range charge separation and transport at the in-plane organic heterojunction, **S. Dev Verma, R. Chen, J. Gorman, F. Auras, A. Rao,** *University of Cambridge, United Kingdom*

Concerted proton-electron transfer reactions in the marcus inverted region, **G. Parada, Z, Goldsmith, S. Kolmar, B. Pettersson Rimgard, B. Q. Mercado, L. Hammarström, S. Hammes-Schiffer, J. Mayer,** *Yale University, U.S.A.*

Photon correlation: A definitive test of 'bright' or 'brightened' exciton fission, **K. Morenz, M. W. B. Wilson,** *University of Toronto, Canada*

Donor-acceptor pre-association and excited state solvation threshold and optical energy cost of photobases, **C. Tseng, J. R. Hunt, J. Dawlaty,** *University of Southern California, U.S.A.*

Two-dimensional electronic spectroscopy reveals the spectral dynamics of forster resonance energy transfer in a cyanine dye heterodyad, **B. Petkov, T. A. Gellen, C. A. Farfan, W. Carbery, B. Hetzler, D. Trauner, X. Li, W. J. Glover, D. Ulness, D. B. Turner**, *New York University, U.S.A.*

Energetics and escape of interchain-delocalized ion pairs in nonpolar media, **J. Burke, M. Bird**, *Brookhaven National Laboratory, U.S.A.*

Engineering of high efficiency light-harvesters and energy converters, **A. Pieper, M. Hohgardt, M. Willich, D. A. Gacek, N. Hafi, D. Pfennig, A. Albrecht, P. J. Walla**, *University of Braunschweig and Max-Planck-Institute, Germanry/Kavli Energy NanoScience Institute, Berkeley, U.S.A.*

Femtosecond infrared spectroscopy of chlorophyll f-containing photosystem I, **N. Zamzam, M. Kaucikas, D. J. Nürnberg, A. W. Rutherford, J. J. van Thor**, *Imperial College London, United Kingdom*

Ultrafast dynamics of mycosporine-like amino acids, **J. Woolley, N. Auckloo, M. Staniforth, C. Corre, V. Stavros**, *University of Warwick, United Kingdom*

Metal-oxide electrode-electrolyte interfaces investigated by operando photoelectron spectroscopy, **G. Wartner, D. Hein, M. Schellenberger, A. Bergmann, R. Seidel**, *Helmholtz Center for Materials and Energy, Germany*

Transient synchronisation in bio-inspired vibronic dimers with non-Markovian environments, **S. Siwiak-Jaszek, A. Olaya-Castro**, *University College London, United Kingdom*

Modulating singlet fission in tetracene dimers, **J. Joy, N. V. Korovina, M. E. Thompson, S. E. Bradforth**, *University of Southern California, U.S.A.*

Monitoring photoinduced charge carrier dynamics in Au nanoparticle–sensitized TiO with time-resolved X-ray photoelectron spectroscopy, **M. Borgwardt, J. Mahl, N. Huse, F. Roth, S. Molodtsov, L. Wenthaus, W. Wurth, W. Eberhardt, O. Gessner**, *Lawrence Berkeley National Laboratory, U.S.A.*

Unravelling the dynamics of individual quantum systems in an ensemble, **A. Dodin, A. P. Willard**, *MIT, U.S.A.*

Pulling protons with thermodynamics: kinetics and applications of quinoline photobases, Photodriven deprotonation of proton donor solvents by N-heterocyclic aromatics, **J. R. Hunt, C. Tseng, I. Demianets, T. Williams, J. Dawlaty**, *University of Southern California, U.S.A.*

Probing intrinsic photodeactivation pathways of nucleobases and dyes in a molecular beam, **G. Gate, M. Haggmark, J. Berenbeim, R. Szabala, A. Sobolowski, M. Rode, M. de Vries**, *University of California Santa Barbara, U.S.A.*

Photoinduced charge separation in model BODIPY based complexes, **R. K. Swedin, Y. V. Zatsikha, A. T. Healy, T. S. Blesener, V. N. Nemykin, D. A. Blank**, *University of Minnesota, U.S.A.*

Photoinduced intramolecular electron transfer in the azathioprine prodrug, **G. Ortiz-Zayas, M. Pollum, L. A. Ortiz-Rodríguez, S. Jockusch, C. E. Crespo-Hernández**, *Case Western Reserve University, U.S.A.*

Pump-probe anisotropy on coupled chromophores; a tool to study photoinduced symmetry breaking charge transfer, **L. Estergreen, J. A. Golden, M. Thompson, S. Bradforth,** *University of Southern California, U.S.A.*

Rotational barrier of methyl groups in methyl terminated silicone(III) surface, **D. Bhattacharyya, A. Montenegro, A. V. Benderskii,** *University of Southern California, U.S.A.*

The optimized energy landscapes of photosynthetic proteins, **S. Irgen-Gioro,** *Northwestern University, U.S.A.*

Photodynamics of carotenoids on multiple time scales: Lessons on ultrafast energy dissipation in large organic molecules, **V. Balevicius, T. Wei, D. Di Tommaso, D. Abramavicius, J. Hauer, T. Polivka, C. Duffy,** *Queen Mary University of London, United Kingdom*

Wave-packet dynamics in energy transfer models, **A. Kiessling, J. Cina,** *University of Oregon, U.S.A.*

Long-lived ^{2}MLCT states in vanadium(II) polypyridines: A Hund's rule exception, **R. Dill, S. Shepard, R. Portillo, A. Rappé, M. Shores, N. Damrauer,** *University of Colorado Boulder, U.S.A.*

Tunable biomimetic systems to elucidate chromophore-protein and chromophore-chromophore interactions that facilitate photosynthetic light harvesting, **T. Roberts, M. Delor, J. Dai, J. Rogers, S. Hamed, L. Hamerlynck, A. Bischoff, R. Noriega, S. Penwell, L. Ginsberg, D. Finley, J. Haberstroh, J. B. Neaton, P. L. Geissler, M. B. Francis, N. S. Ginsberg,** *University of California -Berkeley, U.S.A.*

Two-color two dimensional electronic spectroscopy of the bacterial reaction center, **R. Sechrist, A. Konar, Y. Song, V. R. Policht, A. Niedringhaus, J. P. Ogilvie,** *University of Michigan, U.S.A.*

Two-dimensional electronic spectroscopy reveals energy transfer pathways in cyanobacterial photosystem I complexes, **Y. Lee, M. Gorka, J. H. Golbeck, J. M. Anna,** *University of Pennsylvania, U.S.A.*

Uncovering the mechanism of charge separation in the photosystem II reaction center using multispectral multidimensional spectroscopy, **Y. Song, E. Maret, A. Konar, J. Ogilvie,** *University of Michigan, U.S.A.*

Understanding the role of vibronic couplings in photoinduced energy transfer processes using multi-dimensional electronic-vibrational spectroscopy, **J. Gaynor, J. Sandwisch, A. Petrone, X. Li, M. Khalil,** *University of Washington, U.S.A.*

Charge carrier trapping at surface defects and optical properties of perovskites, **K. Yamashita,** *The University of Tokyo, Japan*

An experimental setup for time- and angle resolved photoelectron spectroscopy of organic photovoltaic heterojunctions, **E. Marsell, A. Tully, A. K. Mills, E. Ostroumov, M. Riede, D. J. Jones, S. A Burke,** *University of British Columbia, Canada*

The photochemical ring-opening of 1,3-cyclohexadiene imaged by ultrafast electron diffraction, **T. J. A. Wolf, D. M. Sanchez, J. Yang, R. M. Parrish, J. P. F. Nunes, M. Centurion, R. Coffee, J. P. Cryan, M. Gühr, K. Hegazy, A. Kirrander, R. K. Li, J. Ruddock, X. Shen, T. Vecchione, S. P. Weathersby, P. M. Weber, K. Wilkin, H. Yong, Q. Zheng, X. J. Wang, M. P. Minitti, T. J. Martínez,** *Stanford University, U.S.A.*

An effective model for energy transfer dynamics in the dimeric photosystem II core complex, **S.-T. Hsieh, L. Zhang, D.-W. Ye, X. Huang, Y.-C. Cheng,** *National Taiwan University, Chinese Taipei*

Intramolecular energy transfer facilitates endothermic singlet fission, **N. Korovina, J. Johnson,** *NREL, U.S.A.*

Membrane environment as a gear towards photoprotection in green plants, **M. Son, A. Pinnola, S. C. Gordon, G. Schlau-Coben,** *Massachusetts Institute of Technology, U.S.A.*

Synchrotron hard X-ray spectroscopic investigation of the photoaquation reaction mechanism in hexacyanoferrate(II) with sub-pulse temporal resolution, **A. M. March, G. Doumy, A. Al Haddad, J. Bang, S. Southworth, Y. Kumagai, M.-F. Tu, L. Young, A. Anderson, N. Govind,** *Argonne National Lab, U.S.A.*

Ultrafast imaging of energy transport in CdSe/CdS lateral heterostructures, **D. Blach, L. Huang,** *Purdue University, U.S.A.*

Excitation energy transfer in a subphthalocyanine–Zn porphyrin dimer studied by nonlinear ultrafast spectroscopy, **G. Bressan, A. N. Cambridge, G. Jones, I. Heisler, D. Gonzalez-Lucas, S. Ramiro-Buenamãnana S. Meech,** *University of East Anglia, United Kingdom.*

Ultrafast dynamics of biomolecular systems, **M. Duchi, M. P. O'Hagan, R. Kumar, S. J. Bennie, B. F. E. Curchod, M. C. Galan, T. A. A. Oliver,** *University of Bristol, United Kingdom*

An unexpected role of composition in perovskite diffusion, **K. Williams, M. Saidaminov, M. Wei, A. Johnson, R. Quintero-Bermudez, A. H. Proppe, Y. Hou, G. Walters, S. O. Kelley, W. A. Tisdale, E. H. Sargent,** *Massachusetts Institute of Technology, U.S.A.*

Ultrafast electron transfer in a Cu-Corrole - BODIPY array observed by femtosecond pump-probe-spectroscopy, **C. Zahn, T. Stensitzki, H. R. A. Golf, A. Wiehe, K. Heyne,** *Free University Berlin, Germany*

Dissociative electron attachment via resonances: Insights from theory, **S. Gulania, A. I. Krylov,** *University of Southern California, Los Angeles, U.S.A.*

Effect of T65 mutation on brightness and fluorescence lifetime in EGFP, **T. Sen, A. Krylov, A. M. Boganor, E. Maksimov, K. A. Lukyanor,** *University of Southern California, U.S.A.*

A consistent model of ultrafast energy transfer in peridinin chlorophyll-a protein using two dimensional electronic spectroscopy and Förster theory, **Z. S. D. Toa, M. H. deGolian, C. C. Jumper, R. G. Hiller, G. D. Scholes,** *Princeton University, U.S.A.*

Ultrafast [2+2] cycloaddition reaction upon photoactivation of 4-thiothymidine in single-stranded DNA, **L. Ortiz-Rodríguez, C. Crespo-Hernández,** *Case Western Reserve University, U.S.A.*

Hydrogen-bonding interactions mediate the ultrafast dynamics of tripyrrindiones, **A. Swain, B. Cho, R. Gautam, C. Curtis, E. Tomat, V. Huxter,** *University of Arizona, U.S.A.*

Ultrafast photodynamics of homosalate, a sunscreen molecule, **K. Krokidi, E. Holt, M. Turner, N. Das Neves Rodrigues, V. Stavros,** *University of Warwick, United Kingdom*

The Faraday Discussions poster prize for the best poster was awarded to Mr Trevor Roberts from the University of California – Berkeley, U.S.A. for his poster on tuneable biomimetic systems to elucidate chromophore-protein and chromophore-chromophore interactions that facilitate photosynthetic light harvesting.

List of participants

Mrs Najla Aldossari, *Hampton University, U.S.A.*
Professor Michael Ashfold, *University of Bristol, United Kingdom*
Dr Shuming Bai, *Duke University, U.S.A.*
Dr Vytautas Balevicius Jr., *Queen Mary University of London, United Kingdom*
Mr Jon Bender, *The University of Texas at Austin, U.S.A.*
Professor David Beratan, *Duke University, U.S.A.*
Mr Dhritiman Bhattacharyya, *University of Southern California, U.S.A.*
Dr Matthew Bird, *Brookhaven National Laboratory, U.S.A.*
Professor Eric Bittner, *University of Houston, U.S.A.*
Ms Daria Blach, *Purdue University, U.S.A.*
Professor David Blank, *University of Minnesota, U.S.A*
Dr Mario Borgwardt, *Lawrence Berkeley National Laboratory, U.S.A.*
Stephen Bradforth, *University of Southern California, U.S.A.*
Mr Giovanni Bressan, *University of East Anglia, United Kingdom*
Mr William Carbery, *New York University, United Kingdom*
Dr Pavel Chabera, *Lund University, Sweden*
Professor Yuan-Chung Cheng, *National Taiwan University, Taiwan*
Professor Majed Chergui, *Ecole Polytechnique Fédérale de Lausanne, Switzerland*
Dr Jeffrey Cina, *University of Oregon, U.S.A.*
Mr Nicholas D. Cooper, *The University of Georgia, U.S.A.*
Professor Dr Carlos E. Crespo-Hernández, *Case Western Reserve University, U.S.A*
Dr Natercia das Neves Rodrigues, *University of Warwick, United Kingdom*
Dr Animesh Datta, *University of Warwick, United Kingdom*
Professor Jahan Dawlaty, *University of Southern California, U.S.A.*
Mr Ryan Dill, *University of Colorado Boulder, U.S.A.*
Mr Amro Dodin, *MIT, U.S.A.*
Ms Marta Duchi, *University of Bristol, United Kingdom*
Ms Laura Estergreen, *University of Southern California, U.S.A.*
Dr Christopher Fecko, *U.S. Department of Energy, Office of Basic Energy Sciences, U.S.A*
Dr Benjamin Fingerhut, *Max Born Institute, Germany*
Professor Graham Fleming, *University of California - Berkeley, U.S.A.*
Dr Terry Frankcombe, *UNSW Canberra, Australia*
Mr Gregory Gate, *University of California, Santa Barbara, U.S.A.*
Mr James Gaynor, *University of Washington, U.S.A.*
Dr Oliver Gessner, *Berkeley Lab, U.S.A.*
Professor Dr Naomi Ginsberg, *University of California, Berkeley, U.S.A.*
Dr Christopher Grieco, *The Ohio State University, U.S.A.*
Mr Sahil Gulania, *University of Southern California, U.S. A.*
Dr Scott Habershon, *University of Warwick, United Kingdom*
Mr Michael Haggmark, *University of California, Santa Barbara, U.S.A.*
Professor Sharon Hammes-Schiffer, *Yale University, U.S.A.*
Mr Lucian Hand, *Light Conversion USA, U.S.A*

Mr Dennis Hein, *Helmholtz Center for Materials and Energy, Germany*
Dr Michael Horbury, *University of Warwick, United Kingdom*
Ms Shou-Ting Hsieh, *Department of Chemistry, National Taiwan University, Taiwan*
Mr Jonathan Hunt, *University of Southern California, U.S.A.*
Dr Vanessa Huxter, *University of Arizona, U.S.A*
Ms Nha Uyen Huynh, *DelMar Photonics, U.S.A.*
Mr Osamu Ide, *Clean Energy Research Laboratory/Chief, Japan*
Mrs Yuriko Ide, *Clean Energy Research Laboratory/Secretary, Japan*
Mr Shawn Irgen-Gioro, *Northwestern University, U.S.A.*
Mr Jimmy Joy, *University of Southern California, U.S.A.*
Mr Michael Kellogg, *University of Southern California, U.S.A.*
Ms Alexis Kiessling, *University of Oregon, U.S.A.*
Professor Bern Kohler, *The Ohio State University, U.S.A.*
Dr Nadia Korovina, *NREL, U.S.A.*
Dr Thomas Kraft, *Layertec GmbH, U.S.A.*
Miss Konstantina Krokidi, *University of Warwick, United Kingdom*
Ms Yumin Lee, *University of Pennsylvania, U.S.A.*
Dr Jenny Lee, *Royal Society of Chemistry, U.S.A.*
Professor Graham Leggett, *University of Sheffield, United Kingdom*
Dr Heiko Lokstein, *Charles University, Prague*
Mr Johannes Mahl, *Lawrence Berkeley National Laboratory, U.S.A.*
Dr Anne Marie March, *Argonne National Lab, U.S.A.*
Professor Andrew Marcus, *University of Oregon, U.S.A.*
Dr Erik Marsell, *University of British Columbia, Canada*
Dr Todd Martinez, *Stanford University, U.S.A.*
Professor Spiridoula Matsika, *Temple University, U.S.A.*
Dr Alex Metherell, *Royal Society of Chemistry, United Kingdom*
Miss Karen Morenz, *University of Toronto, Canada*
Dr Amanda Morris, *Virginia Tech, U.S.A.*
Professor Jennifer Ogilvie, *University of Michigan, U.S.A.*
Professor Alexandra Olaya-Castro, *University College London, United Kingdom*
Dr Tom Oliver, *University of Bristol, United Kingdom*
Mr Luis Ortiz-Rodríguez, *Case Western Reserve University, U.S.A*
Dr Giovanny Parada, *Yale University, U.S.A.*
Dr Thomas Penfold, *Newcastle University, United Kingdom*
Dr Petter Persson, *Lund University, Sweden*
Professor Tomas Polivka, *University of South Bohemia, Czech Republic*
Mr Jamie Riley, *University College London, United Kingdom*
Mr Trevor Roberts, *University of California - Berkeley, U.S.A.*
Dr Eve Rooks, *Royal Society of Chemistry, United Kingdom*
Mr David Sanchez, *Stanford University, U.S.A.*
Professor Benjamin Schwartz, *UCLA, U.S.A.*
Other Riley Sechrist, *University of Michigan, U.S.A.*
Mr Tirthendu Sen, *University of Southern California, U.S.A.*
Mr Stefan Siwiak-Jaszek, *UCL (University College London), United Kingdom*
Miss Minjung Son, *Massachusetts Institute of Technology, U.S.A.*
Dr Yin Song, *University of Michigan, U.S.A.*
Dr Mark Spitler, *University of Wyoming Center for Photoconversion and Catalysis, U.S.A*
Professor Vasilios Stavros, *University of Warwick, United Kingdom*
Mr Collin Steen, *University of California, Berkeley, U.S.A.*
Ms Alicia Swain, *University of Arizona, U.S.A*
Miss Rachel Swedin, *University of Minnesota, U.S.A*

Mrs Sarah Thirkell, *Royal Society of Chemistry, United Kingdom*
Professor Mark Thompson, *University of Southern California, U.S.A.*
Mr Zi Siang Desmond Toa, *Princeton University, U.S.A*
Ms Cindy Tseng, *University of Southern California, U.S.A.*
Mr Ming-Feng Tu, *Argonne National Laboratory, U.S.A.*
Mr Matthew Turner, *University of Warwick, United Kingdom*
Professor Claire Vallance, *University of Oxford, United Kingdom*
Dr Sachin Dev Verma, *University of Cambridge, United Kingdom*
Professor Dr Peter Jomo Walla, *University of Braunschweig/Max-Planck-Insitute/ Kavli EnergyNanoScience Institute, Germany*
Mr Garlef Wartner, *Helmholtz Center for Materials and Energy, Germany*
Professor Michael Wasielewski, *Northwestern University, U.S.A.*
Professor Emily Weiss, *Northwestern University, U.S.A.*
Dr Kristopher Williams, *Massachusetts Institute of Technology, U.S.A.*
Mr Jack Woolley, *University of Warwick, United Kingdom*
Emer Prof Koichi Yamashita, *The University of Tokyo, Japan*
Mr Clark Zahn, *Free University Berlin, Germany*
Mrs Noura Zamzam, *Imperial College London, United Kingdom*